T0235306

Lecture Notes in Artificial Intelligence 11499

Subseries of Lecture Notes in Computer Science

Series Editors

Randy Goebel
University of Alberta, Edmonton, Canada
Yuzuru Tanaka
Hokkaido University, Sapporo, Japan
Wolfgang Wahlster
DFKI and Saarland University, Saarbrücken, Germany

Founding Editor

Jörg Siekmann
DFKI and Saarland University, Saarbrücken, Germany

More information about this series at http://www.springer.com/series/1244

Tamás Mihálydeák · Fan Min ·
Guoyin Wang · Mohua Banerjee ·
Ivo Düntsch · Zbigniew Suraj ·
Davide Ciucci (Eds.)

Rough Sets

International Joint Conference, IJCRS 2019
Debrecen, Hungary, June 17–21, 2019
Proceedings

Editors
Tamás Mihálydeák
University of Debrecen
Debrecen, Hungary

Guoyin Wang
Chongqing University of Posts and Telecommunications
Chongqing, China

Ivo Düntsch
Fujian Normal University
Fuzhou, China

Davide Ciucci
University of Milano-Bicocca
Milan, Italy

Fan Min
Southwest Petroleum University
Chengdu, China

Mohua Banerjee
Indian Institute of Technology Kanpur
Kanpur, India

Zbigniew Suraj
University of Rzeszów
Rzeszów, Poland

ISSN 0302-9743 ISSN 1611-3349 (electronic)
Lecture Notes in Artificial Intelligence
ISBN 978-3-030-22814-9 ISBN 978-3-030-22815-6 (eBook)
https://doi.org/10.1007/978-3-030-22815-6

LNCS Sublibrary: SL7 – Artificial Intelligence

© Springer Nature Switzerland AG 2019
This work is subject to copyright. All rights are reserved by the Publisher, whether the whole or part of the material is concerned, specifically the rights of translation, reprinting, reuse of illustrations, recitation, broadcasting, reproduction on microfilms or in any other physical way, and transmission or information storage and retrieval, electronic adaptation, computer software, or by similar or dissimilar methodology now known or hereafter developed.
The use of general descriptive names, registered names, trademarks, service marks, etc. in this publication does not imply, even in the absence of a specific statement, that such names are exempt from the relevant protective laws and regulations and therefore free for general use.
The publisher, the authors and the editors are safe to assume that the advice and information in this book are believed to be true and accurate at the date of publication. Neither the publisher nor the authors or the editors give a warranty, expressed or implied, with respect to the material contained herein or for any errors or omissions that may have been made. The publisher remains neutral with regard to jurisdictional claims in published maps and institutional affiliations.

This Springer imprint is published by the registered company Springer Nature Switzerland AG
The registered company address is: Gewerbestrasse 11, 6330 Cham, Switzerland

Preface

Professor Zdzisław Pawlak's fundamental papers connected with rough sets theory were published in 1982. Roman Słowiński highlighted the crucial importance of the original rough set theory:

> "This theory helps to find answers to many basic questions in mathematics, computer science, artificial intelligence, decision theory, conflict theory, machine learning, knowledge discovery and control theory. This theory is founded on an observation that knowledge about objects from a real or abstract world is granular. Indeed, objects described by the same information are indiscernible and create elementary sets, which are knowledge granules for that world. When willing to express a concept, referring to a given set of objects, in terms of knowledge about the world the objects come from, one encounters a situation in which in general, the concept is not expressible exactly by the available granules; in other words, the union of elementary sets having non-empty intersection with our set, does not coincide with the set. This set – a concept – may, however, be expressed roughly, using sets called lower and upper approximations – lower approximation containing elementary sets (granules) which are wholly included in our set, and upper approximation containing also those sets which are partly included in our set. The difference between those approximations is called a boundary of a set, and contains ambiguous objects, for which one cannot claim with certainty, whether they do or do not belong to our set. Differentiating between definite knowledge represented by lower approximation and approximate knowledge represented by the boundary of a set has a fundamental impact on the deduction process. Rough set theory complements fuzzy set theory and soft computing, with which it now delivers the best tools for reasoning about data bearing different types of "imperfections", such as ambiguity, inaccuracy, inconsistency, incompleteness, and uncertainty."[1]

The International Joint Conference on Rough Sets (IJCRS) is a major international forum that brings researchers and industry practitioners together to discuss and deliberate on fundamental issues of rough sets and practical solutions relying on different versions of rough set theory. The objective of the conference is to investigate rough set theory, which has been receiving more and more attention in varied hybrid approaches in different practical fields, with a special emphasis on fostering interaction between academia and industry. The IJCRS conferences aim at gathering experts from academia and industry representing fields of research in which theoretical and practical aspects of rough set theory already find or may potentially find usage. They also provide opportunities for researchers to present their ideas before the rough set community, or for those who would like to learn about rough sets and find out whether the rough set approach could be useful for their problems.

The proceedings of IJCRS 2019 contain the papers selected for presentation at the meeting of the International Rough Sets Society, held at the University of Debrecen, Hungary, during June 17–21, 2019.

---

[1] See in Słowiński, R.: Laudatio dedicated to Mr Professor Ph.D. hab. M.Eng. Zdzisław I. Pawlak. In: Długosz, K. (ed.) Zdzisław Pawlak. Doctor Honoris Causa of Poznań University of Technology, pp. 7–11. Poznań University of Technology, Poznań (2002) (in Polish); English translation in Skowron, A., Suraj, Z. (Eds.): Rough Sets and Intelligent Systems – Professor Zdzisław Pawlak in Memoriam, Vol. 1, Springer-Verlag Berlin Heidelberg, 2013, pp. 11.

Conferences in the IJCRS series are held annually and incorporate four main tracks (and conferences) relating to the theory of rough sets and its connection with other paradigms:

- Rough sets and data analysis: RSCTC conference series from 1998
- Rough sets and granular computing: RSFDGrC from 1999
- Rough sets and knowledge technology: RSKT from 2006
- Rough sets and intelligent systems: RSEISP from 2007

The main topics of IJCRS 2019 consisted of three groups:

- Core Rough Set Models and Methods (e.g., covering rough set models, decision-theoretic rough set methods, dominance-based rough set methods, rough clustering, rough computing, rough mereology, partial rough set models, game-theoretic rough set methods)
- Related Methods and Hybridization (e.g., artificial intelligence, machine learning, pattern recognition, decision support systems, fuzzy sets and near sets, uncertain and approximate reasoning, information granulation, formal concept analysis, Petri nets, nature-inspired computation models)
- Areas of Application (e.g., medicine and health, bioinformatics, business intelligence, smart cities, Semantic Web, computer vision and image processing, cybernetics and robotics, knowledge discovery)

IJCRS 2019 received 71 papers from 17 countries. Following the tradition of the previous IJCRS conferences, all submissions underwent a very rigorous reviewing process. Every submission was reviewed by at least two Program Committee (PC) members; on average, each submission received 2.54 reviews. Finally, the PC chairs selected 41 regular papers, based on their originality, significance, correctness, relevance, and clarity of presentation to be included in the proceedings of IJCRS 2019. We would like to thank all authors for submitting their papers. We also wish to congratulate those authors whose papers were selected for presentation and publication in the proceedings.

IJCRS 2019 would not have been successful without the support of many colleagues and organizations. We acknowledge the acceptance of our proposal of organizing IJCRS 2019 at the Faculty of Informatics, University of Debrecen, in Debrecen, Hungary, by the authorities of the International Rough Set Society, the owner of the rights to the series. We wish to express our gratitude to the following for their invaluable suggestions, support, and excellent work throughout the organization process:

- Andrzej Skowron, Mihir Chakraborty, and Attila Pethő, the honorary chairs of IJCRS 2019
- Yiyu Yao, Nguyen Hung Son, and Dominik Ślęzak, the members of the Steering Committee of IJCRS 2019
- The members of the Program Committee of IJCRS 2019

We are very grateful to Chris Cornelis, Eyke Hüllermeier, Sergei Kuznetsov, Wojciech Ziarko, and Mihir Chakraborty, the invited and plenary speakers, for accepting our invitations.

We are also grateful to László Aszalós, Tamás Kádek, Dávid Nagy, Ildikó Vecsei, Ernőné Kása, Rita Koroknai, and Nóra Bende of the Faculty of Informatics, University of Debrecen, whose great efforts ensured the success of the conference.

We greatly appreciate the support of the International Rough Set Society, the Faculty of Informatics, University of Debrecen, and IT Services Hungary Ltd.

This conference was supported by the EFOP–3.6.3–VEKOP–16–2017–00002. The project was supported by the European Union, co-financed by the European Social Fund.

Special thanks go to Alfred Hofmann of Springer, for accepting to publish the proceedings of IJCRS 2019 in the LNCS/LNAI series, and to Anna Kramer and the excellent LNCS team for their help with the proceedings. We are grateful to Springer for the grant of 1,000 EUR for the best conference papers.

May 2019

Tamás Mihálydeák
Fan Min
Guoyin Wang
Mohua Banerjee
Ivo Düntsch
Zbigniew Suraj
Davide Ciucci

Organization

Honorary Chairs

Andrzej Skowron	University of Warsaw, Poland
Mihir Chakraborty	Jadavpur University, India
Attila Pethő	University of Debrecen, Hungary

Conference Chairs

Guoyin Wang	Chongqing University of Posts and Telecommunications, China
Tamás Mihálydeák	University of Debrecen, Hungary

Program Committee Chairs

Fan Min	Southwest Petroleum University, China
Mohua Banerjee	Indian Institute of Technology Kanpur, India
Ivo Düntsch	Fujian Normal University, China
Zbigniew Suraj	University of Rzeszów, Poland

Steering Committee

Davide Ciucci	Università di Milano-Bicocca, Italy
Yiyu Yao	University of Regina, Canada
Hung Son Nguyen	University of Warsaw, Poland
Dominik Ślęzak	University of Warsaw, Poland

Program Committee

Mani A.	Kolkata, India
Piotr Artiemjew	University of Warmia and Mazury, Poland
Jaume Baixeries	Universitat Politècnica de Catalunya, Spain
Rafael Bello	Universidad Central de Las Villas, Cuba
Nizar Bouguila	Concordia University, Canada
Shampa Chakraverty	Netaji Subhas Institute of Technology, India
Chien-Chung Chan	University of Akron, USA
Mu-Chen Chen	National Chiao Tung University, Taiwan
Costin-Gabriel Chiru	Politehnica University of Bucharest, Romania
Víctor Codocedo	Universidad Técnica Federico Santa María, Chile
Zoltán Ernő Csajbók	University of Debrecen, Hungary
Dayong Deng	Zhejiang Normal University, China
Thierry Denoeux	Université de Technologie de Compiègne, France

Murat Diker	Hacettepe University, Turkey
Paweł Drozda	University of Warmia and Mazury, Poland
Soma Dutta	Vistula University, Poland
Zied Elouedi	Institut Supéur de Gestion de Tunis, Tunisia
Victor Flores	Universidad Católica del Norte, Chile
Anna Gomolińska	University of Białystok, Poland
Rafał Gruszczyński	Nicolaus Copernicus University, Poland
Jerzy Grzymała-Busse	University of Kansas, USA
Christopher Hinde	Loughborough University, UK
Dmitry Ignatov	National Research University Higher School of Economics, Russia
Masahiro Inuiguchi	Osaka University, Japan
Ryszard Janicki	McMaster University, Canada
Jouni Jarvinen	University of Turku, Finland
Richard Jensen	Aberystwyth University, UK
Xiuyi Jia	Nanjing University of Science and Technology, China
Michał Kępski	University of Rzeszów, Poland
Md. Aquil Khan	Indian Institute of Technology Indore, India
Yoo-Sung Kim	Inha University, South Korea
Marzena Kryszkiewicz	Warsaw University of Technology, Poland
Yasuo Kudo	Muroran Institute of Technology, Japan
Huaxiong Li	Nanjing University, China
Tianrui Li	Southwest Jiaotong University, China
Jiye Liang	Shanxi University, China
Churn-Jung Liau	Academia Sinica, Taiwan
Caihui Liu	Gannan Normal University, China
Guilong Liu	Beijing Language and Culture University, China
Nguyen Long Giang	Vien Công nghe thông tin, Vietnam
Pradipta Maji	Indian Statistical Institute, India
Jesús Medina	University of Cádiz, Spain
Claudio Meneses	Universidad Católica del Norte, Chile
Marcin Michalak	Silesian University of Technology, Poland
Michinori Nakata	Josai International University, Japan
Amedeo Napoli	Université de Lorraine, France
Loan T. T. Nguyen	University of Warsaw, Poland
M. C. Nicoletti	Federal University of São Carlos, Brazil
Vilem Novak	University of Ostrava, Czech Republic
Krzysztof Pancerz	University of Rzeszów, Poland
Vladimir Parkhomenko	Peter the Great St. Petersburg Polytechnic University, Russia
Andrei Paun	University of Bucharest, Romania
Georg Peters	Munich University of Applied Sciences and Australian Catholic University, Germany/Australia
Alberto Pettorossi	Università di Roma Tor Vergata, Italy
Lech Polkowski	Polish-Japanese Academy of Information Technology, Poland

Mohamed Quafafou	Aix-Marseille University, France
Sándor Radeleczki	University of Miskolc, Hungary
Anna Radzikowska	Warsaw University of Technology, Poland
Sheela Ramanna	University of Winnipeg, Canada
Zbigniew W. Raś	University of North Carolina at Charlotte, USA
Henryk Rybiński	Warsaw University of Technology, Poland
Wojciech Rząsa	University of Rzeszów, Poland
Hiroshi Sakai	Kyushu Institute of Technology, Japan
Gerald Schaefer	Loughborough University, UK
B. Uma Shankar	Indian Statistical Institute, India
Marek Sikora	Silesian University of Technology, Poland
Roman Słowiński	Poznań University of Technology, Poland
Jarosław Stepaniuk	Białystok University of Technology, Poland
Paul Sushmita	Indian Institute of Technology Jodhpur, India
Andrzej Szałas	University of Warsaw, Poland
Marcin Szczuka	University of Warsaw, Poland
Ryszard Tadeusiewicz	AGH University of Science and Technology, Poland
Li-Shiang Tsay	North Carolina A&T State University, USA
Bay Vo	Ho Chi Minh City University of Technology, Vietnam
Arkadiusz Wojna	Security On-Demand, Inc., Poland
Marcin Wolski	Maria Curie-Skłodowska University, Poland
Yan Yang	Southwest Jiaotong University, China
Jingtao Yao	University of Regina, Canada
Dongyi Ye	Fuzhou University, China
Bing Zhou	Sam Houston State University, USA
Wojciech Ziarko	University of Regina, Canada
Beata Zielosko	University of Sielsia, Poland

Additional Reviewers

Andrea Campagner
María Eugenia Cornejo
Michał Czołombitko
Anna Formica
Dariusz Jankowski
Vahid Khorasani Ghassab
Tomasz Krzeszowski
Narges Manouchehri
Amaldev Manuel
Luca Manzoni
Eloísa Ramírez-Poussa
Giuseppe Vizzari
Nuha Zamzami

Organizing Committee

Tamás Mihálydeák	University of Debrecen, Hungary
László Aszalós	University of Debrecen, Hungary
Tamás Kádek	University of Debrecen, Hungary
Ildikó Vecsei	University of Debrecen, Hungary
Ernőné Kása	University of Debrecen, Hungary
Rita Koroknai	University of Debrecen, Hungary
Nóra Bende	University of Debrecen, Hungary

Invited Speakers

Chris Cornelis	Ghent University, Belgium
Eyke Hüllermeier	University of Paderborn, Germany
Sergei Kuznetsov	National Research University Higher School of Economics, Russia
Wojciech Ziarko	University of Regina, Canada
Mihir Chakraborty	Jadavpur University, India

Abstracts of Invited Talks

Fuzzy Rough Sets: Achievements and Opportunities

Chris Cornelis

Computational Web Intelligence, Department of Applied Mathematics, Computer Science and Statistics, Ghent University, Belgium
Chris.Cornelis@UGent.be

Fuzzy logic, introduced by Zadeh [3] in 1965, caters to the idea that for many logical propositions, it is not possible to determine in a black-or-white fashion whether they are true or false. Think of a sentence like "today is a sunny day". For this reason, graded degrees of truth are drawn from a continuous scale, usually the unit interval [0,1], with 0 representing absolute falsehood and 1 representing complete truth, and the intermediate degrees corresponding to partial truth. Fuzzy logic can, in a sense, be seen as the culmination of the tradition of many-valued logics initiated in the first half of the twentieth century by eminent logicians like Łukasiewicz, Gödel and Kleene. In a completely analogous fashion, fuzzy sets embody the notion that membership of objects to a set, category or class is often a matter of degree. Fuzzy set theory is also involved with the expression of gradual relationships between objects, and the well-known concepts of equivalence relation, dominance relation, order relation, etc. have all been adequately generalized to this setting.

Rough sets, introduced by Pawlak [2] in 1982, provide approximations of concepts based on incomplete and possibly inconsistent information about objects and their relationships. Specifically, given a subset A of X, an object $x \in X$ belongs to the lower approximation of A if all objects related to it belong to A, and to the upper approximation if at least one object related to x belongs to A. In Pawlak's original model, object relationships are represented using an equivalence relation over the universe of discourse X (or equivalently, a partition of X) to express object indiscernibility. Subsequent research generalized this assumption to consider various types of binary relations R over X to replace the equivalence relation, including tolerance and dominance relations, or to work with a covering, i.e., a set of possibly overlapping subsets of X whose union equals X, to replace the partition. The different rough set models have found widespread application in data analysis, where they are used e.g. to infer data dependencies that can be exploited in feature selection and decision model construction.

Fuzzy sets and rough sets share a long common history. In 1990, Dubois and Prade [1] proposed the first fuzzy rough set model, in which fuzzy sets are approximated from below and above using a fuzzy relation. Since then, many researchers have focused on the refinement of this model using constructive approaches, involving fuzzy logic

Supported by the Odysseus Programme of the Science Foundation–Flanders.

operations to shape the approximations, and axiomatic ones, proposing a set of desirable properties that approximation operators are expected to satisfy. During the past two decades, practical interest in fuzzy rough sets has also been steadily rising by their application potential in various data analysis tasks, including data reduction, classification and clustering. These applications also raised new challenges for the fuzzy-rough hybridization process, which led amongst others to the introduction of various robust alternatives to the classical fuzzy rough set definitions. In this presentation, I will discuss some of the most prominent machine learning approaches using fuzzy rough sets, and identify some current challenges and directions for the hybrid theory.

References

1. Dubois, D., Prade, H.: Rough fuzzy sets and fuzzy rough sets. Int. J. Gen. Syst. **17**, 91–209 (1990)
2. Pawlak, Z.: Rough sets. Int. J. Comput. Inf. Sci. **11**(5), 341–356 (1982)
3. Zadeh, L.A.: Fuzzy sets. Inf. Control **8**, 338–353 (1965)

Pattern Structures and Pattern Setups for Mining Complex Data

Sergei O. Kuznetsov

National Research University Higher School of Economics, Moscow, Russia

Abstract. Pattern mining started with mining itemset patterns, however many applied problems of data mining make researchers face more complex data like numerical intervals, strings, graphs, geometric figures, etc. Like in itemset mining closed patterns proved to be very important for concise representations of association rules and other types of dependencies. An acknowledged approach to representing closed patterns was formulated in terms of Pattern Structures [3, 5], which were implemented for various description spaces, among them tuples of intervals [7], convex polygons [2], partitions [4], graphs [6], and strings [1]. Pattern structures, however, require that the description space makes a complete semilattice. Pattern setups is a generalization of pattern structures that allows for a partially ordered description space. We consider various examples of pattern structures and pattern setups arising in different applied domains, together with approximation schemes based on kernel operators and efficient algorithms for computing closed patterns and dependencies based on them.

References

1. Buzmakov, A., Egho, E., Jay, N., Kuznetsov, S., Napoli, A., Rassi, C.: On mining complex sequential data by means of FCA and pattern structures. Int. J. Gen. Syst. **45**(2), 135–159 (2016)
2. Belfodil, A., Kuznetsov, S., Robardet, C., Kaytoue, M.: Mining convex polygon patterns with formal concept analysis. In: IJCAI, pp. 1425–1432 (2017)
3. Ganter, B., Kuznetsov, S.: Pattern structures and their projections. In: ICCS, pp. 129–142 (2001)
4. Baixeries, J., Kaytoue, M., Napoli, A.: Characterizing functional dependencies in formal concept analysis with pattern structures. Ann. Math. Artif. Intell. **72**(1–2), 129–149 (2014)
5. Kuznetsov, S.: Pattern structures for analyzing complex data. In: RSFDGrC (2009)
6. Kuznetsov, S.: Fitting pattern structures to knowledge discovery in big data. In: ICFCA 2013, pp. 254–266 (2013)
7. Kaytoue, M., Kuznetsov, S., Napoli, A.: Revisiting numerical pattern mining with formal concept analysis. In: IJCAI, pp. 1342–1347 (2011)

Contents

Core Rough Set Models and Methods

Related Methods and Hybridization

Areas of Application

Core Rough Set Models and Methods

An Application of Bayesian Confirmation Theory for Three-Way Decision

Mengjun Hu[(✉)], Xiaofei Deng, and Yiyu Yao

Department of Computer Science, University of Regina,
Regina, SK S4S 0A2, Canada
{hu258,deng200x,yyao}@cs.uregina.ca

Abstract. Bayesian confirmation theory studies how a piece of evidence confirms a hypothesis. In a qualitative approach, a piece of evidence may confirm, disconfirm, or be neutral with respect to a hypothesis. A quantitative approach uses Bayesian confirmation measures to evaluate the degree to which a piece of evidence confirms a hypothesis. In both approaches, we may perform a three-way classification of a set of pieces of evidence for a given hypothesis. The set of evidence is divided into three regions of positive evidence that confirms the hypothesis, negative evidence that disconfirms the hypothesis, and neutral evidence that neither confirms nor disconfirms the hypothesis. In this paper, we investigate three-way classification models in both qualitative and quantitative Bayesian confirmation approaches and explore their relationships to three-way classification models in rough set theory.

Keywords: Three-way decision · Bayesian confirmation · Rough set · Attribute reduct

1 Introduction

The integration of Bayesian confirmation theory into rough s et theory [17,18] has been studied by several researchers [6–8,26,30]. Rough sets may be viewed as a model that employs three-way decision. This paper focuses on relationships between three-way decision and Bayesian confirmation.

A theory of three-way decision is originally developed from rough sets and has been applied and generalized by researchers in a variety of topics beyond rough sets, such as three-way classifications [13,22,25], three-way clusterings [1,27,28], throo way rccommendations [2,29], and three-way concept analysis [19,21]. In a recent paper [23], Yao proposes a Trisecting-Acting-Outcome (TAO) model for modelling three-way decision in a wide sense. The model includes three steps, that is, a trisecting step of dividing a whole into three parts, an acting step of devising and applying strategies to process the three parts, and an outcome evaluation step to evaluate the results of trisecting and acting steps. Yao also

Y.Y. Yao—This work is partially supported by a Discovery Grant from NSERC, Canada.

© Springer Nature Switzerland AG 2019
T. Mihálydeák et al. (Eds.): IJCRS 2019, LNAI 11499, pp. 3–15, 2019.
https://doi.org/10.1007/978-3-030-22815-6_1

demonstrates that the idea of three-way decision is a common human practice and is widely practiced in many disciplines. In this paper, we investigate the ideas of three-way decision in Bayesian confirmation theory.

Bayesian confirmation theory studies how a piece of evidence e confirms a hypothesis h. Intuitively, there are three possible relationships between e and h, that is, e confirms h, e disconfirms h, and e is neutral with respect to h. These three relationships naturally imply a trisecting of all available evidence with respect to a given hypothesis h. That is, we can divide the set of all pieces of evidence into three parts of evidence confirming h, disconfirming h, and neutral with respect to h. The conditions for these three parts depend on the determination of the three relationships in a specific Bayesian confirmation approach. We investigate the formulation of these three parts in both qualitative and quantitative Bayesian confirmation approaches, which results in a three-way classification model of evidence. To illustrate the application of this model, we explore three-way classification of evidence in rough set theory from two views. The first view takes equivalence classes in rough set theory as evidence, which is a commonly used view in existing related studies [6–8,26,30]. The second view takes attributes that are used to describe objects as evidence. This view leads to a three-way classification of attributes in rough set theory, and enables us to define the concept of class-specific attribute reducts [12,15,18,20] in rough set theory based on Bayesian confirmation.

The remaining part of this paper is arranged as follows. Section 2 provides a brief overview of Bayesian confirmation approaches. In Sect. 3, we propose a three-way classification model of evidence by using Bayesian confirmation approaches. The proposed model is examined with respect to rough set theory in Sect. 4. The examination results in a new definition of class-specific attribute reducts, which is presented in Sect. 5. Section 6 concludes the paper and discusses possible directions for future work.

2 An Overview of Bayesian Confirmation

Bayesian confirmation theory [4,5] studies how a piece of evidence e confirms a hypothesis h. A basic and commonly used idea is to compare the a priori probability $Pr(h)$ and the a posteriori probability $Pr(h|e)$. By employing qualitative and quantitative comparisons, Bayesian confirmation can be categorized into qualitative and quantitative approaches, respectively.

In a qualitative approach, a piece of evidence e confirms a hypothesis h if the a posteriori probability $Pr(h|e)$ increases from the a priori probability $Pr(h)$, that is, the observation of e increases the probability of h. Similarly, e disconfirms h if $Pr(h|e)$ decreases from $Pr(h)$, that is, the observation of e decreases the probability of h. Otherwise, if $Pr(h|e)$ is unchanged from $Pr(h)$, then e is considered to be neutral with respect to h, that is, the observation of e neither increases nor decreases the probability of h. This approach is referred to as P-incremental confirmation [4], which can be formally expressed as:

$$\begin{cases} e \text{ confirms } h, & \text{iff } Pr(h|e) > Pr(h), \\ e \text{ is neutral with respect to } h, & \text{iff } Pr(h|e) = Pr(h), \\ e \text{ disconfirms } h, & \text{iff } Pr(h|e) < Pr(h). \end{cases}$$

The three conditions can be equivalently expressed in several forms [30]. Take the condition $Pr(h|e) > Pr(h)$ as an example. With an assumption $Pr(h) \neq 0$, we have:

$$Pr(h|e) > Pr(h) \Longleftrightarrow \frac{Pr(h|e)}{Pr(h)} > 1. \tag{1}$$

According to the Bayes' theorem, one can compute the probability $Pr(h|e)$ as:

$$Pr(h|e) = \frac{Pr(e|h)}{Pr(e)} Pr(h), \tag{2}$$

which implies that:

$$\frac{Pr(h|e)}{Pr(h)} = \frac{Pr(e|h)}{Pr(e)}. \tag{3}$$

Thus, we have:

$$Pr(h|e) > Pr(h) \Longleftrightarrow \frac{Pr(h|e)}{Pr(h)} > 1 \Longleftrightarrow \frac{Pr(e|h)}{Pr(e)} > 1. \tag{4}$$

The probability $Pr(e)$ can be computed as:

$$Pr(e) = Pr(e|h)Pr(h) + Pr(e|\neg h)Pr(\neg h), \tag{5}$$

where $\neg h$ denotes the negation of hypothesis h. Accordingly, we have:

$$\begin{aligned} \frac{Pr(e|h)}{Pr(e)} > 1 &\Longleftrightarrow \frac{Pr(e|h)}{Pr(e|h)Pr(h) + Pr(e|\neg h)Pr(\neg h)} > 1 \\ &\Longleftrightarrow Pr(e|h) > Pr(e|h)Pr(h) + Pr(e|\neg h)Pr(\neg h) \\ &\Longleftrightarrow (1 - Pr(h))Pr(e|h) > Pr(e|\neg h)Pr(\neg h) \\ &\Longleftrightarrow \frac{Pr(e|h)}{Pr(e|\neg h)} > 1. \end{aligned} \tag{6}$$

To sum up, we have the following equivalent expressions of the condition for e confirming h in a qualitative approach:

$$Pr(h|e) > Pr(h) \Longleftrightarrow \frac{Pr(h|e)}{Pr(h)} > 1 \Longleftrightarrow \frac{Pr(e|h)}{Pr(e)} > 1 \Longleftrightarrow \frac{Pr(e|h)}{Pr(e|\neg h)} > 1. \tag{7}$$

One may similarly get equivalent expressions of the two conditions for e disconfirming h and e being neutral with respect to h.

Although the four conditions in Eq. (7) are mathematically equivalent, they provide very different semantics. The two conditions $Pr(h|e) > Pr(h)$ and

$\frac{Pr(h|e)}{Pr(h)} > 1$ compare the a posteriori probability $Pr(h|e)$ and the a priori probability $Pr(h)$. The former considers the difference between the two probabilities and the latter considers their ratio. The other two conditions focus on the likelihood of e regarding the hypothesis h. The condition $\frac{Pr(e|h)}{Pr(e)} > 1$ compares the likelihood of e given h (i.e., $Pr(e|h)$) and the likelihood of e without the given hypothesis h (i.e., $Pr(e)$). The condition $\frac{Pr(e|h)}{Pr(e|\neg h)} > 1$ compares the likelihood of e given h (i.e., $Pr(e|h)$) and the likelihood of e given the negation of h (i.e., $Pr(e|\neg h)$).

The quantitative Bayesian confirmation approach uses quantitative Bayesian confirmation measures to evaluate the degree to which a piece of evidence e confirms a hypothesis h. The equivalent expressions in Eq. (7) inspire the following quantitative confirmation measures:

$$\begin{aligned} c_d(e,h) &= Pr(h|e) - Pr(h), \\ c_r(e,h) &= \frac{Pr(h|e)}{Pr(h)} = \frac{Pr(e|h)}{Pr(e)}, \\ c_r^+(e,h) &= \frac{Pr(e|h)}{Pr(e|\neg h)}, \end{aligned} \tag{8}$$

which are called P-incremental confirmation measures [4]. By requiring additional properties, many confirmation measures have been proposed and studied in the literature, such as [4–6,9,10]:

$$\begin{aligned} c_{nr}(e,h) &= \frac{Pr(h|e)}{Pr(h)} - 1 = \frac{Pr(e|h)}{Pr(e)} - 1, \\ c_{nr}^+(e,h) &= \frac{Pr(e|h)}{Pr(e|\neg h)} - 1, \\ c_{lr} &= \log\frac{Pr(h|e)}{Pr(h)} = \log\frac{Pr(e|h)}{Pr(e)}, \\ c_{lr}^+ &= \log\frac{Pr(e|h)}{Pr(e|\neg h)}. \end{aligned} \tag{9}$$

3 Three-Way Classification of Evidence

The Bayesian confirmation approaches focus on evaluating how a single piece of evidence confirms a hypothesis. In real-world applications, we often have a set of pieces of evidence observed from a dataset and are interested in which part can be used to confirm or disconfirm a given hypothesis. Accordingly, we desire to divide the set into three parts or regions: a positive region of evidence confirming the hypothesis; a negative region of evidence disconfirming the hypothesis; and a boundary region of evidence that is neutral with respect to the hypothesis. This leads to a three-way classification [23] of evidence.

The formal definition of the three regions is straightforward in the qualitative Bayesian confirmation approach, which is given in the following definition.

Definition 1. *Given a set of evidence E and a hypothesis h, the qualitative positive* POS*, negative* NEG*, and boundary* BND *regions of E given h are defined as:*

$$\begin{aligned}\text{POS}(E,h) &= \{e \in E \mid Pr(h|e) > Pr(h)\},\\ \text{NEG}(E,h) &= \{e \in E \mid Pr(h|e) < Pr(h)\},\\ \text{BND}(E,h) &= \{e \in E \mid Pr(h|e) = Pr(h)\}.\end{aligned} \tag{10}$$

One may also equivalently formulate the three qualitative regions by using equivalent expressions as given in Eq. (7).

To construct the three regions based on a quantitative Bayesian confirmation approach, we may apply two thresholds on the quantitative values given by a confirmation measure.

Definition 2. *Given a set of evidence E, a hypothesis h, and a confirmation measure c, the quantitative positive* POS*, negative* NEG*, and boundary* BND *regions of E given h are defined as:*

$$\begin{aligned}\text{POS}_{(t,s)}(E,h) &= \{e \in E \mid c(e,h) > s\},\\ \text{NEG}_{(t,s)}(E,h) &= \{e \in E \mid c(e,h) < t\},\\ \text{BND}_{(t,s)}(E,h) &= \{e \in E \mid t \leq c(e,h) \leq s\},\end{aligned} \tag{11}$$

where t and s are two thresholds satisfying $t < s$.

The construction of three quantitative regions can be illustrated by Fig. 1. If a piece of evidence e confirms h to a degree greater than s, then e is in the positive region $\text{POS}_{(t,s)}(E,h)$ and we consider that e confirms h. If e confirms h to a degree less than t, e is in the negative region $\text{NEG}_{(t,s)}(E,h)$ and we consider that e disconfirms h. Otherwise, e is in the boundary region $\text{BND}_{(t,s)}(E,h)$ and we consider that e is neutral with respect to h, that is, e neither confirms nor disconfirms h. Equation (11) can also be applied to formulate the three qualitative regions by using the confirmation measure $c_d(e,h) = Pr(h|e) - Pr(h)$ and two thresholds $t = s = 0$. In this sense, it can be considered as a general formulation of the three regions in both qualitative and quantitative approaches, which will be used in our following discussions.

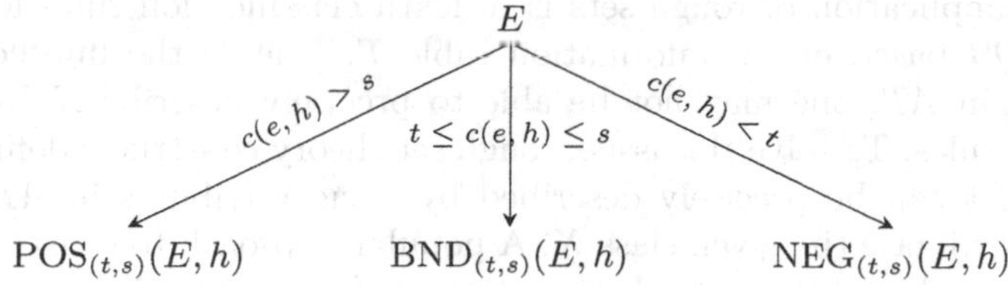

Fig. 1. Three-way classification of evidence based on quantitative Bayesian confirmation

It should be noted that a confirmation measure c actually evaluates both the degree to which e confirms h and the degree to which e disconfirms h. For example, the measure c_d may give both positive and negative values. A greater positive value indicates that a piece of evidence confirms h to a greater degree, and a less negative value indicates that a piece of evidence disconfirms h to a greater degree. Thus, it is meaningful to define the quantitative negative region by applying a threshold t. For example, if the measure c_d is used, one may choose a negative value as the threshold t. Accordingly, $c_d(e, h) < t$ means that the degree to which e disconfirms h is greater than a certain degree of disconfirmation represented by t. Thus, it is reasonable to use $c_d(e, h) < t$ as the condition for the negative region.

4 Three-Way Classification of Evidence in Rough Set Theory

Based on the general formulation of three-way classification of evidence presented in the last section, this section examines specific three-way classification of evidence in rough set theory. Specifically, we discuss two views of evidence in rough sets. The first view takes equivalence classes as evidence, which is adopted in existing confirmation theoretic rough set models [6–8,31]. We propose a second view that takes attributes as evidence. This view leads to a definition of class-specific attribute reduct based on Bayesian confirmation, which will be discussed in Sect. 5.

4.1 Equivalence Classes as Evidence

In rough set theory [17,18], a dataset is formally represented by an information table. There are two types of information tables studied in the literature, namely, complete and incomplete information tables. In this paper, we restrict our discussion to complete information tables. A complete table can be formally represented as the following tuple:

$$T = (OB, AT, \{V_a \mid a \in AT\}, \{I_a : OB \to V_a \mid a \in AT\}), \tag{12}$$

where OB is a set of objects as rows, AT is a set of attributes as columns, V_a is the domain of an attribute $a \in AT$, and I_a is an information function that maps each object to a unique value in V_a.

A major application of rough sets is to learn classification rules for a given class $X \subseteq OB$ based on an information table T. Due to the limited number of attributes in AT, one may not be able to precisely describe X by a set of classification rules. To solve this issue, rough set theory constructs definable sets of objects that can be precisely described by using attributes in AT and use them to approximate the given class X. A popular approach to constructing the definable sets is based on equivalence relations. Suppose $Q \subseteq OB \times OB$ is an equivalence relation (e.g., Q is reflexive, symmetric, and transitive) defined as:

$$Q = \{(x, y) \in OB \times OB \mid \forall a \in AT, I_a(x) = I_a(y)\}. \tag{13}$$

That is, $(x, y) \in Q$ if and only if x and y have the same values on all attributes in AT. Given an object $x \in OB$, its equivalence class $[x] = \{y \in OB \mid (x, y) \in Q\}$ is a definable set since it can be precisely described by a formula $\bigwedge_{a \in AT} a = I_a(x)$ where $\wedge$ denotes the logic AND operator. The family of equivalence classes $OB/Q = \{[x] \mid x \in OB\}$ is used to approximate a given class $X \subseteq OB$, that is, to construct the rough set approximations of X.

From the view of Bayesian confirmation, the fact that an object $o \in OB$ is included in an equivalence class $[x]$ (i.e., $o \in [x]$) can be considered as a piece of evidence. The statement that an object o is a positive instance of the given class X (i.e., $o \in X$) is considered as a hypothesis. For simplicity, we denote such a piece of evidence as $[x]$ and the hypothesis as X. Following Definition 2, one may divide the set of evidence OB/Q into three regions given the hypothesis X.

Definition 3. *The positive, negative, and boundary regions of OB/Q given a class $X \subseteq OB$ are defined as:*

$$\begin{aligned}
\mathrm{POS}_{(t,s)}(OB/Q, X) &= \bigcup\{[x] \in OB/Q \mid c([x], X) > s\},\\
\mathrm{NEG}_{(t,s)}(OB/Q, X) &= \bigcup\{[x] \in OB/Q \mid c([x], X) < t\},\\
\mathrm{BND}_{(t,s)}(OB/Q, X) &= \bigcup\{[x] \in OB/Q \mid t \leq c([x], X) \leq s\},
\end{aligned} \tag{14}$$

where c is a confirmation measure and $c([x], X)$ is the degree to which a piece of evidence $o \in [x]$ confirms the hypothesis $o \in X$.

The three regions in Eq. (14) form a three-way rough set approximation [24] of X. If a piece of evidence $[x]$ confirms X to a degree greater than s, then $[x]$ is a piece of positive evidence. In other words, for an object $o \in OB$, $o \in [x]$ confirms $o \in X$. Similarly, if $[x]$ confirms X to a degree less than t, then $[x]$ is a piece of negative evidence, that is, $o \in [x]$ disconfirms $o \in X$. Otherwise, $[x]$ is a piece of neutral evidence and cannot be used to confirm or disconfirm X. It should be noted that by taking unions in Eq. (14), the three regions are defined as sets of objects instead of sets of equivalence classes. This formulation is consistent with the formulations used in the mainstream of research in the literature, which is referred to as unstructured approximations. A few researchers [3,11,16] have studied structured approximations that are defined as sets of equivalence classes or other building blocks derived from various approaches.

Definition 3 provides a general formulation of certain quantitative rough set models. By taking c_d as the confirmation measure, one may immediately get the following three regions:

$$\begin{aligned}
\mathrm{POS}_{(t,s)}(OB/Q, X) &= \{[x] \in OB/Q \mid Pr(X|[x]) > s + Pr(X)\},\\
\mathrm{NEG}_{(t,s)}(OB/Q, X) &= \{[x] \in OB/Q \mid Pr(X|[x]) < t + Pr(X)\},\\
\mathrm{BND}_{(t,s)}(OB/Q, X) &= \{[x] \in OB/Q \mid t + Pr(X) \leq Pr(X|[x]) \leq s + Pr(X)\}.
\end{aligned} \tag{15}$$

where the probabilities can be estimated as:

$$Pr(X) = \frac{|X|}{|OB|},\ Pr(X|[x]) = \frac{|X \cap [x]|}{|[x]|}. \tag{16}$$

Since $Pr(X|[x]) - Pr(X) \in [-1, 1]$, it is reasonable to require the condition $-1 \leq t \leq 0 \leq s \leq 1$ for the two thresholds. By substitutions $\alpha = s + Pr(X)$ and $\beta = t + Pr(X)$, one may immediately get the well-known probabilistic rough set approximations [14,24]. These three regions provide a new interpretation of the two thresholds (α, β) used in probabilistic rough set approximations from the view of Bayesian confirmation. That is, the interval $[\beta, \alpha]$ represents an interval around the a priori probability $Pr(X)$ determined by a designated level of confirmation s and a designated level of disconfirmation t.

A few researchers have considered Bayesian confirmation in the context of rough sets by considering equivalence classes as evidence. For example, Greco, Matarazzo, and Słowiński [6–8] propose the parameterized rough set model by using both the a posteriori probability $Pr(X|[x])$ and a confirmation measure $c([x], X)$ in formulating the approximations. Yao and Zhou [26] consider the a posteriori probability $Pr(X|[x])$ and a confirmation measure $c([x], X)$ separately and study two Bayesian approaches to rough sets.

4.2 Attributes as Evidence

The majority of existing studies on rough sets and Bayesian confirmation takes a row-wise view, that is, they consider an equivalence class of objects as a piece of evidence. From the column-wise view, an attribute can also be considered as a piece of evidence. A confirmation measure evaluates the degree to which an attribute can be used to confirm or disconfirm a hypothesis. Accordingly, one can perform a three-way classification of attributes based on Bayesian confirmation.

In Eq. (13), we define the equivalence relation Q with respect to all the attributes in AT. In a similar manner, one may also define an equivalence relation with respect to an arbitrary subset $A \subseteq AT$:

$$Q_A = \{(x, y) \in OB \times OB \mid \forall a \in A, I_a(x) = I_a(y)\}. \tag{17}$$

By using the family of equivalence classes OB/Q_A, one may construct the three positive, negative, and boundary regions using the formulation given by Eq. (14) or any other existing three-way rough set models. Let $\text{POS}(OB/Q_A, X)$, $\text{NEG}(OB/Q_A, X)$, and $\text{BND}(OB/Q_A, X)$ denote the three regions of OB/Q_A constructed with respect to a given class $X \subseteq OB$. One may evaluate the performance of the set of attributes A in classifying instances of X by developing quantitative measures based on these three regions. From the view of Bayesian confirmation, we desire a quantitative confirmation measure that reflects both how A confirms X (i.e., how A classifies the positive instances of X) and how A disconfirms X (i.e., how A classifies the negative instances of X). These two sides correspond with the two regions $\text{POS}(OB/Q_A, X)$ and $\text{NEG}(OB/Q_A, X)$, respectively. Thus, such a quantitative confirmation measure $c(A, X)$ is desired to be an increasing function of the size of $\text{POS}(OB/Q_A, X)$ and a decreasing function of the size of $\text{NEG}(OB/Q_A, X)$, which can be formally represented as:

$$c(A, X) = f(|\text{POS}(OB/Q_A, X)|_{\uparrow}, |\text{NEG}(OB/Q_A, X)|_{\downarrow}), \tag{18}$$

where $|\cdot|$ denotes the cardinality of a set, and $\uparrow$ and $\downarrow$ denote the increasing and decreasing functions, respectively. For example, following similar ideas of c_d and c_r given in Eq. (8), one may define the following two measures that use difference and ratio, respectively:

$$c_d(A,X) = Pr(\text{POS}(OB/Q_A,X)) - Pr(\text{NEG}(OB/Q_A,X)),$$
$$c_r(A,X) = \frac{Pr(\text{POS}(OB/Q_A,X))}{Pr(\text{NEG}(OB/Q_A,X))}. \tag{19}$$

The probabilities can be estimated as follows:

$$Pr(\text{POS}(OB/Q_A,X)) = \frac{|\text{POS}(OB/Q_A,X)|}{|OB|},$$
$$Pr(\text{NEG}(OB/Q_A,X)) = \frac{|\text{NEG}(OB/Q_A,X)|}{|OB|}. \tag{20}$$

Following c_{nr} and c_{lr} given in Eq. (9), one may also consider the following two measures:

$$c_{nr}(A,X) = \frac{Pr(\text{POS}(OB/Q_A,X))}{Pr(\text{NEG}(OB/Q_A,X))} - 1,$$
$$c_{lr}(A,X) = \log\frac{Pr(\text{POS}(OB/Q_A,X))}{Pr(\text{NEG}(OB/Q_A,X))}. \tag{21}$$

By applying a specific measure $c(A,X)$ to Eq. (11), one may immediately construct the three regions of the set of evidence AT.

Definition 4. *The positive, negative, and boundary regions of AT given a class $X \subseteq OB$ are defined as:*

$$\text{POS}_{(t,s)}(AT,X) = \{a \in AT \mid c(a,X) > s\},$$
$$\text{NEG}_{(t,s)}(AT,X) = \{a \in AT \mid c(a,X) < t\},$$
$$\text{BND}_{(t,s)}(AT,X) = \{a \in AT \mid t \le c(a,X) \le s\}, \tag{22}$$

where, for simplicity, we use $c(a,X)$ to denote $c(\{a\},X)$.

If an attribute a confirms X to a degree greater than s, then a is a positive attribute with respect to X. In other words, the value of an object on a may help us confirm the object as a positive instance of X. Similarly, if a confirms X to a degree less than t, then a is a negative attribute with respect to X. That is, a may help us confirm an object as a negative instance of X, or equivalently, disconfirm an object as a positive instance of X. Otherwise, a is a neutral attribute with respect to X, which means the values on a may not be quite helpful in classifying instances of X. The selection and determination of thresholds t and s depend on the specific quantitative measures used.

5 Class-Specific Attribute Reduct Based on Bayesian Confirmation

A consideration of attributes as evidence in rough sets relates Bayesian confirmation to the topic of attribute reduction [12,15,18,20] in rough sets. Suppose we have the following sequence of subsets of attributes:

$$A_1 \subset A_2 \subset \cdots \subset A_n \subset AT. \tag{23}$$

By Eq. (17), one may easily verify that:

$$Q_{A_1} \supseteq Q_{A_2} \supseteq \cdots \supseteq Q_{A_n} \supseteq Q_{AT}, \tag{24}$$

or equivalently, for any $x \in OB$, we have:

$$[x]_{A_1} \supseteq [x]_{A_2} \supseteq \cdots \supseteq [x]_{A_n} \supseteq [x]_{AT}. \tag{25}$$

That is, a larger subset of attributes gives smaller equivalence classes as building blocks of rough set approximations. Consequently, the three regions of the corresponding families of equivalence classes satisfy the following properties:

$$\begin{aligned} &\mathrm{POS}(OB/Q_{A_1}, X) \subseteq \mathrm{POS}(OB/Q_{A_2}, X) \subseteq \cdots \subseteq \mathrm{POS}(OB/Q_{AT}, X),\\ &\mathrm{NEG}(OB/Q_{A_1}, X) \subseteq \mathrm{NEG}(OB/Q_{A_2}, X) \subseteq \cdots \subseteq \mathrm{NEG}(OB/Q_{AT}, X),\\ &\mathrm{BND}(OB/Q_{A_1}, X) \supseteq \mathrm{BND}(OB/Q_{A_2}, X) \supseteq \cdots \supseteq \mathrm{BND}(OB/Q_{AT}, X). \end{aligned} \tag{26}$$

In classifications, especially when there are multiple classes considered, confirming an object as a negative instance of a class might not be quite informative and useful. It provides very limited information about which class the object belongs to, with so many remaining classes as possibilities. In this sense, we usually focus more on classifying positive instances of a specific class. Thus, in rough sets, the performance of a subset of attributes A is usually measured based on the positive region $\mathrm{POS}(OB/Q_A, X)$. In this context, we may consider a special case of the confirmation measure $c(A, X)$ as:

$$c(A, X) = f(|\mathrm{POS}(OB/Q_A, X)|_{\uparrow}), \tag{27}$$

which is an increasing function of the size of $\mathrm{POS}(OB/Q_A, X)$. Consequently, we have:

$$c(A_1, X) \leq c(A_2, X) \leq \cdots \leq c(A_n, X) \leq c(AT, X). \tag{28}$$

It can be interpreted as: by considering more attributes, we can obtain more detailed information and confirm more positive instances of a given class X.

An intuitive question is whether it is sufficient to use a subset of AT instead of all the attributes in AT in classifying positive instances of a given class. This leads to the topic of class-specific attribute reduction in rough sets [12,15,20]. Qualitatively, such an attribute reduct is a minimal subset of AT that derives the same positive region as the set AT with respect to a given class X. Quantitatively, one may define quantitative measures to evaluate the performance of a set of attributes. Based on it, an attribute reduct can be defined as a minimal subset of AT that has the same performance as AT with respect to a given class. By using Bayesian confirmation measures to evaluate the performance of a set of attributes, we present the following definition of a class-specific attribute reduct.

Definition 5. *Given a class $X \subseteq OB$, a subset of attributes $R \subseteq AT$ is an attribute reduct with respect to X if it satisfies the following two conditions:*

$$\begin{aligned} &(1)\ c(R, X) = c(AT, X), \\ &(2)\ \forall R' \subset R, c(R', X) < c(AT, X). \end{aligned} \tag{29}$$

The first condition in Definition 5 states that R confirms X to the same degree as AT. The second condition states that any proper subset of R confirms X to a less degree than AT. Thus, R is a minimal set that has the same performance as AT. In the case that the measure c satisfies the property $c(A, X) \leq c(A', X)$ for $A \subseteq A' \subseteq AT$, the second condition can be equivalently expressed as:

$$(2')\ \forall a \in R, c(R - \{a\}, X) < c(AT, X), \tag{30}$$

which indicates that removing any attribute in R will decrease the degree to which R confirms X.

6 Conclusions and Future Work

Bayesian confirmation theory is closely related to three-way decision. We propose a general formulation of three-way classification of evidence by using qualitative and quantitative Bayesian confirmation approaches. This formulation is examined and applied with respect to rough set theory from two views. A first view considers equivalence classes as evidence, which leads to a three-way classification of objects based on quantitative Bayesian confirmation measures. This three-way classification model provides a new interpretation of the two thresholds used in probabilistic rough set models from the view of Bayesian confirmation. A second view considers attributes as evidence, which gives a three-way classification of attributes based on quantitative Bayesian confirmation measures. This view inspires a new definition of class-specific attribute reducts using Bayesian confirmation.

This work considers three-way classification of evidence with respect to only one given hypothesis. A first direction of future work is to consider a set of hypotheses and build the three-way classification model of evidence. Since one hypothesis relates to one given class in rough set theory, such a new model can be applied in rough set theory with respect to multiple classes, which is a second direction of future work.

Acknowledgement. The authors thank reviewers for their valuable comments and constructive suggestions.

References

1. Afridi, M.K., Azam, N., Yao, J.T., Alanazi, E.: A three-way clustering approach for handling missing data using GTRS. Int. J. Approx. Reason. **98**, 11–24 (2018). https://doi.org/10.1016/j.ijar.2018.04.001
2. Azam, N., Yao, J.T.: Game-theoretic rough sets for recommender systems. Knowl.-Based Syst. **72**, 96–107 (2014). https://doi.org/10.1016/j.knosys.2014.08.030
3. Bryniarski, E.: A calculus of rough sets of the first order. Bull. Pol. Acad. Sci. Math. **37**, 71–78 (1989)
4. Festa, R.: Bayesian confirmation. In: Galavotti, M.C., Pagnini, A. (eds.) Experience, Reality, and Scientific Explanation. WONS, vol. 61, pp. 55–87. Springer, Dordrecht (1999). https://doi.org/10.1007/978-94-015-9191-1_4
5. Fitelson, B.: Studies in Bayesian confirmation theory. Ph.D. dissertation, University of Wisconsin (2001). http://fitelson.org/thesis.pdf
6. Greco, S., Matarazzo, B., Słowiński, R.: Parameterized rough set model using rough membership and Bayesian confirmation measures. Int. J. Approx. Reason. **49**, 285–300 (2008). https://doi.org/10.1016/j.ijar.2007.05.018
7. Greco, S., Matarazzo, B., Słowiński, R.: Rough membership and Bayesian confirmation measures for parameterized rough sets. In: Ślęzak, D., Wang, G., Szczuka, M., Düntsch, I., Yao, Y.Y. (eds.) RSFDGrC 2005. LNCS (LNAI), vol. 3641, pp. 314–324. Springer, Heidelberg (2005). https://doi.org/10.1007/11548669_33
8. Greco, S., Pawlak, Z., Słowiński, R.: Can Bayesian confirmation measures be useful for rough set decision rules? Eng. Appl. Artif. Intell. **17**, 345–361 (2004). https://doi.org/10.1016/j.engappai.2004.04.008
9. Greco, S., Słowiński, R., Szczęch, I.: Measures of rule interestingness in various perspectives of confirmation. Inf. Sci. **346–347**, 216–235 (2016). https://doi.org/10.1016/j.ins.2016.01.056
10. Greco, S., Słowiński, R., Szczęch, I.: Finding meaningful Bayesian confirmation measures. Fundam. Inf. **127**, 161–176 (2013). https://doi.org/10.3233/FI-2013-902
11. Hu, M.J., Yao, Y.Y.: Structured approximations as a basis for three-way decisions with rough sets. Knowl.-Based Syst. **165**, 92–109 (2019). https://doi.org/10.1016/j.knosys.2018.11.022
12. Jia, X.Y., Shang, L., Zhou, B., Yao, Y.Y.: Generalized attribute reduct in rough set theory. Knowl.-Based Syst. **91**, 204–218 (2016). https://doi.org/10.1016/j.knosys.2015.05.017
13. Li, H.X., Zhang, L.B., Huang, B., Zhou, X.Z.: Sequential three-way decision and granulation for cost-sensitive face recognition. Knowl.-Based Syst. **91**, 241–251 (2016). https://doi.org/10.1016/j.knosys.2015.07.040
14. Ma, W.M., Sun, B.Z.: Probabilistic rough set over two universes and rough entropy. Int. J. Approx. Reason. **53**, 608–619 (2012). https://doi.org/10.1016/j.ijar.2011.12.010
15. Ma, X., Yao, Y.Y.: Three-way decision perspectives on class-specific attribute reducts. Inf. Sci. **450**, 227–245 (2018). https://doi.org/10.1016/j.ins.2018.03.049
16. Ma, J.M., Zou, C.J., Pan, X.C.: Structured probabilistic rough set approximations. Int. J. Approx. Reason. **90**, 319–332 (2017). https://doi.org/10.1016/j.ijar.2017.08.004
17. Pawlak, Z.: Rough Sets: Theoretical Aspects of Reasoning about Data. Kluwer Academic Publishers, Boston (1991)
18. Pawlak, Z.: Rough sets. Int. J. Comput. Inf. Sci. **11**, 341–356 (1982)

19. Qi, J.J., Qian, T., Wei, L.: The connections between three-way and classical concept analysis. Knowl.-Based Syst. **91**, 143–151 (2016). https://doi.org/10.1016/j.knosys.2015.08.006
20. Qian, Y.H., Liang, J.Y., Pedrycz, W., Dang, C.: Positive approximation: An accelerator for attribute reduction in rough set theory. Artif. Intell. **174**, 597–618 (2010). https://doi.org/10.1016/j.artint.2010.04.018
21. Ren, R.S., Wei, L.: The attribute reductions of three-way concept lattices. Knowl.-Based Syst. **99**, 92–102 (2016). https://doi.org/10.1016/j.knosys.2016.01.045
22. Sun, B.Z., Ma, W.M., Xiao, X.: Three-way group decision making based on multigranulation fuzzy decision-theoretic rough set over two universes. Int. J. Approx. Reason. **81**, 87–102 (2017). https://doi.org/10.1016/j.ijar.2016.11.001
23. Yao, Y.Y.: Three-way decision and granular computing. Int. J. Approx. Reason. **103**, 107–123 (2018). https://doi.org/10.1016/j.ijar.2018.09.005
24. Yao, Y.Y.: Probabilistic rough set approximations. Int. J. Approx. Reason. **49**, 255–271 (2008). https://doi.org/10.1016/j.ijar.2007.05.019
25. Yao, Y.Y., Hu, M.J., Deng, X.F.: Modes of sequential three-way classifications. In: Medina, J., Ojeda-Aciego, M., Verdegay, J.L., Pelta, D.A., Cabrera, I.P., Bouchon-Meunier, B., Yager, R.R. (eds.) IPMU 2018. CCIS, vol. 854, pp. 724–735. Springer, Cham (2018). https://doi.org/10.1007/978-3-319-91476-3_59
26. Yao, Y.Y., Zhou, B.: Two Bayesian approaches to rough sets. Eur. J. Oper. Res. **251**, 904–917 (2016). https://doi.org/10.1016/j.ejor.2015.08.053
27. Yu, H.: A framework of three-way cluster analysis. In: Polkowski, L., Yao, Y.Y., Artiemjew, P., Ciucci, D., Liu, D., Ślęzak, D., Zielosko, B. (eds.) IJCRS 2017. LNCS (LNAI), vol. 10314, pp. 300–312. Springer, Cham (2017). https://doi.org/10.1007/978-3-319-60840-2_22
28. Yu, H., Zhang, C., Wang, G.Y.: A tree-based incremental overlapping clustering method using the three-way decision theory. Knowl.-Based Syst. **91**, 189–203 (2016). https://doi.org/10.1016/j.knosys.2015.05.028
29. Zhang, H.R., Min, F.: Three-way recommender systems based on random forests. Knowl.-Based Syst. **91**, 275–286 (2016). https://doi.org/10.1016/j.knosys.2015.06.019
30. Zhou, B.: A cost-sensitive approach to ternary classification. Ph.D. dissertation, University of Regina (2012)
31. Zhou, B., Yao, Y.Y.: Comparison of two models of probabilistic rough sets. In: Lingras, P., Wolski, M., Cornelis, C., Mitra, S., Wasilewski, P. (eds.) RSKT 2013. LNCS (LNAI), vol. 8171, pp. 121–132. Springer, Heidelberg (2013). https://doi.org/10.1007/978-3-642-41299-8_12

Concept Approximation Based on Rough Sets and Judgment

Jaroslaw Stepaniuk[1], Grzegorz Góra[2], and Andrzej Skowron[3,4(✉)]

[1] Faculty of Computer Science, Bialystok University of Technology, Wiejska 45A, 15-351 Bialystok, Poland
j.stepaniuk@pb.edu.pl

[2] Faculty of Mathematics, Informatics and Mechanics, University of Warsaw, Banacha 2, 02-097 Warsaw, Poland
ggora@mimuw.edu.pl

[3] Systems Research Institute, Polish Academy of Sciences, Newelska 6, 01-447 Warsaw, Poland
skowron@mimuw.edu.pl

[4] Digital Science and Technology Centre, UKSW, Dewajtis 5, 01-815 Warsaw, Poland

Abstract. We discuss an approach of concept approximation based on judgment rather than on partial containment of sets only. This approach seems to be much more general than the traditional one. However, it requires developing some new logical tools for reasoning based on judgment, which is often expressed in natural language.

Keywords: Rough set · Approximation · Judgment

1 Introduction

In this paper, we discuss an approach of concept approximation based on a kind of reasoning, called judgment, rather than on partial inclusion of sets. The former approach seems to be much more general than the latter one. The approach based on judgment is especially relevant in data analysis, where it is required in order to have a deeper judgment about the perceived complex situation related to classification of complex vague concepts. First, we present a short introduction to the approximation of concepts used in the rough set approach. This presentation is based on the definition from [1]. Other existing approaches to concept approximation [2,3] are based on partial containment of sets. We claim that this approach is not satisfactory for dealing with many real-life applications, where more advanced judgment should be made to identify the perceived situation and classify it relative to the complex vague concepts. So, we present an introductory discussion on the need of new logical tools for reasoning toward approximation of complex vague concepts.

We illustrate this using the case of classification of imbalanced data [4–6]. For example (see, *e.g.*, [4]), in neighborhoods types of objects from the minority class

© Springer Nature Switzerland AG 2019
T. Mihálydeák et al. (Eds.): IJCRS 2019, LNAI 11499, pp. 16–27, 2019.
https://doi.org/10.1007/978-3-030-22815-6_2

such as *safe* and *unsafe* (among them *borderline*, *rare examples*, and *outliers*) are considered. In particular, it is important to distinguish the outliers from the noise. In [4] it is emphasized that the results of the noise identification by filters are often identified by medical experts as valid outliers. Hence, it is visible that to provide a decision support system some more advanced reasoning tools, which we call judgment are required. These reasoning tools should help the system to judge properly about such perceived cases. It should be noted that this judgment should be supported by the relevant information about perceived cases extracted from knowledge bases representing experience. As the result of perception of the perceived cases the system should be able to 'derive' a text, in a fragment of natural language, that can be next used in further judgment about these cases [7]. The system should generate relevant model for performing judgment using the derived text [8].

Here, it is worthwhile mentioning, in more detail, two views from [7] and [8]. The first one is by Zadeh, the founder of fuzzy sets and the computing with words paradigm (see also http://www.cs.berkeley.edu/~zadeh/presentations.html):

> *Manipulation of perceptions plays a key role in human recognition, decision and execution processes. As a methodology, computing with words provides a foundation for a computational theory of perceptions - a theory which may have an important bearing on how humans make- and machines might make - perception-based rational decisions in an environment of imprecision, uncertainty and partial truth. [...] computing with words, or CW for short, is a methodology in which the objects of computation are words and propositions drawn from a natural language.*

Another view is by Pearl (the 2011 winner of the ACM Turing Award, "for fundamental contributions to artificial intelligence through the development of a calculus for probabilistic and causal reasoning") [8]:

> *Traditional statistics is strong in devising ways of describing data and inferring distributional parameters from sample. Causal inference requires two additional ingredients: a science-friendly language for articulating causal knowledge, and a mathematical machinery for processing that knowledge, combining it with data and drawing new causal conclusions about a phenomenon.*

In the judgment process, the arguments *for* and *against* the hypothesis about membership of the perceived case to a given concept are collected. In this way, the results of judgment clearly indicating that a given case belongs to one of the regions viz., lower approximation, boundary region, or complement to the upper approximation are obtained.

The question arises about the logic, which is relevant for the above mentioned tasks [9]. First let us observe that the satisfiability relations in the granular framework can be treated as tools for constructing new information granules. In fact, for a given satisfiability relation, the semantics of formulas relative to this relation is defined. In this way, the candidates for new relevant information granules are obtained. We would like to emphasize that the relevant satisfiability relation for the considered problems is not given but it should be induced (discovered) on the basis of a partial information encoded in the information (decision) systems. For real-life problems, it is often necessary to discover a hierarchy of satisfiability relations before the relevant target level is reached. Information granules, constructed at different levels of this hierarchy, finally lead to the relevant ones for approximating the concerned complex vague concepts expressed in natural language (see Fig. 1). The reasoning should also concern about how to derive relevant information granules for solving the target tasks. This kind of reasoning is called adaptive judgment. Deduction, induction, abduction as well as analogy based reasoning all are involved in adaptive judgment. Among the different aspects, the following ones are a few which one needs to address in order to do reasoning with adaptive judgment.

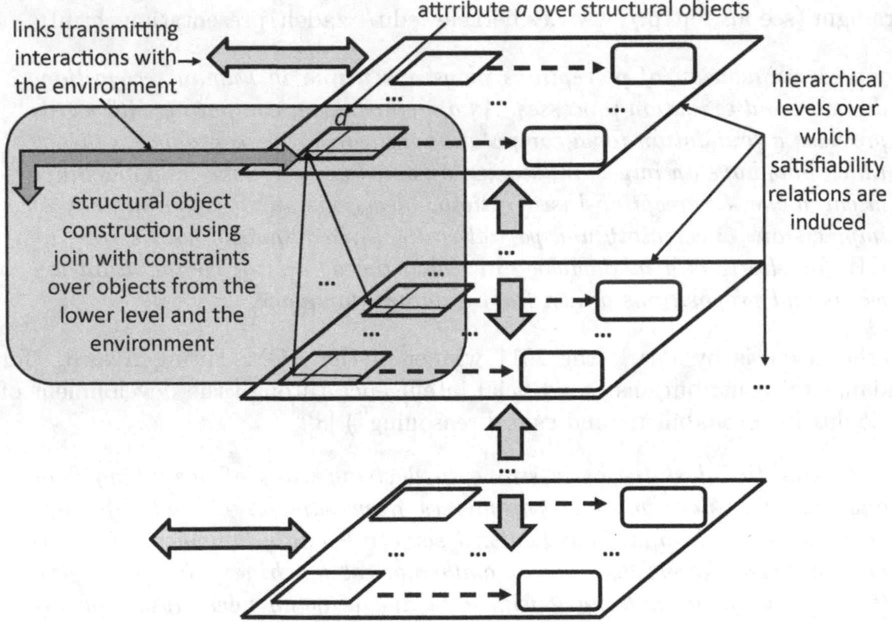

Fig. 1. Interactive hierarchical structures (gray arrows show interactions between hierarchical levels and the environment, arrows at hierarchical levels point from information (decision) systems representing partial specifications of satisfiability relations to theories, induced from them, which consist of rule sets)

- searching for relevant approximation spaces,
- discovery of new features,
- selection of relevant features,
- rule induction,

- discovery of inclusion measures,
- strategies for conflict resolution,
- adaptation of measures based on the minimum description length principle,
- reasoning about changes,
- perception (action and sensory) attributes selection,
- adaptation of quality measures over computations relative to agents,
- adaptation of object structures,
- discovery of relevant contexts,
- strategies for knowledge representation and interaction with knowledge bases,
- ontology acquisition and approximation,
- learning in dialogue of inclusion measures between information granules from different languages (*e.g.*, the formal language of the system and the user natural language),
- strategies for adaptation of existing models,
- strategies for development and evolution of communication language among agents in distributed environments,
- strategies for risk management in distributed computational systems.

One should note that judgment is not only based on deduction, induction or abduction. It has roots not only in logic, but also in psychology and phenomenology [10] (see Fig. 2).

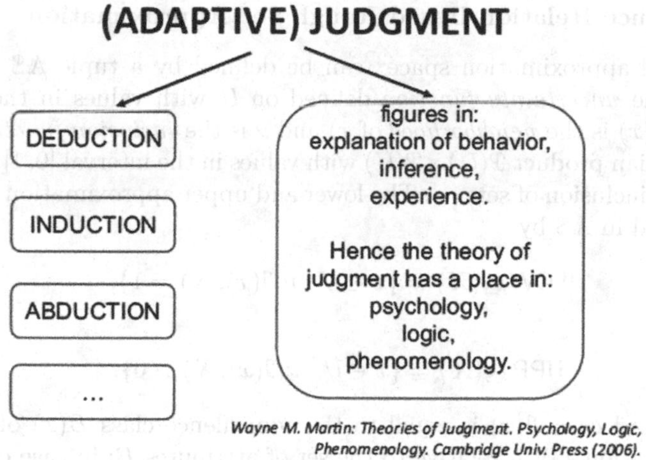

Fig. 2. Judgment has its roots in psychology and phenomenology

Our approach is consistent with the opinion of Valiant[1]:

[1] The 2011 winner of the ACM Turing Award, the highest distinction in computer science, "for his fundamental contributions to the development of computational learning theory and to the broader theory of computer science" (http://people.seas.harvard.edu/~valiant/researchinterests.htm).

A fundamental question for artificial intelligence is to characterize the computational building blocks that are necessary for cognition. A specific challenge is to build on the success of machine learning so as to cover broader issues in intelligence. [...] This requires, in particular a reconciliation between two contradictory characteristics – the apparent logical nature of reasoning and the statistical nature of learning.

The paper is structured as follows. In Sect. 2 two illustrative examples related to judgment in decision making are included. In Subsect. 2.1, we recall the rough set approach towards concept approximation in the context of tolerance relation. In Subsect. 2.2, we discuss some aspects of judgment necessary for the deeper reasoning about imbalanced data. In Sect. 3, we discuss computations based on judgment.

2 Judgment in Decision Making

In the following two sections, we illustrate two judgment strategies related to classification. The first one is related to judgment based on partial inclusion of sets, widely used in rule-based classifiers. The aim of the second one, related to imbalanced data, is to illustrate the need for more advanced judgment strategies in decision making.

2.1 Tolerance Relation Based Rough Set Approximation

A generalized approximation space[2] can be defined by a tuple $\mathbb{AS} = (U, \mathcal{I}, \nu)$ where $\mathcal{I}$ is the *uncertainty function* defined on U with values in the powerset $\mathcal{P}(U)$ of U ($\mathcal{I}(x)$ is the *neighborhood* of x) and ν is the *inclusion function* defined on the Cartesian product $\mathcal{P}(U) \times \mathcal{P}(U)$ with values in the interval $[0, 1]$ measuring the degree of inclusion of sets [1]. The lower and upper approximation operations can be defined in $\mathbb{AS}$ by

$$\mathsf{LOW}_{\mathbb{AS}}(X) = \{x \in U : \nu(\mathcal{I}(x), X) = 1\}, \tag{1}$$

$$\mathsf{UPP}_{\mathbb{AS}}(X) = \{x \in U : \nu(\mathcal{I}(x), X) > 0\}. \tag{2}$$

In the standard case, $\mathcal{I}(x)$ is equal to the equivalence class $B(x)$ of the indiscernibility relation IND_B relative to the set of attributes B; in case of tolerance (similarity) relation $\mathcal{T} \subseteq U \times U$ we take $\mathcal{I}(x) = [x]_{\mathcal{T}} = \{y \in U : x \; \mathcal{T} \; y\}$, *i.e.*, $\mathcal{I}(x)$ is equal to the tolerance class of x defined by $\mathcal{T}$. For $X, Y \subseteq U$ the standard rough inclusion relation ν_{SRI} is defined by

$$\nu_{SRI}(X, Y) = \begin{cases} \dfrac{card(X \cap Y)}{card(X)}, & \text{if } X \text{ is non} - \text{empty}, \\ 1, & \text{otherwise}. \end{cases} \tag{3}$$

[2] More general cases are considered, *e.g.*, in articles [11,12].

For applications it is important to have some constructive definitions of $\mathfrak{I}$ and ν.

The recalled above definition is a formalisation of a simple judgment strategy based on set containment. This judgment can be described in natural language as follows. Due to uncertainty, it is not possible to perceive objects exactly. The objects are perceived using information about them represented by vectors of attribute values. These vectors define some neighborhoods of objects (indiscernibility or tolerance classes). In making decision we use these neighborhoods of objects to judge membership of a perceived object into a concept. The result of judgment is based on the degree to which the neighborhoods are included into the concepts. The above formulas present the result of modeling this judgment in the mathematical language (see also the above citations from [7,8]).

Example 1. Let d be a decision attribute with $\{+,-\}$ as the set of values. For two decision classes $X_{d=+} = \{x \in U: d(x) = +\}$, $X_{d=-} = \{x \in U: d(x) = -\}$, we can label objects $x \in U$ as in Table 1.

Table 1. The judgment about which label to assign to the object.

Label(x)	*argument "for"* $\nu(\mathfrak{I}(x), X_{d=+})$	*argument "against"* $\nu(\mathfrak{I}(x), X_{d=-})$
LOWER	$=1$	$\neq 1$
BOUNDARY	$\in (0,1)$	$\notin (0,1)$
NEGATIVE	$=0$	$\neq 0$

An analogous judgment strategy, enhanced by conflict resolution, is also widely used in inducing classifiers from decision tables (training samples). In the next section, we present an illustrative simple example to emphasize the necessity for developing of more advanced judgment strategies for classifying objects.

2.2 Judgment in Classification of Imbalanced Data

First let us recall a strategy of labelling of objects from the minority class introduced in [13].

The idea of formation of information granule facilitates splitting the problem into more feasible subtasks. Then, they can be easily managed by applying appropriate approaches, dedicated to specific types of entities. After defining groups of similar instances $NN_k(x)$ (namely minority class instances $x \in X_{d=+}$ and their k nearest neighbors $NN_k(x)$), the inclusion degree of each information granule $NN_k(x)$ in $X_{d=+}$ is examined. Based on this analysis, *Label*(x), the labels are assigned to all positive examples $x \in X_{d=+}$.

We assume that evaluation of information granules is crucial for further processing. Before applying oversampling mechanism, each information granule, defined by $NN_k(x)$ having positive instance x as the anchor point, is labelled with one of the following etiquettes: $SAFE$, $BOUNDARY$ and $NOISE$. The category of the individual entity is determined by the inclusion degree of $NN_k(x)$ in the information granule $X_{d=+}$ (the whole minority class). Details of the proposed technique are presented in Definitions 1, 2 and 3.

Definition 1. *Etiquette* $Label(x) = SAFE$ *for* $x \in X_{d=+}$
High inclusion degree indicates that the information granule $NN_k(x)$ *is placed in a homogeneous area and therefore* x *can be considered as* $SAFE$*. The inclusion level is obtained by the analysis of granule characteristics, especially cardinalities of instances from both classes. The number of positive class representatives belonging to the analysed entity (except the anchor example),* i.e., $card(NN_k(x) \cap X_{d=+})$ *is compared to the number of negative class instances,* i.e., $card(NN_k(x) \cap X_{d=-})$*. More than a half of minority class instances belonging to the positive class implies that* x *should be labelled as* $SAFE$ *(see Table 2 for* $k = 5$*).*

Definition 2. *Etiquette* $Label(x) = BOUNDARY$ *for* $x \in X_{d=+}$
Low inclusion degree is determined by the large representation of majority class $X_{d=-}$ *in the information granule* $NN_k(x)$*. When half or more than a half of instances belong to the negative class, the label* $BOUNDARY$ *is chosen for* x*. These kind of entities are placed in the area surrounding class boundaries, where examples from both classes overlap (see Table 2 for* $k = 5$*).*

Definition 3. *Etiquette* $Label(x) = NOISE$ *for* $x \in X_{d=+}$
Noninclusion of the information granule $NN_k(x)$ *in the minority class* $X_{d=+}$ *is identified with the situation when no instances belong to the minority class (except the anchor instance),* i.e., $NN_k(x) \cap X_{d=+} = \emptyset$*. Since only one of the analysed instances is the positive example (namely the core instance* x*), it means that the information granule is created around the rare individual placed in the area occupied by the representatives of the negative class* $X_{d=-}$*. This case is considered as* $NOISE$ *(see Table 2 for* $k = 5$*).*

Table 2. Identification of the type of the minority class instance x in the case of $k = 5$ nearest neighbours.

$Label(x)$	$card(NN_5(x) \cap X_{d=+})$	$card(NN_5(x) \cap X_{d=-})$
SAFE	5	0
	4	1
	3	2
BOUNDARY	2	3
	1	4
NOISE	0	5

The example of labelling instances from the minority class is presented in the Table 2. It shows all possible cases of the type of the minority class instance. Assuming that the parameter k is equal to 5, the second column presents the number of nearest neighbours belonging to the same class as the instance under consideration and the third column shows the number of nearest neighbours representing the opposite class.

Example 2. Let $x \in X_{d=+}$ be an instance from the minority class, *i.e.*, the decision $d(x) = +$. Let x_1, x_2, x_3, x_4, x_5 be five nearest neighbors of x. Let us assume that $d(x_1) = -, d(x_2) = +, d(x_3) = +, d(x_4) = -$ and $d(x_5) = -$. We obtain $card(NN_5(x) \cap X_{d=+}) = card(\{x_1, x_2, x_3, x_4, x_5\} \cap X_{d=+}) = card(\{x_2, x_3\}) = 2$ and $card(NN_5(x) \cap X_{d=-}) = card(\{x_1, x_2, x_3, x_4, x_5\} \cap X_{d=-}) = card(\{x_1, x_4, x_5\}) = 3$. Hence, we conclude that the correct label $Label(x) = BOUNDARY$.

After categorizing information granules (instances from minority class $X_{d=+}$), the mode of algorithm for oversampling is obtained. Three methods are proposed in [13] to deal with various real–life data characteristics. They mainly depend on the number of information granules labelled as BOUNDARY. Assuming that a certain threshold value is one of the parameters of the algorithm, the complexity of the problem is defined based on this value and the number of granules recognized as BOUNDARY. The threshold indicates how many instances of the entire minority class should be placed in boundary regions to treat the problem as a complex one.

Having less *BOUNDARY* entities, *i.e.*,

$$\frac{card(\{x \in X_{d=+} : Label(x) = BOUNDARY\})}{card(X_{d=+})} < complexity\_threshold$$

means that the problem is not complex and the following method of creating new instances can be applied:

Definition 4. [13] *LowComplexity mode for obtaining of the decision table* $DT_{balanced}$ *from the table* DT: $DT \longmapsto^{LowComplexity} DT_{balanced}$

- $Label(x) = SAFE$: *there is no need to significantly increase the number of instances in these safe areas. Only one new instance per existing minority SAFE instance is generated. Numeric attributes are handled by the interpolation with one of the k nearest neighbours. For the nominal features, new sample has the same values of attributes as the instance under consideration.*
- $Label(x) = BOUNDARY$: *the most of synthetic samples are generated in these borderline areas, since numerous majority class representatives may have greater impact on the classifier learning, when there are not enough minority examples. Hence, many new examples are created closer to the instance x under consideration. One of the k nearest neighbours is chosen for each new sample when determining the value of numeric feature. Values of nominal attributes are obtained by the majority vote of k nearest neighbours' features.*

– *Label*(x) = *NOISE: no new samples are created.*

On the other hand, prevalence of *BOUNDARY* information granules, *i.e.*,

$$\frac{card(\{x \in X_{d=+} : Label(x) = BOUNDARY\})}{card(X_{d=+})} \geq complexity\_threshold$$

involves more complications during the learning process. Therefore, dedicated approach (described below) is chosen:

Definition 5. [13] *HighComplexity mode for obtaining* $DT_{balanced}$ *table from DT table:* $DT \longmapsto^{HighComplexity} DT_{balanced}$

– *Label*(x) = *SAFE: assuming that these concentrated instances provide specific and easy to learn patterns that enable proper recognition of minority samples, plenty of new data is created by interpolation between SAFE instance and one of its k nearest neighbours. Nominal attributes are determined by majority vote of k nearest neighbours' features.*
– *Label*(x) = *BOUNDARY: the number of instances is doubled by creating one new example along the line segment between half of the distance from BOUNDARY instance and one of its k nearest neighbours. For nominal attributes values describing the instance under consideration are replicated.*
– *Label*(x) = *NOISE: new examples are not created.*

The last option is the special case, when any information granule is recognized as SAFE, *i.e.*, $\{x \in X_{d=+} : Label(x) = SAFE\} = \emptyset$. Hence, the following method is applied:

Definition 6. [13] *noSAFE mode:* $DT \longmapsto^{noSAFE} DT_{balanced}$

– *Label*(x) = *BOUNDARY: all of the synthetic instances are created in the area surrounding class boundaries. This particular solution is selected in case of especially complex data distribution, which does not include any SAFE samples. Missing SAFE elements indicate that most of the examples are labelled as BOUNDARY (there are no homogeneous regions). Since only BOUNDARY and NOISE examples are available, only generating new instances in the neighbourhood of BOUNDARY instances would provide sufficient number of minority samples.*
– *Label*(x) = *NOISE: no new instances are created.*

NOISE granules are completely excluded from the preprocessing phase, since their anchor instances are erroneous examples or outliers. Therefore, they should not be removed, but they also should not be taken into consideration when creating new synthetic instances to avoid more inconsistencies.

Looking on the above formalisation of labelling of objects from the minority class, one can find possible judgment strategy which can be easily expressed in a fragment of natural language. However, in real-life applications, the situation may be much more complex. For example, one can ask about the risk of making decision based on such modeling. For example, labels SAFE or NOISE are related to complex vague concepts. In the real-life medical applications, the judgment leading to decision about labelling by SAFE and NOISE may require interactions with domain knowledge, which can be the experience or recent discoveries in medicine reported in the literature, or can be additional testing of objects or performing some other actions on them. Some advanced judgments strategies should be used to obtain relevant information for making proper decision.

3 Computations Based on Judgment

From the above discussion it follows that judgment is performed on the basis of information (more precisely on complex granules (see, *e.g.*, [14]) representing the current result of perception of the situation. On the basis of this result, the judgment resulting in selection of the most relevant action(s) (with respect to the current goals) in the perceived situation is performed. By performing actions the complex granule about the current situation is updated and again the judgment leads to the selection of the next action(s). In this way computations on complex granules are controlled by actions aiming to preserve the required constraints on the computation trajectory. For example, the constraints related to classification tasks may be seen as reaching conclusion about the membership of the perceived situation in the considered concept(s) with the additional constraints related, *e.g.*, to time or/and space complexity of the derivation process. It should be noted that this process of transitions from the current granule to the next one is based on judgment.

One of the main aim of the judgment performed by complex granules (e.g., agents) is to derive conclusions for selection of action(s) which should be currently initiated (or terminated). The actions are activated on the basis of satisfiability of some complex vague concepts labelled by actions. It should be noted that these concepts are drifting with time. Adaptive learning of such concepts based on judgment is a great challenge. The whole process towards inducing approximation of these vague concepts labelled by actions, which are initiated on the basis of satisfiability of these concepts, may be treated as a process of discovery of complex game (see Fig. 3). In such a game the concepts (together with assigned relevant judgment mechanisms to them) can be treated as players who by using judgment are deriving arguments *for* and *against* the satisfiability of these concepts on the basis of information about the perceived situation. Next, comes other judgment mechanisms which are used to resolve conflicts among the collected arguments to select the winning player, (concept) and thus the action labelling the winning concept is initiated.

For simplicity of reasoning let us now concentrate on the case of classification with one concept and its complement, and let us consider the issue of *concept approximation* based on judgment. Any perceived situation can be classified on the basis of the complex game between the concept and its complement leading, by using judgment, to arguments *for* and *against* the satisfiability of these concepts on the basis of information about the perceived situation. The judge of the game assigns the perceived situation to (i) the lower approximation of the concept in case the result of her/his judgment points to the concept, (ii) the lower approximation of the concept's complement in case the result of her/his judgment points to the concept complement, (iii) the boundary region of the concept in case she/he is not able to discern between the arguments *for* and *against*. The reader can generalize this definition to the case of multiclass classification. In this way it is possible to generalize the approximation of concepts in the case of approximation based on judgment.

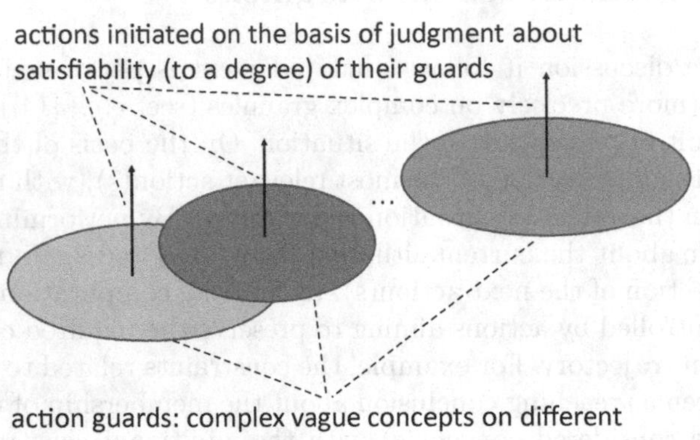

Fig. 3. Actions initiated on the basis of judgement

One can consider judgment as a binary relation over perceived information (or more precisely, complex granules) for deriving conclusions about the current situation on the basis of existing information. One should note that this relation is evolving with time, and new relations should be adaptively learned from the data to guarantee derivation of the relevant conclusions about the perceived situations influenced often by unpredictable interactions caused by the environment. Hence, adaptive learning strategies for modeling judgment corresponding to such relations should be developed. In modeling such relations, usually on the basis of the perceived situation, one can model the next expected situation (after performing the selected action). However, this should be compared with the real perception of the situation after performing the action. The 'difference' can be used for modifying models of the current complex game and/or the current judgment relation.

4 Conclusions

We discussed the approach to concept approximation based on judgment. Two illustrative examples are included. The approach is pointing out to the necessity of developing a new approach of reasoning based on judgment (often represented in natural language) together with the methods for modeling of judgment in intelligent systems. The paper realizes the first step toward this goal. In the future we plan to investigate in more detail the foundations of machine learning based on judgment.

Acknowledgments. The work of Jaroslaw Stepaniuk was supported by the grant S/WI/1/2018 from Bialystok University of Technology and funded with resources for research by the Ministry of Science and Higher Education in Poland. The research of Andrzej Skowron was partially supported by the NCBiR grant POIR.01.02.00-00-0184/17-01.

References

1. Skowron, A., Stepaniuk, J.: Tolerance approximation spaces. Fundam. Inform. **27**(2–3), 245–253 (1996). https://doi.org/10.3233/FI-1996-272311
2. Skowron, A., Jankowski, A., Swiniarski, R.W.: Foundations of rough sets, pp. 331–348. [3]
3. Kacprzyk, J., Pedrycz, W. (eds.): Springer Handbook of Computational Intelligence. Springer, Heidelberg (2015). https://doi.org/10.1007/978-3-662-43505-2
4. Napierala, K., Stefanowski, J.: Types of minority class examples and their influence on learning classifiers from imbalanced data. J. Intell. Inf. Syst. **46**, 563–597 (2016). https://doi.org/10.1007/s10844-015-0368-1
5. Paula Branco, L.T., Ribeiro, R.P.: A survey of predictive modeling on imbalanced domains. ACM Comput. Surv. **49**, 1–50 (2016). https://doi.org/10.1145/2907070
6. He, H., Ma, Y. (eds.): Imbalanced Learning: Foundations, Algorithms, and Applications. Wiley, Hoboken (2013)
7. Zadeh, L.A.: From computing with numbers to computing with words - from manipulation of measurements to manipulation of perceptions. IEEE Trans. Circ. Syst. **45**, 105–119 (1999)
8. Pearl, J.: Causal inference in statistics: an overview. Stat. Surv. **3**, 96–146 (2009). https://doi.org/10.1214/09-SS057
9. Skowron, A., Dutta, S.: Rough sets: past, present, and future. Nat. Comput. **17**, 855–876 (2018). https://doi.org/10.1007/s11047-018-9700-3
10. Martin, W.M. (ed.): Theories of Judgment. Psychology, Logic, Phenomenology. Cambridge University Press, New York (2006). https://doi.org/10.1017/CBO9780511487613
11. Skowron, A., Stepaniuk, J.: Approximation spaces in rough-granular computing. Fundam. Inform. **100**, 141–157 (2010). https://doi.org/10.3233/FI-2010-267
12. Skowron, A., Stepaniuk, J., Swiniarski, R.: Modeling rough granular computing based on approximation spaces. Inf. Sci. **184**, 20–43 (2012). https://doi.org/10.1016/j.ins.2011.08.001
13. Borowska, K., Stepaniuk, J.: Granular computing and parameters tuning in imbalanced data preprocessing. In: Saeed, K., Homenda, W. (eds.) CISIM 2018. LNCS, vol. 11127, pp. 233–245. Springer, Cham (2018). https://doi.org/10.1007/978-3-319-99954-8_20
14. Skowron, A., Jankowski, A.: Interactive computations: toward risk management in interactive intelligent systems. Nat. Comput. **15**(3), 465–476 (2016). https://doi.org/10.1007/s11047-015-9486-5

Rough Sets and the Algebra of Conditional Logic

Gayatri Panicker(✉) and Mohua Banerjee

Department of Mathematics and Statistics, Indian Institute of Technology Kanpur, Kanpur, India
{gayatri,mohua}@iitk.ac.in

Abstract. In this paper, we consider McCarthy's three-valued logic and its corresponding algebra, a C-algebra, and introduce a connection of C-algebras with rough sets. Rough sets, with suitable operations, are shown to form a C-algebra. Further, we present a representation theorem for the class of C-algebras, establishing that every C-algebra can be embedded in a family of rough sets. The results are illustrated with examples.

Keywords: Rough sets · C-algebra · Three-valued logic

1 Introduction

The concept of rough sets was introduced by Pawlak in [12] after which it has been studied in much detail under various contexts, both in theory and applications. We are interested in the algebraic study of rough sets. Investigations in algebras stemming from studies of rough sets have been carried out extensively, since the first proposal of a 'rough algebra' by Iwiński [7]. Different (but equivalent) definitions of rough sets have led to several algebraic structures, some well-known and some new (cf. e.g. [3]). The former include three-valued Łukasiewicz algebras, Stone algebras, Nelson algebras and Kleene algebras, while the latter include topological quasi-Boolean algebras, pre-rough and rough algebras. These investigations have assumed special significance in that, the well-known algebras (such as those in the former list), proposed in different contexts altogether, get a rough set interpretation. In particular, representation results for these classes of algebras have been studied till as recently as [8]. Our work is a continuation of studies in this direction, and introduces a connection of rough sets with *C-algebras* [5], structures that capture McCarthy's three-valued logic.

The notion of set membership in the context of rough sets naturally leads to interpretations through three-valued logics, viz., assigning values `true`, `false` and `undefined` when an element x certainly belongs to set A, certainly does not belong to A, and when it is in the boundary of A respectively (cf. [1,2,6,8,10]). McCarthy in [9] had proposed certain truth-value conditions on

The first author gratefully acknowledges the support provided under the post-doctoral fellowship at the Indian Institute of Technology Kanpur.

© Springer Nature Switzerland AG 2019
T. Mihálydeák et al. (Eds.): IJCRS 2019, LNAI 11499, pp. 28–39, 2019.
https://doi.org/10.1007/978-3-030-22815-6_3

the set {`true, false, undefined`}, resulting in what is referred to in the literature as McCarthy's three-valued logic. The truth value conditions proposed by McCarthy in [9] are an extension of those in two-valued Boolean logic, and model the lazy evaluation of expressions in programming languages. Indeed, McCarthy's three-valued logic is the non-commutative regular extension of Boolean logic to three truth values, `true, false` and `undefined` denoted by $\mathbf{T}, \mathbf{F}, \mathbf{U}$ respectively. The evaluation of expressions in this logic occurs sequentially, from left to right; $\mathbf{F} \wedge \mathbf{U} = \mathbf{F}$ while $\mathbf{U} \wedge \mathbf{F} = \mathbf{U}$. Note that a programming language following lazy evaluation would evaluate the truth value of an expression $P \wedge Q$ where P is `false` to be `false`. If instead P were `undefined`, since the control would not reach statement Q, the composite statement $P \wedge Q$ would be labelled `undefined`. A complete axiomatization of this three-valued logic was given in [5] by Guzmán and Squier who named its corresponding algebra a C-*algebra*, or the algebra of conditional logic.

In this work we consider the definition of rough sets adopted by Pagliani in [10]. Under the C-algebraic interpretation of operations mentioned above, on considering whether an element x belongs to the 'conjunction' $A \wedge B$ of sets A and B, this operation $\wedge$ assumes a certain hierarchy, viz., that 'belongingness' of x in A takes precedence over that in B. If x belongs to A, only then need we check the membership of x with respect to B. If x certainly does not belong to A, then it will not belong to $A \wedge B$. However if x is in the boundary of A then its membership in $A \wedge B$ is undecided and takes the value `undefined`. It follows that the concept of $A \wedge B$ differs from that of $B \wedge A$. With this interpretation we show that every family of rough sets forms a C-algebra. Using this, we also present an example of the connection between rough sets and C-algebras. Furthermore, we establish a representation theorem for the class of C-algebras, proving that every C-algebra can be embedded in a family of rough sets.

In Sect. 2 we list fundamental notions and results pertaining to rough sets and C-algebras. In particular, we give several examples of C-algebras to elucidate the notion. In Sect. 3 we give a C-algebraic interpretation of rough sets along with an example of the connections between these two concepts in Sect. 3.1. In Sect. 4 we arrive at the main representation result of this paper, Theorem 5, and in Sect. 4.1 we give a simple example of an embedding of definable sets within a C-algebra of rough sets. We conclude in Sect. 5.

2 Preliminaries

Consider an approximation space $\langle U, E\rangle$, where U is the *universe* and E is an equivalence (*indiscernibility*) relation on U. For each $A \subseteq U$, we have an associated pair of elements in the power set $\wp(U)$ of U, viz., the *lower approximation* and *upper approximation* of A, denoted by $\underline{A}$ and $\overline{A}$ respectively:

$$\underline{A} := \bigcup\{[x] \in U/E : [x] \subseteq A\}, \text{ and } \overline{A} := \bigcup\{[x] \in U/E : [x] \cap A \neq \emptyset\}.$$

The set $\overline{A} \setminus \underline{A}$ is called the *boundary* of A. The lower approximation of A is also termed its *positive region* with respect to the approximation space $\langle U, E\rangle$, in that it consists of all the elements of U that positively belong to A. In contrast, $(\overline{A})^c$ is the *negative region* of A, consisting of elements of U that certainly do not belong to A. In this work we consider the following definition of rough sets, adopted by Pagliani in [10]: it is given by the above-mentioned pair of definite regions in the approximation space $\langle U, E\rangle$.

Definition 1. *Given an approximation space $\langle U, E\rangle$, a* rough set *is a pair $(\underline{A}, (\overline{A})^c)$ where $A \subseteq U$.*

Thus the family of rough sets for a given approximation space $\langle U, E\rangle$ that we consider henceforth, is given by the following:

$$\mathcal{RS} := \{(\underline{A}, (\overline{A})^c) : A \subseteq U\}.$$

Definition 2. *A set $A \subseteq U$ is said to be* definable *if $\underline{A} = A = \overline{A}$.*

We list the following results which help in establishing the main result of this paper, viz. Theorem 5. We omit the proofs of the results as they are well-known in the literature. In the remainder of this section, consider the pair $\langle U, E\rangle$ to be an approximation space, and A, B, C and so on, to be subsets of U, unless mentioned otherwise.

Proposition 1. *Let $a \in U$ such that $[a] = \{a\}$. Then $a \in A \Leftrightarrow a \in \overline{A} \Leftrightarrow a \in \underline{A}$.*

Proposition 2.

(i) $(\underline{A})^c = \overline{(A^c)}$, $(\overline{A})^c = \underline{(A^c)}$.

(ii) $\underline{A} \cap \underline{B} = \underline{A \cap B}$, $\overline{A} \cup \overline{B} = \overline{A \cup B}$.

(iii) If A, B are definable sets then $A \cup B$, $A \cap B$, A^c are all definable.

McCarthy first studied the three-valued non-commutative logic in the context of programming languages in [9]. The `undefined` state is denoted by the truth value **U**. The following complete axiomatization is due to Guzmán and Squier (cf. [5]) and they called the corresponding algebraic structure, a C-algebra.

Definition 3. *A* C-algebra *is an algebra $\langle M, \vee, \wedge, \neg\rangle$ of type $(2, 2, 1)$, which satisfies the following axioms for all $\alpha, \beta, \gamma \in M$:*

$$\neg\neg\alpha = \alpha \tag{1}$$
$$\neg(\alpha \wedge \beta) = \neg\alpha \vee \neg\beta \tag{2}$$
$$(\alpha \wedge \beta) \wedge \gamma = \alpha \wedge (\beta \wedge \gamma) \tag{3}$$
$$\alpha \wedge (\beta \vee \gamma) = (\alpha \wedge \beta) \vee (\alpha \wedge \gamma) \tag{4}$$
$$(\alpha \vee \beta) \wedge \gamma = (\alpha \wedge \gamma) \vee (\neg\alpha \wedge \beta \wedge \gamma) \tag{5}$$
$$\alpha \vee (\alpha \wedge \beta) = \alpha \tag{6}$$
$$(\alpha \wedge \beta) \vee (\beta \wedge \alpha) = (\beta \wedge \alpha) \vee (\alpha \wedge \beta) \tag{7}$$

Example 1. Every Boolean algebra is a C-algebra. In particular, the two element Boolean algebra $\mathbb{2}$ is a C-algebra.

Example 2. The three element set $\{\mathbf{T}, \mathbf{F}, \mathbf{U}\}$ with McCarthy's truth value conditions is denoted by $\mathbb{3}$. This is a C-algebra with the following operations:

$\neg$	
T	**F**
F	**T**
U	**U**

$\wedge$	**T**	**F**	**U**
T	**T**	**F**	**U**
F	**F**	**F**	**F**
U	**U**	**U**	**U**

$\vee$	**T**	**F**	**U**
T	**T**	**T**	**T**
F	**T**	**F**	**U**
U	**U**	**U**	**U**

Example 3. Consider the four element-structure M with universe $\{a, b, c, d\}$ and the following operations:

$\neg$	
a	b
b	a
c	d
d	c

$\wedge$	a	b	c	d
a	a	b	c	d
b	b	b	b	b
c	c	d	c	d
d	d	d	d	d

$\vee$	a	b	c	d
a	a	a	a	a
b	a	b	c	d
c	c	c	c	c
d	c	d	c	d

This is a C-algebra. M shows that even in the finite case, there may be non-isomorphic C-algebras of the same cardinality, unlike what is known for Boolean algebras.

Example 4. Consider the set $\mathcal{F}$ of eventually constant binary sequences with the following operations defined pointwise:

$$\begin{aligned} \neg x_n &:= (x_n + 1) \pmod 2 \\ x_n \wedge y_n &:= \min\{x_n, y_n\} \\ x_n \vee y_n &:= \max\{x_n, y_n\} \end{aligned}$$

This forms a Boolean algebra isomorphic to the finite-cofinite Boolean algebra on $\mathbb{N}$. Consider $\mathcal{M} := \mathcal{F} \cup \{a_n, b_n\}$ where a_n and b_n are defined as follows:

$$a_n := \begin{cases} 1, & \text{if } n \text{ is odd,} \\ 0, & \text{if } n \text{ is even.} \end{cases} \qquad b_n := \begin{cases} 0, & \text{if } n \text{ is odd,} \\ 1, & \text{if } n \text{ is even.} \end{cases}$$

Extend the operations of $\mathcal{F}$ to the set $\mathcal{M}$ by defining $\neg a_n := b_n$ and $\neg b_n := a_n$. Also for every $x_n \in \mathcal{F}$ set $a_n \wedge x_n := x_n$ and $x_n \wedge a_n := x_n$, while $b_n \wedge x_n := b_n$ and $x_n \wedge b_n =: x_n \wedge \neg x_n$. Set $a_n \wedge b_n = b_n = b_n \wedge a_n$. Finally, extend the operation $\vee$ to $\mathcal{M}$ by defining $z_n \vee w_n := \neg(\neg z_n \wedge \neg w_n)$ for all $z_n, w_n \in \mathcal{M}$. The appended set $\mathcal{M}$ with the aforementioned operations is a C-algebra which is not isomorphic to any Boolean algebra.

Remark 1. The C-algebra $\mathbb{3}$ is the smallest non-trivial C-algebra which is not a Boolean algebra. Note that the operations of $\wedge$ and $\vee$ induce a poset structure on the C-algebra (cf. [11]), but this poset does not form a lattice, in general. For instance, in the C-algebra $\mathbb{3}$ defining $x \leq y$ if $x \vee y = y$ gives $\mathbf{F} \leq \mathbf{T}$ and $\mathbf{F} \leq \mathbf{U}$ while $\mathbf{U}$ and $\mathbf{T}$ are incomparable and have no least upper bound.

Remark 2. In [5] Guzmán and Squier showed that given any C-algebra M, one could append to it two new elements $\mathbf{T}$ and $\mathbf{F}$, i.e., say $M' := M \cup \{\mathbf{T}, \mathbf{F}\}$, and extend the operations of M to M' in such a way that M' becomes a C-algebra as well. The operation $\neg$ is extended to M' by defining $\neg\mathbf{T} := \mathbf{F}$ and $\neg\mathbf{F} := \mathbf{T}$. The operation $\wedge$ of M is extended to M' as follows for each $a \in M'$:

$$\mathbf{T} \wedge a := a$$
$$a \wedge \mathbf{T} := a$$
$$\mathbf{F} \wedge a := \mathbf{F}$$
$$a \wedge \mathbf{F} := a \wedge \neg a$$

By virtue of axioms (1) and (2) in Definition 3, we extend $\vee$ to M' by simply defining $a \vee b := \neg(\neg a \wedge \neg b)$ for each $a, b \in M'$. Then M' with these newly defined operations is a C-algebra. Example 4 is a special case of the above. We also make use of the above feature in Sect. 4.1 to give a simple example of a C-algebra obtained from rough sets.

Since the class of C-algebras is a variety, for any set X, $\mathfrak{Z}^X$ is a C-algebra with the operations defined pointwise. Guzmán and Squier in [5] gave a *pairs of sets* representation of elements of $\mathfrak{Z}^X$. This is a pair (A, B) where $A, B \subseteq X$ and $A \cap B = \emptyset$. Similar to the correlation between 2^X and the power set $\wp(X)$ of X, for each $\alpha \in \mathfrak{Z}^X$, associate the pair of sets (A, B) where $A := \{x \in X : \alpha(x) = \mathbf{T}\}$ and $B := \{x \in X : \alpha(x) = \mathbf{F}\}$. Conversely, for any pair of sets (A, B) where $A, B \subseteq X$ and $A \cap B = \emptyset$ consider the function α where $\alpha(x) := \mathbf{T}$ if $x \in A$, $\alpha(x) := \mathbf{F}$ if $x \in B$ and $\alpha(x) := \mathbf{U}$ otherwise. Thus the operations on $\mathfrak{Z}^X$ (or on the associated family of pairs of sets) can be given as follows:

$$\neg(A_1, A_2) = (A_2, A_1)$$
$$(A_1, A_2) \wedge (B_1, B_2) = (A_1 \cap B_1, A_2 \cup (A_1 \cap B_2))$$
$$(A_1, A_2) \vee (B_1, B_2) = ((A_1 \cup (A_2 \cap B_1), A_2 \cap B_2)$$

Further, Guzmán and Squier showed that every C-algebra is a subalgebra of $\mathfrak{Z}^X$ for some X as stated below. For further details on subdirectly irreducible algebras refer to [4].

Theorem 1 ([5]). *$\mathfrak{Z}$ and $\mathbf{2}$ are the only subdirectly irreducible C-algebras. Hence, every C-algebra is a subalgebra of a product of copies of $\mathfrak{Z}$.*

Notation 2 *A C-algebra with $\mathbf{T}, \mathbf{F}, \mathbf{U}$ is a C-algebra with nullary operations $\mathbf{T}, \mathbf{F}, \mathbf{U}$, where $\mathbf{T}$ is the (unique) left-identity (and right-identity) for $\wedge$, $\mathbf{F}$ is the (unique) left-identity (and right-identity) for $\vee$ and $\mathbf{U}$ is the (unique) fixed point for $\neg$. Note that $\mathbf{U}$ is also a left-zero for both $\wedge$ and $\vee$ while $\mathbf{F}$ is a left-zero for $\wedge$ and $\mathbf{T}$ is a left-zero for $\vee$.*

Remark 3. The pairs of sets representation in C-algebras also leads us to the usual three-valued interpretation of rough sets. If we consider the rough set

$(\underline{A}, (\overline{A})^c)$ for some $A \subseteq U$, we can associate the function $\alpha \in 3^U$ given as: $\alpha(x) := \mathbf{T}$ when $x \in \underline{A}$, the positive region of A, $\alpha(x) := \mathbf{F}$ when $x \in (\overline{A})^c$, the negative region of A, and $\alpha(x) := \mathbf{U}$ otherwise, i.e., when x is in the boundary region of A. However, under the C-algebra operations, while evaluating $\wedge$ and $\vee$, precedence is given to the first rough set occurring on the left. This leads to the natural question of whether the family of rough sets forms a C-algebra under the given operations.

In the following sections, we detail the connections between the notions of C-algebras and rough sets.

3 A C-algebraic Interpretation of $\mathcal{RS}$

In this section we show that for every approximation space, under a specific interpretation, the associated family $\mathcal{RS}$ of rough sets forms a C-algebra. The following well-known statement is instrumental in achieving the main result. We omit the proof of this result; see e.g. [3].

Lemma 1. *Let $A, B \subseteq U$ such that A and B are definable, $A \subseteq B$ and for each $b \in B \setminus A$ we have $[b] \neq \{b\}$. Then there exists $C \subseteq U$ such that*

$$\underline{C} = A \text{ and } \overline{C} = B.$$

Theorem 3. *Given an approximation space $\langle U, E\rangle$, the collection of sets $\mathcal{RS} := \{(\underline{A}, (\overline{A})^c) : A \subseteq U\}$ forms a C-algebra under the following operations:*

$$\neg(\underline{A}, (\overline{A})^c) := ((\overline{A})^c, \underline{A})$$

$$(\underline{A}, (\overline{A})^c) \wedge (\underline{B}, (\overline{B})^c) := (\underline{A} \cap \underline{B}, (\overline{A})^c \cup (\underline{A} \cap (\overline{B})^c))$$

$$(\underline{A}, (\overline{A})^c) \vee (\underline{B}, (\overline{B})^c) := (\underline{A} \cup ((\overline{A})^c \cap \underline{B}), (\overline{A})^c \cap (\overline{B})^c)$$

Proof. It suffices to show that $\mathcal{RS}$ under the given interpretation is a subalgebra of the functional C-algebra 3^U. Note that $\underline{A} \cap (\overline{A})^c = \emptyset$ for each $A \subseteq U$. It follows that one only need check for closure with respect to operations $\neg$ and $\wedge$, since closure for $\vee$ follows from axioms (1) and (2) of Definition 3.

Closure under $\neg$: Let $(\underline{A}, (\overline{A})^c) \in \mathcal{RS}$. We must find $B \subseteq U$ such that $\underline{B} = (\overline{A})^c$ and $(\overline{B})^c = \underline{A}$. We show that $B = A^c$ is the required set. This follows from Proposition 2(i) since we have $\underline{B} = \underline{A^c} = (\overline{A})^c$. Along similar lines we have $\overline{B} = \overline{A^c} = (\underline{A})^c$ from which it follows that $(\overline{B})^c = \underline{A}$. This completes the case for $\neg$.

Closure under $\wedge$: Let $(\underline{A}, (\overline{A})^c)$ and $(\underline{B}, (\overline{B})^c) \in \mathcal{RS}$. In 3^U we have

$$(\underline{A}, (\overline{A})^c) \wedge (\underline{B}, (\overline{B})^c) = (\underline{A} \cap \underline{B}, (\overline{A})^c \cup (\underline{A} \cap (\overline{B})^c)).$$

We aim to find $C \subseteq U$ such that $\underline{C} = \underline{A} \cap \underline{B}$ and $(\overline{C})^c = (\overline{A})^c \cup (\underline{A} \cap (\overline{B})^c)$. Using elementary set-theoretic operations we have $((\overline{A})^c \cup (\underline{A} \cap (\overline{B})^c))^c = \overline{A} \cap ((\underline{A})^c \cup \overline{B})$. Clearly $\underline{A} \cap \underline{B} \subseteq \underline{A} \subseteq \overline{A}$. Along similar lines $\underline{A} \cap \underline{B} \subseteq \underline{B} \subseteq \overline{B} \subseteq (\underline{A})^c \cup \overline{B}$ so that $\underline{A} \cap \underline{B} \subseteq \overline{A} \cap ((\underline{A})^c \cup \overline{B})$. Further, consider $z \in \overline{A} \cap ((\underline{A})^c \cup \overline{B}) \setminus (\underline{A} \cap \underline{B})$ such that $[z] = \{z\}$. Using Proposition 1 we have $z \in A$ and $z \in A^c \cup B$ so that $z \in A \cap B$, a contradiction since $z \notin \underline{A} \cap \underline{B}$. Thus, using Proposition 2(iii) and Lemma 1 we obtain the required result. □

3.1 An Illustration

We mention a connection between the notions of C-algebra and rough sets by way of the following hypothetical situation, using a computer equipped with a program which prompts the observer. Consider a universe of possible real-valued inputs to the computer from 0 to 10 till 5 decimal places given as follows:

$$U := \{a_0.a_1a_2a_3a_4a_5 : \text{ for each } 0 \leq i \leq 5,\ a_i \in \{0, 1, 2, \ldots 9\}\}.$$

Note that the inputs range from 0.00000 to 9.99999. Over this finite universe U define the relation E as follows. For $a = a_0.a_1a_2a_3a_4a_5, b = b_0.b_1b_2b_3b_4b_5 \in U$,

$$(a, b) \in E \text{ if and only if } a_i = b_i \text{ for each } 0 \leq i \leq 3.$$

It can easily be seen that E is an equivalence relation on U. E models the physical limitations of the computer with regards to memory in storing the input, since the computer must necessarily truncate the real-valued input. Consider the approximation space $\langle U, E\rangle$. Using Theorem 3 we can say that the family $\mathcal{RS}$ of rough sets forms a C-algebra. We assume that in this scenario, along with the computer there exists an observer. Indeed, the pair $(\underline{A}, (\overline{A})^c)$ where $A \subseteq U$ can be seen to symbolize the following:

(i) The first component, viz., $\underline{A}$, signifies the elements that the *observer knows definitely belong to A*.
(ii) The second component, viz., $(\overline{A})^c$, comprises the elements that the *computer knows definitely do not belong to A*.

Note that the observer is aware of subsets of U only via the equivalence E. In other words, the observer is aware of those elements that definitely belong to A and those that definitely do not belong to A. The observer is not able to discern when elements in the boundary of A, i.e., $\overline{A} \setminus \underline{A}$, belong to A. In the case when there is any uncertainty in the membership of the element, the observer simply does not make any commitment and ends the process of finding the membership of the given element.

On the other hand, the computer is unable to distinguish between the sets $\underline{A}$ and $\overline{A}$ owing to its physical limitations. However it is able to ascertain when an element certainly does not belong to A. In fact, the computer stores only the representative elements of A under the relation E. For instance, inputs 2.12345 and 2.12346 would be stored as simply 2.123. Hence the computer is only able to assert the elements of $(\overline{A})^c$ with certainty. We make an additional assumption that the computer prompts the observer for further information, which we detail below. The computer only proceeds when given a definite answer from the observer, and therefore in case the observer is unable to do so, the computer gets stuck in a loop.

Note that the non-commutativity of $\wedge$ and $\vee$ play an important role. Under this connotation, the operations $\neg$, $\wedge$ and $\vee$ on $\mathcal{RS}$ can be interpreted as follows:

(i) $\neg(\underline{A}, (\overline{A})^c) = ((\overline{A})^c, \underline{A})$: The first component $(\overline{A})^c$ contains all elements that the observer knows definitely belong to A^c while $\underline{A}$ comprises the elements that the computer knows definitely do not belong to A^c.

(ii) $(\underline{A}, (\overline{A})^c) \wedge (\underline{B}, (\overline{B})^c) = (\underline{A} \cap \underline{B}, (\overline{A})^c \cup (\underline{A} \cap (\overline{B})^c))$: Using Proposition 2(ii) we see that the first component $\underline{A} \cap \underline{B}$ contains all elements that the observer knows definitely belong to $A \cap B$. We claim that, using the prompting of the observer, the second component $(\overline{A})^c \cup (\underline{A} \cap (\overline{B})^c)$ contains all the elements that the computer knows definitely do not belong to $A \cap B$. If the element is in $(\overline{A})^c$ then the computer can decide that this element is not in $A \cap B$. However if the element is in $\overline{A}$ then the computer prompts the observer to state whether this element is in $\underline{A}$ or not. If the observer inputs 'yes' then the computer proceeds to check whether this is in $(\overline{B})^c$. If true, it can then decide that this element does not belong to $A \cap B$. If, however, the element is in $\overline{B}$, then the computer cannot proceed further. On the other hand consider the case where the element is in $\overline{A} \setminus \underline{A}$. Here, the computer, even with the help of the observer, is unable to come to a decisive conclusion about the membership of the element in $A \cap B$ and is stuck in a loop.

(iii) $(\underline{A}, (\overline{A})^c) \vee (\underline{B}, (\overline{B})^c) = (\underline{A} \cup ((\overline{A})^c \cap \underline{B}), (\overline{A})^c \cap (\overline{B})^c)$: The second component, i.e., $(\overline{A})^c \cap (\overline{B})^c$ can easily be seen to contain those elements that the computer knows definitely do not belong to $A \cup B$. Consider the first component, $\underline{A} \cup ((\overline{A})^c \cap \underline{B})$: we argue that it contains all those elements which the observer can ascribe definitely to $A \cup B$. Indeed, if the element is in $\underline{A}$, the observer can say that it definitely belongs to $A \cup B$. If the element is in $\overline{A} \setminus \underline{A}$ the observer cannot be certain if the element is within A or not, and ends the procedure here. On the other hand, if the element is in $(\overline{A})^c$ then the observer knows that it definitely does not belong to A. Therefore she can check if it belongs to $\underline{B}$, in which case she knows that it definitely belongs to $A \cup B$. If the element is in $(\overline{A})^c \cap (\overline{B})^c$ then the observer knows that it definitely does not belong to $A \cup B$. Finally, in case the element belongs to $(\overline{A})^c \cap (\overline{B} \setminus \underline{B})$, the observer cannot say that it definitely belongs to $A \cup B$ – so it will not be in the first component of the pair of rough sets corresponding to $A \cup B$.

The combination of the interpretation of the operations $\neg$, $\wedge$ and $\vee$ through the observer and the computer leads to an interesting observation. Under this interpretation, while computing the $\vee$ (or $\wedge$) of rough sets $(\underline{A}, (A)^c)$ and $(\underline{B}, (\overline{B})^c)$, both the observer and the computer must necessarily give precedence to the underlying set A over set B. This follows from the fact that if there is any uncertainty in the membership of the element in A, the observer simply discards it from consideration in the $\vee$ (or $\wedge$), and the computer is unable to leave the infinite loop.

4 Embedding C-algebras in $\mathcal{RS}$

We now prove a representation theorem for the class of C-algebras: given any C-algebra M there exists an approximation space $\langle U, E\rangle$ such that M can be embedded in $\mathcal{RS}$. For this, we show that given any set X, the C-algebra $\mathfrak{Z}^X$ is isomorphic to the C-algebra formed by the family $\mathcal{RS}$ of rough sets for some approximation space $\langle U, E\rangle$. We make use of the pairs of sets representation of elements of $\mathfrak{Z}^X$ throughout this section. The symbols $\simeq$ and $\hookrightarrow$ will denote isomorphism and embedding of C-algebras respectively.

Theorem 4. *Given any set X, there exists an approximation space $\langle U, E\rangle$ such that $\mathfrak{Z}^X \simeq \mathcal{RS}$.*

Proof. If $X = \emptyset$ then $\mathfrak{Z}^X$ is the trivial algebra. If we take the trivial approximation space $\langle U, E\rangle$ where $U = \emptyset = E$, $\mathcal{RS}$ is a singleton and is isomorphic to $\mathfrak{Z}^X$.

Let $X \neq \emptyset$. For each $x \in X$ define the element x' as distinct from each element of X. Further for $a \neq b$ assume that $a' \neq b'$. Finally for any $A \subseteq X$ define $A' := \{a' : a \in A\}$. Define the universe of the required approximation space $U := X \cup X'$, and the equivalence relation E corresponding to the partition $\{\{a, a'\} : a \in X\}$. We shall show that $\mathfrak{Z}^X \simeq \mathcal{RS}$.

Note that for all $A, B, C, D \subseteq X$ we have the following relations:

$$A^{\mathsf{c}} = (X \setminus A) \cup X',\ (A')^{\mathsf{c}} = (X' \setminus A') \cup X \tag{8}$$
$$(X \setminus A)' = X' \setminus A' \tag{9}$$
$$(A \cap B)' = A' \cap B' \tag{10}$$
$$(A \cup B)' = A' \cup B' \tag{11}$$
$$(A \cup B') \cap (C \cup D') = (A \cap C) \cup (B' \cap D') \tag{12}$$

Given a pair (A, B) of sets, where $A, B \subseteq X$ such that $A \cap B = \emptyset$, define the following subset of U:

$$C_{A,B} := A' \cup (X \setminus B).$$

Consider the function $\varphi : \mathfrak{Z}^X \to \mathcal{RS}$ defined as follows:

$$\varphi((A, B)) := (\underline{C_{A,B}}, (\overline{C_{A,B}})^{\mathsf{c}}).$$

Since $A \cap B = \emptyset$, we have $A \subseteq (X \setminus B)$ so that $A \subseteq B^{\mathsf{c}}$. We observe the following:

(i) $\underline{C_{A,B}} = A \cup A'$.

(ii) $\overline{C_{A,B}} = (X \setminus B) \cup (X' \setminus B')$.

It follows that $(\overline{C_{A,B}})^{\mathsf{c}} = (X \setminus B)^{\mathsf{c}} \cap (X' \setminus B')^{\mathsf{c}}$. Using (8) and (12) we have $(\overline{C_{A,B}})^{\mathsf{c}} = (B \cup X') \cap (B' \cup X) = B \cup B'$. Consequently we have the following:

$$\varphi((A, B)) = (A \cup A', B \cup B').$$

Claim: $\varphi(\neg(A,B)) = \neg\varphi((A,B))$.
We see that $\varphi(\neg(A,B)) = \varphi((B,A)) = (B \cup B', A \cup A') = \neg\varphi((A,B))$.

Claim: $\varphi((A_1,B_1) \wedge (A_2,B_2)) = \varphi((A_1,B_1)) \wedge \varphi((A_2,B_2))$.
Note that $(A_1,B_1) \wedge (A_2,B_2) = (A_1 \cap A_2, B_1 \cup (A_1 \cap B_2))$. Thus, using (10) and (11) we have $\varphi((A_1,B_1) \wedge (A_2,B_2)) = \varphi((A_1 \cap A_2, B_1 \cup (A_1 \cap B_2)) = ((A_1 \cap A_2) \cup (A_1' \cap A_2'), (B_1 \cup (A_1 \cap B_2)) \cup (B_1' \cup (A_1' \cap B_2'))$. On the other hand, since $\varphi((A_1,B_1)) = (A_1 \cup A_1', B_1 \cup B_1')$ and $\varphi((A_2,B_2)) = (A_2 \cup A_2', B_2 \cup B_2')$ we have $\varphi((A_1,B_1)) \wedge \varphi((A_2,B_2)) = (A_1 \cup A_1', B_1 \cup B_1') \wedge (A_2 \cup A_2', B_2 \cup B_2') = ((A_1 \cup A_1') \cap (A_2 \cup A_2'), (B_1 \cup B_1') \cup ((A_1 \cup A_1') \cap (B_2 \cup B_2')))$. Using (12) we have $\varphi((A_1,B_1)) \wedge \varphi((A_2,B_2)) = ((A_1 \cap A_2) \cup (A_1' \cap A_2'), (B_1 \cup (A_1 \cap B_2)) \cup (B_1' \cup (A_1' \cap B_2')))$. Note that for the operation $\vee$ we simply use axioms (1) and (2) from Definition 3. Consequently, φ is a C-algebra homomorphism.

Claim: φ is injective.
Let (A_1,B_1) and (A_2,B_2) be two pairs of sets in 3^X such that $\varphi((A_1,B_1)) = \varphi((A_2,B_2))$. Hence $(A_1 \cup A_1', B_1 \cup B_1') = (A_2 \cup A_2', B_2 \cup B_2')$. Since U is the disjoint union of X and X' it follows that $A_1 = A_2$ and $B_1 = B_2$.

Claim: φ is surjective.
Let $(\underline{C}, (\overline{C})^c) \in \mathcal{RS}$ for some $C \subseteq U$. Since U is the disjoint union of X and X' we can say that $C = Y \cup Z'$ for some $Y, Z \subseteq X$. Hence, using (10) we have $(\underline{C}, \overline{C}) = ((Y \cap Z) \cup (Y' \cap Z'), (Y \cup Y') \cup (Z \cup Z'))$. Using (8) and (12) we have $(\overline{C})^c = (Y \cup Z)^c \cap (Y' \cup Z')^c = ((X \setminus (Y \cup Z)) \cup X') \cap (X' \setminus (Y' \cup Z')) \cup X) = (X \setminus (Y \cup Z)) \cup (X' \setminus (Y' \cup Z'))$.

Define the pair (A,B) as $A := Y \cap Z$ and $B := X \setminus (Y \cup Z)$. Clearly, since $Y \cap Z \subseteq (Y \cup Z)$ we have $(Y \cap Z) \cap (X \setminus (Y \cup Z)) = \emptyset$. Also $\varphi((A,B)) = (A \cup A', B \cup B') = ((Y \cap Z) \cup (Y' \cap Z'), (X \setminus (Y \cup Z)) \cup (X' \setminus (Y' \cup Z'))) = (\underline{C}, (\overline{C})^c)$. The result follows. □

Remark 4. Note that as a consequence, φ also preserves constants $\mathbf{T}, \mathbf{F}, \mathbf{U}$ and is therefore an isomorphism of C-algebras with $\mathbf{T}, \mathbf{F}, \mathbf{U}$.

Theorem 5. (Representation) *Given any C-algebra M, there exists an approximation space $\langle U, E\rangle$ such that $M \hookrightarrow \mathcal{RS}$.*

Proof. The proof follows in a straighforward manner from Theorems 1 and 4. □

4.1 An Illustration of Theorem 5

It is well-known that given an approximation space $\langle U, E\rangle$, the family $\mathcal{D}$ of definable sets forms a Boolean algebra. Using Remark 2 we see that by appending two new elements $\mathbf{T}, \mathbf{F}$ to $\mathcal{D}$ with the requisite operations, we obtain a new C-algebra, say $\mathcal{D}'$. As a consequence of Theorem 5, there exists an approximation space $\langle W, R\rangle$ such that $\mathcal{D}' \hookrightarrow \mathcal{RS}$. A natural question would be what this approximation space is, for certain simple examples of families of definable sets. Consider $U := \{a, b\}$ and E as the purely reflexive equivalence on U corresponding to the partition $\{\{a\}, \{b\}\}$. Thus $\mathcal{D} = \{\emptyset, \{a\}, \{b\}, U\}$. We present a candidate for the approximation space $\langle W, R\rangle$ below.

Note that $\mathcal{D}' \hookrightarrow \mathfrak{Z}^X$ where $X := \{x, y, z\}$. The map $\psi : \mathcal{D}' \to \mathfrak{Z}^X$ given below is a C-algebra embedding:

$$\begin{aligned}
\psi(\mathbf{F}) &:= (\mathbf{F}, \mathbf{F}, \mathbf{F}) \\
\psi(\mathbf{T}) &:= (\mathbf{T}, \mathbf{T}, \mathbf{T}) \\
\psi(\emptyset) &:= (\mathbf{F}, \mathbf{F}, \mathbf{U}) \\
\psi(\{a\}) &:= (\mathbf{T}, \mathbf{F}, \mathbf{U}) \\
\psi(\{b\}) &:= (\mathbf{F}, \mathbf{T}, \mathbf{U}) \\
\psi(U) &:= (\mathbf{T}, \mathbf{T}, \mathbf{U})
\end{aligned}$$

On utilizing the pairs of sets representation of elements in $\mathfrak{Z}^X$, we can rewrite the embedding $\psi : \mathcal{D}' \to \mathfrak{Z}^X$ as follows:

$$\begin{aligned}
\psi(\mathbf{F}) &= (\emptyset, \{x, y, z\}) \\
\psi(\mathbf{T}) &= (\{x, y, z\}, \emptyset) \\
\psi(\emptyset) &= (\emptyset, \{x, y\}) \\
\psi(\{a\}) &= (\{x\}, \{y\}) \\
\psi(\{b\}) &= (\{y\}, \{x\}) \\
\psi(U) &= (\{x, y\}, \emptyset)
\end{aligned}$$

We use the procedure delineated in Theorem 4 to obtain the required universe $W := \{x, y, z, x', y', z'\}$ and equivalence $R := \{\{x, x'\}, \{y, y'\}, \{z, z'\}\}$. Thus the embedding $\varphi : \mathcal{D}' \to \mathcal{RS}$ is given by $\varphi(A) := (\underline{C}, (\overline{C})^c)$, where $A \subseteq U$ or $A \in \{\mathbf{T}, \mathbf{F}\}$ and $C \subseteq W$. The required subsets C of W along with their correspondence with elements of $\mathcal{D}'$ are given below:

$$\begin{aligned}
A &\mapsto C : \\
\mathbf{F} &\mapsto \emptyset \\
\mathbf{T} &\mapsto W \\
\emptyset &\mapsto \{z\} \\
\{a\} &\mapsto \{x, x', z\} \\
\{b\} &\mapsto \{y, y', z\} \\
U &\mapsto \{x, x', y, y', z\}
\end{aligned}$$

Thus, given any approximation space, the collection of its definable sets can be embedded in a C-algebra of rough sets. This gives a family of examples of C-algebras that have direct connections to rough sets.

5 Conclusion

The interconnectivity between three-valued logics and rough sets has been delved into deeply in the literature. Most of the studies thus far have focussed on commutative operations of $\wedge$ and $\vee$. In this work, we attempt to give an interpretation of rough sets using the non-commutative $\wedge$ and $\vee$ of C-algebras or the algebra of conditional logic. We show that this interpretation is meaningful by ascertaining that, in fact, every family of rough sets forms a C-algebra. On the other hand, we are also able to obtain a representation theorem: every C-algebra is shown to be embedded in a family of rough sets. Further explorations of the interplay between these two notions appear worthwhile, and could possibly also throw light on connections between rough sets and programming languages.

References

1. Avron, A., Konikowska, B.: Rough sets and 3-valued logics. Stud. Logica **90**(1), 69–92 (2008)
2. Banerjee, M.: Rough sets and 3-valued Łukasiewicz logic. Fundamenta Informaticae **31**(3, 4), 213–220 (1997)
3. Banerjee, M., Chakraborty, M.K.: Algebras from rough sets. In: Pal, S., Polkowski, L., Skowron, A. (eds.) Rough-Neural Computing. Cognitive Technologies, pp. 157–184. Springer, Heidelberg (2004). https://doi.org/10.1007/978-3-642-18859-6_7
4. Burris, S., Sankappanavar, H.P.: A Course in Universal Algebra. Graduate Texts in Mathematics, vol. 78. Springer, New York (1981)
5. Guzmán, F., Squier, C.C.: The algebra of conditional logic. Algebra Universalis **27**(1), 88–110 (1990)
6. Iturrioz, L.: Rough sets and three-valued structures. In: Orłowska, E. (ed.) Logic at Work, Heidelberg, pp. 24–596 (1999)
7. Iwiński, T.B.: Algebraic approach to rough sets. Bull. Pol. Acad. Sci. **35**(3,4), 673–683 (1987)
8. Kumar, A., Banerjee, M.: Kleene algebras and logic: boolean and rough set representations, 3-valued, rough set and perp semantics. Stud. Logica **105**(3), 439–469 (2017)
9. McCarthy, J.: A basis for a mathematical theory of computation. In: Computer Programming and Formal Systems, North-Holland, Amsterdam, pp. 33–70 (1963)
10. Pagliani, P.: Rough set theory and logic-algebraic structures. In: Orłowska, E. (ed.) Incomplete Information: Rough Set Analysis. Studies in Fuzziness and Soft Computing, vol. 13, pp. 109–190. Physica, Heidelberg (1998). https://doi.org/10.1007/978-3-7908-1888-8_6
11. Panicker, G.: If-then-else over the Algebra of Conditional Logic. Ph.D. thesis, Indian Institute of Technology Guwahati, India (2018). http://gyan.iitg.ernet.in/handle/123456789/922
12. Pawlak, Z.: Rough sets. Int. J. Comput. Inf. Sci. **11**(5), 341–356 (1982)

Rough Sets Defined by Multiple Relations

Jouni Järvinen[1(✉)], László Kovács[2(✉)], and Sándor Radeleczki[3(✉)]

[1] Department of Mathematics and Statistics, University of Turku, 20014 Turku, Finland
jjarvine@utu.fi

[2] Institute of Information Science, University of Miskolc, 3515 Miskolc-Egyetemváros, Hungary
kovacs@iit.uni-miskolc.hu

[3] Institute of Mathematics, University of Miskolc, 3515 Miskolc-Egyetemváros, Hungary
matradi@uni-miskolc.hu

Abstract. We generalize the standard rough set pair induced by an equivalence E on U in such a way that the upper approximation defined by E is replaced by the upper approximations determined by tolerances $T_1, \ldots, T_n$ on U. Using this kind of multiple upper approximations we can express "softer" uncertainties of different kinds. We can order the set $RS(E, T_1, \ldots, T_n)$ of the multiple approximations of all subsets of the universe U by the coordinatewise inclusion. We show that whenever the tolerances $T_1, \ldots, T_n$ are E-compatible, this ordered set forms a complete lattice. As a special case we show how this complete lattice can be reduced to the complete lattice of the traditional rough sets defined by the equivalence E.

Keywords: Lower and upper approximation · Rough set · Compatibility condition · Tolerance relation · Multiple borders

1 Compatibility Condition and Multiple Approximations

The aim of this paper is to extend the "traditional" rough set model to be able to represent different levels of uncertainty. Rough sets were introduced by Pawlak in [8]. He assumed that our knowledge about the objects of a universe U is given in the terms of an information relation R reflecting their indiscernibility.

For any relation $R \subseteq U \times U$ and $x \in U$, denote $R(x) = \{y \in U \mid (x, y) \in R\}$. Then for any subset $X \subseteq U$ its *lower approximation* is defined as

$$X_R = \{x \in U \mid R(x) \subseteq X\},$$

and the *upper approximation* of X is given by

$$X^R = \{x \in U \mid R(x) \cap X \neq \emptyset\}.$$

If R is a reflexive relation, then $X_R \subseteq X \subseteq X^R$ and the elements of U may be divided into three disjoint classes:

© Springer Nature Switzerland AG 2019
T. Mihálydeák et al. (Eds.): IJCRS 2019, LNAI 11499, pp. 40–51, 2019.
https://doi.org/10.1007/978-3-030-22815-6_4

(C1) The elements which are certainly *in* X. These are the elements in X_R, because if $x \in X_R$, then all the elements to which x is R-related are in X.
(C2) The elements which certainly are *not in* X. These are the elements x such that all the elements to which x is R-related are outside X.
(C3) The elements which are *possibly in* X. These are the elements x which are R-related at least to one element from X and also at least to one element outside X. In other words, $x \in X^R \setminus X_R$.

Initially, Pawlak assumed that R is an equivalence, that is, a reflexive, symmetric and transitive relation. There are many generalizations of Pawlak's construction based on non-equivalence relations, and replacing equivalence classes by coverings; see [13,14], for instance. A natural variant is to assume that our information is given by a *tolerance relation*, that is, a reflexive and symmetric binary relation, being not transitive in general. Authors of this paper have considered lattice-theoretical properties of rough sets defined by tolerances, for example, in [3,5,6].

In [4], we used both equivalences and tolerances to form approximations. As a motivation for this kind of setting consider the case in which U consists of a set of patients of a hospital and $x\,E\,y$ means that all the attributes of x and y representing some medical information are the same. Let X be a set of patients with a certain disease. If $x \in X^E$, then X contains a patient y such that x cannot be distinguished from y in terms of any attribute. On the other hand, sometimes it would be useful to know also those patients who have a risk to have the disease in the near future or who are at an initial phase of the disease. These persons may have different symptoms as the patients with illness have. But they may have, for instance, similar symptoms. Thus, we can use a tolerance relation T to represent this similarity. The upper approximation X^T consists of persons who are similar to patients with disease, thus they may have some risk to get the disease. It may be reasonable to introduce several tolerance relations to represent different types of risks and different types of similarity, and therefore in this paper we consider also multiple tolerances.

In [4] we considered tolerances compatible with equivalences, which turned to be closely related to "similarity relations extending equivalences" studied in [11]. In this work, we slightly generalize the notion of compatibility to be used also between tolerances.

Definition 1. *Let R and T be two tolerances on U. If $R \circ T = T$, then T is **R-compatible**.*

If T is R-compatible, then $R \subseteq T$ and $R^2 \subseteq R \circ T = T$, so R is "transitive" inside T. Since $T^{-1} = T$ and $(R \circ T)^{-1} = T^{-1} \circ R^{-1} = T \circ R$ we get

$$R \circ T = T \iff (R \circ T)^{-1} = T^{-1} \iff T \circ R = T. \tag{1.1}$$

Hence, $R \circ T = T$ and $T \circ R = T$ are equivalent conditions.

For a tolerance T, the *kernel* of T is defined by

$$\ker T = \{(x, y) \mid T(x) = T(y)\}.$$

Proposition 2. *Let R and T be tolerances on U. The tolerance T is R-compatible if and only if $R \subseteq \ker T$.*

Proof. ($\Rightarrow$) Suppose that T is R-compatible. We show that $R \subseteq \ker T$. Assume $(x, y) \in R$. Let $z \in T(x)$. Then $z\,T\,x$ and $x\,R\,y$, that is, $(z, y) \in T \circ R = T$. Thus, $z \in T(y)$ and $T(x) \subseteq T(y)$. Similarly, we can show that $T(y) \subseteq T(x)$: if $z \in T(y)$, then $(x, z) \in R \circ T = T$ and $z \in T(x)$. Thus, $T(x) = T(y)$ and $(x, y) \in \ker T$. Therefore, $R \subseteq \ker T$.

($\Leftarrow$) Assume that $R \subseteq \ker T$. Let $(x, y) \in R \circ T$. Then, there is z such that $x\,R\,z$ and $z\,T\,y$. Because $(x, z) \in \ker T$, $y \in T(z) = T(x)$. Thus, $(x, y) \in T$ and $R \circ T \subseteq T$. Because $T \subseteq R \circ T$ holds always, we have $T = R \circ T$ and T is R-compatible. □

We can also present the following characterization.

Proposition 3. *Suppose R and T are tolerances on U. The tolerance T is R-compatible if and only if*

$$T(x) = \bigcup\{R(y) \mid y \in T(x)\} \tag{1.2}$$

for all $x \in U$.

Proof. ($\Rightarrow$) Assume that T is R-compatible. Let $z \in T(x)$. Then $z \in R(z)$ gives $z \in \bigcup\{R(y) \mid y \in T(x)\}$. On the other hand, if $z \in \bigcup\{R(y) \mid y \in T(x)\}$, then $z\,R\,y$ and $y\,T\,x$ give $(z, x) \in R \circ T = T$, that is, $z \in T(z)$. So, (1.2) holds.

($\Leftarrow$) Suppose (1.2) is true for any $x \in U$. If $(x, z) \in T \circ R$, then there is y such that $y \in T(x)$ and $z \in R(y)$. By (1.2), these give $z \in T(x)$. Thus, $(x, z) \in T$ and $T \circ R \subseteq T$. Since, $T \subseteq T \circ R$ holds always, T is R-compatible. □

Let $X \subseteq U$ be arbitrary and let T be an R-compatible tolerance. The following properties can be proved:

$$(X^T)^R = X^{T \circ R} = X^T = X^{R \circ T} = (X^R)^T; \tag{1.3}$$

$$(X_T)_R = X_{T \circ R} = X_T = X_{R \circ T} = (X_R)_T. \tag{1.4}$$

Indeed, $X^{T \circ R} = X^T = X^{R \circ T}$ is clear by (1.1). Let us check $(X^T)^R = X^{R \circ T}$ as an example:

$$\begin{aligned} x \in (X^T)^R &\iff (\exists z)\, x\,R\,z \text{ and } z \in X^T \\ &\iff (\exists z)(\exists y)\, x\,R\,z \text{ and } z\,T\,y \text{ and } y \in X \\ &\iff (\exists y)\, x\,(R \circ T)\,y \text{ and } y \in X \\ &\iff x \in X^{R \circ T} \end{aligned}$$

Hence (1.3) is satisfied. Equalities (1.4) are proved analogously.

If our knowledge about the attributes of the elements is incomplete, then classification (C1)–(C3) of the elements of U into three disjoint subsets

$$X_E \cup (X^E \setminus X_E) \cup (U \setminus X^E)$$

may be insufficient [2]. For instance, beside those elements which are in the *boundary* $X^E \setminus X_E$ of X, there may exist other elements in U whose attributes are not enough known to exclude that they are somehow related to X. Hence a division of the elements of U in four, or even more classes might be more convenient. In this work, we will consider several tolerances $T_1, \ldots, T_n$ on U. This enables us to define multiple borders and consider cases in which there are several degrees of possibility. Our work is related to a multi-granulation rough set model (MGRS), where the set approximations are defined by using multi equivalence relations on the universe [10].

The tolerances $T_1, \ldots, T_n$ are assumed to be E-compatible. This means that if x is T_i-similar to y, then any element E-indistinguishable with x must also be T_i-similar to y. The obtained tuples $(X_E, X^{T_1}, \ldots, X^{T_n})$ can be considered as generalizations of rough sets.

2 Rough Sets of Multiple Approximations

For a binary relation R on U, the "traditional" *R-rough set of X* is defined as the pair (X_R, X^R). We denote by

$$RS(R) = \{(X_R, X^R) \mid X \subseteq U\}$$

the set of all *R-rough sets*. The set $RS(R)$ can be ordered *coordinatewise inclusion* by

$$(X_R, X^R) \leq (Y_R, Y^R) \iff X_R \subseteq Y_R \text{ and } X^R \subseteq Y^R,$$

obtaining a partially ordered set $(RS(R), \leq)$, which we denote simply by $RS(R)$. If E is an equivalence relation, then $RS(E)$ is a complete lattice such that

$$\bigvee_{X \in \mathcal{H}} (X_E, X^E) = \Big(\bigcup_{X \in \mathcal{H}} X_E, \bigcup_{X \in \mathcal{H}} X^E \Big) \tag{2.1}$$

and

$$\bigwedge_{X \in \mathcal{H}} (X_E, X^E) = \Big(\bigcap_{X \in \mathcal{H}} X_E, \bigcap_{X \in \mathcal{H}} X^E \Big) \tag{2.2}$$

for all $\mathcal{H} \subseteq \wp(U)$, where $\wp(U)$ the *powerset* of U, that is, the set of all subsets of U. It is also known that a so-called regular double Stone algebra can be defined on $RS(E)$ [1,9]. If T is a tolerance, then in [3] it is proved that $RS(T)$ is not necessarily even a semilattice.

In [4] we considered the following generalization

$$RS(E,T) = \{(X_E, X^T) \mid X \subseteq U\}$$

of the traditional rough set system. The idea behind studying such pairs (X_E, X^T) is that the equivalence E represents "strict" information (*indistinguishability*) and the information represented by T is "soft" (*similarity*). Hence X_E is defined as it is usual in rough set theory, but X^T is now more permissible, because $E \subseteq T$ and thus $X \subseteq X^E \subseteq X^T$. We proved several results about the structure of $RS(E,T)$, particularly that it always forms a complete lattice.

First we generalize our setting to multiple E-compatible tolerances. If E is an equivalence on U and $T_1, \ldots, T_n$ are tolerances on U, then

$$X^{T_1} \setminus X_E,\ X^{T_2} \setminus X_E,\ \ldots, X^{T_n} \setminus X_E$$

may express uncertainties of different kinds. We denote

$$RS(E, T_1, \ldots, T_n) = \{(X_E, X^{T_1}, \ldots, X^{T_n}) \mid X \subseteq U\}.$$

As earlier, $RS(E, T_1, \ldots, T_n)$ is ordered coordinatewise.

Proposition 4. *Let E be an equivalence on U and $T_1, \ldots, T_n$ be E-compatible tolerances. Then $RS(E, T_1, \ldots, T_n)$ is a complete lattice.*

Proof. Because $\underbrace{(\emptyset, \emptyset, \ldots, \emptyset)}_{n+1}$ is the least element of $\mathbf{RS} := RS(E, T_1, \ldots, T_n)$, it suffices to show that for any $\emptyset \neq \mathcal{H} \subseteq \wp(U)$, the set $\{(X_E, X^{T_1}, \ldots, X^{T_n}) \mid X \in \mathcal{H}\}$ has a supremum in $\mathbf{RS}$. Since $\left(\bigcup_{X \in \mathcal{H}} X_E, \bigcup_{X \in \mathcal{H}} X^E\right)$ is an E-rough set by (2.1), there exists a set $Y \subseteq U$ with

$$Y_E = \bigcup_{X \in \mathcal{H}} X_E \quad \text{and} \quad Y^E = \bigcup_{X \in \mathcal{H}} X^E.$$

By Property (1.3) we have that for $1 \leq i \leq n$,

$$Y^{T_i} = (Y^E)^{T_i} = \Big(\bigcup_{X \in \mathcal{H}} X^E\Big)^{T_i} = \bigcup_{X \in \mathcal{H}} (X^E)^{T_i} = \bigcup_{X \in \mathcal{H}} X^{T_i}.$$

This implies that

$$\Big(\bigcup_{X \in \mathcal{H}} X_E, \bigcup_{X \in \mathcal{H}} X^{T_1}, \ldots, \bigcup_{X \in \mathcal{H}} X^{T_n}\Big) = (Y_E, Y^{T_1}, \ldots, Y^{T_n})$$

belongs to $\mathbf{RS}$.

Now $(Y_E, Y^{T_1}, \ldots, Y^{T_n})$ is an upper bound of $(X_E, X^{T_1}, \ldots, X^{T_n})$ for all $X \in \mathcal{H}$. It is also clear that if

$$(Z_E, Z^{T_1}, \ldots, Z^{T_n})$$

is an upper bound of $\{(X_E, X^{T_1}, \ldots, X^{T_n}) \mid X \in \mathcal{H}\}$, then $X_E \subseteq Z_E$ and $X^{T_i} \subseteq Z^{T_i}$ for all $X \in \mathcal{H}$ and $1 \leq i \leq n$. This gives

$$\bigcup_{X \in \mathcal{H}} X_E \subseteq Z_E \quad \text{and} \quad \bigcup_{X \in \mathcal{H}} X^{T_i} \subseteq Z^{T_i}$$

for $1 \leq i \leq n$. Therefore,

$$(Y_E, Y^{T_1}, \ldots, Y^{T_n}) \leq (Z_E, Z^{T_1}, \ldots, Z^{T_n})$$

and $(Y_E, Y^{T_1}, \ldots, Y^{T_n})$ is the supremum of $\{(X_E, X^{T_1}, \ldots, X^{T_n}) \mid X \in \mathcal{H}\}$. □

Example 5. Let $U = \{1, 2, 3, 4\}$ and E be an equivalence on U such that $U/E = \{\{1\}, \{2, 3\}, \{4\}\}$. Assume T_1 is an equivalence (and thus a tolerance) such that

$$T_1(1) = T_1(2) = T_1(3) = \{1, 2, 3\} \quad \text{and} \quad T_1(4) = \{4\}.$$

In addition, let T_2 be a tolerance such that

$$T_2(1) = U, \quad T_2(2) = T_2(3) = \{1, 2, 3\} \quad \text{and} \quad T_2(4) = \{1, 4\}.$$

Because $E \subseteq \ker T_1 = T_1$ and $E = \ker T_2$, T_1 and T_2 are E-compatible.

We have also $T_1 \subseteq T_2$, but T_2 is not T_1-compatible, since $T_1 \nsubseteq \ker T_2 = E$. The elements of

$$RS(E, T_1, T_2) = \{(X_E, X^{T_1}, X^{T_2}) \mid X \subseteq U\}$$

are given in Table 1. Note that here we denote sets just by sequences of their elements, the set $\{1, 2, 4\}$ is written 124, for instance. The Hasse diagram of $RS(E, T_1, T_2)$ can be found in Fig. 1.

Table 1. The 3-tuple approximations of subsets of U

X	(X_E, X^{T_1}, X^{T_2})	X	(X_E, X^{T_1}, X^{T_2})
$\emptyset$	$(\emptyset, \emptyset, \emptyset)$	23	$(23, 123, 123)$
1	$(1, 123, U)$	24	$(4, U, U)$
2	$(\emptyset, 123, 123)$	34	$(4, U, U)$
3	$(\emptyset, 123, 123)$	123	$(123, 123, U)$
4	$(4, 4, 14)$	124	$(14, U, U)$
12	$(1, 123, U)$	134	$(14, U, U)$
13	$(1, 123, U)$	234	$(234, U, U)$
14	$(14, U, U)$	U	(U, U, U)

Let us note that if $n = 1$ and $T_1 = T$, we obtain the complete lattice $RS(E, T) = \{(X_E, X^T) \mid X \subseteq U\}$ investigated in [4]. Our next theorem shows that adding T-compatible tolerances $S_1, \ldots, S_n$ to $RS(E, T)$ does not change the lattice-theoretical structure. Notice that if T is an E-compatible tolerance and a tolerance S is compatible with T, then S is also E-compatible because

$$E \circ S \subseteq T \circ S \subseteq S,$$

which implies $E \circ S = S$, since $S \subseteq E \circ S$.

Theorem 6. *Let E be an equivalence on U and let T be an E-compatible tolerance. If $S_1, \ldots, S_n$ are tolerances which are T-compatible, then*

$$RS(E, T) \cong RS(E, T, S_1, \ldots, S_n).$$

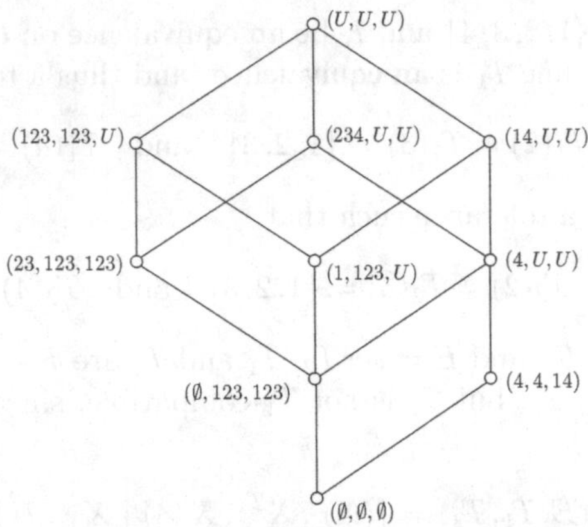

Fig. 1. The lattice $RS(E, T_1, T_2)$

Proof. Note first that each $S_1, \ldots, S_n$ is E-compatible. This means that

$$RS(E, T, S_1, \ldots, S_n)$$

is a complete lattice by Proposition 4. We define a map

$$\varphi\colon RS(E,T) \to RS(E,T,S_1,\ldots,S_n),\ (X_E, X^T) \mapsto (X_E, X^T, X^{S_1}, \ldots, X^{S_n}).$$

The map φ is well defined, because if $(X_E, X^T) = (Y_E, Y^T)$, then by (1.3),

$$X^{S_k} = (X^T)^{S_k} = (Y^T)^{S_k} = Y^{S_k}$$

for any $1 \leq k \leq n$, which yields $\varphi(X_E, X^T) = \varphi(Y_E, Y^T)$. Next we prove that φ is an order-embedding, that is,

$$(X_E, X^T) \leq (Y_E, Y^T) \iff \varphi(X_E, X^T) \leq \varphi(Y_E, Y^T).$$

Suppose $(X_E, X^T) \leq (Y_E, Y^T)$. Then $X^T \subseteq Y^T$ and for any $1 \leq k \leq n$,

$$X^{S_k} = (X^T)^{S_k} \leq (Y^T)^{S_k} = Y^{S_k}.$$

Hence, $\varphi(X_E, X^T) \leq \varphi(Y_E, Y^T)$. It is trivial that if $\varphi(X_E, X^T) \leq \varphi(Y_E, Y^T)$, then $(X_E, X^T) \leq (Y_E, Y^T)$. The mapping φ is obviously surjective, because if $(X_E, X^T, X^{S_1}, \ldots, X^{S_n})$ belongs to $RS(E, T, S_1, \ldots, S_n)$, then $\varphi(X_E, X^T) = (X_E, X^T, X^{S_1}, \ldots, X^{S_n})$. □

The following consequence is immediate. Notice that each equivalence E is compatible with itself, that is $E \circ E = E$.

Corollary 7. *Let E be an equivalence relation on U and $T_1, \ldots, T_n$ be E-compatible tolerances. If $T_1 = E$, then*

$$RS(E) \cong RS(E, T_1, \ldots, T_n).$$

Let $E_0 \subseteq E_1 \subseteq E_2 \subseteq \cdots \subseteq E_n$ be equivalences on U. Note that the kernel of an equivalence is the equivalence itself. Therefore, E_1 is E_0-compatible and $E_2, \ldots, E_n$ are E_1-compatible. By Theorem 6 we can write the following corollary.

Corollary 8. *Let $E_0 \subseteq E_1 \subseteq \cdots \subseteq E_n$ be equivalences on U. Then*

$$RS(E_0, E_1, \ldots, E_n) \cong RS(E_0, E_1).$$

We end this section by presenting a couple of examples where multiple rough sets can be defined in a natural way.

Example 9. Let R be a *fuzzy equivalence* on U. This means that for all $x, y \in U$, $R(x, y) \in [0, 1]$ and that R is

- *reflexive*: $R(x, x) = 1$ for each $x \in U$,
- *symmetric*: $R(x, y) = R(y, x)$ for all $x, y \in U$, and
- *transitive*: $R(x, z) \geq \min\{R(x, y), R(y, z)\}$ for any $x, y, z \in U$.

It is known that for any $\alpha \in [0, 1]$ the *α-cut*

$$R_\alpha = \{(x, y) \in U \times U \mid R(x, y) \geq \alpha\}$$

of R is a "crisp" equivalence on U. Let $0 \leq \alpha_0 \leq \alpha_1 \leq \cdots \leq \alpha_n \leq 1$. Then $R_{\alpha_0} \subseteq R_{\alpha_1} \subseteq \cdots \subseteq R_{\alpha_n}$ are equivalences on U. By Corollary 8 we get

$$RS(R_{\alpha_0}, R_{\alpha_1}, \ldots, R_{\alpha_n}) \cong RS(R_{\alpha_0}, R_{\alpha_1}).$$

Example 10. An *information system* in the sense of Pawlak [7] is a triple

$$(U, A, \{V\}_{a \in A}),$$

where U is a set of objects, A is a set of attributes and V_a is the value set of $a \in A$. Each attribute is a mapping $a \colon U \to V_a$. For any $\emptyset \neq B \subseteq A$, the *strong indiscernibility relation* of B is defined by

$$\text{ind}(B) = \{(x, y) \mid a(x) = a(y) \text{ for all } a \in B\}.$$

The *weak indiscenibility relation of B* is given by

$$\text{wind}(B) = \{(x, y) \mid a(x) = a(y) \text{ for some } a \in B\}.$$

Clearly, $\text{ind}(B)$ is an equivalence and $\text{wind}(B)$ is a tolerance.

Let $\emptyset \neq C \subseteq B \subseteq A$. It is easy to see that $\text{wind}(C)$ is $\text{ind}(B)$-compatible. Indeed, the inclusion $\text{wind}(C) \subseteq \text{ind}(B) \circ \text{wind}(C)$ is clear. In order to prove the converse inclusion, let $(x, y) \in \text{ind}(B) \circ \text{wind}(C)$. Then $(x, z) \in \text{ind}(B)$ and $(z, y) \in \text{wind}(C)$ for some $z \in U$. As $C \subseteq B$, $(x, z) \in \text{ind}(B)$ yields $a(x) = a(z)$ for all $a \in C$. Because $(z, y) \in \text{wind}(C)$, we have $b(y) = b(z) = b(x)$ for some $b \in C$. Thus, $(x, y) \in \text{wind}(C)$. This means $\text{ind}(B) \circ \text{wind}(C) \subseteq \text{wind}(C)$, completing the proof.

Suppose $\emptyset \neq C_1, \ldots, C_n \subseteq B$. Since $\text{wind}(C_i)$ is $\text{ind}(B)$-compatible for any $1 \leq i \leq n$, we can form the generalized rough set complete lattice

$$RS(\text{ind}(B), \text{wind}(C_1), \ldots, \text{wind}(C_n)).$$

3 Comparison with the Fuzzy Set Approach

The relationship between rough set theory and fuzzy set theory is widely discussed in the literature. One of the key differences between these approaches is the fact that in fuzzy set theory the membership value does not depend on other elements. In contrast, the rough approximations and rough membership functions are defined in terms of a relation on the object set [15]. According to [12], one may treat rough set in set-oriented view as a special class of fuzzy sets. In this section, we argue that from the viewpoint of set approximation, rough sets with multiple borders significantly increase the functionality of the standard rough set model and it provides a more general model of uncertainty than the fuzzy model.

In the fuzzy set theory [16], a *fuzzy set* A on U is defined by a membership function

$$f_A \colon U \to [0,1],$$

where the value $f_A(x)$ for any $x \in U$ denotes the "grade of membership" of x in A. For any $\alpha \in [0,1]$, the closed alpha-cut set A_α and the open alpha-cut set $A_{>\alpha}$ are crisp sets, where

$$A_\alpha = \{x \in U \mid f_A(x) \geq \alpha\}$$

and

$$A_{>\alpha} = \{x \in U \mid f_A(x) > \alpha\}.$$

Let $X \subseteq U$ be a (crisp) set. A fuzzy set A can be considered as a "rough approximation" of X, if

$$A_1 \subseteq X \subseteq A_{>0}.$$

The set A_1 denotes the elements which are certainly in X and the elements which may belong to X are contained in $A_{>0}$. In "fuzzy terminology", A_1 is called the *core* of A and $A_{>0}$ is the *support* of A.

Similarly as in case of multiple tolerances, we may use several cut sets to approximate X. More precisely, let $X \subseteq U$ and suppose that there exists a fuzzy set A on U and $1 > \alpha_1 > \alpha_2 > \ldots > \alpha_n > 0$ such that

$$A_1 \subseteq X \subseteq A_{\alpha_1} \subseteq A_{\alpha_2} \subseteq \cdots \subseteq A_{\alpha_n}.$$

Our next proposition shows that we can always construct the same tuple

$$(A_1, A_{\alpha_1}, \ldots A_{\alpha_n})$$

using multiple rough sets.

Proposition 11. *Let A be a fuzzy set U and $1 > \alpha_1 > \alpha_2 > \ldots > \alpha_n > 0$. Then there exist a set $X \subseteq U$, an equivalence E on U, and E-compatible tolerances $T_1, \ldots, T_n$ satisfying*

$$(A_1, A_{\alpha_1}, \ldots, A_{\alpha_n}) = (X_E, X^{T_1}, \ldots, X^{T_n}).$$

Proof. Having $(A_1, A_{\alpha_1}, \ldots, A_{\alpha_n})$, we define the equivalences:

$$\begin{aligned} E &= A_1 \times A_1 \cup \{(x,x) \mid x \in U\}, \\ T_1 &= A_{\alpha_1} \times A_{\alpha_1} \cup (U \setminus A_{\alpha_1}) \times (U \setminus A_{\alpha_1}), \\ T_2 &= A_{\alpha_2} \times A_{\alpha_2} \cup (U \setminus A_{\alpha_2}) \times (U \setminus A_{\alpha_2}), \\ &\vdots \\ T_n &= A_{\alpha_n} \times A_{\alpha_n} \cup (U \setminus A_{\alpha_n}) \times (U \setminus A_{\alpha_n}). \end{aligned}$$

It is clear that $E \subseteq T_i$ for any $1 \le i \le n$, so each $T_1, \ldots, T_n$ is E-compatible. We have that

$$\begin{aligned} X_E &= X = A_1, \\ X^{T_1} &= (A_1)^{T_1} = A_{\alpha_1}, \\ X^{T_2} &= (A_1)^{T_2} = A_{\alpha_2}, \\ &\vdots \\ X^{T_n} &= (A_1)^{T_n} = A_{\alpha_n}. \end{aligned}$$

Thus, $(A_1, A_{\alpha_1}, \ldots, A_{\alpha_n}) = (X_E, X^{T_1}, \ldots, X^{T_n})$. □

We end this section by showing that the converse is not true.

Proposition 12. *Let U be a set with at least 3 elements. There exists an equivalence E on U, E-compatible tolerances T_1 and T_2, and a set $X \subseteq U$, such that (X_E, X^{T_1}, X^{T_2}) cannot be given in terms of α-cut sets of some fuzzy set A on U.*

Proof. If $|U| \ge 3$, we may define tolerances T_1 and T_2 on U such that neither $T_1 \subseteq T_2$ nor $T_2 \subseteq T_1$ hold. In addition, let $E = \{(x,x) \mid x \in U\}$. Then trivially T_1 and T_2 are E-compatible. Let us consider the case $T_1 \nsubseteq T_2$ only, because $T_2 \nsubseteq T_1$ can be treated similarly. Now $T_1 \nsubseteq T_2$ means that there is $(x,y) \in T_1$ such that $(x,y) \notin T_2$. We get that $\{x\}^{T_1} \nsubseteq \{x\}^{T_2}$.

Next consider the rough set 3-tuple $(\{x\}_E, \{x\}^{T_1}, \{x\}^{T_2})$. Suppose that there exists a fuzzy set A on U and α_1 and α_2 such that

$$(A_1, A_{\alpha_1}, A_{\alpha_2}) = (\{x\}_E, \{x\}^{T_1}, \{x\}^{T_2}).$$

Because $\alpha_1, \alpha_2 \in [0,1]$, without loss of generality we may assume that $\alpha_1 \ge \alpha_2$. Then $A_{\alpha_1} \subseteq A_{\alpha_2}$ would imply $\{x\}^{T_1} \subseteq \{x\}^{T_2}$, a contradiction. □

These properties mean that every multiple alpha-cuts fuzzy model can be given using multiple rough set model, but not every multiple rough set model can be obtained with some alpha-cuts of a fuzzy set. From this point of view, the multiple rough set model is a more general model of uncertainty than the fuzzy set model with multiple cuts.

4 Conclusions

The paper presented an extension of the traditional rough set model introducing multiple upper approximations using more tolerance relations where the tolerance relations are compatible with the inner equivalence relation. Regarding the main properties of the proposed model, it can be proven that the set of multiple upper approximations rough sets form a complete lattice. In special cases, this lattice is isomorphic with the lattice generated from the base rough set pairs. The proposed model can be used to represent a novel multi-level uncertainty-based approximation of selected base sets. It is shown in the paper that for presenting multiple borders, this approximation model is more general than the widely used fuzzy approximation model.

References

1. Comer, S.D.: On connections between information systems, rough sets, and algebraic logic. In: Algebraic Methods in Logic and Computer Science, pp. 117–124. No. 28 in Banach Center Publications (1993)
2. Grzymala-Busse, J.W.: Rough set strategies to data with missing attribute values. In: Young Lin, T., Ohsuga, S., Liau, C.J., Hu, X. (eds.) Foundations and Novel Approaches in Data Mining. Studies in Computational Intelligence, vol. 9, pp. 197–212. Springer, Heidelberg (2006). https://doi.org/10.1007/11539827_11
3. Järvinen, J.: Knowledge representation and rough sets. Ph.D. dissertation, Department of Mathematics, University of Turku, Finland (1999). TUCS Dissertations 14
4. Järvinen, J., Kovács, L., Radeleczki, S.: Defining rough sets using tolerances compatible with an equivalence. Inf. Sci. **496**, 264–283 (2019)
5. Järvinen, J., Radeleczki, S.: Rough sets determined by tolerances. Int. J. Approximate Reasoning **55**, 1419–1438 (2014)
6. Järvinen, J., Radeleczki, S.: Representing regular pseudocomplemented Kleene algebras by tolerance-based rough sets. J. Aust. Math. Soc. **105**, 57–78 (2018)
7. Pawlak, Z.: Information systems theoretical foundations. Inf. Syst. **6**, 205–218 (1981)
8. Pawlak, Z.: Rough sets. Int. J. Comput. Inf. Sci. **11**, 341–356 (1982)
9. Pomykała, J., Pomykała, J.A.: The Stone algebra of rough sets. Bull. Pol. Acad. Sci. Math. **36**, 495–512 (1988)
10. Qian, Y., Liang, J., Yao, Y., Dang, C.: MGRS: a multi-granulation rough set. Inf. Sci. **180**, 949–970 (2010)
11. Słowiński, R., Vanderpooten, D.: Similarity relation as a basis for rough approximations. ICS Research Report 53/95, Warsaw University of Technology (1995). Also in: Wang, P.P. (ed.) Advances in Machine Intelligence & Soft-Computing, vol. IV, pp. 17–33. Duke University Press, Durham, NC (1997)
12. Wong, S., Ziarko, W.: Comparison of the probabilistic approximate classification and the fuzzy set model. Fuzzy Sets Syst. **21**, 357–362 (1987)
13. Yao, Y.Y.: Generalized rough set models. In: Polkowski, L., Skowron, A. (eds.) Rough Sets in Knowledge Discovery, pp. 286–318. Physica-Verlag, Heidelberg (1998)

14. Yao, Y.Y.: On generalizing rough set theory. In: Wang, G., Liu, Q., Yao, Y., Skowron, A. (eds.) Rough Sets, Fuzzy Sets, Data Mining, and Granular Computing, pp. 44–51. Springer, Berlin, Heidelberg (2003)
15. Yao, Y.: A comparative study of fuzzy sets and rough sets. Information Sciences **109**, 227–242 (1998)
16. Zadeh, L.: Fuzzy sets. Information and Control **8**, 338–353 (1965)

On the Roughly Continuous Real Functions

Zoltán Ernő Csajbók(✉)

Department of Health Informatics, Faculty of Health, University of Debrecen, Sóstói út 2-4, Nyíregyháza 4406, Hungary
csajbok.zoltan@foh.unideb.hu

Abstract. Studying rough calculus was originated by Pawlak in many papers. In this paper, fundamental features of roughly continuous–discontinuous real functions are presented in a systematic manner.

1 Introduction

In the mid 1990s Pawlak relying on the rough set theory (RST) [7,8,14] in many papers initiated the study of rough calculus, see mainly [9,10,12,13]. He invented the investigation of its different subfields such as rough continuity–discontinuity, derivatives–integrals, differential equations, etc. Since then, however, to the best knowledge of the author, relatively little progress has been made in this area.

This paper, after an adequate preparation, systematically summarizes the fundamental features of rough continuity–discontinuity concerning rough real functions. The world of rough real functions is a strange but an interesting one.

In Sect. 2, some important notations are summarized for the sake of fully clarity. Section 3 presents rough numbers. The main part of the paper is Sects. 5–8 which basically deal with the most significant issues of rough continuity–discontinuity.

2 Preliminaries

Let U, V be two classical nonempty sets. A *function* f with domain U and co-domain V is denoted by $f : U \to V$, $u \mapsto f(u)$, where $u \mapsto f(u)$ is the assignment or mapping rule of f. Usually, V^U denotes the set of all functions with domain U and co-domain V. In particular, $f \in V^U$ means that f maps U to V, but its assignment rule is not specified.

If $f, g \in V^U$, the operation $f \odot g$, $\odot \in \{+, -, \cdot, /\}$ and the relation $f \boxdot g$, $\boxdot \in \{=, \neq, \leq, <, \geq, >\}$ are understood by pointwise.

For any $S \subseteq U$, $f(S) = \{f(u) \mid u \in S\} \subseteq V$ is the direct image of S. Especially, $f(U) \subseteq V$ is the *range* of f.

If $a, b \in \mathbb{R}$ $(a \leq b)$, $[a, b] = \{x \in \mathbb{R} \mid a \leq x \leq b\}$ and $]a, b[= \{x \in \mathbb{R} \mid a < x < b\}$ denote *closed* and *open* intervals, respectively. $[a, a] = \{a\}$ is identified with the real number $a \in \mathbb{R}$. In addition, it is easy to interpret the open-closed $]a, b]$ and closed-open $[a, b[$ intervals.

© Springer Nature Switzerland AG 2019
T. Mihálydeák et al. (Eds.): IJCRS 2019, LNAI 11499, pp. 52–65, 2019.
https://doi.org/10.1007/978-3-030-22815-6_5

(a, b) means an ordered pair of real numbers.

Let $\mathbb{R}^+$ be the set of nonnegative real numbers, and $[n] = \{0, 1, \ldots, n\} \subseteq \mathbb{N}$ be a finite set of natural numbers.

3 Rough Numbers

Let I denote a closed interval $I = [0, a]$ ($a \in \mathbb{R}^+$, $a > 0$). A *categorization* of I is a sequence $S_I = \{x_i\}_{i \in [n]} \subseteq \mathbb{R}^+$ where $n \geq 1$ and $0 = x_0 < x_1 < \cdots < x_n = a$.

Remark 1. (i) S_I is often called the *discretization* of I as well.

(ii) A categorization has two components, an interval and a sequence. In the literature, many different forms of these two components exist. For instance, I may be $]a, b[$ with $a, b \in \mathbb{R}$, $a < b$, and $0 \in]a, b[$ [13]; $[a, b]$ with $a, b \in \mathbb{R}^+$, $0 < a < b$; $]-\infty, \infty[$ [13]; $[0, \infty[$ [12]; etc. Accordingly, S_I may be finite or infinite, and may contain negative real numbers. However, for the sake of simplicity, such more general cases are not considered in this paper. In spite of the simplicity of the framework, it is general enough to study the fundamental properties of roughly continuous–discontinuous real functions.

(iii) The notion of categorization referring to as *landmark* is also used in a somewhat different manner in qualitative reasoning [4,5].

(iv) The categorization S_I can be interpreted as a *scale* by which the real numbers can be approximated [12]. With the help of a special scale on $\mathbb{R}^+$, a *measurement system* can be constructed to approximate the accuracy of measurement results ([7], Example 1). □

Let I_S denote an equivalence relation which is generated by the categorization S_I and is defined as follows. If $x, y \in I$, $x I_S y$ if and only if $x = y = x_i \in S_I$ for some $i \in [n]$ or $x, y \in]x_i, x_{i+1}[$ for some $i \in [n-1]$. Hence, the partition I/I_S associated with the equivalence relation I_S is the following:

$$I/I_S = \{\{x_0\},]x_0, x_1[, \{x_1\}, \ldots, \{x_{n-1}\},]x_{n-1}, x_n[, \{x_n\}\}$$

where $[x_i, x_i] = \{x_i\}$ ($i \in [n]$).

The block of the partition I/I_S containing $x \in I$ is denoted by $[x]_{I_S}$. In particular, if $x \in S_I$, $[x]_{I_S} = \{x\}$. If $x \in [x]_{I_S} =]x_i, x_{i+1}[$, $\overline{[x]}_{I_S} = [x_i, x_{i+1}]$ is the *closure* of $[x]_{I_S}$. Of course, when $x \in S_I$, $[x]_{I_S} = \overline{[x]}_{I_S} = \{x\}$. Hence, any $x \in S_I$ is called a *roughly isolated point* in I/I_S.

Evidently I_S is an indiscernibility relation on the real interval I, thus the naming of the following notion is consistent with the standard terminology of rough set theory. The ordered pair (I, I_S) is an I_S*-approximation space.*

In the approximation space (I, I_S), closed intervals of the form $[0, x]$ ($x \in I$) will be approximated.

According to the standard process of rough set theory, I_S-lower and I_S-upper approximations of $[0, x]$ are defined as

$$\underline{I}_S([0,x]) = \{y \in I \mid [y]_{I_S} \subseteq [0,x]\} = \bigcup\{[y]_{I_S} \in I/I_S \mid [y]_{I_S} \subseteq [0,x]\},$$
$$\overline{I}_S([0,x]) = \{y \in I \mid, [y]_{I_S} \cap [0,x] \neq \emptyset\} = \bigcup\{[y]_{I_S} \in I/I_S \mid [y]_{I_S} \cap [0,x] \neq \emptyset\}.$$

With a slight abuse of notation, let us define the following numbers:

$$\underline{I}_S(x) = \sup\{y \in S_I \mid y \leq x\}, \quad \overline{I}_S(x) = \inf\{y \in S_I \mid y \geq x\}.$$

The following lemma is straightforward.

Lemma 1. *Let $x \in I$. With the above notations,*

$$\underline{I}_S([0,x]) = [0, \underline{I}_S(x)],\ \overline{I}_S([0,x]) = [0, \overline{I}_S(x)],\ \textit{and}\ \underline{I}_S(x) \leq x \leq \overline{I}_S(x).$$

Thus, $[x]_{I_S} = [\underline{I}_S(x), \overline{I}_S(x)] = \{x\}$, *if* $x \in S_I$, *and* $[x]_{I_S} =]\underline{I}_S(x), \overline{I}_S(x)[$, *if* $x \notin S_I$.

It is said that the number $x \in I$ is *exact* with respect to the approximation space (I, I_S), if $\underline{I}_S(x) = \overline{I}_S(x)$, otherwise x is *inexact* or *rough* [12]. Of course, $x \in I$ is exact if and only if $x \in S_I$. Hence, the members of I/I_S are called *rough numbers* with respect to the approximation space (I, I_S).

In pursuance of Lemma 1, any inexact number $x \in I$ can be represented by the interval $[x]_{I_S}$ or, equivalently, by the pair of exact numbers $\underline{I}_S(x)$ and $\overline{I}_S(x)$.

Remark 2. In [6], rough real number is studied on the whole real line. The paper [1] presents many similar number constructions, e.g., interval, fuzzy, grey, vague, etc. numbers. □

4 Roughly Constant and Monotone Real Functions

Let $I = [0, a_I]$ and $J = [0, a_J]$ be two closed intervals with $a_I, a_J \in \mathbb{R}^+$, $a_I, a_J > 0$. Let S_I and P_J be the categorizations of I and J, where $S_I = \{x_i\}_{i \in [n]}$, $P_J = \{y_j\}_{j \in [m]} \subseteq \mathbb{R}^+$ with $m, n \geq 1$, $0 = x_0 < x_1 < \cdots < x_n = a_I$ and $0 = y_0 < y_1 < \cdots < y_m = a_J$. The corresponding I_S and J_P-approximation spaces are (I, I_S) and (J, J_P).

To make the blocks of the partition I/I_S easier to handle, they are enumerated as follows.

$$N_I : I/I_S \to [2n], \quad [x]_{I_S} \mapsto \begin{cases} B_{2i} = 2i, & \text{if } \exists i \in [n]([x]_{I_S} = \{x_i\} \subset S_I), \\ B_{2i+1} = 2i+1, & \text{if } \exists i \in [n-1](x \in]x_i, x_{i+1}[). \end{cases}$$

The inverse of N_I is:

$$N_I^{-1} : [2n] \to I/I_S, \quad B_i \mapsto \begin{cases} \{x_{i/2}\}, & \text{if } i \equiv 0 \pmod 2 \\ ]x_{\frac{i-1}{2}}, x_{\frac{i+1}{2}}[, & \text{if } i \equiv 1 \pmod 2 \end{cases}.$$

Evidently, there is a one-to-one correspondence between the equivalent classes of I/I_S and [2n], where the elements of [2n] are referred to as B_i's. Therefore, equivalent classes of I/I_S and B_i's can and *will* be used interchangeably. In particular, $B_i = [x]_{I_S}$ will be written for an appropriate $i \in [2n]$, and then $\overline{[x]}_{I_S}$ and $\overline{B}_i$ will also be used interchangeably.

The equivalent classes of J/J_P can be enumerated in the same way by the help of enumeration function N_J. They are referred to as C_j's ($j \in [2m]$).

Before studying different features of rough real functions, it is important to note that all of them occur as the result of the *superposition* of two scales, S_I and P_J. Altering of S_I and/or P_J may change the nature of the considered features. Nevertheless, it may permit one way or another to "improve" the features of rough real functions.

Definition 1 ([13]). A function $f \in J^I$ is (S_I, P_J)*–constant* or *roughly constant*, if for all $i \in [2n]$, $f(B_i) \subseteq C_j$ for some $j \in [2m]$.

Example 1. In the running example, let $I = [0, x_5]$, $J = [0, y_4]$, and $S_I = \{x_0, x_1, x_2, x_3, x_4, x_5\}$, $P_J = \{y_0, y_1, y_2, y_3, y_4\}$. Accordingly,

$$I/I_S = \{B_0 = \{x_0\}, B_1 =]x_0, x_1[, B_2 = \{x_1\}, B_3 =]x_1, x_2[, B_4 = \{x_2\},$$
$$B_5 =]x_2, x_3[, B_6 = \{x_3\}, B_7 =]x_3, x_4[, B_8 = \{x_4\}, B_9 =]x_4, x_5[, B_{10} = \{x_5\}\},$$
$$J/J_P = \{C_0 = \{y_0\}, C_1 =]y_0, y_1[, C_2 = \{y_1\}, C_3 =]y_1, y_2[, C_4 = \{y_2\},$$
$$C_5 =]y_2, y_3[, C_6 = \{y_3\}, C_7 =]y_3, y_4[, C_8 = \{y_4\}\}.$$

Figure 1 depicts roughly constant functions. □

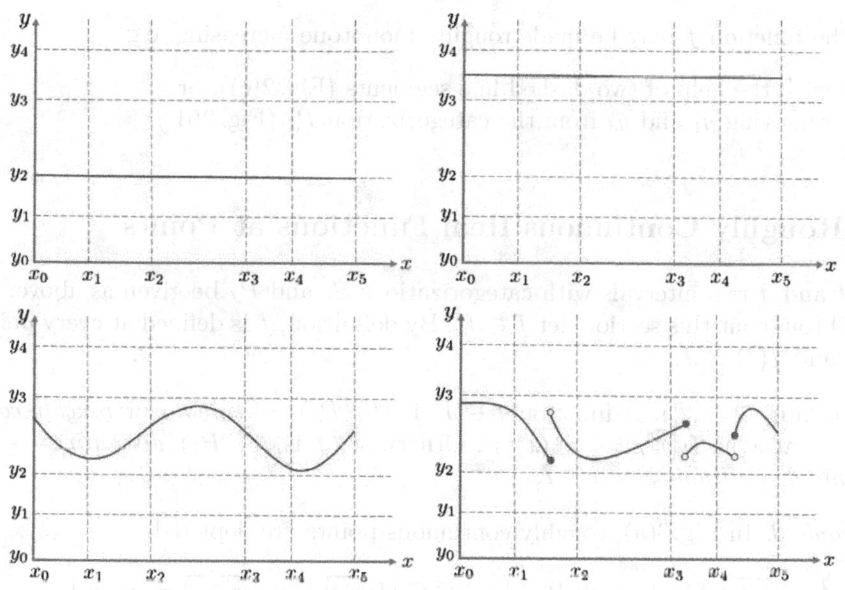

Fig. 1. Roughly constant functions

Definition 2 ([13]). A function $f \in J^I$ is (S_I, P_J)*–monotone increasing* or *roughly monotone increasing*, if

– $N_J([f(B_0)]_{J_P}) = j_0$ for some $j_0 \in [2m]$;

- $j_{i-1} \leq N_J([\inf f(B_i)]_{J_P}) \leq N_J([\sup f(B_i)]_{J_P}) = j_i$, where $j_{i-1}, j_i \in [2m]$ $(i = 1, \ldots, 2n)$.

Roughly monotone decreasing functions can be defined similarly.

Example 2 In Fig. 2(a), the function is roughly monotone increasing, because

- $N_J([f(B_0)]_{J_P}) = 3$
- $3 \leq N_J([\inf f(B_1)]_{J_P}) = 3 \leq N_J([\sup f(B_1)]_{J_P}) = 7$
- $7 \leq N_J([\inf f(B_2)]_{J_P}) = N_J([\sup f(B_2)]_{J_P}) = 7$
- $7 \leq N_J([\inf f(B_3)]_{J_P}) = N_J([\sup f(B_3)]_{J_P}) = N_J([\inf f(B_4)]_{J_P})$ $= N_J([\sup f(B_4)]_{J_P}) = \ldots = N_J([\inf f(B_8)]_{J_P}) = N_J([\sup f(B_8)]_{J_P}) = 7$
- $7 \leq N_J([\inf f(B_9)]_{J_P}) = 7 \leq N_J([\sup f(B_9)]_{J_P}) = 8$
- $8 \leq N_J([\inf f(B_{10})]_{J_P}) = N_J([\sup f(B_{10})]_{J_P}) = 8$

In Fig. 2(b), the function is not roughly monotone increasing, because

- $N_J([f(B_0)]_{J_P}) = 3$, however, $3 > N_J([\inf f(B_1)]_{J_P}) = 1$ (horizontally shaded area);
- $N_J([\inf f(B_5)]_{J_P}) = 5 \leq N_J([\sup f(B_5)]_{J_P}) = 7$ (vertically shaded area), however, $7 > N_J([\inf f(B_6)]_{J_P}) = N_J([\sup f(B_6)]_{J_P}) = 5$.

The function f may be made roughly monotone increasing, e.g.,

(1) with the help of two dashed line segments (Fig. 2(c)), or
(2) removing y_1 and y_3 from the categorization P_J (Fig. 2(d)). □

5 Roughly Continuous Real Functions at Points

Let I and J two intervals with categorizations S_I and P_J be given as above.

Throughout this section, let $f \in J^I$. By definition, f is defined at every point of I, and $f(I) \subseteq J$.

Definition 3 ([12]). A function $f \in J^I$ is (S_I, P_J)*–continuous* or *roughly continuous* at x, if $f(\overline{[x]}_{I_S}) \subseteq \overline{[f(x)]}_{J_p}$. Otherwise, f is (S_I, P_J)*–discontinuous* or *roughly discontinuous* at $x \in I$.

Example 3. In Fig. 3(a), roughly continuous points are depicted.

$$f(\overline{[x^i]}_{I_S}) = f([x_1, x_2]) \subseteq]y_3, y_4[\subseteq \overline{[f(x^i)]}_{J_P} = \overline{]y_3, y_4[} = [y_3, y_4]$$
$$f(\overline{[x^{ii}]}_{I_S}) = f([x_2, x_3]) \subseteq]y_3, y_4[\subseteq \overline{[f(x^{ii})]}_{J_P} = \overline{]y_3, y_4[} = [y_3, y_4]$$

that is, f is roughly continuous at x^i and x^{ii} by definition.

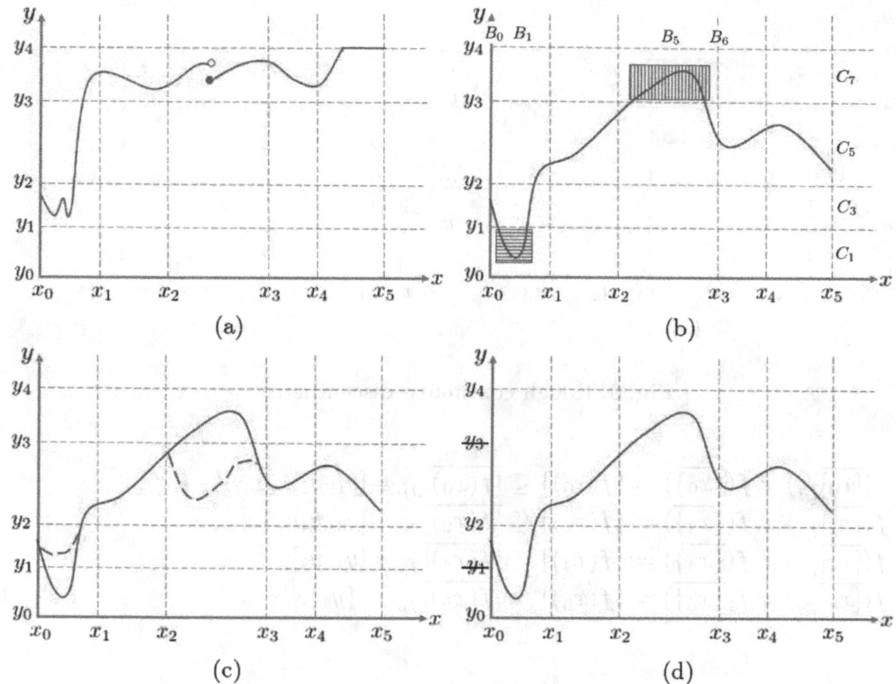

Fig. 2. Rough monotonicity

Figure 3(b) shows roughly discontinuous points.

$$f(\overline{[x^{iii}]}_{I_S}) = f([x_0, x_1]) \subseteq]y_1, y_4[\nsubseteq \overline{[f(x^{iii})]}_{J_P} = \overline{]y_1, y_2[} = [y_1, y_2]$$
$$f(\overline{[x^{iv}]}_{I_S}) = f([x_2, x_3]) \subseteq]y_3, y_4[\cup \{f(x^{iv})\} \nsubseteq \overline{[f(x^{iv})]}_{J_P} = \overline{]y_2, y_3[} = [y_2, y_3],$$
$$f(\overline{[x^{v}]}_{I_S}) = f([x_3, x_4]) \subseteq]y_3, y_4[\cup \{f(x_4)\} \nsubseteq \overline{[f(x^{v})]}_{J_P} = \overline{]y_3, y_4[} = [y_3, y_4],$$
$$f(\overline{[x^{vi}]}_{I_S}) = f([x_4, x_5]) \subseteq]y_3, y_4] \cup \{f(x_4)\} \nsubseteq \overline{[f(x^{vi})]}_{J_P} = \overline{\{y_4\}} = \{y_4\},$$

that is, f is roughly discontinuous at points x^{iii}, x^{iv}, x^{v}, and x^{vi} by definition.

This example also shows that a function which is continuous at a point in the classical sense, it is not necessary roughly continuous at the same point (e.g., x^{iii}, x^{v}, x^{vi}). On the contrary, if a function is discontinuous at a point in the classical sense, it may be roughly continuous at the same point (e.g., x^{ii}, x_4). □

Proposition 1. *A function $f \in J^I$ is (S_I, P_J)–continuous at every $x \in S_I$ roughly isolated point.*

Proof. For any $x \in S_I$, $[x]_{I_S} = \{x\}$. And so, $f(\overline{[x]}_{I_S}) = f(\overline{\{x\}}) = \{f(x)\} \subseteq \overline{[f(x)]}_{J_p}$ □

Example 4. Figure 4 demonstrates that the function f is roughly continuous at various roughly isolated points.

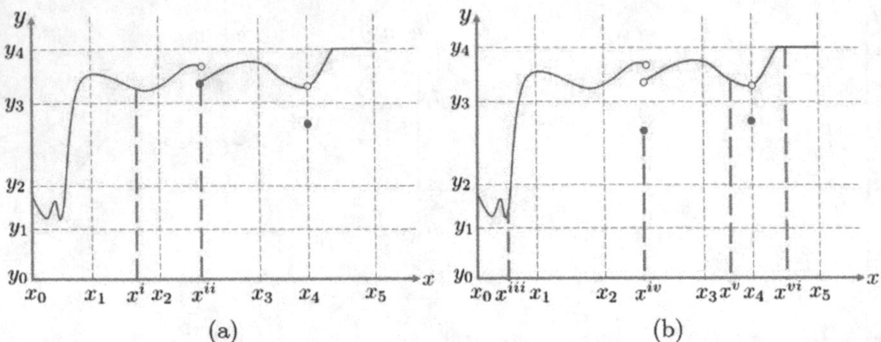

Fig. 3. Rough continuity–discontinuity

- $f(\overline{[x_0]}_{I_S}) = f(\overline{\{x_0\}}) = \{f(x_0)\} \subseteq \overline{[f(x_0)]}_{J_P} = [y_1, y_2]$;
- $f(\overline{[x_2]}_{I_S}) = f(\overline{\{x_2\}}) = \{f(x_2)\} \subseteq \overline{[f(x_2)]}_{J_P} = [y_3, y_4]$;
- $f(\overline{[x_4]}_{I_S}) = f(\overline{\{x_4\}}) = \{f(x_4)\} \subseteq \overline{[f(x_4)]}_{J_P} = [y_2, y_3]$;
- $f(\overline{[x_5]}_{I_S}) = f(\overline{\{x_5\}}) = \{f(x_5)\} = \overline{[f(x_5)]}_{J_P} = \{y_4\}$. □

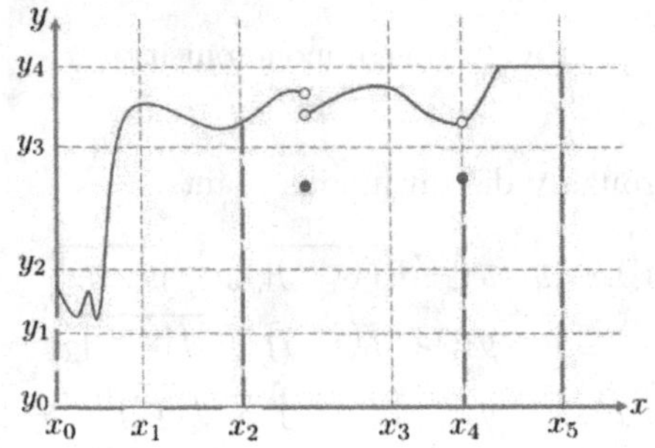

Fig. 4. Rough continuity at roughly isolated points

6 Roughly Continuous Real Functions on Sets

Definition 4. A function $f \in J^I$, is (S_I, P_J)*–continuous* on $I' \subseteq I$ or *roughly continuous* on $I' \subseteq I$, if f is (S_I, P_J)–continuous at every point of I'. Otherwise, f is *not roughly continuous on* I'.

Function f is (S_I, P_J)*–discontinuous* on $I' \subseteq I$ or *roughly discontinuous* on $I' \subseteq I$, if f is (S_I, P_J)–discontinuous at every point of I'.

Remark 3. Definition 4 differentiates between not rough continuity and rough discontinuity on sets. f is not roughly continuous, if it has at least one roughly discontinuous point, but roughly discontinuous, if it is discontinuous at every point. □

Definition 5. Let $x \in I$, but $x \notin S_I$. Let I_i^{PC} and I_i^{CP} denote the following proper subsets of $B_i = [x]_{I_S}$ ($i \in [2n]$, $i \equiv 1 \pmod 2$) which are defined as

$$I_i^{PC} = \{x' \in [x]_{I_S} \mid f(x') = y_j \text{ for any } j \in [m], \text{where } f \text{ touches } y = y_j\} \subsetneq B_i$$
$$I_i^{CP} = \{x' \in [x]_{I_S} \mid f(x') = y_j, \text{ for any } j \in [m], \text{where } f \text{ intersects } y = y_j\} \subsetneq B_i.$$

Let $I_i^{PC,CP} = I_i^{PC} \cup I_i^{CP}$.

Remark 4. The acronyms "PC" and "CP" refer to "Point of Contact" and "Cross Point", respectively. □

Lemma 2. *With the notations of Definition 5, if $I_i^{PC,CP} \neq \emptyset$, $f(\overline{B}_i)$ contains at least one open interval from J/J_P whose intersection with $f(\overline{B}_i)$ is nonempty.*

Proof. If $I_i^{CP} \neq \emptyset$ and f intersects, e.g., the straight line $y = y_j$, $f(\overline{B}_i) \cap C_{2j-1} \neq \emptyset$ and $f(\overline{B}_i) \cap C_{2j+1} \neq \emptyset$.

If $I_i^{PC} \neq \emptyset$ and f touches, e.g., the straight line $y = y_{j'}$, $f(\overline{B}_i) \cap C_{2j'-1} \neq \emptyset$ or $f(\overline{B}_i) \cap C_{2j'+1} \neq \emptyset$ depending on f. □

Proposition 2. *Let $\bigcup_i I_i^{PC,CP} \neq \emptyset$. Function f is (S_I, P_J)-discontinuous on $\bigcup_i I_i^{PC,CP}$.*

Proof. For some $I_i^{PC,CP} \neq \emptyset$, let $x \in I_i^{PC,CP}$ be an arbitrary point in $\bigcup_i I_i^{PC,CP}$. Then, $x \in B_i$ ($i \in [2n]$, $i \equiv 1 \pmod 2$) and $f(x) = y_j$ for some $j \in [m]$, i.e., f touches or intersects the straight line $y = y_j$ at $x \in B_i$. Applying Lemma 2, $f(\overline{B}_i)$ contains at least one open interval from J/J_P whose intersection with $f(\overline{B}_i)$ is nonempty. Consequently, $f(\overline{B}_i) = f(\overline{[x]_{I_S}}) \nsubseteq \overline{[f(x)]_{J_P}} = \{y_j\}$. □

It is easy to see the following important corollary.

Corollary 1. *If f is (S_I, P_J)-continuous on I, $\bigcup_i I_i^{PC,CP} = \emptyset$.*

Proof. On the contrary, let us assume that $\bigcup_i I_i^{PC,CP} \neq \emptyset$. Then, applying Proposition 2, there exists at least one point in I at which f is roughly discontinuous. However, it contradicts the condition that f is roughly continuous on I. □

7 Rough Jump Discontinuity

Corollary 1 means geometrically that a roughly continuous function neither touches nor intersects any straight line $y = y_j$ ($j \in [m]$) on every open interval of I/I_S. Nevertheless, the converse statement is not true. Namely, if a function neither touches nor intersects any straight line $y = y_j$ ($j \in [m]$) on every open interval of I/I_S, it may be not roughly continuous on I. That is, there must be a third kind of discontinuity which may damage the rough continuity of a function.

Definition 6. The rough discontinuity of f is called

- the *rough jump discontinuity of the first kind*, if it is derived from touching a straight line $y = y_j$ for some $j \in [m]$;
- the *rough jump discontinuity of the second kind*, if it is derived from intersecting a straight line $y = y_j$ for some $j \in \{1, 2, \ldots, m-1\}$;
- any other type of discontinuity is called the *rough jump discontinuity of the third kind.*

It should be noted that when f touches or intersects a straight line $y = y_j$ for some $j \in [m]$ at a roughly isolated point, f is roughly continuous at this point automatically by Proposition 1.

The following proposition shows that the rough jump discontinuity of the third kind actually exists.

Proposition 3. *With the notations of Definition 5, let $I_i^{PC,CP} = \emptyset$. Then, one can construct a function f which is (S_I, P_J)–discontinuous on B_i.*

Proof. The first variant of f. Let $I_i^{PC,CP} = \emptyset$ and $f(\overline{B}_i) \subseteq \overline{C}_j$ for some $j \in [2m]$ ($j \equiv 1 \pmod 2$) except only one point $x' \in B_i$ in such a way that $f(x') \notin \overline{C}_j$. Since $f(x') \notin \overline{C}_j$, so $f(x') \in \overline{C}_{j'}$ for some $j' \in [2m]$ with $j' \neq j-1, j, j+1$. Then, $f(\overline{[x']_{I_S}}) = f(\overline{B}_i) \subseteq \overline{C}_j \cup \{f(x')\} \not\subseteq \overline{[f(x')]}_{J_P} = \overline{C}_{j'}$. In addition, for any $x'' \in B_i \setminus \{x'\}$, $f(\overline{[x'']_{I_S}}) = f(\overline{B}_i) \subseteq \overline{C}_j \cup \{f(x')\} \not\subseteq \overline{[f(x'')]}_{J_P} = \overline{C}_j$.

The last two statements mean that f is roughly discontinuous on B_i.

The second variant of f. Let $f(B_i) \subseteq C_j$ for some $j \in [2m]$ ($j \equiv 1 \pmod 2$), and $f(x_{\frac{i-1}{2}}) \notin \overline{C}_j$ and/or $f(x_{\frac{i+1}{2}}) \notin \overline{C}_j$. Then, similarly to the first variant of f, it can be proved that f is also roughly discontinuous on B_i. □

Example 5. Figure 5(a) depicts rough jump discontinuity of the third kind in order to illustrate Proposition 3.

The first variant. $I_5^{PC,CP} = \emptyset$ and $f(\overline{B}_5) = f([x_2, x_3]) \subseteq \overline{C}_7 = [y_3, y_4]$ except only one point $x' \in B_5$ in such a way that $f(x') \in \overline{C}_5 = [y_2, y_3]$. Then,

$$f(\overline{[x']_{I_S}}) = f([x_2, x_3]) \subseteq [y_3, y_4] \cup \{f(x')\} \not\subseteq \overline{[f(x')]}_{J_P} = [y_2, y_3].$$

Let $x'' \in B_5 \setminus \{x'\}$. Then,

$$f(\overline{[x'']_{I_S}}) = f(\overline{B}_5) \subseteq [y_3, y_4] \cup \{f(x')\} \not\subseteq \overline{[f(x'')]}_{J_P} = [y_3, y_4].$$

In conclusion, f is roughly discontinuous on B_5. □

The second variant. $I_7^{PC,CP} = \emptyset$ and $f(B_7) = f(]x_3, x_4[) \subseteq C_7 =]y_3, y_4[$ and $f(x_4) \notin \overline{C}_7 = [y_3, y_4]$. Although, f is roughly continuous at x_4 because it is a roughly isolated point, x_4 makes f roughly discontinuous on B_7.

In order to show it, let $x \in B_7 =]x_3, x_4[$. Then,

$$f(\overline{[x]}_{I_S}) = f([x_3, x_4]) \subseteq]y_3, y_4[\cup \{f(x_4)\} \not\subseteq \overline{[f(x)]}_{J_P} = [y_3, y_4]. \qquad \square$$

Rough jump discontinuity of the third kind may also arise, but not necessarily, when $I_i^{PC} = B_i$. (In this case, the notion of I_i^{PC} is temporarily used in an extended way.) In Fig. 5(b), $I_1^{PC} = B_1$, $I_3^{PC} = B_3$, $I_9^{PC} = B_9$. Nevertheless, function f is roughly continuous on B_1, but roughly discontinuous on B_3 and B_9 owing to the rough jump discontinuity of the third kind. In conclusion, f is

- roughly continuous on B_1, because
  $f(\overline{[x]}_{I_S}) = \{y_1\} = \overline{[f(x)]}_{J_P}$ $(x \in B_1)$;
- roughly discontinuous on B_3, because
  $f(\overline{[x]}_{I_S}) = \{y_1\} \cup \{y_2\} \not\subseteq \overline{[f(x)]}_{J_P} = \{y_2\}$ $(x \in B_3)$;
- roughly continuous on B_5, because
  $f(\overline{[x]}_{I_S}) = \{y_2\} \cup \{f(x)\} \subseteq \overline{[f(x)]}_{J_P} = [y_2, y_3]$ $(x \in B_5)$;
- roughly continuous on B_7, because
  $f(\overline{[x]}_{I_S}) = \{f(x_3)\} \cup \{f(x)\} \subseteq \overline{[f(x)]}_{J_P} = [y_2, y_3]$ $(x \in B_7)$;
- roughly discontinuous on B_9, because
  $f(\overline{[x]}_{I_S}) = \{f(x_4)\} \cup \{y_4\} \not\subseteq \overline{[f(x)]}_{J_P} = \{y_4\}$ $(x \in B_9)$.

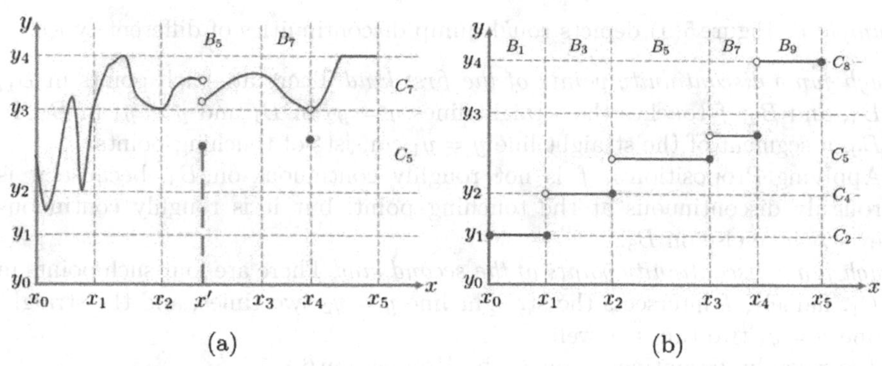

Fig. 5. Rough jump discontinuities

Proposition 4. *A function $f \in J^I$ is (S_I, P_J)–continuous on I if and only if f does not have jump discontinuity of any kind.*

Proof. It follows from Definition 6 and Corollary 1, Proposition 3. $\square$

Proposition 5. *With the notations of Definition 5, let $I_i^{PC} \neq \emptyset$ but $I_i^{CP} = \emptyset$. In addition, let us assume that f does not have any rough jump discontinuity of the third kind on $\overline{B}_i$. Then, f is (S_J, P_J)–discontinuous on I_i^{PC}, but (S_J, P_J)–continuous on $B_i \setminus I_i^{PC}$.*

Proof. The rough discontinuity of f on I_i^{PC} follows directly from Proposition 2.

Turning to the second statement of the proposition, let $I_i^{CP} = \emptyset$ but $I_i^{PC} \neq \emptyset$ and assume that f does not have any rough jump discontinuity of the third kind on $\overline{B}_i$. Taken together, these conditions mean that f does not intersect any straight line $y = y_j$, but touches at least one but at most two straight lines, maybe more times, on B_i.

Let us assume that f touches the straight lines $y = y_j$ and/or $y = y_{j+1}$ for some $j = 0, 1, \ldots, m-1$. Then, $f(\overline{[x]}_{I_S})$ is the subset of one of the intervals $[y_j, y_{j+1}[,]y_j, y_{j+1}]$, $[y_j, y_{j+1}]$ depending on whether f touches either $y = y_j$ or $y = y_{j+1}$, or both of them.

Let $x \in B_i \setminus \underline{I_i^{PC}}$. In this case, f touches neither y_j, nor $\underline{y_{j+1}}$, i.e., $f(x) \in]y_j, y_{j+1}[$, and so $\overline{[f(x)]}_{J_P} = [y_j, y_{j+1}]$. Therefore, $f(\overline{[x]}_{I_S}) \subseteq \overline{[f(x)]}_{J_P}$. □

Proposition 6. *With the notations of Definition 5, if $I_i^{CP} \neq \emptyset$, function f is (S_J, P_J)-discontinuous on B_i.*

Proof. If $I_i^{CP} \neq \emptyset$, f intersects at least one straight line $y = y_j$ for some $j \in \{1, \ldots, m-1\}$. Then, the intersections of $f(\overline{[x]}_{I_S})$ with at least the following two intervals C_{2j-1}, C_{2j+1} are nonempty. However, $\overline{[f(x)]}_{J_P}$ forms only exactly one interval from J/J_P. Consequently, $f(\overline{[x]}_{I_S}) \subseteq \overline{[f(x)]}_{J_P}$ cannot hold for any $x \in B_i$. □

Example 6. Figure 5(a) depicts rough jump discontinuities of different types.

Rough jump discontinuity points of the first kind There are such points in B_1, B_3, and B_9: f touches the straight lines $y = y_2$ in B_1 and $y = y_4$ in B_3. In B_9, a segment of the straight line $y = y_4$ consists of touching points.
Applying Proposition 5, f is not roughly continuous on B_3, because it is roughly discontinuous at the touching point, but it is roughly continuous everywhere else on B_3.

Rough jump discontinuity points of the second kind There are four such points in B_1, namely, f intersects the straight line $y = y_2$ two times, and the straight line $y = y_3$ two times as well.
f is roughly discontinuous on B_1 by Proposition 6.

Rough jump discontinuity of the third kind There is a jump discontinuity of the third kind on B_5 owing to x'.
Although, f is roughly continuous at x_4 because it is a roughly isolated point, it causes rough jump discontinuities of the third kind on B_7 and B_9. □

8 Rough Darboux Property

Definition 7 ([11])**.** A function $f \in J^I$ has the (S_I, P_J)*–Darboux property* or *rough Darboux property,* if for all $i \in [2n]$ $f(B_i) \subseteq C_j$ for some $j \in [2m]$ in such a way that for any interval pair (B_i, B_{i+1}) $(i = 0, 1, 2, \ldots, 2n-1)$,

$$N_J(f(B_{i+1})) = N_J(f(B_i)) + \alpha \text{ with } \alpha \in \{-1, 0, 1\}.$$

Remark 5. In the classical real analysis, the Darboux property means that if $f : [a, b] \to \mathbb{R}$ is real value function defined on a closed bounded interval and k is a number between $f(a)$ and $f(b)$, then there is at least one point $c \in]a, b[$ in such a way that $f(k) = c$. This property is also known as the Intermediate Value Property (IVP). Both the Darboux property and IVP have many other formulations and generalizations, in some cases they are not equivalent.

> Until the work of Darboux in 1875 some mathematicians believed that this property actually implied continuity of $f(x)$. Darboux showed that there are discontinuous functions with the property of Darboux. ([3], p. 111)

According to Darboux's famous theorem, Intermediate Value Property holds for every derivative function independently of whether it is continuous or not (see, e.g., [2], Theorem 6.2.12, p. 178).

Intermediate Value Property expresses an intuitive property of continuous functions. Pawlak's notion of Darboux property also captures this intuitive property of roughly continuous functions, of course, in the roughly real function context. □

Proposition 7. *If f (S_I, P_J)-continuous on I, f has (S_I, P_J)-Darboux property.*

Proof. Applying Proposition 4, f does not have rough jump discontinuity of any kind. Hence, for all $i \in [2n]$, $f(B_i) \subseteq C_j$ holds for some $j \in [2m]$. Moreover, $f(\overline{B}_i) \subseteq \overline{C}_j$ also holds by the definition of rough continuity (Definition 3). (It is noted, that $f(\overline{B}_i) \subseteq \overline{C}_j$ does not imply necessarily the inclusion $f(B_i) \subseteq C_j$.)

- Let $f(B_i) = \{y_j\} = C_{2j}$ for some $i = 0, 1, \ldots, 2n-1$, where $\{y_j\} \in P_J$ $(j \in [m])$ is a roughly isolated point. The function f can leave this straight line segment only through its endpoint $\left(x_{\frac{i+1}{2}}, y_j\right)$, otherwise f would not be roughly continuous on B_i. It means that only the following three cases are possible:

  (1) $f(B_{i+1}) \subseteq C_{2j+1}$, i.e., $\alpha = 1$;
  (2) $f(B_{i+1}) \subseteq C_{2j}$, i.e., $\alpha = 0$;
  (3) $f(B_{i+1}) \subseteq C_{2j-1}$, i.e., $\alpha = -1$.

- Let $f(B_i) = C_j$ for some $i = 0, 1, \ldots, 2n-1$, where $C_j \in P_J$ is an open interval. In this case $j \in [2m]$ with $j \equiv 1 \pmod 2$. The function f can leave the open interval C_j only through

  (1) its endpoint $\left(x_{\frac{i+1}{2}}, y_{\frac{j+1}{2}}\right)$, in which case $f(B_{i+1}) \subseteq C_{j+1}$, i.e., $\alpha = 1$;
  (2) the open interval $\left]y_{\frac{j-1}{2}}, y_{\frac{j+1}{2}}\right[$, in which case $f(B_{i+1}) \subseteq C_j$, i.e., $\alpha = 0$;
  (3) its endpoint $\left(x_{\frac{i+1}{2}}, y_{\frac{j-1}{2}}\right)$, in which case $f(B_{i+1}) \subseteq C_{j-1}$, i.e., $\alpha = -1$. □

Example 7. In Fig. 6(a), f is roughly continuous on I and possesses the rough Darboux property:

- $N_J(f(B_0)) = 2$, $N_J(f(B_1)) = 2 + 1 = 3$, $N_J(f(B_2)) = 3 + 1 = 4$,

– $N_J(f(B_3)) = 4+1 = 5$, $N_J(f(B_4)) = 5+1 = 6$, $N_J(f(B_5)) = 6+0 = 6$,
– $N_J(f(B_6)) = 6+0 = 6$, $N_J(f(B_7)) = 6-1 = 5$, $N_J(f(B_8)) = 5+0 = 5$,
– $N_J(f(B_9)) = 5+0 = 5$, $N_J(f(B_{10})) = 5-1 = 4$. □

Proposition 8. *If f has the (S_I, P_J)-Darboux property, f is not necessarily roughly continuous.*

Proof. This statement is proved by an example in which the function f has the rough Darboux property, but it is not roughly continuous.

Let $f(x) = y_2$ on I except that $f(x_3) \in C_3$ (see Fig. 6(b)).

f has the rough Darboux property because

– $N_J(f(B_0))=4$,
– $N_J(f(B_1))=N_J(f(B_2))=N_J(f(B_3))=N_J(f(B_4))=N_J(f(B_5))=4+0=4$,
– $N_J(f(B_6))=4-1=3$,
– $N_J(f(B_7))=N_J(f(B_8))=N_J(f(B_9))=N_J(f(B_{10}))=3+1=4$,

but f is not roughly continuous on I because

– for any $x' \in B_5$, $f(\overline{[x']_{J_P}}) = \{y_2\} \cup \{f(x_3)\} \not\subseteq \overline{[f(x')]_{J_P}} = \{y_2\}$,
– for any $x'' \in B_7$, $f(\overline{[x'']_{J_P}}) = \{y_2\} \cup \{f(x_3)\} \not\subseteq \overline{[f(x'')]_{J_P}} = \{y_2\}$. □

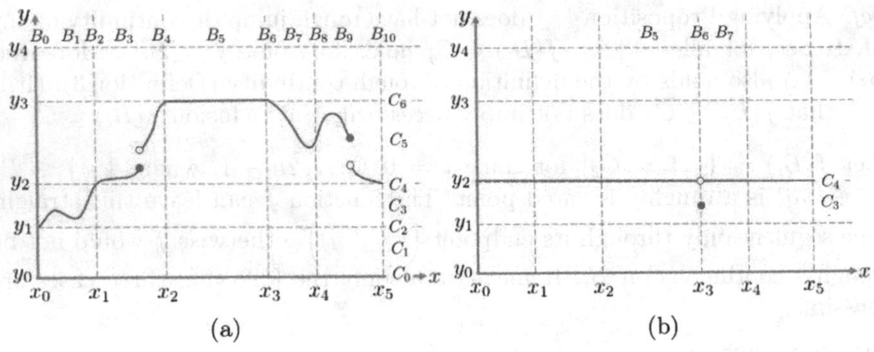

Fig. 6. Rough Darboux property

9 Conclusions

In this paper, the most fundamental properties of rough continuity of rough real functions have been studied systematically. It is a subfield of the rough calculus which was originated by Pawlak in the mid 1990s. Among other things, such a notion of rough Darboux property has been proposed which preserves its classical feature: rough continuous functions have rough Darboux property, but the rough Darboux property does not imply the rough continuity of rough functions.

Acknowledgement. The authors would like to thank the anonymous referees for their useful comments and suggestions.

References

1. Alsawy, A.A., Hefny, H.A.: On uncertain granular numbers. Int. J. Comput. Appl. **62**(18), 20–27 (2013)
2. Bartle, R., Sherbert, D.: Introduction to Real Analysis, 4th edn. Wiley, Hoboken (2011). Incorporated
3. Halperin, I.: Discontinuous functions with the darboux property. Can. Math. Bull. **2**(2), 111–118 (1959)
4. Kuipers, B.: Qualitative simulation. Artif. Intell. **29**(3), 289–338 (1986)
5. Kuipers, B.: Qualitative reasoning: modeling and simulation with incomplete knowledge. Automatica **25**(4), 571–585 (1989)
6. Michalak, M.: Rough numbers and rough regression. In: Kuznetsov, S.O., Ślęzak, D., Hepting, D.H., Mirkin, B.G. (eds.) RSFDGrC 2011. LNCS (LNAI), vol. 6743, pp. 68–71. Springer, Heidelberg (2011). https://doi.org/10.1007/978-3-642-21881-1_12
7. Pawlak, Z.: Rough sets. Int. J. Comput. Inf. Sci. **11**(5), 341–356 (1982)
8. Pawlak, Z.: Rough Sets: Theoretical Aspects of Reasoning about Data. Kluwer Academic Publishers, Dordrecht (1991)
9. Pawlak, Z.: Rough real functions, vol. 50. Institute of Computer Science Report, Warsaw University of Technology, Warsaw (1994)
10. Pawlak, Z.: On some issues connected with roughly continuous functions, vol. 21. Institute of Computer Science Report, Warsaw University of Technology, Warsaw (1995)
11. Pawlak, Z.: Rough calculus, vol. 58. Institute of Computer Science Report, Warsaw University of Technology, Warsaw (1995)
12. Pawlak, Z.: Rough sets, rough relations and rough functions. Fundam. Inform. **27**(2/3), 103–108 (1996)
13. Pawlak, Z.: Rough real functions and rough controllers. In: Lin, T., Cercone, N. (eds.) Rough Sets and Data Mining: Analysis of Imprecise Data, pp. 139–147. Kluwer Academic Publishers, Boston (1997)
14. Pawlak, Z., Skowron, A.: Rudiments of rough sets. Inf. Sci. **177**(1), 3–27 (2007)

On Topologies Defined by Binary Relations in Rough Sets

Michiro Kondo(✉)

Tokyo Denki University, Tokyo 120-8551, Japan
mkondo@mail.dendai.ac.jp

Abstract. We consider relationship between binary relations in approximation spaces and topologies defined by them. In any approximation space (X, R), a reflexive closure R_ω determines an Alexandrov topology $\mathcal{T}_{(R_\omega)}$ and, for any Alexandrov topology $\mathcal{T}$ on X, there exists a reflexive relation $R_\mathcal{T}$ such that $\mathcal{T} = \mathcal{T}_R$. From the result, we also obtain that any Alexandrov topology satisfying (clop), A is open if and only if A is closed, can be characterized by reflexive and symmetric relation.

Moreover, we provide a negative answer to the problem left open in [1].

Keywords: Approximation space · (Alexandrov) Topology · Residuated lattice

1 Introduction

Since Pawlak [6] introduced a notion of *rough sets* in 1982, many papers about rough sets are published and extended to more general cases. One of the most important concept in rough sets is an approximation space (X, R), where X is a finite non-empty set and R is an equivalence relation on X. Now, different kinds of generalizations of approximation spaces are obtained by replacing X to be an infinite set and the equivalence relation R to be an arbitrary binary relation on X. In this paper, we treat an approximation space (X, R) of generalized rough sets, that is, we do not restrict X to be finite and moreover R is not always an equivalence relation. We consider fundamental topological properties of approximation spaces induced by binary relations and prove that

1. For any binary relation R on X, a reflexive closure R_ω of R forms an Alexandrov topology $\mathcal{T}_{(R_\omega)} = \{A \subseteq X \mid (R_\omega)_-(A) = A\}$.
2. A topology $\mathcal{T}^* = \{A \subseteq X \mid A \; is \; R-open\}$ defined by a binary relation R is identical with the topology $\mathcal{T}_{(R_\omega)}$.

Moreover, we consider another topology (uniform topology) on residuated lattices which are algebraic semantics of fuzzy logics and solve an open problem [1] left open by providing a counterexample.

This work was supported by Tokyo Denki University Science Promotion Fund (Q18K-01).

© Springer Nature Switzerland AG 2019
T. Mihálydeák et al. (Eds.): IJCRS 2019, LNAI 11499, pp. 66–77, 2019.
https://doi.org/10.1007/978-3-030-22815-6_6

2 Preliminaries

Let (X, R) be an approximation space of a generalized rough set, that is, X is a non-empty set and R is a binary relation on X. We define three operators R (we use the same symbol as the approximation space for the sake of simplicity), R_- and R_+ as follows: For every $x \in X$ and $A \subseteq X$,

$$\begin{aligned} R(x) &= \{y \in X \mid xRy\} \\ R(A) &= \{y \in X \mid \exists y \in A \text{ s.t. } xRy\} = \cup_{x\in A} R(x) \\ R_-(A) &= \{x \in X \mid R(x) \subseteq A\} = \{x \in X \mid \forall y\,(xRy \to y \in A)\} \\ R_+(A) &= \{x \in X \mid R(x) \cap A \neq \emptyset\} = \{x \in X \mid \exists y \in A \text{ s.t. } xRy\} \end{aligned}$$

It is clear that

$$R_+(A) = (R_-(A^c))^c \quad \text{and} \quad R_-(A) = (R_+(A^c))^c$$

Moreover, we have

Proposition 1 *For all $A, B \subseteq X$,*

$$R(A) \subseteq B \ \Leftrightarrow \ A \subseteq R_-(B),$$

that is, the operator R is a left adjoint operator of R_-.

Proof. Suppose $R(A) \subseteq B$ and $x \in A$. For all $y \in X$, if xRy, since $x \in A$, then we have $y \in R(A) \subseteq B$, that is, $x \in R_-(B)$. This means that $A \subseteq R_-(B)$.

Conversely, we assume that $A \subseteq R_-(B)$ and $y \in R(A)$. There exists $x \in A$ such that xRy. This implies that $x \in A \subseteq R_-(B)$ and hence $y \in B$. We get $R(A) \subseteq B$. □

We note that $R(A) = (R^{-1})_+(A)$ for all $A \subseteq X$. Since R_- and R_+ are dual operators, we mainly treat R_- in this paper. With respect to the operator R_-, we have

Proposition 2 *For any relation R on X,*

1. $A \subseteq B \implies R_-(A) \subseteq R_-(B)$;
2. $R_-(\bigcap_\lambda A_\lambda) = \bigcap_\lambda R_-(A_\lambda)$;
3. $\bigcup_\lambda R_-(A_\lambda) \subseteq R_-(\bigcup_\lambda A_\lambda)$;
4. $xRy \Leftrightarrow x \in (R_-(\{y\}^c))^c$ *for all* $x, y \in X$.

Corollary 1 *Let R and S be binary relations on X. If the operators R_-, S_- are identical, then so the two relations are, that is,*

$$R_- = S_- \iff R = S.$$

Proof. Suppose that $R \neq S$. Since there exists $(x, y) \in X \times X$ such that xRy but *not* xSy. Since xRy, we have

$$x \in (R_-(\{y\}^c))^c = (S_-(\{y\}^c))^c.$$

This means that xSy. But this is a contradiction. Thus, if $R_- = S_-$ then we have $R = S$. □

For a binary relation R on X,

R is *serial* $\Leftrightarrow \forall x\, \exists y\, xRy$.
R is *reflexive* $\Leftrightarrow \forall x\, xRx$.
R is *symmetric* $\Leftrightarrow \forall x\, \forall y\, (xRy \to yRx)$.
R is *transitive* $\Leftrightarrow \forall x\, \forall y\, \forall z\, (xRy \wedge yRz \to xRz)$.
R is *weakly dense* $\Leftrightarrow \forall x\, \forall y\, (xRy \to \exists z\, (xRz \wedge zRy))$.
R is *Euclidean* $\Leftrightarrow \forall x\, \forall y\, \forall z\, (xRy \wedge xRz \to yRz)$.

We have following results which are reminiscent of correspondence between R_- (R_+) operators and modal operators $\Box$ ($\Diamond$) in modal logics:

Proposition 3 *For any binary relation R on X,*

1. *R is serial $\Longleftrightarrow R_-(A) \subseteq R_+(A)$ for all $A \subseteq X$.*
2. *R is reflexive $\Longleftrightarrow R_-(A) \subseteq A$ for all $A \subseteq X$;*
3. *R is symmetric $\Longleftrightarrow A \subseteq R_-((R_-(A^c))^c) = R_+(A)$ for all $A \subseteq X$;*
4. *R is transitive $\Longleftrightarrow R_-(A) \subseteq R_-(R_-(A))$ for all $A \subseteq X$.*
5. *R is weakly dense $\Longleftrightarrow R_+(A) \subseteq R_+(R_+(A))$ for all $A \subseteq X$.*
6. *R is Euclidean $\Longleftrightarrow R_+(A) \subseteq R_-(R_+(A))$ for all $A \subseteq X$.*

The results above makes us to introduce topologies on approximation spaces of generalized rough sets, which is an analogy to do for modal logics.

Now we consider a following family of subsets constructed by the operator R_-:

$$\mathcal{T}_R = \{A \subseteq X \mid R_-(A) = A\}$$

Then, it naturally occurs questions:

Q1: Under what conditions, does a family $\mathcal{T}_R$ of subsets form a topology on an approximation space (X, R)?
Q2: If we consider an approximation space (X, R) based on other algebras, then what properties does have the topology $\mathcal{T}_R$ on (X, R)?

With respect to the second question Q2, there is a problem left open in [1]:

Let (X, τ) be a topological residuated lattice, that is, $X = (X, \wedge, \vee, \odot, \to, 0, 1)$ is a residuated lattice, τ is a topology on X and two operations $\odot$, $\to$ are continuous with respect to τ. Then, does the following hold?

$$\odot \text{ is continuous if and only if } \to \text{ is continuous.}$$

We provide a negative answer to the question by indicating a counterexample.

3 Topologies Induced by Relations

Let X be a non-empty set and R be a relation on X. A subset $A \subseteq X$ is called *R-open* if $x \in A$ and xRy then $y \in A$. By $\mathcal{T}^*$, we mean the class of all R-open subsets of X. We also define a family $\mathcal{T}_R$ of subsets induced by the relation R: $\mathcal{T}^* = \{A \subseteq X \mid A \text{ is } R\text{-open}\}$ and $\mathcal{T}_R = \{A \subseteq X \mid R_-(A) = A\}$.

Proposition 4 *A is R-open if and only if $A \subseteq R_-(A)$ if and only if $R(A) \subseteq A$.*

Moreover, we have the following result.

Proposition 5 *For arbitrary binary relation R on X, the family $\mathcal{T}^*$ forms an Alexandrov topology, that is, a topology closed under intersection.*

Proof. We only show that $\cap_\lambda A_\lambda \in \mathcal{T}^*$ for all $A_\lambda \in \mathcal{T}^*$. Let $x \in \cap_\lambda A_\lambda$ and xRy. Since $x \in A_\lambda$ and A_λ is R-open for all λ, we get $y \in A_\lambda$ and thus $y \in \cap_\lambda A_\lambda$. This means that $\cap_\lambda A_\lambda$ is R-open, that is, $\cap_\lambda A_\lambda \in \mathcal{T}^*$. □

In [4], it was also proved that if R is reflexive then $\mathcal{T}_R$ is an Alexandrov topology on X. Now we have a naive question whether $\mathcal{T}^*$ is identical with $\mathcal{T}_R$ for a reflexive relation R.

Proposition 6 *If R is reflexive then $\mathcal{T}^* = \mathcal{T}_R$.*

Proof. Taking into account of the fact $R_-(A) \subseteq A$ for a reflexive relation R, we have that $A \in \mathcal{T}^*$ iff A is R-open iff $A \subseteq R_-(A)$ and $R_-(A) \subseteq A$ iff $R_-(A) = A$ iff $A \in \mathcal{T}_R$. Therefore, we get $\mathcal{T}^* = \mathcal{T}_R$. □

We have a little bit generalization of the result above. Let R be an arbitrary binary relation on X. We define $R_\omega = R \cup \omega$, where $\omega = \{(x, x) \,|\, x \in X\}$. It is obvious that R_ω is the smallest reflexive relation containing R, that is, R_ω is the reflexive closure of R. It follows from the above that $\mathcal{T}_{(R_\omega)}$ is an Alexandrov topology.

Lemma 1 *For any relation R, $\mathcal{T}_{(R_\omega)}$ is an Alexandrov topology.*

Theorem 1 *Let (X, R) be an approximation space. The topology $\mathcal{T}^*$ of all R-open subsets is identical with the topology $\mathcal{T}_{(R_\omega)}$ induced by reflexive closure of R, that is, $\mathcal{T}^* = \mathcal{T}_{(R_\omega)}$.*

Proof. Let $A \in \mathcal{T}^*$. Since R_ω is reflexive, it is sufficient to show $A \subseteq (R_\omega)_-(A)$. Let $x \in A$. For all $y \in X$, if $xR_\omega y$ then we have $x = y$ or xRy. If $x = y$ then it is obvious that $y = x \in A$. If xRy, since A is R-open and $x \in A$, then we get $y \in A$. In any case $y \in A$ for all y such that $xR_\omega y$. This means that $x \in (R_\omega)_-(A)$ and $A \subseteq (R_\omega)_-(A)$. Thus we have $\mathcal{T}^* \subseteq \mathcal{T}_{(R_\omega)}$.

Conversely, suppose that $A \in \mathcal{T}_{(R_\omega)}$. Let $x \in A$ and xRy. It follows from $xR_\omega y$ and $x \in A \subseteq (R_\omega)_-(A)$ that $y \in A$ and A is R-open. This implies that $A \in \mathcal{T}^*$ and $\mathcal{T}_{(R_\omega)} \subseteq \mathcal{T}^*$.

Therefore $\mathcal{T}^* = \mathcal{T}_{(R_\omega)}$. □

Moreover, as proved in [5], if R is reflexive the $\mathcal{T}_R = \mathcal{T}_{(R^*)}$, where R^* is a transitive closure of R, that is, $R^* = \bigcup_{n \geq 1} R^n$. It follows that

Theorem 2 *For any binary relation R on X, the families $\mathcal{T}^*, \mathcal{T}_{(R_\omega)}$ and $\mathcal{T}_{(R_\omega)^*}$ form a same Alexandrov topology.*

Example 1. Let $X = \{0,1,2\}$ and R be a relation defined by

$$R = \{(0,0),(2,2),(0,1),(1,2)\}.$$

Then it is clear that

$$R_\omega = \{(0,0),(1,1),(2,2),(0,1),(1,2)\}$$

and its transitive closure $(R_\omega)^*$ is

$$(R_\omega)^* = \{(0,0),(1,1),(2,2),(0,1),(1,2),(0,2)\}.$$

For example, $\{1\}$ is not R-open, because $1 \in \{1\}$ and $1R2$ but $2 \notin \{1\}$. On the other hand, $\{2\}$ is R-open. All R-open sets are $\emptyset, \{2\}, \{1,2\}$ and X, that is,

$$\mathcal{T}^* = \{\emptyset, \{2\}, \{1,2\}, X\}.$$

Since, $(R_\omega)_-(\{0\}) = (R_\omega)_-(\{1\}) = \emptyset, (R_\omega)_-(\{2\}) = \{2\}, (R_\omega)_-(\{0,1\}) = \{0\}, (R_\omega)_-(\{0,2\}) = \{2\}$ and $(R_\omega)_-(\{1,2\}) = \{1,2\}$, we get

$$\mathcal{T}_{(R_\omega)} = \{\emptyset, \{2\}, \{1,2\}, X\}$$

On the other hand, since $((R_\omega)^*)_-(\{0\}) = ((R_\omega)^*)_-(\{1\}) = \emptyset, ((R_\omega)^*)_-(\{2\}) = \{2\}, ((R_\omega)^*)_-(\{0,1\}) = ((R_\omega)^*)_-(\{0,2\}) = \emptyset$ and $((R_\omega)^*)_-(\{1,2\}) = \{1,2\}$, we also have

$$\mathcal{T}_{(R_\omega)^*} = \{\emptyset, \{2\}, \{1,2\}, X\} = \mathcal{T}_{(R_\omega)}.$$

We note that $((R_\omega)^*)_-(\{0,1\}) = \emptyset \in \mathcal{T}_{(R_\omega)^*} = \mathcal{T}_{(R_\omega)}$ but $(R_\omega)_-(\{0,1\}) = \{0\}$ is not an open set, because of $\{0\} \notin \mathcal{T}_{(R_\omega)} = \mathcal{T}_{(R_\omega)^*}$. Therefore, two different operators $(R_\omega)_- \neq (R_\omega)^*_-$ construct the same topology.

Conversely we can show that

Lemma 2 *For any Alexandrov topology $\mathcal{T}$, there exists a reflexive relation relation $R_\mathcal{T}$ on X such that $\mathcal{T} = \mathcal{T}_{(R_\mathcal{T})}$*

Proof. Let $\mathcal{T}$ be an Alexandrov topology. We define an operator I as follows: For every subset $A \subseteq X$, the operator I is defined by

$$I(A) = \bigcup\{O \in \mathcal{T} \mid O \subseteq A\}.$$

We note $I(A) \in \mathcal{T}$. Using this operator, we also define a relation R on X as follows: For all $x, y \in X$,

$$xR_\mathcal{T}y \iff \forall B \subseteq X\ (x \in I(B) \to y \in B).$$

Then, it is obvious that $R_\mathcal{T}$ is a reflexive relation. We show that $(R_\mathcal{T})_-(A) = I(A)$ for all $A \subseteq X$. If $x \in I(A)$, then we have $y \in A$ for any y such that $xR_\mathcal{T}y$ and thus $x \in (R_\mathcal{T})_-(A)$. This means $I(A) \subseteq (R_\mathcal{T})_-(A)$. Conversely, let $x \notin I(A)$. We take

$$\Gamma = \{B \subseteq X \mid x \in I(B)\} \cup \{A^c\}.$$

Then we claim

$$\bigcap \Gamma \neq \emptyset.$$

Otherwise, we have $\bigcap \Gamma = \emptyset$, that is,

$$\bigcap \{B \mid x \in I(B)\} \cap A^c = \emptyset.$$

This means that

$$\bigcap \{B \mid x \in I(B)\} \subseteq A$$

and hence that

$$I(\bigcap \{B \mid x \in I(B)\}) \subseteq I(A).$$

Since $\mathcal{T}$ satisfies the condition (IP), it follows

$$\bigcap \{I(B) \mid x \in I(B)\} \subseteq I(A).$$

This implies $x \in I(A)$, but this is a contradiction. Thus we conclude that $\bigcap \Gamma \neq \emptyset$. Since $y \in \bigcap \Gamma$ for some y, we have $xR_{\mathcal{T}}y$ and $y \notin A$. Hence

$$x \notin (R_{\mathcal{T}})_-(A).$$

That is, $(R_{\mathcal{T}})_-(A) \subseteq I(A)$ and

$$(R_{\mathcal{T}})_-(A) = I(A).$$

Now we prove $\mathcal{T} = \mathcal{T}_{(R_{\mathcal{T}})}$. Suppose that $O \in \mathcal{T}$. We have $O = I(O) = (R_{\mathcal{T}})_-(O)$ and thus $O \in \mathcal{T}_{(R_{\mathcal{T}})}$, that is, $\mathcal{T} \subseteq \mathcal{T}_{(R_{\mathcal{T}})}$. Conversely, if $O \in \mathcal{T}_{(R_{\mathcal{T}})}$ then $O = (R_{\mathcal{T}})_-(O) = I(O) \in \mathcal{T}$, that is, $\mathcal{T}_{(R_{\mathcal{T}})} \subseteq \mathcal{T}$. Therefore, $\mathcal{T} = \mathcal{T}_{(R_{\mathcal{T}})}$. □

Remark 1. The relation $R_{\mathcal{T}}$ defined above by an Alexandrov topology $\mathcal{T}$ is not only reflexive but also transitive. It is easy to prove that $xR_{\mathcal{T}}y$ if and only $x \in \{y\}^-$.

It follows from the above that

Theorem 3 *Every Alexandrov topology can be constructed by a reflexive relation.*

Moreover, we have following characterization theorems of Alexandrov topologies with some proportioo (cf. [4]). Wc consider an interesting topological property called (*clop*) here. We say that a topology has the property (clop) if it satisfies the following property.

For every subset A, A is open if and only if A is closed.

For instance, the discrete topology has the property (clop). Any Alexandrov topology with (clop) can be characterized as follows.

Theorem 4 ([4])

1. *Every Alexandrov topology satisfying (clop) can be characterized by a reflexive and symmetric relation.*
2. *Every Alexandrov topology induced by an interior operator can be characterized by a reflexive and transitive relation.*

Proof. We only show the second case. It is obvious that if a relation R is reflexive and transitive then R_- is an interior operator and hence the topology $\mathcal{T}_R$ induced by the interior operator R_- is the Alexandrov topology. Conversely, let $\mathcal{T}$ be an Alexandrov topology induced by an interior operator I, that is,

(1) $I(A) \subseteq A$ for all $A \subseteq X$;
(2) $I(A) = I(I(A))$ for all $A \subseteq X$;
(3) $I(A \cap B) = I(A) \cap I(B)$ for all $A, B \subseteq X$;
(IP) $I(\bigcap_\lambda A_\lambda) = \bigcap_\lambda I(A_\lambda)$ for all $A_\lambda \in \mathcal{T}$.

Then the topology $\mathcal{T}$ can be represented by

$$\mathcal{T} = \{I(A) \mid A \subseteq X\}.$$

We take the relation $R_\mathcal{T}$ defined by

$$xR_\mathcal{T}y \Leftrightarrow \forall B \subseteq X\,(x \in I(B) \rightarrow y \in B).$$

It follows from the proof of Lemma 2 that $(R_\mathcal{T})_-(A) = I(A)$ for all $A \subseteq X$. Thus, we obtain $\mathcal{T} = \mathcal{T}_{(R_\mathcal{T})}$. This means that for any Alexandrov topology $\mathcal{T}$ induced by an interior operator, there exists a reflexive and transitive relation R such that $\mathcal{T} = \mathcal{T}_{(R_\mathcal{T})}$.

Therefore, every Alexandrov topology induced by an interior operator is characterized by a reflexive and transitive relation. □

We also consider topologies of direct products of approximation spaces. Let (X_λ, R_λ) be an approximation space for all $\lambda \in \Lambda$. A binary relation R on the direct product $\Pi_{\lambda\in\Lambda}X_\lambda$ (simply denoted by $\Pi_\lambda X_\lambda$) of X_λ is defined: For all $x, y \in \Pi_\lambda X_\lambda$,

$$xRy \Leftrightarrow \forall\lambda \in \Lambda\; x(\lambda)R_\lambda y(\lambda).$$

Proposition 7 $R_-(\Pi_\lambda A_\lambda) = \Pi_\lambda (R_\lambda)_-(A_\lambda)$

Proof. Suppose that $x \in R_-(\Pi_\lambda A_\lambda)$. For all $\lambda \in \Lambda$, it is sufficient to show $x(\lambda) \in (R_\lambda)_-(A_\lambda)$. For any $y_\lambda \in X_\lambda$, if $x(\lambda)R_\lambda y_\lambda$, by Axiom Choice (AC), then there exists $y \in \Pi_\lambda X_\lambda$ such that $y_\lambda = y(\lambda)$ for all $\lambda \in \Lambda$. Since $x(\lambda)R_\lambda y(\lambda)$ for all $\lambda \in \Lambda$ and thus xRy, we have $y \in \Pi_\lambda A_\lambda$ by $x \in R_-(\Pi_\lambda A_\lambda)$. This means that $y_\lambda = y(\lambda) \in A_\lambda$, that is, $x(\lambda) \in (R_\lambda)_-(A_\lambda)$ for all $\lambda \in \Lambda$. Therefore, $x \in \Pi_\lambda(R_\lambda)_-(A_\lambda)$ and $R_-(\Pi_\lambda A_\lambda) \subseteq \Pi_\lambda(R_\lambda)_-(A_\lambda)$.

Conversely, we assume $x \in \Pi_\lambda(R_\lambda)_-(A_\lambda)$. For any $y \in \Pi_\lambda X_\lambda$, if xRy, since $x(\lambda)R_\lambda y(\lambda)$ and $x(\lambda) \in (R_\lambda)_-(A_\lambda)$ for all $\lambda \in \Lambda$, then we have $y(\lambda) \in A_\lambda$ and thus $y \in \Pi_\lambda A_\lambda$. This implies $x \in R_-(\Pi_\lambda A_\lambda)$ and $\Pi_\lambda(R_\lambda)_-(A_\lambda) \subseteq R_-(\Pi_\lambda A_\lambda)$.

Therefore, we obtain the result $R_-(\Pi_\lambda A_\lambda) = \Pi_\lambda(R_\lambda)_-(A_\lambda)$. □

4 Uniform Topology

In this section, we consider an open problem [1] and solve it by providing a counterexample. In general, to introduce topologies on algebras is to define open sets in the algebras, that is, to define a family of subsets satisfying the axiom of topology. A uniform topology, introduced by A. Weil and so on, is a topology defined by equivalence relations and is researched by many people. We define uniform topologies on (commutative) residuated lattices and prove some fundamental results, moreover, we solve the problem left open in [1].

At first, we define a residuated lattice. An algebraic structure $X = (X, \wedge, \vee, \odot, \rightarrow, 0, 1)$ is called a *residuated lattice* if

(1) $(X, \wedge, \vee, 0, 1)$ is a bounded lattice;
(2) $(X, \odot, 1)$ is a commutative monoid;
(3) For all $x, y, z \in X$,

$$x \odot y \leq z \Leftrightarrow x \leq y \rightarrow z.$$

The class of all residuated lattices is an important algebraic semantics for studying properties of fuzzy logics. A binary relation R on X is called a *congruence relation* if it is an equivalence relation and satisfies the compatibility property (CP):

(CP) If $(x, y), (u, v) \in R$ then $(x * u, y * v) \in R$, where $* \in \{\wedge, \vee, \odot, \rightarrow\}$.

A non-empty subset $F(\subseteq X)$ is called a *filter* if

(F1) If $x, y \in F$ then $x \odot y \in F$;
(F2) If $x \in F$ and $x \leq y$ then $y \in F$.

By $Fil(X)$, we mean the set of all filters of X. We denote the set of all congruences on X by $Con(X)$. Then it is well-known that

Proposition 8

$$Fil(X) \cong Con(X)$$

It follows from the above that the quotient structure $X/F = \{x/F \,|\, x \in X\}$ is also a residuated lattice under the following operations: For all $x/F, y/F \in X/F$,

$$x/F \circ y/F = (x \circ y)/F, \text{ where } \circ \in \{\wedge, \vee, \rightarrow, \odot\}.$$

Let $F \in Fil(X)$. A binary relation R on X is called *compatible with a filter* F if xRy and $x \leftrightarrow a, y \leftrightarrow b \in F$ then aRb, where $x \leftrightarrow a$ is an abbreviation of $(x \rightarrow a) \wedge (a \rightarrow x)$.

It is easy to show that if R is compatible with F then a binary relation R/F is well-defined on the quotient residuated lattice X/F, where R/F is defined by

$$x/F(R/F)y/F \Leftrightarrow xRy.$$

The following result can be proved directly from the definition of compatibility.

Proposition 9 *Let R be a binary relation compatible with a filter F. Then*

1. *R is reflexive $\Leftrightarrow R/F$ is reflexive;*
2. *R is symmetric $\Leftrightarrow R/F$ is symmetric;*
3. *R is transitive $\Leftrightarrow R/F$ is transitive;*

Let R be a relation on a residuated lattice X, which is compatible with F. Since R/F is the relation on the residuated lattice X/F, we define an operator $(R/F)_-$ on X/F and consider its properties.

Lemma 3 *For every subset $A \subseteq X$,*

$$R_-(A)/F \subseteq (R/F)_-(A/F).$$

Proof. Suppose $x/F \in R_-(A)/F$. There exists $a \in R_-(A)$ such that $x/F = a/F$, that is, $x \leftrightarrow a \in F$. For every $y/F \in X/F$, since R is compatible with F, if $x/F(R/F)y/F$ then xRy and hence aRy. This means $y \in A$ from $a \in R_-(A)$ and $y/F \in A/F$. Therefore, we get $x/F \in (R/F)_-(A/F)$ and $R_-(A)/F \subseteq (R/F)_-(A/F)$. □

Conversely,

Lemma 4 *If $F \subseteq A$ and $A \in Fil(X)$, then we have*

$$(R/F)_-(A/F) \subseteq R_-(A)/F.$$

Proof. Suppose that $x/F \in (R/F)_-(A/F)$. If xRy, since $x/F(R/F)y/F$, then $y/F \in A/F$ and hence there exists an element $a \in A$ such that $y/F = a/F$. Since $y \leftrightarrow y \in F \subseteq A$ and A is a filter, we have $y \in A$ and thus $x \in R_-(A)$. This means $x/F \in R_-(A)/F$ and $(R/F)_-(A/F) \subseteq R_-(A)/F$. □

It follows from the above that

Theorem 5 *Let R be a binary relation compatible with F and $F \subseteq A$ for $A \in Fil(X)$. Then we have*

$$R_-(A)/F = (R/F)_-(A/F).$$

We define a topology on a residuated lattice by using congruence relations. Let $\mathcal{K}^* \subseteq Con(X)$ be closed under intersection, that is, if $\theta, \varphi \in \mathcal{K}^*$ then $\theta \cap \varphi \in \mathcal{K}^*$.

Proposition 10 *We have the following results in $\mathcal{K}^*$.*

1. *$\varphi \in \mathcal{K}^* \;\Rightarrow \omega \subseteq \varphi$, where $\omega = \{(x, x) \mid x \in X\}$;*
2. *$\varphi \in \mathcal{K}^* \;\Rightarrow \varphi^{-1} \in \mathcal{K}^*$;*
3. *$\varphi \in \mathcal{K}^* \;\Rightarrow \exists \psi \in \mathcal{K}^*$ s.t. $\psi \circ \psi \subseteq \varphi$.*

We define a class $\mathcal{K}$ of binary relations on X by

$$\mathcal{K} = \{\varphi \subseteq X \times X \mid \exists \theta \in \mathcal{K}^* \text{ s.t. } \theta \subseteq \varphi\}.$$

Then it is easy to show that $\mathcal{K}$ is a *uniformity*, that is, it satisfies the following.

Proposition 11 *$\mathcal{K}$ is a uniformity, that is, it satisfies*

(U1) $\varphi \in \mathcal{K} \Rightarrow \omega \subseteq \varphi$;
(U2) $\varphi \in \mathcal{K} \Rightarrow \varphi^{-1} \in \mathcal{K}$;
(U3) $\varphi \in \mathcal{K} \Rightarrow \exists \psi \in \mathcal{K}$ s.t. $\psi \circ \psi \subseteq \varphi$;
(U4) $\varphi, \psi \in \mathcal{K} \Rightarrow \varphi \cap \psi \in \mathcal{K}$;
(U5) $\varphi \in \mathcal{K}, \varphi \subseteq \psi \Rightarrow \psi \in \mathcal{K}$.

From the general theory of uniformity, a topology $\mathcal{T}_{\mathcal{K}}$ is introduced on the residuated lattice X as follows.

$$\mathcal{T}_{\mathcal{K}} = \{O \subseteq X \mid \forall x \in O\, \exists \varphi \in \mathcal{K}\, \text{s.t.}\, \varphi[x] \subseteq O\},$$

where $\varphi[x] = \{y \in X \mid (x, y) \in \varphi\}$. Then it is obvious to show that

Proposition 12 $\mathcal{T}_{\mathcal{K}} = \{O \subseteq X \mid \forall x \in O\, \exists \theta \in \mathcal{K}^*\, \text{s.t.}\, \theta[x] \subseteq O\}$.

Therefore, on a residuated lattice X, we have

$$O \subseteq X \text{ is an open set } \Leftrightarrow O \text{ is a join of equivalence classes.}$$

We also have the fundamental result with respect to the topology $\mathcal{T}_{\mathcal{K}}$.

Theorem 6 *For any $\theta \in \mathcal{K}^*$, $x \in X$, a subset $\theta[x]$ is a closed and open set.*

Proof. Let $\theta \in \mathcal{K}^*$. For every $y \in \theta[x]$, since $\theta \in \mathcal{K}^*$ and $\theta[y] = \theta[x]$, $\theta[x]$ is an open set.

If $y \in (\theta[x])^c$, since $y \notin \theta[x]$ and $\theta[y] \neq \theta[x]$, then we have $\theta[x] \cap \theta[y] = \emptyset$ and thus $y \in \theta[y] \subseteq (\theta[x])^c$. This means that $(\theta[x])^c$ is an open set, that is, $\theta[x]$ is a closed set. □

Corollary 2 *For every filter F of X, F is a closed and open set in $(X, \mathcal{T}_{\mathcal{K}})$.*

We simply denote $\mathcal{T}_{\mathcal{K}}$ for $\mathcal{K}^* = \{\theta\}$ by $\mathcal{T}_\theta$.

Proposition 13 *For $\theta, \varphi \in Con(X)$, we have*

$$\theta \subseteq \varphi \Leftrightarrow \mathcal{T}_\theta \subseteq \mathcal{T}_\varphi$$

Proof. Suppose $\theta \subseteq \varphi$. Let $O \in \mathcal{T}_\varphi$. For every $x \in O$, there exists $\psi \subseteq X \times X$ such that $\varphi \subseteq \psi$ and $\psi[x] \subseteq O$. Since $\theta[x] \subseteq \varphi[x] \subseteq \psi[x]$, we get $\theta[x] \subseteq O$. This means that $O \in \mathcal{T}_\theta$ and $\mathcal{T}_\varphi \subseteq \mathcal{T}_\theta$.

Conversely, we assume $\mathcal{T}_\varphi \subseteq \mathcal{T}_\theta$. For each $(a, b) \in \theta$, since $\varphi[a] \in \mathcal{T}_\varphi \subseteq \mathcal{T}_\theta$ and $a \in \varphi[a]$, we have $\theta[a] \subseteq \varphi[a]$. It follows from $b \in \theta[a]$ that $b \in \varphi[a]$ and hence that $(a, b) \in \varphi$. Therefore, $\theta \subseteq \varphi$. □

A topological space $(X, \mathcal{T}_{\mathcal{K}})$ is called *totally bounded* if for each $\varphi \in \mathcal{K}$ there exist $x_1, \cdots, x_n \in X$ such that

$$X = \bigcup_{i=1}^{n} \varphi[x_i].$$

It follows from the above that

Theorem 7 ([2]) *$(X, \mathcal{T}_\theta)$ is a compact space $\iff$ $(X, \mathcal{T}_\theta)$ is totally bounded.*

Proof. Suppose that $(X, \mathcal{T}_\theta)$ is totally bounded. Let $\{O_\lambda\}$ $(\lambda \in \Lambda)$ be an open covering of X in $(X, \mathcal{T}_\theta)$. Since X is totally bounded, for $\theta \in \mathcal{K}$, there exists $x_1, \cdots, x_n \in X$ such that $X = \bigcup_{i=1}^n \theta[x_i]$. The fact $x_i \in X = \bigcup_\lambda O_\lambda$ implies $x_i \in O_{\lambda_i} \in \mathcal{T}_\theta$ for some λ_i and thus $\theta[x_i] \subseteq O_{\lambda_i}$. We get $X = \bigcup_{i=1}^n \theta[x_i] \subseteq \bigcup_{i=1}^n O_{\lambda_i}$ and $X = \bigcup_{i=1}^n O_{\lambda_i}$. This means that X is compact.

Conversely, for each $\varphi \in \mathcal{K}$ there exist $\theta \in \mathcal{K}^*$ such that $\theta \subseteq \varphi$. Since θ is a congruence, we have $x \in \theta[x] \subseteq \varphi[x] \in \mathcal{T}_\theta$ for any $x \in X$ and hence $X = \bigcup_{x\in X} \theta[x] \subseteq \bigcup_{x\in X} \varphi[x]$. That is, $\{\varphi[x]\}_{x\in X}$ is an open covering of X. Since X is compact, there exist finite number of open sets $\varphi[x_i]$ such that $X = \bigcup_{i=1}^n \varphi[x_i]$. Therefore, X is totally bounded. □

5 Topological Residuated Lattices

We consider a topology on a residuated lattice, for which operations of the residuated lattice are continuous. Let $X = (X, \wedge, \vee, \odot, \rightarrow, 0, 1)$ be a residuated lattice and τ be a topology on X. A structure (X, τ) is called a *topological residuated lattice* [1] if the operations $\odot$ and $\rightarrow$ are continuous with respect to τ. Namely, For all $A, B \subseteq X$, if we set

$$A \odot B = \{x \odot y \,|\, x \in A, y \in B\},\ A \rightarrow B = \{x \rightarrow y \,|\, x \in A, y \in B\},$$

then the following results hold for all $O \in \tau, a, b \in X$:

(1) $a \odot b \in O \ \Rightarrow\ \exists O_a, O_b \in \tau\,\text{s.t.}\, a \in O_a, b \in O_b\,\text{and}\, O_a \odot O_b \subseteq O$
(2) $a \rightarrow b \in O \ \Rightarrow\ \exists O_a, O_b \in \tau\,\text{s.t.}\, a \in O_a, b \in O_b\,\text{and}\, O_a \rightarrow O_b \subseteq O$

Theorem 8 *Let X be a residuated lattice and $\mathcal{K}^*$ be the set of all congruences on X. Then $(X, \mathcal{T}_\mathcal{K})$ is a topological residuated lattice.*

There is a problem left open in [1] with respect to topologies on residuated lattices:

Let X be a residuated lattice and τ be a topology on X. Then,

$$\odot \text{ is continuous } \Leftrightarrow\ \rightarrow \text{ is continuous ?}$$

We give a negative answer to the problem by providing a following counterexample. Let $X = \{0, a, 1\}$ with $0 < a < 1$. We define operations on X: For all $x, y \in X$

$$x \wedge y = x \odot y = \min\{x, y\}$$
$$x \vee y = \max\{x, y\}$$
$$x \rightarrow y = \begin{cases} 1 & \text{if } x \leq \text{y} \\ y & \text{otherwise} \end{cases}$$

It is obvious that

$$\tau = \{\emptyset, X, \{a, 1\}\}$$

is a topology on X and the operator $\odot$ is continuous with respect to τ. However, the operator $\rightarrow$ is not continuous. Because, $0 \rightarrow 0 = 1 \in \{a, 1\}$ but

$$X \rightarrow X = X \not\subseteq \{a, 1\}.$$

This means that $\rightarrow$ is not continuous with respect to the topology τ.

References

1. Ghorbani, S., Hasankhani, A.: Implicative topology on residuated lattices. J. Mult.-valued Log. Soft Comput. **17**, 521–535 (2011)
2. Haveshki, M., Eslami, E., Saeido, A.B.: A topology induced by uniformity on BL-algebras. Math. Log. Q. **53**, 162–169 (2007)
3. James, I.M.: Introduction to Uniform Topology. Cambridge University Press, New York (1990)
4. Kondo, M.: On the structure of generalized rough sets. Inf. Sci. **176**, 589–600 (2006)
5. Li, Z.: Topological properties of generalized rough sets. In: FKSD 2010 Seventh International Conference on Fuzzy and Knowledge Discovery, pp. 2067–2070 (2010)
6. Pawlak, Z.: Rough sets. Int. J. Comput. Inf. Sci. **11**, 341–356 (1982)

Iterative Set Approximations Based on Tolerance Relation

László Aszalós and Dávid Nagy[(✉)]

Faculty of Informatics, University of Debrecen, Debrecen, Hungary
{aszalos.laszlo,nagy.david}@inf.unideb.hu
https://inf.unideb.hu/en/aszalos.laszlo,
https://inf.unideb.hu/en/nagy.david

Abstract. We introduce two covering approximation spaces which utilise a ranking method to reduce the number of base sets used at approximation of a set. The ranking method aggregates all the information embedded in the tolerance relation and selects the most promising representatives. We present the method in the context of its process and describe some interesting features of our approximation pairs.

Keywords: Set approximation · Representatives of a set · Power method

1 Introduction

One of the popular tools for data mining and statistics is classification, which assigns a new object into one of several categories. There are various methods for classifying data, out of which—if the categories cannot be linearly separated—perhaps the k-NN method is the most preferred. The basic question here is *how far the new object is from the already tagged objects*, and what tags the k closest objects have. The more objects we have, the more computation, or the more data storage for objects may be needed.

Big data can cause both of these problems. The solution is not to store all the similar objects, but only some of them. Then the question arises *which of the many elements should be a representative?* On the one hand, our goal would be to minimise the number of representatives which would reduce the storage needed and speed up the calculations required by the classification. On the other hand, we would like to have a representative of all objects in a category, i.e. at least one object similar to it would be a representative. In most cases, similarity means that the values describing the objects are close to each other or, in other words, their distance is small.

The word *representative* has several related meanings, one of which is *a typical example*, *sample*. So we are interested in the representatives of a set of objects

The work/publication is supported by the EFOP-3.6.1-16-2016-00022 project and the *ÚNKP-18-3 New National Excellence Program* of the Ministry of Human Capacities. The project is co-financed by the European Union and the European Social Fund.

© Springer Nature Switzerland AG 2019
T. Mihálydeák et al. (Eds.): IJCRS 2019, LNAI 11499, pp. 78–90, 2019.
https://doi.org/10.1007/978-3-030-22815-6_7

in a given category. In relation to sets, the relation *element* is a fundamental tool of mathematics. However, since it is a binary relation, by only using this relation we cannot distinguish between the elements of the set. Therefore, we would like to extend this relationship by introducing different degrees. We want to say e.g. that element a is more/better part of the set than element b. This is achieved by assigning a number to each object, and whichever object has a larger number, a better/more important component it is of the set.

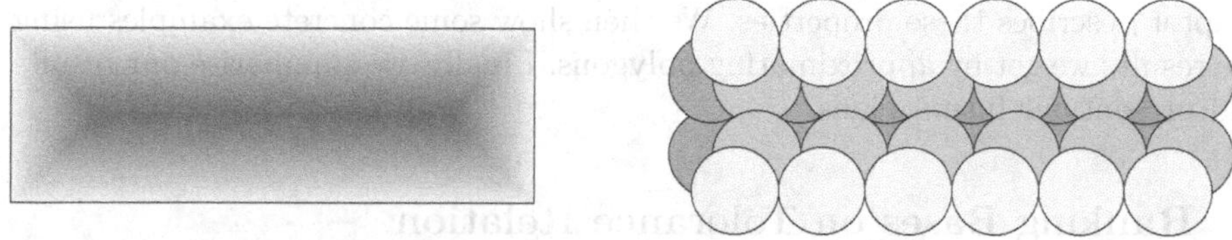

Fig. 1. The "centres" of a rectangle and the approximation of this rectangle

The origin of our motivation is shown in Fig. 1. Here, the inner points of the rectangle on the left are coloured according to how far they are from the edges of the rectangle. If we are looking for the characteristic points of a polygon or other 2D shapes, for some people these are clearly defined as the vertices (or points of their edges), because they determine the shape of the polygons. On the other hand, if we think in areas, e.g. regions or countries, then only a few would say that the border characterises a county or country. After all, the border is not much more than a place of transition, with the exact positions of the borders shaped by history; and therefore we consider the interior as the characteristic region of a country. Therefore, we believe that the darkest grey points in the figure are the most characteristic of the rectangle, and the increasingly lighter colours give an increasingly weaker characterisation. (The edges of the rectangle marked with black is only for the representation of the boundaries of the rectangle.) Therefore, if a polygon nodes should be characterised by a few points, and if the similarity of the points ends with the distance r (i.e. two points are similar if and only if their distance is less than r) then we cover the polygon with circles of radius r. Here, the first point of approximation is our darkest point. This point is the centre of the first circle. The next point will be the darkest point that is not yet covered by a circle. Following this method, you can get the coverage shown on the right side of the Fig. 1, where the colours of the circles indicate the colour of their centres. (That is, the darker circle is positioned first, and the lighter at the end.)

It is easy to provide a coverage of this rectangle that contains fewer circles than our greedy algorithm outlined above; but we believe that the universality and complexity of our method makes it suitable for solving large-scale real-life tasks.

Our method will use a similarity relationship that can be generated from arbitrary metrics or can be given independently. Based on this relationship, we define a rank between our objects to find the central element(s) of any set. As the central objects are given, we can use some variant of our algorithm presented in Sect. 3 to choose the representatives.

The structure of the article is as follows: first, we show how a ranking can be created based on similarity and membership, assigning a real number to each object. Next we present how we can select our representatives with these rankings. Different selection methods have different characteristics and the fourth chapter describes these properties. We then show some concrete examples using the results we get by approximating polygons. Finally, we summarise our results and present our future plans.

2 Ranking Bases on Tolerance Relation

There is a more or less serious ranking in the mathematical community: the Erdős number. Here, mathematicians (and often other scholars) correspond to a vertex of a graph where two vertices are connected if and only if the corresponding scholars are co-authors. The centre of the graph is Pál Erdős, who was a highly productive author. The numbers belonging to the mathematicians are given by the distances from the centre. imilar rankings exist between actors (The oracle of Bacon) and even chess players. Of course, the smaller number/distance here constitutes to a higher rank.

Generally however, the larger number usually means a higher rank, and this will be true in our case too. Here the number of steps needed to reach the *centre* of the set is not important. We wish to know how quickly we can reach the boundary of the set from any arbitrary element. Phrasing it differently: how far is this element from the complement of the set?

These kind of rankings can be calculated easily, but this simpleness prevents the development of a sophisticated rankings. In case of the Erdős number, it does not matter how many articles have been written by the two authors, or the extent to which the article was accepted by the scientific community. However, our ranking considers the finer details; whilst the traditional distance only takes the single shortest path into consideration, in our case all possible paths are counted. We were influenced by the design of the PageRank algorithm [1], but we could not directly apply it here, as it is made for directed graphs. The tolerance relation—giving the similarity of the objects—is symmetric by definition, so the corresponding graph we use is not directed.

Let's look at the notations used in this article. Let indicate the set of objects by V, where for the sake of simplicity $V = \{1, 2, \dots, n\}$. Denote the—possibly partial—tolerance relation by T. Let the set of objects similar to x be given by $C_x = \{y|xTy\}$. Finally, let the set we wish to approximate be $Q \subset V$. As only the set Q and the relation T are given, these are the only inputs of out methods which produce the representatives—the objects that characterise the set Q the most.

The PageRank algorithm is formulated in the literature in various ways. One of them uses the synonym of voting: initially each vertex has one vote, which is distributed equally along the outgoing edges, and is then repeated for each incoming vote too. For getting a limit on the distribution of votes, we will need a damping factor.

Let's use that synonym. Since our graph of similarity is not directed, we make alternations in a number of places. Each vertex votes for each vertex, not just the similar ones. A vote can be one of two kinds: to support or to oppose. That is, the vote supports or opposes the other vertex being representative of its own set. A vertex supports the similar vertices in its own set, and the different vertices outside of its own set. Moreover a vertex opposes the dissimilar vertices in its own set, and the similar vertices outside its own set. In other words to be similar to mates and to be different from aliens is rang-raising property, while similarity to aliens and dissimilarity from mates is a rang-losing one. In the cases when the tolerance relation is partial, incomparable vertices do not vote on each other (hence are neither similar nor dissimilar). The same voting rules apply to the next rounds too. However, the votes of every vertex shall be weighted by the sum of the weights of its received votes. These rounds are repeated until an equilibrium is reached. To detect the equilibrium in a uniform way, the weights are normalised in each round.

Based on these voting rules we can construct the support matrix S of size $n \times n$, where

$$s_{ij} = \begin{cases} 1, \text{if } iTj \text{ holds and } (i, j \in Q, \text{ or } i, j \notin Q) \\ 1, \text{if } iTj \text{ does not hold and } (i \in Q, \ j \notin Q \text{ or } i \notin Q, \ j \in Q) \\ -1, \text{if } iTj \text{ holds and } (i \in Q, \ j \notin Q \text{ or } i \notin Q, \ j \in Q) \\ -1, \text{if } iTj \text{ does not hold and } (i, j \in Q, \text{ or } i, j \notin Q) \\ 0, \text{otherwise} \end{cases} \tag{1}$$

The column vector R_k contains the ranks of each object in the kth round. These ranks are updated in an iterative way by multiplication with the support matrix S. We follow the von Mises' algorithm [4] by using matrix operations as shown in (2). If $R_k \approx R_{k+1}$—i.e. we get close to the equilibrium—our algorithm stops.

$$\begin{aligned} R_1 &= (1, \ldots, 1)^T \\ R_{k+1} &= \frac{SR_k}{||SR_k||} \end{aligned} \tag{2}$$

Of course, successive multiplications can be triggered by the successive powering of the support matrix S, giving us the rank values of thousands of objects in little time. Table 1 describes the algorithm of this faster variant, where $B2$ corresponds to matrix B^2, `Ones` to vector $(1, \ldots, 1)^T$.

3 Selecting Representatives

Based on the above, we already have a vector R_k which assigns a number r_i to each object i. Higher values mean higher rank. To be aware of our task, consider

Table 1. Power method implemented in Python+Numpy

```
def powers(S, eps=1e−8):
    B = S
    Ones = np.ones((len(S),))
    B2 = B @ B
    B2 /= np.linalg.norm(B2 @ Ones, np.inf)
    while np.linalg.norm(B @ Ones − B2 @ Ones) > eps:
        B, B2 = B2, B2 @ B2
        B2 /= np.linalg.norm(B2 @ Ones, np.inf)
    return B2
```

the ranks associated with the square grid in the unit square. In Fig. 2, the set Q forms a triangle where the similarities of the grid points are measured by the Manhattan distance. Here the limit of similarity r is 0.1. The limit of difference R is 0.2 on the picture on the left and is 0.3 on the right. (Two objects are different from each other if and only if their distance is greater than R.) It is easy to see that the larger incomparability zone (where $r \leq d \leq R$, i.e. at distance d where the objects neither similar nor dissimilar) blends the original set Q better.

After normalising the ranks obtained by the power method onto the interval $[-1, 1]$, their values fell to the interval $[-0.95, 0.85] \cup [0.9, 1]$ for the left image, while for the other case the same became $[-0.97, -0.81] \cup [0.88, 1]$, so the ranks are also getting more and more diverse. The situation is similar when the set Q forms a square (Fig. 3). By observing the figures, we can see that the boundaries of both the square and the triangle—that are organised around 0.3 and 0.7—are blurred, so the similarity or proximity to aliens (to the border of the set) reduces the ranks.

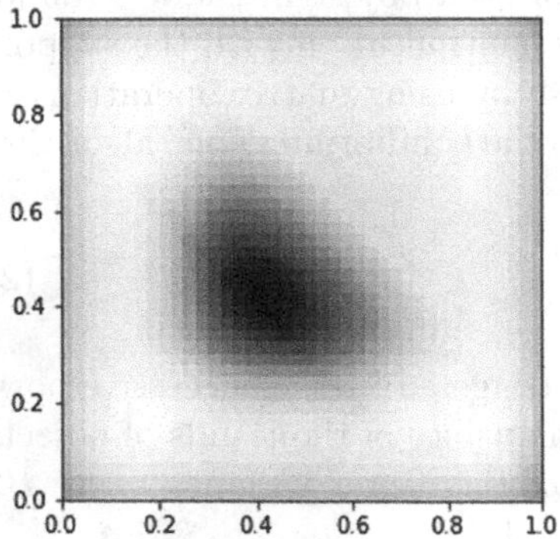

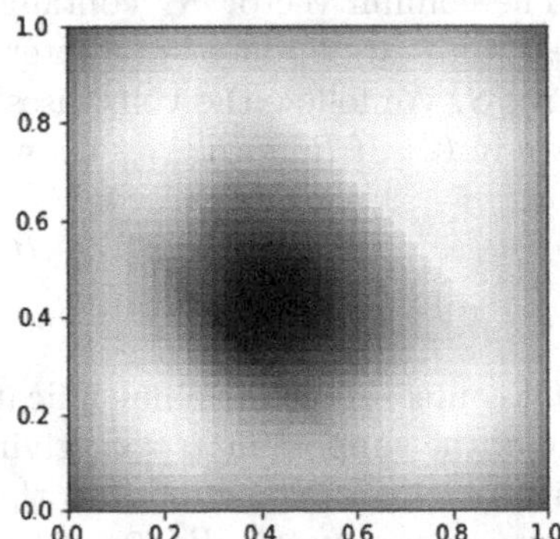

Fig. 2. Rank on a grid at in a triangular case

For an efficient and universal implementation we constructed a greedy algorithm. Therefore it is relatively easy—but not necessarily faster—to create an approximation that contains less representatives, but each object of the set also has a similar representative.

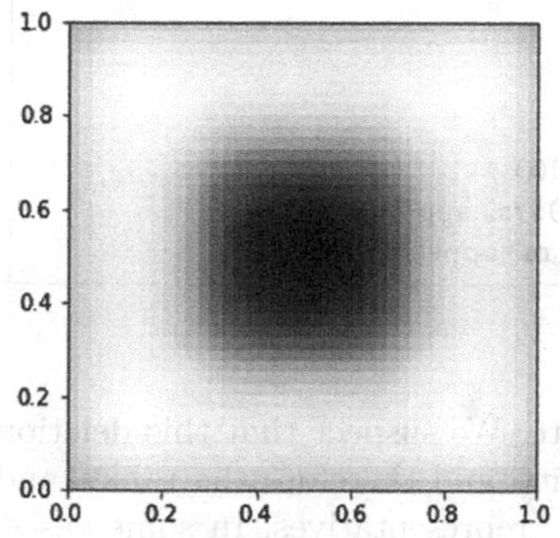

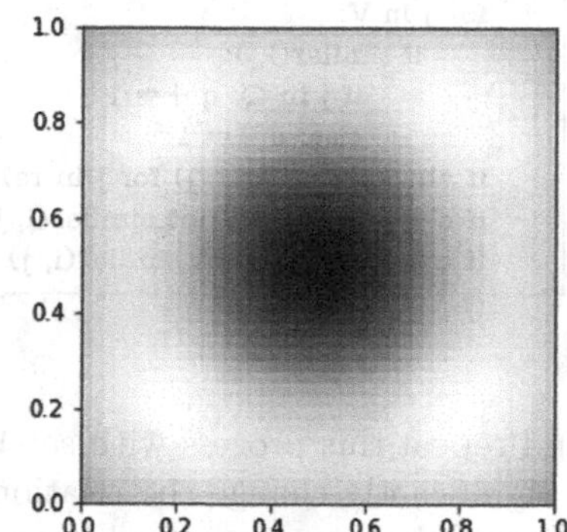

Fig. 3. Rank on a grid at square case

Table 2 shows the essence of our method. In Line 1 we sort the objects of the set Q in descending order according to the ranks $\mathtt{r}$ in the vector R_k. In Line 3 we process the objects in this order. In Line 4 we initialise two variables, one of our variants observes the environment C_i of the object i and counts how many objects are from the set Q and how many are from outside it. These numbers are stored in variables q and c, respectively. The similarity of two objects (or, in some cases, their distance) can be stored in a matrix which needs $O(V^2)$ space. In Line 5 the scanning of environment C_i needs $O(n)$ time because we check each object whether they are similar (close enough) or not. By using k-dimensional trees this could be done with $O(\ln n)$ complexity. In Line 7 and 8 we note that the next object is in the set of object i, or in its complementer set. Finally the algorithm judges whether the object i would be a representative or not.We have the following variations:

(1) if there is no representative j yet in the environment C_i—i.e. due symmetry $i \notin C_j$, (Line 9)
(2) if there is no representative j yet in the environment C_i (as before) and C_i does not contain any object not in set Q—$V \setminus Q \cap C_i = \emptyset$ (Line 10)
(3) if there is no representative j yet in the environment C_i (like at first) and more then half of this environment is member of set Q—$||V \setminus Q \cap C_i|| < ||Q \cap C_i||$, (Line 11)

then object i should be a representative of set Q.

We repeat these steps until all of the objects in Q are reviewed. Finally, the lists `rs1`, `rs2`, `rs3` will contain the representatives for the various methods.

As an alternative to these methods, after finding the first representative i, we reassemble the ranks of objects in $V \setminus C_i$, then search for the next representative

Table 2. Three set approximation methods implemented in Python+Numpy

```
ps = sorted([(r,i) for i,r in enumerate(Rk) if i in Q], reverse=True)
rs1, rs2, rs3 = [], [], []
for r, i in ps:
    q, c = 0, 0
    for j in V:
        if similar(i,j):
            if j in Q: q += 1
            else: c += 1
    if all(not_similar(i, j) for j in rs1): rs1.append(i)
    if c == 0 and all(not_similar(i, j) for j in rs2): rs2.append(i)
    if c < q and all(not_similar(i, j) for j in rs3): rs3.append(i)
```

j, and repeat this process with set $V\backslash C_i\backslash C_j$, etc. We suspect that this deletion will significantly change the relation of similarity, and therefore the ranks, and hence not the most typical objects will become representatives. In some cases, this method produced less representatives, so we plan to compare the resulting representations with the standard classification benchmarks, examining their effectiveness.

4 Properties of Different Set Approximations

From the theoretical point of view a Pawlakian approximation space [5–7] can be characterised by an ordered pair $\langle U, \mathcal{R}\rangle$ where U is a nonempty set of objects and $\mathcal{R}$ is an equivalence relation on U. In order to approximate an arbitrary subset S of U the followings have to be introduced:

- *the set of base sets*: $\mathbf{B} = \{B \mid B \subseteq U, \text{ and } x, y \in B \text{ if } x\mathcal{R}y\}$, the partition of U generated by the equivalence relation $\mathcal{R}$;
- *the functions* l, u form a Pawlakian approximation pair $\langle \mathsf{l}, \mathsf{u}\rangle$, i.e.
  1. $\mathrm{Dom}(\mathsf{l}) = \mathrm{Dom}(\mathsf{u}) = 2^U$
  2. $\mathsf{l}(S) = \bigcup\{B \mid B \in \mathbf{B} \text{ and } B \subseteq S\}$;
  3. $\mathsf{u}(S) = \bigcup\{B \mid B \in \mathbf{B} \text{ and } B \cap S \neq \emptyset\}$.

The base sets represent the background knowledge (or its limit). In a Pawlakian system two objects are treated as indiscernible if all of their known attribute values are the same. The indiscernibility relation defines an equivalence relation.

In this paper, we propose two new possible approximation pairs based on the representatives. We recall, the concept of C_x—the set of objects covered (similar to) by x—as introduced in Sect. 2. Let R_S denote the set of representatives of any arbitrary set S, obtained as described in the previous section.

The two proposed approximation pairs can be given as $\langle \mathsf{l}, \mathsf{u}_1 \rangle$ and $\langle \mathsf{l}, \mathsf{u}_2 \rangle$, where

$$\mathsf{l}(S) = \bigcup_{\substack{x \in R_S \\ C_x \subseteq S}} \{C_x\}$$

$$\mathsf{u}_1(S) = \bigcup_{x \in R_S} \{C_x\}$$

$$\mathsf{u}_2(S) = \bigcup_{\substack{x \in R_S \\ \|C_x \cap S\| > \|C_x \setminus S\|}} \{C_x\}$$

The following properties of approximation pairs are examined (full description can be seen in [2]): $\langle \mathsf{l}, \mathsf{u} \rangle$ denotes an arbitrary approximation pair.

Monotonicity
l and u are said to be monotone if $S \subset S'$ then $\mathsf{l}(S) \subset \mathsf{l}(S')$ and $\mathsf{u}(S) \subset \mathsf{u}(S')$
Weak Approximation Property
$\forall S \in 2^U : \mathsf{l}(S) \subseteq \mathsf{u}(S)$
Strong Approximation Property
$\forall S \in 2^U : \mathsf{l}(S) \subseteq S \subset \mathsf{u}(S)$
Normality of l
$\mathsf{l}(\emptyset) = \emptyset$
Normality of u
$\mathsf{u}(\emptyset) = \emptyset$

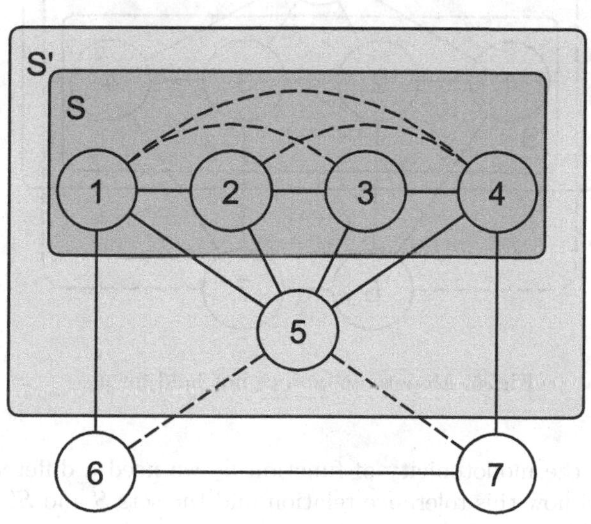

Fig. 4. Monotonicity does not hold for u_1.

Monotonicity

None of the aforementioned approximation pairs are monotone. The first counterexample is presented in Fig. 4. In our figures the solid/dashed edge between nodes i and j denotes that the relation iTj holds/does not hold, respectively. If there is no edge between some nodes, then the given relation is partial [3,8], and the two nodes that aren't connected are neither similar nor dissimilar, i.e. they are incomparable or not have been compared yet. The ranks of the objects are for the two given sets:

S: $\langle 0.126, 0.626, 0.626, 0.126, -1.000, 0.547, 0.547\rangle$,
S': $\langle 0.126, 0.626, 0.626, 0.126, 1.000, 0.547, 0.547\rangle$.

In case of the set S we have $r_1 = r_4 < r_2 = r_3$, henceforth one possible R_S can be $\{2, 4\}$, where $C_2 = \{1, 2, 3, 5\}$ and $C_4 = \{3, 4, 5, 7\}$. Object 4 is similar to 5 and 7, and 2 is similar to 5, hence $\mathsf{l}(S) = \emptyset$.

In case of the set S' we have $r_1 = r_4 < r_2 = r_3 < r_5$, and object 5 which covers every member of S', so object 5 becomes the only representative. Hence $\mathsf{l}(S') = \{1, 2, 3, 4, 5\}$, thus $\mathsf{l}(S) \subset \mathsf{l}(S')$ holds.

However, although $S \subset S'$ we have $\mathsf{u}_1(S) \not\subset \mathsf{u}_1(S')$, where $\mathsf{u}_1(S) = \{1, 2, 3, 4, 5, 7\}$ and $\mathsf{u}_1(S') = \{1, 2, 3, 4, 5\}$. This proves that the approximation pair $\langle \mathsf{l}, \mathsf{u}_1\rangle$ is not monotone.

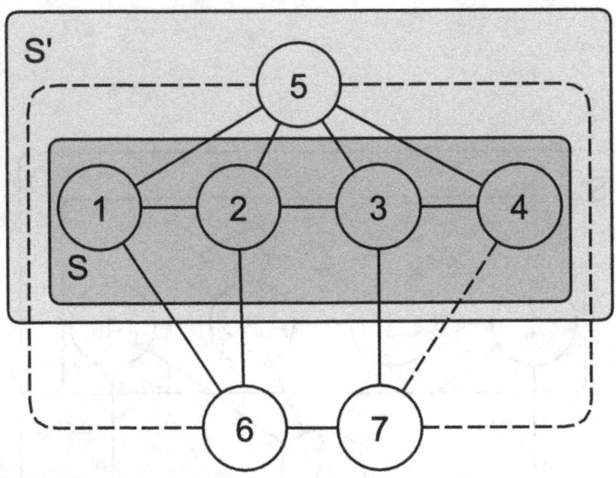

Fig. 5. Monotonicity does not hold for u_2.

To disprove the monotonicity of function u_2 we need a different counterexample. Figure 5 show this tolerance relation and the sets S and S'. The ranks of the objects in this case are:

S: $\langle 0.585, 0.759, 0.698, 0.668, -1.000, -0.009, 0.318\rangle$,
S': $\langle 0.585, 0.759, 0.698, 0.668, 1.000, -0.009, 0.318\rangle$.

For ranks in S we get $r_2 > r_3 > r_4 > r_1$, so at first we need to check the size of sets $C_2 \cap S = \{1, 2, 3\}$ and $C_2 \setminus S = \{5, 6\}$. The first of the two sets is bigger, so $2 \in R_S$. The set $S \setminus C_2$ only contains object 4, so we need to check the sizes of the sets $C_4 \cap S = \{3, 4\}$ and $C_4 \setminus S = \{5\}$. Again, the first is the bigger set, so $4 \in R_S$. Therefore $\mathsf{u}_2(S) = C_2 \cup C_4 = \{1, 2, 3, 4, 5, 6\}$. Summarising this, we have: $S \subset S'$ but $\mathsf{u}_2(S) \not\subset \mathsf{u}_2(S')$. Hence the approximation pair $\langle \mathsf{l}, \mathsf{u}_2\rangle$ is not monotone, too.

Weak Approximation
The weak approximation property holds for both approximation pairs. In both cases, the functions use the same representatives based on the same order of ranks. Therefore the lower approximation cannot be a larger set than the upper approximation.

Strong Approximation
For $\langle \mathsf{l}, \mathsf{u}_1\rangle$ the strong approximation also holds. By definition $\mathsf{l}(S) \subseteq S$ is always true. In case of u_1, every member of the set S is covered by at least one of the representatives. Thus $S \subseteq \mathsf{u}_1(S)$ is also true.

However, for $\langle \mathsf{l}, \mathsf{u}_2\rangle$ the strong approximation does not hold. In Fig. 6 an example can be seen that contradicts this property. The ranks of the objects are the following:

$$r = \langle 1.00, 0.85, 0.36, 0.36, 0.65, 0.55, -0.32, -0.80, -0.11, -0.11, -0.40\rangle.$$

Hence the possible representatives of the set S are the objects 1 and 7, where $C_1 = \{1, 2, 3, 4, 5, 6\}$ and $C_7 = \{6, 7, 8, 9, 10, 11\}$. The upper approximation of the set S is $\mathsf{u}_2(S) = \{1, 2, 3, 4, 5, 6\}$, because $C_7 \cap S = \{6, 7, 8\}$ and $C_7 \setminus S = \{9, 10, 11\}$, therefore object 7 cannot be a representative of S. So $S \not\subseteq \mathsf{u}_2(S)$, meaning that the strong approximation is not necessarily true.

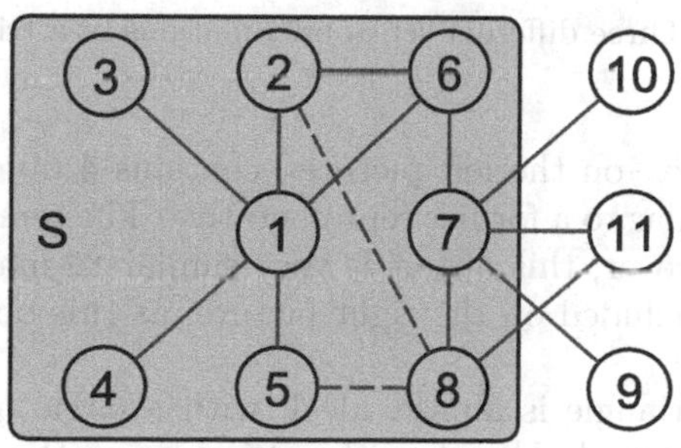

Fig. 6. Strong approximation property does not hold for the second approximation pair

Normality of l and u

The representatives are selected from the members of the set. If $S = \emptyset$ then we have no representatives, so $\mathsf{l}(S) = \mathsf{u}_1(S) = \mathsf{u}_2(S) = \emptyset$. In both cases, the normality of the lower approximation holds because an empty set does not have a representative member. The same holds for the normality of the upper approximation.

5 Experimental Results

The previously tested grids have a very special structure and they provide ideal results. Let's also look at some cases where chance has significant role in selecting the representatives. We randomly generated 500 points in the unit square. The idealised concept Q, the set to approximate is the same triangle as before. The randomly selected points of this triangle are marked with larger squares. Previously we have seen that the vertices of the triangle are challenging for our method, their rankings were rather low before. We chose the limit of similarity r to be 0.1 again. The radius of environments C_i—denoted with circles on Fig. 7—is the same. The limit of difference R is 0.2. The colours of the circles refer to the process of approximation,the environments of the first representatives are darker, and environments of the later representatives are brighter circles.

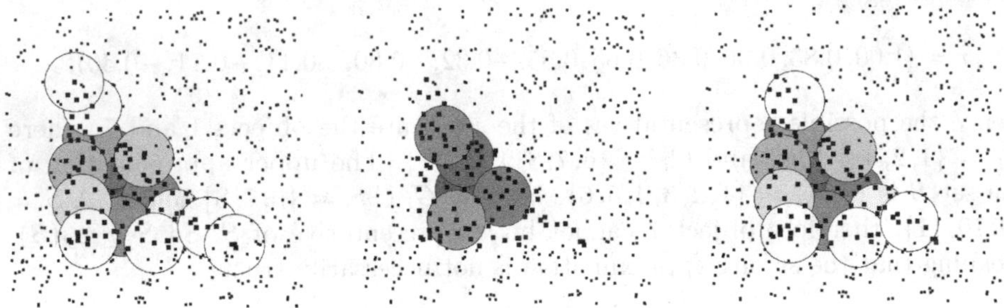

Fig. 7. Three different set approximations of a triangle

The topmost circle—on the left picture—contains 4 object from the set Q, but 3 of them are similar to a former representative. The remaining one becomes a representative. However, this object is very similar to many non-set objects, so this circle is not included on the right picture, as this object is not a representative in this case.

The size of this triangle is almost ideal: with a circle of radius R most of the triangle can be covered. Therefore the objects near the centre of gravity of the triangle hardly receive any opposing votes from their own set,so this part is really the *centre of the triangle*.

In Fig. 8 only the idealised set Q has changed, it is a square. The middle image shows that the firstly selected representative is far from the centre of

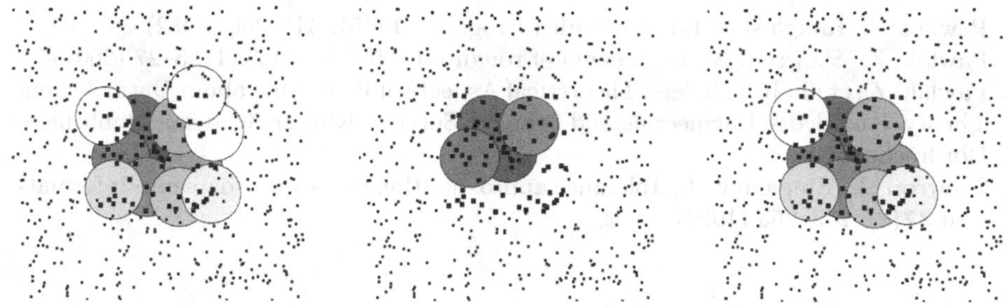

Fig. 8. Three different set approximations of a square

gravity of the square. Here we would not be able to cover a large part of the square with a circle of radius R, so there is no similar, small *centre of the square*. For larger central area the exact location of the other objects determines the place of the initial representative earned by our methods.

6 Conclusion and Further Work

In this article, we have described a method—more specifically 3 variants—which specifies the representatives of a set of objects Q by using a tolerance relation on this set. From these representatives we reach the others in one step, i.e. each object is similar to one of the representatives. Through these representatives different classification tasks can be solved with much less complexity.

Our method is based on the power method which uses matrix multiplications. This operation has very effective implementation in various libraries, so our method can be used widely, even for larger sets.

The ranks assigned to the elements of Q can be used in several ways to make approximations, three of which have been suggested in the paper. We reviewed the features of these 3 approximations and presented them in practice.

Although the method has been used to approximate one set only, it can be used to approximate disjoint sets without significant change. We plan to analyse the applicability of the method for a system of non-disjoint sets.

References

1. Altman, A., Tennenholtz, M.: Ranking systems: the PageRank axioms. In: Proceedings of the 6th ACM Conference on Electronic Commerce, pp. 1–8. ACM (2005)
2. Csajbók, Z., Mihálydeák, T.: Partial approximative set theory: a generalization of the rough set theory. Int. J. Comput. Inf. Syst. Ind. Manag. Appl. **4**, 437–444 (2012)
3. Mani, A.: Choice inclusive general rough semantics. Inf. Sci. **181**(6), 1097–1115 (2011)
4. Mises, R.V., Pollaczek-Geiringer, H.: Praktische verfahren der gleichungsauflösung. ZAMM - J. Appl. Math. Mech./Zeitschrift für Angewandte Mathematik und Mechanik **9**(2), 152–164 (1929). https://doi.org/10.1002/zamm.19290090206, https://onlinelibrary.wiley.com/doi/abs/10.1002/zamm.19290090206

5. Pawlak, Z.: Rough sets. Int. J. Parallel Program. **11**(5), 341–356 (1982)
6. Pawlak, Z., Skowron, A.: Rudiments of rough sets. Inf. Sci. **177**(1), 3–27 (2007)
7. Pawlak, Z., et al.: Rough Sets: Theoretical Aspects of Reasoning about Data. System Theory, Knowledge Engineering and Problem Solving. Kluwer Academic Publishers, Dordrecht (1991)
8. Skowron, A., Stepaniuk, J.: Tolerance approximation spaces. Fundamenta Informaticae **27**(2), 245–253 (1996)

Approximation Based on Representatives

Dávid Nagy[(✉)] and László Aszalós

Department of Computer Science, Faculty of Informatics, University of Debrecen, Egyetem tér 1, Debrecen 4010, Hungary
{nagy.david,aszalos.laszlo}@inf.unideb.hu

Abstract. In the authors' previous research, a possible usage of the correlation clustering in rough set theory was investigated. Correlation clustering relies on a tolerance relation. Its output is a partition. The system of base sets can be derived from the partition and a new approximation space appears. This space focuses on the similarity (the tolerance relation) itself and it is different from the covering type approximation space relying on a tolerance relation. In real-world applications, the number of objects is very high. So it can be effective only if a portion of the data points is used. In this paper, a possible method is provided to choose the necessary number of objects that represent the data set. These members are called representatives and it can be useful to use them in the approximation of an arbitrary set.

Keywords: Rough set theory · Correlation clustering · Set approximation · Representatives

1 Introduction

In our previous study, we examined whether the clusters, generated by correlation clustering, can be understood as a system of base sets. Correlation clustering is a clustering method in data mining which creates a partition of the input data set. The groups, defined by this partition, contain similar objects. In our previous paper [9,10] we showed that it is worth to generate the system of base sets from the partition. This way the base sets contain objects that are typically similar to each other and they are pairwise disjoint. Data sampling is a technique used to select a representative subset of data points to identify patterns in the larger data set being examined. Sampling can be particularly important when data sets are so large that it could be inefficient to analyse them in full. Finding and analysing a sample is more cost-effective than surveying the entirety of the population. However, it must be representative. This means that the sample points must be as similar in the sample as they are in the entire set. In this paper, a possible way is shown to choose objects from a set that can be used as representatives of the given set. Using the representatives, the execution time of the set approximations can be notably reduced.

The structure of the paper is the following: we begin with introducing the theoretical background and discussing rough set theory. In Sect. 4 we present our

© Springer Nature Switzerland AG 2019
T. Mihálydeák et al. (Eds.): IJCRS 2019, LNAI 11499, pp. 91–101, 2019.
https://doi.org/10.1007/978-3-030-22815-6_8

previous work. In Sect. 3 correlation clustering is defined. In the next section, we demonstrate our method for selecting representative members of a data set. In Sect. 6 the aforementioned method is used in the approximation process. Finally we conclude the results.

2 Theoretical Background

From the theoretical point of view a Pawlakian approximation space [11–13] can be characterised by an ordered pair $\langle U, \mathcal{R} \rangle$ where U is a non-empty set of objects and $\mathcal{R}$ is an equivalence relation on U. In order to approximate an arbitrary subset S of U the following have to be introduced:

- *the set of base sets*: $\mathfrak{B} = \{B \mid B \subseteq U, \text{ and } x, y \in B \text{ if } x\mathcal{R}y\}$, the partition of U generated by the equivalence relation $\mathcal{R}$;
- *the set of definable sets*: $\mathfrak{D}_{\mathfrak{B}}$ is an extension of $\mathfrak{B}$, and it is given by the following inductive definition:
  1. $\mathfrak{B} \subseteq \mathfrak{D}_{\mathfrak{B}}$;
  2. $\emptyset \in \mathfrak{D}_{\mathfrak{B}}$;
  3. if $D_1, D_2 \in \mathfrak{D}_{\mathfrak{B}}$, then $D_1 \cup D_2 \in \mathfrak{D}_{\mathfrak{B}}$.
- *the functions* l, u form a Pawlakian approximation pair $\langle \mathsf{l}, \mathsf{u} \rangle$, i.e.
  1. $Dom(\mathsf{l}) = Dom(\mathsf{u}) = 2^U$
  2. $\mathsf{l}(S) = \bigcup\{B \mid B \in \mathfrak{B} \text{ and } B \subseteq S\}$;
  3. $\mathsf{u}(S) = \bigcup\{B \mid B \in \mathfrak{B} \text{ and } B \cap S \neq \emptyset\}$.

3 Correlation Clustering

Cluster analysis is an unsupervised learning method in data mining. The goal is to group the objects so that the objects in the same group are more similar to each other than to those which are in other groups. In many cases, the similarity is based on the attribute values of the objects. Although, there are some cases when these values are not numbers, but we can still say something about their similarity or dissimilarity. For example, let's consider humans. We cannot describe someone's looks using only a number, but we can make simple statements on whether two people are similar or dissimilar. These opinions are dependent on the person making the statements. Someone can say that two people are similar while others treat them as dissimilar. If we want to formulate the similarity and dissimilarity using mathematics, we need a tolerance relation (i.e. a reflexive and symmetric relation). If this relation holds for two objects, we can say that they are similar. If this relation does not hold, then they are dissimilar. This relation is reflexive because every object is similar to itself. It is also symmetric because if some object is similar to another one, then the similarity is equivalent the other way round. However transitivity does not necessarily hold. If we take a human and a mouse, then due to their inner structure they are considered similar. This is the reason mice are used in many drug experiments. A human and a mannequin are also similar, this time according to their shape.

This is why these dolls are used in display windows. However, a mouse and a mannequin are dissimilar (except that both are similar to the same object). Correlation clustering is a clustering technique based on a tolerance relation [5,6,15].

The task is to find an $R \subseteq V \times V$ equivalence relation which is *closest* to the tolerance relation. A (partial) tolerance relation $\mathcal{R}$ [7,14] can be represented by a matrix M. Let matrix $M = (m_{ij})$ be the matrix of the partial relation $\mathcal{R}$ of similarity: $m_{ij} = 1$ if objects i and j are similar, $m_{ij} = -1$ if objects i and j are dissimilar, and $m_{ij} = 0$ otherwise.

A relation is called partial if there exist two elements (i, j) such that $m_{ij} = 0$. It means that if we have an arbitrary relation $R \subseteq V \times V$ we have two sets of pairs. Let R_{true} be the set of those pairs of elements for which R holds and R_{false} be the one for which R does not hold. If R is partial, then $R_{true} \cup R_{false}$ is a proper subset of $V \times V$. If R is total, then $R_{true} \cup R_{false} = V \times V$.

A partition of a set S is a function $p : S \to \mathbb{N}$. Objects $x, y \in S$ are in the same cluster at partitioning p, if $p(x) = p(y)$. For a conflict one of the following two cases holds:

- Two dissimilar objects end up in the same cluster
- Two similar objects end up in different clusters

The cost function is the number of these disagreements. The formal definition can be seen in [9]. For a relation the partition with the minimal cost function value is called *optimal.* Solving a correlation clustering problem is equivalent to minimising its cost function for the fixed relation. If the cost function's value is 0, the partition is called *perfect.* Given the $\mathcal{R}$ and R we call the value f the distance of the two relations. With this definition, the partition generates an equivalence relation. This relation can be considered to be the closest to the tolerance relation.

It is easy to check that we cannot necessarily find a perfect partition for an arbitrary similarity relation. Consider the simplest such case, given three objects A, B and C, and A is similar to both B and C, but B and C are dissimilar. In this situation, the following 5 partitions can be given:

$$\{\{A, B, C\}, \{\{A, B\}, \{C\}\}, \{\{A, C\}, \{B\}\}, \{\{B, C\}, \{A\}\}, \{\{A\}, \{B\}, \{C\}\}.$$

It is easy to see that in every of one them there is at least 1 conflict. The number of partitions can be given by the Bell number [1], which grows exponentially. So the optimal partition cannot be determined in reasonable time. In a practical case a quasi optimal partition can be sufficient, so a search algorithm can be used.

The main advantage of the correlation clustering is that the number of clusters does not need to be specified in advance like in many clustering algorithms, and this number is optimal based on the similarity. However, as the number of partitions grows exponentially it is an NP hard problem.

4 Similarity Based Rough Sets

The system of base sets is based on the background knowledge embedded in an information system. The base sets represent the background knowledge (or its limit). In a Pawlakian system two objects are treated as indiscernible if all of their known attribute values are identical. The indiscernibility relation defines an equivalence relation. In some cases we only have a similarity (tolerance) relation. If we change the negativity of indiscernible relations to positivity of similarity (based on background knowledge), then we may rely on a tolerance relation. Some covering systems are also based on a tolerance relation. However, in our case the emphasis is on the similarity to a given object and not the similarity of objects in the general sense. With correlation clustering, a quasi optimal partition of the universe can be obtained [2–4]. The members of a partition are called clusters. They contain elements that are typically similar to each other and not just to a distinguished member. In our previous research, we investigated if the partition can be understood as a system of base sets [8–10]. According to our results, it is worth to generate a partition with correlation clustering. The system of base sets can defined as:

$$\mathfrak{B} = \{B \mid B \subseteq U, \text{ and } x, y \in B \text{ if } p(x) = p(y)\},$$

where p is the partition gained from the correlation clustering. The base sets have several useful properties:

- the similarity of objects relying on their properties (and not the similarity to a distinguished object) plays an important role in the definition of base sets;
- the system of base sets consists of disjoint sets, so the lower and upper approximations are closed in the following sense: Let S be a set and $x \in U$. If $x \in \mathsf{l}(S)$, then we can say, that every object $y \in U$ which is in the same cluster as x is in $\mathsf{l}(S)$. If $x \in \mathsf{u}(S)$, then we can say, that every object $y \in U$ which is in the same cluster as x is in $\mathsf{u}(S)$.
- the number of clusters is not set by the user because the algorithm finds the optimal number. This way, only the necessary number of base sets appear (in applications we have to use an acceptable number of base sets);
- the size of the base sets is not too small, nor too big.

5 Representative Member

In data mining, to reduce the execution time of an algorithm, it is common to use samples. A sample contains points from the original data set. There are numerous ways to choose a part of the input data set which can be treated as a sample. However, in every method it is crucial that the chosen objects must represent the entire population. In this case, representativeness means that the specific properties are as similar in the sample as in the entire set. Without this property, important information might be disregarded.

Imagine that a product is needed to be sold, for example a toy to a group of children. In almost every group of youngsters, there is at least one member whose decision has the most influence on the group's life. In this case, one child is enough to be found and convinced to buy the toy. The rest of the group will follow them.

In [10] the authors of this article provided a method to calculate the representative of a given set.

A member is called a representative if it is similar to most, and different from the least of the members in the group. For any member m two values have been stored:

- α - the number of elements that are similar to m.
- β - the number of elements that are different from m.

Figure 1 shows a very simple example to the method. For the member A these two values are:

- $\alpha = 4$. Because there are four members (B,C,E,F) that are similar to A.
- $\beta = 2$. Because there are two members (H and D) that are different from A.

Here, similarity is denoted by a solid line and dissimilarity by a dashed line.

A member can be considered a possible representative if the following fraction is maximal:

$$r = \frac{\alpha^w - \beta^v}{\alpha + \beta + 1} \; v, w \in \mathbb{R}, v, w > 1, w > v \tag{1}$$

v and w are weights. By default their values can be set to 1. In this case, both the similarity and dissimilarity have equal importance. Using default weights in the example shown in Fig. 1 the object F has the maximal r value. Naturally, any other types of methods can used to determine the representatives. The aforementioned formula is a simple way to do it.

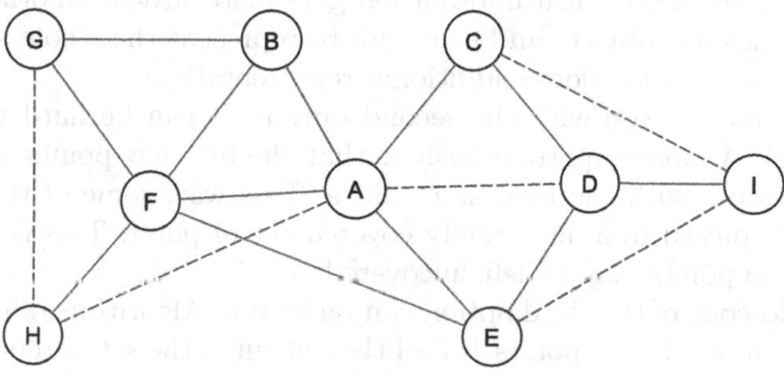

Fig. 1. α and β values for the member A

In many applications, however, it might not be enough to have only one representative in each set of objects. Figure 2 shows a very simple example to this problem. Clearly object A has the highest r value so it is the most representative object of the set. However, it is only similar to objects B, C, D, E, F and G and does not have any kind of connections with the rest of the objects. So the above mentioned property for samples is not satisfied as object A alone cannot represent the entire set.

We offer three possible ways to generate more than one representative from which only the third option proves to be appropriate for real-world applications. A representative member X is said to cover the member Y if X is similar to Y.

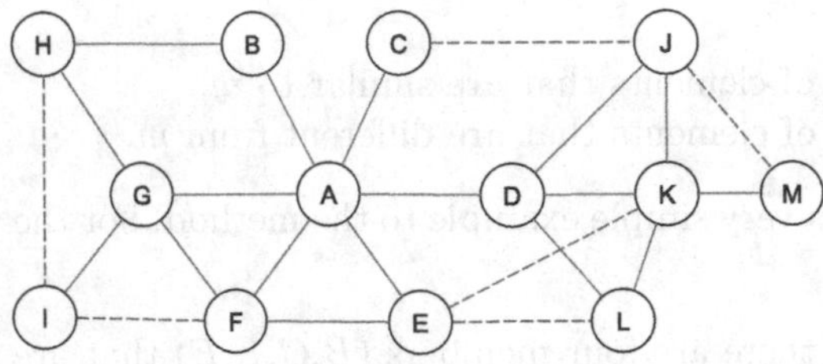

Fig. 2. Multiple representatives are needed

1. The user gives a threshold value k. Then k percent of random objects are treated as representatives.
2. The user gives an interval for the r values. If the r value of an object is in this interval, then it is considered as a representative.
3. Use an algorithm based on similarity to generate the necessary number of representatives.

The main issue with the first option is that the user must have some knowledge about the given set in order to choose an optimal k value. Due to randomness, it is possible that critical information gets disregarded. Another problem is that sometimes one object can be enough to represent the whole set, but the user forces the system to choose additional representatives.

Similar problems arise with the second option. It can be hard to choose a proper interval. A more important issue is that the first few points selected will always be the ones with the highest r values. This way, some of the representatives may be picked from an already covered set of point. This is redundant, and some of the points may be left uncovered.

The pseudo-code of the third option can be seen in Algorithm 1. The input of the function is a set of data points D and the output is the set of representatives REP. It is an iterative method, in each step the algorithm keeps a record of the covered objects (i.e. the objects that are similar to one of the representatives) which is empty in the first iteration (line 6). In every iteration, the object with the highest r value is selected from the uncovered objects (line 10–15). In line 16–20, the data points, that are covered by the currently selected representative, are inserted into the set C. At the end of each step, the chosen representative is moved into the set *REP*. The algorithm stops when there are no uncovered members left.

The strength of the method is that it uses the similarity between objects, and so it generates the optimal number of representatives. The other two methods can create too few or too many representatives. Another advantage is that it does not need any user-defined parameters. The algorithm can be treated as a directed sampling method which can be a very powerful tool in many applications.

A political party contains members that share a common political ideology. However, in some parties it can happen that even though the members follow the same vision, there are some disagreements. So the group can be divided into smaller groups. In this case, one politician is not enough to represent the entire party. The above mentioned algorithm could be a solution as it takes into account the variety of the members.

Algorithm 1. Selecting representatives

```
function SELECT REPRESENTATIVES(D)
    REP ← ∅
    for each p ∈ D do
        calculate the r value of point p
    end for
    C ← ∅
    while C ≠ D do
        max ← −inf
        max_p ← None
        for each p ∈ (D \ C) do
            if r value of point p > max then
                max ← r value of point p
                max_p ← p
            end if
        end for
        for each p ∈ (D \ C) do
            if max_p covers p then
                C ← C ∪ {p}
            end if
        end for
        REP ← REP ∪ {max_p}
    end while
    return REP
end function
```

Figure 3 presents the steps of the algorithm for the data set shown in Fig. 2. The grey ellipses contain the covered objects by a chosen representative member. In the first step, object A is chosen. In the second step, objects B, C, D, E, F and G are not considered as they are covered by A. The second method, mentioned in the beginning of this section, could have chosen B or G as possible representatives because they have the second highest r value. Naturally, it is pointless to select them because object A makes it redundant (both of them are similar to A). After four steps, the algorithm finishes and the four representatives are objects A, K, H, I. It can be easily seen that these 4 members share the diversity of the original data set.

6 Approximation Based on Representatives

The lower approximation of a set S is the union of those base sets that are subsets of S. In order to get these base sets, every point in each base set must be considered. It can be a time consuming task if the number of points is high. The effectiveness of the representatives lies in situations when the number of objects is very large. It can be practical to use the power of representatives in the approximation process. For each base set, let us consider only its representatives. Let $B \in \mathfrak{B}$ be a base set, and $REP(B)$ be the set of its representatives. The approximation functions are defined as the following:

- $\mathsf{l}(S) = \bigcup\{B \mid B \in \mathfrak{B} \text{ and } \forall x \in REP(B) : x \in S\}$;
- $\mathsf{u}(S) = \bigcup\{B \mid B \in \mathfrak{B} \text{ and } \exists x \in REP(B) : x \in S\}$.

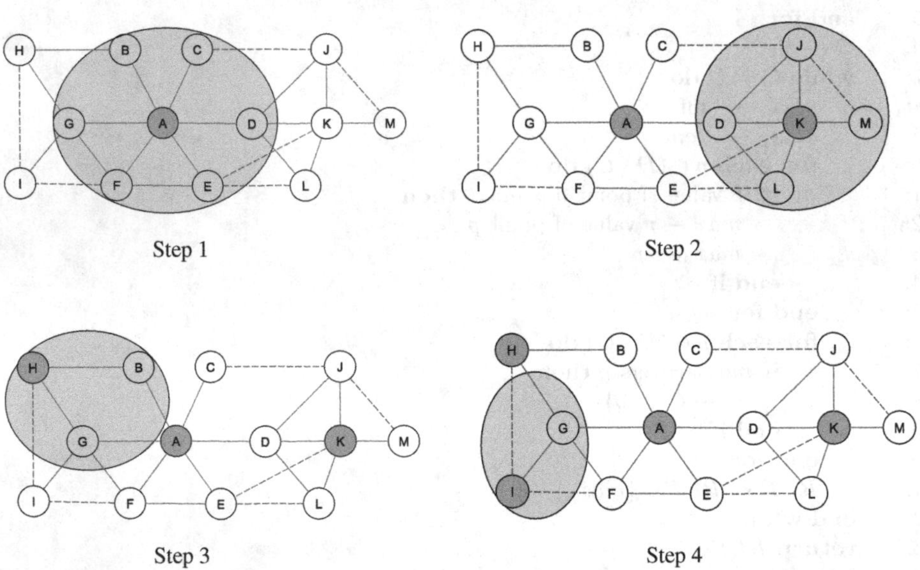

Fig. 3. The execution of the algorithm

This way, the lower approximation of a set S becomes the union of those base sets for which every representative is a member of S. A base set belongs to the upper approximation if at least one of its representatives is in the set S. Naturally, the certainty of the lower approximation might be lost, but as the number of points are increasing, it can be very useful.

In Fig. 4 a simple example is provided for the method. The base sets are denoted by solid-line rectangles, and the set we wish to approximate (S) is denoted by a grey ellipse. For each base set, the black circles symbolise the representatives.

The approximation of the set S is the following based on the representatives:

- $l(S) = \{B_2, B_6\}$
- $u(S) = \{B_1, B_2, B_3, B_6\}$

The approximation of the set S is the following based on the classical approximation pair:

- $l(S) = \{B_2, B_6\}$
- $u(S) = \{B_1, B_2, B_3, B_5, B_6\}$

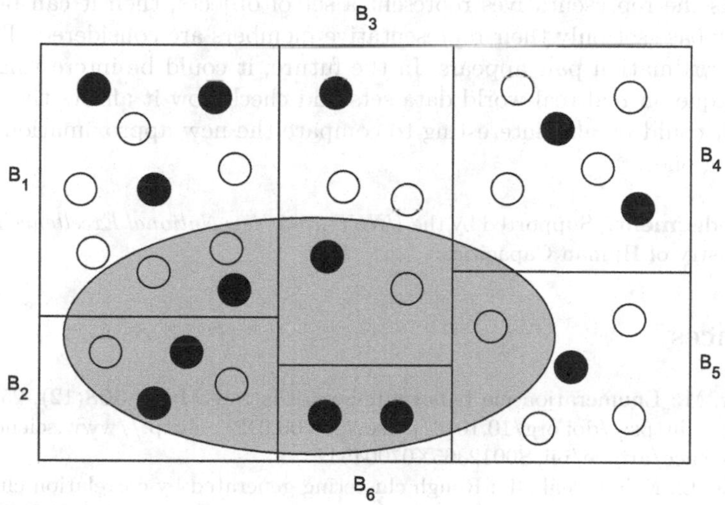

Fig. 4. Approximation based on representatives

The lower approximation is the same in both cases. The upper approximation differs in one base set (B_5). When there is a huge number of points and there are several sets to be approximated, we recommend approximation using representatives. In this case, the method can reduce the run-time of the approximation significantly. Determining the approximation with the classical functions 32 objects needed to be considered. Using the proposed method, only 13 of them had to be tested, so almost 60% of the original points were left-out. Of course, with 32 to 13 points is not a significant change, but in case of millions of objects it can be very useful.

7 Conclusion and Future Work

In [9,10] the authors introduced a partial approximation space relying on a similarity relation (a tolerance relation). The genuine novelty of this new approximation spaces is the way in which the systems of base sets is defined: it is the result of correlation clustering, and so the similarity is taken into consideration generally. Singleton clusters does not have real information in approximation process, these clusters cannot be taken as base sets, therefore the approximation spaces are partial in general cases (the unions of base sets are proper subsets of universes). In data mining and also in statistics, it is very common to use only a subset of the original data set instead of the entire collection. The members of this subset can be called as representatives. A very important criteria is that these objects must have the same properties as the whole data set. In this paper, a possible way is provided to choose the necessary number of representatives of a set. The authors also showed how this method can be applied in set approximation. As the representatives represent a set of objects, then it can be useful if for every base set only their representative members are considered. This way a new approximation pair appears. In the future, it could be interesting to use this technique on real real-world data sets and check how it affects the approximation. It could be also interesting to compare the new approximation pair to the existing ones.

Acknowledgement. Supported by the *ÚNKP-18-3 New National Excellence Program* of the Ministry of Human Capacities.

References

1. Aigner, M.: Enumeration via ballot numbers. Discrete Math. **308**(12), 2544–2563 (2008). https://doi.org/10.1016/j.disc.2007.06.012, http://www.sciencedirect.com/science/article/pii/S0012365X07004542
2. Aszalós, L., Mihálydeák, T.: Rough clustering generated by correlation clustering. In: Ciucci, D., Inuiguchi, M., Yao, Y., Ślęzak, D., Wang, G. (eds.) RSFDGrC 2013. LNCS (LNAI), vol. 8170, pp. 315–324. Springer, Heidelberg (2013). https://doi.org/10.1007/978-3-642-41218-9_34
3. Aszalós, L., Mihálydeák, T.: Rough classification based on correlation clustering. In: Miao, D., Pedrycz, W., Ślęzak, D., Peters, G., Hu, Q., Wang, R. (eds.) RSKT 2014. LNCS (LNAI), vol. 8818, pp. 399–410. Springer, Cham (2014). https://doi.org/10.1007/978-3-319-11740-9_37
4. Aszalós, L., Mihálydeák, T.: Correlation clustering by contraction. In: 2015 Federated Conference on Computer Science and Information Systems (FedCSIS), pp. 425–434. IEEE (2015)
5. Bansal, N., Blum, A., Chawla, S.: Correlation clustering. Mach. Learn. **56**(1–3), 89–113 (2004)
6. Becker, H.: A survey of correlation clustering. Advanced Topics in Computational Learning Theory, pp. 1–10 (2005)
7. Mani, A.: Choice inclusive general rough semantics. Inf. Sci. **181**(6), 1097–1115 (2011)

8. Mihálydeák, T.: Logic on similarity based rough sets. In: Nguyen, H.S., Ha, Q.-T., Li, T., Przybyła-Kasperek, M. (eds.) IJCRS 2018. LNCS (LNAI), vol. 11103, pp. 270–283. Springer, Cham (2018). https://doi.org/10.1007/978-3-319-99368-3_21
9. Nagy, D., Mihálydeák, T., Aszalós, L.: Similarity based rough sets. In: Polkowski, L., et al. (eds.) IJCRS 2017. LNCS (LNAI), vol. 10314, pp. 94–107. Springer, Cham (2017). https://doi.org/10.1007/978-3-319-60840-2_7
10. Nagy, D., Mihálydeák, T., Aszalós, L.: Similarity based rough sets with annotation. In: Nguyen, H.S., Ha, Q.-T., Li, T., Przybyła-Kasperek, M. (eds.) IJCRS 2018. LNCS (LNAI), vol. 11103, pp. 88–100. Springer, Cham (2018). https://doi.org/10.1007/978-3-319-99368-3_7
11. Pawlak, Z.: Rough sets. Int. J. Parallel Prog. **11**(5), 341–356 (1982)
12. Pawlak, Z., Skowron, A.: Rudiments of rough sets. Inf. Sci. **177**(1), 3–27 (2007)
13. Pawlak, Z., et al.: Rough Sets: Theoretical Aspects of Reasoning About Data. System Theory, Knowledge Engineering and Problem Solving, vol. 9. Kluwer Academic Publishers, Dordrecht (1991)
14. Skowron, A., Stepaniuk, J.: Tolerance approximation spaces. Fundamenta Informaticae **27**(2), 245–253 (1996)
15. Zimek, A.: Correlation clustering. ACM SIGKDD Explor. Newsl. **11**(1), 53–54 (2009)

Local Search for Attribute Reduction

Xiaojun Xie[1,2], Ryszard Janicki[2], Xiaolin Qin[1(✉)], Wei Zhao[2], and Guangmei Huang[3]

[1] College of Computer Science and Technology, Nanjing University of Aeronautics and Astronautics, Nanjing 211106, China
{xiexj,qinxcs}@nuaa.edu.cn

[2] Department of Computing and Software, McMaster University, Hamilton L8S 4K1, Canada
{janicki,zhaow9}@mcmaster.ca

[3] Faculty of Education, Guangxi Normal University, Guilin 541001, China
guangmeihuang@126.com

Abstract. Two new attribute reduction algorithms based on iterated local search and rough sets are proposed. Both algorithms start with a greedy construction of a relative reduct. Then attempts to remove some attributes to make the reduct smaller. Process of attributes selection is the main difference between the algorithms. It is random for the first one, and a sophisticated selection procedure is used for the second algorithm. Moreover a fixed number of iterations is assumed for the first algorithms whereas the second stops when a local optimum is reached. Various experiments using eight well-known data sets from UCI have been made and they show substantial superiority of our algorithms.

Keywords: Rough set · Attribute reduction · Local search · Positive region

1 Introduction

Feature selection, or attribute reduction, is a process of finding a minimal subset of attributes that still provides the same, or similar information as the set of all original attributes. Rough set theory has been very successful as a theoretical base used in filter-based feature selection algorithms in many fields, such as data mining, machine learning, pattern recognition and many others [1–7].

Attribute reduction methods can be divided into four categories: exact algorithms, approximation algorithms, general heuristic algorithms and meta-heuristic algorithms.

Exact algorithms can find all reducts and an optimal reduct. The classical exact algorithm [8], consists in finding the discernibility matrix first, then deriving the discernibility function in its conjunctive normal form (CNF) from it, and at the end transforming CNF into DNF i.e. disjunctive normal form. Then, each prime implicant of the DNF corresponds to a reduct, and each minimal prime implicant of the DNF corresponds to an optimal reduct. Unfortunately, finding

© Springer Nature Switzerland AG 2019
T. Mihálydeák et al. (Eds.): IJCRS 2019, LNAI 11499, pp. 102–117, 2019.
https://doi.org/10.1007/978-3-030-22815-6_9

all reducts or an optimal reduct has been proven to be in general an NP-hard problem [8,9], which is a problem for big data sets with many attributes and objects.

Several efficient approximation algorithms have been proposed in recent years. Yang et al. [12] provided a new efficient method based on related family for computing all attribute reducts and relative attribute reducts. Tan et al. [10] proposed very time efficient matrix based approximation algorithm by introducing the concepts of minimal and maximal descriptions. Hacibeyoglu et al. [11] analyzed the main shortcoming of this algorithm, namely is its excessively high space complexity, and proposed a substantial improvement with the worst case space complexity of $\binom{N}{N/2}/2$, where N is the number of attributes.

For many big real-world applications, efficiency of approximation algorithms is still not enough. Frequently it is also not necessary to find all reducts, on contrary, quite often finding one reduct is enough, which leads to the idea of looking for heuristic algorithms.

The general heuristic algorithm normally starts with the core attribute set or an empty attribute set, then gradually adds an attribute with the maximal significance into the attribute reduct until the attribute reduct satisfies the stopping criterion. Different models have been used for stopping criteria, namely positive region [13], information entropy [14], knowledge granularity [15], and other models [16,17].

General heuristic algorithms usually fail to obtain an optimal reduct, so many meta-heuristic algorithms have been proposed such as genetic algorithms, tabu search, ant colony optimization, particle swarm optimization and artificial fish swarm algorithm, and so on. In [18], Xu et al. illustrated the shortcomings of the previous genetic algorithm-based methods and designed new fitness function, which resulted in more efficient genetic algorithm. Chen et al. [19] provided a novel rough set based method to feature selection using fish swarm algorithm. Inbarani et al. [20] proposed a supervised feature selection method based on quick reduct and improved harmony search. Luan et al. [21] developed a novel attribute reduction algorithm based on rough set and improved artificial fish swarm algorithm. Aziz and Hassanien [22] proposed an improved social spider algorithm for the minimal reduction problem. Xie et al. [23] designed a test-cost-sensitive rough set-based algorithm for the minimum weight vertex cover problem, which can also be used to solve attribute reduction problem in rough sets.

Nevertheless, for big data sets with huge number of attributes and objects, meta-heuristic algorithms are often still not sufficiently efficient. In recent years, *local search* has been shown to be an effective and promising approach to solve many NP-hard problems, such as, for example, the minimum vertex cover problem [24,25]. In this paper we will design, discuss and test two new algorithms for attribute reduction that is based on local search paradigm. The main ideas of these two algorithms can be described as follows (Fig. 1).

If a reduct has been obtained, then an upper bound of the target problem has also been found. Then, we decrease the upper bound by removing an attribute

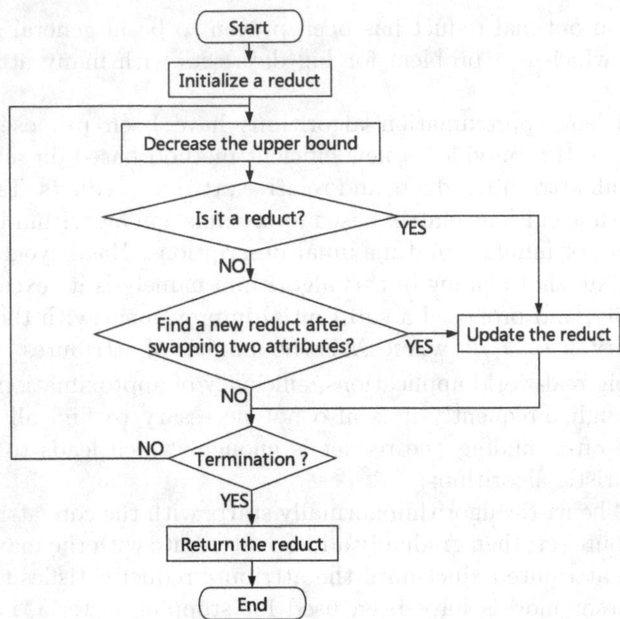

Fig. 1. Basic flowchart of our two algorithms. Procedures for termination and finding new reducts are different in each algorithm.

from the current reduct. The outcome may or may not be a reduct. If it is not a reduct, we swap attributes, one attribute from the current candidate reduct and the other that does not belong to the current candidate reduct. If the result is a reduct, it has smaller, i.e. better, upper bound. We continue this process as long as it is possible, the outcome is a relatively small reduct or even an optimal reduct.

Finding a new relative reduct after swapping two attributes is the key process in each iteration and the difference between our two algorithms. To make this efficient, the second algorithm uses the reverse incremental verification to check if a swapping results in a reduct. The second algorithm also uses a set of *removed attributes* to adjust the iteration process, which additionally improves the efficiency of our algorithm. Moreover the second algorithm stops when a local optimum is found while the first one performs given in advance number of iterations.

The rest of the paper is organized as follow. In Sect. 2, basic concepts about rough sets are introduced. Section 3 exposes the local search-based algorithms for attribute reduction. Experimental results on UCI data sets are presented in Sect. 4. Some conclusions and further researches are drawn in Sect. 5.

2 Preliminaries

This section recalls some basic concepts, definitions and notation used in this paper.

For any equivalence relation $R \subseteq U \times U$, where U is a set, $[x]_R$ denotes the equivalence class containing $x \in U$, i.e. $[x]_R = \{y \mid (x, y) \in R\}$, and U/R denotes the partition of U defined by R, i.e. $U/R = \{[x]_R \mid x \in U\}$.

A *decision table* is the 5-tuple: $S = (U, C, D, V, f)$, where U is a finite nonempty set of objects, called *universe*, C is a set of *conditional attributes*, D is a set of *decision attributes*, V is *domain of attributes* $C \cup D$ and $f : U \times (C \cup D) \rightarrow V$ is an *information function*.

Table 1 is a simple example of a decision table, where $U = \{x_1, x_2, x_3, x_4, x_5, x_6, x_7\}$, $C = \{a_1, a_2, a_3, a_4, a_5, a_6\}$, and $D = \{Flu\}$ (or $D = \{a_7\}$).

Table 1. An example of a decision table.

	a_1	a_2	a_3	a_4	a_5	a_6	a_7
Patient	Headache	Temperature	Lymphocyte	Leukocyte	Eosinophil	Heartbeat	Flu
x_1	Yes	High	High	High	High	Normal	Yes
x_2	Yes	High	Normal	High	High	Abnormal	Yes
x_3	Yes	High	High	High	Normal	Abnormal	Yes
x_4	No	High	Normal	Normal	Normal	Normal	No
x_5	Yes	Normal	Normal	Low	High	Abnormal	No
x_6	Yes	Normal	Low	High	Normal	Abnormal	No
x_7	Yes	Low	Low	High	Normal	Normal	Yes

Let $S = (U, C, D, V, f)$ be a decision table. For each nonempty $B \subseteq C$ or $B = D$ we define a ***indiscernibility relation induced by*** B, denoted $\mathsf{ind}(B)$, as:

$$\mathsf{ind}(B) = \{(x, y) \mid x, y \in U \wedge \forall a \in B, f(x, a) = f(y, a)\}.$$

The relation $\mathsf{ind}(B)$ is clearly an equivalence relation on U.

When $B = D$, $\mathsf{ind}(B)$ is called ***classification relation induced by*** D and denoted by $\mathcal{D}$. In this case a partition $U/\mathcal{D}$ is called *classification defined by the decision attributes* D.

For every nonempty $B \subseteq C$, and every $X \subseteq U$, we define $B_-(X)$, the *B-lower approximation* of X, as $B_-(X) = \{x \in U \mid [x]_{\mathsf{ind}(B)} \subseteq X\}$. For every for every nonempty $B \subseteq C$ and every $U' \subseteq U$, we define the ***positive region*** (or ***lower approximation***) ***of*** D ***over*** U' ***with respect to*** B as:

$$\mathsf{POS}_B^{U'}(D) = \bigcup_{X \in U'/\mathcal{D}} B_-(X).$$

When $U' = U$, the most popular case, we will just write $\mathsf{POS}_B(D)$. Note that we always have: $\mathsf{POS}_B^{U'}(D) \subseteq U'$.

Definition 1. *Let $S = (U, C, D, V, f)$ be a decision table and let $B \subseteq C$.*

1. *A set B is called a* ***relative attribute reduct*** *if and only if $\mathsf{POS}_B(D) = \mathsf{POS}_C(D)$, and*
2. *a set B is called an* ***attribute reduct*** *if and only if it is a relative reduct and for each $B' \subsetneq B$, we have $\mathsf{POS}_{B'}(D) \neq \mathsf{POS}_B(D)$,*
3. *a set B is called an* ***optimal attribute reduct*** *if and only if it is a reduct and for any other reduct B', we have $|B| \leq |B'|$.* ◇

In other words, reducts are minimal relative reducts and optimal reduct is a reduct with smallest cardinality.

3 Local Search for Attribute Reduction

This section describes in detail our local search method for solving the attribute reduction problem.

3.1 A Plain Local Search Algorithm for Attribute Reduction

Our method stems from the following simple result.

Suppose $S = (U, C, D, V, f)$ is a decision table, $Red \subseteq C$ is a relative reduct, we randomly select $a \in Red$. If $Red \setminus \{a\}$ is also a relative reduct, then we update $Red \setminus \{a\}$ as new Red and jump into the next iteration. If $Red \setminus \{a\}$ is *not* a relative reduct, we randomly choose $u \in Red \setminus \{a\}$ and $v \in C \setminus Red$ and verify if $Red_{auv} = (Red \setminus \{a, u\}) \cup \{v\}$ is a relative reduct. If it is, we update Red_{auv} as new Red, and go to the next iteration. Since $|Red_{auv}| = |Red| - 1$, Red_{auv} is better relative reduct than Red. If Red_{auv} is not a relative reduct, Red is not changed and we continue with the next iteration. The algorithm stops when it iterates T times, where T is a parameter given in advance.

The process always returns a relative reduct and the bigger value of T, the smaller, i.e. better, the solution is. Algorithm 1 represents the procedure described above. The algorithm starts with a construction some relative reduct Red (steps 1 and 2). The computation is greedy, the set Red is initially empty and then, in each iteration we choose an attribute $a \in C \setminus Red$ at random and add it to Red. The computation process stops when $\mathsf{POS}_{Red}(D) = \mathsf{POS}_C(D)$. The process always converges, the worst case is when $Red = C$, so no reduct exists. The worst case time complexity is $O(\sum_{i=1}^{|Red|} i|U|) = O(|Red|^2|U|) = O(|C|^2|U|)$. Steps 3–14 represent T iterations that result in a derivation of a reduct Red_T from a relative reduct Red. Clearly $|Red_T| \leq |Red|$. The worst case time complexity of the ith iteration is $O(|Red_i||U|)$, where Red_i is Red from i^{th} iteration, so the worst case time complexity of lines 3–14 is $O((\sum_{i=1}^{T} |Red_i|)|U|) = O(T|Red||U|) = O(T|C||U|)$ as clearly $|Red_i| \leq |Red| \leq |C|$ for all $i = 1, \ldots, T$. For the entire Algorithm 1 we have $O(\max(T, |Red|)|Red||U|) = O(\max(T, |C|)|C||U|) = O(T|C||U|)$ as usually $T > |C|$.

Algorithm 1 always finds some reducts but not necessarily an optimal reduct. The quality of solution clearly depends on the size of T, but also on smart selection of pairs (u, v). Foundations of such selection process are presented in the next section. We would also like to get rid of this arbitrary limit T and just stop when a local minimum is found.

Algorithm 1. (LSAR) Local search algorithm for attribute reduction

Input: A decision table $S = (U, C, D, V, f)$, the maximum number of iterations T

Output: The attribute reduction *Red*.

$t = 0$, $Red = \emptyset$;
construct a relative reduct *Red* using greedy algorithm;
while $t < T$ **do**
    remove an attribute a from *Red* randomly;
    if $\mathsf{POS}_{Red\setminus\{a\}}(D) = \mathsf{POS}_C(D)$ **then**
        $Red = Red \setminus \{a\}$;
    else
        select randomly the deleting attribute $u \in Red \setminus \{a\}$ and the adding attribute $v \in C \setminus Red$;
        if $\mathsf{POS}_{(Red\setminus\{a,u\})\cup\{v\}}(D) = \mathsf{POS}_C(D)$ **then**
            $Red = (Red \setminus \{a, u\}) \cup \{v\}$;
        end
    end
    $t = t + 1$;
end
return *Red*;

3.2 Attribute Pair Selection Mechanism

In principle, the basic problem we have to deal with in Algorithm 1 can be formulated as follows. Suppose that $\mathsf{POS}_B(D) \neq \mathsf{POS}_C(D)$. How to select attributes u and v such that $\mathsf{POS}_{(B\setminus\{u\})\cup\{v\}}(D) = \mathsf{POS}_C(D)$? We will use a reverse incremental verification approach to solve this problem and start with two useful lemmas.

Lemma 1. *Let $S = (U, C, D, V, f)$ be a decision table. For each $B \subseteq C$, we have:* $\mathsf{POS}_B(D) = \mathsf{POS}_B^{\mathsf{POS}_B(D)}(D)$.

Proof. Clearly $\mathsf{POS}_B^{\mathsf{POS}_B(D)}(D) \subseteq \mathsf{POS}_B(D)$. Let $x \in \mathsf{POS}_B(D)$ and $\mathsf{POS}_B(D)/\mathcal{D}$ be the partition of $\mathsf{POS}_B(D)$ defined by D. Then $x \in X_x \in \mathsf{POS}_B(D)/\mathcal{D}$. But by the definition: $\mathsf{POS}_B^{\mathsf{POS}_B(D)}(D) = \bigcup_{X\in\mathsf{POS}_B(D)/\mathcal{D}} B_-(X)$, hence $x \in \mathsf{POS}_B^{\mathsf{POS}_B(D)}(D)$. □

Before formulating our next result we need to introduce one more concept.

Let $S = (U, C, D, V, f)$ be a decision table. For each nonempty $B \subseteq C$ we define the ***inconsistent objects pairs***, denoted $\mathsf{iop}(B)$, as:

$$\mathsf{iop}(B) = \{(x, y) \mid x, y \in U \wedge (\forall a \in B. f(x, a) = f(y, a)) \wedge (\exists d \in D. f(x, d) \neq f(y, d))\}.$$

If (x, y) forms an inconsistent object pair, then the value of all conditional attribute are the same and the values of some decision attributes are different.

Lemma 2. *Let $S = (U, C, D, V, f)$ be a decision table and $B \subseteq C$. Then we have:*

1. *For each attribute $v \in C \setminus B$,*

$$\mathsf{POS}_{B \cup \{v\}}(D) = \mathsf{POS}_B(D) \cup \mathsf{POS}^{U'}_{\{v\}}(D),$$

where $U' = \mathsf{POS}_{B \cup \{v\}}(D) \setminus \mathsf{POS}_B(D)$.

2. *For each attribute $u \in B$,*

$$\mathsf{POS}_{B \setminus \{u\}}(D) = \mathsf{POS}_B(D) \setminus \bigcup_{X \in \mathcal{X}_u} X,$$

where $\mathcal{X}_u = \{X \mid X \in \mathsf{POS}_B(D)/\mathsf{ind}(B) \wedge (X \times U) \cap \mathsf{iop}(B \setminus \{u\}) \neq \emptyset\}$.

Proof. (sketch) (1) First note that $U' \cap \mathsf{POS}_B(D) = \emptyset$ and $\mathsf{POS}^{U'}_{\{v\}}(D) \subseteq U'$, so $\mathsf{POS}_{B \cup \{v\}}(D) = \mathsf{POS}_B(D) \cup \mathsf{POS}^{U'}_{\{v\}}(D) \iff U' = \mathsf{POS}^{U'}_{\{v\}}(D)$. Suppose that $x \in U' \setminus \mathsf{POS}^{U'}_{\{v\}}(D)$, i.e. $x \in \mathsf{POS}_{B \cup \{v\}}(D)$, $x \notin \mathsf{POS}_B(D)$ and $x \notin \mathsf{POS}^{U'}_{\{v\}}(D)$, which clearly implies $[x]_{\mathsf{ind}(B \cup \{v\})} \subseteq \mathsf{POS}_{B \cup \{v\}}(D)$, $[x]_{\mathsf{ind}(B)} \cap \mathsf{POS}_B(D) = \emptyset$ and $[x]_{\mathsf{ind}(\{v\})} \cap \mathsf{POS}^{U'}_{\{v\}}(D) = \emptyset$. However, since $v \notin B$, we also have $\mathsf{ind}(B \cup \{v\}) = \mathsf{ind}(B) \cap \mathsf{ind}(\{v\})$, which means $[x]_{\mathsf{ind}(B \cup \{v\})} \subseteq [x]_{\mathsf{ind}(B)} \cap [x]_{\mathsf{ind}(\{v\})}$, a contradiction.
(2) Since $B \setminus \{u\} \subsetneq B$ then $\mathsf{POS}_{B \setminus \{u\}}(D) \subseteq \mathsf{POS}_B(D)$. Consider $X \in \mathcal{X}_u$. Since $X \in \mathsf{POS}_B(D)/\mathsf{ind}(B)$ then $X \subseteq \mathsf{POS}_B(D)$ and since $(X \times U) \cap \mathsf{iop}(B \setminus \{u\}) \neq \emptyset$ then $X \cap \mathsf{POS}_{B \setminus \{u\}}(D) = \emptyset$. Hence $\mathsf{POS}_{B \setminus \{u\}}(D) \subseteq \mathsf{POS}_B(D) \setminus \bigcup_{X \in \mathcal{X}_u} X$. Let $x \in \mathsf{POS}_B(D) \setminus \bigcup_{X \in \mathcal{X}_u} X$. Hence $x \in \mathsf{POS}_B(D)$ and there is $y \in U$ such that $(x, y) \notin \mathsf{iop}(B \setminus \{u\})$, i.e. $x \in \mathsf{POS}_{B \setminus \{u\}}(D)$. □

Lemma 2 shows the results of adding and deleting attributes to and from a positive region $\mathsf{POS}_B(D)$. We will use them to provide a pair selection mechanism described in Algorithm 1. More precise rules are given by the next result.

Proposition 1. *Let $S = (U, C, D, V, f)$ be a decision table, $B \subsetneq C$, $u \in B$ and $v \in C \setminus B$ such that*

- $\mathsf{POS}_B(D) \neq \mathsf{POS}_C(D)$,
- $\mathsf{POS}_{(B \setminus \{u\}) \cup \{v\}}(D) = \mathsf{POS}_C(D)$ *and*
- $\mathsf{POS}_C(B)/\mathsf{ind}(B \cup \{v\}) = \{X_1, \ldots, X_n\}$.

Then the following properties hold.

1. $\mathsf{POS}^{U'}_{\{v\}}(D) = U'$, *where* $U' = \mathsf{POS}_C(D) \setminus \mathsf{POS}_B(D)$.
2. $\mathsf{POS}^{\widehat{U}}_{(B\setminus\{u\})\cup\{v\}}(D) = \widehat{U}$, *for every* $\widehat{U} = \{x_1, \ldots, x_n\} \subseteq U$ *such that* $\widehat{U} \cap X_i = \{x_i\}$ *for* $i = 1, \ldots, n$.

Proof. (sketch) (1) Since $\mathsf{POS}_{(B\setminus\{u\})\cup\{v\}}(D) = \mathsf{POS}_C(D)$, then directly from the definition of positive region we have: $\mathsf{POS}_{B\cup\{v\}}(D) = \mathsf{POS}_C(D)$. By Lemma 2(1) we have: $\mathsf{POS}_C(D) = \mathsf{POS}_{B\cup\{v\}}(D) = \mathsf{POS}_B(D) \cup \mathsf{POS}^{\mathsf{POS}_C(D)\setminus\mathsf{POS}_B(D)}_{\{v\}}(D)$, i.e. $\mathsf{POS}^{\mathsf{POS}_C(D)\setminus\mathsf{POS}_B(D)}_{\{v\}}(D) = \mathsf{POS}_C(D) \setminus \mathsf{POS}_B(D)$.
(2) From Lemma 1 it follows $\mathsf{POS}_{(B\setminus\{u\})\cup\{v\}}(D) = \mathsf{POS}^{\mathsf{POS}_{(B\setminus\{u\})\cup\{v\}}(D)}_{(B\setminus\{u\})\cup\{v\}}(D)$ and $\mathsf{POS}_{B\cup\{v\}}(D) = \mathsf{POS}^{\mathsf{POS}_{B\cup\{v\}}(D)}_{B\cup\{v\}}(D)$. However $\mathsf{POS}_{(B\setminus\{u\})\cup\{v\}}(D) = \mathsf{POS}_{B\cup\{v\}}(D) = \mathsf{POS}_C(D)$, so $\mathsf{POS}^{\mathsf{POS}_C(D)}_{(B\setminus\{u\})\cup\{v\}}(D) = \mathsf{POS}^{\mathsf{POS}_C(D)}_{B\cup\{v\}}(D)$. On the other hand, since $(B \setminus \{u\}) \cup \{v\} = (B \cup \{v\}) \setminus \{u\}$, by Lemma 2(2) we have $\mathsf{POS}^{\mathsf{POS}_C(D)}_{(B\setminus\{u\})\cup\{v\}}(D) = \mathsf{POS}^{\mathsf{POS}_C(D)}_{B\cup\{v\}}(D) \setminus \bigcup_{X\in\mathcal{X}^v_u} X$, where $\mathcal{X}^v_u = \{X \mid X \in \mathsf{POS}_C(D)/\mathsf{ind}(B\cup\{v\}) \wedge (X\times\mathsf{POS}_C(D)) \cap \mathsf{iop}((B\cup\{v\})\setminus\{u\}) \neq \emptyset\}$. This means that $\bigcup_{X\in\mathcal{X}^v_u} X = \emptyset$, i.e. $\mathcal{X}^v_u = \emptyset$, or, equivalently, $X \in \mathsf{POS}_C(D)/\mathsf{ind}(B \cup \{v\})$ implies $(X\times\mathsf{POS}_C(D)) \cap \mathsf{iop}((B\cup\{v\})\setminus\{u\}) = \emptyset$. But this also means that $X \in \mathsf{POS}_C(D)/\mathsf{ind}(B \cup \{v\}) = \{X_1, \ldots, X_n\}$ implies $X \subseteq \mathsf{POS}^{\mathsf{POS}_C(D)}_{(B\setminus\{u\})\cup\{v\}}(D)$. For each $i = 1, \ldots, n$, let x_i be an arbitrary element of X_i and set $\widehat{U} = \{x_1, \ldots, x_n\}$. If $i \neq j$, then we now have $(x_i, x_j) \in \mathsf{ind}((B \setminus \{u\}) \cup \{v\})$ and $f(x_i, d) = f(x_j, d)$ for each $d \in D$. But this means that we have $\mathsf{POS}^{\widehat{U}}_{(B\setminus\{u\})\cup\{v\}}(D) = \widehat{U}$. □

Proposition 1 suggests the following useful definition. Let $S = (U, C, D, V, f)$ be a decision table, $B \subsetneq C$ and $U' = \mathsf{POS}_C(D) \setminus \mathsf{POS}_B(D)$. We define $C^*_B \subseteq C$, a ***set of attributes filtered by*** B as:

$$C^*_B = \{v \mid v \in C \setminus B \wedge \mathsf{POS}^{U'}_{\{v\}}(D) = U'\}.$$

We will now show a sample application of the results stated above.

Example 1. Take the decision table Table 1, where $U = \{x_1, x_2, \ldots, x_7\}$, $C = \{a_1, a_2, \ldots, a_6\}$, and $D = \{Flu\}$. Consider $B = \{a_1, a_4\}$. In this case $\mathsf{POS}_{\{a_1,a_4\}}(D) = \{x_4, x_5\}$ and $\mathsf{POS}_C(D) = U$. We want to find such $u \in B = \{a_1, a_4\}$ and $v \in C \setminus B = \{a_2, a_3, a_5, a_6\}$ that $\mathsf{POS}_{(\{a_1,a_4\}\setminus\{u\})\cup\{v\}}(D) = \mathsf{POS}_C(D)$. We have to perform the following steps.

1. First we compute U' as defined in Proposition 1(1). In this case $U' = \mathsf{POS}_C(D) \setminus \mathsf{POS}_{\{a_1,a_4\}}(D) = \{x_1, x_2, x_3, x_6, x_7\}$.
2. For each $v \in \{a_2, a_3, a_5, a_6\}$, we compute $\mathsf{POS}^{U'}_{\{v\}}(D)$ and for this case we have: $\mathsf{POS}^{U'}_{\{a_2\}}(D) = \{x_1, x_2, x_3, x_6, x_7\}$, $\mathsf{POS}^{U'}_{\{a_3\}}(D) = \{x_1, x_2, x_3\}$, $\mathsf{POS}^{U'}_{\{a_5\}}(D) = \{x_1, x_2\}$ and $\mathsf{POS}^{U'}_{\{a_6\}}(D) = \{x_1, x_7\}$.

3. We now can calculate $C^*_{\{a_1,a_4\}}$. Only $\mathsf{POS}^{U'}_{\{a_2\}}(D) = U'$, so $C^*_{\{a_1,a_4\}} = \{a_2\}$, i.e. we set $v = a_2$.
4. We calculate $\mathsf{POS}_{B\cup\{v\}}(D) = \mathsf{POS}_{\{a_1,a_4\}\cup\{a_2\}}(D) = \mathsf{POS}_{\{a_1,a_2,a_4\}}(D) = U$.
5. We calculate that $\mathsf{POS}_{\{a_1,a_2,a_4\}}(D)/\mathsf{ind}(\{a_1,a_2,a_4\}) = \{\{x_1,x_2,x_3\}, \{x_4\},\{x_5\},\{x_6\},\{x_7\}\}$, and construct $\widehat{U}$ as $\widehat{U} = \{x_1,x_4,x_5,x_6,x_7\}$.
6. We will now use Proposition 1(2) to find proper u. Since we have $\mathsf{POS}^{\widehat{U}}_{\{a_1,a_2\}}(D) = \widehat{U}$ and $\mathsf{POS}^{\widehat{U}}_{\{a_2,a_4\}}(D) = \widehat{U}$, we set either $u = a_1$ or $u = a_4$.
7. Finally we set $(u,v) = (a_1,a_2)$ or $(u,v) = (a_4,a_2)$. ◇

3.3 A Local Search Algorithm with the Attribute Pair Selection Mechanism for Attribute Reduction

In step 4 of Algorithm 1, some element a is randomly removed from *Red*. Next we try to find appropriate u and v, but we may not succeed. In such a case a should not be used in next iteration. To implement this we use a set of *removed attributes* denoted by *RemoveSet* in Algorithm 2. Moreover at some point we will reach some local optimum so no more iteration is needed as we have just got our result. Local optimum means that we cannot remove any attribute a from the current reduct *Red*, all elements of *Red* have been tried but none has worked so they all have been put into *RemoveSet*, i.e. a local optimum is reached when *Red* = *RemoveRed*. Therefore we have designed the following four adjustment rules.

Adjustment rule 1: In each iteration, the randomly deleted attribute a must not belong to *RemoveSet*.

Adjustment rule 2: If a pair of attributes (u,v) cannot be found in the current iteration, the randomly deleted attribute a is added to the set *RemoveSet*.

Adjustment rule 3: *RemoveSet* is initialized to empty set. If a pair of attributes (u,v) is found, the search of current reduct is stopped, *RemoveSet* is reset to empty set again and the new iteration begins.

Adjustment rule 4: If the current attribute reduct *Red* equals *RemoveSet*, the algorithm stops and returns *Red*. Since *RemoveSet* $\subseteq$ *Red*, we can replace equality *Red* = *RemoveSet* with computationally simpler $|Red| = |RemoveSet|$.

Algorithm 2 applies all the above four rules and techniques described in Sect. 3.2. As opposed to Algorithm 1, it does not have an arbitrary limit of iterations T.

The analysis of its time complexity is similar to that for Algorithm 1. Algorithm 2 also starts with construction of a relative reduct using the same greedy procedure, so the worst case time complexity of this step (i.e. step 2) is $O(|Red|^2|U|) = O(|C|^2|U|)$.

Algorithm 2. (LSAR-APS) Local search algorithm with the attribute pair selection mechanism for attribute reduction

```
Input: A decision table $S = (U, C, D, V, f)$.
Output: The attribute reduction $Red$.
1  $t = 0$, $Red = \emptyset$ and $RemoveSet = \emptyset$;
2  construct a relative reduct $Red$ using greedy algorithm;    /* the same as in Algorithm 1 */
3  while $|Red| \neq |RemoveSet|$                      /* Adjustment rule 4 */ do
4      remove at random an attribute $a$ from $Red \setminus RemoveSet$;    /* Adjustment rule 1 */
5      if $\mathsf{POS}_{Red}(D) = \mathsf{POS}_{Red \setminus \{a\}}(D)$ then
6          $Red = Red \setminus \{a\}$;
7      else
8          calculate $C^*_{Red}$;
9          $flag = 0$;    /* the tag $flag$ is used to mark whether or not the attribute pair $(u, v)$ can be found */
10         for each $v \in C^*_{Red}$ and each $u \in Red \setminus \{a\}$ when $flag = 0$ do
11             compute $\mathsf{POS}_C(D)/\mathsf{ind}((Red \setminus \{a\}) \cup \{v\}) = \{X_1, \ldots, X_n\}$;
12             construct a set $\widehat{U} = \{x_1, \ldots, x_n\}$, where $x_i \in X_i$;
13             if $\mathsf{POS}^{\widehat{U}}_{(Red \setminus \{a,u\}) \cup \{v\}}(D) = \widehat{U}$ then
14                 $Red = (Red \setminus \{u\}) \cup \{v\}$;
15                 $flag = 1$; /* $flag = 1$ means finding an attribute pair and it causes exit from the loop, as by Adjustment rule 3 */
16             end
17         end
18         if $flag = 0$ then
19             $RemoveSet = RemoveSet \cup \{a\}$;              /* Adjustment rule 2 */
20         else
21             $RemoveSet = \emptyset$;                         /* Adjustment rule 3 */
22         end
23     end
24 end
25 return $Red$;
```

For the time essential steps inside the loop **while do** (step 3) we have the following worst case time complexities. Let Red_i represents the relative reduct used in the ith iteration. Step 5 is $O(|Red_i||U|) = O(|C||U|)$. Time complexity of step 8, i.e. finding $C^*_{Red_i}$, is $O(|C \setminus Red_i||\mathsf{POS}_C(D) \setminus \mathsf{POS}_{Red_i}(D)|) = O(|C||U|)$. Steps 11–12 construct $\widehat{U}$ and their time complexity is $O(|Red_i||\mathsf{POS}_C(D)|) = O(|C||U|)$, while steps 13–16 verify if a pair (u, v) fixes Red_i, and they are $O(|Red_i||\mathsf{POS}_{Red_i}(D)) = O(|C||U|)$ as well. The remaining steps inside **while do** have complexity $O(1)$. Hence the entire worst case time complexity of the ith iteration is $O(|C||U|)$, or more precisely $O(|Red_i||U|)$.

As far as the worst case time complexity is concerned, the ith iteration of Algorithm 1 and the ith iteration of Algorithm 2, have the same upper approximation $O(|Red_i||U|) = O(|C||U|)$. However, because $|\mathsf{POS}_C(D) \setminus \mathsf{POS}_{Red_i \setminus \{a\}}(D)| \ll |U$, $|\widehat{U} \leq |\mathsf{POS}_C(D)| \leq |U|$ and, usually, $|C^*_{Red_i}| \ll |Red_i|$, an average case time complexity of Algorithm 2 is usually much smaller than $O(|Red_i||U|)$ for the ith iteration.

The loop **while do** executes $O(|Red|) = O(|C|)$ times, so the overall worst case time complexity of Algorithm 2 is $O(|C|^2|U|)$. In reality, Algorithm 2 (LSAR-APS) is usually much faster than Algorithm 1 (LSAR), however there might be some exceptions (for example see Table 4, data set CNAE-9).

4 Experiments

In this section, we will present the results of experiments conducted to evaluate the performance of Algorithms 1 and 2, also named as LSAR and LSAR-APS, on eight well-known UCI data sets [26]. The characteristics of these data sets are given in Table 2. We compare our two algorithms with the positive region-based heuristic algorithm POSR [13], the backward search strategy-based quick heuristic algorithm GARA-BS [16], and the immune quantum-behaved particle swarm attribute reduction algorithm IQPOSR [23]. All the experiments have been ran on a personal computer with Inter(R) Core(TM) i5-7300HQ CPU, 2.50 GHz and 16 GB memory. The programming language is Matlab R2016a.

Table 2. Description of data sets.

Data sets	Names	No. of objects	No. of attributes	No. of classes
S1	Soybean (small)	47	35	4
S2	Zoo	101	16	7
S3	Dermatology	366	33	6
S4	Mushroom	8124	22	2
S5	Letter	20000	16	26
S6	CNAE-9	1080	856	9
S7	Musk (Ver.2)	6598	166	2
S8	Connect-4	67557	42	3

4.1 Reduct Size and Computation Time

We evaluate the feasibility and effectiveness of our two algorithms according to two aspects: the reduct size and the computation time. The algorithms POSR, GARA-BS and LSAR-APS have no parameters. For IQPOSR, the parameters use the settings on small-scale problem instances in [23], and the specific settings are as follows: the particle size $M = 50$, the total number of iterations $T = 200$, the particle protection period $K = 10$, the accuracy error $\varepsilon_0 = 0.01$, and the test cost of each attribute $c(a) = 1$. LSAR is a single candidate solution-based stochastic local search algorithm, and it requires more iterations than population-based iterated algorithms. Hence the maximum iterations of LSAR is 10 times that of IQPOSR, i.e., $T = 2000$. However the time complexity of LSAR is much less than that of IQPOSR. Each algorithm runs 10 times on each data set, and we record the best reduct and the average computation time of the 10 runs. The experiment results shown in Tables 3 and 4.

Table 3. Comparison of reduct size on eight data sets

Data set	Reduct size				
	POSR	GARA-BS	IQPSOR	LSAR	LSAR-APS
Soybean (small)	2	2	2	2	2
Zoo	5	5	5	5	5
Dermatology	10	9	9	8	8
Mushroom	4	4	4	4	4
Letter	11	12	11	11	11
CNAE-9	81	75	84	80	71
Musk (Ver.2)	4	4	4	4	4
Connect-4	34	34	35	34	34

Table 3 shows that the reduct sizes obtained by LSAR and LSAR-APS are the same on all data sets, except for the data set CNAE-9. From all five algorithms, LSAR-APS is the best one in terms of the reduct size, especially for the data set CNAE-9. The reduct size of these five algorithms are the same on data sets Soybean (small), Zoo, Mushroom, and Musk (Ver.2). POSR obtains the worst reduct size on data sets Dermatology, and the reduct size of GARA-BS is the worst one on data set Letter. On data sets CNAE-9 and Connect-4, IQPSOR performs the worst in terms of the reduct size.

From Table 4 we have that GARA-BS is the fastest algorithm on data sets Soybean (small) and Zoo. On data sets Dermatology, Mushroom, Letter, Musk (Ver.2), and Connect-4, the algorithm LSAR-APS performs the best in terms of the computational time. On data set CNAE-9, the computational time of LSAR is the best one. *This is one of these rare cases when LSAR performed better than LSAR-APS.* The algorithm POSR is very complex, so its computational time grows dramatically as the data set increases. IQPSOR is a population-based meta-heuristic algorithm, and its computational times are stable. Among three previous algorithms, GARA-BS obtains the smallest computational time, but its computational time is still far greater than that of LSAR-APS.

In summary, especially when large data sets are concerned, our algorithm LSAR-APS can achieve a better reduct in a much shorter time. For example,

Table 4. Comparison of computational time on eight data sets

Data set	Computational time/s				
	POSR	GARA-BS	IQPSOR	LSAR	LSAR-APS
Soybean (small)	0.138	0.011	2.357	1.130	0.017
Zoo	0.184	0.014	3.387	1.339	0.033
Dermatology	3.186	0.132	12.751	2.817	0.105
Mushroom	19.773	1.022	324.006	16.930	0.617
Letter	245.631	3.813	737.317	72.026	2.239
CNAE-9	2064.218	220.204	718.453	23.013	74.643
Musk (Ver.2)	365.983	10.319	449.029	17.112	1.695
Connect-4	12417.113	175.689	2665.508	614.399	56.953

the algorithm LSAR-APS only takes an average of 74.643 s to find a reduct with a smallest size 71, and this is definitely the best results among these five algorithms. To the best of our knowledge, the reduct size 71 on data set CNAE-9 is also the best solution obtained so far.

4.2 Classification Accuracy Analysis

The classification accuracy was conducted on the selected attribute reducts found by all five algorithms with classifier 3NN (k-Nearest Neighbor algorithm and $k = 3$), which is a popular classifier for testing the attribute reduction algorithms. All of the classification accuracies are obtained with 10-fold cross validation. In 10-fold cross validation, a given data set is randomly divided into 10 nearly equally sized subsets, of these 10 subsets, 9 subsets are used as training set, a single subset is retained as testing set to assess the classification accuracy. The average performance results in terms of the classification accuracy are summarized in Table 5, where the column "Raw" depicts the classification accuracies with the original data and the boldface highlights the highest accuracy among these five algorithms.

Table 5. Classification accuracy on different data sets.

Data set	Classification accuracy/%					
	Raw	POSR	GARA-BS	IQPSOR	LSAR	LSAR-APS
S1	100.00±0.00	**100.00±0.00**	**100.00±0.00**	**100.00±0.00**	**100.00±0.00**	**100.00±0.00**
S2	93.18 ±7.93	90.09±8.17	89.51 ±10.54	89.09±8.77	90.18±10.36	**91.00±11.01**
S3	96.72±2.50	**92.64±3.62**	73.75± 8.56	90.16 ±5.51	76.26± 7.66	76.52± 5.24
S4	100.00±0.00	**100.00±0.00**	**100.00±0.00**	**100.00±0.00**	**100.00±0.00**	**100.00±0.00**
S5	95.63±0.41	**94.61±0.31**	94.23±0.55	93.68±0.30	93.38±0.51	94.36±0.50
S6	85.83±2.73	85.74±3.30	85.83±2.66	85.93±3.17	85.28±4.37	**86.11±2.99**
S7	96.79±0.58	90.85±1.29	91.60±0.59	**92.83±0.89**	91.54±0.99	91.71±1.05
S8	66.60±1.23	67.31±1.53	67.30±0.72	67.03±0.86	67.21±0.98	**67.68±0.86**

Table 5 shows that the algorithm LSAR-APS achieves the best classification performance as its number of the highest classification accuracy is five times out of eight data sets. For POSR this number is four times among eight data sets, QIPSOR matched the best classification accuracies for 3 out of 8 cases while and LSAR and GARA-BS only obtain the best classification performance on data sets S1 and S4. Hence, LSAR-APS can achieve better or comparable classification accuracy in comparison with other four algorithms.

5 Conclusion

In this paper, we studied local search approach for attribute reduction problem in rough set theory that has a wide range of applications. We introduced a local search framework for this problem and proposed two advanced strategies to improve the iteration process of the local search-based algorithm, i.e., attribute pair selection mechanism and adjustment rules. The results of the experiment on the broadly used data set indicated that our proposed algorithm LSAR-ASP significantly outperforms other state-of-the-art algorithms.

We are surprised to find that the reduct found by LSAR-APS on data set CNAE-9 is actually an optimal reduct (see Appendix A). In this sense, this work provides a new idea for solving the optimal reduct of large data sets. In the future work, we will test our proposed algorithm on high-dimensional large data sets and propose some additional improved strategies to enhance the efficiency of the local search-based attribute reduction algorithm.

Acknowledgment. The authors gratefully acknowledge three anonymous referees for their helpful comments. The research was supported by The National Natural Science Foundation of China (grant nos. 61373015, 61728204), China Scholarship Council (grant no. 201806830058), State Key Laboratory for smart grid protection and operation control Foundation, Science and Technology Funds from National State Grid Ltd(The Research on Key Technologies of Distributed Parallel Database Storage and Processing based on Big Data), and NSERC of Canada (Discovery grant no. 6466-15).

Appendix A

Here we report the optimal solution found by LSAR-APS on data set CNAE-9. The optimal reduct is: 7 20 63 68 73 75 77 105 118 119 133 150 151 183 191 194 199 201 202 207 211 246 247 258 272 276 328 333 334 338 345 350 359 360 373 382 390 403 415 417 421 423 424 443 476 483 499 518 519 539 546 555 581 607 608 614 615 618 619 631 648 650 673 684 705 726 731 815 823 824 832.

References

1. Pawlak, Z.: Rough sets. Int. J. Comput. Inf. Sci. **11**(5), 341–356 (1982)
2. Swiniarski, R.W., Skowron, A.: Rough set methods in feature selection and recognition. Pattern Recogn. Lett. **24**(6), 833–849 (2003)
3. Lingras, P.J., Yao, Y.Y.: Data mining using extensions of the rough set model. J. Am. Soc. Inf. Sci. **49**(5), 415–422 (1998)
4. Herawan, T., Deris, M.M., Abawajy, J.H.: A rough set approach for selecting clustering attribute. Knowl. Based Syst. **23**(3), 220–231 (2010)
5. Janicki, R., Lenarčič, A.: Optimal approximations with rough sets and similarities in measure spaces. Int. J. Approximate Reasoning **71**, 1–14 (2016)
6. Janicki, R.: Approximations of arbitrary relations by partial orders. Int. J. Approximate Reasoning **98**, 177–195 (2018)
7. Xie, X., Qin, X.: Dynamic feature selection algorithm based on minimum vertex cover of hypergraph. In: Phung, D., Tseng, V.S., Webb, G.I., Ho, B., Ganji, M., Rashidi, L. (eds.) PAKDD 2018. LNCS (LNAI), vol. 10939, pp. 40–51. Springer, Cham (2018). https://doi.org/10.1007/978-3-319-93040-4_4
8. Skowron, A., Rauszer, C.: The discernibility matrices and functions in information systems. In: Słowiński, R. (ed.) Intelligent Decision Support. Handbook of Applications and Advances of the Rough Sets Theory, Dordrecht, Kluwer (1992)
9. Nguyen, H.S.: Approximate boolean reasoning approach to rough sets and data mining. In: Ślęzak, D., Yao, J.T., Peters, J.F., Ziarko, W., Hu, X. (eds.) RSFDGrC 2005. LNCS (LNAI), vol. 3642, pp. 12–22. Springer, Heidelberg (2005). https://doi.org/10.1007/11548706_2
10. Tan, A., Li, J., Lin, Y., Lin, G.: Matrix-based set approximations and reductions in covering decision information systems. Int. J. Approximate Reasoning **59**, 68–80 (2015)
11. Hacibeyoglu, M., Salman, M.S., Selek, M., Kahramanli, S.: The logic transformations for reducing the complexity of the discernibility function-based attribute reduction problem. Knowl. Inf. Syst. **46**(3), 599–628 (2016)
12. Yang, T., Li, Q., Zhou, B.: Related family: a new method for attribute reduction of covering information systems. Inf. Sci. **228**, 175–191 (2013)
13. Xu, Z., Liu, Z., Yang, B.: A quick attribute reduction algorithm with complexity of $\max(O(|C||U|), O(|C|^2|U/C|))$. Chin. J. Comput. **29**(3), 391–399 (2006)
14. Jiang, F., Sha-sha, W., Du, J.W., Yue-Fei, S.: Attribute reduction based on approximation decision entropy. Control Decis. **30**(1), 65–70 (2015)
15. Deng, T., Yang, C., Hu, Q.: Feature selection in decision systems based on conditional knowledge granularity. Int. J. Comput. Intell. Syst. **4**(4), 655–671 (2011)
16. Ge, H., Li, L., Xu, Y., Yang, C.: Quick general reduction algorithms for inconsistent decision tables. Int. J. Approximate Reasoning **82**, 56–80 (2017)
17. Xie, X., Qin, X.: A novel incremental attribute reduction approach for dynamic incomplete decision systems. Int. J. Approximate Reasoning **93**, 443–462 (2018)
18. Xu, Z., Gu, D., Yang, B.: Attribute reduction algorithm based on genetic algorithm. In: Proceedings of International Conference on Intelligent Computation Technology and Automation, Zhangjiajie, China, pp. 169–172 (2009)
19. Chen, Y., Zhu, Q., Xu, H.: Finding rough set reducts with fish swarm algorithm. Knowl. Based Syst. **81**, 22–29 (2015)
20. Inbarani, H.H., Bagyamathi, M., Azar, A.T.: A novel hybrid feature selection method based on rough set and improved harmony search. Neural Comput. Appl. **26**(8), 1859–1880 (2015)

21. Luan, X.Y., Li, Z.P., Liu, T.Z.: A novel attribute reduction algorithm based on rough set and improved artificial fish swarm algorithm. Neurocomputing **174**, 522–529 (2016)
22. Abd El Aziz, M., Hassanien, A.E.: An improved social spider optimization algorithm based on rough sets for solving minimum number attribute reduction problem. Neural Comput. Appl. **30**(8), 2441–2452 (2018)
23. Xie, X., Qin, X., Yu, C., Xu, X.: Test-cost-sensitive rough set based approach for minimum weight vertex cover problem. Appl. Soft Comput. **64**, 423–435 (2018)
24. Cai, S., Su, K., Sattar, A.: Local search with edge weighting and configuration checking heuristics for minimum vertex cover. Artif. Intell. **175**(9), 1672–1696 (2011)
25. Cai, S., Hou, W., Lin, J., Li, Y.: Improving local search for minimum weight vertex cover by dynamic strategies. In: Proceedings of International Joint Conferences on Artificial Intelligence, Stockholm, Sweden, pp. 1412–1418 (2018)
26. UCI machine learning repository. http://www.ics.uci.edu/mlearn/MLRepository.html

Rough Matroids Based on Dual Approximation Operators

Mauricio Restrepo[1(✉)] and Chris Cornelis[2]

[1] Universidad Militar Nueva Granada, Bogotá, Colombia
mauricio.restrepo@unimilitar.edu.co
[2] Ghent University, Ghent, Belgium
http://www.unimilitar.edu.co

Abstract. This paper presents the concept of lower and upper rough matroids based on approximation operators for covering-based rough sets. This concept is a generalization of lower and upper rough matroids based on coverings. A new definition of lower and upper definable sets related with an approximation operator is presented and these definable sets are used for defining rough matroids based on an approximation operator. Finally, an order relation for a special type of rough matroids is established from the order relation among approximation operators.

Keywords: Covering rough sets · Rough matroid · Order relation

1 Introduction

Covering-based rough sets are a generalization of rough set theory, which was developed by many authors and had applications in other contexts [6,9,12,22, 23]. Yao and Yao introduced a general framework for the study of dual pairs of covering-based approximation operators, distinguishing between element-based, granule-based and subsystem-based definitions [21]. Other pairs of approximation operators have been studied in literature; for instance, in [18], Yang and Li present a summary of seven non-dual pairs of approximation operators used by Żakowski [22], Pomykala [9], Tsang et al. [12], Zhu [24], Zhu and Wang [26], Xu and Wang [17]. Restrepo et al. present a general framework of pairs of dual operators and established a partial order relation among these operators [10,11]. Matroids were introduced in 1935 by Whitney as a generalization of independence in linear algebra. Matroids have been used in combinatorial optimization and algorithm design.

Many works have shown interesting connections between matroids and rough sets [4–7,12,13,16]. Zhu et al. presented the concept of rough matroid based on a relation [27,28] and rough matroids based on coverings [19], using a particular approximation operator. The concepts of rough matroids based on coverings and binary relations were presented in [19] and [27], respectively. In both cases, the elements in the matroids are definable sets. In [19] it was stated that the lower

© Springer Nature Switzerland AG 2019
T. Mihálydeák et al. (Eds.): IJCRS 2019, LNAI 11499, pp. 118–129, 2019.
https://doi.org/10.1007/978-3-030-22815-6_10

approximation of a set X is equal to X itself if and only if X is also equal to its upper approximation, which is false, as this paper will show. Because of this, it is necessary to separately consider the definable sets for the operators $\underline{apr}$ and $\overline{apr}$. The integration of matroids with covering-based rough sets can bring new theories and practical significance in important problems such as the reduction of attributes as shown in [16]. This article presents new definitions for lower and upper rough matroids, using a dual pair of approximation operators.

The paper is organized as follows: Sect. 2 presents preliminary concepts regarding covering-based rough sets, such as lower and upper approximations, the main neighborhood operators, definable sets and matroids. Section 3 presents the concept of lower and upper rough matroids based on an approximation operator. This section also presents a generalization of rough matroid based on coverings. Section 4 presents an order relation among rough matroids. Finally, Sect. 5 presents the main conclusions of the paper and describes future work.

2 Preliminaries

In Pawlak's rough set model, an approximation space is an ordered pair $apr = (U, E)$, where E is an equivalence relation defined on a non-empty set U [8]. In this paper U is considered as a finite set.

According to Yao and Yao [20,21], there are three different, but equivalent ways to define lower and upper approximation operators: element-based definition, granule-based definition and subsystem-based definition. For each $A \subseteq U$, the granule-based lower and upper approximations are defined by:

$$\underline{apr}(A) = \bigcup\{[x]_E \in U/E : [x]_E \subseteq A\} \tag{1}$$

$$\overline{apr}(A) = \bigcup\{[x]_E \in U/E : [x]_E \cap A \neq \emptyset\} \tag{2}$$

The set $[x]_E$ represents the equivalence class of x and U/E the partition obtained from the equivalence relation. $\mathscr{P}(U)$ represents the set of parts of U.

Other equivalent element-based and sub-system based definitions for approximation in covering-based rough sets can be found in [21].

2.1 Covering-Based Rough Sets

Covering-based rough sets were proposed to extend the range of applications of rough set theory. In covering-based rough sets an element $x \in U$ can belong to many sets, so we have to consider the sets K in $\mathbb{C}$ such that $x \in K$.

Definition 1 *[23]. Let $\mathbb{C} = \{K_i\}$ be a family of nonempty subsets of U. $\mathbb{C}$ is called a covering of U if $\bigcup K_i = U$. The ordered pair $(U, \mathbb{C})$ is called a covering approximation space.*

Duality

Definition 2 *[3]. Let $f, g : B \to B$ be two self-maps on a complete Boolean lattice B. We say that g is the dual of f, if for all $x \in B$,*

$$g(-x) = -f(x),$$

where $-x$ represents the complement of $x \in B$.

Meet and Join Morphisms

Definition 3 *[3]. Let L be a finite lattice. A meet-morphism f is a morphism that satisfies $f(a \wedge b) = f(a) \wedge f(b)$ for a and b in L. Dually, a join-morphism f is a morphism that satisfies $f(a \vee b) = f(a) \vee f(b)$ for a and b in L.*

Minimal and Maximal Description. In a covering approximation space for each $x \in U$, it is very important to take into account the collection of sets in $K \in \mathbb{C}$ such that $x \in K$.

$$\mathscr{C}(\mathbb{C}, x) = \{K \in \mathbb{C} : x \in K\}$$

Definition 4. *Let $(U, \mathbb{C})$ be a covering approximation space and x in U. The set*

$$md(\mathbb{C}, x) = \{K \in \mathscr{C}(\mathbb{C}, x) : (\forall S \in \mathscr{C}(\mathbb{C}, x), S \subseteq K) \Rightarrow K = S)\} \quad (3)$$

is called the minimal description of x [1]. On the other hand, the set

$$MD(\mathbb{C}, x) = \{K \in \mathscr{C}(\mathbb{C}, x) : (\forall S \in \mathscr{C}(\mathbb{C}, x), S \supseteq K) \Rightarrow K = S\} \quad (4)$$

is called the maximal description of x [25].

Approximation Operators. A first definition of approximation operator can derive from neighborhood operators.

Definition 5. *A neighborhood operator is a function $N : U \to \mathscr{P}(U)$. In general we consider functions N such that $x \in N(x)$.*

The element-based definitions of approximation operators based on a neighborhood operator N are defined as:

$$\underline{apr}_N(A) = \{x \in U : N(x) \subseteq A\} \quad (5)$$

$$\overline{apr}_N(A) = \{x \in U : N(x) \cap A \neq \emptyset\} \quad (6)$$

From $md(\mathbb{C}, x)$ and $MD(\mathbb{C}, x)$, Yao and Yao define the following neighborhood operators [21]:

1. $N_1(x) = \bigcap\{K : K \in md(\mathbb{C}, x)\}$
2. $N_2(x) = \bigcup\{K : K \in md(\mathbb{C}, x)\}$

3. $N_3(x) = \bigcap\{K : K \in MD(\mathbb{C}, x)\}$
4. $N_4(x) = \bigcup\{K : K \in MD(\mathbb{C}, x)\}$

Therefore, four different pairs of dual approximation operators can be defined in a covering space: $(\underline{apr}_{N_i}, \overline{apr}_{N_i})$.

The granule-based definitions of approximation operators based on a covering $\mathbb{C}$ were considered before, see Table 2 in [21]:

$$\underline{apr}'_{\mathbb{C}}(A) = \bigcup\{K \in \mathbb{C} : K \subseteq A\} \tag{7}$$

$$\overline{apr}''_{\mathbb{C}}(A) = \bigcup\{K \in \mathbb{C} : K \cap A \neq \emptyset\} \tag{8}$$

The approximation operators shown above do not satisfy the dual relation:

$$\overline{apr}''_{\mathbb{C}}(-A) = -\underline{apr}'_{\mathbb{C}}(A) \tag{9}$$

Therefore, it is possible to define a dual operator for each one and get two different pairs of dual approximation operators in a covering space.

This paper considers the following two properties for any operator:

1. $\underline{apr}(A) = -\overline{apr}(-A)$.
2. $\underline{apr}(A) \subseteq A \subseteq \overline{apr}(A)$.

Other coverings obtained from a covering $\mathbb{C}$ have been used for new definitions of approximation operators.

From a covering $\mathbb{C}$ of U, the following new coverings have been defined:

1. $\mathbb{C}_1 = \bigcup\{md(\mathbb{C}, x) : x \in U\}$
2. $\mathbb{C}_2 = \bigcup\{MD(\mathbb{C}, x) : x \in U\}$
3. $\mathbb{C}_3 = \{\bigcap(md(\mathbb{C}, x)) : x \in U\}$
4. $\mathbb{C}_4 = \{\bigcup(MD(\mathbb{C}, x)) : x \in U\}$
5. $\mathbb{C}_\cap = \mathbb{C} \setminus \{K \in \mathbb{C} : (\exists \mathbb{K} \subseteq \mathbb{C} \setminus \{K\})\,(K = \bigcap \mathbb{K})\}$

Covering $\mathbb{C}_\cap$ is called the $\cap$-reduction of $\mathbb{C}$. The main idea is to eliminate the elements K in $\mathbb{C}$ that can be expressed as the intersection of other sets in the covering.

The dual pairs of approximation operators which satisfy the meet/join property, according to the results established in [10] are:

1. $(\underline{apr}_{N_1}, \overline{apr}_{N_1}) = (\underline{apr}'_{\mathbb{C}_3}, \overline{apr}'_{\mathbb{C}_3})$
2. $(\underline{apr}_{N_2}, \overline{apr}_{N_2})$
3. $(\underline{apr}_{N_3}, \overline{apr}_{N_3})$
4. $(\underline{apr}_{N_4}, \overline{apr}_{N_4}) = (\underline{apr}''_{\mathbb{C}}, \overline{apr}''_{\mathbb{C}}) = (\underline{apr}''_{\mathbb{C}_2}, \overline{apr}''_{\mathbb{C}_2}) = (\underline{apr}''_{\mathbb{C}_\cap}, \overline{apr}''_{\mathbb{C}_\cap})$
5. $(\underline{apr}''_{\mathbb{C}_1}, \overline{apr}''_{\mathbb{C}_1}) = (\underline{apr}''_{\mathbb{C}_\cup}, \overline{apr}''_{\mathbb{C}_\cup})$
6. $(\underline{apr}''_{\mathbb{C}_3}, \overline{apr}''_{\mathbb{C}_3})$
7. $(\underline{apr}''_{\mathbb{C}_4}, \overline{apr}''_{\mathbb{C}_4})$

2.2 Definable Sets Based on an Approximation Operator

Different notations for approximation operators have been used. For example, XL for $\underline{apr}_{N_1}$ and XH for $\overline{apr}_{N_1}$ where used in [18].

The duality of the XL and XH operators is not enough to show that $XL(X) = X$ if and only if $XH(X) = X$; therefore the Corollary 1 in [19] is wrong. A counterexample can be seen in Example 1 below, demonstrating that it is necessary to consider the definable sets separately, and to understand them as the fixed points of the $\underline{apr}$ and $\overline{apr}$ operators respectively.

Definition 6. *Let $(U, \mathbb{C})$ be a covering approximation space and $\underline{apr}_{\mathbb{C}}$ any lower approximation operator from the list above. The family of lower definable sets for the approximation operator is defined as follows:*

$$\mathbb{D}^{\mathbb{C}}_{\underline{apr}} = \{X \subseteq U : \underline{apr}_{\mathbb{C}}(X) = X\}.$$

A similar set can be defined from an upper approximation operator. We denote with $\mathbb{D}^{\mathbb{C}}_{\overline{apr}}$ the families of upper definable sets for $\overline{apr}$, respectively. We can leave out $\mathbb{C}$ when there is no confusion. In general, $\mathbb{D}_{\underline{apr}} \neq \mathbb{D}_{\overline{apr}}$.

$\mathbb{D}_{apr} = \mathbb{D}_{\underline{apr}} \cap \mathbb{D}_{\overline{apr}}$ is called the set of definable sets for the pair of approximation operators.

Example 1. For the covering $\mathbb{C} = \{\{1,2\}, \{3,4\}, \{1,4\}, \{2,3,4\}, \{1,2,4\}\}$ of the set $U = \{1,2,3,4\}$ and the approximation operators $\underline{apr}_{N_1}$ and $\overline{apr}_{N_1}$, we have that $N_1(1) = \{1\}$, $N_1(2) = \{2\}$, $N_1(3) = \{3,4\}$ and $N_1(4) = \{4\}$. The result of the approximations are shown in Table 1:

Table 1. Approximations for operators $\underline{apr}_{N_1}$ and $\overline{apr}_{N_1}$.

A	$\underline{apr}_{N_1}(A)$	$\overline{apr}_{N_1}(A)$	A	$\underline{apr}_{N_1}(A)$	$\overline{apr}_{N_1}(A)$
$\{1\}$	$\{1\}$	$\{1\}$	$\{2,3\}$	$\{2\}$	$\{2,3\}$
$\{2\}$	$\{2\}$	$\{2\}$	$\{2,4\}$	$\{2,4\}$	$\{2,3,4\}$
$\{3\}$	$\emptyset$	$\{3\}$	$\{3,4\}$	$\{3,4\}$	$\{3,4\}$
$\{4\}$	$\{4\}$	$\{3,4\}$	$\{1,2,3\}$	$\{1,2\}$	$\{1,2,3\}$
$\{1,2\}$	$\{1,2\}$	$\{1,2\}$	$\{1,2,4\}$	$\{1,2,4\}$	$\{1,2,3,4\}$
$\{1,3\}$	$\{1\}$	$\{1,3\}$	$\{1,3,4\}$	$\{1,3,4\}$	$\{1,3,4\}$
$\{1,4\}$	$\{1,4\}$	$\{1,3,4\}$	$\{2,3,4\}$	$\{2,3,4\}$	$\{2,3,4\}$

In this case, the lower and upper definable sets are, respectively: $\mathbb{D}_{\underline{apr}_{N_1}} = \{\emptyset, \{1\}, \{2\}, \{4\}, \{1,2\}, \{1,4\}, \{2,4\}, \{3,4\}, \{1,2,4\}, \{1,3,4\}, \{2,3,4\}, \{1,2,3,4\}\}$ and $\mathbb{D}_{\overline{apr}_{N_1}} = \{\emptyset, \{1\}, \{2\}, \{3\}, \{1,2\}, \{1,3\}, \{2,3\}, \{3,4\}, \{1,2,3\}, \{1,3,4\}, \{2,3,4\}, \{1,2,3,4\}\}$.

Thus, in this case $\mathbb{D}_{apr_{N_1}} = \{\emptyset, \{1\}, \{2\}, \{1,2\}, \{3,4\}, \{1,3,4\}, \{2,3,4\}, \{1,2,3,4\}\}$.

Proposition 1. *If* $(\underline{apr}_{\mathbb{C}}, \overline{apr}_{\mathbb{C}})$ *is a dual pair of approximation operators,* $\underline{apr}_{\mathbb{C}}$ *is a meet-morphism and* $A, B \in \mathbb{D}_{apr}$*, then* $A \cap B \in \mathbb{D}_{\underline{apr}}$ *and* $A \cup B \in \mathbb{D}_{\overline{apr}}$*.*

Proof. It follows from the definition of meet-morphism: $\underline{apr}(A \cap B) = \underline{apr}(A) \cap \underline{apr}(B) = A \cap B$. Thus, $A \cap B \in \mathbb{D}_{\underline{apr}}$. Using the duality property, it is easy to show that $\overline{apr}_{\mathbb{C}}$ is a join-morphism, therefore: $\overline{apr}(A \cup B) = \overline{apr}(A) \cup \overline{apr}(B) = A \cup B$, and so, $A \cup B \in \mathbb{D}_{\overline{apr}}$.

The order relation among approximation operators can be used for establishing an order relation among definable sets.

Proposition 2. *If* $\underline{apr}_1 \leq \underline{apr}_2$*, then* $\mathbb{D}_{\underline{apr}_1} \subseteq \mathbb{D}_{\underline{apr}_2}$*.*

Proof. If $\underline{apr}_1(X) \subseteq \underline{apr}_2(X)$ and $X \in \mathbb{D}(U, \underline{apr}_1)$, then $\underline{apr}_1(X) = X$. We will show that $\underline{apr}_2(X) = X$. Obviously $\underline{apr}_2(X) \subseteq X$. Now, if $x \in X$, then $x \in \underline{apr}_1(X) \subseteq \underline{apr}_2(X)$ and $X \subseteq \underline{apr}_2(X)$. Therefore, $\underline{apr}_2(X) = X$ and $X \in \mathbb{D}(U, \underline{apr}_2)$. So, $\mathbb{D}_{\underline{apr}_1} \subseteq \mathbb{D}_{\underline{apr}_2}$.

Corollary 1. *If* $\overline{apr}_1 \leq \overline{apr}_2$*, then* $\mathbb{D}_{\overline{apr}_1} \supseteq \mathbb{D}_{\overline{apr}_2}$*.*

Proof. The proof is similar to the one above.

Example 2. According to the order relation established in [11] we have, for example, the relations shown in Fig. 1.

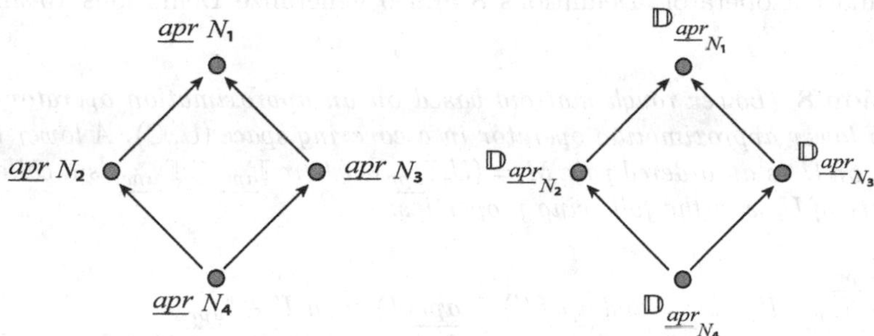

Fig. 1. Order relation for lower approximation operators and lower definable sets.

2.3 Matroids

Matroids can be introduced from an elementary point of view as a collection of sets of linearly independent vectors. Let us suppose that $\{a_1, a_2, a_3, a_4\}$ represents the column vectors of matrix A:

$$A = \begin{pmatrix} 1 & 0 & 2 & 1 \\ 1 & 0 & 2 & 2 \\ 2 & -1 & 0 & 0 \end{pmatrix} \rightsquigarrow \cdots \rightsquigarrow E_A = \begin{pmatrix} 1 & 0 & 2 & 0 \\ 0 & 1 & 4 & 0 \\ 0 & 0 & 0 & 1 \end{pmatrix} \tag{10}$$

Then, according to the reduced row echelon form E_A, $\{a_1, a_2, a_4\}$ is a set of linearly independent vectors. Additionally, we know that any subset of linearly independent vectors is also linearly independent. In this case, the collection of independent sets is: $\mathbb{I} = \{\emptyset, \{a_1\}, \{a_2\}, \{a_4\}, \{a_1, a_2\}, \{a_1, a_4\} \{a_2, a_4\}, \{a_1, a_2, a_4\}\}$.

There are different definitions of a matroid. In this case, we consider the following definition in terms of independence.

Definition 7 *[27]. Let U be a finite set. A matroid on U is an ordered pair $M = (U, \mathbb{I})$, where $\mathbb{I}$ is a collection of subsets of U with the following properties:*

1. *$\emptyset \in \mathbb{I}$.*
2. *If $I \in \mathbb{I}$, $I' \subseteq I$ then $I' \in \mathbb{I}$.*
3. *If $I_1, I_2 \in \mathbb{I}$ and $|I_1| < |I_2|$, then there exists $I \in \mathbb{I}$ such that $I_1 \subset I \subseteq I_1 \cup I_2$.*

The members of $\mathbb{I}$ are called independent sets of U. A **base** for the matroid M is any maximal set in $\mathbb{I}$. The sets not contained in $\mathbb{I}$ are called dependent. A minimal dependent subset of U is called a **circuit** of M.

3 Rough Matroids Based on Approximation Operators

The concept of rough matroid based on a covering $\mathbb{C}$ was proposed by Yang et al. [27], using a particular pair of approximation operators. The following definitions present the notion of lower and upper rough matroids based on an approximation operator. Definitions 8 and 9 generalize Definitions 15 and 16 in [27].

Definition 8 *(Lower rough matroid based on an approximation operator). Let $\underline{apr}$ be a lower approximation operator in a covering space $(U, \mathbb{C})$. A lower rough matroid on U is an ordered pair $M = (U, \mathbb{I}_{\underline{apr}})$, where $\mathbb{I}_{\underline{apr}} \subseteq \mathbb{D}_{\underline{apr}}$ is a collection of subsets of U with the following properties:*

1. *$\emptyset \in \mathbb{I}_{\underline{apr}}$.*
2. *If $I \in \mathbb{I}_{\underline{apr}}$, $I' \in \mathbb{D}_{\underline{apr}}$ and $\underline{apr}(I') \subseteq \underline{apr}(I)$ then $I' \in \mathbb{I}_{\underline{apr}}$.*
3. *If $I_1, I_2 \in \mathbb{I}_{\underline{apr}}$ and $|\underline{apr}(I_1)| < |\underline{apr}(I_2)|$, then there exists $I \in \mathbb{I}_{\underline{apr}}$ such that $\underline{apr}(I_1) \subset \underline{apr}(I) \subseteq \underline{apr}(I_1) \cup \underline{apr}(I_2)$.*

Similarly, the definition of upper rough matroid based on an upper approximation operator is presented.

Definition 9 *(Upper rough matroid based on an approximation operator). Let $\overline{apr}$ be an upper approximation operator in a covering space $(U, \mathbb{C})$. An upper rough matroid on U is an ordered pair $M = (U, \mathbb{I}_{\overline{apr}})$, where $\mathbb{I}_{\overline{apr}} \subseteq \mathbb{D}_{\overline{apr}}$ is a collection of subsets of U with the following properties:*

1. $\emptyset \in \mathbb{I}_{\overline{apr}}$.
2. *If $I \in \mathbb{I}_{\overline{apr}}$, $I' \in \mathbb{D}_{\overline{apr}}$ and $\overline{apr}(I') \subseteq \overline{apr}(I)$ then $I' \in \mathbb{I}_{\overline{apr}}$.*
3. *If $I_1, I_2 \in \mathbb{I}_{\overline{apr}}$ and $|\overline{apr}(I_1)| < |\overline{apr}(I_2)|$, then there exists $I \in \mathbb{I}_{\overline{apr}}$ such that $\overline{apr}(I_1) \subset \overline{apr}(I) \subseteq \overline{apr}(I_1) \cup \overline{apr}(I_2)$.*

Example 3. For the covering $\mathbb{C} = \{\{1,2\},\{3,4\},\{1,4\},\{2,3,4\},\{1,2,4\}\}$ and the approximations $\underline{apr}_{N_1}$ and $\overline{apr}_{N_1}$, we have that:

1. $(U, \mathbb{I}) = \{\emptyset, \{1\}\}$ is a lower and an upper rough matroid based on the approximation operator.
2. $(U, \mathbb{I}) = \{\emptyset, \{1\}, \{3\}, \{1,3\}\}$ is an upper rough matroid, but it is not a lower rough matroid, because $\{3\} \notin \mathbb{D}_{\underline{apr}}$.

Definition 10 *(Rough matroid based on an approximation operator). Let $\mathbb{I}$ be a family of subsets of U in a covering space $(U, \mathbb{C})$. If there exists a pair of approximation operators $(\underline{apr}, \overline{apr})$ such that M is a lower and an upper rough matroid, then M is called a rough matroid based on an approximation operator.*

Proposition 3. *If $M = (U, \mathbb{I})$ is a matroid, then M is a rough matroid based on approximation operators.*

Proof. Let us consider the approximation operators $\underline{apr}(X) = \overline{apr}(X) = X$, for all $X \subseteq U$. Thus, M is a lower and an upper rough matroid based on approximation operators.

Proposition 4. *If $M = (U, \mathbb{I}_{\underline{apr}})$ is a lower rough matroid and $M' = (U, \mathbb{I}_{\overline{apr}})$ is an upper rough matroid, then $\underline{M} = (U, \mathbb{I})$ is a rough matroid where $\mathbb{I} = \mathbb{I}_{\underline{apr}} \cap \mathbb{I}_{\overline{apr}}$.*

Proof. 1. $\emptyset \in \mathbb{I}_{apr}$, because $\emptyset \in \mathbb{I}_{\underline{apr}}$ and $\emptyset \in \mathbb{I}_{\overline{apr}}$.
2. If $I \in \mathbb{I}_{apr}$, $I' \in \mathbb{D}_{apr}$ and $\underline{apr}(I') \subseteq \underline{apr}(I)$, $(\overline{apr}(I') \subseteq \overline{apr}(I))$ then, $I' \in \mathbb{I}_{\underline{apr}}$ $(I' \in \mathbb{I}_{\overline{apr}})$. So, $I' \in \mathbb{I}_{apr}$.
3. If $I_1, I_2 \in \mathbb{I}_{apr}$ and $|\overline{apr}(I_1)| < |\overline{apr}(I_2)|$, then there exists $I \in \mathbb{I}_{apr}$ such that $\underline{apr}(I_1) \subset \underline{apr}(I) \subseteq \underline{apr}(I_1) \cup \underline{apr}(I_2)$. Since $\mathbb{I} = \mathbb{I}_{\underline{apr}} \cap \mathbb{I}_{\overline{apr}}$ and $\mathbb{D}_{apr} = \mathbb{D}_{\underline{apr}} \cap \mathbb{D}_{\overline{apr}}$, we have that $I \in \mathbb{I}_{apr}$ such that $\overline{apr}(I_1) \subset \overline{apr}(I) \subseteq \overline{apr}(I_1) \cup \overline{apr}(I_2)$.

3.1 Some Types of Rough Matroids Based on Approximation Operators

In this section we consider a dual pair $(\underline{apr}, \overline{apr})$ of approximation operators, where $\underline{apr}$ is a meet-morphism and therefore, $\overline{apr}$ is a join-morphism and a special type of rough matroid.

Proposition 5. *Let $0 < r \leq n$ an integer, where $n = |U|$ for a covering space $(U, \mathbb{C})$ and $(\underline{apr}, \overline{apr})$ is a dual pair of approximation operators. For*

$$\mathbb{I}^r_{apr} = \{I \in \mathbb{D}_{apr} : |I| \leq r\}$$

we have that: if $\{x\} \in \mathbb{D}_{apr}$ for all $x \in U$, then $(U, \mathbb{I}^r_{apr})$ is a lower and an upper rough matroid.

Proof. 1. $\emptyset \in \mathbb{I}^r_{apr}$.
2. If $I \in \mathbb{I}^r_{apr}$, $I' \subseteq I$ and $I' \in \mathbb{D}_{apr}$, then $|I'| \leq |I| \leq r$, and so $I' \in \mathbb{I}^r_{apr}$.
3. If $I_1, I_2 \in \mathbb{I}^r_{apr}$, with $|I_1| < |I_2|$ and $\overline{apr}(I_1) = I_1$ and $\overline{apr}(I_2) = I_2$. Let $w \in I_2 - I_1$ and $I = I_1 \cup \{w\}$. We have that $\overline{apr}(I) = \overline{apr}(I_1 \cup \{w\}) = \overline{apr}(I_1) \cup \overline{apr}(\{w\}) = I_1 \cup \{w\} = I$. So, $I \in \mathbb{I}^r_{apr}$.
Similarly, it is possible to see that $\mathbb{I}^r_{apr}$ is a lower rough matroid.

Proposition 6. *Let $(U, \mathbb{C})$ be a covering space, $X \subseteq U$ and $(\underline{apr}, \overline{apr})$ a dual pair of approximation operators. For*

$$\mathbb{I}_{apr}(X) = \{\underline{apr}(Y) : Y \subseteq U, Y \in \mathbb{D}_{apr}, \underline{apr}(Y) \subseteq \underline{apr}(X)\}$$

we have that $(U, \mathbb{I}_{apr}(X))$ is a lower rough matroid.

Proof. 1. $\emptyset \in \mathbb{I}_{apr}(X)$, because $\emptyset = \underline{apr}(\emptyset) \subseteq \underline{apr}(X)$.
2. If $I \in \mathbb{I}_{apr}(X)$, $I' \subseteq I$ and $I' \in \mathbb{D}_{apr}$, then $\underline{apr}(I') \subseteq \underline{apr}(I) \subseteq \underline{apr}(X)$, and so $I' \in \mathbb{I}_{apr}(X)$.
3. If $I_1, I_2 \in \mathbb{I}_{apr}(X)$, with $|I_1| < |I_2|$. Let $I = I_1 \cup I_2$, we have that $\underline{apr}(I) \subseteq \underline{apr}(I_1) \cup \underline{apr}(I_2) \subseteq \underline{apr}(X)$. Therefore, $I \in \mathbb{I}_{apr}(X)$.

Proposition 7. *If $\underline{apr}$ is order-preserving and $X \subseteq Y$, then $\mathbb{I}_{apr}(X) \subseteq \mathbb{I}_{apr}(Y)$.*

Proof. If $W \in \mathbb{I}_{apr}(X)$, $W = \underline{apr}(W) \subseteq \underline{apr}(X) \subseteq \underline{apr}(Y)$, then $W \in \mathbb{I}_{apr}(Y)$.

4 Order Relation Among Rough Matroids

The order relation among approximation operators can be used for defining an order relation among rough matroids $\mathbb{I}^r_{apr}$ and $\mathbb{I}_{apr}(X)$ for each $X \subseteq U$.

Proposition 8. *If $\underline{apr}_1 \leq \underline{apr}_2$, then $\mathbb{I}^r_{\underline{apr}_1} \subseteq \mathbb{I}^r_{\underline{apr}_2}$*

Proof. According to Proposition 2, $\mathbb{D}(U, \underline{apr}_1) \subseteq \mathbb{D}(U, \underline{apr}_2)$. If $I \in \mathbb{I}^r_{\underline{apr}_1}$, then $I \in \mathbb{D}(U, \underline{apr}_1) \subseteq \mathbb{D}(U, \underline{apr}_2)$ and $|I| \leq r$. So, $I \in \mathbb{I}^r_{\underline{apr}_2}$.

According to the order relation established among approximation operators in [11] and the list in Sect. 2.1, we have that:

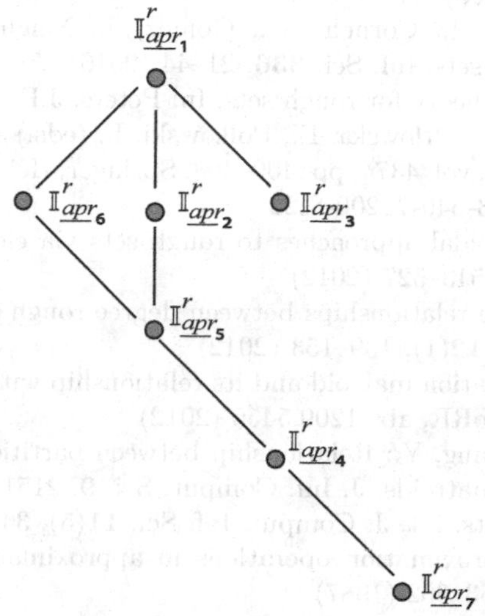

Fig. 2. Order relation for lower rough matroids based on approximation operators.

The same order relation can be established for $\mathbb{I}_{apr}(X)$.

Proposition 9. *If* $\underline{apr}_1 \leq \underline{apr}_2$*, then* $\mathbb{I}_{apr_1}(X) \subseteq \mathbb{I}_{apr_2}(X)$.

Proof. If $W \in \mathbb{I}_{apr_1}(X)$, $W = \underline{apr_1}(W) \subseteq \underline{apr_1}(X) \subseteq \underline{apr_2}(X)$, then $W \in \mathbb{I}_{apr_2}(X)$.

A similar diagram to Fig. 2 can be obtained for this order relation.

5 Conclusions

This paper proposes a generalization of rough matroids based on a covering: rough matroids based on an approximation operator. We introduce new definitions of definable sets for lower and upper rough matroids. We show that any matroid is a lower and an upper matroid based on approximation operators. Finally, we extended the pre-order relation among approximation operators to these rough matroids.

As future work, we want to study the properties of this class of lower and upper rough matroids for a particular pair of approximation operators. More specifically, properties related with lower and upper definable sets and properties related with duality and adjointness.

Acknowledgement. This work was supported by Universidad Militar Nueva Granada's Special Research Fund, under project CIAS 2948-2019 and by the Odysseus program of the Research Foundation-Flanders.

References

1. Bonikowski, Z., Brynarski, E.: Extensions and intentions in rough set theory. Inf. Sci. **107**, 149–167 (1998)
2. D'Eer, L., Restrepo, M., Cornelis, C., Gómez, J.: Neighborhood operators for covering-based rough sets. Inf. Sci. **336**, 21–44 (2016)
3. Järvinen, J.: Lattice theory for rough sets. In: Peters, J.F., Skowron, A., Düntsch, I., Grzymała-Busse, J., Orłowska, E., Polkowski, L. (eds.) Transactions on Rough Sets VI, Part I. LNCS, vol. 4374, pp. 400–498. Springer, Heidelberg (2007). https://doi.org/10.1007/978-3-540-71200-8_22
4. Li, X., Liu, S.: Matroidal approaches to rough sets via closure operators. Int. J. Approx. Reason. **53**, 513–527 (2012)
5. Li, Y., Wang, Z.: The relationships between degree rough sets and matroids. An. Fuzzy Math. Inform. **12**(1), 139–153 (2012)
6. Liu, Y., Zhu, W.: Relation matroid and its relationship with generalized rough set based on relations. CoRR, abs 1209.5456 (2012)
7. Liu, Y., Zhu, W., Zhang, Y.: Relationship between partition matroids and rough sets through k-rank matroids. J. Inf. Comput. Sci. **9**, 2151–2163 (2012)
8. Pawlak, Z.: Rough sets. Int. J. Comput. Inf. Sci. **11**(5), 341–356 (1982)
9. Pomykala, J.A.: Approximation operations in approximation space. Bull. Acad. Pol. Sci. **35**(9–10), 653–662 (1987)
10. Restrepo, M., Cornelis, C., Gómez, J.: Duality, conjugacy and adjointness of approximation operators in covering-based rough sets. Int. J. Approx. Reason. **55**, 469–485 (2014)
11. Restrepo, M., Cornelis, C., Gómez, J.: Partial order relation for approximation operators in covering-based rough sets. Inf. Sci. **284**, 44–59 (2014)
12. Tsang, E., Chen, D., Lee J., Yeung, D.S.: On the upper approximations of covering generalized rough sets. In: Proceedings of the 3rd International Conference on Machine Learning and Cybernetics, pp. 4200–4203 (2004)
13. Wang, S., Zhu, W., Min, F.: Transversal and function matroidal structures of covering-based rough sets. In: Yao, J.T., Ramanna, S., Wang, G., Suraj, Z. (eds.) RSKT 2011. LNCS (LNAI), vol. 6954, pp. 146–155. Springer, Heidelberg (2011). https://doi.org/10.1007/978-3-642-24425-4_21
14. Wybraniec-Skardowska, U.: On a generalization of approximation space. Bull. Pol. Acad. Sci. Math. **37**, 51–61 (1989)
15. Wu, M., Wu, X., Shen, T.: A new type of covering approximation operators. In: IEEE International Conference on Electronic Computer Technology, pp. 334–338 (2009)
16. Wang, S., Zhu, Q., Zhu, W., Min, F.: Matroidal structure of rough sets and its characterization to attribute reduction. Knowl. Based Syst. **54**, 155–161 (2012)
17. Xu, Z., Wang, Q.: On the properties of covering rough sets model. J. Henan Normal Univ. (Nat. Sci.) **33**(1), 130–132 (2005)
18. Yang, T., Li, Q.: Reduction about approximation spaces of covering generalized rough sets. Int. J. Approx. Reason. **51**, 335–345 (2010)
19. Yang, B., Zhao, H., Zhu, W.: Rough matroids based on covering. In: Proceedings of Sixth IEEE International Conference on Data Mining - Workshops, pp. 407–411 (2013)
20. Yao, Y.Y.: Constructive and algebraic methods of the theory of rough sets. Inf. Sci. **109**, 21–47 (1998)

21. Yao, Y., Yao, B.: Covering based rough sets approximations. Inf. Sci. **200**, 91–107 (2012)
22. Zakowski, W.: Approximations in the space (u, π). Demonstr. Math. **16**, 761–769 (1983)
23. Zhu, W.: Properties of the first type of covering-based rough sets. In: Proceedings of Sixth IEEE International Conference on Data Mining - Workshops, pp. 407–411 (2006)
24. Zhu, W.: Relationship between generalized rough sets based on binary relation and covering. Inf. Sci. **179**, 210–225 (2009)
25. Zhu, W., Wang, F.: On three types of covering based rough sets. IEEE Trans. Knowl. Data Eng. **19**(8), 1131–1144 (2007)
26. Zhu, W., Wang, F.: A new type of covering rough set. In: Proceedings of Third International IEEE Conference on Intelligence Systems, pp. 444–449 (2006)
27. Zhu, W., Wang, S.: Rough matroids. In: IEEE International Conference on Granular Computing, pp. 817–8221 (2011)
28. Zhu, W., Wang, S.: Rough matroids based on relation. Inf. Sci. **232**, 241–252 (2013)

Studies on Reducing the Necessary Data Size for Rule Induction from the Decision Table by STRIM

Yuichi Kato[1(✉)] and Tetsuro Saeki[2]

[1] Shimane University, 1060 Nishikawatsu-cho, Matsue, Shimane 690-8504, Japan
ykato@cis.shimane-u.ac.jp

[2] Yamaguchi University, 2-16-1 Tokiwadai, Ube, Yamaguchi 755-8611, Japan
tsaeki@yamaguchi-u.ac.jp

Abstract. STRIM (Statistical Test Rule Induction Method) has been proposed for an if-then rule induction method from the decision table. STRIM judges the significance of a trying rule by a statistical test based on the table. The method judging the trying rule has been executed based on the standard normal distribution approximating the distribution of the decision attribute's values so that the judging method needs the proper size dataset satisfying the conditions of the approximation. This paper proposes a new STRIM named minor-STRIM not incorporating the test by the approximating distribution but by the original distribution, which expands the applicable range to cases not satisfying the conditions. Specifically, minor-STRIM uses a binomial distribution for the testing and shows the applicable range expanded and performance evaluation by use of a simulation experiment compared with those by the conventional STRIM. The simulation also shows that it gives discussing and confirming information the validity of the results obtained from applying minor-STRIM to a real-world dataset.

1 Introduction

Rough Set (RS) theory was introduced by Pawlak [1] and used for inducing if-then rules from a dataset called the decision table (DT). To date, various methods and algorithms for inducing rules by the theory have been proposed [2–5] since the inducing rules are useful to simply and clearly express the structure of rating and/or knowledge hiding behind the table. The basic idea to induce rules is to approximate the concept in the DT by use of the lower and/or upper approximation sets which are respectively derived from the equivalence relations and their equivalence sets in the given DT. However, those methods and algorithms by RS paid little attention to the fact that the DT was just a sample set gathered from the population of interest. If resampling the DT from the population or the DT by Bootstrap method for example, the new DT will change equivalence relations,

© Springer Nature Switzerland AG 2019
T. Mihálydeák et al. (Eds.): IJCRS 2019, LNAI 11499, pp. 130–143, 2019.
https://doi.org/10.1007/978-3-030-22815-6_11

their equivalence sets, and the lower and/or upper approximation sets, so the induced rules will change and fluctuate. Those methods and algorithms also had the problem that those induced rules were not arranged from the statistical views.

Then, we proposed a rule induction method named STRIM (Statistical Test Rule Induction Method) taking the above mentioned problems into consideration [6–14]. Specifically, STRIM

(1) Proposed a data generation model for generating a DT. This model recognized the DT as an input-output system which transformed a tuple of the condition attribute's value occurred by chance (the input) into the decision attribute value (the output) through pre-specified if-then rules (generally unknown) under some hypotheses. That is, the input was recognized as an outcome of the random variables and the output was also the outcome of a random variable dependent on the input and the pre-specified rules. Accordingly, the pairs of input and output formed the DT containing rules.

(2) Assumed a trying proper condition part of if-then rules and judged whether it was a candidate of rules by statistically testing whether the condition part caused bias in the distribution of the decision attribute's values.

(3) Arranged the candidates having inclusion relationships by representing them with one of the highest bias and finally induced if-then rules with a statistical significance level after systematically exploring the trying condition part of rules.

The validity and capacity of STRIM have been confirmed by the simulation experiments that STRIM can induce pre-specified if-then rules from the DT proposed in (1). Accordingly, the validity and capacity also secure a certain extent of the confidence of rules induced by STRIM from the DT of real-world datasets. The DT proposed in (1) is also used for confirming the validity and capacity of other rule induction methods proposed previously [10,13].

However, the conventional STRIM executed the statistical test of the bias in (2) based on the standard normal distribution approximating the distribution in order to easily test it so that the testing required a proper data size satisfying the conditions for the approximation. That is, the condition controlled the applicable range of the conventional STRIM. Then, this paper proposes a new STRIM named minor-STRIM which expands the range by incorporating a test not using the approximate distribution, but the original, specifically, a binomial distribution and its validity and capacity is also clarified in the same way as the conventional. Finally, minor-STRIM is applied to the Car Evaluation dataset of UCI [15] and the validity and capacity of the induced rules are discussed based on the information obtained from the simulation studies.

2 Conventional Rough Sets and STRIM

Rough Set theory is used for inducing if-then rules from a decision table S. S is conventionally denoted by $S = (U, A = C \cup \{D\}, V, \rho)$. Here, $U = \{u(i)|i = 1, ..., |U| = N\}$ is a sample set, A is an attribute set, $C = \{C(j)|j = 1, ..., |C|\}$

is a condition attribute set, $C(j)$ is a member of C and a condition attribute, and D is a decision attribute. Moreover, V is a set of attribute values denoted by $V = \cup_{a \in A} V_a$ and is characterized by the information function ρ: $U \times A \rightarrow V$.

The conventional Rough Set theory first focuses on the following equivalence relation and the equivalence set of indiscernibility within the decision table S of interest: $I_B = \{(u(i), u(j)) \in U^2 | \rho(u(i), a) = \rho(u(j), a), \forall a \in B \subseteq C\}$. I_B is an equivalence relation in U and derives the quotient set $U/I_B = \{[u_i]_B | i = 1, 2, ..., |U| = N\}$. Here, $[u_i]_B = \{u(i) \in U | (u(j), u_i) \in I_B, u_i \in U\}$. $[u_i]_B$ is an equivalence set with the representative element u_i.

Let be $\forall X \subseteq U$ then X can be approximated like $B_*(X) \subseteq X \subseteq B^*(X)$ by use of the equivalence set. Here, $B_*(X) = \{u_i \in U | [u_i]_B \subseteq X\}$, and $B^*(X) = \{u_i \in U | [u_i]_B \cap X \neq \phi\}$, $B_*(X)$ and $B^*(X)$ are referred to as the lower and upper approximations of X by B respectively. The pair of $(B_*(X), B^*(X))$ is usually called a rough set of X by B.

Specifically, let be $X = \{u(i) | \rho(u(i), D) = d\} = U(d) = \{u(i) | u^{D=d}(i)\}$ called the concept of $D = d$, and define a set of $u(i)$ as $U(CP) = \{u(i) | u^{C=CP}(i)$, meaning CP satisfies $u^C(i)$, where $u^C(i)$ is the condition attribute values of $u(i)\} = B_*(X)$, then CP can be used as the condition part of the if-then rule of $D = d$, with necessity. That is, the following expression of if-then rules with necessity is obtained: if $CP = \wedge_j(C(j) = v_{jk})$ then $D = d$. In the same way, $B^*(X)$ derives the condition part CP of the if-then rule of $D = d$ with possibility.

However, the approximation of $X = U(d)$ by the lower or upper approximation is respectively too strict or loose so that the rules induced by the approximations are often no use. Then, Ziarko expanded the original RS by introducing an admissible error in two ways [4]: $\underline{B}_\varepsilon(U(d) = \{u(i) | accuracy \geq 1 - \varepsilon\}$, $\overline{B}_\varepsilon(U(d)) = \{u(i) | accuracy > \varepsilon\}$, where $\varepsilon \in [0, 0.5)$. The pair of $(\underline{B}_\varepsilon(U(d)), \overline{B}_\varepsilon(U(d)))$ is called an ε-lower and ε-upper approximation which satisfies the following properties: $B_*(U(d)) \subseteq \underline{B}_\varepsilon(U(d)) \subseteq \overline{B}_\varepsilon(U(d)) \subseteq B^*(U(d))$, $\underline{B}_{\varepsilon=0}(U(d)) = B_*(U(d))$ and $\overline{B}_{\varepsilon=0}(U(d)) = B^*(U(d))$. The ε-lower and/or ε-upper approximation induce if-then rules with admissible errors the same as the lower and/or upper approximation.

As mentioned above, the conventional RS theory basically focuses on the equivalence relation I_B and its equivalence sets U/I_B in U given in advance and induces rules approximating the concept by use of the approximation sets derived from the U/I_B. However, I_B is very dependent on the DT provided. Accordingly, every DT obtained from the same population is different from each other and, I_B, U/I_B and the approximation sets are different from each other for each DT, which leads to inducing different rule sets. That is, the rule induction methods by the conventional RS theory lack statistical views.

Then, STRIM has proposed a data generation model for the DT and a rule induction method based on the model. Specifically, STRIM considers the decision table to be a sample dataset obtained from an input-output system including a rule box, as shown in Fig. 1, and hypotheses regarding the decision attribute values, as shown in Table 1. A sample $u(i)$ consists of its condition attributes values $u^C(i)$ and its decision attribute value $u^D(i)$. $u^C(i)$ is the input for the rule box,

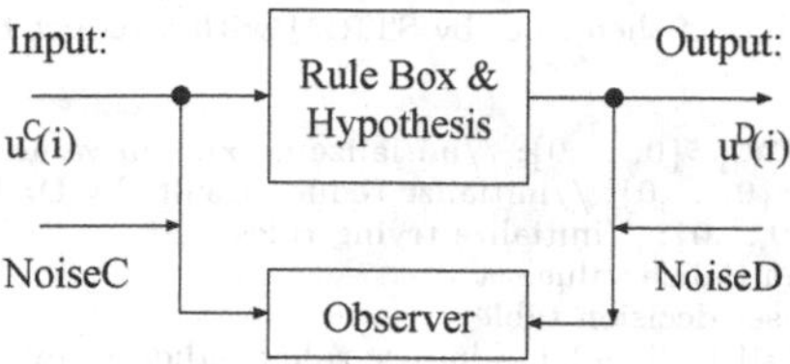

Fig. 1. A data generation mode: Rule box contains if-then rules $R(d,k)$: if $CP(d,k)$ then $D = d$, where $CP(d,k) = \wedge_l(C(l_k) = v_{l_k})$ $(d = 1,2,...,k = 1,2,...)$.

Table 1. Hypotheses with regard to the decision attribute value.

Hypothesis 1	$u^C(i)$ coincides with $CP(d,k)$, and $u^D(i)$ is uniquely determined as $D = d$ (uniquely determined case)
Hypothesis 2	$u^C(i)$ does not coincide with any $CP(d,k)$, and $u^D(i)$ can only be determined randomly (indifferent case).
Hypothesis 3	$u^C(i)$ coincides with several $CP(d,k)$ $(d = d1, d2, ...)$, and their outputs of $u^C(i)$ conflict with each other. Accordingly, the output of $u^C(i)$ must be randomly determined from the conflicted outputs (conflicted case)

and is transformed into the output $u^D(i)$ using the rules (generally unknown) contained in the rule box and the hypotheses. The hypotheses consist of three cases corresponding to the input. They are uniquely determined, indifferent and conflicted cases (see Table 1). In contrast, $u(i) = (u^C(i), u^D(i))$ is measured by an observer, as shown in Fig. 1. The existence of NoiseC and NoiseD makes missing values in $u^C(i)$, and changes $u^D(i)$ to create another value for $u^D(i)$, respectively. Those noises bring the system closer to a real-world system. The data generation model suggests that a pair of $(u^C(i), u^D(i))$ $(i = 1,...,N)$, i.e. a decision table is an outcome of these random variables: $(C, D) = ((C(1),...,C(|C|), D)$ observing the population.

Based on the data generation model, STRIM (1) extracted significant pairs of a condition attribute and its value like $C(j_k) = v_{j_k}$ for rules of $D = d$ by the local reduct [9,10,12], (2) constructed a trying condition part of the rules like $CP = \wedge_j(C(j_k) = v_{j_k})$ by use of the reduct results, and (3) investigated whether $U(CP)$ caused a bias at n_d in the frequency distribution of the decision attribute values $f = (n_1, n_2, ..., n_{M_D})$ or not, where $n_m = |U(CP) \cap U(m)|$ $(m = 1, ..., |V_{a=D}| = M_D)$ and $U(m) = \{u(i)|u^{D=m}(i)\}$, since the $u^C(i)$ coinciding to $CP(d,k)$ in the rule box is transformed into $u^D(i)$ based on Hypotheses 1 or 3. Accordingly, the CP coinciding to one of rules in the rule box produces bias in f. Specifically, STRIM used a statistical test method for the investigation specifying a null hypothesis $H0$: f does not have any bias, that is, CP is not a rule and its alternative hypothesis $H1$: f has a bias, that is, CP is a rule, and a proper significance level, and tested $H0$ by use of the sample dataset, that is, the decision table and the proper test statistics, for example, $z = \frac{(n_d+0.5-np_d)}{(np_d(1-p_d))^{0.5}}$, where $n_d = \max_m f = (n_1, ..., n_m, ..., n_{M_D})$,

```
Line Algorithm to induce if-then rules by STRIM with a reduct function
No.
 1  int main(void) {
 2    int rdct_max[|CV|]={0,. . . ,0}; //initialize maximum value of C(j)
 3    int rdct[|CV|]={0,. . . ,0}; //initialize reduct results by D=l
 4    int rule[|C|]={0,...,0}; //initialize trying rules
 5    int tail=-1; //initialize value set
 6    input data; // set decision table
 7    for (di=1; di<=|D|; di++) {// induce rule candidates every D=l
 8      attribute_reduct(rdct_max)
 9      set rdct[ck] ; // if (rdct_max[ck]==0) {rdct[ck]=0; }else {rdct[ck]=1; }
10      rule_check(rcdct, redct_max, tail, rule); // the first stage process
11    }// end di
12    arrange rule candidates // the second stage
13  }// end main
14  int attribute_reduct(int rdct_max[]) {
15    make contingency table for D=l vs. C(j)
16    Test H0(j,l);
17    if H0(j,l) is rejected then set rdct_max[j,l]=jmax else rdct_max[j,l]=0; //
    jmax:the attribute value of the maximum frequency
18  }// end of attribute_reduct
19  int rule_check(int rdct[], int rdct_max[], int tail,int rule[]) {// the first stage
    process
20    for (ci=tail+1; cj<|C|; ci++) {
21      for (cj=1; cj<=rdct[ci]; cj++) {
22        rule[ci]=rdct_max[cj]; // a trying rule set for test
23        count frequency of the trying rule; // count n1, n2,...
24        if (frequency>=N0) {//sufficient frequency ?
25          if (|z|>3.0) {//sufficient evidence ?
26            add the trying rule as a rule candidate
27          }// end of if |z|
28          rule_check(ci,rule)
29        }// end if frequency
30      }// end cj
31      rule[ci]=0; // trying rules reset
32    }// end ci
33  }// end rule_check
```

Fig. 2. An algorithm for STRIM including a reduct function.

$p_d = P(D = d)$, $n = \sum_{j=1}^{M_D} n_j$. z obeys the standard normal distribution under test conditions: $np_d \geq 5$ and $n(1 - p_d) \geq 5$ [16] and is considered to be an index of the bias of f. (4) If $H0$ is rejected then the assumed CP becomes a candidate for the rules in the rule box. (5) After repeating the processes from (1) to (4) and obtaining the set of rule candidates, STRIM arranged their rule candidates and induced the final results (see literatures [11,12] for details).

Figure 2 shows an algorithm written in C language style for STRIM including a reduct function. Line No. (LN) 8 and 9 are the reduct portion of the above (1), and (2) is executed at LN 10 and the dimension rule[] is used for trying rules, (3) is executed at LN 25 in the function rule_check(), (4) is executed at LN 26 and (5) is LN12.

To summarize, STRIM directly induces rules with statistical significance level assuming the condition part of rules: $CP = \wedge_j(C(j_k) = v_{j_k})$ and statistically testing it by use of U. STRIM does not require the basic concept of the approximation which is an essence for the rule induction by the conventional RS theory. Conversely, the RS theory has nothing directly to do with statistical significance.

Table 2. An example of pre-specified rules $R(d,k)$ in the rule box: if $CP(d,k)$ then $D = d$ $(d = 1, ..., 4, k = 1, 2)$.

$R(d,k)$	$CP(d,k)$	$D = d$
$R(1,1)$	110000	$D = 1$
$R(1,2)$	001100	$D = 1$
$R(2,1)$	220000	$D = 2$
$R(2,2)$	002200	$D = 2$
$R(3,1)$	330000	$D = 3$
$R(3,2)$	003300	$D = 3$
$R(4,1)$	440000	$D = 4$
$R(4,2)$	004400	$D = 4$

3 Studies on the Conventional STRIM by Simulation Experiment

We implemented the data generation process and verified the capacity of inducing the rules by the conventional STRIM as follows: (1) Specified rules by eight $(d = 1, ..., 4, k = 1, 2$, the number of rules $(N_{rule}) = 8)$ as shown in Table 2 corresponding to the rule box in Fig. 1, where $|C| = 6$, $V_a = \{1, 2, 3, 4\}$ $(a = C(j)$ $(j = 1, ..., |C|)$, $a = D)$, and $CP(1,1) = 110000$ denoted $CP(1,1) = (C(1) = 1) \wedge (C(2) = 1)$ and was called a rule of the rule length 2 $(RL = 2)$, having two conditions. (2) Generated $v_{C(j)}(i)$ $(j = 1, ..., |C| = 6)$ with a uniform distribution and formed $u^C(i) = (v_{C(1)}(i), ..., v_{C(6)}(i))$ $(i = 1, ..., N = 10,000)$. (3) Transformed $u^C(i)$ into $u^D(i)$ using the pre-specified rules in Table 2 and hypotheses in Table 1, without generating NoiseC and NoiseD for a plain experiment and then generated the decision table.

After randomly selecting samples by $N_B = 1,000$ from N samples, newly forming the DT and applying STRIM to the DT, Table 3 was obtained. The table shows us the following: For example, the estimated Rule No. 1 "1100001" denotes if $(C(1) = 1) \wedge (C(2) = 1)$ then $D = 1$, has $f = (n_1, n_2, n_3, n_4) = (57, 1, 1, 1)$ and the bias at $D = 1$. The outcome probability to cause such a bias is around 1.59E-36, which leads to rejecting $H0$ and adopting $H1$. As the result, "1100001" was adopted as a rule. It should be noted that the reason it was adopted as the rule was not the high accuracy $= 57/60 = 0.950$. STRIM induced all the pre-specified rules in Table 2 and the two extra rules.

In order to examine the performance for the rule induction at N_B, Fig. 3 shows (a): the number of all the induced rule N_{rule} and (b): the number of the induced pre-specified N_{rule} at $N_B = 200, 300, 500, 1000, 2000$. Those were plotted by the average of ten times' experiments in the same way as the one shown in Table 3. Figure 3 shows us the following: $N_B = 2,000$ was the sufficient data size since STRIM induced the same rules of $N_{rule} = 8$ in (a) and (b) corresponding to

Table 3. An example of estimated rules for the dataset with $N_B = 1,000$ generated by the data generation model in Fig. 1 with the pre-specified rules in Table 2.

Rule No.	Estimated rules $(C(1), ..., C(6), D)$	$f = (n_1, n_2, n_3, n_4)$	p-value(z)	Accuracy	Coverage
1	(1100001)	(57, 1, 1, 1)	1.59E-36(12.57)	0.95	0.23
2	(0033003)	(4, 2, 56, 3)	1.32E-35(12.40)	0.86	0.25
3	(0011001)	(56, 1, 1, 2)	6.52E-35(2.27)	0.93	0.22
4	(0022002)	(1, 56, 2, 2)	2.83E-33(11.96)	0.92	0.22
5	(0044004)	(3, 2, 1, 56)	5.01E-33(11.91)	0.91	0.22
6	(2200004)	(5, 56, 1, 1)	1.60E-31(11.62)	0.89	0.22
7	(3300003)	(3, 3, 44, 1)	1.23E-28(11.04)	0.86	0.19
8	(4400004)	(1, 2, 3, 51)	3.20E-28(10.95)	0.89	0.19
9	(3003003)	(4, 7, 40, 8)	1.38E-17(8.46)	0.68	0.18
10	(0404004)	(12, 9, 8, 42)	7.98E-11(6.40)	0.59	0.16

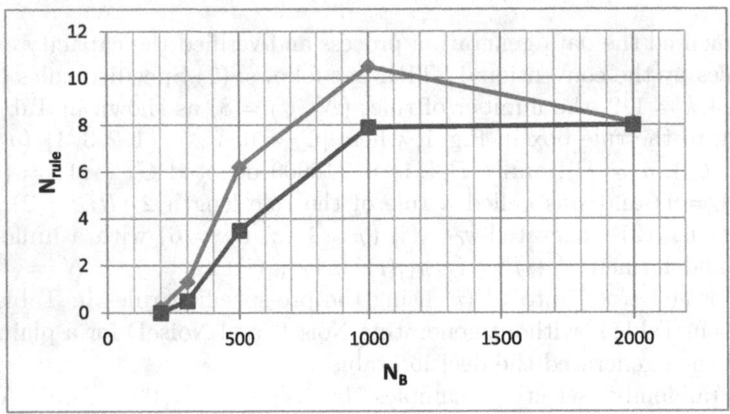

Fig. 3. Studies on the number of induced rules at N_B: (a) the number of all the induced rules (♦), (b) the number of induced pre-specified rules (■).

the pre-specified rules in spite of different DTs. At $N_B = 1,000$, STRIM almost induced all the pre-specified rules although there were some differences between (a) and (b), the same as Table 3. Less than $N_B = 1,000$, STRIM could not abruptly induce rules and the N_{rule} of (a) and (b) was almost 0 at $N_B = 200$. That is why the small size dataset could not properly execute local reducts at LN = 7 − 11 and satisfy the test condition: $np_d \geq 5 \rightarrow n \geq \frac{5}{p_d} = \frac{5}{\frac{1}{4}} = 20 = N0$ at LN = 24 respectively in Fig. 2.

Just for reference, Table 4 shows N_{rule} by RL of the rules induced by ROSE II [17] implementing the conventional RS method for the dataset in Table 3. Table 4 shows us the following: ROSE II could not induce any pre-specified rules with $RL = 2$ but a lot of rules with larger RL than that of the pre-specified.

Table 4. The number of rules by rule length induced by ROSE II for the dataset in Table 3.

RL	1	2	3	4	5	6
N_{rule}	0	0	35	168	312	8
$N_{subrule}$	0	0	35	22	5	0

The rules with $RL = 4, 5$ and 6 seems to be almost meaningless from the statistical views. $N_{subrule}$ denotes the number of sub-rules of the pre-specified rules. Here, for example, the rule "1100021" is called the sub-rule of "1100001" since $U(\text{"1100001"}) \supseteq U(\text{"1100021"})$. Accordingly, ROSE II induced some part of the sub-rules of the pre-specified although no one knows about such situations for real-world datasets.

Generally, the statistical test problems are divided into two cases: One is the case of large data size and the other is that of small data size. The former is usually studied in the standard normal distribution and the latter is studied in its individual distribution. This paper experimentally studies the problems inducing if-then rules from the DT with $N_B = 200, 300, 500$ as the small data size problem. Specifically, the conventional STRIM is developed into minor-STRIM, which incorporates a test method for the small data size into the conventional STRIM, and improves it to be able to be used even for small sized datasets.

4 Studies on Minor-STRIM by Simulation Experiment

As was experimentally shown in Sect. 3, the conventional STRIM for large data sizes could not induce all of the pre-specified rules around less than $N_B = 1,000$ since it did not properly execute the local reduct and the statistical test due to the small data size. This Section improves its procedure without the reduct and by substituting the test with one based on the original distribution. Specifically, n_d in principle obeys a Binomial distribution $B_n(n, p_d)$ and the p-value is given as follows: $p-value = \sum_{l=n_d}^{n} {}_nC_l p_d^l (1-p_d)^{n-l} = P(F_{2n_d}^{2(n-n_d+1)} \geq \frac{n_d(1-p_d)}{(n-n_d+1)})$, where $F_{2n_d}^{2(n-n_d+1)}$ is a random variable and obeys F-Distribution with degrees of freedom $(2(n-n_d+1), 2n_d)$ [18]. The algorithm in Fig. 2 should also be modified as follows:

(1) Rearrange LN = 7-11 as follows: rule_check(tail, rule); // the first stage process
(2) Rearrange LN = 24-25 as follows: if (p-value $< \alpha$) { // p-value is less than the pre-specified significance level α, and delete LN = 29.

The improved STRIM incorporating (1) and (2) is named minor-STRIM to distinguish from the conventional version.

Figure 4 shows the results of the simulation experiment by minor-STRIM corresponding with Fig. 3 for each significance level: α= 1.0E-2, 1.0E-3, 1.0E-4,

Table 5. An arrangement of Car Evaluation dataset of UCI.

Unified attribute value	$C(1)$: buying	$C(2)$: maint	$C(3)$: doors	$C(4)$: person	$C(5)$: lug boot	$C(6)$: safety	D: class (freq.)
1	vhigh	vhigh	2	2	small	low	unacc (1210)
2	high	high	3	4	med	med	acc (383)
3	med	med	4	more	big	high	good (69)
4	low	low	5more	–	–	–	vgood (65)

1.0E-5. (a) N_{rule} of all the induced rules and (b) N_{rule} of all the induced pre-specified rules plotted by the average of ten times' experiments in the same way as Fig. 3 show us the following:

(1) Around less than $N_B = 500$ minor-STRIM could not induce all the pre-specified rules regardless of α. However, minor-STRIM greatly improves the experimental results compared with those in Fig. 3.
(2) The tendency of (1) will be somewhat improved by use of $\alpha = 1.0$E-2 or 1.0E-3 (see (b)) while the use of them increases extra rules at $N_B = 200, 300$ (see (a)).
(3) Conversely, the use of $\alpha = 1.0$E-5 is sever to squeeze N_{rule}.
(4) The information (2) and (3) recommend to use $\alpha = 1.0$E-4 when inducing rules from real-world datasets.

The data generation model in Fig. 1 appears useful to gain the information from (1) to (4).

This experimental study may suggest that minor-STRIM can be applied to a DT with a large data size. However, the conventional STRIM should be used in such a situation since the conventional version use the function of reduct and of stopping exploration for sub-rules in the case once not satisfying the testing condition at LN = 24-29 in Fig. 2, while minor-STRIM, in principle, explores all the rule patterns. Just for reference, the run time at $N_B = 1,000$ by the conventional STRIM was 1.09 [s], and minor-STRIM needed 49.39 [s] and 3.75 [s] at $N_B = 300$ on a PC with Intel(R) Core(TM) i5-6500 CPU at 3.20 GHz.

5 Application to Car Evaluation Dataset

The literature [15] provides a lot of datasets for machine learning. This paper applied minor-STRIM to the "Car Evaluation" dataset included in them. Table 5 shows the summaries and specifications of the dataset: $|C| = 6$, $|V_{C(i)}| = 4$ $(i = 1, 2, 3)$, $|V_{C(i)}| = 3$ $(i = 4, 5, 6)$, $|V_D| = 4$, $N = |U| = \prod_{i=1}^{6} |V_{C(i)}| = 1728$ which consists of every combination of condition attributes' values, and there is not any conflicted or identical samples. The frequencies of the decision attribute values extremely incline toward $D = 1$ as shown in Table 5.

The literature [9] shows some examples of only trying rules for $D = 1$ and 4. In order to compare and discuss those with the rule induction results by minor-STRIM, this paper repeatedly shows them as Table 6. Table 6 shows χ^2-values

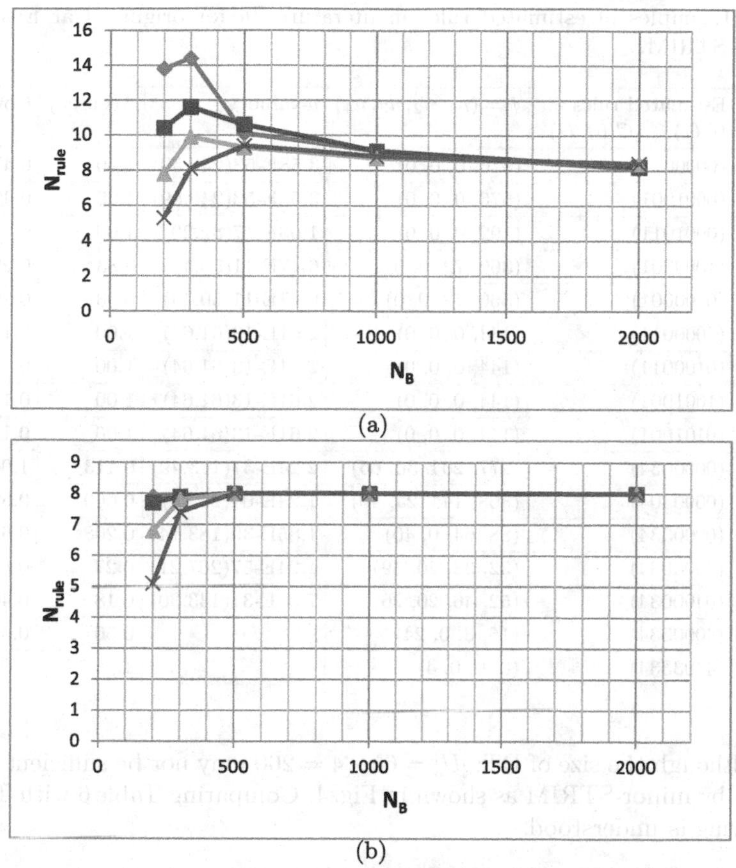

Fig. 4. Studies on the number of induced rules at N_B by significance levels $\alpha = 1.0\text{E-}2$ (♦), 1.0E-3 (▲), 1.0E-4 (■), 1.0E-5 (×): (a) the number of all the induced rules, (b) the number of induced pre-specified rules.

having the same testing condition as z-values, and its corresponding p-values which were used for the statistical test due to heavy inclination of the outcome frequencies between the decision attribute' values and the rest of the surface caput is the same as Table 3. The literature [9] judged Rule No. = 1 and 2 valid for $D = 1$ and Rule No. = 13 for $D = 4$ since the p-value was the smallest.

The conventional STRIM needed a sufficient data size satisfying the test condition so that the rule induction was executed neglecting the unbalance of the frequencies. Then, samples for $D = 1, 2$ and 3 were randomly selected by 65 which corresponded to the frequency of $D = 4$, and a new DT was formed for analyzing it by minor-STRIM. An example of rule induction results is shown in

Table 6. Examples of estimated rules in literature [9] for original Car Evaluation dataset by STRIM.

Rule No.	Estimated rules $(C(1),...,C(6),D)$	$f=(n_1,n_2,n_3,n_4)$	p-value(χ^2)	Accuracy	Coverage
1	(0000011)	(576, 0, 0, 0)	3.58E-53(246.59)	1.00	0.476
2	(0001001)	(576, 0, 0, 0)	3.58E-53(246.59)	1.00	0.476
3	(0001011)	(192, 0, 0, 0)	1.03E-17(82.20)	1.00	0.157
4	(1000001)	(360, 72, 0, 0)	6.47E-11(50.43)	0.83	0.296
5	(0100001)	(360, 72, 0, 0)	6.47E-11(50.43)	0.83	0.296
6	(1000011)	(144, 0, 0, 0)	2.61E-13(61.64)	1.00	0.119
7	(0100011)	(144, 0, 0, 0)	2.61E-13(61.64)	1.00	0.119
8	(1001001)	(144, 0, 0, 0)	2.61E-13(61.64)	1.00	0.119
9	(0101001)	(144, 0, 0, 0)	2.61E-13(61.64)	1.00	0.12
10	(0000034)	(277, 204, 30, 65)	2.24E-37(173.49)	0.113	1.00
11	(0000304)	(368, 144, 24, 40)	1.24E-04(20.65)	0.069	0.62
12	(0000334)	(88, 64, 0, 40)	1.85E-39(183.14)	0.208	0.62
13	(4000034)	(52, 33, 20, 39)	1.24E-57(267.21)	0.27	0.60
14	(0400034)	(52, 46, 20, 26)	7.32E-31(143.30)	0.18	0.40
15	(4000334)	(16, 8, 0, 24)	–	0.50	0.37
16	(4403334)	(0, 0, 0, 4)	–	1.00	0.062

Table 7 although the size of DT: $|U| = 65 \times 4 = 260$ may not be sufficient for the induction by minor-STRIM as shown in Fig. 4. Comparing Table 6 with Table 7, the following is understood:

(1) With regard to $D = 1$, both results coincide with each other in spite of the different sample size.

(2) With regard to $D = 4$, minor-STRIM induced the rule "0000334" in No. 3 corresponding to No. 12 in Table 6 which was not adopted in the literature [9] and was covered with the unbalance of the frequency distribution due to the test condition: $np_4 \geq 5 \rightarrow n \geq \frac{5}{p_4} = \frac{5}{\frac{65}{1728}} \cong 133$. minor-STRIM seems to unveil the cover.

(3) minor-STRIM also discovered No. 4, 5, 6 and 7 which could not have been induced by the conventional STRIM due to the unbalance of the frequency distribution.

To arrange Table 7 by the original rating words, the following expressions are obtained:

If $person = 2 \vee safety = low$ then $class = unac$ $(D = 1)$
If $maint = med \wedge safety = med$ then $class = acc$ $(D = 2)$
If $(buying = vhigh \wedge lug\ boot = big \vee maint = vhigh) \wedge safety = med$ then $class = good$ $(D = 3)$
If $(lug\ boot = big \vee buying = high \wedge maint = high) \wedge safety = high$ then $class = vgood$ $(D = 4)$

Table 7. An example of estimated rules for the newly constructed Car Evaluation dataset ($|U| = 65 \times 4 = 260$) by minor-STRIM with $\alpha = 1.0\text{E-}4$.

Rule No.	Estimated rules $(C(1),...,C(6),D)$	$f = (n_1, n_2, n_3, n_4)$	*p*-value	Accuracy	Coverage
1	(0001001)	(38, 0, 0, 0)	1.32E-23	1.00	0.58
2	(0000011)	(26, 0, 0, 0)	2.22E-16	1.00	0.40
3	(0000334)	(4, 8, 0, 40)	5.99E-15	0.77	0.62
4	(1000323)	(0, 2, 16, 0)	2.08E-08	0.89	0.25
5	(0100023)	(4, 8, 26, 0)	2.23E-08	0.68	0.40
6	(2200034)	(1, 0, 0, 13)	1.60E-07	0.93	0.2
7	(0300022)	(2, 11, 0, 0)	1.11E-05	0.85	0.17

Taking into account that each attribute in the Car dataset can be recognized on an ordinal scale, there may be some conflicts between Rule No. 5 and No. 7. That is, there are relationships No. 5 > No. 7 in "*Class*" and No. 5 < No. 7 in "*maint*" although they commonly contain $safety = med$ factor. The reason is supposed to be the shortage of data size: $|U| = 260$, as shown in Fig. 4. The same experiments repeated several times showed that rules of middle rating such as $D = 2$ and/or $D = 3$ slightly and delicately changed.

The dataset is a kind of evidence for rule induction, which is better the more evidence there is. It should be noted that the preliminary experiment by use of the data generation model in Fig. 1 gives us various guiding principles such as certificate and reproducibility of experiment results against N_B.

6 Conclusion

The rule induction methods by the conventional RS [2–5] basically approximate the concept in the DT given as one of samples by the lower and/or upper approximation. The concept and both approximations are constructed by the equivalence relation and its equivalence sets in the given DT. Accordingly, the rules induced from the sample will change and fluctuate every resampling from the population and/or the given DT. That is, the conventional method has had defects not considering the rule induction problem from the statistical views. Then, STRIM which does not use the concept of the approximation and directly and systematically explores rules by use of a statistical test has been proposed and confirmed its validity in simulation experiments and real-world datasets [6–14].

However, the conventional STRIM used the test method applicable in the case of the large data size so that it was constrained based on the data size. Generally, the test problem is discussed in two cases, a large and a small data size. The conventional STRIM was applicable with the former case and used the statistic obeying the standard normal distribution for the test. This paper developed it into the latter case and proposed minor-STRIM which used the statistic obeying a binomial distribution. The validity and capacity of minor-STRIM have been confirmed in a simulation experiment and by applying it to Car Evaluation dataset of UCI [15]. Generally, the classification of the large or small data size depends on the specification of the DT such as $|A| = |C \cup \{D\}|$ and $|V| = |\cup_{a \in A} V_a|$. Accordingly, the process firmly conducting the simulation experiments, and obtaining knowledge from the experimental results and their certificates shown in Sects. 3 and 4 are very important before applying the conventional STRIM or minor-STRIM to real-world datasets of interest. It should be noted that the data generation model in Fig. 1 enables this process.

References

1. Pawlak, Z.: Rough sets. Int. J. Inf. Comput. Sci. **11**(5), 341–356 (1982)
2. Skowron, A., Rauser, C.M.: The discernibility matrix and functions in information systems. In: Słowiński, R. (ed.) Intelligent Decision Support, Handbook of Application and Advances of Rough Set Theory, pp. 331–362. Kluwer Academic Publishers, Dordrecht (1992)
3. Grzymala-Busse, J.W.: LERS—a system for learning from examples based on rough sets. In: Słowiński, R. (ed.) Intelligent Decision Support, Handbook of Applications and Advances of the Rough Sets Theory, pp. 3–18. Kluwer Academic Publishers, Dordrecht (1992)
4. Ziarko, W.: Variable precision rough set model. J. Comput. Syst. Sci. **46**, 39–59 (1993)
5. Shan, N., Ziarko, W.: Data-based acquisition and incremental modification of classification rules. Comput. Intell. **11**(2), 357–370 (1995)
6. Matsubayashi, T., Kato, Y., Saeki, T.: A new rule induction method from a decision table using a statistical test. In: Li, T., et al. (eds.) RSKT 2012. LNCS (LNAI), vol. 7414, pp. 81–90. Springer, Heidelberg (2012). https://doi.org/10.1007/978-3-642-31900-6_11
7. Kato, Y., Saeki, T., Mizuno, S.: Studies on the necessary data size for rule induction by STRIM. In: Lingras, P., Wolski, M., Cornelis, C., Mitra, S., Wasilewski, P. (eds.) RSKT 2013. LNCS (LNAI), vol. 8171, pp. 213–220. Springer, Heidelberg (2013). https://doi.org/10.1007/978-3-642-41299-8_20
8. Kato, Y., Saeki, T., Mizuno, S.: Considerations on rule induction procedures by STRIM and their relationship to VPRS. In: Kryszkiewicz, M., Cornelis, C., Ciucci, D., Medina-Moreno, J., Motoda, H., Raś, Z.W. (eds.) RSEISP 2014. LNCS (LNAI), vol. 8537, pp. 198–208. Springer, Cham (2014). https://doi.org/10.1007/978-3-319-08729-0_19
9. Kato, Y., Saeki, T., Mizuno, S.: Proposal of a statistical test rule induction method by use of the decision table. Appl. Soft Comput. **28**, 160–166 (2015)
10. Kitazaki, Y., Saeki, T., Kato, Y.: Performance Comparison to a classification problem by the second method of quantification and STRIM. In: Flores, V., et al. (eds.) IJCRS 2016. LNCS (LNAI), vol. 9920, pp. 406–415. Springer, Cham (2016). https://doi.org/10.1007/978-3-319-47160-0_37
11. Fei, J., Saeki, T., Kato, Y.: Proposal for a new reduct method for decision tables and an improved STRIM. In: Tan, Y., Takagi, H., Shi, Y. (eds.) DMBD 2017. LNCS, vol. 10387, pp. 366–378. Springer, Cham (2017). https://doi.org/10.1007/978-3-319-61845-6_37

12. Kato, Y., Itsuno, T., Saeki, T.: Proposal of dominance-based rough set approach by STRIM and its applied example. In: Polkowski, L., et al. (eds.) IJCRS 2017, Part I. LNCS (LNAI), vol. 10313, pp. 418–431. Springer, Cham (2017). https://doi.org/10.1007/978-3-319-60837-2_35
13. Kato, Y., Kawaguchi, S., Saeki, T.: Studies on CART's performance in rule induction and comparisons by STRIM. In: Nguyen, H.S., Ha, Q.-T., Li, T., Przybyła-Kasperek, M. (eds.) IJCRS 2018. LNCS (LNAI), vol. 11103, pp. 148–161. Springer, Cham (2018). https://doi.org/10.1007/978-3-319-99368-3_12
14. Kato, Y., Saeki, T., Mizuno, S.: Considerations on the principle of rule induction by STRIM and its relationship to the conventional rough sets methods. Appl. Soft Comput. J. **73**, 933–942 (2018)
15. Asunction, A., Newman, D.J.: UCI Machine Learning Repository, University of California, School of Information and Computer Science, Irvine (2007). http://www.ics.edu/~mlearn/MlRepository.html
16. Walpole, R.E., Myers, R.H., Myers, S.L., Ye, K.: Probability and Statistics for Engineers and Scientists, 8th edn., pp. 187–191. Pearson Prentice Hall, Upper Saddle River (2007)
17. Laboratory of Intelligent Decision Support System (IDSS). http://idss.cs.put.poznan.pl/site/139.html
18. Fleiss, J.L., Levin, B., Paik, M.C.: Statistical Methods for Rates and Proportions, 3rd edn., p. 25. Wiley, New York (2003)

Rough Approximations on Two Universes Under a Mapping

Tong-Jun Li[1,3](✉), Wei-Zhi Wu[1,3], and Xiuyi Jia[2]

[1] School of Mathematics, Physics and Information Science, Zhejiang Ocean University, Zhoushan 316022, Zhejiang, China
{litj,wuwz}@zjou.edu.cn

[2] School of Computer Science and Engineering, Nanjing University of Science and Technology, Nanjing 210094, Jiangsu, China
jiaxy@njust.edu.cn

[3] Key Laboratory of Oceanographic Big Data Mining and Application of Zhejiang Province, Zhejiang Ocean University, Zhoushan 316022, Zhejiang, China

Abstract. In this paper, the focus is on the relation-based rough sets on two universes. Two universes are connected with a mapping, by which a relation on one universe is constructed based on the relation on the another universe, so two relations on the different universes are induced. The relationships between the rough approximations based on the induced relations and the original relations are examined in detail.

Keywords: Rough sets · Rough approximations · Binary relations · Mappings

1 Introduction

Rough set theory [1], which was first proposed by Pawlak in 1982, is an important mathematic approach, which has been successfully applied in many fields, such as, artificial intelligence, decision making, knowledge representation, pattern recognition, etc. In the classical rough sets, equivalence relations are used to construct lower and upper approximation operators. In order to deal with complex practical problems, Pawlak rough set model is extended, and various generalized rough set models have been established, for example, generalized rough set based on relation [2–4], covering rough set model [4–7], rough set models in multigranulation spaces [8–10], etc.

Relation-based rough sets are natural extensions of the classical Pawlak rough set model. Skowron and Stepaniuk [11] proposed rough approximation operators based on tolerance relation, and examined attribute reduction of tolerance information systems. Slowinski and Vanderpooten [3] defined lower and upper approximation operators by means of similarity relation. Reflexive and transitive relation based rough sets have rich topological properties [12,13]. Yao [2,14] investigated rough approximations based on binary relation by constructive and algebraic methods, and compared some rough set models.

© Springer Nature Switzerland AG 2019
T. Mihálydeák et al. (Eds.): IJCRS 2019, LNAI 11499, pp. 144–154, 2019.
https://doi.org/10.1007/978-3-030-22815-6_12

Rough set models on two universes of discourse have been studied in the literature [15,16,18]. In generalized rough set models, the approximated sets and the approximating sets usually locate at two different universes [15], Li and Zhang [16] proposed one kind of rough set model on two universes, in which they included in the same universe. Sun and Ma extended multigranulation rough sets in the framework of two universes [17,18].

In the above rough set models, a relation connects two universes. In another case, two universes are connected with a mapping. For example, the communication between two information systems is a very important topic in the field of artificial intelligence. In mathematics, it can be explained as a mapping between two universes [19], which can maintain the knowledge structures of two universes unchanging. On the other hand, one can use a mapping to transform a big database into a small database, so that target on the big database can be obtained by dealing with the small database. In this aspect, Wang et al. [20–22] studied the communication between information systems, in which the notions of homomorphisms of relation or covering information systems, and attribute reduction of information systems was investigated. All these motivate us to study the relationship between two universes under a mapping.

In this paper, we study the relationships of the rough approximations on the two universes with a mapping. The rest of this paper is organized as follows. We briefly review in the next section some basic notions. In Sect. 3 we discuss of induced binary relations, and examine the rough approximations based on the induced relations. The paper is then concluded with a brief summary.

2 Preliminaries

In this section, a lot of basic knowledge about relation-based rough approximations and mapping are reviewed briefly.

Let U be a nonempty and finite set, and called the universe of discourse. A *binary relation* R on U means a subset of U^2, that is, $R \subseteq U^2$, and $(x, y) \in R$ is often denoted as xRy. For any $x \in U$, the set, $R(x) = \{y \in U | xRy\}$, is called the *successor neighborhood* of x, and the *predecessor neighborhood* of y $(y \in U)$ means the set $R^{-1}(y) = \{x \in U | xRy\}$.

Some types of binary relations on the universe U are often mentioned in the literature:

- R is *reflexive* if $x \in R(x)$, $\forall x \in U$;
- R is *symmetric* if xRy implies yRx, $\forall x, y \in U$;
- R is *transitive* if xRy, yRz implies xRz, $\forall x, y, z \in U$.

Let R be a binary relation on U. The tuple (U, R) is called an *approximation space.* For $A \subseteq U$, *the lower and upper rough approximations* of A with respect to (U, R), denoted by $\underline{R}(A)$ and $\overline{R}(A)$ respectively, are defined by

$$\underline{R}(A) = \{x \in U | R(x) \subseteq A\}, \quad \overline{R}(A) = \{x \in U | R(x) \cap A \neq \emptyset\}.$$

The basic properties of the rough approximation operators, $\underline{R}$ and $\overline{R}$, are enumerated as follows [2]: $\forall A, B \subseteq U$,

(L1) $\underline{R}(A) = \sim(\overline{R}(\sim A))$, (U1) $\overline{R}(A) = \sim(\underline{R}(\sim A))$;
(L2) $\underline{R}(U) = U$, (U2) $\overline{R}(\emptyset) = \emptyset$;
(L3) $\underline{R}(A \cap B) = \underline{R}(A) \cap \underline{R}(B)$, (U3) $\overline{R}(A \cup B) = \overline{R}(A) \cup \overline{R}(B)$;
(L4) $A \subseteq B \Rightarrow \underline{R}(A) \subseteq \underline{R}(B)$, (U4) $A \subseteq B \Rightarrow \overline{R}(A) \subseteq \overline{R}(B)$.

Properties (L1) and (U1) show that $\underline{R}$ and $\overline{R}$ are dual to each other. The rough approximation operators based on a variety of binary relations have different properties, conversely some kinds of binary relations can be characterized by corresponding rough approximation operators [23,24].

A *mapping* f from U to W, denoted as $f: U \to W$, assigns each element $x \in U$ an element $f(x) \in W$. Here U is called the *domain* of f, and the *image set* of f is the set $\{f(x) | x \in U\} \subseteq W$. For $x \in U$ the element $f(x)$ is called the *image* of x under f. For $y \in W$ the *preimage* of y under f is a subset of U, denoted by $f^{-1}(y)$ and defined by $f^{-1}(y) = \{x \in U | f(x) = y\}$.

It should be pointed out that the family, $U/f = \{f^{-1}(y) \neq \emptyset | y \in W\}$, forms a partition of U, that is, $f^{-1}(y_1) \cap f^{-1}(y_2) \neq \emptyset$ implies $f^{-1}(y_1) = f^{-1}(y_2)$, and $\bigcup_{y \in W} f^{-1}(y) = U$.

Let f be a mapping from U to W. The notions of the image and the preimage of elements can be extend to the image set and the preimage set of subsets, respectively. For $X \subseteq U$, the *image set* of X under f is a subset $f(X)$ of W, and defined by $f(X) = \{f(x) \in W | x \in X\}$. For $Y \subseteq W$, the *preimage set* of Y is a subset $f^{-1}(Y)$ of U, and defined by $f^{-1}(Y) = \{x \in U | \exists y \in Y (f(x) = y)\}$.

For any $x \in U$, in the following we denote $f^{-1}(f(x))$ as $[x]_R^f$.

Then the following properties of $f: U \to W$ can be checked directly: for any $X, X_1, X_2 \subseteq U$, and $Y, Y_1, Y_2 \subseteq W$, we have

(1) $X_1 \subseteq X_2 \Rightarrow f(X_1) \subseteq f(X_2)$;
(2) $f(X_1 \cap X_2) \subseteq f(X_1) \cap f(X_2)$;
(3) $f(X_1 \cup X_2) = f(X_1) \cup f(X_2)$;
(4) $\sim f(X) \subseteq f(\sim X)$;
(5) $Y_1 \subseteq Y_2 \Rightarrow f^{-1}(Y_1) \subseteq f^{-1}(Y_2)$;
(6) $f^{-1}(Y_1 \cap Y_2) = f^{-1}(Y_1) \cap f^{-1}(Y_2)$;
(7) $f^{-1}(Y_1 \cup Y_2) = f^{-1}(Y_1) \cup f^{-1}(Y_2)$;
(8) $\sim f^{-1}(Y) = f^{-1}(\sim Y)$;
(9) $X \subseteq f^{-1}(f(X))$, denoted $f^{-1}(f(X))$ as $\overline{X}$;
(10) $f(f^{-1}(Y)) = Y$.

3 Rough Approximations Based on Induced Relations

In the following, we assume that U and W are two finite and nonempty sets, f is a mapping from U to W, and f is surjective, that is, $f(U) = W$.

In this section, we examine rough approximation operators based on two relations. One relation is defined on U, which is induced by a relation on W via the mapping f, and another relation is defined on W, which is constructed by using of a relation on U and the mapping f.

3.1 Two Relations Induced by the Mapping

Given a relation R on U, by which and based on f, a relation on W can be induced as follows.

Definition 1. *Let R be a relation on U. Then a relation R_f on W can be defined as follows:*

$$R_f = \{(y_1, y_2) \in W^2 | \exists x_1, x_2 \in U (y_1 = f(x_1), y_2 = f(x_2), x_1 R x_2)\}.$$

From Definition 1 we can see that $y_1 R_f y_2$ if and only if

$$[f^{-1}(y_1) \times f^{-1}(y_2)] \cap R \neq \emptyset.$$

Example 1. Let $U = \{1, 2, 3, 4, 5\}$, $W = \{a, b, c\}$, and $f : U \to W$. The mapping f satisfies

$$f(1) = f(2) = a, f(3) = f(4) = b, f(5) = c.$$

Taking

$$R = \{(1,3), (2,2), (2,5), (3,1), (3,4), (3,5), (4,3), (5,2), (5,3)\},$$

in terms of Definition 1, we can figure out the relation R_f on W, i.e.,

$$R_f = \{(a,b), (a,a), (a,c), (b,a), (b,b), (b,c), (c,a), (c,b)\}.$$

Proposition 1. *Let R be a relation on U. For any $x \in U$, $y \in W$, if $y = f(x)$, then $f(R(x)) \subseteq R_f(y)$.*

Proof. According to Definition 1, for any $x' \in R(x)$, we have $y R_f f(x')$, that is, $f(x') \in R_f(y)$. Therefore $f(R(x)) \subseteq R_f(y)$.

It is easy to check that for any $y = f(x)$, $f(R(x)) \subseteq R_f(y)$ if and only if $R(x) \subseteq f^{-1}(R_f(y))$. It should be noted that the reversed inequality of $f(R(x)) \subseteq R_f(y)$ may not be satisfied.

Example 2. Let R and R_f be the two relations of Example 1. Noticing $a = f(1)$, from Example 1 we can check that $R(1) = \{3\}$, $f(R(1)) = f(3) = \{b\}$, and $R_f(a) = \{a, b, c\}$. Thus $f(R(1)) \subseteq R_f(a)$, but $f(R(1)) \neq R_f(a)$.

Definition 2. *Let R be a relation on U. Then R is said to be compatible with f if $[x]_R^f \times [x']_R^f \subseteq R$ for xRx'.*

The below proposition shows that the inequality in Proposition 1 becomes an equality under the compatibility of R.

Proposition 2. *Let R be a relation on U. If R is compatible with f, then for any $x \in U$ and $y \in W$ with $y = f(x)$, we have $f(R(x)) = R_f(y)$.*

Proof. According to Proposition 1, it is only needed to prove that if $y = f(x)$, then $f(R(x)) \supseteq R_f(y)$.

For $x \in U$, $y \in W$, if $y = f(x)$, then for any $y' \in R_f(y)$, there exists $x^*, x' \in U$ such that $y = f(x^*)$, $y' = f(x')$, and $x^* R x'$, where x^* and x may not be the same one, but $x \in [x^*]_R^f$. Since R is compatible with f, it follows from $x^* R x'$ that $[x^*]_R^f \times [x']_R^f \subseteq R$, thus xRx', that is, $x' \in R(x)$, so $y' \in f(R(x))$. Therefore $R_f(y) \subseteq f(R(x))$.

Similarly we can prove that if $y = f(x)$, then $R(x) = f^{-1}(R_f(y))$.

On the other hand, if a relation R on W is known, then by the mapping f, a relation on U can be got.

Definition 3. *Let R be a relation on W. Then a relation $R_{f^{-1}}$ on U can be defined as follows:*

$$R_{f^{-1}} = \{(x_1, x_2) \in U^2 | f(x_1) R f(x_2)\}.$$

Example 3. Let U, W, and f be the same as those of Example 1. If we put

$$R = \{(a, a), (a, c), (c, b), (b, c), (c, c)\},$$

then R is a relation on W. According to Definition 3, we can calculate the relation $R_{f^{-1}}$ as

$$R_{f^{-1}} = \{(1, 2), (2, 1), (1, 5), (2, 5), (5, 3), (5, 4), (3, 5), (4, 5), (5, 5)\}.$$

Proposition 3. *Let R be a relation on W. Then*

(1) *for any $y_1, y_2 \in W$, if $y_1 R y_2$ then $f^{-1}(y_1) \times f^{-1}(y_2) \subseteq R_{f^{-1}}$;*

(2) *for any $x \in U$, $y \in W$, if $y = f(x)$, then*

$$R(y) = f(R_{f^{-1}}(x)), R_{f^{-1}}(x) = f^{-1}(R(y)).$$

Proof. (1) It follows from Definition 3 directly.

(2) For any $x \in U$, $y \in W$, if $y = f(x)$ and $y' \in R(y)$, by Definition 3 we have $f^{-1}(y') \subseteq R_{f^{-1}}(x)$, so $y' \in f(R_{f^{-1}}(x))$. Thus $R(y) \subseteq f(R_{f^{-1}}(x))$. Conversely, if $y' \in f(R_{f^{-1}}(x))$, then there is an $x' \in R_{f^{-1}}(x)$ such that $f(x') = y'$, so $(y, y') \in R$, that is, $y' \in R(y)$. Thus $f(R_{f^{-1}}(x)) \subseteq R(y)$, so we can conclude that $R(y) = f(R_{f^{-1}}(x))$.

It follow from $R(y) = f(R_{f^{-1}}(x))$ that $f^{-1}(R(y)) = f^{-1}(f(R_{f^{-1}}(x)))$, by $R_{f^{-1}}(x) \subseteq f^{-1}(f(R_{f^{-1}}(x)))$ we get $R_{f^{-1}}(x) \subseteq f^{-1}(R(y))$. On the other hand, if $x' \in f^{-1}(R(y))$, then $f(x') \in R(y)$, so $(x, x') \in R_{f^{-1}}$, that is, $x' \in R_{f^{-1}}(x)$. Thus $f^{-1}(R(y)) \subseteq R_{f^{-1}}(x)$. It is proved that $R_{f^{-1}}(x) = f^{-1}(R(y))$.

Applying Definitions 2 and 3, Proposition 2, we know that the relation $R_{f^{-1}}$ is compatible with f^{-1}.

About the transformation between the relations on U and W, we have the following results.

Proposition 4. *Let R be a relation on U. Then $R \subseteq (R_f)_{f^{-1}}$. Specially, if R is compatible with f, then $R = (R_f)_{f^{-1}}$.*

Proof. By Definitions 1 and 3, the inequality $R \subseteq (R_f)_{f^{-1}}$ can be proved directly.

If R be compatible with f, then for any $(x_1, x_2) \in (R_f)_{f^{-1}}$, by the definition of $(R_f)_{f^{-1}}$ we have $(y_1, y_2) = (f(x_1), f(x_2)) \in R_f$, subsequently by the definition of R_f, there are $x_1^* \in f^{-1}(y_1), x_2^* \in f^{-1}(y_2)$ such that $(x_1^*, x_2^*) \in R$. Since R is compatible with f, we have $[x_1^*]_R^f \times [x_2^*]_R^f \subseteq R$. It follows from $x_1 \in [x_1^*]_R^f$ and $x_2 \in [x_2^*]_R^f$ that $(x_1, x_2) \in R$. Thus $(R_f)_{f^{-1}} \subseteq R$, combining $R \subseteq (R_f)_{f^{-1}}$ we have $R = (R_f)_{f^{-1}}$.

For the relation R on W we have the below conclusion.

Proposition 5. *Let R be a relation on W. Then $R = (R_{f^{-1}})_f$.*

Proof. It directly follows from Definitions 1 and 3.

3.2 Rough Approximations Based on the Induced Relation R_f

Let R be a relation on U. For any $X \subseteq U$, by the mapping f, it turns into $f(X)$. The rough approximations of X on (U, R) and the rough approximations of $f(X)$ of (W, R_f) have the following relationships under f.

Theorem 1. *Let R be a relation on U. Then for any $X \subseteq U$,*

$$\underline{R_f}(f(X)) \subseteq f(\underline{R}(\overline{X})),\ f^{-1}(\underline{R_f}(f(X))) \subseteq \underline{R}(\overline{X}).$$

Specially, if R is compatible with f, then

$$f(\underline{R}(X)) \subseteq \underline{R_f}(f(X)) \subseteq f(\underline{R}(\overline{X})),\ \underline{R}(X) \subseteq f^{-1}(\underline{R_f}(f(X))) \subseteq \underline{R}(\overline{X}).$$

Proof. For any $X \subseteq U$, if $y \in \underline{R_f}(f(X))$, then $R_f(y) \subseteq f(X)$, so we have $f^{-1}(R_f(y)) \subseteq f^{-1}(f(X)) = \overline{X}$. Since $R(x) \subseteq f^{-1}(R_f(y))$, where $f(x) = y$, we have $R(x) \subseteq \overline{X}$, so $x \in \underline{R}(\overline{X})$, thus $y \in f(\underline{R}(\overline{X}))$. It can be concluded that $\underline{R_f}(f(X)) \subseteq f(\underline{R}(\overline{X}))$.

For any $x \in f^{-1}(\underline{R_f}(f(X)))$, we have $y = f(x) \in \underline{R_f}(f(X))$, from which it follows that $R_f(y) \subseteq f(X)$. Then $f^{-1}(R_f(y)) \subseteq f^{-1}(f(X)) = \overline{X}$, according to $R(x) \subseteq f^{-1}(R_f(y))$ we have $x \in \underline{R}(\overline{X})$. Therefore $f^{-1}(\underline{R_f}(f(X))) \subseteq \underline{R}(\overline{X})$.

If R be compatible with f, then for any $y \in f(\underline{R}(X))$, there is an $x \in \underline{R}(X)$ such that $y = f(x)$. By $x \in \underline{R}(X)$ we have $R(x) \subseteq X$, so $f(R(x)) \subseteq f(X)$. By Proposition 2 we get $f(R(x)) = R_f(y)$, thus $R_f(y) \subseteq f(X)$, i.e., $y \in \underline{R_f}(f(X))$. It is proved that $f(\underline{R}(X)) \subseteq \underline{R_f}(f(X)) \subseteq f(\underline{R}(\overline{X}))$.

From $f(\underline{R}(X)) \subseteq \underline{R_f}(f(X))$, it follows that $f^{-1}(f(\underline{R}(X))) \subseteq f^{-1}(\underline{R_f}(f(X)))$, by $\underline{R}(X) \subseteq f^{-1}(f(\underline{R}(X)))$ we have $\underline{R}(X) \subseteq f^{-1}(\underline{R_f}(f(X)))$. It can be concluded that $\underline{R}(X) \subseteq f^{-1}(\underline{R_f}(f(X))) \subseteq \underline{R}(\overline{X})$.

Theorem 1 shows that, when the relation R on U is compatible with f, for any $X \subseteq U$, if $X = \overline{X}$, then $f(\underline{R}(X)) = \underline{R_f}(f(X))$, $\underline{R}(X) = f^{-1}(\underline{R_f}(f(X)))$.

Theorem 2. *Let R be a relation on U. Then for any $X \subseteq U$,*

$$f(\overline{R}(X)) \subseteq \overline{R_f}(f(X)),\ \overline{R}(X) \subseteq f^{-1}(\overline{R_f}(f(X))).$$

Specially, if R is compatible with f, then

$$f(\overline{R}(X)) \subseteq \overline{R_f}(f(X)) \subseteq f(\overline{R}(\overline{X})),\ \overline{R}(X) \subseteq f^{-1}(\overline{R_f}(f(X))) \subseteq \overline{R}(\overline{X}).$$

Proof. For any $y \in f(\overline{R}(X))$, there is an $x \in \overline{R}(X))$ such that $y = f(x)$, so $R(x) \cap X \neq \emptyset$. There exists an $x' \in U$ such that $x' \in R(x)$, and $x' \in X$, that is, $y' = f(x') \in f(X)$. By the definition of R_f we have $y' \in R_f(y)$, so $R_f(y) \cap f(X) \neq \emptyset$, that is, $y \in \overline{R_f}(f(X))$. Thus $f(\overline{R}(X)) \subseteq \overline{R_f}(f(X))$.

From $f(\overline{R}(X)) \subseteq \overline{R_f}(f(X))$, it follows that $f^{-1}(f(\overline{R}(X))) \subseteq f^{-1}(\overline{R_f}(f(X)))$. Noticing $\overline{R}(X) \subseteq f^{-1}(f(\overline{R}(X)))$ we have $\overline{R}(X) \subseteq f^{-1}(\overline{R_f}(f(X)))$.

Assume that R is compatible with f. If $y \in \overline{R_f}(f(X))$, then $R_f(y) \cap f(X) \neq \emptyset$, so $f^{-1}(R_f(y)) \cap f^{-1}(f(X)) = f^{-1}(R_f(y) \cap f(X)) \neq \emptyset$. If we take $x \in f^{-1}(y)$, by Proposition 2 we have $R(x) = f^{-1}(R_f(y))$. Thus $R(x) \cap \overline{X} \neq \emptyset$, i.e., $x \in \overline{R}(\overline{X})$, so $y = f(x) \in f(\overline{R}(\overline{X}))$. We get $f(\overline{R}(X)) \subseteq \overline{R_f}(f(X)) \subseteq f(\overline{R}(\overline{X}))$.

If $x \in f^{-1}(\overline{R_f}(f(X)))$, we have $y = f(x) \in \overline{R_f}(f(X))$, so $R_f(y) \cap f(X) \neq \emptyset$. Similarly, we have $x \in \overline{R}(\overline{X})$. Thus $\overline{R}(X) \subseteq f^{-1}(\overline{R_f}(f(X))) \subseteq \overline{R}(\overline{X})$.

From Theorem 2 we can see that, when R is compatible with f, for any $X \subseteq U$, if $X = \overline{X}$, then $f(\overline{R}(X)) = \overline{R_f}(f(X))$, $\overline{R}(X) = f^{-1}(\overline{R_f}(f(X)))$.

On the other hand, for any $Y \subseteq W$, we have $f^{-1}(Y) \subseteq U$, about the rough approximations of Y and $f^{-1}(Y)$ we have the following conclusions.

Theorem 3. *Let R be a relation on U. Then for any $Y \subseteq W$,*

$$\overline{R}(f^{-1}(Y)) \subseteq f^{-1}(\overline{R_f}(Y)),\ f(\overline{R}(f^{-1}(Y))) \subseteq \overline{R_f}(Y).$$

Specially, if R is compatible with f, then

$$\overline{R}(f^{-1}(Y)) = f^{-1}(\overline{R_f}(Y)),\ f(\overline{R}(f^{-1}(Y))) = \overline{R_f}(Y).$$

Proof. The inequalities $\overline{R}(f^{-1}(Y)) \subseteq f^{-1}(\overline{R_f}(Y))$ and $f(\overline{R}(f^{-1}(Y))) \subseteq \overline{R_f}(Y)$ and the equation $f(\overline{R}(f^{-1}(Y))) = \overline{R_f}(Y)$ can be deduced from Theorem 2 directly. As an example, in the following we prove $\overline{R}(f^{-1}(Y)) \subseteq f^{-1}(\overline{R_f}(Y))$.

For any $Y \subseteq W$, putting $X = f^{-1}(Y)$, we obtain $f(X) = Y$ and $X = \overline{X}$. By Theorem 2 we have $f(\overline{R}(X)) \subseteq \overline{R_f}(f(X))$, i.e. $f(\overline{R}(f^{-1}(Y))) \subseteq \overline{R_f}(Y)$, so $f^{-1}(f(\overline{R}(f^{-1}(Y)))) \subseteq f^{-1}(\overline{R_f}(Y))$, and $\overline{R}(f^{-1}(Y)) \subseteq f^{-1}(\overline{R_f}(Y))$ follows from $\overline{R}(f^{-1}(Y)) \subseteq f^{-1}(f(\overline{R}(f^{-1}(Y))))$.

As for $\overline{R}(f^{-1}(Y)) = f^{-1}(\overline{R_f}(Y))$, by $\overline{R}(f^{-1}(Y)) \subseteq f^{-1}(\overline{R_f}(Y))$ we know that it is needed only to prove $f^{-1}(\overline{R_f}(Y)) \subseteq \overline{R}(f^{-1}(Y))$. If $x \in f^{-1}(\overline{R_f}(Y))$, we have $y = f(x) \in \overline{R_f}(Y)$, that is, $R_f(y) \cap Y \neq \emptyset$, so $f^{-1}(R_f(y)) \cap f^{-1}(Y) \neq \emptyset$. Since R is compatible with f, we have $R(x) = f^{-1}(R_f(y))$, so $x \in \overline{R}(f^{-1}(Y))$. Thus $f^{-1}(\overline{R_f}(Y)) \subseteq \overline{R}(f^{-1}(Y))$.

Theorem 4. *Let R be a relation on U. Then for any $Y \subseteq W$,*

$$f^{-1}(\underline{R_f}(Y)) \subseteq \underline{R}(f^{-1}(Y)),\ \underline{R_f}(Y) \subseteq f(\underline{R}(f^{-1}(Y))).$$

Specially, if R is compatible with f, then

$$f^{-1}(\underline{R_f}(Y)) = \underline{R}(f^{-1}(Y)),\ \underline{R_f}(Y) = f(\underline{R}(f^{-1}(Y))).$$

Proof. From Theorem 1 it can be proved similarly that $\underline{R_f}(Y) \subseteq f(\underline{R}(f^{-1}(Y)))$, $f(\underline{R}(f^{-1}(Y))) = \underline{R_f}(Y)$. And from Theorem 3 we can directly prove that

$$f^{-1}(\underline{R_f}(Y)) \subseteq \underline{R}(f^{-1}(Y)), f^{-1}(\underline{R_f}(Y)) = \underline{R}(f^{-1}(Y)).$$

As an example, in the following we prove $\underline{R}(f^{-1}(Y)) \supseteq f^{-1}(\underline{R_f}(Y))$: for any $Y \subseteq W$,

$$\begin{aligned} f^{-1}(\underline{R_f}(Y)) \subseteq \underline{R}(f^{-1}(Y)) &\Longleftrightarrow \sim f^{-1}(\underline{R_f}(Y)) \supseteq \sim \underline{R}(f^{-1}(Y)) \\ &\Longleftrightarrow f^{-1}(\sim \underline{R_f}(Y)) \supseteq \overline{R}(\sim f^{-1}(Y)) \\ &\Longleftrightarrow f^{-1}(\overline{R_f}(\sim Y)) \supseteq \overline{R}(f^{-1}(\sim Y)) \\ &\Longleftrightarrow f^{-1}(\overline{R_f}(Y)) \supseteq \overline{R}(f^{-1}(Y)). \end{aligned}$$

3.3 Rough Approximations Based on the Induced Relation $R_{f^{-1}}$

Let R be a relation on W. Analogously, in the following we investigate the relationships between the rough approximations on $(U, R_{f^{-1}})$ and (W, R).

Firstly, by $X \subseteq U$ and $f(X)$ we obtain the following results.

Theorem 5. *Let R be a relation on W. Then for any $X \subseteq U$,*

$$\begin{aligned} f(\underline{R_{f^{-1}}}(X)) \subseteq \underline{R}(f(X)) \subseteq f(\underline{R_{f^{-1}}}(\overline{X})), \\ \underline{R_{f^{-1}}}(X) \subseteq f^{-1}(\underline{R}(f(X)) \subseteq \underline{R_{f^{-1}}}(\overline{X}). \end{aligned}$$

Proof. For any $y \in f(\underline{R_{f^{-1}}}(X))$, there exists an $x \in \underline{R_{f^{-1}}}(X)$ such that $y = f(x)$, thus $R_{f^{-1}}(x) \subseteq X$. Then $f(R_{f^{-1}}(x)) \subseteq f(X)$, by Proposition 3 we have $R(y) = f(R_{f^{-1}}(x))$, so $y \in \underline{R}(f(X))$. We get $f(\underline{R_{f^{-1}}}(X)) \subseteq \underline{R}(f(X))$. From $R(y) \subseteq f(X)$ we have $f^{-1}(R(y)) \subseteq f^{-1}(f(X)) = \overline{X}$, by $R_{f^{-1}}(x) = f^{-1}(R(y))$ we get $R_{f^{-1}}(x) \subseteq \overline{X}$, i.e., $x \in \underline{R_{f^{-1}}}(\overline{X})$, so $y = f(x) \in f(\underline{R_{f^{-1}}}(\overline{X}))$. Consequently, we conclude that $f(\underline{R_{f^{-1}}}(X)) \subseteq \underline{R}(f(X)) \subseteq f(\underline{R_{f^{-1}}}(\overline{X}))$.

By $f(\underline{R_{f^{-1}}}(X)) \subseteq \underline{R}(f(X))$ we have $f^{-1}(f(\underline{R_{f^{-1}}}(X))) \subseteq f^{-1}(\underline{R}(f(X)))$, from $\underline{R_{f^{-1}}}(X) \subseteq f^{-1}(f(\underline{R_{f^{-1}}}(X)))$ it follows that $\underline{R_{f^{-1}}}(X) \subseteq f^{-1}(\underline{R}(f(X)))$. If $x \in f^{-1}(\underline{R}(f(X)))$, then $y = f(x) \in \underline{R}(f(X))$, that is, $R(y) \subseteq f(X)$, so $f^{-1}(R(y)) \subseteq f^{-1}(f(X)) = \overline{X}$ is deduced, by $R_{f^{-1}}(x) = f^{-1}(R(y))$ we get $x \in \underline{R_{f^{-1}}}(\overline{X})$. It can be concluded that $\underline{R_{f^{-1}}}(X) \subseteq f^{-1}(\underline{R}(f(X))) \subseteq \underline{R_{f^{-1}}}(\overline{X})$.

Theorem 5 shows that when $X = \overline{X}$, we have $f(\underline{R_{f^{-1}}}(X)) = \underline{R}(f(X))$ and $\underline{R_{f^{-1}}}(X) = f^{-1}(\underline{R}(f(X)))$.

Theorem 6. *Let R be a relation on W. Then for any $X \subseteq U$,*

$$f(\overline{R_{f^{-1}}}(X)) \subseteq \overline{R}(f(X)) \subseteq f(\overline{R_{f^{-1}}}(\overline{X})),$$
$$\overline{R_{f^{-1}}}(X) \subseteq f^{-1}(\overline{R}(f(X))) \subseteq \overline{R_{f^{-1}}}(\overline{X}).$$

Proof. For any $y \in f(\overline{R_{f^{-1}}}(X))$, there exists an $x \in \overline{R_{f^{-1}}}(X)$ such that $y = f(x)$, so $R_{f^{-1}}(x) \cap X \neq \emptyset$. Thus $f(R_{f^{-1}}(x)) \cap f(X) \supseteq f(R_{f^{-1}}(x) \cap X) \neq \emptyset$, by $f(R_{f^{-1}}(x)) = R(y)$, we have $y \in \overline{R}(f(X))$. It follows from $R(y) \cap f(X) \neq \emptyset$ that $f^{-1}(R(y)) \cap f^{-1}(f(X)) \neq \emptyset$, by $f^{-1}(R(y)) = R_{f^{-1}}(x)$, we get $R_{f^{-1}}(x) \cap \overline{X} \neq \emptyset$, i.e., $x \in \overline{R_{f^{-1}}}(\overline{X})$, which means $y \in f(\overline{R_{f^{-1}}}(\overline{X}))$. Consequently, we conclude $f(\overline{R_{f^{-1}}}(X)) \subseteq \overline{R}(f(X)) \subseteq f(\overline{R_{f^{-1}}}(\overline{X}))$.

From $f(\overline{R_{f^{-1}}}(X)) \subseteq \overline{R}(f(X))$ we get $f^{-1}(f(\overline{R_{f^{-1}}}(X))) \subseteq f^{-1}(\overline{R}(f(X)))$, again by $\overline{R_{f^{-1}}}(X) \subseteq f^{-1}(f(\overline{R_{f^{-1}}}(X)))$, we have $\overline{R_{f^{-1}}}(X) \subseteq f^{-1}(\overline{R}(f(X)))$. For $x \in f^{-1}(\overline{R}(f(X)))$, we have $y = f(x) \in \overline{R}(f(X))$, that is, $R(y) \cap f(X) \neq \emptyset$. Thus $f^{-1}(R(y)) \cap f^{-1}(f(X)) = f^{-1}(R(y) \cap f(X)) \neq \emptyset$, by $f^{-1}(R(y)) = R_{f^{-1}}(x)$, we get $R_{f^{-1}}(x) \cap \overline{X} \neq \emptyset$, that is, $x \in \overline{R_{f^{-1}}}(\overline{X})$. Therefore it can be concluded that $\overline{R_{f^{-1}}}(X) \subseteq f^{-1}(\overline{R}(f(X))) \subseteq \overline{R_{f^{-1}}}(\overline{X})$.

From Theorem 6 we can see that for $X \subseteq U$, if $X = \overline{X}$, then $f(\overline{R_{f^{-1}}}(X)) = \overline{R}(f(X))$, $\overline{R_{f^{-1}}}(X) = f^{-1}(\overline{R}(f(X)))$.

Similarly, by $Y \subseteq W$ and $f^{-1}(Y) \subseteq U$, the below conclusions are gained.

Theorem 7. *Let R be a relation on W. Then for any $Y \subseteq W$,*

$$f^{-1}(\underline{R}(Y)) = \underline{R_{f^{-1}}}(f^{-1}(Y)).$$

Proof. For any $x \in U$, we have

$$\begin{aligned} x \in f^{-1}(\underline{R}(Y)) &\Longleftrightarrow f(x) \in \underline{R}(Y) \\ &\Longleftrightarrow R(f(x)) \subseteq Y \\ &\Longleftrightarrow f^{-1}(R(f(x))) \subseteq f^{-1}(Y) \\ &\Longleftrightarrow R_{f^{-1}}(x) \subseteq f^{-1}(Y) \\ &\Longleftrightarrow x \in \underline{R_{f^{-1}}}(f^{-1}(Y)). \end{aligned}$$

The proof is completed.

It follows from Theorem 7 that for any $Y \subseteq W$, $\underline{R}(Y) = f(\underline{R_{f^{-1}}}(f^{-1}(Y)))$.

Theorem 8. *Let R be a relation on W. Then for any $Y \subseteq W$,*

$$f^{-1}(\overline{R}(Y)) = \overline{R_{f^{-1}}}(f^{-1}(Y)).$$

Proof. For any $Y \subseteq W$, we have

$$\begin{aligned} f^{-1}(\overline{R}(Y)) = \overline{R_{f^{-1}}}(f^{-1}(Y)) &\Longleftrightarrow \sim f^{-1}(\overline{R}(Y)) = \sim \overline{R_{f^{-1}}}(f^{-1}(Y)) \\ &\Longleftrightarrow f^{-1}(\sim \overline{R}(Y)) = \underline{R_{f^{-1}}}(\sim f^{-1}(Y)) \\ &\Longleftrightarrow f^{-1}(\underline{R}(\sim Y)) = \underline{R_{f^{-1}}}(f^{-1}(\sim Y)). \end{aligned}$$

According to Theorem 7, we conclude that $f^{-1}(\overline{R}(Y)) = \overline{R_{f^{-1}}}(f^{-1}(Y))$.

From Theorem 8 we know that for any $Y \subseteq W$, $\overline{R}(Y) = f(\overline{R_{f^{-1}}}(f^{-1}(Y)))$.

4 Summaries

Much attention has been paid on the rough sets on two universes, in many cases two universes are connected by a relation between them. In this paper, two universes are linked with a mapping, a binary relation on one universe can be induced according to the given relation on the other universe, the properties of induced relations are investigated. The rough approximations based on the original relation and induced relation are compared, as a result, it can found that under some conditions, the rough approximations on the two universes can be transformed to each other.

Based on the obtained results, the knowledge discovery on the relation information systems linked with a mapping can be studied further in the framework of rough set theory.

Acknowledgements. This work was supported by grants from the National Natural Science Foundation of China (Nos. 61773349, 61573321, 61773208).

References

1. Pawlak, Z.: Rough sets. Int. J. Comput. Inf. Sci. **11**, 341–356 (1982)
2. Yao, Y.Y.: Constructive and algebraic methods of theory of rough sets. Inf. Sci. **109**, 21–47 (1998)
3. Slowinski, R., Vanderpooten, D.: A generalized definition of rough approximations based on similarity. IEEE Trans. Knowl. Data Eng. **12**, 331–336 (2000)
4. Zhang, Y.L., Luo, M.K.: Relationships between covering-based rough sets and relation-based rough sets. Inf. Sci. **225**, 57–71 (2013)
5. Yao, Y.Y., Yao, B.: Covering based rough sets approximations. Inf. Sci. **200**, 91–107 (2012)
6. Zhu, W., Wang, F.: The fourth type of covering-based rough sets. Inf. Sci. **201**, 80–92 (2012)
7. Deer, L., Restrepo, M., Cornelis, C., Gomez, J.: Neighborhood operators for covering-based rough sets. Inf. Sci. **336**, 21–44 (2016)
8. Lin, G., Liang, J., Qian, Y.: Multigranulation rough sets: from partition to covering. Inf. Sci. **241**, 101–118 (2013)
9. Wu, W., Leung, Y.: Optimal scale selection for multi-scale decision tables. Int. J. Approx. Reason. **54**, 1107–1129 (2013)
10. Zhang, X., Miao, D., Liu, C., Le, M.: Constructive methods of rough approximation operators and multigranulation rough sets. Knowl. Based Syst. **91**, 114–125 (2016)
11. Skowron, A., Stepaniuk, J.: Tolerance approximation spaces. Fundam. Inform. **27**, 245–253 (1996)
12. Qin, K., Yang, J., Pei, Z.: Generalized rough sets based on reflexive and transitive relations. Inf. Sci. **178**, 4138–4141 (2008)
13. Li, Z.W., Xie, T.S., Li, Q.G.: Topological structure of generalized rough sets. Comput. Math. Appl. **63**, 1066–1071 (2012)
14. Yao, Y.Y.: Relational interpretations of neighborhood operators and rough set approximation operators. Inf. Sci. **111**, 239–259 (1998)
15. Wu, W., Zhang, W.: Constructive and axiomatic approaches of fuzzy approximation operators. Inf. Sci. **159**, 233–254 (2004)

16. Li, T., Zhang, W.X.: Rough fuzzy approximations on two universes of discourse. Inf. Sci. **178**, 892–906 (2008)
17. Qian, Y., Liang, J., Yao, Y., Dang, C.: MGRS: a multi-granulation rough set. Inf. Sci. **180**, 949–970 (2010)
18. Sun, B.Z., Ma, W.M.: Multigranulation rough set theory over two universes. J. Intell. Fuzzy Syst. **28**, 1251–1269 (2015)
19. Dick, S., Schenker, A., Pedrycz, W., Kandel, A.: Regranulation: a granular algorithm enabling communication between granular worlds. Inf. Sci. **177**, 408–435 (2007)
20. Wang, C., Wu, C.X., Chen, D., Hu, Q., Wu, C.: Communicating between information systems. Inf. Sci. **178**, 3228–3239 (2008)
21. Wanga, C., Wu, C.X., Chen, D., Du, W.: Some properties of relation information systems under homomorphisms. Appl. Math. Lett. **21**, 940–945 (2008)
22. Wang, C., Chen, D., Sun, B., Hu, Q.: Communication between information systems with covering based rough sets. Inf. Sci. **216**, 17–33 (2012)
23. Wu, W.Z., Zhang, W.X.: Constructive and axiomatic approaches of fuzzy approximation operators. Inf. Sci. **159**, 233–254 (2004)
24. Yao, Y.Y.: Constructive and algebraic methods of the theory of rough sets. Inf. Sci. **109**, 21–47 (1998)

Rough Sets Based on Possible Indiscernibility Relations in Incomplete Information Tables with Continuous Values

Michinori Nakata[1(✉)], Hiroshi Sakai[2], and Keitarou Hara[3]

[1] Faculty of Management and Information Science, Josai International University, 1 Gumyo, Togane, Chiba 283-8555, Japan
nakatam@ieee.org

[2] Department of Mathematics and Computer Aided Sciences, Faculty of Engineering, Kyushu Institute of Technology, Tobata, Kitakyushu 804-8550, Japan
sakai@mns.kyutech.ac.jp

[3] Department of Informatics, Tokyo University of Information Sciences, 4-1 Onaridai, Wakaba-ku, Chiba 265-8501, Japan
hara@rsch.tuis.ac.jp

Abstract. Rough sets under incomplete information with continuous domains are examined on the basis of possible world semantics. We show an approach under possible indiscernibility relations, although the traditional approaches are done under possible tables. This is because the number of possible indiscernibility relations is finite, even if the number of possible tables is infinite. First, lower and upper approximations are described using the indiscernibility relation on an attribute in a complete information table. Second, these are addressed in an incomplete information table under possible world semantics. Two types of indiscernibility relations; namely, certain and possible ones, are obtained on an attribute in an information table. The actual indiscernibility relation is one of possible ones. The family of indiscernibility relations is a lattice for inclusion. The minimal element is the certain indiscernibility relation while the maximal one is the maximal possible indiscernibility relation. By using certain and possible indiscernibility relations, we obtain four types of approximations: certain lower, certain upper, possible lower, and possible upper approximations. The approach based on possible world semantics gives the same approximations as ones obtained from our extended approach, which is proposed in the previous work directly using indiscernibility relations.

Keywords: Neighborhood rough sets · Possible world semantics · Incomplete information · Possible indiscernibility relations · Lower and upper approximations · Continuous values

© Springer Nature Switzerland AG 2019
T. Mihálydeák et al. (Eds.): IJCRS 2019, LNAI 11499, pp. 155–165, 2019.
https://doi.org/10.1007/978-3-030-22815-6_13

1 Introduction

Big data consists of various types of data. When we focus on string data, data is broadly classified into discrete and continuous data.

Rough sets, constructed by Pawlak [18], are used as an effective method for data mining. The framework is usually applied to complete information tables with nominal attributes and creates fruitful results in various fields. However, attributes taking continuous values frequently appear, when we describe properties of an object in our daily life. Furthermore, incomplete information ubiquitously exists in daily life. We cannot sufficiently utilize information obtained from our daily life unless we deal with continuous and incomplete information. Therefore, extended versions of rough sets are proposed to deal with incomplete information in continuous domains.

An approach, which is most frequently used [7,20–22], is to use the way applied to nominal attributes by Kryszkiewicz [8]. The approach fixes the indiscernibility of an object with incomplete information with another object. However, it is natural that an object characterized by incomplete information has two possibilities; namely, the object is indiscernible with another object and not so. To fix the indiscernibility corresponds to taking into consideration only one of the two possibilities. Therefore, the approach creates poor results and information loss occurs [12,19]. Furthermore, the fixing is not compatible with the approach by Lipski in the field of incomplete databases [9,10], because Lipski handles all possibilities of objects with incomplete information.

Another is to directly use indiscernibility relations that are extended to deal with incomplete information [15]. Yet another is to use possible classes obtained from the indiscernibility relation on a set of attributes [16]. These approaches have the same order of computational complexity as the one in complete information. However, no justification is shown, although it is known in discrete data that these give the same results as the approach based on possible world semantics [13]. To give these approaches a correctness criterion, it is required to develop an approach based on possible world semantics. The approaches so far under possible world semantics use possible tables derived from an incomplete information table. Unfortunately, infinite possible tables can be derived from an incomplete information table with continuous values. Possible world semantics is unavailable as long as to use possible tables.

Rough sets are based on the indiscernibility relation on a set of attributes. The number of possible indiscernibility relations is finite, even if the number of possible tables is infinite. We focus on possible indiscernibility relations, not possible tables. In this work, we develop an approach based on possible world semantics by using possible indiscernibility relations in an incomplete information table with continuous values.

The paper is organized as follows. In Sect. 2, an approach directly using indiscernibility relations is described in a complete information table. In Sect. 3, we develop an approach in an incomplete information table under possible world semantics. In Sect. 4, conclusions are addressed.

2 Rough Sets by Using Indiscernibility Relations in Complete Information Systems with Continuous Values

A data set is represented as a two-dimensional table, called an information table. In the information table, each row and each column represent an object and an attribute, respectively. A mathematical model of an information table with complete information is called a complete information system. The complete information system is a triplet expressed by $(U, AT, \{D(a_i) \mid a_i \in AT\})$. U is a non-empty finite set of objects, which is called the universe. AT is a non-empty finite set of attributes such that $a_i : U \to D(a_i)$ for every $a_i \in AT$ where $D(a_i)$ is the continuous domain of attribute a_i.

We have two approaches for dealing with attributes taking continuous values. One approach is to discretize a continuous domain into disjunctive intervals in which objects are regarded as indiscernible [4]. The discretization has a heavy influence over results. The other approach is to use neighborhood [11]. The indiscernibility of two objects is derived from the distance of them. When the distance of the two objects is less than or equal to a given threshold, they are regarded as indiscernible. Results gradually change as the threshold changes. Thus, we adopt the latter approach.

Binary relation R_{a_i} expressing indiscernibility of objects on attribute $a_i \in AT$ is called the indiscernibility relation for a_i:

$$R_{a_i} = \{(o, o') \in U \times U \mid |a_i(o) - a_i(o')| \leq \delta\}, \tag{1}$$

where $a_i(o)$ is the value for attribute a_i of object o and δ is a threshold that denotes a range in which $a_i(o)$ is indiscernible with $a_i(o')$.

Proposition 1
If $\delta 1 \leq \delta 2$, then $R^{\delta 1}_{a_i} \subseteq R^{\delta 2}_{a_i}$, where $R^{\delta 1}_{a_i}$ and $R^{\delta 2}_{a_i}$ are the indiscernibility relations on attribute a_i with thresholds $\delta 1$ and $\delta 2$, respectively.

From the indiscernibility relation, indiscernible class $[o]_{a_i}$ for object o is obtained:

$$[o]_{a_i} = \{o' \mid (o, o') \in R_{a_i}\}. \tag{2}$$

Directly using indiscernibility relation R_{a_i}, lower approximation $\underline{apr}_{a_i}(\mathcal{O})$ and upper approximation $\overline{apr}_{a_i}(\mathcal{O})$ for a_i of set $\mathcal{O}$ of objects are:

$$\underline{apr}_{a_i}(\mathcal{O}) = \{o \mid \forall o' \in U\ (o, o') \notin R_{a_i} \vee o' \in \mathcal{O}\}, \tag{3}$$

$$\overline{apr}_{a_i}(\mathcal{O}) = \{o \mid \exists o' \in U\ (o, o') \in R_{a_i} \wedge o' \in \mathcal{O}\}. \tag{4}$$

Proposition 2 [15]
If $\delta 1 \leq \delta 2$, then $\underline{apr}^{\delta 1}_{a_i}(\mathcal{O}) \supseteq \underline{apr}^{\delta 2}_{a_i}(\mathcal{O})$ and $\overline{apr}^{\delta 1}_{a_i}(\mathcal{O}) \subseteq \overline{apr}^{\delta 2}_{a_i}(\mathcal{O})$, where $\underline{apr}^{\delta 1}_{a_i}(\mathcal{O})$ and $\overline{apr}^{\delta 1}_{a_i}(\mathcal{O})$ are lower and upper approximations under threshold $\delta 1$ and $\underline{apr}^{\delta 2}_{a_i}(\mathcal{O})$ and $\overline{apr}^{\delta 2}_{a_i}(\mathcal{O})$ are lower and upper approximations under threshold $\delta 2$.

For object o in the lower approximation of $\mathcal{O}$, all objects with which o is indiscernible are included in $\mathcal{O}$; namely, $[o]_{a_i} \subseteq \mathcal{O}$. On the other hand, for an object in the upper approximation of $\mathcal{O}$, some objects with which o is indiscernible are in $\mathcal{O}$; namely, $[o]_{a_i} \cap \mathcal{O} \neq \emptyset$. Thus, $\underline{apr}_{a_i}(\mathcal{O}) \subseteq \overline{apr}_{a_i}(\mathcal{O})$.

3 Rough Sets by Possible Indiscernibility Relations in Incomplete Information Systems with Continuous Domains

An information table with incomplete information is called an incomplete information system. In incomplete information systems, $a_i : U \rightarrow s_{a_i}$ for every $a_i \in AT$ where s_{a_i} is the set of values over domain $D(a_i)$ of attribute a_i or the set of intervals on $D(a_i)$. Single value v with $v \in a_i(o)$ or $v \subseteq a_i(o)$ is a possible value that may be the actual one as the value of attribute a_i in object o. The possible value is the actual one if $a_i(o)$ is a single value.

We have two kinds of indiscernibility relations from an incomplete information table[1]. One is the certain indiscernibility relation. The others are possible indiscernibility relations. Certain indiscernibility relation CR_{a_i} is:

$$CR_{a_i} = \{(o, o') \in U \times U \mid (o = o') \vee (\forall u \in a_i(o) \forall v \in a_i(o') |u - v| \leq \delta)\}. \quad (5)$$

In this binary relation that is unique, two objects o and o' of $(o, o') \in CR_{a_i}$ are certainly indiscernible with each other on a_i. Such a pair is called a certain pair. On the other hand, we have lots of possible indiscernibility relations. The number of possible indiscernibility relations grows exponentially as the number of values with incomplete information increases. Family $\mathcal{F}(R_{a_i})$ of possible indiscernibility relations is:

$$\mathcal{F}(R_{a_i}) = \{e \mid e = CR_{a_i} \cup e' \wedge e' \in \mathcal{P}(MPPR_{a_i})\}, \quad (6)$$

where each element is a possible indiscernibility relation and $\mathcal{P}(MPPR_{a_i})$ is the power set of $MPPR_{a_i}$ and $MPPR_{a_i}$ is:

$$\begin{aligned} MPPR_{a_i} &= \{\{(o', o), (o, o')\} | (o', o) \in MPR_{a_i}\}, \\ MPR_{a_i} &= \{(o, o') \in U \times U \mid \exists u \in a_i(o) \exists v \in a_i(o') |u - v| \leq \delta)\} \backslash CR_{a_i}. \quad (7) \end{aligned}$$

A pair of objects that is included in MPR_{a_i} is called a possible pair. $\mathcal{F}(R_{a_i})$ is a lattice for set inclusion. CR_{a_i} is the minimum possible indiscernibility relation in $\mathcal{F}(R_{a_i})$, which is the minimal element, whereas $CR_{a_i} \cup MPR_{a_i}$ is the maximum possible indiscernibility relation, which is the maximal element. One of possible indiscernibility relations is the actual indiscernibility relation, although we cannot know it without additional information.

[1] For the sake of simplicity and space limitation, We describe the case of an attribute, although our approach can be easily extended to the case of more than one attribute.

Example 1

T

U	a_1	a_2
o_1	1.74	{6.21,6.27}
o_2	[1.77,1.81]	[6.43,6.49]
o_3	1.84	6.47
o_4	[1.87,1.97]	6.52
o_5	{1.69,1.71}	{6.28,6.35}

In incomplete information table T, let threshold δ be 0.05 on attribute a_1. The set of certain pairs of indiscernible objects on a_1 is:

$$\{(o_1,o_1),(o_1,o_5),(o_2,o_2),(o_3,o_3),(o_4,o_4),(o_5,o_5),(o_5,o_1)\}.$$

The set of possible pairs of indiscernible objects is:

$$\{(o_1,o_2),(o_2,o_1),(o_2,o_3),(o_3,o_2),(o_3,o_4),(o_4,o_3)\}.$$

Using formulae (5)–(7), the family of possible indiscernibility relations and each possible indiscernibility relation pr_i with $i = 1, \dots, 8$ are:

$$\begin{aligned}
\mathcal{F}(R_{a_1}) &= \{pr_1, \cdots, pr_8\},\\
pr_1 &= \{(o_1,o_1),(o_1,o_5),(o_2,o_2),(o_3,o_3),(o_4,o_4),(o_5,o_5),(o_5,o_1)\},\\
pr_2 &= \{(o_1,o_1),(o_1,o_5),(o_2,o_2),(o_3,o_3),(o_4,o_4),(o_5,o_5),(o_5,o_1),\\
&\quad (o_1,o_2),(o_2,o_1)\},\\
pr_3 &= \{(o_1,o_1),(o_1,o_5),(o_2,o_2),(o_3,o_3),(o_4,o_4),(o_5,o_5),(o_5,o_1),\\
&\quad (o_2,o_3),(o_3,o_2)\},\\
pr_4 &= \{(o_1,o_1),(o_1,o_5),(o_2,o_2),(o_3,o_3),(o_4,o_4),(o_5,o_5),(o_5,o_1),\\
&\quad (o_3,o_4),(o_4,o_3)\},\\
pr_5 &= \{(o_1,o_1),(o_1,o_5),(o_2,o_2),(o_3,o_3),(o_4,o_4),(o_5,o_5),(o_5,o_1),\\
&\quad (o_1,o_2),(o_2,o_1),(o_2,o_3),(o_3,o_2)\},\\
pr_6 &= \{(o_1,o_1),(o_1,o_5),(o_2,o_2),(o_3,o_3),(o_4,o_4),(o_5,o_5),(o_5,o_1),\\
&\quad (o_1,o_2),(o_2,o_1),(o_3,o_4),(o_4,o_3)\},\\
pr_7 &= \{(o_1,o_1),(o_1,o_5),(o_2,o_2),(o_3,o_3),(o_4,o_4),(o_5,o_5),(o_5,o_1),\\
&\quad (o_2,o_3),(o_3,o_2),(o_3,o_4),(o_4,o_3)\},\\
pr_8 &= \{(o_1,o_1),(o_1,o_5),(o_2,o_2),(o_3,o_3),(o_4,o_4),(o_5,o_5),(o_5,o_1),\\
&\quad (o_1,o_2),(o_2,o_1),(o_2,o_3),(o_3,o_2),(o_3,o_4),(o_4,o_3)\}.
\end{aligned}$$

These possible indiscernibility relations have the following lattice structure for inclusion:

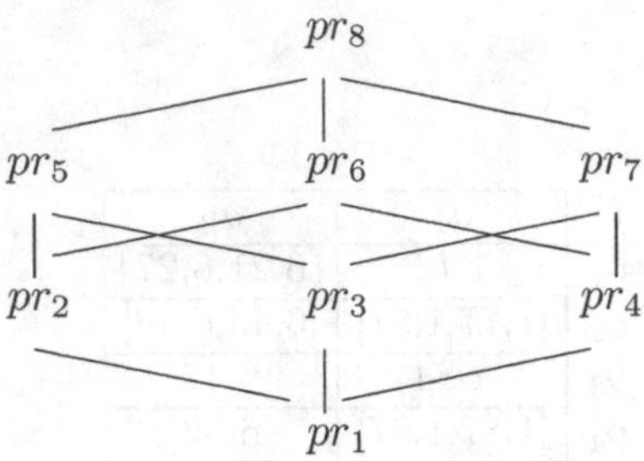

pr_1 is the minimal element, whereas pr_8 is the maximal element.

We develop an approach based on possible indiscernibility relations in an incomplete information table with continuous domains. We apply the formulae (3) and (4) of lower and upper approximations to every possible indiscernibility relations.

Proposition 3
If $pr_k \subseteq pr_l$ for possible indiscernibility relations $pr_k, pr_l \in \mathcal{F}(R_{a_i})$, then $\underline{apr}_{a_i}(\mathcal{O})^{pr_k} \supseteq \underline{apr}_{a_i}(\mathcal{O})^{pr_l}$ and $\overline{apr}_{a_i}(\mathcal{O})^{pr_k} \subseteq \overline{apr}_{a_i}(\mathcal{O})^{pr_l}$.

This proposition shows that the sets of lower and upper approximations under possible indiscernibility relations are also lattices for set inclusion.

We aggregate the lower and upper approximations under possible indiscernibility relations. Let $\mathcal{O}$ be a set of objects. Certain lower approximation $C\underline{apr}_{a_i}(\mathcal{O})$ is:

$$C\underline{apr}_{a_i}(\mathcal{O}) = \{o \mid \forall pr \in \mathcal{F}(R_{a_i}) o \in \underline{apr}_{a_i}(\mathcal{O})^{pr}\}, \tag{8}$$

where $\underline{apr}_{a_i}(\mathcal{O})^{pr}$ is the lower approximation derived from possible indiscernibility relation pr. Certain upper approximation $C\underline{apr}_{a_i}(\mathcal{O})$ is:

$$C\overline{apr}_{a_i}(\mathcal{O}) = \{o \mid \forall pr \in \mathcal{F}(R_{a_i}) o \in \overline{apr}_{a_i}(\mathcal{O})^{pr}\}, \tag{9}$$

where $\overline{apr}_{a_i}(\mathcal{O})^{pr}$ is the upper approximation derived from possible indiscernibility relation pr. Possible lower approximation $P\underline{apr}_{a_i}(\mathcal{O})$ is:

$$P\underline{apr}_{a_i}(\mathcal{O}) = \{o \mid \exists pr \in \mathcal{F}(R_{a_i}) o \in \underline{apr}_{a_i}(\mathcal{O})^{pr}\}. \tag{10}$$

Possible upper approximation $P\overline{apr}_{a_i}(\mathcal{O})$ is:

$$P\overline{apr}_{a_i}(\mathcal{O}) = \{o \mid \exists pr \in \mathcal{F}(R_{a_i}) o \in \overline{apr}_{a_i}(\mathcal{O})^{pr}\}. \tag{11}$$

From Proposition 3, these approximations are transformed into the following formulae:

$$C\underline{apr}_{a_i}(\mathcal{O}) = \underline{apr}_{a_i}(\mathcal{O})^{pr_{max}}, \tag{12}$$

$$C\overline{apr}_{a_i}(\mathcal{O}) = \overline{apr}_{a_i}(\mathcal{O})^{pr_{min}}, \tag{13}$$
$$P\underline{apr}_{a_i}(\mathcal{O}) = \underline{apr}_{a_i}(\mathcal{O})^{pr_{min}}, \tag{14}$$
$$P\overline{apr}_{a_i}(\mathcal{O}) = \overline{apr}_{a_i}(\mathcal{O})^{pr_{max}}, \tag{15}$$

where pr_{min} and pr_{max} are the minimal and maximal possible indiscernibility relations. These formulae show that we can obtain the four approximations without the computational complexity in the number of possible indiscernibility relations.

Example 2
We go back to Example 1. Let set $\mathcal{O}$ of objects be $\{o_2, o_3, o_4\}$. Using formulae (3) and (4), lower and upper approximations from each possible indiscernibility relation are:

$$\underline{apr}_{a_1}(\mathcal{O})^{pr_1} = \{o_2, o_3, o_4\}, \overline{apr}_{a_1}(\mathcal{O})^{pr_1} = \{o_2, o_3, o_4\},$$
$$\underline{apr}_{a_1}(\mathcal{O})^{pr_2} = \{o_3, o_4\}, \overline{apr}_{a_1}(\mathcal{O})^{pr_2} = \{o_1, o_2, o_3, o_4\},$$
$$\underline{apr}_{a_1}(\mathcal{O})^{pr_3} = \{o_2, o_3, o_4\}, \overline{apr}_{a_1}(\mathcal{O})^{pr_3} = \{o_2, o_3, o_4\},$$
$$\underline{apr}_{a_1}(\mathcal{O})^{pr_4} = \{o_2, o_3, o_4\}, \overline{apr}_{a_1}(\mathcal{O})^{pr_4} = \{o_2, o_3, o_4\},$$
$$\underline{apr}_{a_1}(\mathcal{O})^{pr_5} = \{o_3, o_4\}, \overline{apr}_{a_1}(\mathcal{O})^{pr_5} = \{o_1, o_2, o_3, o_4\},$$
$$\underline{apr}_{a_1}(\mathcal{O})^{pr_6} = \{o_3, o_4\}, \overline{apr}_{a_1}(\mathcal{O})^{pr_6} = \{o_1, o_2, o_3, o_4\},$$
$$\underline{apr}_{a_1}(\mathcal{O})^{pr_7} = \{o_2, o_3, o_4\}, \overline{apr}_{a_1}(\mathcal{O})^{pr_7} = \{o_2, o_3, o_4\},$$
$$\underline{apr}_{a_1}(\mathcal{O})^{pr_8} = \{o_3, o_4\}, \overline{apr}_{a_1}(\mathcal{O})^{pr_8} = \{o_1, o_2, o_3, o_4\}.$$

By using formulae (12)–(15),

$$C\underline{apr}_{a_1}(\mathcal{O}) = \{o_3, o_4\},$$
$$C\overline{apr}_{a_1}(\mathcal{O}) = \{o_2, o_3, o_4\},$$
$$P\underline{apr}_{a_1}(\mathcal{O}) = \{o_2, o_3, o_4\},$$
$$P\overline{apr}_{a_1}(\mathcal{O}) = \{o_1, o_2, o_3, o_4\}.$$

As with the case of nominal attributes [13], the following proposition holds.

Proposition 4
$C\underline{apr}_{a_i}(\mathcal{O}) \subseteq P\underline{apr}_{a_i}(\mathcal{O}) \subseteq \mathcal{O} \subseteq C\overline{apr}_{a_i}(\mathcal{O}) \subseteq P\overline{apr}_{a_i}(\mathcal{O})$.

Using the four approximations denoted by formulae (12)–(15), lower and upper approximations are expressed in interval sets, as is described in [14][2]:

$$\underline{apr}^{\bullet}_{a_i}(\mathcal{O}) = [C\underline{apr}_{a_i}(\mathcal{O}), P\underline{apr}_{a_i}(\mathcal{O})], \tag{16}$$
$$\overline{apr}^{\bullet}_{a_i}(\mathcal{O}) = [C\overline{apr}_{a_i}(\mathcal{O}), P\overline{apr}_{a_i}(\mathcal{O})]. \tag{17}$$

[2] Hu and Yao also say that approximations are described by using an interval set in information tables with incomplete information [5].

Certain and possible approximations are the lower and upper bounds of the actual approximation. The two approximations $\underline{apr}^{\bullet}_{a_i}(\mathcal{O})$ and $\overline{apr}^{\bullet}_{a_i}(\mathcal{O})$ depend on each other; namely, the complementarity property $\underline{apr}^{\bullet}_{a_i}(\mathcal{O}) = U - \overline{apr}^{\bullet}_{a_i}(U - \mathcal{O})$ linked with them holds, as is so in complete information systems.

Example 3
Using four approximations in Example 2, from formulae (16) and (17),

$$\underline{apr}^{\bullet}_{a_1}(\mathcal{O}) = [\{o_3, o_4\}, \{o_2, o_3, o_4\}],$$
$$\overline{apr}^{\bullet}_{a_1}(\mathcal{O}) = [\{o_2, o_3, o_4\}, \{o_1, o_2, o_3, o_4\}].$$

Furthermore, the following proposition is valid from formulae (12)–(15).

Proposition 5

$$C\underline{apr}_{a_i}(\mathcal{O}) = \{o \mid \forall o' \in U \; (o, o') \notin (CR_{a_i} \cup MPR_{a_i}) \vee o' \in \mathcal{O}\},$$
$$C\overline{apr}_{a_i}(\mathcal{O}) = \{o \mid \exists o' \in U \; (o, o') \in CR_{a_i} \wedge o' \in \mathcal{O}\},$$
$$P\underline{apr}_{a_i}(\mathcal{O}) = \{o \mid \forall o' \in U \; (o, o') \notin CR_{a_i} \vee o' \in \mathcal{O}\},$$
$$P\overline{apr}_{a_i}(\mathcal{O}) = \{o \mid \exists o' \in U \; (o, o') \in (CR_{a_i} \cup MPR_{a_i}) \wedge o' \in \mathcal{O}\}.$$

This proposition shows that our extended approach directly using indiscernibility relations [15] is justified. Namely, results from the extended approach directly using indiscernibility relations are the same as the ones from possible world semantics. A criterion for justification is formally represented as

$$q(R_{a_i}) = \bigodot q'(\mathcal{F}(R_{a_i})),$$

where q' is the approach for complete information, which is described in Sect. 2, and q is an extended approach of q', which directly deals with incomplete information, $\bigodot$ is an aggregate operator, and $\mathcal{F}(R_{a_i})$ is the set of possible indiscernibility relations from the original indiscernibility relation R_{a_i} under possible world semantics. This is also schematized in Fig. 1.

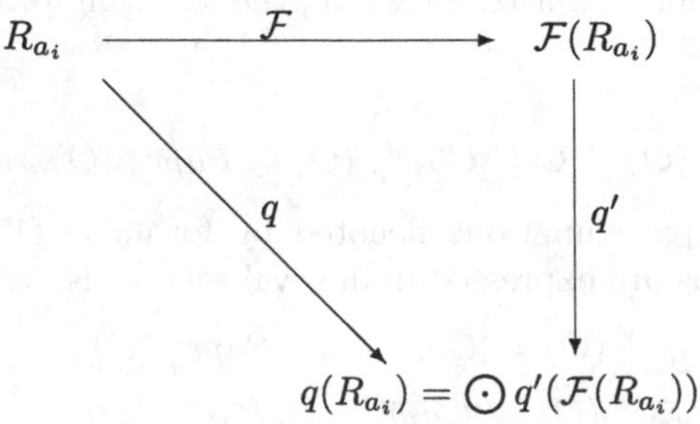

Fig. 1. Correctness criterion of extended method q

This type of correctness criterion is usually used in the field of databases dealing with incomplete information [1–3,6,17,23].

When objects in $\mathcal{O}$ are specified by nominal attribute a_j with incomplete information, $\mathcal{O}$ is specified by using an element in domain $D(a_j)$. In the case where $\mathcal{O}$ is specified by restriction $a_j = x$ with $x \in D(a_j)$, four approximations: certain lower, certain upper, possible lower, and possible upper ones, are:

$$Capr_{\underline{a_i}}(\mathcal{O}) = \underline{apr}_{a_i}(C\mathcal{O}_{a_j=x})^{pr_{max}}, \tag{18}$$
$$C\overline{apr}_{a_i}(\mathcal{O}) = \overline{apr}_{a_i}(C\mathcal{O}_{a_j=x})^{pr_{min}}, \tag{19}$$
$$Papr_{\underline{a_i}}(\mathcal{O}) = \underline{apr}_{a_i}(P\mathcal{O}_{a_j=x})^{pr_{min}}, \tag{20}$$
$$P\overline{apr}_{a_i}(\mathcal{O}) = \overline{apr}_{a_i}(P\mathcal{O}_{a_j=x})^{pr_{max}}. \tag{21}$$

where

$$C\mathcal{O}_{a_j=x} = \{o \in \mathcal{O} \mid a_j(o) = x\}, \tag{22}$$
$$P\mathcal{O}_{a_j=x} = \{o \in \mathcal{O} \mid a_j(o) \supseteq x\}. \tag{23}$$

When we describe the case where $o \in \mathcal{O}$ is specified by numerical attribute a_j with incomplete information. Set $\mathcal{O}$ is specified by an interval where precise values of a_j are used.

$$Capr_{\underline{a_i}}(\mathcal{O}) = \underline{apr}_{a_i}(C\mathcal{O}_{[a_j(o_m),a_j(o_n)]})^{pr_{max}}, \tag{24}$$
$$C\overline{apr}_{a_i}(\mathcal{O}) = \overline{apr}_{a_i}(C\mathcal{O}_{[a_j(o_m),a_j(o_n)]})^{pr_{min}}, \tag{25}$$
$$Papr_{\underline{a_i}}(\mathcal{O}) = \underline{apr}_{a_i}(P\mathcal{O}_{[a_j(o_m),a_j(o_n)]})^{pr_{min}}, \tag{26}$$
$$P\overline{apr}_{a_i}(\mathcal{O}) = \overline{apr}_{a_i}(P\mathcal{O}_{[a_j(o_m),a_j(o_n)]})^{pr_{max}}, \tag{27}$$

where

$$C\mathcal{O}_{[a_j(o_m),a_j(o_n)]} = \{o \in \mathcal{O} \mid a_j(o) \subseteq [a_j(o_m), a_j(o_n)]\}, \tag{28}$$
$$P\mathcal{O}_{[a_j(o_m),a_j(o_n)]} = \{o \in \mathcal{O} \mid a_j(o) \cap [a_j(o_m), a_j(o_n)] \neq \emptyset\}, \tag{29}$$

where $a_j(o_m)$ and $a_j(o_n)$ are precise and $a_j(o_m) \leq a_j(o_n)$.

Example 4
In incomplete information table T of Example 1, let $\mathcal{O}$ be specified by values $a_2(o_3)$ and $a_2(o_4)$. Using formulae (28) and (29),

$$C\mathcal{O}_{[a_2(o_3),a_2(o_4)]} = \{o_3, o_4\},$$
$$P\mathcal{O}_{[a_2(o_3),a_2(o_4)]} = \{o_2, o_3, o_4\}.$$

Possible indiscernibility relations pr_{min} and pr_{max} on a_1 is pr_1 and pr_8 in Example 1. Using formulae (24)–(27),

$$Capr_{\underline{a_i}}(\mathcal{O}) = \{o_4\},$$
$$C\overline{apr}_{a_i}(\mathcal{O}) = \{o_3, o_4\},$$
$$Papr_{\underline{a_i}}(\mathcal{O}) = \{o_2, o_3, o_4\},$$
$$P\overline{apr}_{a_i}(\mathcal{O}) = \{o_1, o_2, o_3, o_4\}.$$

4 Conclusions

We have described rough sets based on possible world semantics in information tables with continuous domains. First, we have dealt with complete information tables. Rough sets are obtained from directly using the indiscernibility relation on an attribute. Second, we have dealt with incomplete information tables under possible world semantics.

In incomplete information tables, we focus on that the number of possible indiscernibility relations is finite, although the number of possible tables is infinite. The family of possible indiscernibility relations is expressed by a lattice having the minimal and maximal elements. The families of lower and upper approximations that are derived from each possible indiscernibility relation are also a lattice for inclusion. The number of possible indiscernibility relations increases exponentially as the number of attribute values with incomplete information grows. However, approximations are obtained by using the minimal and the maximal possible indiscernibility relations. Therefore, we have no difficulty of computational complexity. By using the minimal and the maximal possible indiscernibility relations, four types approximations: certain lower, certain upper, possible lower, and possible upper approximations are obtained, as is so in incomplete information tables with nominal attributes. These approximations are the same as those obtained from an extended approach directly using indiscernibility relations. Therefore, the approach based on possible world semantics gives justification of the extended approach.

Acknowledgment. The authors wish to thank the anonymous reviewers for their valuable comments.

References

1. Abiteboul, S., Hull, R., Vianu, V.: Foundations of Databases. Addison-Wesley Publishing Company, Reading (1995)
2. Bosc, P., Duval, L., Pivert, O.: An initial approach to the evaluation of possibilistic queries addressed to possibilistic databases. Fuzzy Sets Syst. **140**, 151–166 (2003)
3. Grahne, G. (ed.): The Problem of Incomplete Information in Relational Databases. LNCS, vol. 554. Springer, Heidelberg (1991). https://doi.org/10.1007/3-540-54919-6
4. Grzymala-Busse, J.W.: Mining numerical data – a rough set approach. In: Peters, J.F., Skowron, A. (eds.) Transactions on Rough Sets XI. LNCS, vol. 5946, pp. 1–13. Springer, Heidelberg (2010). https://doi.org/10.1007/978-3-642-11479-3_1
5. Hu, M.J., Yao, Y.Y.: Rough set approximations in an incomplete information table. In: Polkowski, L., et al. (eds.) IJCRS 2017, Part II. LNCS (LNAI), vol. 10314, pp. 200–215. Springer, Cham (2017). https://doi.org/10.1007/978-3-319-60840-2_14
6. Imielinski, T., Lipski, W.: Incomplete information in relational databases. J. ACM **31**, 761–791 (1984)
7. Jing, S., She, K., Ali, S.: A universal neighborhood rough sets model for knowledge discovering from incomplete heterogeneous data. Expert Syst. **30**(1), 89–96 (2013). https://doi.org/10.1111/j.1468-0394.2012.00633_x

8. Kryszkiewicz, M.: Rules in incomplete information systems. Inf. Sci. **113**, 271–292 (1999)
9. Lipski, W.: On semantics issues connected with incomplete information databases. ACM Trans. Database Syst. **4**, 262–296 (1979)
10. Lipski, W.: On databases with incomplete information. J. ACM **28**, 41–70 (1981)
11. Lin, T.Y.: Neighborhood systems: a qualitative theory for fuzzy and rough sets. In: Wang, P. (ed.) Advances in Machine Intelligence and Soft Computing, vol. IV, pp. 132–155. Duke University (1997). https://doi.org/10.1007/11548669_34
12. Nakata, M., Sakai, H.: Applying rough sets to information tables containing missing values. In: Proceedings of 39th International Symposium on Multiple-Valued Logic, pp. 286–291. IEEE Press (2009). https://doi.org/10.1109/ISMVL.2009.1
13. Nakata, M., Sakai, H.: Twofold rough approximations under incomplete information. Int. J. Gen. Syst. **42**, 546–571 (2013). https://doi.org/10.1080/17451000.2013.798898
14. Nakata, M., Sakai, H.: Describing rough approximations by indiscernibility relations in information tables with incomplete information. In: Carvalho, J.P., et al. (eds.) IPMU 2016, Part II. CCIS, vol. 611, pp. 355–366. Springer, Cham (2016). https://doi.org/10.1007/978-3-319-40581-0_29
15. Nakata, M., Sakai, H., Hara, K.: Rules induced from rough sets in information tables with continuous values. In: Medina, J., et al. (eds.) IPMU 2018, Part II. CCIS, vol. 854, pp. 490–502. Springer, Cham (2018). https://doi.org/10.1007/978-3-319-91476-3_41
16. Nakata, M., Sakai, H., Hara, K.: Rule induction based on indiscernible classes from rough sets in information tables with continuous values. In: Nguyen, H.S., Ha, Q.-T., Li, T., Przybyła-Kasperek, M. (eds.) IJCRS 2018. LNCS (LNAI), vol. 11103, pp. 323–336. Springer, Cham (2018). https://doi.org/10.1007/978-3-319-99368-3_25
17. Paredaens, J., De Bra, P., Gyssens, M., Van Gucht, D.: The Structure of the Relational Database Model. Springer, Heidelberg (1989). https://doi.org/10.1007/978-3-642-69956-6
18. Pawlak, Z.: Rough Sets: Theoretical Aspects of Reasoning About Data. Kluwer Academic Publishers, Dordrecht (1991). https://doi.org/10.1007/978-94-011-3534-4
19. Stefanowski, J., Tsoukiàs, A.: Incomplete information tables and rough classification. Comput. Intell. **17**, 545–566 (2001)
20. Yang, X., Zhang, M., Dou, H., Yang, Y.: Neighborhood systems-based rough sets in incomplete information system. Inf. Sci. **24**, 858–867 (2011). https://doi.org/10.1016/j.knosys.2011.03.007
21. Zenga, A., Lia, T., Liuc, D., Zhanga, J., Chena, H.: A fuzzy rough set approach for incremental feature selection on hybrid information systems. Fuzzy Sets Syst. **258**, 39–60 (2015). https://doi.org/10.1016/j.fss.2014.08.014
22. Zhao, B., Chen, X., Zeng, Q.: Incomplete hybrid attributes reduction based on neighborhood granulation and approximation. In: 2009 International Conference on Mechatronics and Automation, pp. 2066–2071. IEEE Press (2009)
23. Zimányi, E., Pirotte, A.: Imperfect information in relational databases. In: Motro, A., Smets, P. (eds.) Uncertainty Management in Information Systems: From Needs to Solutions, pp. 35–87. Kluwer Academic Publishers (1997)

The Prototype View of Concepts

Ruisi Ren and Ling Wei[(✉)]

School of Mathematics, Northwest University,
Xi'an 710069, People's Republic of China
ruisiren_rose@163.com, wl@nwu.edu.cn

Abstract. Concepts are important and basic elements in human's cognition process. The formal concept gives a mathematical format of the classical view of concepts in which all instances of a concept share common properties. But in some situation this view is not consistent with human's understanding of concepts. The prototype view of concepts is more appropriate in our daily life. This view characters some analog categories as internally structured into a prototype (clearest cases, best examples of the category) and non-prototype members, with non-prototype members tending toward an order from better to poorer examples. The objective of this paper is to give a mathematical description of prototype view of concepts. Firstly, we give a similarity measurement of an object to another object in a formal context. Then based on this similarity measurement, the mathematical format of prototype view of concepts, named k-cutting concept, induced by one typical object is obtained. Finally, the properties of k-cutting concepts are studied. In addition to presenting theorems to summarize our results, we use some examples to illustrate the main ideas.

Keywords: Prototype view of concepts · Similarity measurement · k-cutting concepts · Object concepts

1 Introduction

Concepts are important and basic constituents in human's cognition process. Consequently, they are crucial in many psychological processes, such as categorization, inference, memory, learning, and decision-making. In philosophy, there are different views or structures of concepts. In classical view, a concept contains two parts, extension and intension. The extension is a group of objects belonging to the concept and the intension is a family of attributes characterizing the properties of the concept. The classical view holds that all instances of a concept share common properties, which are necessary and sufficient conditions for defining the concept. In order to apply the philosophical concept into data processing, Wille [22] proposed a new field, formal concept analysis (FCA), giving a mathematical format of the classical view of concepts.

FCA [22] shows a mathematical format of classical view of concepts, named formal concepts. A formal concept consists of a pair of an object set (extent) and

© Springer Nature Switzerland AG 2019
T. Mihálydeák et al. (Eds.): IJCRS 2019, LNAI 11499, pp. 166–178, 2019.
https://doi.org/10.1007/978-3-030-22815-6_14

an attribute set (intent). The objects in extent possess all the attributes in intent and the attributes in intent are possessed by all the objects in extent. Based on the partial order theory, Wille and Ganter [8] presented a lattice structure of formal concepts named a concept lattice which reveals hierarchical structure of concepts with respect to the generalization and the specialization of concepts. However, the formal concept is an all-or-none phenomenon. That is, if an object possesses all the attributes in the intent of a formal concept, it is definitely in the extent of this formal concept, but if an object does not possess all the attributes in the intent, even though it possesses most attributes in the intent, this object is definitely not in this formal concept. In other words, if two objects are in the extent of same concept, they must have same degree of typicality in this formal concept. That is, the objects in the extent of a concept are equally important in people's understanding of the concept. This view of concepts is mostly used in machine-oriented concept learning [1,10,12,25,27], but not always consistent with human's understanding of concepts. Classical formal concepts have been extended to other types, such as preconcepts [23], semiconcepts [24], protoconcepts [21], property oriented concepts [4], object oriented concepts [26], dual concepts [2,13], monotone concepts [3], RS-definable concepts [28] and three-way concepts [15–17].

There is increasing evidence that memberships of objects in semantic categories which are expressed by words of natural languages can be graded rather than all-or-none. Lakoff [11], Rosch [19] and Zadeh [29] argued that some natural categories are analog and must be represented logically in a manner which reflects their analog structure. Rosch [19] has further characterized some natural analog categories as internally structured into a prototype (clearest cased, best examples of the category) and non-prototype members, with non-prototype members tending toward an order from better to poorer examples. For example, *chair* is a more reasonable exemplar than *radio* of the concept *furniture*, or we can say that the *chair* has a larger membership than *radio* of the concept *furniture*. When we talk about color, *vermilion*, *fuchsia*, *pink*, *cerise*, *peach*, *garnet*, *cardinal*, *rose*, *wine* all belong to concept *red*. However, *rose* is more typical than *pink*. This kind of view of concepts are called prototype view of concepts.

In this paper, we try to give a mathematical representation of the prototype view of concepts [6,7]. Considering the cognitive process of recognizing concepts, we firstly choose an object as the prototype of a concept, which is the most typical object and can be a representative of this concept. Then the similarities between other objects and prototype are given according to a similarity measurement. Since the prototype is described by a group of attributes [9], the similarity measurement is defined based on the description of objects. The objects with high similarity to the prototype can be put into the concept. In order to quantitatively define high similarity, we preset a threshold k and the corresponding prototype view concepts are called k-cutting concepts. Since prototype o is the most typical object of this concept, the description of prototype o is regarded as the intent of this k-cutting concept and the objects whose similarity to prototype o is bigger than k are put into the extent of this k-cutting concept. Furthermore we study the properties of k-cutting concepts.

The rest of the paper is organized as follows. Section 2 gives the basic notions in formal concept analysis. Then Sect. 3 presents the similarity measurement between two different objects and defines the k-cutting concept. Furthermore we show the properties of k-cutting concepts. Finally, this paper is concluded in Sect. 4.

2 Formal Concept Analysis

This section reviews basic notions in FCA. FCA, proposed by Wille in 1982 [22], gives a mathematical way to represent a concept with a pair of objects set (called the extent) and attributes set (called the intent). The data source of FCA is called formal context defined as follows [8,22].

Definition 1. *A formal context* $(OB, AT, \mathbf{I})$ *consists of two sets* OB *and* AT*, and a relation* $\mathbf{I}$ *between* OB *and* AT*. The elements of* OB *are called the objects and the elements of* AT *are called the attributes of the context. In order to express that an object* o *is in a relation* $\mathbf{I}$ *with an attribute* a*, we write* $o\mathbf{I}a$ *or* $(o, a) \in \mathbf{I}$ *and read it as "the object* o *has the attribute* a*".*

Based on the formal context, the set of attributes possessed by an object o and the set of objects possessing an attribute a are given as

$$\begin{aligned} o\mathbf{I}. &= \{a \in AT \mid o\mathbf{I}a\} \subseteq AT, \\ .\mathbf{I}a &= \{o \in OB \mid o\mathbf{I}a\} \subseteq OB. \end{aligned} \tag{1}$$

Actually, $o\mathbf{I}.$ can be regarded as the description of object o and $.\mathbf{I}a$ can be understood as a set of objects which can be described by attribute a or a set of representatives of description $\{a\}$. Given a formal context $(OB, AT, \mathbf{I})$, if for any $o \in OB$, we have $o\mathbf{I}. \neq \emptyset$, $o\mathbf{I}. \neq AT$, and for any $a \in AT$, we have $.\mathbf{I}a \neq \emptyset$, $.\mathbf{I}a \neq OB$, then the formal context $(OB, AT, \mathbf{I})$ is called canonical. If for any objects $o_1, o_2 \in OB$, from $o_1\mathbf{I}. = o_2\mathbf{I}.$, it always follows that $o_1 = o_2$ and, consequently, $.\mathbf{I}a_1 = .\mathbf{I}a_2$ implies $a_1 = a_2$ for all $a_1, a_2 \in AT$. We call this context a clarified formal context. In this paper, we suppose all formal contexts are canonical, clarified and finite. Based on the description of an object and the representatives of an attribute, a pair of operators called derivation operators are defined on an objects set $O \subseteq OB$ and an attributes set $A \subseteq AT$, respectively, in $(OB, AT, \mathbf{I})$ [8]:

$$\begin{aligned} O^* &= \{a \in AT \mid \forall o \in O(o\mathbf{I}a)\} = \{a \in AT \mid O \subseteq .\mathbf{I}a\} = \bigcap\{o\mathbf{I}. \mid o \in O\}, \\ A^* &= \{o \in OB \mid \forall a \in A(o\mathbf{I}a)\} = \{o \in OB \mid A \subseteq o\mathbf{I}.\} = \bigcap\{.\mathbf{I}a \mid a \in A\}. \end{aligned} \tag{2}$$

It is obvious to see that, for any object $o \in OB$ and any attribute $a \in AT$, it always follows $o\mathbf{I}. = \{o\}^*$ and $.\mathbf{I}a = \{a\}^*$. Then based on above derivation operators, a formal concept is obtained [8].

Definition 2. *A formal concept of the context* $(OB, AT, \mathbf{I})$ *is a pair* (O, A) *with* $O^* = A$ *and* $O = A^*$ $(O \subseteq OB, A \subseteq AT)$. *We call* O *the extent and* A *the intent of the formal concept* (O, A).

The formal concepts of a formal context $(OB, AT, \mathbf{I})$ are ordered by

$$(O_1, A_1) \leq (O_2, A_2) \Leftrightarrow O_1 \subseteq O_2 \ (\Leftrightarrow A_1 \supseteq A_2). \tag{3}$$

All formal concepts of $(OB, AT, \mathbf{I})$ can form a complete lattice called the formal concept lattice of $(OB, AT, \mathbf{I})$, denoted by $L(OB, AT, \mathbf{I})$. The infimum and supremum are given by

$$(O_1, A_1) \wedge (O_2, A_2) = (O_1 \cap O_2, (A_1 \cup A_2)^{**}),$$
$$(O_1, A_1) \vee (O_2, A_2) = ((O_1 \cup O_2)^{**}, A_1 \cap A_2) \tag{4}$$

In a formal context, there is a kind of important concept, named object concept [8].

Definition 3. *Let* $(OB, AT, \mathbf{I})$ *be a formal context,* (o^{**}, o^*) *is a formal concept for all* $o \in OB$*, which is called an object concept. Here, for convenience, we write* o^* *instead of* $\{o\}^*$ *for any* $o \in OB$.

The object concept (o^{**}, o^*) can be understood as a concept induced by object o, which means the object o is a typical object (prototype) of concept (o^{**}, o^*). Specifically, the description (intent) of concept (o^{**}, o^*) is the description of object o and the extent of this concept is a set of objects which can be described by the description of object o. In order to show the importance of the object concept, the notion of join-dense is recalled in next definition [8].

Definition 4. *Let* P *be an ordered set and let* $Q \subseteq P$. *Then* Q *is called join-dense in* P *if for every element* $a \in P$ *there is a subset* A *of* Q *such that* $a = \vee_P A$.

Following theorem shows that any formal concept can be constructed based on a set of object concepts, so the object concepts can be regarded as the fundamental elements in concept construction [8].

Theorem 1. *Let* $(OB, AT, \mathbf{I})$ *be a formal context and* $L(OB, AT, \mathbf{I})$ *the associated complete lattice of concepts. Then the set of all the object concepts is join-dense in* $L(OB, AT, \mathbf{I})$. *Specifically, for a formal concept* (O, A),

$$\bigvee \{(o^{**}, o^*) \mid o \in O\} = (O, A) \tag{5}$$

holds.

Finally, we give an example to illustrate the definitions and theorems presented in this section.

Table 1. A formal context $(OB, AT, \mathbf{I})$

OB	a	b	c	d
o_1	1	0	0	1
o_2	0	1	0	1
o_3	1	1	1	0
o_4	0	1	1	0

Example 1. Table 1 is a formal context $(OB, AT, \mathbf{I})$ with four objects $OB = \{o_1, o_2, o_3, o_4\}$ and four attributes $AT = \{a, b, c, d\}$. The description of every object and the representatives of every attribute are as follows:

$$o_1\mathbf{I}. = \{a, d\},\ o_2\mathbf{I}. = \{b, d\},\ o_3\mathbf{I}. = \{a, b, c\},\ o_4\mathbf{I}. = \{b, c\}.$$
$$.\mathbf{I}a = \{o_1, o_3\},\ .\mathbf{I}b = \{o_2, o_3, o_4\},\ .\mathbf{I}c = \{o_3, o_4\},\ .\mathbf{I}d = \{o_1, o_2\}.$$

We can see that for any object $o_i \in OB$, its description is neither whole attribute set AT nor the empty set. Also, for any attribute in AT, its representatives set is neither whole object set OB nor the empty set. Thus the formal context $(OB, AT, \mathbf{I})$ is canonical. Moreover, for any two different objects, their descriptions are different, and for any two different attributes, their representatives sets are different. Thus the formal context $(OB, AT, \mathbf{I})$ is clarified.

The formal concept lattice of context $(OB, AT, \mathbf{I})$ is shown in Fig. 1. The object concepts are: $(o_1^{**}, o_1^*) = (o_1, ad)$, $(o_2^{**}, o_2^*) = (o_2, bd)$, $(o_3^{**}, o_3^*) = (o_3, abc)$, $(o_4^{**}, o_4^*) = (o_3o_4, bc)$. After calculation, we have

$$\begin{aligned}
&(o_2o_3o_4, b) = (o_2, bd) \vee (o_3, abc) \vee (o_3o_4, bc),\\
&(o_1o_3, a) = (o_1, ad) \vee (o_3, abc),\\
&(o_1o_2, d) = (o_1, ad) \vee (o_2, bd),\\
&(OB, \emptyset) = (o_1, ad) \vee (o_2, bd) \vee (o_3, abc) \vee (o_3o_4, bc).
\end{aligned}$$

That is, any formal concept can be constructed by joining a set of object concepts. Thus, the set of all object concepts is join-dense in formal concept lattice.

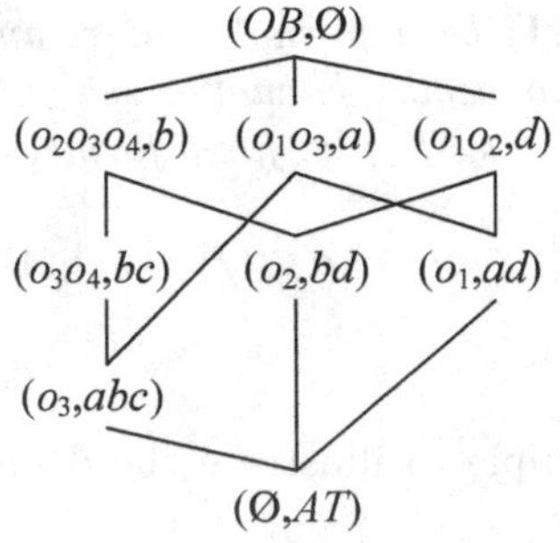

Fig. 1. The formal concept lattice $L(OB, AT, \mathbf{I})$

3 The Prototype View of Concept

Section 2 shows the importance of object concepts in concept construction. However, the definition of object concepts is too strict. According to Definition 3, the extent of an object concept (o^{**}, o^*) is a set of objects which can be fully described by the description of object o. Actually, the semantic concept in our daily life is based on a typical object (prototype), but the extent of semantic concept is not required to be fully described by the description of the typical object. Based on the similarity [5,18,20] to the typical object, the typicality of objects in extent can be defined. In this section, we give a mathematical way to represent the semantic concept and discuss its properties. Firstly, the similarity measurement of one object to another is shown in Sect. 3.1.

3.1 Similarity Measurement Between Two Objects

In a formal context $(OB, AT, \mathbf{I})$, an object o can be described by a set of attributes $o\mathbf{I}.$ (called description of object o). And if two objects have same description, they can be regarded as same one [8]. Thus, in order to measure the similarity of object o_i to object o, we only need to measure the similarity between the descriptions of these two objects. The more similar the descriptions of two objects are, the more similar the two objects are.

Definition 5. *Let $(OB, AT, \mathbf{I})$ be a formal context and o be a reference object. For any object $o_i \in OB$, the similarity measurement of o_i to o is defined as*

$$Sim(o_i, o) = \frac{|o_i\mathbf{I}. \cap o\mathbf{I}.|}{|o\mathbf{I}.|}. \tag{6}$$

The range of the value of this similarity measurement is $0 \leq Sim(o_i, o) \leq 1$. The closer the similarity is to 1, the more similar object o_i is to object o; the closer the similarity is to 0, the less similar object o_i is to object o. In other words, the more attributes in description of object o can be used to describe object o_i, the more similar object o_i is to object o. Now let us consider the similarity of objects in the extent of an object concept (o^{**}, o^*) to the object o.

Proposition 1. *Let $(OB, AT, \mathbf{I})$ be a formal context and (o^{**}, o^*) be an object concept induced by object o. Then the value of similarity measurement of any object in o^{**} to object o is* 1.

Proof. Suppose (o^{**}, o^*) is an object concept of formal context $(OB, AT, \mathbf{I})$ and $o_i \in o^{**}$ is an object from the extent of this concept. Since $o_i \in o^{**}$, according to the properties of operator $*$, we have $o^{***} \subseteq o_i^*$ and $o^* = o^{***}$. Thus, we obtain $o^* \subseteq o_i^*$. That is, $o\mathbf{I}. \subseteq o_i\mathbf{I}$. Hence we have $o_i\mathbf{I}. \cap o\mathbf{I}. = o\mathbf{I}$. Thus, $Sim(o_i, o) = \frac{|o_i\mathbf{I}. \cap o\mathbf{I}.|}{|o\mathbf{I}.|} = \frac{|o\mathbf{I}.|}{|o\mathbf{I}.|} = 1$.

By Proposition 1, the extent o^{**} of object concept (o^{**}, o^*) consists of objects whose value of similarity measurement to object o is 1. If we regard object o as a typical object (prototype) of an semantic concept, in order to get all the objects in this prototype view of concepts, we should not only consider objects with similarity value of 1, but also objects with similarity value less than 1.

We use a simple example to illustrate the basic notions and ideas introduced so far.

Example 2. (Continued with Example 1) We set object o_3 as a reference object. In the following, we will compute the similarity of each object in OB to reference object o_3.

$$Sim(o_1, o_3) = \frac{|o_1\mathbf{I}. \cap o_3\mathbf{I}.|}{|o_3\mathbf{I}.|} = \frac{|\{a,d\} \cap \{a,b,c\}|}{|\{a,b,c\}|} = \frac{1}{3},$$
$$Sim(o_2, o_3) = \frac{|o_2\mathbf{I}. \cap o_3\mathbf{I}.|}{|o_3\mathbf{I}.|} = \frac{|\{b,d\} \cap \{a,b,c\}|}{|\{a,b,c\}|} = \frac{1}{3},$$
$$Sim(o_3, o_3) = \frac{|o_3\mathbf{I}. \cap o_3\mathbf{I}.|}{|o_3\mathbf{I}.|} = \frac{|\{a,b,c\} \cap \{a,b,c\}|}{|\{a,b,c\}|} = 1,$$
$$Sim(o_4, o_3) = \frac{|o_4\mathbf{I}. \cap o_3\mathbf{I}.|}{|o_3\mathbf{I}.|} = \frac{|\{b,c\} \cap \{a,b,c\}|}{|\{a,b,c\}|} = \frac{2}{3}.$$

Then, we check the correctness of Proposition 1. In Example 1, we get four object concepts (o_1, ad), (o_2, bd), (o_3, abc) and (o_3o_4, bc) inducing by objects o_1, o_2, o_3 and o_4, respectively. The similarities of objects in extent of object concepts to the objects inducing these concepts are computed as follows:

$$Sim(o_1, o_1) = \frac{|o_1\mathbf{I}. \cap o_1\mathbf{I}.|}{|o_1\mathbf{I}.|} = \frac{|\{a,d\} \cap \{a,d\}|}{|\{a,d\}|} = 1,$$
$$Sim(o_2, o_2) = \frac{|o_2\mathbf{I}. \cap o_2\mathbf{I}.|}{|o_2\mathbf{I}.|} = \frac{|\{b,d\} \cap \{b,d\}|}{|\{b,d\}|} = 1,$$
$$Sim(o_3, o_3) = \frac{|o_3\mathbf{I}. \cap o_3\mathbf{I}.|}{|o_3\mathbf{I}.|} = \frac{|\{a,b,c\} \cap \{a,b,c\}|}{|\{a,b,c\}|} = 1,$$
$$\begin{cases} Sim(o_3, o_4) = \frac{|o_3\mathbf{I}. \cap o_4\mathbf{I}.|}{|o_4\mathbf{I}.|} = \frac{|\{a,b,c\} \cap \{b,c\}|}{|\{b,c\}|} = \frac{|\{b,c\}|}{|\{b,c\}|} = 1, \\ Sim(o_4, o_4) = \frac{|o_4\mathbf{I}. \cap o_4\mathbf{I}.|}{|o_4\mathbf{I}.|} = \frac{|\{b,c\} \cap \{b,c\}|}{|\{b,c\}|} = 1. \end{cases}$$

The computation results are consistent to Proposition 1. That is, the value of similarity of objects in extent of object concept to the object inducing this concept is 1.

3.2 The *k*-cutting Concept Induced by One Typical Object

The classical view of concepts holds that all instances of a concept share common properties that are necessary and sufficient conditions for defining the concept [14]. However, in our daily life, semantic category in our nature language is not an all-or-none phenomenon. For example, we know that a *chair* is a more reasonable exemplar of the category *furniture* than a *radio*. In other words, the *chair* is more typical than *radio* in the category *furniture*. This is contrary to the assumption that categories are necessarily logical, bounded entities, Rosch [19] has characterized some natural analog categories as internally structured into a prototype (clearest cases, best example of the category) and nonprototype members, with nonprototype members tending toward an order from better to poorer example. Based on the results of Rosch's study, we summarize the process of human to recognize a semantic concept as follows:

step 1: *Pick up the typical object (prototype) of the concept*;

step 2: *Calculate the characterized attributes (description) of the typical object*;

step 3: *Calculate the similarity of each object to the typical object*;

step 4: *Put the objects with high similarity into the extent of concept.*

The above steps are just a qualitative description of process to obtain semantic concepts. If we want to express this process in a mathematical way, some quantitative index is needed. For example, in step 4, *high* is a qualitative description. In order to determine which object has high similarity to the typical object, we give a preset threshold k. If the similarity measurement of an object to typical object is bigger than k, this object can be regarded as being highly similar to typical object. Thus, this object can be put into the extent and the corresponding concept is called the k-cutting concept. Since the objects in one extent belong to the same concept, they should possess some common attributes with each other. Thus we strict that k should satisfy $k > \frac{1}{2}$. The following use of k satisfies these settings.

The above process is easy for us to understand, but it is hard to give a mathematical definition of k-cutting concept directly. In the following, we show the mathematical definition of the k-cutting concept. Firstly, a pair of k-cutting derivation operators are given as follows.

Definition 6. *Let* $(OB, AT, \mathbf{I})$ *be a formal context. A pair of* k*-cutting derivation operators* ($k > \frac{1}{2}$) *for objects set* $O \subseteq OB$ *and attributes set* $A \subseteq AT$ *are defined as:*

$$O^{*k} = \{a \in AT \mid |a^* \cap O| \geq k \cdot |O|\},$$
$$A^{*k} = \{o \in OB \mid |o^* \cap A| \geq k \cdot |A|\} \tag{7}$$

In Definition 6, the attribute shared by more than $k \cdot |O|$ objects in O belongs to attributes set O^{*k}; the object possessing more than $k \cdot |A|$ attributes in A belongs to objects set A^{*k}. In the following, we present the properties of k-cutting derivation operators.

Property 1. Let $(OB, AT, \mathbf{I})$ be a formal context. The following properties hold for any objects sets $O, O_1, O_2 \subseteq OB$ and attributes sets $A, A_1, A_2 \subseteq AT$:

$$
\begin{aligned}
&G1.\ \emptyset^{*k} = AT, \text{ when } \emptyset \subseteq OB,\\
&\quad\ \emptyset^{*k} = OB, \text{ when } \emptyset \subseteq AT;\\
&G2.\ O^{*1} = O^*, \quad A^{*1} = A^*;\\
&G3.\ O^{*k} = O^* \text{ when } O \text{ is a singleton set},\\
&\quad\ A^{*k} = A^* \text{ when } A \text{ is a singleton set};\\
&G4.\ k \leq h \Rightarrow O^{*h} \subseteq O^{*k},\\
&\quad\ k \leq h \Rightarrow A^{*h} \subseteq A^{*k};\\
&G5.\ O^{*k} = \cup_{k \leq k_i} O^{*k_i} = \cap_{k_j \leq k} O^{*k_j},\\
&\quad\ A^{*k} = \cup_{k \leq k_i} A^{*k_i} = \cap_{k_j \leq k} A^{*k_j};\\
&G6.\ O \subseteq O^{**k}, \quad A \subseteq A^{**k}.
\end{aligned}
$$

Proof. The results in $G1$ and $G2$ are obvious.

$G3$. If object set O is a singleton set, then there exists an object $o_i \in OB$ satisfying $O = \{o_i\}$. Since $\{o_i\}$ is a singleton set, the result of $|a^* \cap \{o_i\}|$, for any $a \in AT$, is either 0 or 1. According to Definition 6, for any $k > \frac{1}{2}$, we have $O^{*k} = \{a \in AT \mid |a^* \cap O| \geq k \cdot |O|\} = \{a \in AT \mid |a^* \cap \{o_i\}| \geq k|\{o_i\}|\} = \{a \in AT \mid |a^* \cap \{o_i\}| \geq k\}$. That is, if attribute $a \in O^{*k}$, then $|a^* \cap \{o_i\}| \geq k$. That means $|a^* \cap \{o_i\}| = 1$. Thus $O^{*k} = \{a \in AT \mid |a^* \cap \{o_i\}| = 1\} = \{a \in AT \mid \{o_i\} \subseteq a^*\} = \{o_i\}^* = O^*$. The rest part can be proved similarly.
$G4$. For any $a \in O^{*h}$, according to Definition 6, we have $|a^* \cap O| \geq h \cdot |O|$. Since $k \leq h$, we have $h \cdot |O| \geq k \cdot |O|$. Thus, $|a^* \cap O| \geq h \cdot |O| \geq k \cdot |O|$. That is, $a \in O^{*k}$. Because of the arbitrariness of attribute a, we obtain $O^{*h} \subseteq O^{*k}$. The formula $k \leq h \Rightarrow A^{*h} \subseteq A^{*k}$ can be proved similarly.
$G5$. From property $G4$, for any $k_i \geq k$, we have $O^{*k_i} \subseteq O^{*k}$. Hence, we obtain $\cup_{k \leq k_i} O^{*k_i} \subseteq O^{*k}$. Also, since $k \leq k$, we can get $O^{*k} \subseteq \cup_{k \leq k_i} O^{*k_i}$. Thus, we obtain $O^{*k} = \cup_{k \leq k_i} O^{*k_i}$. Analogously, from property $G4$, for any $k_j \leq k$, we have $O^{*k} \subseteq O^{*k_j}$. Hence, we obtain $O^{*k} \subseteq \cap_{k_j \leq k} O^{*k_j}$. Also, since $k \leq k$, we can get $\cap_{k_j \leq k} O^{*k_j} \subseteq O^{*k}$. Thus, we obtain $O^{*k} = \cap_{k_j \leq k} O^{*k_j}$. The rest part $A^{*k} = \cup_{k \leq k_i} A^{*k_i} = \cap_{k_j \leq k} A^{*k_j}$ can be proved similarly.
$G6$. For any $o_i \in O$, from property of operator $*$, we have $O^* \subseteq o_i^*$. Consequently, we obtain $o_i^* \cap O^* = O^*$. Thus, $|o_i^* \cap O^*| = |O^*| \geq k|O^*|$ holds no matter what value k has. According to Definition 6, we can get $o_i \in O^{**k}$. The rest part $A \subseteq A^{**k}$ can be proved similarly.

Then based on the k-cutting derivation operators, the definition of k-cutting concept induced by one prototype (we will simply call it k-cutting concept if there is no confusion) is given as follows.

Definition 7. *Let $(OB, AT, \mathbf{I})$ be a formal context. The k-cutting concept induced by one typical object o is defined as $(\hat{o}^{**k}, o^*)$. $\hat{o}^{**k}$ and o^* are called extent and intent of k-cutting concept $(\hat{o}^{**k}, o^*)$. Here, $\hat{o}^{**k} = (o_i, m(o_i))$, $o_i \in o^{**k}$ is a set of objects in o^{**k} accompanied with a membership value.*

Specifically, the element of $\hat{o}^{**k}$ is an object-membership pair $(o_i, m(o_i))$, in which $o_i \in o^{**k}$ and $m(o_i)$ is the membership of object o_i belonging to the k-cutting concept $(\hat{o}^{**k}, o^*)$. The membership can be measured in different ways, and the most common way is using the similarity measurement value of object o_i to typical object o. In following analysis, for convenience, we can regard o^{**k} instead of $\hat{o}^{**k}$ as an extent of k-cutting concept $(\hat{o}^{**k}, o^*)$. That is, we can rewrite k-cutting concept $(\hat{o}^{**k}, o^*)$ as (o^{**k}, o^*).

From Definition 7, the intent of the k-cutting concept (o^{**k}, o^*) is a set of attributes which is the description of typical object o and the extent of the concept is a set of object-membership pairs in which the description of object contains more than $k \cdot |o^*|$ attributes in description of o. The set of all k-cutting concepts induced by one typical object in formal context $(OB, AT, \mathbf{I})$ is denoted by $OCC_k(OB, AT, \mathbf{I})$. Now we check the similarity of any object in k-cutting concept given in Definition 7 to verify its rationality.

Theorem 2. *Let $(OB, AT, \mathbf{I})$ be a formal context and (o^{**k}, o^*) is a k-cutting concept. An object $o_i \in OB$ belongs to o^{**k} if and only if the similarity of o_i to o is bigger than k. That is, $Sim(o_i, o) \geq k$.*

Proof. According to Definition 6, we have $o_i \in o^{**k}$ is equivalent to $|o_i^* \cap o^*| \geq k \cdot |o^*|$. Since we assumed that the formal context in this paper is canonical, we have $o^* \neq \emptyset$, that is, $|o^*| \neq 0$. Thus, both sides of the inequality $|o_i^* \cap o^*| \geq k \cdot |o^*|$ can be divided by $|o^*|$. The result is $\frac{|o_i^* \cap o^*|}{|o^*|} \geq k$. That is, $Sim(o_i, o) \geq k$. Thus, $o_i \in o^{**k}$ holds if and only if $Sim(o_i, o) \geq k$ holds.

Remark 1. The higher the value of similarity the object has, the closer it is to the typical object. However, the similarity measure can not be used to decide whether or not the object is a prototype or typical object. That is, for some object, the value of similarity measurement is 1, but it is not a prototype of this concept, since it has more attributes than the attributes in intent.

At the beginning of Sect. 3.2, we discussed that since all the objects in one extent belong to a same concept, they should possess some common attributes with each other. Hence, we restrict the value of $k > \frac{1}{2}$. The following proposition shows that the restriction of k guarantees the existence of common attributes of a concept.

Proposition 2. *Let (o^{**k}, o^*) be a k-cutting concept. If $k > \frac{1}{2}$, then, for any $o_1, o_2 \in o^{**k}$, we have $o_1^* \cap o_2^* \cap o^* \neq \emptyset$.*

Proof. Because of $o_1, o_2 \in o^{**k}$, based on Definition 6, we have $|o_1^* \cap o^*| \geq k \cdot |o^*|$ and $|o_2^* \cap o^*| \geq k \cdot |o^*|$. Since we suppose $k > \frac{1}{2}$, the formulas $|o_1^* \cap o^*| > \frac{|o^*|}{2}$ and $|o_2^* \cap o^*| > \frac{|o^*|}{2}$ can be obtained. Thus, $|o_1^* \cap (o_2^* \cap o^*)| = |(o_1^* \cap o^*) \cap (o_2^* \cap o^*)| = |(o_1^* \cap o^*)| + |(o_2^* \cap o^*)| - |(o_1^* \cap o^*) \cup (o_2^* \cap o^*)| > \frac{|o^*|}{2} + \frac{|o^*|}{2} - |(o_1^* \cap o^*) \cup (o_2^* \cap o^*)| > |o^*| - |o^*| = 0$. That is, $o_1^* \cap o_2^* \cap o^* \neq \emptyset$.

The result $o_1^* \cap o_2^* \cap o^* \neq \emptyset$ in Proposition 2 can be rewritten as $(o_1^* \cap o^*) \cap (o_2^* \cap o^*) \neq \emptyset$. This proposition shows that in order to let objects in the same concept have common attributes, the value of k should satisfy $k > \frac{1}{2}$. These common attributes are the most important characters of the concept, since they can reflect the commonness of objects in extent.

We continue with Example 2 to demonstrate the ideas of k-cutting concept induced by one typical object and to verify the correctness of Theorem 2 and Proposition 2.

Example 3. According to Definition 7, the $\frac{2}{3}$-cutting concepts induced by typical object o_1, o_2, o_3, and o_4 are $(\{(o_1, 1)\}, ad)$, $(\{(o_2, 1)\}, bd)$, $(\{(o_3, 1), (o_4, \frac{2}{3})\}, abc)$, and $(\{(o_3, 1), (o_4, 1)\}, bc)$. Compared with the classical object concept induced by object o_3 whose extent only contains object o_3, the $\frac{2}{3}$-cutting concepts induced by object o_3 contains objects o_3 and o_4. And these two objects have different memberships. Since object o_3 is the prototype of this concept, its membership is 1. The membership of object o_4 is $\frac{2}{3}$. The $\frac{2}{3}$-cutting concept can be regarded as a more general concept than the classical concept. The object in k-cutting concept induced by object o does not need to possess all the attributes in description of object o. We only restrict that the description of any object in extent of k-cutting concept contains more than $k \cdot |o|$ attributes in the description of prototype o.

We use $(\{(o_3, 1), (o_4, \frac{2}{3})\}, abc)$, the $\frac{2}{3}$-cutting concept induced by typical object o_3, as an example to show the correctness of Theorem 2. According to the similarity measurement calculated in Example 2, we have $Sim(o_1, o_3) = \frac{1}{3} < \frac{2}{3}$, $Sim(o_2, o_3) = \frac{1}{3} < \frac{2}{3}$, $Sim(o_3, o_3) = 1 \geq \frac{2}{3}$ and $Sim(o_4, o_3) = \frac{2}{3} \geq \frac{2}{3}$. Based on Theorem 2, only objects o_3 and o_4 belong to the extent of $\frac{2}{3}$-cutting concept $(\{(o_3, 1), (o_4, \frac{2}{3})\}, abc)$, which is consistent with our calculation by Definition 7. Also, from $(\{(o_3, 1), (o_4, 1)\}, bc)$, the $\frac{2}{3}$-cutting concept induced by typical object o_4, we can see that the similarity of object o_3 to typical object o_4 is 1, but object o_3 is not the typical object of this concept, since its description is $\{a, b, c\}$ which is bigger than $\{b, c\}$, description of typical object o_4.

We will check the correctness of Proposition 2 in the following. Since $k = \frac{2}{3} > \frac{1}{2}$, every two objects in $o_3^{**\frac{2}{3}}$ should have common attributes. Based on Table 1, we have $o_3^{**\frac{2}{3}} = \{o_3, o_4\}$, and we can calculate $(o_3^* \cap o_3^*) \cap (o_4^* \cap o_3^*) = \{a, b, c\} \cap \{b, c\} = \{b, c\} \neq \emptyset$. The results are consistent to Proposition 2.

4 Conclusion

In formal concept analysis, the formal concept is a mathematical formation of the classical view of concept and reflects a semantic meaning "commonly possessing". However, in our daily life, the prototype view of concepts is more common and just reflects the meaning of "mostly possessing". In this paper, we discussed the similarity between two objects and defined the mathematical formation of the prototype of concepts, named k-cutting concepts. Moreover, the properties of this newly proposed concept are studied and its rationality is discussed.

The results of this paper suggest several future research topics. It is interesting to investigate the structure of k-cutting concepts and the k-cutting concepts can be generalized as the k-cutting concepts induced by a group of typical objects.

Acknowledgement. The authors gratefully acknowledge the support of the Natural Science Foundation of China (No.61772021 and No.11371014).

References

1. Boucher-Ryan, P., Bridge, D.: Collaborative recommending using formal concept analysis. Knowl.-Based Syst. **19**(5), 309–315 (2006)
2. Chen, Y.H., Yao, Y.Y.: A multiview approach for intelligent data analysis based on data operators. Inf. Sci. **178**(1), 1–20 (2008)
3. Deogun, J.S., Saquer, J.: Monotone concepts for formal concept analysis. Discrete Appl. Math. **144**(1), 70–78 (2004)
4. Düntsch, I., Gediga, G.: Modal-style operators in qualitative data analysis. In: Proceedings of the 2002 IEEE International Conference on Data Mining, pp. 155–162. IEEE Computer Society, Washington, D.C. (2002)
5. Formica, A.: Concept similarity in formal concept analysis: an information content approach. Knowl.-Based Syst. **21**(1), 80–87 (2008)
6. Frixione, M., Lieto, A.: Representing concepts in formal ontologies: compositionality vs. typicality effects. Log. Log. Philos. **21**, 391–414 (2012)
7. Frixione, M., Lieto, A.: Towards an extended model of conceptual representations in formal ontologies: a typicality-based proposal. J. Univ. Comput. Sci. **20**(3), 257–276 (2014)
8. Ganter, B., Wille, R.: Formal Concept Analysis: Mathematical Foundations. Springer, Heidelberg (1999). https://doi.org/10.1007/978-3-642-59830-2
9. Hu, M., Yao, Y.: Definability in incomplete information tables. In: Flores, V., et al. (eds.) IJCRS 2016. LNCS (LNAI), vol. 9920, pp. 177–186. Springer, Cham (2016). https://doi.org/10.1007/978-3-319-47160-0_16
10. Jiang, G.Q., Chute, C.G.: Auditing the semantic completeness of SNOMED CT using formal concept analysis. J. Am. Med. Inform. Assoc. **16**(1), 89–102 (2009)
11. Lakoff, G.: Hedges: a study in meaning criteria and the logic of fuzzy concepts. In: Hockney, D., Harper, W., Freed, B. (eds.) Contemporary Research in Philosophical Logic and Linguistic Semantics, vol. 4, pp. 221–27. Springer, Netherlands (1975). https://doi.org/10.1007/978-94-010-1756-5_9
12. Li, W., Wei, L.: Data dimension reduction based on concept lattices in image mining. In: Proceedings of the Sixth IEEE International Conference on Fuzzy Systems and Knowledge Discovery, vol. 5, pp. 369–373 (2009)
13. Ma, J.M., Zhang, W.X.: Axiomatic characterizations of dual concept lattices. Int. J. Approx. Reason. **54**(5), 690–697 (2013)
14. Medin, D.L., Smith, E.E.: Concepts and concept formation. Ann. Rev. Psychol. **35**(1), 113–138 (1984)
15. Qi, J.J., Qian, T., Wei, L.: The connections between three-way and classical concept lattices. Knowl.-Based Syst. **91**, 143–151 (2016)
16. Qi, J., Wei, L., Yao, Y.: Three-way formal concept analysis. In: Miao, D., Pedrycz, W., Ślęzak, D., Peters, G., Hu, Q., Wang, R. (eds.) RSKT 2014. LNCS (LNAI), vol. 8818, pp. 732–741. Springer, Cham (2014). https://doi.org/10.1007/978-3-319-11740-9_67

17. Ren, R.S., Wei, L.: The attribute reductions of three-way concept lattices. Knowl.-Based Syst. **99**, 92–102 (2016)
18. Resnik, P.: Using information content to evaluate semantic similarity in a taxonomy. In: Proceedings of the International Joint Conference on Artificial Intelligence, pp. 448–453. Morgan Kaufmann, San Francisco (1995)
19. Rosch, E.H.: On the internal structure of perceptual and semantic categories. In: Moore, T.E. (ed.) Cognitive Development and the Aquisition of Language, pp. 111–144. Academic Press, New York (1973)
20. Tversky, A.: Features of similarity. Psychol. Rev. **84**(4), 327–352 (1977)
21. Vormbrock, B., Wille, R.: Semiconcept and protoconcept algebras: the basic theorems. In: Ganter, B., Stumme, G., Wille, R. (eds.) Formal Concept Analysis. LNCS (LNAI), vol. 3626, pp. 34–48. Springer, Heidelberg (2005). https://doi.org/10.1007/11528784_2
22. Wille, R.: Restructuring lattice theory: an approach based on hierarchies of concepts. In: Rival, I. (ed.) Ordered Sets, pp. 445–470. Reidel Publishing Company, Dordrecht (1982)
23. Wille, R.: Preconcept algebras and generalized double boolean algebras. In: Eklund, P. (ed.) ICFCA 2004. LNCS (LNAI), vol. 2961, pp. 1–13. Springer, Heidelberg (2004). https://doi.org/10.1007/978-3-540-24651-0_1
24. Wille, R.: Concept lattices and conceptual knowledge systems. Comput. Math. Appl. **23**(6–9), 493–515 (1992)
25. Xiao, Q.Z., Qin, K., Guan, Z.Q., Wu, T.: Image mining for robot vision based on concept analysis. In: IEEE International Conference on Robotics and Biomimetics (ROBIO 2007), pp. 207–212 (2007)
26. Yao, Y.: A comparative study of formal concept analysis and rough set theory in data analysis. In: Tsumoto, S., Słowiński, R., Komorowski, J., Grzymała-Busse, J.W. (eds.) RSCTC 2004. LNCS (LNAI), vol. 3066, pp. 59–68. Springer, Heidelberg (2004). https://doi.org/10.1007/978-3-540-25929-9_6
27. Yao, Y.Y.: Interpreting concept learning in cognitive informatics and granular computing. IEEE Trans. Syst. Man Cybern. Part B (Cybern.) **39**(4), 855–866 (2009)
28. Yao, Y.Y.: Rough-set concept analysis: interpreting RS-definable concepts based on ideas from formal concept analysis. Inf. Sci. **346**, 442–462 (2016)
29. Zadeh, L.A.: Fuzzy sets. Inf. Control **8**(3), 338–353 (1965)

A Three-Way Clustering Algorithm via Decomposing Similarity Matrices for Multi-view Data with Noise

Jing Xiong and Hong Yu(✉)

Chongqing Key Laboratory of Computational Intelligence,
Chongqing University of Posts and Telecommunications,
Chongqing 400065, People's Republic of China
xiongjing99@qq.com, yuhong@cqupt.edu.cn

Abstract. The multiple views of data can provide complementary information to each other, a large number of studies have demonstrated that one can achieve the better clustering performance by integrating information from multiple views than using only a single view. However, identifying the explicit cluster structure in the multi-view data with noise and reflecting uncertain relationships between objects and clusters is still a problem that has not been satisfactorily solved. To address the problem, this paper propose a three-way clustering algorithm for multi-view data with noise. The algorithm is mainly divided into two stages. In the first stage, we decompose the similarity matrix of each view into the good data and the corruptions to eliminate the noise contained in the multi-view data. In the second stage, only the clean data of each view is used to obtain the consistency information, and the final three-way clustering results are generated based on the theory of three-way decisions. The experimental results show that the proposed algorithm has better clustering performance in dealing with multi-view data with noise.

Keywords: Noisy data · Multi-view clustering · Three-way decision · Similarity matrix decomposition · Co-regularization

1 Introduction

Multi-view data are captured from heterogeneous sources or views, where the different view represents distinct information of the same objects [1,4,34]. Many real-world data sets can be represented by multiple views. Although each view can be used separately for learning, the views can provide complementary information to each other and improve learning performance, which makes the analysis and learning of multi-view data attract more and more attention [23,27]. The clustering problem is to use a certain similarity measure for the given data set, so that the similarity of objects of the same class is as large as possible, and the similarity of objects in different classes is as small as possible [6,10,29]. Multi-view clustering approaches attempt to mine valuable information underlying different views of data and integrate them to improve clustering performance [23,26].

© Springer Nature Switzerland AG 2019
T. Mihálydeák et al. (Eds.): IJCRS 2019, LNAI 11499, pp. 179–193, 2019.
https://doi.org/10.1007/978-3-030-22815-6_15

In the past few years, many multi-view clustering have been proposed, including subspace learning [7], multiple kernel learning [9], and co-training [1,13]. Many of these methods are based on spectral clustering, since spectral clustering uses graphical structures to represent multi-view data, reveals complex structures between objects, handles the distribution of data in arbitrary shapes including non-convex structures, and it has clear mathematical principles [17,28]. The main difference among these multi-view spectral clustering methods is that the processing of the similarity matrixes between views. Generally speaking, there are three categories methods for multi-view clustering. One reconstructs a similarity matrix containing the consensus information as an input of the corresponding spectral clustering algorithm by a certain projection transformation method [12]. For example, the multi-view subspace learning methods [18] usually assume that the existence of a shared latent representation for reconstructing all views. A common subspace representation of the data shared across multiple views is first learned. Then standard spectral clustering is applied on the learned subspace representation matrix to generate the clustering result [26]. The second category merges the similarity matrices to obtain a common similarity matrix that minimizes the differences between the input similarity matrices, such as co-training [13], co-regularization [14], feature selection [24] or graph fusion [16]. The third one is to independently cluster each view and then obtain the consistent clustering result through the designed weighting strategies [3,8].

In this paper, we focus on the second category multi-view clustering method, especially on the co-training multi-view spectral clustering methods. Co-training multi-view spectral clustering is a method that can effectively deal with multi-view clustering. The idea is to minimize the inconsistency between views. The co-regularization framework can be regarded as a regularized version of co-training algorithm. Kumar et al. [14] proposed co-regularized multi-view spectral clustering method (CMSC), in which two methods were proposed. The pairwise disagreement term and centroid based disagreement term for different views are added into the objective function of spectral clustering. The clustering results which are consistent across the views are achieved after the optimization process.

In the real world applications, the input data may be noisy, which results in the corresponding similarity matrices being corrupted by noise data. However, in the conventional multi-view clustering method of co-regularization, in addition to considering the fusion of view information, more attention is paid to the problem of weight setting between views [11,26,34]. It often combines multiple representations of data with possibly noise data, which may often degrade the clustering performance. In multi-view noise processing, Xia et al. [28] used the Markov chain method to handle the possible noise in the transition probability matrices associated with different views. Ren et al. [21] learned a graph containing K connected components containing consistency information, where K is the number of clusters, and the l_1 norm is used to constrain the interference of noise on the cluster.

On the other hand, the real world data sets may not be well separated for uncertainty. The above multi-view clustering approaches are all based on two-way clustering, in which there exist two relationships between an object and a cluster. However, there might be three relationships between an object and a cluster, namely, belong-to definitely, not belong-to definitely and uncertain. To address the problem, Yu et al. [32,33] proposed a three-way clustering method to solve the problem of uncertainty clustering, inspired by the theory three-way decisions [30,31]. That is, we use a pair of sets to represent a cluster instead of using a single set, so there are three regions such as the core region, fringe region and trivial region. Objects in the core region are typical elements of the cluster and objects in the fringe region might belong to the cluster, and the objects in the trivial region certainly does not belong to the cluster definitely. The three-way representation intuitively shows which objects are fringe to the cluster. Thus, this paper uses three-way clustering method to obtain the final clustering results, which adapts to the hard clustering as well as to the soft clustering.

In this paper, a novel method is proposed for multi-view clustering problem with noisy data. We assumed that the similarity matrix to be decomposed into two latent factors: the clean (good) data and the corruptions. Only the clean data is used for subsequent multi-view information fusion (Fig. 1). The similarity matrix of each view based on the corresponding good data is used as the input of the co-regularized multi-view spectral clustering framework, and finally the clustering result is obtained.

The remainder of this paper is organized as follows. Section 2 briefly reviews the relevant preliminary concepts. Section 3 describes the proposed method in detail. Section 4 reports the results of comparative experiments and conclusions are provided in Sect. 5.

2 Preliminaries

2.1 Representation of Three-Way Clustering

The purpose of clustering is to divide the objects in the data set $\mathbf{X}$ into corresponding clusters. If there are K clusters, the cluster set is represented as $\mathbf{C} = \{C_1, \cdots, C_k, \cdots, C_K\}$. In the existing works, a cluster is usually represented by a single set, namely, $C_k = \{\mathbf{x}_1, \cdots, \mathbf{x}_i, \cdots, \mathbf{x}_{|C_k|}\}$. In contrast to the general crisp representation of a cluster, we represent a three-way cluster C as a pair of sets [32]:

$$C = (Co(C), Fr(C)). \tag{1}$$

Here, $Co(C) \subseteq \mathbf{X}$ and $Fr(C) \subseteq \mathbf{X}$. Let $Tr(C) = \mathbf{X} - Co(C) - Fr(C)$. Then, $Co(C)$, $Fr(C)$ and $Tr(C)$ naturally form the three regions of a cluster as Core Region, Fringe Region and Trivial Region respectively. That is:

$$\begin{aligned} &CoreRegion(C) = Co(C), \\ &FringeRegion(C) = Fr(C), \\ &TrivialRegion(C) = \mathbf{X} - Co(C) - Fr(C). \end{aligned} \tag{2}$$

If $\mathbf{x} \in CoreRegion(C)$, the object $\mathbf{x}$ belongs to the cluster C definitely; if $\mathbf{x} \in FringeRegion(C)$, the object $\mathbf{x}$ might belong to C; if $\mathbf{x} \in TrivialRegion(C)$, the object $\mathbf{x}$ does not belong to C definitely.

These subsets have the following properties.

$$\begin{aligned}&\mathbf{X} = Co(C) \cup Fr(C) \cup Tr(C),\\ &Co(C) \cap Fr(C) = \emptyset,\\ &Fr(C) \cap Tr(C) = \emptyset,\\ &Tr(C) \cap Co(C) = \emptyset.\end{aligned} \tag{3}$$

If $Fr(C) = \emptyset$, the representation of C in Eq. (1) turns into $C = Co(C)$; it is a single set and $Tr(C) = \mathbf{X} - Co(C)$. This is a representation of two-way decisions. In other words, the representation of a single set is a special case of the representation of three-way cluster.

Furthermore, according to Formula (3), we know that it is enough to represent a cluster expediently by the core region and the fringe region.

In another way, we can define a cluster by the following properties:

$$\begin{aligned}&(i)\ Co(C_k) \neq \emptyset, 1 \leq k \leq K;\\ &(ii)\ \bigcup Co(C_k) \bigcup Fr(C_k) = \mathbf{X}, 1 \leq k \leq K.\end{aligned} \tag{4}$$

Property (i) implies that a cluster cannot be empty. This makes sure that a cluster is physically meaningful. Property (ii) states that any object of $\mathbf{X}$ must definitely belong to or might belong to a cluster, which ensures that every object is properly clustered.

With respect to the family of clusters $\mathbf{C}$, we have the following family of clusters formulated by three-way decisions as:

$$\mathbf{C} = \{(Co(C_1), Fr(C_1)), \cdots, (Co(C_k), Fr(C_k)), \cdots, (Co(C_K), Fr(C_K))\}. \tag{5}$$

Obviously, we have the following family of clusters formulated by two-way decisions as:

$$\mathbf{C} = \{Co(C_1), \cdots, Co(C_k), \cdots, Co(C_K)\}. \tag{6}$$

In the approaches based on the theory of three-way decisions, an evaluation function $v(\mathbf{x}, C_k)$ is usually designed and where two thresholds α and β are set in advance; then, the three-way decision rules can be constructed as Eq. (7).

$$\begin{aligned}&if\ v(\mathbf{x}, C_k) \geq \alpha, && decide\ \mathbf{x}\ to\ Co(C_k);\\ &if\ \beta \leq v(\mathbf{x}, C_k) < \alpha, && decide\ \mathbf{x}\ to\ Fr(C_k);\\ &if\ v(\mathbf{x}, C_k) < \beta, && decide\ \mathbf{x}\ to\ Tr(C_k).\end{aligned} \tag{7}$$

In fact, the evaluation function $v(\mathbf{x}, C_k)$ can be a risk decision function, a similarity function, a distance function and so on.

2.2 Review of Spectral Clustering

Spectral clustering is a technique that exploits the properties of the Laplacian of the graph, whose vertices denote the data points, edges denote the similarities between the data points. The basic idea is to divide the weighted undirected graph into K optimal subgraphs, so that the interior of the subgraphs are as similar as possible, and the distance between subgraphs is as far as possible to achieve the purpose of clustering [6,20]. Here we briefly outline the spectral clustering algorithm.

First, the similarity matrix $\mathbf{A} \in \mathbb{R}^{n \times n}$ is constructed by using a similarity measure for objects. The common similarity function generally has a Gaussian kernel function. In this work, we focus on the symmetric KNN method [19], the matrix $\mathbf{A}$ is given by $a_{i,j} = 1$ if i is the NN nearest neighbor of j or vice versa, and $a_{i,j} = 0$ else.

Then, the graph Laplacian $\mathbf{L} = \mathbf{D} - \mathbf{A}$ is computed and the Laplacian matrix is normalized as needed, where $\mathbf{D}$ is a diagonal matrix with $\mathbf{D}_{ii} = \sum_j \mathbf{A}_{i,j}$. Let $\mathbf{H}$ denote a matrix with columns as the smallest K eigenvectors of $\mathbf{L}$.

$$\min_{\mathbf{H} \in \mathbb{R}^{n \times K}} tr(\mathbf{H}^T \mathbf{L} \mathbf{H}), \quad s.t.\ \mathbf{H}^T \mathbf{H} = I. \tag{8}$$

Finally, the object i is assigned to the cluster C if the i-th row of $\mathbf{H}$ is assigned to cluster C by the k-means algorithm.

2.3 Co-regularized Multi-view Spectral Clustering

Let $\chi = \{\mathbf{X}^{(1)}, \cdots, \mathbf{X}^{(v)}, \cdots, \mathbf{X}^{(m)}\}$ denote a set of data consisting of m views, $\mathbf{X}^{(v)} = \{\mathbf{x}_1^{(v)}, \mathbf{x}_2^{(v)}, \cdots, \mathbf{x}_n^{(v)}\} \in \mathbb{R}^{n \times d^{(v)}}$ denotes the examples in the view v, $\mathbf{x}_i^{(v)}$ denotes the i-th data object of the v-th view, $d^{(v)}$ is the feature dimension of the v-th view, and n represents the number of objects in a view. We use $\mathbf{A}^{(v)}$ to represent the similarity matrix of the v-th view, and the symmetrically normalized $\mathbf{L}_{sym}^{(v)} = I - {\mathbf{D}^{(v)}}^{-1/2} \mathbf{A}^{(v)} {\mathbf{D}^{(v)}}^{-1/2}$ to represent the Laplacian matrix of the view.

The objective function of the spectral clustering in a single view is as described in Eq. (8). The core idea of co-regularized multi-view spectral clustering (CMSC) [14] is to minimize the differences between the input similarity matrices. The disagreement term for different views are added into the objective function of spectral clustering. $D(\mathbf{H}^{(v)}, \mathbf{H}^{(w)}) = -tr(\mathbf{H}^{(v)} {\mathbf{H}^{(v)}}^T \mathbf{H}^{(w)} {\mathbf{H}^{(w)}}^T)$ is a measure of disagreement between clusterings of two views [14]. The CMSC objective function based on pairwise constraints is as follows:

$$\begin{aligned} \min_{\mathbf{H}^{(1)}, \cdots, \mathbf{H}^{m} \in \mathbb{R}^{n \times K}} \quad & tr({\mathbf{H}^{(v)}}^T \mathbf{L}_{sym}^{(v)} \mathbf{H}^{(v)}) + \lambda \sum_{1 \leqslant v, w \leqslant m, v \neq m} D(\mathbf{H}^{(v)}, \mathbf{H}^{(w)}), \\ & s.t. {\mathbf{H}^{(v)}}^T \mathbf{H}^{(v)} = I, \forall 1 \le v \le m. \end{aligned} \tag{9}$$

The hyperparameter λ trades-off the spectral clustering objectives and the spectral embedding disagreement term. Let $\mathbf{L}^{(v)} = {\mathbf{D}^{(v)}}^{-1/2} \mathbf{A}^{(v)} {\mathbf{D}^{(v)}}^{-1/2}$ be used to represent the Laplacian matrix, then, $\mathbf{H}^{(v)}$ denote a matrix with columns as the maximal K eigenvectors of $\mathbf{L}^{(v)}$, the objective function can be written in the following form of the maximum value:

$$\max_{\mathbf{H}^{(1)},\cdots,\mathbf{H}^{m}\in\mathbb{R}^{n\times K}} tr(\mathbf{H}^{(v)^T}\mathbf{L}^{(v)}\mathbf{H}^{(v)}) + \lambda \sum_{1\leqslant v,w\leqslant m, v\neq m} tr(\mathbf{H}^{(v)}\mathbf{H}^{(v)^T}\mathbf{H}^{(w)}\mathbf{H}^{(w)^T}), \quad s.t. \mathbf{H}^{(v)^T}\mathbf{H}^{(v)} = I, \forall 1 \leq v \leq m. \tag{10}$$

The Eq. (10) can be solved using alternating maximization. For a given $\mathbf{H}^{(v)}$, we get the following optimization problem in $\mathbf{H}^{(v)}$:

$$\max_{\mathbf{H}^{(v)}} tr(\mathbf{H}^{(v)^T}(\mathbf{L}^{(v)} + \lambda \sum_{1\leqslant v,w\leqslant m, v\neq m} \mathbf{H}^{(w)}\mathbf{H}^{(w)^T})\mathbf{H}^{(v)}), \quad s.t. \mathbf{H}^{(v)^T}\mathbf{H}^{(v)} = I. \tag{11}$$

3 The Proposed Method

In this section, we present the robust multi-view three-way clustering algorithm which effectively handles noisy data and obtains three-way representation of clusters.

3.1 The Framework

Figure 1 shows the main framework of the proposed method, which consists of two stages. The first stage is to decompose the similarity matrix of each view into two parts, namely the clean (good) data $\mathbf{A}^{g(v)}$ and the corruptions $\mathbf{E}^{(v)}$. In the second part, we use the co-regularized multi-view spectral clustering framework and three-way k-means method to obtain the final three-way clustering results.

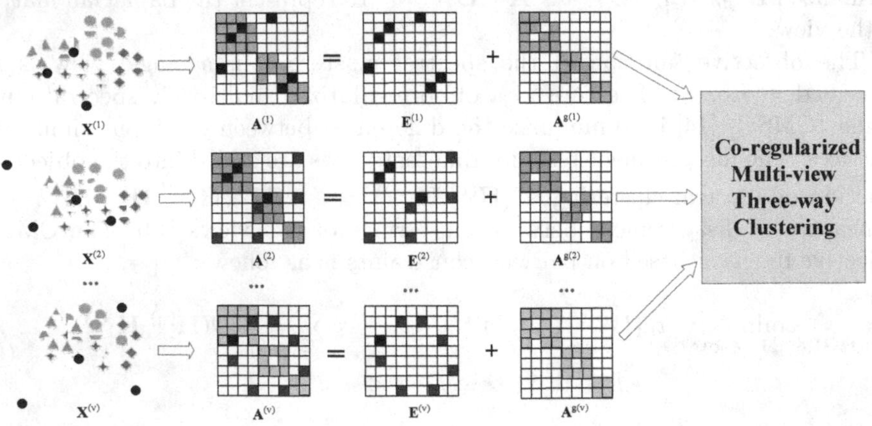

Fig. 1. The framework of the proposed method

First, we construct the similarity matrix of each view by the KNN method [19]. The matrix $\mathbf{A}^{(v)}$ is given by $\mathbf{a}_{i,j}^{(v)} = 1$ if i is the NN nearest neighbor of j or vice versa, and $\mathbf{a}_{i,j}^{(v)} = 0$ else. Due to the existence of noise data, the corresponding similarity matrices may be corrupted. So we assume the similarity matrix $\mathbf{A}^{(v)}$ to be decomposed into two latent factors: the clean(good) data $\mathbf{A}^{g(v)}$ and the corruptions $\mathbf{E}^{(v)}$. Namely,

$$\mathbf{A}^{(v)} = \mathbf{E}^{(v)} + \mathbf{A}^{g(v)} \quad \mathbf{E}^{(v)}, \mathbf{A}^{g(v)} \in \mathbb{R}_{\geq 0}^{n\times n}, \text{ both are symmetric.} \tag{12}$$

Then, the similarity matrix based on clean(good) data of each view is used as the input of the co-regularized multi-view spectral clustering framework and the clustering result is obtained. Subsect. 3.2 introduces how to find the matrices $\mathbf{E}^{(v)}$ and $\mathbf{A}^{g(v)}$, and Subsect. 3.3 presents the co-regularized multi-view three-way clustering process.

3.2 Similarity Matrix Decomposition

Generally speaking, corruptions are relatively rare, if they were not rare, i.e. the majority of the data is corrupted, a reasonable clustering structure can not be expected [2]. Thus, it is reasonable that we assume that $\mathbf{E}^{(v)}$ is sparse, use $\theta^{(v)}$ to limit the degree to which each view is contaminated by noise, and use the l_0 norm to constrain the sparsity, i.e. $\|\mathbf{E}^{(v)}\|_0 \leq 2\theta^{(v)}$, $\|\cdot\|_0$ represents the number of non-zero elements in the matrix, $2\theta^{(v)}$ indicates the symmetry of $\mathbf{E}^{(v)}$. To prevent extreme situations, such as $\theta^v = \frac{1}{2}\|\mathbf{A}^{(v)}\|_0$, we use $\left\|a_{i,:}^{g(v)}\right\|_0 \geq m$ to constrain each object $\mathbf{x}_i^{(v)}$ in the matrix to be connected to at least m objects. Usually, we set $m = \left\lceil \frac{1}{2}\left\|a_{i,:}^{g(v)}\right\|_0 \right\rceil$.

We jointly perform spectral clustering and decomposition of matrix $\mathbf{A}^{(v)}$, and add constraints on $\mathbf{E}^{(v)}$ in the objective function of the spectral clustering Eq. (8). The problem is transformed as follows:

$$\begin{aligned}
(\mathbf{H}^{(v)}, \mathbf{A}^{g(v)}) = & \underset{\mathbf{H}^{(v)},\mathbf{A}^{g(v)}}{\arg\min} \;\; tr(\mathbf{H}^{(v)T}\mathbf{L}(\mathbf{A}^{g(v)})\mathbf{H}^{(v)}), \\
& s.t. \mathbf{H}^T\mathbf{H} = I, \mathbf{A}^{g(v)} = \mathbf{A}^{g(v)^T}, \|\mathbf{A}^{(v)} - \mathbf{A}^{g(v)}\|_0 \leq 2\theta^v, \left\|a_{i,:}^{g(v)}\right\|_0 \geq m, \\
& \forall i \in \{1, \cdots, n\}, \forall 1 \leq v \leq m.
\end{aligned} \tag{13}$$

where $\mathbf{L}(\mathbf{A}^{g(v)})$ represents the Laplacian of the clean data $\mathbf{A}^{g(v)}$, $\mathbf{H}^{(v)}$ represents the indicator matrix of the v-th view, and $\mathbf{H}^{(v)} \in \mathbb{R}^{n\times K}$, K represents the number of clusters.

Equation (13) is hard to optimize (in particular due to the $\|\cdot\|_0$ constraints the problem becomes NP-hard in general). The $\|\cdot\|_0$ norm is simply handled by relaxation to the $\|\cdot\|_1$ norm. In this work, we aim to preserve the interpretability of the $\|\cdot\|_0$ norm. Aleksandar et al. [2] proposes a block coordinate-descent (alternating) optimization scheme to approximate it. That is, given $\mathbf{H}^{(v)}$, update $\mathbf{A}^{g(v)}$ or vice versa. Since $\mathbf{A}^{g(v)}$ determines $\mathbf{E}^{(v)}$ and vice versa, we just focus on the update of one, e.g., $\mathbf{E}^{(v)}$.

Update of $\mathbf{H}^{(v)}$ when $\mathbf{E}^{(v)}$ fixed: since $\mathbf{A}^{g(v)} = \mathbf{A}^{(v)} - \mathbf{E}^{(v)}$, $\mathbf{L}(\mathbf{A}^{g(v)})$ are now constant, the problem is transformed into a traditional spectral clustering problem. That is, the solution of $\mathbf{H}^{(v)}$ are the K first eigenvectors of $\mathbf{L}(\mathbf{A}^{g(v)})$.

Update of $\mathbf{E}^{(v)}$ when $\mathbf{H}^{(v)}$ fixed: since $\mathbf{L}^{(v)} = \mathbf{D}^{(v)} - \mathbf{A}^{(v)}$ and $\mathbf{A}^{g(v)} = \mathbf{A}^{(v)} - \mathbf{E}^{(v)}$, we have $\mathbf{L}(\mathbf{A}^{g(v)}) = \mathbf{L}(\mathbf{A}^{(v)}) - \mathbf{L}(\mathbf{E}^{(v)})$, and $tr(\mathbf{H}^{(v)^T}\mathbf{L}(\mathbf{A}^{g(v)})\mathbf{H}^{(v)}) = tr(\mathbf{H}^{(v)^T}\mathbf{L}(\mathbf{A}^{(v)})\mathbf{H}^{(v)}) - tr(\mathbf{H}^{(v)^T}\mathbf{L}(\mathbf{E}^{(v)})\mathbf{H}^{(v)})$. The term $tr(\mathbf{H}^{(v)^T}\mathbf{L}(\mathbf{A}^{(v)})\mathbf{H}^{(v)})$ is constant, so the problem of minimizing the previous term is equivalent to maximizing $tr(\mathbf{H}^{(v)^T}\mathbf{L}(\mathbf{E}^{(v)})\mathbf{H}^{(v)})$. Let $e_{i,j}^{(v)}$ denotes the element in $\mathbf{E}^{(v)}$ and $\mathbf{h}_i^{(v)}$ denotes the i-th row in $\mathbf{H}^{(v)}$, $tr(\mathbf{H}^{(v)^T}\mathbf{L}(\mathbf{E}^{(v)})\mathbf{H}^{(v)}) = \sum_{i,j} \frac{1}{2} e_{i,j}^{(v)} \left\|\mathbf{h}_i^{(v)} - \mathbf{h}_j^{(v)}\right\|_2^2$ [25]. The Eq. (13) can be transformed into solving the following objective function:

$$f([e_{i,j}^{(v)}]) := \sum\nolimits_{i,j} e_{i,j}^{(v)} \left\|\mathbf{h}_i^{(v)} - \mathbf{h}_j^{(v)}\right\|_2^2. \tag{14}$$

Algorithm 1. Similarity Matrix Decomposition

```
Input: Similarity matrix of each view A^(1),···,A^(v),···,A^(m).
Output: Good similarity matrix of each view A^g(1),···,A^g(v),···,A^g(m).
\\Initialize A^g(v);
for v=1 to m do
  A^g(v) ← A^(v);
for v=1 to m do
  while true do
    \\Updata of H^(v);
    Compute Laplacian L(A^g(v)) = D^g(v) − A^g(v), matrix H^(v) ∈ R^(n×K) ,
    and trace;
    if trace could not be lowered then
      break;
    \\Updata of A^g(v);
    Calculate the maximum number of edges that each object can delete
    and count_i^(v) ← |a_{i,:}^(v)| − m;
    Calculate the value p_{a_{i,j}^(v)} of each edge of A^g(v) lower triangle according
    to Eq.(15) and sort;
    if count_i^(v) > 0 and count_j^(v) > 0 then
      let let e_{i,j}^(v) = a_{i,j}^(v);
      count_i^(v) − −, count_j^(v) − −;
      if |e_{i,j}^(v)| > θ^(v) then
        break;
    construct E^(v) according to e_{i,j}^(v); A^g(v) = A^g(v)−E^(v);
return A^g(1),···,A^g(v),···,A^g(m).
```

Our problem is to find a set $[e_{i,j}^{(v)}]$ containing the edge $e_{i,j}^{(v)}$ affected by the noise data, where $\mathbf{A}^{(v)}$ contains the elements of $\mathbf{E}^{(v)}$, so the problem of Eq. (14) is equivalent to solving the maximum of the following problem:

$$p_{e_{i,j}^{(v)}} = p_{a_{i,j}^{(v)}} = a_{i,j}^{(v)} \left\| \mathbf{h}_i^{(v)} - \mathbf{h}_j^{(v)} \right\|_2^2. \tag{15}$$

This function result matches the intuition of corrupted edges: the higher the $p_{e_{i,j}^{(v)}}$ value means that object i and object j do not belong to the same cluster, but there is still an edge connection between them. So we only need to find the largest $\theta^{(v)}$ of $p_{a_{i,j}^{(v)}}$ and remove it to eliminate the effect of noise. Since the matrix is symmetrical, we only need to sort each edge of $\mathbf{A}^{(v)}$'s upper or lower triangle according to Eq. (15), and find the $\theta^{(v)}$ largest edges, let $e_{i,j}^{(v)} = a_{i,j}^{(v)}$, if the edge to which the object is connected is less than m, skip this next substitute. The rest of the elements in $\mathbf{E}^{(v)}$ are set to 0. Algorithm 1 gives the process of decomposing the similarity matrix.

3.3 Co-regularized Multi-view Three-Way Clustering Process

According to Algorithm 1, we get $\mathbf{A}^{g(1)},\cdots,\mathbf{A}^{g(v)},\cdots,\mathbf{A}^{g(m)}$. According to Eq. (10), the indicator matrix $\mathbf{H}^{(v)}$ containing the view consistency information

---

Algorithm 2. Co-regularized Multi-View Three-Way Clustering

Input: The good similarity matrix of each view $\mathbf{A}^{g(1)},\cdots,\mathbf{A}^{g(v)},\cdots,\mathbf{A}^{g(m)}$.
Output: Three-way clustering results $\mathbf{C} =$
$\{(Co(C_1), Fr(C_1)),\cdots,(Co(C_k), Fr(C_k)),\cdots,(Co(C_K), Fr(C_K))\}$.
Compute Laplacian of each view $\mathbf{L}^{(v)} = \mathbf{D}^{(v)^{-1/2}}\mathbf{A}^{g(v)}\mathbf{D}^{(v)^{-1/2}}$, indicator matrix of each view $\mathbf{H}^{(v)} \in \mathbb{R}^{n\times K}$ and the sum of $Trace\_sum$;
while *true* **do**
  for *v=1 to m* **do**
    if $w \neq v$ **then**
      Calculate the updated indicator matrix $\mathbf{H}^{(v)}$ and *trace* for each view according to Eq.(10);
  if *trace could not be higher* **then**
    break;
Choose the indicator matrix $\mathbf{H}^{(v)}$ of the richest information and apply k-means on it, get the hard clustering results $\{C_1, C_k,\cdots,C_K\}$ and the cluster center points $\{cen_1, cen_k,\cdots,cen_K\}$;
Calculate the average distance $dist_{aver}^k$ and the farthest distance $dist_{far}^k$ from each cluster C_k to the cluster center cen_k.
Let $\alpha_k = dist_{aver}^k$ and $\beta_k = dist_{aver}^k + dist_{far}^k$;
for *k=1 to K* **do**
  for *i=1 to n* **do**
    if $dist(\mathbf{h}_i, cen_k) \leq \alpha_k$, decide to Co(C$_k$);
    if $\alpha_k \leq dist(\mathbf{h}_i, cen_k) \leq \beta_k$, decide to Fr(C$_k$).
return
$C = \{(Co(C_1), Fr(C_1)),\cdots,(Co(C_k), Fr(C_k)),\cdots,(Co(C_K), Fr(C_K))\}$.

---

can be obtained. Then we use three-way k-means clusters to obtain the final clustering results, which is described in Algorithm 2.

4 Experiments

In this section, in order to verify the effectiveness of the proposed method, we use several representative multi-view clustering algorithms as compared methods to test on the four real data sets. The general clustering accuracy (ACC) [33], normalized mutual information (NMI) [5] and adjusted rand index (ARI) [22] are used to evaluate the clustering performance. The experimental setup is described in detail in Sect. 4.1, and the experimental results are analyzed in Sect. 4.2.

4.1 Experiments Setup

4.1.1 Datasets

We conduct experiments on five real-world datasets: Wine, SensIT, 3sources, Digits. A detailed summarization of these datasets is in Table 1. Since we want to verify the robustness to the noise data, after the data is normalized by column, 5% of the data is randomly extracted, and the attribute values of these objects are set to random values.

Table 1. Information about the datasets

Datasets	Objects	Dimensions	View	Cluster
Wine	178	{6,7}	2	3
3sources	169	{3068,3631,3560}	3	6
SensIT	300	{50,50}	2	3
Digits	2000	{240,76}	2	10

- Wine[1]: Wine is the standard data set in UCI. We split the feature vectors into 2 subsets, each subset is considered as one data view as the reference [15].
- 3sources[2]: It is collected from three online news sources: BBC, Reuters and Guardian. In total it consists of 416 distinct news manually categorized into six classes. Among them, 169 are reported in all three sources and each story was manually annotated with one of the six topical labels.
- SensIT dataset[3]: It uses two sensors to classify three types of vehicle. We randomly sample 100 data for each class, and then conduct experiments on 2 views and three classes.

[1] http://archive.ics.uci.edu/ml/.
[2] http://mlg.ucd.ie/datasets/3sources.html.
[3] https://www.csie.ntu.edu.tw/~cjlin/libsvmtools/datasets/multiclass.html.

- Digits[4]: It consists of features of hand-written digits (0–9). The dataset is represented by 6 feature sets and contains 2000 samples with 200 in each category. We choose 76 Fourier coefficients of the character shapes and the 240 pixel averages in 2×3 windows as two views.

4.1.2 Compared Methods

We compare the proposed method with some representative multi-view clustering algorithms.

- Best Single View (BSV): running the proposed methods on each input view, and then reporting the results of the view that achieves the best performance.
- Feature Concatenation (FeatCon): concatenating the features of all views to form a single representation, and then applying the proposed method on the concatenated view.
- Co-regularized Multi-view Spectral Clustering (CMSC) [14]: adopting the co-regularization framework in spectral clustering, and we use KNN method to construct similarity matrix on each view.
- Multi-view Spectral Clustering by Common Eigenvectors (MVSC-CEV) [12] The strategy of this method is different from our method. In this method, each view is projected to obtain a matrix containing consistency information, and then standard spectral clustering is performed.

4.2 Experiments Results

In the proposed method, the similarity matrix of the view needs to be calculated, and all of them are calculated by the KNN. For the co-regularized multi-view clustering, the parameter λ used in the literature has been set to 0.09 after many experiments. Each data set is tested after randomly adding 5% of the noise. Our method needs to set the amount of noise $\theta^{(v)}$ of the view according to experience, where the value of $\theta^{(v)}$ should not be set too large to prevent excessive loss of information. In this experiment, we set $\theta^{(v)} = 10$, and the experimental results are shown in Table 2.

From the above experimental results, it is obvious that the proposed method performs better in most cases than the compared algorithms in all indices, which shows that the proposed method is effective and feasible. The best NMI value and ARI value of the Wine data set are in the FeatCon method. Because the wine data set is not a multi-view data, the result on the FeatCon method is better than our method is explicable.

Figure 2 intuitively shows the comparison of NMI values of CMSC (no noise), CMSC (5% noise), MVSC-CEV (5% noise) and the proposed method(5% noise). It can be seen that the effectiveness of the CMSC method is reduced when it processes noise-containing data. In contrast, the performance of the proposed method is better when dealing with noise-containing data. The results shows that the proposed method has a certain effect to deal with noisy data.

[4] https://archive.ics.uci.edu/ml/datasets.html.

Table 2. Comparison of experimental results

Datasets	Methods	NMI	ACC	ARI
Wine	BSV	0.6663	0.3438	0.6596
	FeatCon	**0.8120**	0.5299	**0.8305**
	CMSC [14]	0.7114	0.6273	0.7391
	MVSC-CEV [12]	0.6363	0.38764	0.6216
	Proposed	0.8055	**0.6488**	0.8299
SensIT	BSV	0.225	0.3591	0.2682
	FeatCon	0.2806	0.3596	0.2306
	CMSC [14]	0.2727	0.4857	0.1995
	MVSC-CEV [12]	0.2336	0.3166	0.2117
	Proposed	**0.3001**	**0.5454**	**0.2848**
3sources	BSV	0.3823	0.5147	0.1575
	FeatCon	0.3087	0.5029	0.1205
	CMSC [14]	0.3186	0.5384	0.1488
	MVSC-CEV [12]	0.3052	0.4201	**0.2008**
	Proposed	**0.4002**	**0.5976**	0.1834
Digits	BSV	0.7751	0.854	0.6796
	FeatCon	0.7731	0.855	0.681
	CMSC [14]	0.8057	0.8815	0.8903
	MVSC-CEV [12]	0.8463	0.9185	0.7791
	Proposed	**0.9051**	**0.928**	**0.9089**

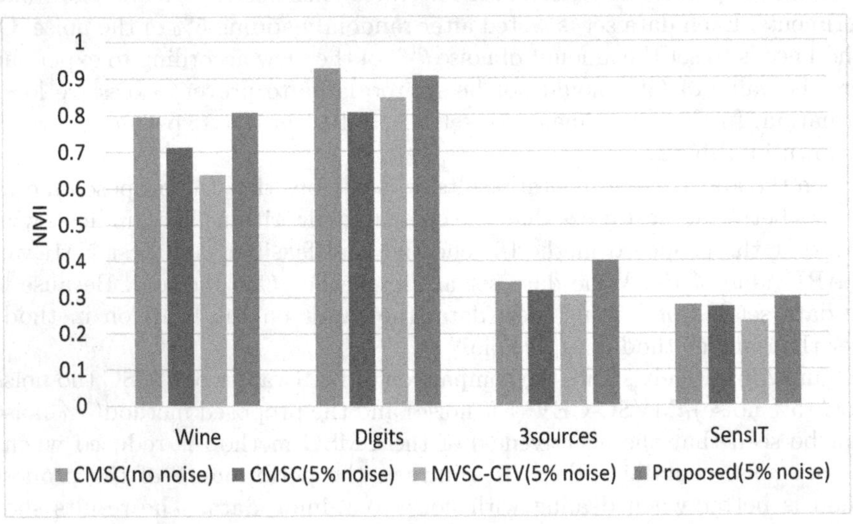

Fig. 2. Comparison of NMI values

5 Conclusions

In this paper, we have addressed the problem of clustering on the multi-view data with noise. Based on the idea of matrix decomposition, we decompose the similarity matrix of each view into two latent factors, namely, the good data and the corruptions. Then, the good similarity matrix of each view is used as the input of the co-regularized multi-view spectral clustering framework. At the same time, the idea of three-way clustering is also applied in this paper, which makes the clustering result reflects all the three relationships between an object and a cluster. Experimental results have demonstrated the effectiveness of the proposed method.

Acknowledgments. This work was supported in part by the National Natural Science Foundation of China under grant No. 61876027, 61672120 and 61533020.

References

1. Blum, A., Mitchell, T.: Combining labeled and unlabeled data with co-training. In: Proceeding of the Eleventh Annual Conference on Computational Learning Theory, pp. 92–100. ACM (1998)
2. Bojchevski, A., Matkovic, Y., Gnnemann, S.: Robust spectral clustering for noisy data: modeling sparse corruptions improves latent embeddings. In: Proceedings of the 23rd ACM SIGKDD International Conference on Knowledge Discovery and Data Mining, pp. 737–746. ACM (2017)
3. Bruno, E., Marchand-Maillet, S.: Multiview clustering: a late fusion approach using latent models. In: Proceedings of the 32nd International ACM SIGIR Conference on Research and Development in Information Retrieval, pp. 736–737. ACM (2009)
4. Chaudhuri, K., Kakade, S.M., Livescu, K., Sridharan, K.: Multi-view clustering via canonical correlation analysis. In: Proceedings of the 26th Annual International Conference on Machine Learning, pp. 129–136. ACM (2009)
5. Danon, L., Daz-Guilera, A., Duch, J., Arenas, A.: Comparing community structure identification. J. Stat. Mech.: Theory Exp. **2005**(09), P09008 (2005)
6. Filippone, M., Camastra, F., Masulli, F., Rovetta, S.: A survey of kernel and spectral methods for clustering. Pattern Recognit. **41**(1), 176–190 (2008)
7. Gao, H.C., Nie, F.P., Li, X.L., Huang, H.: Multi-view subspace clustering. In: IEEE international Conference on Computer Vision, pp. 4238–4246. IEEE (2015)
8. Greene, D., Cunningham, P.: A matrix factorization approach for integrating multiple data views. In: Buntine, W., Grobelnik, M., Mladenić, D., Shawe-Taylor, J. (eds.) ECML PKDD 2009. LNCS (LNAI), vol. 5781, pp. 423–438. Springer, Heidelberg (2009). https://doi.org/10.1007/978-3-642-04180-8_45
9. Houthuys, L., Langone, R., Suykens, A.K.J.: Multi-view kernel spectral clustering. Inf. Fusion **44**, 46–56 (2018)
10. Jain, A.K.: Data clustering: 50 years beyond k-means. Pattern Recognit. Lett. **31**(8), 651–666 (2010)
11. Jiang, Y.Z., Chung, F.L., Wang, S.T., Deng, Z.H., Wang, J., Qian, P.J.: Collaborative fuzzy clustering from multiple weighted views. IEEE Trans. Cybern. **45**(4), 688–701 (2015)

12. Kanaan-Izquierdo, S., Ziyatdinov, A., Perera-Lluna, A.: Multiview and multifeature spectral clustering using common eigenvectors. Pattern Recognit. Lett. **102**, 30–36 (2018)
13. Kumar, A., Daum, H.: A co-training approach for multi-view spectral clustering. In: Proceedings of the 28th International Conference on Machine Learning (ICML 2011), pp. 393–400 (2011)
14. Kumar, A., Rai, P., Daume, H.: Co-regularized multi-view spectral clustering. In: Advances in Neural Information Processing Systems, pp. 1413–1421 (2011)
15. Li, K., Li, S., Ding, Z.M., Zhang, W.D., Fu, Y.: Latent discriminant subspace representations for multi-view outlier detection. In: Proceedings of the 32th AAAI Conference on Artificial Intelligence, pp. 3522–3529 (2018)
16. Li, Y., Nie, F., Huang, H., Huang, J.: Large-scale multi-view spectral clustering via bipartite graph. In: Proceedings of the 29th AAAI Conference on Artificial Intelligence, pp. 2750–2756 (2015)
17. Lee, T.L., Liu, T.L.: Guided co-training for large-scale multi-view spectral clustering. arXiv preprint arXiv:1707.09866 (2017)
18. Luo, S.R., Zhang, C.Q., Zhang, W., Cao, X.C.: Consistent and specific multi-view subspace clustering. In: Proceedings of the 32th AAAI Conference on Artificial Intelligence, pp. 3730–3737 (2018)
19. Luxburg, U.V.: A tutorial on spectral clustering. Stat. Comput. **17**(4), 395–416 (2007)
20. Ng, A.Y., Jordan, M.I., Weiss, Y.: On spectral clustering: analysis and an algorithm. In: Advances in Neural Information Processing Systems, pp. 849–856 (2001)
21. Ren, P.Z., et al.: Robust auto-weighted multi-view clustering. In: 27th International Joint Conference on Artificial Intelligence (IJCAI), pp. 2644–2650 (2018)
22. Steinley, D.: Properties of the hubert-arable adjusted rand index. Psychol. Methods **9**(3), 386 (2004)
23. Sun, S.: A survey of multi-view machine learning. Neural Comput. Appl. **23**(7–8), 2031–2038 (2013)
24. Wang, H., Nie, F., Huang, H.: Multi-view clustering and feature learning via structured sparsity, In: Proceedings of the 30th International Conference on Machine Learning (ICML 2013), pp. 352–360 (2013)
25. Wang, X., Zhang, T., Gao, X.: Multiview clustering based on non-negative matrix factorization and pairwise measurements. IEEE Trans. Cybern. **99**, 1–14 (2018)
26. Wang, Y., Chen, L.: Multi-view fuzzy clustering with minimax optimization for effective clustering of data from multiple sources. Expert Syst. Appl. **72**, 457–466 (2017)
27. Wen, J., Xu, Y., Liu, H.: Incomplete multiview spectral clustering with adaptive graph learning. IEEE Trans. Cybern. 1–12 (2018)
28. Xia, R., Pan, Y., Du, L., Jin, Y.: Robust multi-view spectral clustering via low-rank and sparse decomposition. In: Proceedings of the 28th AAAI Conference on Artificial Intelligence, pp. 2149–2155 (2014)
29. Xu, R., Wunsch, D.: Survey of clustering algorithms. IEEE Trans. Neural Netw. **16**(3), 645–678 (2005)
30. Yao, Y.: An outline of a theory of three-way decisions. In: Yao, J.T., et al. (eds.) RSCTC 2012. LNCS (LNAI), vol. 7413, pp. 1–17. Springer, Heidelberg (2012). https://doi.org/10.1007/978-3-642-32115-3_1
31. Yao, Y.Y.: Three-way decisions and cognitive computing. Cogn. Comput. **8**(4), 543–554 (2016)

32. Yu, H.: A framework of three-way cluster analysis. In: Polkowski, L., et al. (eds.) IJCRS 2017. LNCS (LNAI), vol. 10314, pp. 300–312. Springer, Cham (2017). https://doi.org/10.1007/978-3-319-60840-2_22
33. Yu, H., Wang, X.C., Wang, G.Y., Zeng, X.H.: An active three-way clustering method via low-rank matrices for multi-view data. Inf. Sci. **000**, 1–17 (2018)
34. Zhang, G.Y., Wang, C.D., Huang, D., Zheng, W.X., Zhou, Y.R.: TW-Co-k-means: two-level weighted collaborative k-means for multi-view clustering. Knowl.-Based Syst. **150**, 127–138 (2018)

Related Methods and Hybridization

A Scalable Approach to Fuzzy Rough Nearest Neighbour Classification with Ordered Weighted Averaging Operators

Oliver Urs Lenz[1](✉), Daniel Peralta[1,2], and Chris Cornelis[1]

[1] Department of Applied Mathematics, Computer Science and Statistics, Ghent University, Ghent, Belgium
{oliver.lenz,chris.cornelis}@ugent.be
http://www.cwi.ugent.be

[2] Data Mining and Modelling for Biomedicine Group, VIB Center for Inflammation Research, Ghent University, Ghent, Belgium
daniel.peralta@irc.vib-ugent.be
https://www.irc.ugent.be

Abstract. Fuzzy rough sets have been successfully applied in classification tasks, in particular in combination with OWA operators. There has been a lot of research into adapting algorithms for use with Big Data through parallelisation, but no concrete strategy exists to design a Big Data fuzzy rough sets based classifier. Existing Big Data approaches use fuzzy rough sets for feature and prototype selection, and have often not involved very large datasets. We fill this gap by presenting the first Big Data extension of an algorithm that uses fuzzy rough sets directly to classify test instances, a distributed implementation of FRNN-OWA in Apache Spark. Through a series of systematic tests involving generated datasets, we demonstrate that it can achieve a speedup effectively equal to the number of computing cores used, meaning that it can scale to arbitrarily large datasets.

Keywords: Fuzzy rough sets · OWA operators · Big Data · Apache Spark

1 Introduction

Fuzzy rough sets [7] encode two complementary types of uncertainty: degrees of membership, and the approximation of concepts. This expressiveness has led to their adoption in a variety of machine learning contexts. Fuzzy Rough Nearest Neighbours (FRNN), introduced in [9] (as FRNN-FRS), was an attempt to use fuzzy rough sets directly for classification and obtain better results than existing lazy learners like Fuzzy Nearest Neighbours (FNN) and k Nearest Neighbours (kNN). FRNN considers the lower and upper approximation of each class and classifies a test instance based on its membership in these.

Like other lazy learners, FRNN does not require training and so can be applied directly to classify test instances with a training set. FRNN is also

© Springer Nature Switzerland AG 2019
T. Mihálydeák et al. (Eds.): IJCRS 2019, LNAI 11499, pp. 197–209, 2019.
https://doi.org/10.1007/978-3-030-22815-6_16

conceptually attractive because its predictions are directly interpretable. Upper approximation membership encodes to what extent a test instance is similar to the training instances of a class, and so possibly belongs to this class. Lower approximation membership encodes to what extent a test instance is not similar to the training instances of other classes and so necessarily belongs to this class.

However, it was pointed out by [17] that, as originally defined, FRNN makes predictions that are necessarily identical to those of traditional 1NN. This fact, and the more general observation already made in [9] that FRNN is sensitive to noise, motivated a number of revised proposals. In this paper we focus on FRNN-OWA, introduced in [15], which incorporates Ordered Weighted Averaging (OWA) operators into the definition of lower and upper approximation. This involves the application of weight vectors, and the choice of these weight vectors offers a great degree of flexibility. For example, because lower and upper approximations are calculated for each class, it is possible to use different types of weights for different classes. This idea has been applied successfully by [15] and subsequent studies [19] and [20] to imbalanced datasets, where a judicious choice of weights increases the signal of the minority class.

Over the course of the past two decades, ever larger quantities of data have become available as potential inputs for machine learning algorithms, to the point where the performance of machine learning algorithms is often no longer constrained by the availability of training data, but by the capability of the algorithms to handle training data. One popular tactic to increase data processing capacity is to break down the work of an algorithm into a series of parallel tasks, and to execute these tasks on a cluster of computing cores. A number of frameworks exist that automate many of the aspects of parallel cluster computing, including Apache Spark [11], which we use in this paper.

Handling large amounts of data is a particular challenge for lazy learners like FRNN-OWA, which have to process the entire training set when they receive a test instance. Since the application of fuzzy rough sets in machine learning problems is a relatively recent, ongoing endeavour, it is not surprising that while there exist distributed implementations of kNN [13] and Fuzzy kNN [12] classification, no Big Data implementation exists of a fuzzy rough set classifier. The few implementations that do try to extend the use of fuzzy rough sets to a Big Data context focus on preprocessing algorithms like Fuzzy Rough Feature and Prototype Selection, and only one has been applied to a real dataset with more than 1 million instances [8].

This paper seeks to address this absence by presenting the first Big Data implementation of an algorithm that uses fuzzy rough sets directly to classify test instances (FRNN-OWA). By effectively parallelising the FRNN-OWA algorithm, our implementation can be scaled to arbitrarily large datasets by adding additional computing cores. We demonstrate this through a series of systematic tests on generated datasets of up to 2^{24} instances. In addition, we show that our implementation can be used to classify test instances with real datasets containing over 10 million instances.

In Sect. 2 of this paper, we first define and explain the motivation for FRNN-OWA and give an overview of existing attempts at Big Data implementations of algorithms involving fuzzy rough sets. We then formulate our proposal in Sect. 3, describe our experimental setup in Sect. 4 and present the results in Sect. 5. We conclude in Sect. 6 that our implementation demonstrates the viability of using large quantities of available data to classify unseen instances with fuzzy rough sets.

2 Background

2.1 Fuzzy Rough Nearest Neighbour Classification with OWA Operators

Recall the following concepts from fuzzy rough set theory. An information system (X, A) consists of a set of instances X and a set A of attributes $a : X \longrightarrow V_a$. A t-norm $T : [0,1] \times [0,1] \longrightarrow [0,1]$ is an associative, commutative and monotonically increasing binary operation for which 1 is an identity element. An implication $I : [0,1] \times [0,1] \longrightarrow [0,1]$ is a binary operation that is monotonically decreasing in its first argument and monotonically increasing in its second argument, and for which $I(0,0) = I(0,1) = I(1,1) = 1$ and $I(1,0) = 0$. An indiscernibility relation $R : X \times X \longrightarrow [0,1]$ is a fuzzy tolerance relation (i.e. reflexive and symmetric) such that $(\forall a \in A : a(x) = a(y)) \implies R(x,y) = 1$.

Given an information system (X, A) and a choice of indiscernibility relation R on X, t-norm T and implication I, the upper and lower approximations of a fuzzy set C in X are defined as in (1).

$$
\begin{aligned}
\overline{C}(y) &= \max_{x \in X}(T(R(y,x), C(x)) \\
\underline{C}(y) &= \min_{x \in X}(I(R(y,x), C(x))
\end{aligned}
\tag{1}
$$

In FRNN, C can be any of the crisp decision classes, and a test instance y is classified to the class C for which the average of $\overline{C}(y)$ and $\underline{C}(y)$ is highest. For crisp C and the minimum t-norm $\min(\cdot\,,\cdot)$ and Kleene-Dienes implication $\max(1 - \cdot\,,\cdot)$, which [9] uses, (1) simplifies to (2).

$$
\begin{aligned}
\overline{C}(y) &= \max_{x \in C}(R(y,x)) \\
\underline{C}(y) &= \min_{x \notin C}(1 - R(y,x))
\end{aligned}
\tag{2}
$$

It can be seen from (2) that a test instance necessarily has the highest membership degree in the lower and upper approximations of the class of the most indiscernible training instance. Since the indiscernibility relation R corresponds inversely to a generalised metric, the most indiscernible training instance is the nearest neighbour under this metric, meaning that FRNN is indistinguishable from 1NN classification.

To solve this, FRNN-OWA replaces max and min in (2) with Ordered Weighted Averaging (OWA) operators, which were first defined in [21]. For a given k-dimensional weight vector w with values in $[0,1]$ that sum to 1, the OWA operator F_w corresponding to w acts on any k-dimensional vector v by rearranging its coefficients such that they descend, and taking the inner product with w. With abuse of notation, we will also apply F_w to sets of size k.

For the special cases of the basis vectors $e_1 = \langle 1, 0, \ldots, 0\rangle$ and $e_k = \langle 0, \ldots, 0, 1\rangle$, we get $F_{e_1} = \max$ and $F_{e_k} = \min$. While the choice of weights is in principle open, the idea of FRNN-OWA is to use weights that approximate max and min, such that the contribution of training instances to the membership of a test instance to the lower and upper approximations of a class gradually vanishes as the training distances are ranked further away from the test instance.

Thus, FRNN-OWA changes (2) into (3).

$$\begin{aligned} \overline{C}(y) &= F_{w_1}(\{R(y,x) | x \in C\}) \\ \underline{C}(y) &= F_{w_2}(\{1 - R(y,x) | x \in X \setminus C\}) \end{aligned} \tag{3}$$

Note that the use of OWA operators becomes computationally costly as the number of instances in the training set increases, since we need to sort all training instances for each test instance. The computational complexity of FRNN-OWA is $\mathcal{O}(dn + n\log(n))$ per test instance, for $n = |X|$ and $d = |A|$.

2.2 Big Data Implementations of Fuzzy Rough Sets

The existing literature on using fuzzy rough sets in a Big Data context is limited, and has focused on preprocessing algorithms, which reduce the size of training data, improve its quality, or both, by acting on its instances, its attributes, or both.

The first publication to explicitly adapt a fuzzy rough set algorithm for Big Data was by Asfoor et al. [1]. The authors point out that for a given information system (X, A) and fuzzy set C in X, the time complexity of calculating the membership of each instance of X in the lower and upper approximations of C is $\mathcal{O}(dn^2)$. In addition, the resulting indiscernibility matrix has size $\mathcal{O}(dn^2)$, and storing it in memory becomes highly problematic as n grows. They solve these challenges with a distributed implementation in Message Passing Interface (MPI) that avoids calculating and storing the whole matrix. This work was continued by Vluymans et al. [18], who present a distributed implementation in Apache Spark of Fuzzy Rough Prototype Selection (FRPS), a preprocessing algorithm for kNN classification developed in [16] and adapted in [18] for kNN regression. Asfoor [2] also adapts OWA-FRPS, a more robust version of FRPS with OWA operators, into a distributed implementation (POWA-FPRS) that approximates the ordered weighted average by partitioning the data and calculating the ordered weighted average of the ordered weighted averages within these partitions.

Jensen and Mac Parthaláin [10] point out that the calculation of fuzzy rough sets scales badly to large numbers of instances, and that this is further compounded if the feature space is also large. They propose three variants of Fuzzy Rough Feature Selection (FRFS). In nnFRFS and nnFDM (based on FRFS with Fuzzy Discernability Matrices), the indiscernibility relation is modified to only consider the k nearest neighbours of each instance. Fuzzy Rough Feature Grouping (FRFG) introduces a preliminary step in which overlapping groups of correlated features are defined. For each pass, only the most decisive feature from each group is considered, and other features in the same group are then skipped, thus reducing the number of candidates that have to be evaluated.

A number of other authors have presented Big Data implementations of FRFS. Qian et al. [14] propose to reduce the computational cost of FRFS by relaxing the calculations of the lower and upper approximations, potentially reducing the specificity of the resulting feature selection. Zeng et al. [22,23] present a mechanism to incrementally update fuzzy rough approximations in a hybrid information system (HIS) (in which a hybrid metric combines different types of attributes) and apply this to feature selection. Finally, Hu et al. [8] present a distributed implementation of multi-kernel attribute reduction using kernelised fuzzy rough sets, and evaluate the results for Support Vector Machines (SVM) and Classification and Regression Trees (CART).

As can be seen in Table 1, half of these works only use datasets with up to a few thousand instances. The connected studies of [1,2,18] work with generated datasets of up to 10,000,000 instances and only [8] tests on real datasets with more than one million instances.

Table 1. Articles with Big Data implementations of fuzzy rough algorithms—largest numbers of training instances in generated and real datasets

Article		Generated	Real
[1]	Asfoor et al. 2014	10,000,000	—
[18]	Vluymans et al. 2015	10,000,000	320,395
[2]	Asfoor 2015	10,000,000	320,395
[10]	Jensen and Mac Parthaláin 2015	—	832
[14]	Qian et al. 2015	—	2310
[23]	Zeng et al. 2015	—	2800
[22]	Zeng et al. 2017	—	2800
[8]	Hu et al. 2018	—	4,898,431
	Present study	16,777,216	11,000,000

The studies mentioned above have demonstrated the usefulness of scalable implementations of fuzzy rough prototype and feature selection. However, the potential to apply fuzzy rough classification algorithms in a big data context remains untapped, which is what we wish to address.

3 A Scalable Version of FRNN-OWA

We propose a parallel implementation of FRNN-OWA that can classify test instances with arbitrary large datasets in a fixed amount of time if we add sufficient parallel computing power.

FRNN-OWA is a 'nearest neighbour' classifier in the sense that if we use suitable weights, the influence of training instances vanishes as the training distances are ranked further away from a given test instance. So while, as mentioned in Sect. 2.1, sorting the entire set of training instances for each test instances is computationally costly, the precise order among the more distant training instances is actually of little consequence. For this reason, we adapt an idea from [10] (discussed in Sect. 2.2) and restrict the application of OWA weights to the k nearest training instances of a test instance y, within a class C for the upper approximation and without for the lower approximation, for some value k. We denote these by $\mathrm{NN}(y, C)$ and $\mathrm{NN}(y, X \setminus C)$ respectively.

The definitions for the upper and lower approximation which we use are given in (4), and we classify a test instance y to the class C for which the average of $\overline{C}(y)$ and $\underline{C}(y)$ is highest.

$$\begin{aligned} \overline{C}(y) &= F_{w_1}(\{R(y,x) | x \in \mathrm{NN}(y,C)\}) \\ \underline{C}(y) &= F_{w_2}(\{1 - R(y,x) | x \in \mathrm{NN}(y, X \setminus C)\}) \end{aligned} \tag{4}$$

We have chosen to use additive weights in this paper, defined as $w_1 = (\frac{2(k+1-i)}{k(k+1)})_{1 \le i \le k}$ and $w_2 = (\frac{2i}{k(k+1)})_{1 \le i \le k}$, and to set $k = 20$, after initial testing with different types of weights and a range of values for k on datasets of various sizes convinced us that these generally produce good results.

The time complexity of sorting all distances for every class is $\mathcal{O}(n \log(n))$, whereas the time complexity of identifying the k closest distances per class is just $\mathcal{O}(n)$. Since we do need to sort the k smallest distances per class, our proposal reduces the overall time complexity per test instance from $\mathcal{O}(dn + n \log(n))$ to $\mathcal{O}(dn + n + 2ck \log(k))$, where c is the number of classes. Since k and c are kept constant, for large n this further reduces to $\mathcal{O}((d+1)n)$. Thus, this variant of FRNN-OWA scales linearly with training set size.

There exist several different frameworks for parallel computing that provide different trade-offs between ease of use, automated performance optimisation and user control. Since our main objective is to demonstrate the conceptual viability of our approach, rather than to obtain the absolutely fastest run times possible, we have chosen to implement our algorithm in Spark, which offers a relatively straightforward path to parallelisation. We implement FRNN-OWA through the Python API of Spark, using high-level dataframe operations that allow us to express operations as SQL instructions which are automatically distributed across the nodes in the cluster.

Our implementation is structured as follows:

0. Initialise Spark.
1. Read the training set, combine all attributes into a feature vector. If the attributes are numerical, scale the features to $[0, 1]$.

2. Read the test set, combine all attributes into a feature vector. If the attributes are numerical, apply the same scaling as in step 1.
3. Optional: divide the training set from step 1 into a large number of small partitions.
4. Fill a dataframe of length k with additive weights.
5. Broadcast the test set from step 2 to all partitions, cross join with the training set from step 1, calculate the distance between each pair of test and training instances and select the k closest distances per class per test instance.
6. Cache the dataframe from step 5.
7. Join the weights from 4 with the distances from 5, multiply, and sum per class and test instance to get the upper approximations.
8. For every test instance and class, join the weights from step 5 with the k closest training instances from step 5 that do not belong to that class, multiply, and sum to get the lower approximations.
9. Join the upper and lower approximations from steps 7 and 8 and for every test instance, select the class for which the sum of the approximations is highest.
10. Divide the number of test instances from step 9 for which the predicted class matches the actual class by the total number of test instances and report the accuracy.

Step 3 was used only to prevent out-of-memory errors with the largest datasets when using multiple executors per node. Anecdotally, it seemed to increase run times, and so we did not include step 3 with our baseline measurements with only one core, so as not to obtain unduly positive speedups.

Step 5 is the costliest step, because it involves a cross join between training and test instances. Broadcasting the test set makes it available on all partitions, which means that the training set does not have to be replicated across partitions. Ordinarily, Spark would not preserve the resulting dataframe after its use in step 7, and would have to recalculate step 5 for step 8. To prevent this, we cache the dataframe in step 6.

4 Experimental Setup

All experiments were performed on the Golett cluster of the Ghent University Tier-2 of the Flemish Supercomputer Centre (VCS). The computing nodes of the Golett cluster are equipped with 2 × 12-core Intel E5-2680v3 (Haswell-EP @ 2.5 GHz) processors, 64 GB memory and 500 GB hard drives, and connected by FDR-10 InfiniBand. The experiments were run in Spark clusters of up to 64 executors, 4 cores per executor and 16 GB memory per executor. These Spark clusters occupied up to 32 nodes of the Golett cluster, with 8 cores per node. The algorithm was implemented in Spark 2.4.0 and run with the Hadoop Yarn resource manager.

The shared nature of the Golett cluster and the general inavailability of fully free nodes necessitated the choice of using only 8 cores per node, while limiting the number of cores per executor to 4 meant that two executors fit precisely onto one node. During initial testing, increasing the number of nodes per executor far above 4 led to diminishing returns. Of the 64 GB of memory per node, 8 GB was reserved for the operating system. Our cluster was limited to using one third of the remaining 56 GB on the basis of using one third of the number of cores. Thus, we chose 16 GB of memory per executor to maximise this resource, whereas in practice this amount was limited to 9.33 GB per executor.

The scaling of our implementation was tested on a series of generated datasets with varying training set sizes. Each training set had 20 real-valued attributes and 10 classes. Training set size varied from 2^{10} to 2^{24}.

The algorithm was also tested on four real datasets from the UCI Machine Learning Repository [6], summarised in Table 2. SUSY [4], HEPMASS [3] and HIGGS [4] are three large datasets of Monte Carlo simulations of particle physics collisions. The attributes are all real and indiscernibility was defined as the complement of the Manhattan distance, with both attributes and distance scaled to $[0, 1]$. *Poker hand* [5] is a slightly smaller dataset of possible hands of cards in the game of poker. It was included here because its attributes are categorical, necessitating a different indiscernibility relation. We chose the complement of the Hamming distance scaled to $[0, 1]$.

Table 2. Real datasets used in the present study, properties

Name	Number of instances	Attribute type	Number of attributes	Number of classes
Poker hand	1,025,010	Categorical	10	10
SUSY	5,000,000	Real	18	2
HEPMASS	10,500,000	Real	28	2
HIGGS	11,000,000	Real	28	2

Our primary performance measure is $T_{p,n}$, the time it takes using p cores to classify one test instance with n training instances. Time measurement starts with the initialisation of Spark and ends with the calculation of the accuracy. We report the average run time per test instance, derived from running the algorithm with a test set of 100 instances. These were, respectively, generated in addition to the generated training sets, and drawn and subtracted from the real training sets. For the generated training sets, we also report a speedup figure $S_{p,n}$ which is defined as $T_{1,n}/T_{p,n}$.

5 Results

Table 3 summarises the run times of our distributed implementation of FRNN-OWA for various generated training set sizes and various numbers of cores, and Table 4 the resultant speedups with respect to the baseline of using only one core. The speedups are also plotted in Fig. 1.

Table 3. Run times in seconds per test instance of FRNN-OWA applied to generated training sets of different sizes, for different numbers of cores

Cores	Training set size														
	2^{10}	2^{11}	2^{12}	2^{13}	2^{14}	2^{15}	2^{16}	2^{17}	2^{18}	2^{19}	2^{20}	2^{21}	2^{22}	2^{23}	2^{24}
1	0.83	0.83	1.3	1.3	1.9	3.1	6.1	11	21	50	104	201	428	858	1627
2	0.37	0.44	0.63	0.86	1.3	1.8	3.1	5.8	11	27	68	78	202	424	876
4	0.33	0.39	0.55	0.81	1.2	1.0	1.6	3.0	5.4	12	29	39	82	273	356
8	0.54	0.41	0.74	1.0	1.0	1.0	1.3	1.6	3.1	5.9	18	20	39	95	189
16	0.44	0.54	0.59	0.86	1.1	1.1	1.8	1.0	1.5	3.1	6.0	13	27	55	110
32	0.38	0.50	0.65	0.94	1.2	1.1	1.1	1.3	1.1	1.8	3.8	5.9	15	21	42
64	0.55	0.75	0.86	1.4	1.3	1.2	1.2	1.4	1.1	2.2	3.2	6.0	12	11	23
128	0.51	0.63	0.71	1.0	1.2	1.2	1.3	1.2	1.2	1.4	2.0	4.1	6.7	7.2	14
256	0.75	0.77	1.0	1.2	1.5	1.5	1.5	1.4	1.3	1.5	1.5	2.1	7.2	6.4	14

Values rounded for readability to two significant digits (<100) or whole integers (≥100

Table 4. Speedups of FRNN-OWA applied to generated training sets of different sizes, for different numbers of cores

Cores	Training set size														
	2^{10}	2^{11}	2^{12}	2^{13}	2^{14}	2^{15}	2^{16}	2^{17}	2^{18}	2^{19}	2^{20}	2^{21}	2^{22}	2^{23}	2^{24}
1	1	1	1	1	1	1	1	1	1	1	1	1	1	1	1
2	2.2	1.9	2.1	1.5	1.4	1.7	2.0	2.0	1.9	1.9	1.5	2.6	2.1	2.0	1.9
4	2.5	2.1	2.4	1.6	1.6	3.0	3.8	3.7	4.0	4.3	3.6	5.2	5.2	3.1	4.6
8	1.5	2.0	1.8	1.3	1.8	2.9	4.8	6.9	7.0	8.5	5.9	10	11	9.1	8.6
16	1.9	1.5	2.2	1.5	1.7	2.9	3.5	11	14	16	17	15	16	16	15
32	2.2	1.6	2.0	1.3	1.6	2.8	5.6	9.1	20	28	27	34	28	41	38
64	1.5	1.1	1.5	0.89	1.5	2.5	5.1	8.3	19	23	32	33	35	79	72
128	1.6	1.3	1.9	1.3	1.5	2.5	4.8	9.2	18	35	52	49	65	120	118
256	1.1	1.1	1.3	1.0	1.3	2.1	4.2	8.1	16	34	68	95	59	133	115

Values rounded for readability to two significant digits (<100) or whole integers (≥100)

The results show first of all that there is a certain amount of random fluctuation, which is to be expected on shared infrastructure. For training sets with fewer than 2^{11} instances, the overhead of the implementation is the dominating factor, and run time is effectively constant. For training sets with fewer than 2^{13} instances, overhead is still large enough that it negates the effect of adding more cores: speedup is constant. As training set size grows beyond 2^{13} instances, the speedup with p cores starts to climb more or less linearly until it reaches its theoretical maximum, p. This is reflected in the distinct diagonal cluster of lines in Fig. 1. Only the maximal configuration with 256 cores does not reach its full potential speedup within the space of these dataset sizes.

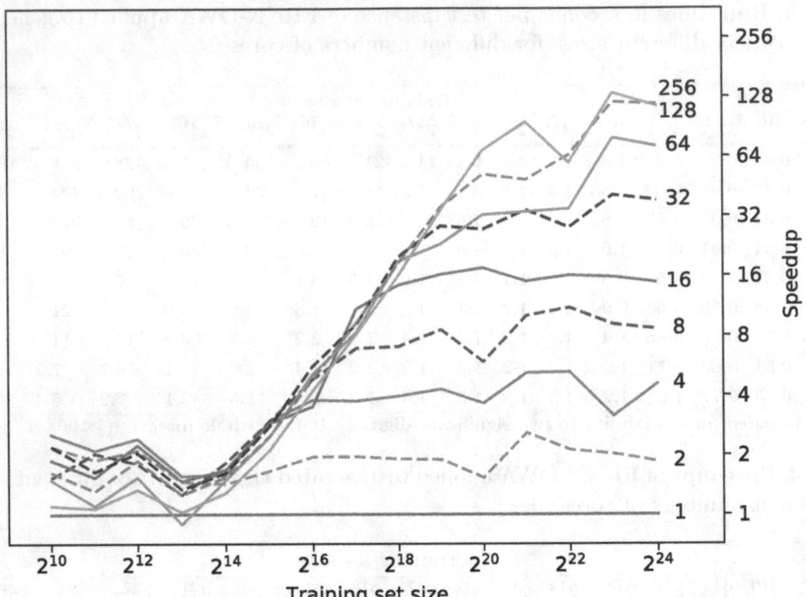

Fig. 1. Speedups for different numbers of cores, with FRNN-OWA applied to generated training sets of different sizes

Table 5. Run times per test instance of FRNN-OWA applied to real datasets, with 256 cores

Name	Time (s)
Poker hand	1.2
SUSY	4.3
HEPMASS	27
HIGGS	30

Table 5 shows the run times of our implementation of FRNN-OWA applied to the real datasets, which demonstrate that our implementation can be used to classify instances using FRNN-OWA with very large training sets.

6 Conclusion and Further Work

In this paper we have argued that until now, classifiers based on fuzzy rough sets have not been fit to handle Big Data, and that other attempts to adapt fuzzy rough sets for use with Big Data have mostly involved demonstrations on not very large datasets. To address this, we have presented the first implementation of a classifier based on fuzzy rough sets that can be scaled to handle arbitrarily large datasets. We have proposed a parallelised version of FRNN-OWA that can divide execution time over as many computing cores as is required.

To evaluate the performance of our implementation, we devised a series of systematic experiments, measuring run time on generated datasets varying in size from 2^{10} to 2^{24} instances, and calculating the speedup obtained by using between 1 and 256 computing cores. The results of these experiments showed that with sufficiently large datasets, the execution time of our implementation is effectively reduced by a factor equal to the number of computing cores. We then demonstrated that our implementation can be used for classifying test instances with a number of large real datasets of up to 11,000,000 instances.

We believe that the work presented in this paper constitutes a necessary first step towards adapting fuzzy rough sets for Big Data, and that it enables both the application of fuzzy rough sets to concrete classification problems, as well as several types of further research.

Having restricted the application of OWA operators to the k nearest neighbours of a test instance, a natural question to ask is what value for k is sufficiently large. In the future we wish to determine whether it is necessary to tune k for each dataset or whether a certain value is always good enough. This question also has to take into account the choice of weights. In fact, restricting the application of OWA operators to the k nearest neighbours opens up for consideration new types of weights whose accuracy reaches a global maximum for value of k and decreases as k approaches the full training set size.

We also want to investigate whether we can further reduce the computational complexity of FRNN-OWA by approximating some of the calculations. It is easy to think of Big Data merely in terms of large datasets that pose computational challenges. However, as data becomes available ever more easily in ever greater quantities, the types of questions that we want to answer change. Traditionally, researchers have asked which machine learning model can produce the best classification results for a given training set. But in a context where the amount of training data is essentially unlimited, it may be more relevant to ask which machine learning model can produce the best classification results in a given amount of time. If the accuracy loss from approximating parts of FRNN-OWA is less than the accuracy gain from the additional training data that can be processed in the same amount of time, this may be a worthwile trade-off.

Acknowledgement. The research reported in this paper was conducted with the financial support of the Odysseus programme of the Research Foundation – Flanders (FWO). D. Peralta is a Postdoctoral Fellow of the Research Foundation – Flanders (FWO).

References

1. Asfoor, H., et al.: Computing fuzzy rough approximations in large scale information systems. In: 2014 IEEE International Conference on Big Data (Big Data), pp. 9–16. IEEE (2014)
2. Asfoor, H.M.: Fuzzy rough set approximations in large scale information systems. Master's thesis, University of Washington (2015)

3. Baldi, P., Cranmer, K., Faucett, T., Sadowski, P., Whiteson, D.: Parameterized neural networks for high-energy physics. Eur. Phys. J. C **76**(5) (2016). Article number 235
4. Baldi, P., Sadowski, P., Whiteson, D.: Searching for exotic particles in high-energy physics with deep learning. Nat. Commun. **5**, 4308 (2014)
5. Cattral, R., Oppacher, F., Deugo, D.: Evolutionary data mining with automatic rule generalization. Recent Adv. Comput. Comput. Commun. **1**(1), 296–300 (2002)
6. Dua, D., Karra Taniskidou, E.: UCI machine learning repository (2017). http://archive.ics.uci.edu/ml
7. Dubois, D., Prade, H.: Rough fuzzy sets and fuzzy rough sets. Int. J. Gen. Syst. **17**(2–3), 191–209 (1990)
8. Hu, Q., Zhang, L., Zhou, Y., Pedrycz, W.: Large-scale multimodality attribute reduction with multi-kernel fuzzy rough sets. IEEE Trans. Fuzzy Syst. **26**(1), 226–238 (2018)
9. Jensen, R., Cornelis, C.: A new approach to fuzzy-rough nearest neighbour classification. In: Chan, C.-C., Grzymala-Busse, J.W., Ziarko, W.P. (eds.) RSCTC 2008. LNCS (LNAI), vol. 5306, pp. 310–319. Springer, Heidelberg (2008). https://doi.org/10.1007/978-3-540-88425-5_32
10. Jensen, R., Mac Parthaláin, N.: Towards scalable fuzzy-rough feature selection. Inf. Sci. **323**, 1–15 (2015)
11. Karau, H., Konwinski, A., Wendell, P., Zaharia, M.: Learning Spark: Lightning-Fast Big Data Analysis. O'Reilly Media Inc, Newton (2015)
12. Maillo, J., Luengo, J., García, S., Herrera, F., Triguero, I.: Exact fuzzy k-nearest neighbor classification for big datasets. In: 2017 IEEE International Conference on Fuzzy Systems (FUZZ-IEEE), pp. 1–6. IEEE (2017)
13. Maillo, J., Ramírez, S., Triguero, I., Herrera, F.: kNN-IS: an iterative spark-based design of the k-nearest neighbors classifier for big data. Knowl.-Based Syst. **117**, 3–15 (2017)
14. Qian, Y., Wang, Q., Cheng, H., Liang, J., Dang, C.: Fuzzy-rough feature selection accelerator. Fuzzy Sets Syst. **258**, 61–78 (2015)
15. Ramentol, E., et al.: IFROWANN: imbalanced fuzzy-rough ordered weighted average nearest neighbor classification. IEEE Trans. Fuzzy Syst. **23**(5), 1622–1637 (2015)
16. Verbiest, N., Cornelis, C., Herrera, F.: OWA-FRPS: a prototype selection method based on ordered weighted average fuzzy rough set theory. In: Ciucci, D., Inuiguchi, M., Yao, Y., Ślęzak, D., Wang, G. (eds.) RSFDGrC 2013. LNCS (LNAI), vol. 8170, pp. 180–190. Springer, Heidelberg (2013). https://doi.org/10.1007/978-3-642-41218-9_19
17. Verbiest, N., Cornelis, C., Jensen, R.: Fuzzy rough positive region based nearest neighbour classification. In: 2012 IEEE International Conference on Fuzzy Systems (FUZZ-IEEE), pp. 1–7. IEEE (2012)
18. Vluymans, S., et al.: Distributed fuzzy rough prototype selection for big data regression. In: 2015 Annual Conference of the North American Fuzzy Information Processing Society (NAFIPS) held jointly with 2015 5th World Conference on Soft Computing (WConSC), pp. 1–6. IEEE (2015)
19. Vluymans, S., Fernández, A., Saeys, Y., Cornelis, C., Herrera, F.: Dynamic affinity-based classification of multi-class imbalanced data with one-versus-one decomposition: a fuzzy rough set approach. Knowl. Inf. Syst. **56**(1), 55–84 (2018)
20. Vluymans, S., Sánchez Tarragó, D., Saeys, Y., Cornelis, C., Herrera, F.: Fuzzy rough classifiers for class imbalanced multi-instance data. Pattern Recognit. **53**, 36–45 (2016)

21. Yager, R.R.: On ordered weighted averaging aggregation operators in multicriteria decisionmaking. IEEE Trans. Syst. Man Cybern. **18**(1), 183–190 (1988)
22. Zeng, A., Li, T., Hu, J., Chen, H., Luo, C.: Dynamical updating fuzzy rough approximations for hybrid data under the variation of attribute values. Inf. Sci. **378**, 363–388 (2017)
23. Zeng, A., Li, T., Liu, D., Zhang, J., Chen, H.: A fuzzy rough set approach for incremental feature selection on hybrid information systems. Fuzzy Sets Syst. **258**, 39–60 (2015)

Learning Multi-granular Features for Harvesting Knowledge from Free Text

Zheng Zhou, Huaming Wang, Zhixing Li, Feng Hu(✉), and Guoyin Wang(✉)

Chongqing Key Laboratory of Computational Intelligence,
Chongqing University of Posts and Telecommunications,
Chongqing 400065, People's Republic of China
{hufeng,wanggy}@cqupt.edu.cn

Abstract. Extracting entities and their relations expressed in free text is essential to correct and populate knowledge graphs. Traditional methods assume that only the information of entities benefits the extraction of relations. They view this task as a two-step task, named entity recognition (NER) and relation classification (RC). However, the inadequate use of information and the error propagation problem constrain methods following this pipeline fashion. Joint extraction methods are proposed to incorporate useful interaction information between the two tasks for improvement, which solve NER and RC simultaneously. Although they have been proved to be superior to pipeline models, their performance is still far from satisfaction. In this paper, we try to combine the idea of data-driven granular cognitive computing and deep learning in joint extraction task. Accordingly, a neural-based joint extraction model named Joint extraction with Multi-granularity Context (JMC) is proposed. It explores the multi-granularity context of natural language sentences and uses neural networks to learn representations of these context automatically. Experiments results on NYT, a large data set produced by the distant supervision technique, show that JMC achieves comparative results to state-of-the-art methods.

Keywords: Knowledge extraction · Joint extraction · Data-driven granular cognitive computing · Deep learning

1 Introduction

There is massive free text containing considerable fragmented knowledge on the Web, which computers can only process with many constraints. With effective extracting methods, knowledge expressed in free text can be organized into structural knowledge bases, such as Knowledge Vault [6], Freebase [2] and Wikidata [28]. Then, the knowledge can be used to build question answering, semantic search and recommendation systems. However, existing knowledge graphs are mostly incomplete and noisy [7], as may lead to wrong decisions in knowledge-based systems. Coping with these problems still counts on knowledge expressed in free text, which is helpful to correct and populate the facts in knowledge graphs.

© Springer Nature Switzerland AG 2019
T. Mihálydeák et al. (Eds.): IJCRS 2019, LNAI 11499, pp. 210–224, 2019.
https://doi.org/10.1007/978-3-030-22815-6_17

An effort of handling knowledge in free text is open information extraction (OpenIE). However, its relation words are picked from the raw text, but it is common that relations are not expressed explicitly in natural language. Relation extraction, aiming to predict semantic relations between named entity pairs, has no such constrains. As a result, semantic relations conveyed implicitly in natural language can be uncovered effectively. Traditional relation extraction methods are often conducted on a pipeline fashion of two separated tasks: named entity recognition (NER) and relation classification (RC) [4,13,14,23]. The main drawback is that the error of entity recognition task may be propagated to relation classification task, limiting the final performance. Moreover, only the result of NER is applied to help RC task in a pipeline fashion.

Actually, entity recognition and relation classification are highly interrelated. Not only the results of NER can help determine the relations among entities, but the results of RC can also help improve the performance of NER. For example, the sentence *"Mrs. Tsuruyama is from Kumamoto Prefecture in Japan."* denotes that the person named *Mrs. Tsuruyama* lives in *Kumamoto Prefecture*. With such prior information that *Mrs. Tsuruyama* is a person and *Kumamoto Prefecture* is a location, the possibility of there is *Live_In* relation between these two entities is high. Besides, given that relation *Live_In* exists in *Mrs. Tsuruyama* and *Kumamoto Prefecture*, one can easily determine that *Mrs. Tsuruyama* is a person and *Kumamoto Prefecture* is a location. Under similar observation, joint extraction methods were designed to make NER and RC benefit from each other by incorporating the interaction information between them. Although joint extraction methods have been proved to be superior to pipeline methods, most of them still rely on millions of lexicalized features and higher-order term features like other natural language processing tasks [10,15,22]. These features are incomplete, sparse and costly in computing [3].

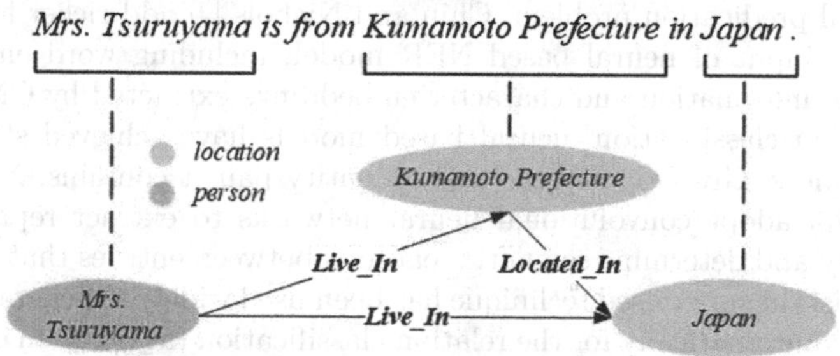

Fig. 1. Illustration of the joint extraction task.

Motivated by data-driven granular cognitive computing model [29], this paper explores multi-granular features for joint extraction task, including word-level features, local context features, segment context features and sentential context features. Moreover, we introduce these multi-granular prior knowledges to neural network architecture and propose a neural-based joint extraction method named

Joint Extraction with Multi-granular Context (JMC). Unlike traditional methods, JMC counts on neural network to learn representations of multi-granular context automatically instead of using hand-crafted features.

The main contributions of this paper are three-fold: (1) A neural model named JMC which extracts entities and relations jointly from unstructured text is proposed. (2) The idea of granular computing is introduced to joint extraction task to find multi-granular context features and design the corresponding neural network. (3) Experiments are conducted to evaluate the effectiveness of the proposed methods. Results imply that multi-granular context features can bring improvement to joint extraction task.

The rest of this paper is organized as follows. Section 2 briefly introduces related works of knowledge extraction. Section 3 states the joint extraction task and gives the multi-task objective. Section 4 depicts the proposed model. Section 5 gives the experiment results on a distant supervision corpus. Conclusions are shown in Sect. 6.

2 Related Works

2.1 Pipeline

Most existing works view relation extraction as a two-step task, where named entity recognition [13] is first conducted to determine the type of entities. Then, the information of entities are taken as input to identify the relations for entity pair [14,23]. Collobert et al. [5] propose a convolutional neural network based model for part-of-speech tagging, chunking, named entity recognition, and semantic role labeling. However, it eliminates the interactions among the predications. Lample et al. [13] modify it by replacing CNNs with bi-directional LSTMs to extract features. A conditional random layer is also adopted to solve the structural predication problem. Chiu and Nichols [4] add richer features for words as the input of neural based NER model, including word embeddings, capitalization information and character embeddings extracted by CNNs.

For relation classification, neural based models have achieved state-of-the-art performances. Given a sentence and an entity pair it contains, Nguyen and Grishman [23] adopt convolutional neural networks to extract representation automatically and determine semantic relations between entities that a sentence expresses. Distant supervised technique has been used widely to generate massive training data automatically for the relation classification task. For an entity pair, there is more than one sentence in distant supervised data set. Only part of them express the considered relation in extract operation, other sentences are noisy samples. To cope with the noise in distant supervised data sets, Lin et al. [17] take a batch of sentences as input and weight them using attention [31] to reduce the influence of noisy sentences. Considering information consistency and complementarity among texts in different languages, Lin et al. [16] generalizes the model to multi-lingual scenario.

2.2 Joint Extraction

Recent studies focus on designing more integrated models to capture the interdependencies between named entity recognition and relation classification tasks. Roth and Yih [27] adopt linear programming formulation to infer entities and relations simultaneously. Kate and Mooney [10] introduce a card-pyramid structure which encodes the entities and relations in a sentence. It adopts dynamic programming to solve the joint extraction task by labeling nodes in a card-pyramid structure jointly. Li and Ji [15] use a segment-based decoder based on the idea of semi-Markov chain to simultaneously extract entity mentions and relations with beam search. Miwa and Sasaki [22] propose the table representation that encodes entities and relations in a sentence. Besides, a history-based structured learning approach is proposed. Miwa and Bansal [21] present a joint model stacking bidirectional tree-structured LSTMs on bidirectional LSTMs to capture word sequence and dependency tree substructure.

Gupta et al. [9] view the entity recognition and relation classification as a table filling problem and design neural models based on multi-task recurrent neural networks to solve it. Zheng et al. [32] transform the joint extraction to a single tagging problem by fusing the relation types with the tags of NER. Ren et al. [26] first embed entity mentions, relation mentions, text features and type labels into two low-dimensional spaces where objects whose types are close also have similar representations. Then, the types of test mentions are estimated based on the learned embeddings. Katiyar and Cardie [11] propose an attention-based recurrent neural network for joint extraction of entity mentions and relations without using dependency trees. Adel and Schütze [1] utilize convolutional neural networks and linear-chain conditional random fields for joint extraction.

In this paper, we design an architecture for the joint extraction task. Different from existing joint extraction methods, it benefits from multi-granular context feature extracted automatically. Experiments results show that the proposed model achieves comparative or better results to state-of-the-art methods.

3 Problem Statement

This paper focuses on extracting facts from single sentence, leaving the integrating of information in multiple sentences for future study. Given a sentence $S = (w_1, w_2, ..., w_n)$, where w_i is the i-th word in the sentence and n is the sentence length. Let R be the set of the predefined semantic relations or the relations in knowledge graph. Set T contains the abstracted types of entities such as `PERSON` and `LOCATION`. Joint extraction is aimed at finding the mentions as well as types of entities and the relations between entities in S. The types of entities and relations are picked from T and R respectively. Challenges are three-fold. First, the extraction of entities and relations are highly related. Second, the assignment for entities are not independent. Third, the results could turn to be a multi-relational graph with the entities and relations in the sentence increasing, as Fig. 1 shows.

Actually, this task can be well represented as a table filling task [22]. As Table 1 shows, the table representation encodes the whole entity and relation structure in a sentence. The diagonal cells are tagged according to the relative position to its corresponding entity and the type of the entity. Other cells are filled with relation types and directions between words ($\rightarrow$ denotes the direction of relations and $\perp$ denotes the non-relation pair). Its relations are defined on word pairs, instead of entities, as enables it extracting relations from raw sentences directly. Besides, that the table structure captures multiple relations in a single sentence comes for free.

Table 1. The table representation of a sentence in joint extraction task.

	Mrs.	Tsuruyama	is	from	Kumamoto	Prefecture	in	Japan	.
Mrs.	B-PER,$\perp$								
Tsuruyama	$\perp$	L-PER,$\perp$							
is	$\perp$	$\perp$	O,$\perp$						
from	$\perp$	$\perp$	$\perp$	O,$\perp$					
Kumamoto	Live_in$\rightarrow$	$\perp$	$\perp$	$\perp$	B-LOC,$\perp$				
Prefecture	$\perp$	$\perp$	$\perp$	$\perp$	$\perp$	L-LOC,$\perp$			
in	$\perp$	$\perp$	$\perp$	$\perp$	$\perp$	$\perp$	O,$\perp$		
Japan	Live_in$\rightarrow$	$\perp$	$\perp$	$\perp$	$\perp$	Located_in$\rightarrow$	$\perp$	U-LOC,$\perp$	
.	$\perp$	$\perp$	$\perp$	$\perp$	$\perp$	$\perp$	$\perp$	$\perp$	O,$\perp$

4 Model

We consider the joint extraction task from granular computing perspective and propose to introduce multi-granular context features. Section 4.1 gives the details of multi-granular context. Section 4.2 introduce the details of the proposed model.

4.1 Multi-granular Features

For table filling tasks, relations are assigned on words. Only taking word itself as features would be very deficient. As a result, capturing rich contextual information is essential for determining the non-diagonal cells. This paper explores information from multi-granular context for the table filling task. For the convenience of statement, word on position i is marked as w_i, its tag, which corresponds with the diagonal cell in the table representation, is marked as t_i. The representation of i-th word is h_i.

Word Feature. Word feature is the representation of tokens. For filling the diagonal cells, only the very basic feature h_i is used. The word feature can be formulated as

$$feat_i^w = h_i \tag{1}$$

When determining other cells, feature h_i as well as its tag t_i is used. The word feature turns to be

$$feat_i^w = [h_i, t_i] \tag{2}$$

where $[\cdot]$ is the concatenation operation.

Local Context Feature. In natural language processing tasks, the surrounding words contribute to the understanding of current word. The local context feature is constituted by the information of surrounding words within the predefined window size. Taking the window size as c, the local context feature is

$$feat_i^{lc} = g(h_{i-c/2}, ..., h_{i+c/2}) \tag{3}$$

where $g(\cdot)$ is the feature extraction function. i is the index of the corresponding word.

Segment Context Feature. Previous works have shown the effectiveness of segment features in dependency parsing task. Table filling and dependency parsing share the characteristic that relations are defined on word pairs. Inspired by the graph-based dependency parsing model [30], we also divide a sentence into three parts (prefix, infix and suffix). The segment context of the dependency word pair is composed of these segments (parts). In this paper, the segment feature is used to produce the relation on word pair. For cell c_{ij} in the table representation, three types of segment feature are considered

$$\begin{aligned} feat_{ij}^{ps} &= k(h_0, ..., h_i) \\ feat_{ij}^{is} &= k(h_{i+1}, ..., h_j) \\ feat_{ij}^{ss} &= k(h_{j+1}, ..., h_n) \end{aligned} \tag{4}$$

where $k(\cdot)$ is the feature extraction function. $feat_{ij}^{ps}$, $feat_{ij}^{is}$ and $feat_{ij}^{ss}$ represent the segments which split by the indexes i and j. The final segment feature is the concatenation of the representations of three segments, formulated as

$$feat_{ij}^{seg} = [feat_{ij}^{ps}, feat_{ij}^{is}, feat_{ij}^{ss}] \tag{5}$$

Sentential Context Feature. The global information can also help the determination of relations. For example, given the prior knowledge that only the `Live_In` relation exists in the given sentence, one could avoid illegal assignments to cells. Sentential context feature captures the global information over the entire sentence, which can be formulated as

$$feat^s = o(h_1, h_2, ..., h_n) \tag{6}$$

where $o(\cdot)$ is the feature extraction function and n is the sentence length.

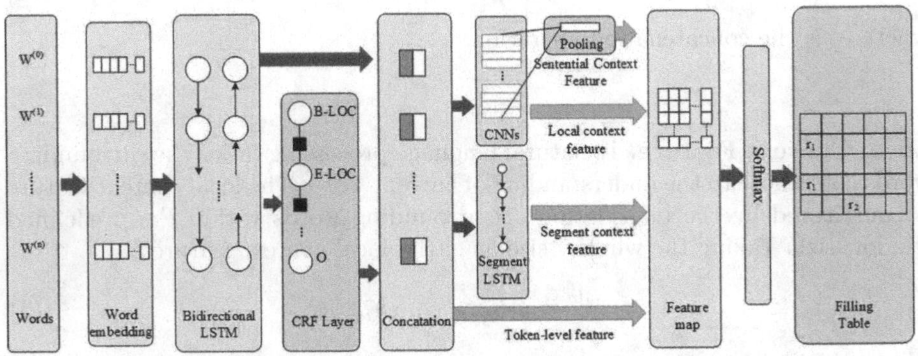

Fig. 2. The JMC architecture. Bidirectional LSTM layer, CNNs&Pooling layer and segment LSTM layer produce multi-granular features, including word feature, local context feature, segment context feature and sentential context feature.

4.2 The Proposed Joint Extraction Model

Different from traditional methods, we propose to learn these features automatically with neural model instead of designing extraction functions by hand. Word feature is generated by feeding the embedding of words into a bi-directional long-short term memory network. Local and sentential context feature are given by convolutions and polling. For segment context feature, a forward LSTM layer is adopted following [30].

Figure 2 depicts the architecture of the proposed joint extraction model. JMC takes only word unigram as input and then leaves the feature combinations learned by the model automatically. First, it embeds words into dense vectors using pre-trained word2vec. Second, following the structure of BiLSTM-CRF (bidirectional long-short term memory network and conditional random field) for NER, dense vectors of words are feed into bi-directional LSTM layer, dense hidden layer and CRF layer sequentially. Then, the NER tags are produced by CRF layer. Third, the outputs of BiLSTM are concatenated with the one-hot vectors of NER tags as word features. They are feed into a forward LSTM and CNNs to generate segment context feature, local context feature and global context feature. The concatenation of these features is taken as the basic representation of cells in feature map.

Word Embedding. Words are discrete and sparse in nature. We adopt a word embedding layer to represent the word. It maps a word to a dense vector of pre-defined dimensionality. The word embedding layer is initialized with the pre-trained 300 dimensional GloVe[1] word vectors trained on Wikipedia corpus.

BiLSTM Layer. Bi-directional LSTM (BiLSTM), presenting each sequence forwards and backwards to two separate hidden states to capture past and future information, has been proved to be effective in sequence labeling tasks. The representation of a word produced by Bi-LSTM is obtained by concatenating its left and right context.

$$h_t = [\overrightarrow{h_t}, \overleftarrow{h_t}] \tag{7}$$

where h_t is the output of the Bi-LSTM layer. $\overrightarrow{h_t}$ and $\overleftarrow{h_t}$ are the output vector of forward and backward LSTM respectively.

Suppose x_t and $\overrightarrow{h_t}$ are the word embedding and the hidden state at time t. The states of forward LSTM unit at time t can be formulated as

$$\begin{aligned} i_t &= \sigma(W_i \overrightarrow{h_{t-1}} + U_i x_t + b_i) \\ f_t &= \sigma(W_f \overrightarrow{h_{t-1}} + U_f x_t + b_f) \\ \tilde{c}_t &= \tanh(W_c \overrightarrow{h_{t-1}} + U_c x_t + b_c) \\ c_t &= f_t \odot c_{t-1} + i_t \odot \tilde{c}_t \\ o_t &= \sigma(W_o \overrightarrow{h_{t-1}} + U_o x_t + b_o) \\ \overrightarrow{h_t} &= o_t \odot \tanh c_t \end{aligned} \tag{8}$$

where $\sigma(\cdot)$ is element-wise sigmoid function and $\odot$ is the element-wise product. U_i, U_f, U_c, U_o and W_i, W_f, W_c, W_o denote the weight matrices of different gates. b_i, b_f, b_c and b_o are the weight matrices and bias vectors. The formulation of the backward LSTM is similar to Eq. 8.

CRF Layer. Conditional Random Field(CRF) layer has been successively used in tagging models. We also use it to model the interdependencies among NER tags. Given an input sentence $X = (x_1, x_2, ..., x_n)$, $P = (p_1, p_2, ..., p_n)$ is considered as the score vectors delivered by the BiLSTM. p_i is a score vector of word x_i whose size is $1 \times k$, where k is the number of distinct tags for NER task.

Given the prediction tags $Y = (y_1, y_2, ..., y_n)$, where y_i is chosen from the tag set $T = \{t_1, t_2, ..., t_k\}$. The score is defined as

$$s(X, Y) = \Sigma_{i=0}^{n} A_{y_i, y_{i+1}} + \Sigma_{i=1}^{n} p_i^{y_i} \tag{9}$$

[1] https://nlp.stanford.edu/projects/glove/.

where A is the transition matrix and $A_{i,j}$ denotes the transition score from tag i to tag j. We predict the output sequence by maximizing the score

$$Y^* = argmax_{y \in \bar{Y}} s(X, y) \tag{10}$$

where $\bar{Y}$ contains all possible output sequences of the input sentence X.

Segment LSTM Layer. We consider three segments described and adopt a forward LSTM layer to learn their representations in Eq. 4. The representation of infix segment is considered as the hidden state of the head word. The representation of inner segment is obtained by subtraction between the hidden vector of the tail word and the head word. For the suffix segment, its representation is the subtraction of the last hidden vector and the hidden state of the tail word. When there has no prefix or suffix, the corresponding embedding is set to zero vector.

Softmax Layer. A Softmax classifier is adopted to determine the relation that the word pair hold. The relation between word i and word j is produced by

$$r_{ij} = softmax(WT'_{ij} + b) \tag{11}$$

where the W and b are weight matrix and bias vector. Besides, instead of feeding the feature table generated by the table convolution layer into the Softmax classifier directly, we add a hidden layer ahead of it, which transforms the representation of each cell into a new feature space with much lower dimensionality.

Objective Function. This paper follows the multi-task framework to avoid the error propagation problem in the pipeline framework. Basic features learned automatically are shared by these two tasks and their objectives are optimized jointly. Let the given sentence be $S = (w_1, w_2, ..., w_n)$. For named entity recognition, the objective function is

$$L^{ner} = \sum_{i=1}^{|D|} \sum_{t=1}^{n_i} (log(p_t^{(i)} = y_t^{(i)} | x^{(i)}, \Theta)) \tag{12}$$

where $|D|$ is the size of training set, n_i is the length of sentence $x^{(i)}$. $y_t^{(i)}$ is the correct tag[2] of word t in sentence $x^{(i)}$ and $p_t^{(i)}$ is the normalized probabilities of tags produced by the model. Besides, Θ is the parameter of the joint model. For relation classification, the objective function is

$$L^{rc} = \sum_{i=1}^{|D|} \sum_{m,n=1}^{n_i^2} (log(c_{mn}^{(i)} = y_{mn}^{(i)} | x^{(i)}, \Theta)) \tag{13}$$

[2] Entity type encoded in BILOU (Begin, Inside, Last, Outside, Unit) scheme.

where $c_{mn}^{(i)}$ is the ground truth relation between m-th word and n-th word in the sentence $x^{(i)}$. The multi-task objective function is

$$L = \alpha L^{ner} + (1-\alpha)L^{rc} \tag{14}$$

where α is the trade-off weight between named entity recognition and relation classification tasks.

5 Experiments

5.1 Implement Details

We use Tensorflow[3] framework to implement our joint extraction model. All hyper-parameters are tuned on the development set. The weights of word embedding are pre-trained by [24] and the dimensionality of embedding vectors is 300. The numbers of hidden units of forward and backward LSTM are both 64. The weights and biases are updated using gradient based optimizer Adam [12] by minimizing crossentropy of the output of CRF layer and softmax layer. The learning rate is initialed to 0.01 and reduced half when there has no decrements of loss. To avoid overfitting, we add dropout operations after the BiLSTM with the dropout rate of 0.2. Early stop technique is also adopted. More detailed setting of parameters can be found in the source code[4].

5.2 Data Set

Distant supervision methods can produce a large amount of training data automatically. With manually labeled test set, its quality can be ensured despite containing noise. Distant supervision has been used in many natural language processing tasks [19,26]. To evaluate the effectiveness of our methods detailedly, we test the proposed method on the public dataset NYT [26], produced by distant supervision technique. There are 353k triplets in the training data and 3,880 triplets in the test set. Besides, the number of valid relations is 24 and `None` is viewed as the undefined relation `UND`.

5.3 Compared Methods

We choose joint extraction methods producing state-of-the-art results on NYT as comparatives. **DS+Logistic** [20] trains a multi-class logistic classifier to predict relations. **DeepWalk** [25] embeds mention-feature co-occurrences and mention-type associations as a homogeneous network. **FCM** [8] adopts neural language model to perform compositional embedding. **Cotype** [26] first runs text segmentation algorithm to extract entity mentions. Then, entity mentions, relation mentions, text features and type labels are embedded into two low-dimensional

[3] www.tensorflow.org.
[4] https://github.com/MingYates/JMC.

spaces. In each space, mentions with close types also have similar representations. **LSTM-LSTM** [32] converts the joint extraction task to a tagging problem and solves it using LSTMs. **REHESSION** [19] benefits from heterogeneous information source, for example, knowledge base and domain heuristics. Besides, state-of-the-art tagging model **BiLSTM+CRF** is also selected as a comparative on the named entity recognition task.

5.4 Results of Named Entity Recognition

We take Strict-F1, Macro-F1 and Micro-F1 proposed in [18] as evaluations for NER. Results are shown in Table 2. BiLSTM+CRF and JMC outperform other methods with more than 0.30 on Strict-F1. The reason might be that **DeepWalk** and **Cotype** have a preprocess step of entity mention detection and the error of entity mention detection will propagate to entity typing. Moreover, the results denote that tagging based NER can also achieve comparative results on distant supervision data set.

Table 2. Performance of named entity recognition on NYT

Methods	Strict-F1	Macro-F1	Micro-F1
DS+Logistic [20]	-	-	-
DeepWalk [25]	0.49	0.54	0.53
FCM [8]	-	-	-
LSTM-LSTM [32]	-	-	-
Cotype [26]	0.60	0.65	0.66
REHESSION [19]	-	-	-
BiLSTM+CRF	0.89	0.91	0.90
JMC (proposed)	**0.94**	**0.93**	**0.91**

5.5 Results of Relation Classification

For a sentence, it is considered correct if the predicted relations are correct without considering the results of entities. Besides, we ignore `BLANK` and `UND` relations and only report the accuracy for valid relations as [26] does. As Fig. 3 shows, JMC produces the best results on relation classification task. It is worth to mention that the proposed method only takes words as input, while **Cotype** and **REHESSION** introduce external knowledge bases.

5.6 Results of Joint Extraction

Performances on the setting of end-to-end relation extraction are also reported in Table 3. A sentence is considered correct if the entities and relations are correct. The results of comparative methods are reported in their original papers

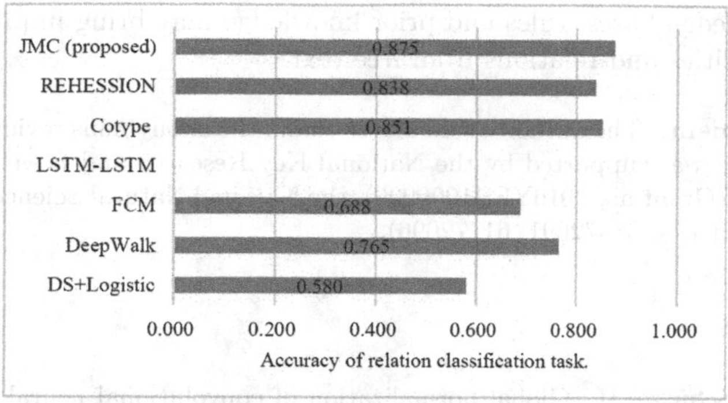

Fig. 3. Accuracy of relation classification on NYT

adopting the same criteria [19,26]. As Table 3 says, JMC produces the highest recall and F1 score compared to other methods. Besides, it gives comparative results evaluated by precision.

Table 3. Performance of joint extraction on NYT

Methods	Precision	Recall	F1
DS+Logistic [20]	0.258	0.393	0.311
DeepWalk [25]	0.176	0.224	0.197
FCM [8]	0.553	0.154	0.240
LSTM-LSTM [32]	**0.615**	0.414	0.495
Cotype [26]	0.423	0.511	0.463
REHESSION [19]	0.412	0.573	0.479
JMC (proposed)	0.524	**0.657**	**0.583**

6 Conclusions

This paper studies joint extraction of entities and relations from free text. Considering that the ground truth of part-of-speech tags and dependency trees are not available in real applications, we design a neural model extracting entities and relations jointly which only takes words as input. Different from existing joint extraction methods, the proposed model needs no hand-designed features and learns representations of multi-granular context among outputs on feature automatically. Results on distant supervision data set show that the proposed method produces comparative performance compared to state-of-the-art methods in the setting of named entity recognition, relation classification and end-to-end joint extraction. For the future works, incorporating heterogeneous source such as knowledge bases, rules and prior knowledge may bring improvement for extraction entities and relations from free text.

Acknowledgment. The author would like to thank the anonymous reviewers for their help. This work was supported by the National Key Research and Development Program of China (Grant no. 2016YFB1000905), the National Natural Science Foundation of China (Grant nos. 61572091, 61772096).

References

1. Adel, H., Schütze, H.: Global normalization of convolutional neural networks for joint entity and relation classification. In: Proceedings of the 2017 Conference on Empirical Methods in Natural Language Processing, EMNLP 2017, Copenhagen, Denmark, 9–11 September 2017, pp. 1723–1729 (2017)
2. Bollacker, K.D., Evans, C., Paritosh, P., Sturge, T., Taylor, J.: Freebase: a collaboratively created graph database for structuring human knowledge. In: Proceedings of the ACM SIGMOD International Conference on Management of Data, SIGMOD 2008, Vancouver, BC, Canada, 10–12 June 2008, pp. 1247–1250 (2008)
3. Chen, D., Manning, C.D.: A fast and accurate dependency parser using neural networks. In: Proceedings of the 2014 Conference on Empirical Methods in Natural Language Processing, EMNLP 2014, A Meeting of SIGDAT, A Special Interest Group of the ACL, 25–29 October 2014, Doha, Qatar, pp. 740–750 (2014)
4. Chiu, J.P.C., Nichols, E.: Named entity recognition with bidirectional LSTM-CNNs. TACL **4**, 357–370 (2016)
5. Collobert, R., Weston, J., Bottou, L., Karlen, M., Kavukcuoglu, K., Kuksa, P.P.: Natural language processing (almost) from scratch. J. Mach. Learn. Res. **12**, 2493–2537 (2011)
6. Dong, X., et al.: Knowledge vault: a web-scale approach to probabilistic knowledge fusion. In: The 20th ACM SIGKDD International Conference on Knowledge Discovery and Data Mining, KDD 2014, New York, NY, USA, 24–27 August 2014, pp. 601–610 (2014)
7. Färber, M., Bartscherer, F., Menne, C., Rettinger, A.: Linked data quality of DBpedia, freebase, OpenCyc, Wikidata, and YAGO. Semant. Web **9**(1), 77–129 (2018)
8. Gormley, M.R., Yu, M., Dredze, M.: Improved relation extraction with feature-rich compositional embedding models. In: Proceedings of the 2015 Conference on Empirical Methods in Natural Language Processing, EMNLP 2015, Lisbon, Portugal, 17–21 September 2015, pp. 1774–1784 (2015)
9. Gupta, P., Schütze, H., Andrassy, B.: Table filling multi-task recurrent neural network for joint entity and relation extraction. In: COLING 2016, 26th International Conference on Computational Linguistics, Proceedings of the Conference: Technical Papers, Osaka, Japan, 11–16 December 2016, pp. 2537–2547 (2016)
10. Kate, R.J., Mooney, R.J.: Joint entity and relation extraction using card-pyramid parsing. In: Proceedings of the Fourteenth Conference on Computational Natural Language Learning, CoNLL 2010, Uppsala, Sweden, 5–16 July 2010, pp. 203–212 (2010)
11. Katiyar, A., Cardie, C.: Going out on a limb: joint extraction of entity mentions and relations without dependency trees. In: Proceedings of the 55th Annual Meeting of the Association for Computational Linguistics, ACL 2017, Vancouver, Canada, 30 July–4 August, vol. 1: Long Papers, pp. 917–928 (2017)

12. Kingma, D.P., Ba, J.: Adam: a method for stochastic optimization. CoRR abs/1412.6980 (2014)
13. Lample, G., Ballesteros, M., Subramanian, S., Kawakami, K., Dyer, C.: Neural architectures for named entity recognition. In: NAACL HLT 2016, The 2016 Conference of the North American Chapter of the Association for Computational Linguistics: Human Language Technologies, San Diego, California, USA, 12–17 June 2016, pp. 260–270 (2016)
14. Lee, J.Y., Dernoncourt, F., Szolovits, P.: MIT at semeval-2017 task 10: relation extraction with convolutional neural networks. In: Proceedings of the 11th International Workshop on Semantic Evaluation, SemEval@ACL 2017, Vancouver, Canada, 3–4 August 2017, pp. 978–984 (2017)
15. Li, Q., Ji, H.: Incremental joint extraction of entity mentions and relations. In: Proceedings of the 52nd Annual Meeting of the Association for Computational Linguistics, ACL 2014, Baltimore, MD, USA, 22–27 June 2014, vol. 1: Long Papers, pp. 402–412 (2014)
16. Lin, Y., Liu, Z., Sun, M.: Neural relation extraction with multi-lingual attention. In: Proceedings of the 55th Annual Meeting of the Association for Computational Linguistics, ACL 2017, Vancouver, Canada, 30 July–4 August, vol. 1: Long Papers, pp. 34–43 (2017)
17. Lin, Y., Shen, S., Liu, Z., Luan, H., Sun, M.: Neural relation extraction with selective attention over instances. In: Proceedings of the 54th Annual Meeting of the Association for Computational Linguistics, ACL 2016, 7–12 August 2016, Berlin, Germany, vol. 1: Long Papers (2016)
18. Ling, X., Weld, D.S.: Fine-grained entity recognition. In: Proceedings of the Twenty-Sixth AAAI Conference on Artificial Intelligence, Toronto, Ontario, Canada, 22–26 July 2012 (2012)
19. Liu, L., et al.: Heterogeneous supervision for relation extraction: a representation learning approach. In: Proceedings of the 2017 Conference on Empirical Methods in Natural Language Processing, EMNLP 2017, Copenhagen, Denmark, 9–11 September 2017. pp. 46–56 (2017)
20. Mintz, M., Bills, S., Snow, R., Jurafsky, D.: Distant supervision for relation extraction without labeled data. In: ACL 2009, Proceedings of the 47th Annual Meeting of the Association for Computational Linguistics and the 4th International Joint Conference on Natural Language Processing of the AFNLP, Singapore, 2–7 August 2009, pp. 1003–1011 (2009)
21. Miwa, M., Bansal, M.: End-to-end relation extraction using LSTMs on sequences and tree structures. In: Proceedings of the 54th Annual Meeting of the Association for Computational Linguistics, ACL 2016, Berlin, Germany, 7–12 August 2016, vol. 1: Long Papers (2016)
22. Miwa, M., Sasaki, Y.: Modeling joint entity and relation extraction with table representation. In: Proceedings of the 2014 Conference on Empirical Methods in Natural Language Processing, EMNLP 2014, A Meeting of SIGDAT, A Special Interest Group of the ACL, Doha, Qatar, 25–29 October 2014, pp. 1858–1869 (2014)
23. Nguyen, T.H., Grishman, R.: Relation extraction: perspective from convolutional neural networks. In: Proceedings of the 1st Workshop on Vector Space Modeling for Natural Language Processing, VS@NAACL-HLT 2015, Denver, Colorado, USA, 5 June 2015, pp. 39–48 (2015)
24. Pennington, J., Socher, R., Manning, C.: Glove: global vectors for word representation. In: Proceedings of the 2014 Conference on Empirical Methods in Natural Language Processing (EMNLP), pp. 1532–1543 (2014)

25. Perozzi, B., Al-Rfou, R., Skiena, S.: DeepWalk: online learning of social representations. In: The 20th ACM SIGKDD International Conference on Knowledge Discovery and Data Mining, KDD 2014, New York, NY, USA, 24–27 August 2014, pp. 701–710 (2014)
26. Ren, X., et al.: CoType: joint extraction of typed entities and relations with knowledge bases. In: Proceedings of the 26th International Conference on World Wide Web, WWW 2017, Perth, Australia, 3–7 April 2017, pp. 1015–1024 (2017)
27. Roth, D., Yih, W.T.: Global inference for entity and relation identification via a linear programming formulation. In: Introduction to Statistical Relational Learning, pp. 553–580 (2007)
28. Vrandecic, D., Krötzsch, M.: Wikidata: a free collaborative knowledgebase. Commun. ACM **57**(10), 78–85 (2014)
29. Wang, G.: Data-driven granular cognitive computing. In: Polkowski, L., et al. (eds.) IJCRS 2017. LNCS (LNAI), vol. 10313, pp. 13–24. Springer, Cham (2017). https://doi.org/10.1007/978-3-319-60837-2_2
30. Wang, W., Chang, B.: Graph-based dependency parsing with bidirectional LSTM. In: Proceedings of the 54th Annual Meeting of the Association for Computational Linguistics, ACL 2016, 7–12 August 2016, Berlin, Germany, vol. 1: Long Papers (2016)
31. Xu, K., et al.: Show, attend and tell: neural image caption generation with visual attention. CoRR abs/1502.03044 (2015)
32. Zheng, S., Wang, F., Bao, H., Hao, Y., Zhou, P., Xu, B.: Joint extraction of entities and relations based on a novel tagging scheme. In: Proceedings of the 55th Annual Meeting of the Association for Computational Linguistics, ACL 2017, Vancouver, Canada, 30 July–4 August, vol. 1: Long Papers, pp. 1227–1236 (2017)

Building a Framework of Rough Inclusion Functions by Means of Computerized Proof Assistant

Adam Grabowski(✉)

Institute of Informatics, University of Białystok, Konstantego Ciołkowskiego 1M, 15-245 Białystok, Poland

adam@math.uwb.edu.pl

Abstract. The paper describes some of the issues concerning the development of automated formal framework for the reasoning about rough inclusion functions, starting with the classical one, and generalizations thereof. We work with the Mizar system; the viewpoint of the rough set theory, and especially mereology by Leśniewski, can allow for the creation of new foundations for the Mizar Mathematical Library, or at least for fresh branch of this formal database, originally based on Tarski-Grothendieck axioms.

Keywords: Rough inclusion function · Rough approximation space · Mizar Mathematical Library · Automated theorem proving

1 Introduction

Rough sets discovered by Pawlak [19] are a tool for knowledge discovery and modelling under imperfect information; this is especially the case nowadays, where we often face large databases of information gathered from various sources. In such situations the use of computer methods in contemporary mathematics seems to be unquestionable. On the other hand, we can also use specialized software to test the correctness of our reasoning and, in the same time, to formulate and prove new hypotheses automatically, or at least with an extensive support from machines.

Rough Inclusion Functions (RIFs for short) seem to be a kind of a bridge between classical set theory and theory of rough sets. They reflect probabilistic nature of rough sets, and focus rather on quantitative than their qualitative nature. At the very first sight, the foundational issues are very important as (rough) inclusions are very fundamental – based on this single predicate one can build the whole theory, as it was done in the case of Leśniewski mereology.

As the proof assistant we have chosen the Mizar system [1], relatively well-known system based on classical logic, together with its repository of texts formally verified by computer [23,26] – the Mizar Mathematical Library (MML). This year MML celebrates thirty years; the first Mizar article, *"Tarski-Grothendieck set theory"* by Andrzej Trybulec, accepted January 1, 1989, made

© Springer Nature Switzerland AG 2019
T. Mihálydeák et al. (Eds.): IJCRS 2019, LNAI 11499, pp. 225–238, 2019.
https://doi.org/10.1007/978-3-030-22815-6_18

axiomatic foundations for the repository; its logical framework was established much earlier. The research on Łukasiewicz inclusion function could be also an interesting addition to this database on its own, regardless of the rough set theory context. Our leading idea however, was to follow the parts of Gomolińska's work [4].

The paper is organized as follows. The next section contains an outline of the notion of rough inclusion functions, starting with the classical $\kappa^{\pounds}$, and two other ones, considered by Gomolińska. Section 3 describes briefly the formalization of rough sets with the use of Mizar proof assistant, and in two following sections we focus on the formal translation of classical rough inclusions and RIFs in general. Section 6 show how this newly created object can really extend the hierarchy of types assuring possibly high level of generalization of theorems and to bridge the gap in already existing formal developments (Sect. 7). Eighth section shows that not all proofs were as simple as we could wish, even based on informally elementary example of concrete subsets of an approximation space, while the last section draws some conclusions and plans for future research.

2 Rough Inclusion Functions

Mereology was authored by S. Leśniewski; it is worth mentioning at this point that for some fifteen years there was a Leśniewski Award (established by the Association of Mizar Users in 1989) granted to authors of Mizar articles with the greatest number of references in the MML – to appreciate the impact of the chosen formalization for the whole computer-checked repository.

The primitive notion, or rather a predicate, "being a part of", with the mixture of axioms and definitions, proposed by Leśniewski in 1916, gained another new life from the works of rough set researchers, with Polkowski and Skowron [22] as the pioneers, focusing rather on "being a part of to degree" concept, at the same time opening the route of granular computing and rough mereology [21]. Potentially then, instead of reusing set theory available in the repository of Mizar texts, one can define a ternary relation $\mu(X, Y, r)$ which could be read as "the object X is a part of Y to a degree at least r" and use this predicate instead of

$$\kappa(X, Y) \leq r.$$

This is still tempting as many theorem provers work with higher-order logic completely abstracting from set theory. However, our aim was to reuse as much set theory as we can, with the future possibility of mereological background for MML.

More generally, for a given universe U, rough inclusion functions (RIFs for short) are the mappings κ from $\wp U \times \wp U$ into unit interval which satisfy two properties:

$$\mathrm{rif}_1(\kappa) \Leftrightarrow \forall_{X,Y \subseteq U} \ (\kappa(X,Y) = 1 \Leftrightarrow X \subseteq Y)$$
$$\mathrm{rif}_2(\kappa) \Leftrightarrow \forall_{X,Y,Z \subseteq U} \ (Y \subseteq Z \Rightarrow \kappa(X,Y) \leq \kappa(X,Z))$$

The formulas mean that they are monotone with respect to the second coordinate and reflect the idea of fuzzy implication (see a footnote in Sect. 6). Essentially, the correspondence with the fuzzy set theory [28], as well as another postulates satisfied by RIFs [2] can be easily seen. Let us recall three basic examples of rough inclusion functions:

$$\kappa^{\pounds}(X,Y) = \begin{cases} \frac{|X\cap Y|}{|X|} & \text{if } X \neq \emptyset \\ 1 & \text{otherwise} \end{cases}$$

$$\kappa_1(X,Y) = \begin{cases} \frac{|Y|}{|X\cup Y|} & \text{if } X \cup Y \neq \emptyset \\ 1 & \text{otherwise} \end{cases}$$

$$\kappa_2(X,Y) = \frac{|(U-X)\cup Y|}{|U|},$$

where $|X|$ denotes cardinality of set X, the universe U is non-empty finite set of objects.

The first one is the most popular – standard RIF, $\kappa^{\pounds}$. A similar idea, closely related to the conditional probability, was explored by J. Łukasiewicz around 1913. The next one, κ_1, was introduced by Gomolińska [4]. The operator κ_2 was considered by G. Drwal and A. Mrózek in 1998. All three RIFs are different but interdefinable:

$$\kappa_1(X,Y) = \kappa^{\pounds}(X\cup Y, Y)$$
$$\kappa_2(X,Y) = \kappa^{\pounds}(U, (U-X)\cup Y)$$
$$\kappa^{\pounds}(X,Y) = \kappa_1(X, X\cap Y).$$

Furthermore, we know that they can be ordered as

$$\kappa^{\pounds}(X,Y) \leq \kappa_1(X,Y) \leq \kappa_2(X,Y).$$

We can construct relatively simple example of subsets of an approximation space where all three are really distinct – the relations are sharp; the discussion will be given in Sect. 8.

3 Rough Sets and Automated Reasoning

Pawlak's early works were devoted to automated reasoning, making it more understandable for ordinary people outside of academia, which is quite unusual nowadays. As far as I know, rough sets are absent in any other popular repositories offered by computer proof assistants: Isabelle/HOL Archive of Formal Proofs, Metamath or Coq. The core idea of the approach available in the Mizar Mathematical Library reflects faithfully original idea by Pawlak: first of all, approximation spaces are defined generally as relational structures, i.e. sets equipped by the indiscernibility binary relation `InternalRel`. Under specific assumptions, we can prove standard properties of lower and upper approximations [14,15].

```
definition
  let A be non empty RelStr;
  let X be Subset of A;
  func LAp X -> Subset of A equals   :: ROUGHS_1:def 4
  { x where x is Element of A : Class (the InternalRel of A, x) c= X };
  correctness;
end;
```

Apart from approximations, the connection of rough sets with conditional probability was already expressed in the following formal definition of a membership function:

```
definition
  let A be finite Tolerance_Space;
  let X be Subset of A;
  func MemberFunc (X, A) -> Function of the carrier of A, REAL means
:: ROUGHS_1:def 9
   for x being Element of A holds
   it.x = card (X /\ Class (the InternalRel of A, x)) /
     (card Class (the InternalRel of A, x));
end;
```

Classical inclusion is practically the core of set-theoretic part of the MML and, of course, it was used already in the aforementioned definition of the lower approximation. As a counterpart, it was used also in the following definition of rough inclusion:

```
definition
  let A be Tolerance_Space, X, Y be Subset of A;
  pred X _c= Y means     :: ROUGHS_1:def 11
  LAp X c= LAp Y;
  reflexivity;
end;
```

Similarly, `c=^` stands for the rough inclusion from the viewpoint of upper approximations while ordinary rough inclusion is just the conjunction of these two conditions.

There is also a lot of code showing algebraic context of rough sets: connection with ordinary set theory [7,10], lattice theory [6], and general topology. Table 1 summarizes main contributions (all five authored by the present author) with some statistical data. Some results, of more general interest, are dispersed over the whole library. To be more explicit, we translated [13] significant parts of J. Järvinen's *Lattice theory for rough sets* [16], studied various generalizations of rough approximations [24] up to pure binary relations [27,29]. By the way, years before, the solution of Robbins conjecture – well-known problem in the world of automated theorem proving – was also translated from Otter proof object into more human-accessible language [6].

Observe that in two last lines the numbers cannot be simply added; it is something like the set-theoretic union of all used files showing overall use of

Table 1. Statistical data about five main submissions

MML Id	ROUGHS_1	ROUGHS_2	ROUGHS_3	ROUGHS_4	ROUGHS_5	Total
lines of code	1686	1791	2392	1605	1386	8860
pages (approx.)	28	30	40	27	23	148
definitions	19	18	12	28	7	84
theorems	61	44	54	23	53	235
TPTP problems	591	557	858	566	860	3432
notations attached	29	19	22	42	21	53
used articles	61	39	55	75	54	83

various theories. TPTP stands for *Thousands of Problems for Theorem Provers* – these problems are automatically generated from the given Mizar source code and can be a testbed for automated theorem provers [18].

4 Rough Inclusion Formalized

As we worked in Tarski-Grothendieck set theory (TG) in the Mizar Mathematical Library (non-conservative extension of ordinary Zermelo-Frænkel with the Axiom of Choice), the only primitive is $\in$, which in the Mizar representation is just `in`. The Tarski's axiom A postulating for each set the existence of a Grothendieck universe it belongs to provides additional features like the existence of arbitrarily large, inaccessible cardinals. This axiom implies the axioms of infinity, choice, and power set and the axiomatics provides a richer ontology than ZFC (in the same time still having many conventional axioms of ZFC at hand), for example supporting work in category theory.

If we recall the definition of $\kappa^{\mathcal{L}}$ from the third section, this definition really reminds us about the probability theory[1]:

```
definition let Omega be set, Sigma be SigmaField of Omega;
  mode Probability of Sigma -> Function of Sigma,REAL means
:: PROB_1:def 8
  (for A being Event of Sigma holds 0 <= it.A) & it.Omega = 1 &
  (for A,B being Event of Sigma st A misses B holds
    it.(A \/ B) = it.A + it.B) &
  for ASeq being SetSequence of Sigma st ASeq is non-ascending holds
    it * ASeq is convergent &
    lim (it * ASeq) = it.Intersection ASeq;
end;
```

[1] All items from the Mizar Mathematical Library which are automatically hyperlinked can be browsed online from the page http://mizar.org/version/current/html/.

The above definition is used in the following notion of conditional probability:

```
definition
  let Omega be set, Sigma be SigmaField of Omega,
      P be Probability of Sigma, B be Event of Sigma;
  assume 0 < P.B;
  func P.|.B -> Probability of Sigma means
:: PROB_2:def 6
  for A being Event of Sigma holds it.A = P.(A /\ B) / P.B;
end;
```

The type of `Probability` reflects its informal counterpart, but it is definitely too complex for our considerations, however it is still possible to revise this approach and reuse the existing formal apparatus.

```
definition let R be finite Approximation_Space;
           let X,Y be Subset of R;
  func kappa (X,Y) -> Element of [.0,1.] equals
     card (X /\ Y) / card X if X <> {}
     otherwise 1;
  correctness;
end;
```

where κ function is defined pointwise as

```
definition let R be finite Approximation_Space;
  func kappa R -> Function of
    [:bool the carrier of R, bool the carrier of R:], [.0,1.] means
    for x,y being Subset of R holds it.(x,y) = kappa (x,y);
end;
```

In the definition above, brackets `[:` and `:]` denote Cartesian binary product of the carrier of an approximation space R. What is the real difference between these two definitions? The first one is a Mizar functor, just the appropriate object (in fact, the real number), which is an element of the unit interval. Another one is set-theoretic function returning for any pair of subsets of R just the earlier one. In the use, as dot `.` stands for the application of a function, it is a difference between `kappa(X,Y)` vs. `(kappa R).(X,Y)` (in the first case, R is a hidden argument reconstructed from the types of subsets X and Y).

```
theorem Prop1a: :: Proposition 1 a)
  kappa (X,Y) = 1 iff X c= Y;
```

We cannot however formulate the theorem that the earlier satisfies rif_1, but in the latter case we can prove *the functorial registration* (essentially expressing that $\kappa^{\pounds}$ is RIF):

```
registration let R be finite Approximation_Space;
  cluster kappa R -> satisfying_RIF1 satisfying_RIF2;
  coherence;
end;
```

and the first part of this proof depends heavily on `Prop1a`.

As we can be interested in much more properties of mappings from $\wp U \times \wp U$ into $[0,1]$ (as [4] lists actually seven of them), we found it useful to introduce a new type – which is a kind of a shortcut for quite complicated radix type.[2] Taking into account MML standards, the name `preRoughInclusionFunction of R` was proposed instead:

```
definition let R be 1-sorted;
  mode preRoughInclusionFunction of R is
    Function of [:bool the carrier of R, bool the carrier of R:], [.0,1.];
end;
```

In order to have even shorter lines, additional type was introduced:

```
definition let R be 1-sorted;
  mode preRIF of R is preRoughInclusionFunction of R;
end;
```

so all three names can be used simultaneously and with these new types, we can proceed further. All properties defining rough inclusion functions were stated in [4] in the form of predicates, but we decided to use adjectives as they enable the proofs to be more straightforward by the more extensive use of the typing hierarchy.

```
definition let R be non empty RelStr;
           let f be preRIF of R;
  attr f is satisfying_RIF1 means
    for X,Y being Subset of R holds f.(X,Y) = 1 iff X c= Y;
end;
```

Many proof assistants are not tightly linked with fixed set theory axioms, as they use rather logic quite extensively; there are however two of them which use just Tarski-Grothendieck as a base: Mizar and Metamath.

5 An Outline of Formalization

In a sense, proving properties of $\kappa^{\mathcal{L}}$ was like building the bridge over the gap between classical set theory and mereology. It is definitely of a more general interest, also for the repository of the Mizar system. We can now focus on more rough set-specific definitions.

$$\kappa_1(X,Y) = \begin{cases} \frac{|Y|}{|X \cup Y|} & \text{if } X \cup Y \neq \emptyset \\ 1 & \text{otherwise} \end{cases}$$

Here, $|X|$ means the cardinality of X, regardless if X is finite or not, but due to Mizar typing hierarchy, for finite X, $|X|$ is a natural number, and the division is well-defined. A short note about undefinedness will be crucial: among many of proof assistants it is quite unfeasible to handle an additional category

[2] Radix type together with the cluster of attributes makes a new type of an object.

of objects: "indefinite". Hence, the result of division by zero is fixed as equal to zero ($x/0 = 0$ for all real numbers; identity $x/x = 1$ can be proved only under the assumption $x \neq 0$ as usual).

It simplifies the typing apparatus, but, on the other hand, it can lead to quite unexpected results, for example, the value of function $f(x)$ is well defined if x is an element of the domain of f, otherwise it returns the empty set. Obviously however, it does not state that $\emptyset$ is always an element of the range of a function f. It is clear that these subtleties are not visible if we work only under standard assumptions.

Similarly to the theorems on $\kappa^{\pounds}$, we have proven the theorems about κ_1 and κ_2, essentially covering most of [4] up to Proposition 4 where all basic properties of rough inclusion functions under consideration are summarized.

To give an impression how the proofs in Mizar look like, we quote here the full proof of Proposition 4e) (stating that $\kappa_1(X,Y) = \kappa^{\pounds}(X \cup Y, Y)$):

```
theorem :: Proposition 4 e)
  kappa_1 (X,Y) = kappa (X \/ Y,Y)
  proof
    per cases;
    suppose
A1:   X \/ Y <> {}; then
      kappa (X \/ Y,Y) = card ((X \/ Y) /\ Y) / card (X \/ Y) by KappaDef
          .= card Y / card (X \/ Y) by XBOOLE_1:21;
      hence thesis by A1,Kappa1;
    end;
    suppose
A1:   X = {}; then
AA:   X c= Y;
A2:   kappa (X \/ Y, Y) = 1 by Prop1a,A1;
      kappa_1 (X,Y) = 1 by Prop11a,AA;
      hence thesis by A2;
    end;
  end;
```

We saved our submission under the name ROUGHIF1. The files (full Mizar script, corresponding vocabulary file and an abstract where the proofs are removed to show what is really developed) are available online.[3] The plans are to submit this for inclusion into the Mizar Mathematical Library; after the acceptance our development will be available under the MML identifier ROUGHIF1 – also in the form of HTML hyperlinked document which could be freely browsed with no need of installing Mizar verification software [25]. Some statistical data about this new submission are as follows: it took 1205 lines of Mizar code, which counts as about 20 pages of ordinary text in LaTeX (which will be also automatically generated). This nearly 40 kBytes contains 15 definitions and 32 proven theorems. Additional 20 registrations of clusters can be also treated as theorems automatically used by the Mizar verifier to enhance the reasoning.

[3] http://mizar.uwb.edu.pl/library/roughif1/.

6 Refining the Hierarchy of Rough Inclusion Functions

As rif_1 is of the form of an equivalence, splitting it into two implications seems quite natural. We can consider the following relaxation of the postulates for RIFs: $\mathrm{rif}_1(\kappa)$ is equivalent to the conjunction of $\mathrm{rif}_0(\kappa)$, $\mathrm{rif}_0^{-1}(\kappa)$ below:

$$\mathrm{rif}_0(\kappa) \Leftrightarrow \forall_{X,Y}(X \subseteq Y \;\Rightarrow\; \kappa(X,Y) = 1)$$
$$\mathrm{rif}_0^{-1}(\kappa) \Leftrightarrow \forall_{X,Y}(\kappa(X,Y) = 1 \Rightarrow X \subseteq Y)$$

In this manner, a mapping κ from $\wp U \times \wp U$ into $[0,1]$ is called

- a quasi-rough inclusion function (q-RIF) over U if it satisfies $\mathrm{rif}_0(\kappa)$ and $\mathrm{rif}_2^*(\kappa)$;
- a weak q-RIF over U if it satisfies $\mathrm{rif}_0(\kappa)$ and $\mathrm{rif}_2(\kappa)$.

Obviously then, every RIF is a q-RIF and every q-RIF is a weak q-RIF [5].

```
definition let R be non empty RelStr;
           let f be preRIF of R;
  attr f is satisfying_RIF0 means
    for X,Y being Subset of R st X c= Y holds f.(X,Y) = 1;
end;
```

To ensure automatic recognizing that RIF0 and RIF01 occurred as a result of equivalence splitting, the following *registration* should be formulated and proved (it can be read as "every function from $\wp U \times \wp U$ into the unit interval (that is, every preRIF of R) which satisfies rif_1 satisfies also rif_0 and rif_0^{-1}"):

```
registration let R be non empty RelStr;
  cluster satisfying_RIF1 -> satisfying_RIF0 satisfying_RIF01
    for preRIF of R;
  coherence;
end;
```

The so-called conditional registrations of clusters improve the work on the generalization of theorems: dedicated software distributed with the Mizar system can automatically discover redundant, unnecessary assumptions. Of course, such registrations should be proven (in the above case – the proof of coherence, but as a rule we do not quote the proofs in this paper, as they are available on the web). Similar situation allowed for stating selected properties of rough approximations in terms of tolerance approximation spaces or more general relational structures, as we noticed in Sect. 3.

Furthermore, such automatization offers also some flexibility in this fully formal approach: RIFs can be defined as preRIFs satisfying, as usual, rif_1 and rif_2^* or, equivalently, those satisfying rif_1 and rif_2, as assuming the condition rif_1 holds, both rif_2^* and rif_2 are equivalent.[4]

[4] The author wishes to thank one of the referees for this valuable remark. Of course, we are interested in the development of quasi RIFs and weak quasi RIFs, so this distinction is important even if it is meaningless in the theory of RIFs.

```
definition let R be non empty RelStr;
          let f be preRIF of R;
  attr f is satisfying_RIF2* means
    for X,Y,Z being Subset of R st f.(Y,Z) = 1 holds f.(X,Y) <= f.(X,Z);
end;

registration let R;
  cluster satisfying_RIF2 -> satisfying_RIF2*
    for satisfying_RIF1 preRIF of R;
  cluster satisfying_RIF2* -> satisfying_RIF2
    for satisfying_RIF1 preRIF of R;
end;
```

Relatively short (but needed!) proof of this equivalence is available in our complete Mizar formalization. Then the core idea of this development, can be defined as follows:

```
definition let R;
  mode RIF of R is satisfying_RIF1 satisfying_RIF2 preRIF of R;
end;
```

(with this naming space chosen similarly to preRIFs, a bit longer but also more meaningful synonym `RoughInclusionFunction of R` was additionally introduced).

7 Comparative Study of Some Generalized Rough Approximations Revisited

At the *Concurrency Specification & Programming'18 Workshop* in Berlin we presented how generalized rough approximations can be automatically studied to support ordinary mathematician in his/her work [11]; as the testbed for the usefulness of the approach we have chosen theorems from another Gomolińska's paper [2], and the proof of one of them failed, namely Theorem 4. All but two items from there were proven: points (d) and (e) involved κ and even defining κ was not very difficult, we tried to avoid mixing approaches. Luckily, this was not the author's fault: until this time, in any of five main Mizar articles about rough sets, rough inclusion κ was not used at all and it was just not defined formally before. Our motivation however for dealing with RIFs, besides the main work, was of course to complete the formalization of the earlier paper.

We will recall briefly two lacking formalization gaps – items of Theorem 4 from earlier Gomolińska paper [2]. She started with general approximation space $\langle U, \rho, \kappa \rangle$, and we tried to avoid the κ ingredient as long as we could. This excerpt reads as follows:

For any set $x \subseteq U$, object $u \in U$ it holds that:
(d) $\forall_{u \in f_1(x)} \kappa(I(u), x) > 0$.
(e) $\forall_{u \in f_1^d(x)} \kappa(I(u), x) = 1$.

One of the mappings considered in this paper was defined as

$$f_1(x) = \{u \in U : I(u) \cap x \neq \emptyset\},$$

where

$$I(u) = \rho^{\leftarrow}(\{u\}) \text{ and } f^d(x) = (f(x^c))^c.$$

But in view of rif_1 property, (e) means that $I(u) \subseteq x$, and (d) means that $I(u)$ is not a negative region w.r.t. rough approximation. After the application of implemented automatization and the unfolding of new definition of κ, we can claim the formalization of the first part of [2] is now really completed.

8 A Formalized Example

Originally, rough approximation spaces proposed by Pawlak were defined as a tuple

$$\langle U, \rho \rangle$$

and we were fixed with a relational structure equipped by an underlying indiscernibility relation.

Until now, even if we faced some hard decisions how to deal with things formally, the informal proofs from [4] were either not very hard to discover (and left by the authors to the readers), or quite nicely sketched in the paper (which is crucial for our formalization work), the formalization was fluent, and we were not forced to abandon Gomolińska's ideas. The prominent exceptions were the proof of Proposition 4(c) before (a) and (b), and, of course, the urgent need for communication between $\kappa(X,Y)$ treated as a Mizar functor and κ's treated as functions between corresponding domains.

In [4], p. 149 Gomolińska constructs a (relatively) simple example of an approximation space (or, to be honest, just the universe with subsets because as it is clear from the definitions of κ's, the indiscernibility relation does not influence RIFs), which shows that all three considered RIFs ($\kappa^{\pounds}, \kappa_1, \kappa_2$) are distinct. She claims that $U = \{0, 1, 2, \ldots, 9\}$, $X = \{0, \ldots, 4\}$, $Y = \{2, \ldots, 6\}$. Then $\kappa^{\pounds}(X,Y) = 3/5$, $\kappa_1(X,Y) = 5/7$, and $\kappa_2(X,Y) = 4/5$. The informal proof was 4 lines long, but in our first attempt after some 50 lines of formulas, we dropped this proving path.

There is quite recent implementation of ellipsis in Mizar [17] and potentially we could use such enumerative set, but we soon realized that this would be the most tedious part of our work (some calculations we still left at the end of the file, under pragma `::$EOF` – the end of the file verified by the Mizar checker). It is quite ordinary that while proofs are obviously important, constructing counterexamples is still underestimated activity in the proof-checking world and definitely an automated tool like MACE (Models and CounterExamples implemented with Prover9 theorem prover) could be really useful. Calculations on real numbers are implemented in the checker via appropriate directive `requirements`, but still the author should put references by appropriate cardinality theorems (and as the number of mutually distinct elements originally approached ten, the combinatorial explosion of 2^{10} appeared on the horizon).

```
definition let X be set;
  func DiscreteApproxSpace X -> strict RelStr equals
    RelStr (# X, id X #);
  coherence;
end;
```

Probably not the simplest one, but our construction depend on the discrete approximation space, where discrete means that the indiscernibility relation is just the identity – essentially all elements of this space are indiscernible.

```
theorem
  for X,Y being Subset of DiscreteApproxSpace {1,2,3,4,5}
    st X = {1,2} & Y = {2,3,4} holds
      kappa (X,Y), kappa_1 (X,Y), kappa_2 (X,Y) are_mutually_distinct;
```

In our Mizar script, we claimed that $U = \{1,2,3,4,5\}, X = \{1,2\}, Y = \{2,3,4\}$ with $\kappa^{£}\ (X,Y) = 1/2, \kappa_1(X,Y) = 3/4$, and $\kappa_2(X,Y) = 4/5$. After 30 lines, which will compress to ca. 20 lines, we finished the proof. In this pessimistic case, the de Bruijn factor (the ratio between full formal proof and its informal counterpart) was approx. 7, which is twice as big as it is claimed as a rule in the field of computerized proof assistants, but there are quite detailed proofs without significant omissions, where it equals 2 (before compression), which is close to the Holy Grail of 1.

9 Conclusions and Further Work

One of the very important aims of the current research was to reflect the concept of rough inclusion functions within the Mizar Mathematical Library. Even though much work on formalization of the building blocks of rough sets in Mizar has been done before [8], no attention has been devoted to rough inclusion functions (including the Łukasiewicz function $\kappa^{£}$) by the Mizar developers till recently. The shortcoming is addressed in this research.

Weak quasi-RIFs could be an interesting source of inspiration; it is quite normal activity to revise the formal approach granted in the MML to drop some assumptions, generalize theorems or lemmas, or make the net of notions more granular [12,20]. Such refinement allows to develop new areas not necessarily by writing new submission from scratch (or, frequently, by copy-and-paste-then-refine technique), but rather improving the originals. By the way, this process of removing of redundant parts of an article is automatized. Furthermore, Gomolińska published later a significant extension of her paper [3], where she focuses on mappings "complementary" to RIFs, semantically close to the fuzzy negation of κ's. Such mappings are also interesting as theory of fuzzy sets is also well represented in the Mizar library and some propositions could be straightforward.

We hope that some researchers will join us formalizing even small parts of their research; in [9] we described how some connections between chosen facts in rough set theory and another areas of mathematics were discovered automatically by means of MML Query search engine: we mean, among others, Isomichi classification of domains and, partially, variants of Kuratowski closure-complement problem. Furthermore, obtaining proper semantic markup of mathematical documents is the core idea of OMDoc (Open Mathematical Documents) community – not primarily presentation-oriented, as, for example, LaTeX source, but allowing also to catch the meaning of represented formulas. Some preliminary works on building a version of the Mizar Mathematical Library partially designed for the mereological reasoning were already done and the results are quite convincing.

References

1. Bancerek, G., et al.: Mizar: state-of-the-art and beyond. In: Kerber, M., Carette, J., Kaliszyk, C., Rabe, F., Sorge, V. (eds.) CICM 2015. LNCS (LNAI), vol. 9150, pp. 261–279. Springer, Cham (2015). https://doi.org/10.1007/978-3-319-20615-8_17
2. Gomolińska, A.: A comparative study of some generalized rough approximations. Fundamenta Informaticae **51**, 103–119 (2002)
3. Gomolińska, A.: On certain rough inclusion functions. In: Peters, J.F., Skowron, A., Rybiński, H. (eds.) Transactions on Rough Sets IX. LNCS, vol. 5390, pp. 35–55. Springer, Heidelberg (2008). https://doi.org/10.1007/978-3-540-89876-4_3
4. Gomolińska, A.: On three closely related rough inclusion functions. In: Kryszkiewicz, M., Peters, J.F., Rybinski, H., Skowron, A. (eds.) RSEISP 2007. LNCS (LNAI), vol. 4585, pp. 142–151. Springer, Heidelberg (2007). https://doi.org/10.1007/978-3-540-73451-2_16
5. Gomolińska, A.: Rough approximation based on weak q-RIFs. In: Peters, J.F., Skowron, A., Wolski, M., Chakraborty, M.K., Wu, W.-Z. (eds.) Transactions on Rough Sets X. LNCS, vol. 5656, pp. 117–135. Springer, Heidelberg (2009). https://doi.org/10.1007/978-3-642-03281-3_4
6. Grabowski, A.: Mechanizing complemented lattices within Mizar type system. J. Autom. Reasoning **55**, 211–221 (2015). https://doi.org/10.1007/s10817-015-9333-5
7. Grabowski, A.: Binary relations-based rough sets - an automated approach. Formalized Math. **24**(2), 143–155 (2016). https://doi.org/10.1515/forma-2016-0011
8. Grabowski, A.: Automated discovery of properties of rough sets. Fundamenta Informaticae **128**(1–2), 65–79 (2013). https://doi.org/10.3233/FI-2013-933
9. Grabowski, A.: Efficient rough set theory merging. Fundamenta Informaticae **135**(4), 371–385 (2014). https://doi.org/10.3233/FI-2014-1129
10. Grabowski, A.: Computer certification of generalized rough sets based on relations. In: Polkowski, L., et al. (eds.) IJCRS 2017. LNCS (LNAI), vol. 10313, pp. 83–94. Springer, Cham (2017). https://doi.org/10.1007/978-3-319-60837-2_7
11. Grabowski, A.: Automated comparative study of some generalized rough approximations. In: Proceedings of CS&P 2018, pp. 39–45 (2018)
12. Grabowski, A., Korniłowicz, A., Naumowicz, A.: Four decades of Mizar. J. Autom. Reasoning **55**, 191 (2015). https://doi.org/10.1007/s10817-015-9345-1
13. Grabowski, A.: Lattice theory for rough sets - a case study with Mizar. Fundamenta Informaticae **147**(2–3), 223–240 (2016). https://doi.org/10.3233/FI-2016-1406

14. Grabowski, A.: On the computer-assisted reasoning about rough sets. In: Dunin-Kęplicz, B., Jankowski, A., Szczuka, M. (eds.) Monitoring, Security and Rescue Techniques in Multiagent Systems. AISC, vol. 28, pp. 215–226 (2005). https://doi.org/10.1007/3-540-32370-8_15
15. Grabowski, A., Jastrzębska, M.: Rough set theory from a math-assistant perspective. In: Kryszkiewicz, M., Peters, J.F., Rybinski, H., Skowron, A. (eds.) RSEISP 2007. LNCS (LNAI), vol. 4585, pp. 152–161. Springer, Heidelberg (2007). https://doi.org/10.1007/978-3-540-73451-2_17
16. Järvinen, J.: Lattice theory for rough sets. In: Peters, J.F., Skowron, A., Düntsch, I., Grzymała-Busse, J., Orłowska, E., Polkowski, L. (eds.) Transactions on Rough Sets VI. LNCS, vol. 4374, pp. 400–498. Springer, Heidelberg (2007). https://doi.org/10.1007/978-3-540-71200-8_22
17. Korniłowicz, A.: Flexary connectives in Mizar. Comput. Lang. Syst. Struct. **44**(Part C), 238–250 (2015). https://doi.org/10.1016/j.cl.2015.07.002
18. Naumowicz, A.: Automating boolean set operations in Mizar proof checking with the aid of an external SAT solver. J. Autom. Reasoning **55**, 285 (2015). https://doi.org/10.1007/s10817-015-9332-6
19. Pawlak, Z.: Rough Sets: Theoretical Aspects of Reasoning about Data. Kluwer, Dordrecht (1991). https://doi.org/10.1007/978-94-011-3534-4
20. Pąk, K.: Improving legibility of formal proofs based on the close reference principle is NP-hard. J. Autom. Reasoning **55**, 295–306 (2015). https://doi.org/10.1007/s10817-015-9337-1
21. Polkowski, L.: Approximate Reasoning by Parts: An Introduction to Rough Mereology. Intelligent Systems Reference Library, vol. 20. Springer, Heidelberg (2011). https://doi.org/10.1007/978-3-642-22279-5
22. Polkowski, L., Skowron, A.: Rough mereology: a new paradigm for approximate reasoning. Int. J. Approximate Reasoning **15**(4), 333–365 (1996). https://doi.org/10.1016/S0888-613X(96)00072-2
23. Rudnicki, P.: Obvious inferences. J. Autom. Reasoning **3**(4), 383–393 (1987). https://doi.org/10.1007/BF00247436
24. Skowron, A., Stepaniuk, J.: Tolerance approximation spaces. Fundamenta Informaticae **27**(2–3), 245–253 (1996). https://doi.org/10.3233/FI-1996-272311
25. Urban, J., Sutcliffe, G.: Automated reasoning and presentation support for formalizing mathematics in Mizar. In: Autexier, S., et al. (eds.) CICM 2010. LNCS (LNAI), vol. 6167, pp. 132–146. Springer, Heidelberg (2010). https://doi.org/10.1007/978-3-642-14128-7_12
26. Wiedijk, F.: Formal proof - getting started. Not. AMS **55**(11), 1408–1414 (2008)
27. Yao, Y.Y.: Two views of the rough set theory in finite universes. Int. J. Approximate Reasoning **15**(4), 291–317 (1996). https://doi.org/10.1016/S0888-613X(96)00071-0
28. Zadeh, L.: Fuzzy sets. Inf. Control **8**(3), 338–353 (1965). https://doi.org/10.1016/S0019-9958(65)90241-X
29. Zhu, W.: Generalized rough sets based on relations. Inf. Sci. **177**(22), 4997–5011 (2007). https://doi.org/10.1016/j.ins.2007.05.037

Membrane Systems and Multiset Approximation: The Cases of Inner and Boundary Rule Application

Péter Battyányi and György Vaszil[(✉)]

Department of Computer Science, Faculty of Informatics, University of Debrecen, Kassai út 26, Debrecen 4028, Hungary
{attyanyi.peter,vaszil.gyorgy}@inf.unideb.hu

Abstract. We continue the study of generalized P systems with dynamically changing structure based on an associated multiset approximation framework. We consider membrane systems where the applicability of the multiset transformation rules is determined by the approximating multisets of the membrane regions. We consider two cases: First, we study systems with inner rules where we allow only rule applications such that the multisets involved in the rules are part of the lower approximation of the respective regions, then we consider systems with boundary rules where rule application is defined on the boundaries, that is, rules can only manipulate the elements outside of the lower approximation. We show that the second variant benefits from the underlying approximation framework by demonstrating an increase in its computational strength. On the other hand, by presenting an appropriate simulating Petri net, we show that the computational power of systems with inner rule application remains weaker than that of Turing machines (as long as the unsynchronized version is considered).

1 Introduction

Membrane systems, introduced in [15], are biologically inspired models of computation: their operation imitates in a sense the functioning of living cells. The computation proceeds in distinct regions, called membranes or compartments. The compartments allow computation with multisets: they accomplish transformations of their contained multisets by various evolution (multiset rewriting) rules. Several variants of P systems have been introduced and studied, see the monograph [16] for a thorough introduction, or the handbook [17] for a summary of notions and results of the area.

In the original symbol object model, the compartments are organized in a tree like structure. Each membrane except for the outermost one, the skin membrane, have a unique parent membrane, the parent-child relationship depicts the connection when one membrane (the parent) contains the other membrane (the child). The rules account for the distributed computational processes in the compartments. In this basic model, the lefthand side of a rule is a multiset of

© Springer Nature Switzerland AG 2019
T. Mihálydeák et al. (Eds.): IJCRS 2019, LNAI 11499, pp. 239–252, 2019.
https://doi.org/10.1007/978-3-030-22815-6_19

objects inside one of the regions and the righthand side of a rule is a multiset of objects labelled with target indications *here*, *out* and in_j indicating the positions the elements should be placed before the next computational step begins. Usually, computation in a region takes place in a maximally parallel manner, this means that a computational step in a region is understood as the simultaneous application of a multiset of rules which is maximal, that is, it cannot be augmented by any applicable rule. The membrane system waits for each of its compartments to finish its maximally parallel computational process, then the objects labelled with the target indications are moved to their correct places and a new computational step of the P system can begin. Objects with target indication *here* remain in the region, objects with *out* move to the parent region, objects with in_j enter the jth child region of the given compartment. The computation proceeds until no rule can be applied in any of the regions. The result is usually formed by the objects of a designated region, the output region, after the computation having come to a halt.

The structure of a membrane system can be represented in various ways, cell-like membrane systems have a membrane structure which can be described by a tree. Systems with graph-like membrane structures called tissue-like P systems were also considered, where the connection between the membranes are established by edges forming the communication routes. Here we study variants of tissue-like systems called generalized P systems (see [3]).

The question of how to define dynamically changing membrane structures using topological spaces, and how the underlying topologies influence the behaviour of P systems was already examined in [4,5]. Multiset approximation spaces were defined in [7,8], which made it possible to talk about lower and upper approximations of the contents of membranes of a P system. This led to various notions of membrane borders, and notions of closeness of membranes. Restricting the interaction to membranes that are close to each other, or permitting only rules that manipulate multisets which are on the boundaries of the membranes can affect the computational strength of the membrane system. The study of this area was initiated in [9], where also an intention to model chemical stability played an important role. The results in [9] were formulated for the so-called symport/antiport P systems, but the investigations were also continued for so called generalized P systems in [2]. In the present paper we also study generalized P systems, but we do not rely on any notion of closeness of membranes. Instead, we focus on the notion of clear observability. We consider lower approximations and boundaries of compartments, and restrict the applicability of the rules accordingly. It will turn out that the use of boundary rules, that is, rules which can only manipulate objects on the boundaries of compartments, results in an increase of the computational power of certain variants of generalized P systems to the level of the power of Turing machines. On the other hand, if we restrict rule applications only to rules that manipulate multisets which lie in the inner approximations of the membranes (inner rules), this restriction is not enough to provide Turing completeness.

The main contributions of the paper are of two kinds: the introduction of the above described variants of generalized P systems with associated multiset approximation spaces, and the presented results about their computational power.

In the following, we first recall the necessary definitions, then take up the examination of the two variants of generalized P systems with dynamically changing communication structure based on multiset approximation spaces. As maximal parallel rule application makes already the basic model of generalized P systems computationally complete, we study the weaker, unsynchronized variants. We first show that generalized P systems with inner rules can be simulated by simple place-transition Petri nets, thus, their computational power is less than that of Turing machines. Then we consider systems with boundary rules and show that they are able to simulate so called register machines, which demonstrates that their computational power is the same as the power of Turing machines. Finally, the paper ends with a few concluding remarks.

2 Preliminaries

Let $\mathbb{N}$ and $\mathbb{N}_{>0}$ be the set of non-negative integers and the set of positive integers, respectively, and let O be a finite nonempty set (the set of object). A *multiset* M over O is a pair $M = (O, f)$, where $f : O \to \mathbb{N}$ is a mapping which gives the *multiplicity* of each object $a \in O$. The set $\text{supp}(M) = \{a \in O \mid f(a) > 0\}$ is called the *support* of M. If $\text{supp}(M) = \emptyset$, then M is the empty multiset. If $a \in \text{supp}(M)$, then $a \in M$, and $a \in^n M$ if $f(a) = n$.

Let $M_1 = (O, f_1), M_2 = (O, f_2)$. Then $(M_1 \sqcap M_2) = (O, f)$ where $f(a) = \min\{f_1(a), f_2(a)\}$; $(M_1 \sqcup M_2) = (O, f')$, where $f'(a) = \max\{f_1(a), f_2(a)\}$; $(M_1 \oplus M_2) = (O, f'')$, where $f''(a) = f_1(a) + f_2(a)$; $(M_1 \ominus M_2) = (O, f''')$ where $f'''(a) = \max\{f_1(a) - f_2(a), 0\}$; and $M_1 \sqsubseteq M_2$, if $f_1(a) \leq f_2(a)$ for all $a \in O$.

For any $n \in \mathbb{N}$, n-times addition of M, denoted by $\oplus_n M$, is given by the following inductive definition:

- $\oplus_0 M = \emptyset$;
- $\oplus_1 M = M$;
- $\oplus_{n+1} M = (\oplus_n M) \oplus M$.

Let $M_1 \neq \emptyset, M_2$ be two multisets. For any $n \in \mathbb{N}$, $M_1 \sqsubseteq^n M_2$, if $\oplus_n M_1 \sqsubseteq M_2$ but $\oplus_{n+1} M_1 \not\sqsubseteq M_2$.

The number of copies of objects in a finite multiset $M = (O, f)$ is its cardinality: $\text{card}(M) = \Sigma_{a \in \text{supp}(M)} f(a)$. Such an M can be represented by any string w over O for which $|w| = \text{card}(M)$, and $|w|_a = f(a)$ where $|w|$ denotes the length of the string w, and $|w|_a$ denotes the number of occurrences of symbol a in w.

We define the $\mathcal{MS}^n(O)$, $n \in \mathbb{N}$, to be the set of all multisets $M = (O, f)$ over O such that $f(a) \leq n$ for all $a \in O$, and we let $\mathcal{MS}^{<\infty}(O) = \bigcup_{n \geq 0} \mathcal{MS}^n(O)$.

2.1 Generalized P Systems

Now we present the notion of generalized P systems, variants of tissue P systems introduced in [3].

An $n+3$-tuple $\Pi = (O, w_1, w_2, \ldots, w_n, R, i_o)$ is a *generalized P system* of degree $n \geq 1$, where

- O is a finite set of objects;
- $w_i \in \mathcal{MS}^{<\infty}(O)$, $1 \leq i \leq n$, is a finite multiset of objects, the initial contents of the ith region of Π;
- R is a finite set of transformation rules of the form $(x_1, \alpha_1) \ldots (x_k, \alpha_k) \to (y_1, \beta_1) \ldots (y_l, \beta_l)$, where $x_i, y_j \in \mathcal{MS}^{<\infty}(O)$, and $1 \leq \alpha_i, \beta_j \leq n$ indicate labels of the regions of the system for all $1 \leq i, j \leq n$;
- $1 \leq i_o \leq n$ is the label of output compartment.

The rules of a generalized P system can be considered to model interactions of objects simultaneously affecting several regions of the membrane system. Thus, the links between participating compartments are defined dynamically, through the applicability of the rules by the functioning of the system.

Given a generalized P system Π as above, a *configuration* of Π is an n-tuple $c = (u_1, u_2, \ldots, u_n)$ with $u_i \in MS^{<\infty}(O)$, $1 \leq i \leq n$, and $c_0 = (w_1, w_2, \ldots, w_n)$ is called its *initial configuration*. The multisets $u_1, u_2, \ldots, u_n$ are the *contents* of the corresponding compartments $1, 2, \ldots, n$, in configuration c.

A generalized P system changes its configurations by applying its rules. In the basic setting, a rule $r \in R$, is *applicable* to a configuration c, if and only if x_i is a submultiset of u_{α_i} for all $1 \leq i \leq k$. As a result of applying r to c, each multiset x_i is removed from the region u_{α_i}, $1 \leq i \leq k$, and each multiset y_j is added to the region u_{β_j}, $1 \leq j \leq l$.

The configuration $c' = (v_1, \ldots, v_n)$ of Π is obtained directly from the configuration $c = (u_1, \ldots, u_n)$ by applying the rules in the *unsynchronized* manner, if there is a multiset R' of rules from R, such that all of them are simultaneously applicable to different copies of objects in configuration c, and the configuration c' is the result of the application of the rules in R'. The configuration c' is obtained from c by applying the rules in the *maximally parallel* manner, if we add the additional requirement that the set R' is maximal, that is, for any $r \in R$, the rules in the rule multiset $\{r\} \oplus R'$ are not simultaneously applicable to c.

A sequence of configurations $c_0, c_1, \ldots$ of Π is called a *computation* if each configuration in the sequence is obtained directly from the previous one, starting from the initial configuration. Computations halt if no rule can be applied, the result of a *halting computation* is the number of objects that are present in the output compartment (compartment i_o) in the halting configuration.

2.2 Multiset Approximation Spaces

There are different ways of set approximations originating in rough set theory proposed in the early 1980's, [11,12]. The theory and its different generalizations uses different kinds of indiscernibility relations to provide lower and upper approximations of sets. An indiscernibility relation on a given set of objects is given by a set of base sets by which lower and upper approximations can be constructed for any set. This way of set approximation was generalized to partial set approximation in [7], giving the possibility to embed available knowledge into an approximation space. The lower and upper approximations also rely on base sets which can be thought of as representants of the available knowledge. Having the concepts of lower and upper approximations, we can also introduce the concept of boundary as the difference between these two.

As membrane systems can be represented by multisets, in order to use the above described concepts in membrane systems theory, we need to generalize the set approximation framework for multisets. With the membrane structure as a background, an underlying multiset approximation space can be defined. The nature of this space is basically determined by its constituents, to a certain extent, independently of the membrane structure. The notion of multiset approximation spaces has been introduced in [7] (see also [8] for more details). Multiset approximations also rely on a set of base multisets given beforehand. By creating the lower and upper approximations using the usual approximation technique, the boundaries of multisets (boundaries of membrane regions) can also be defined, and we will make use of this feature in subsequent parts of the paper.

A multiset approximation space over a finite alphabet O consists of the following:

- A *domain*: in our case it is $\mathcal{MS}^{<\infty}(O)$, the set of finite multisets over some finite set O. The elements of the domain are approximated using the approximation space.
- A *base system*: $\mathfrak{B} \subseteq \mathcal{MS}^{<\infty}(O)$, a nonempty set of finite *base multisets* providing the basis for the approximation process.
- The *approximation functions*: $\mathsf{l}, \mathsf{u}, \mathsf{b} : \mathcal{MS}^{<\infty}(O) \to \mathcal{MS}^{<\infty}(O)$ determining the lower and upper approximations (and the boundaries) of multisets of the domain.

A *multiset approximation space* is a quintuple $(O, \mathfrak{B}, \mathsf{l}, \mathsf{u}, \mathsf{b})$ where O is a finite set, $\mathfrak{B} \subseteq \mathcal{MS}^{<\infty}(O)$ is a base system (a set of base multisets), and $\mathsf{b}, \mathsf{u}, \mathsf{l} : \mathcal{MS}^{<\infty}(O) \to \mathcal{MS}^{<\infty}(O)$ are the approximation functions generated by $\mathfrak{B}$.

For any multiset $M = (O, f) \in \mathcal{MS}^{<\infty}(O)$, we define the *lower approximation function*:

$$\mathsf{l}(M) = \bigsqcup \{\oplus_n B \mid B \in \mathfrak{B},\ B \sqsubseteq M, \text{and } B \sqsubseteq^n M\},$$

the *boundary function*:

$$\mathsf{b}(M) = \bigsqcup \{\oplus_n B \mid B \in \mathfrak{B},\ \text{and } B \sqcap (M \ominus \mathsf{l}(M)) \sqsubseteq^n M \ominus \mathsf{l}(M)\},$$

and the *upper approximation function*:

$$\mathsf{u}(M) = \mathsf{l}(M) \sqcup \mathsf{b}(M).$$

In addition, we also define $\mathsf{b}^{\mathsf{e}}(M) = \mathsf{b}(M) \ominus M$ as the *external part* of the boundary of M, and $\mathsf{b}^{\mathsf{i}}(M) = \mathsf{b}(M) \sqcap M$, the *internal part* of the boundary of M.

Intuitively, we can think of the lower approximation of the multiset M as the collection of elements that can be covered by the base multisets in such a way that the covering is inside M completely. If we also cover those elements of M that are left out of the lower approximation, then the union of the covering base sets contains M, thus, it can be thought of as the upper approximation of M, while the difference between the upper and the lower approximations of M is the boundary.

3 Regulating Rule Application in the Multiset Approximation Framework

In [2] we considered P systems with dynamical structure where the dynamic character of the membrane system was encoded in the reformulation of the region structure regarding a closeness property defined among the membranes based on the actual configuration of the system. Here we examine questions that arise when we require that in order for a rule to be applicable, the multisets on its lefthand side must conform to certain properties defined in the multiset approximation framework associated to the system. We discuss the following two approaches: first we require that a rule to be applied should only work with the lower approximations of the compartments' contents. The second approach demands that the multisets on the lefthand sides of the rules should come from the boundaries of the respective compartments.

Conforming the requirement of clear observability when dealing with rough sets, first we stipulate in the following definition that a rule should be applicable in a P system only if the multisets on its lefthand side come from the inner approximations of the containing regions, this means that we are absolutely sure that the rule application affects elements of the corresponding regions. The second requirement, on the other hand, corresponds to a system where rule application can only alter those elements about which our knowledge is vague, so the configuration changes of these systems might be thought of as steps in the direction of reducing vagueness, obtaining more and more determinate knowledge about the objects distributed in the membranes.

We formalize these notions in the following definition.

Definition 1. Let $\Pi = (O, \mathfrak{B}, w_1, w_2, \ldots, w_n, R, i_o)$ where $\mathfrak{B} \subseteq \mathcal{MS}^{<\infty}(O)$ is a base system and $(O, w_1, w_2, \ldots, w_n, R, i_o)$ is a generalized P system.

We call Π a *generalized P system with an associated multiset approximation space and inner rules*, if the applicability of a rule $r = (x_1, \alpha_1) \ldots (x_k, \alpha_k) \to (y_1, \beta_1) \ldots (y_l, \beta_l) \in R$ in a configuration $c = (u_1, \ldots, u_n)$ is defined by the requirement that x_i is a submultiset of $\mathsf{l}(u_{\alpha_i})$, the inner approximation of the respective region, $1 \leq i \leq k$. If $r \in R$ is applicable to c in this sense, then we call r an *inner rule* (with respect to c).

We call Π a *generalized P system with an associated multiset approximation space and boundary rules*, if the applicability of a rule $r = (x_1, \alpha_1) \dots (x_k, \alpha_k) \to (y_1, \beta_1) \dots (y_l, \beta_l) \in R$ in a configuration $c = (u_1, \dots, u_n)$ is defined by the requirement that x_i is a submultiset of $\mathsf{b}^{\mathsf{i}}(u_{\alpha_i})$, the internal part of the boundary of the respective region, $1 \leq i \leq k$. If $r \in R$ is applicable to c in this sense, then we call r a *boundary rule* (with respect to c).

Example 1. Assume that $C = (w_1, w_2)$ is a configuration of Π, a generalized P system with an associated multiset approximation space for $w_1 = a^3b^3c^2$ and base sets $B_1 = a^2$, $B_2 = bc$. Further, let $r_1 = (ab^2, 1) \to (c, 1)(d^3, 2)$ and $r_2 = (ab, 1) \to (e^2, 1)$ be to rules of Π.

If Π is a system with inner rules, then both rules are applicable in C, as $B_1 \sqcup \oplus_2 B_2 = a^2b^2c^2$ is the lower approximation of w_1.

If Π is a system with boundary rules, then only the rule r_2 is applicable in C, as a^2bc is the boundary of w_1 with inner part ab.

We claim that the use of inner rules do not add much to the computational strength of the P system in the sense that in the non-synchronized mode a generalized P system with an associated multiset approximation space and inner rules is not Turing complete. To show this, we construct a simple place-transition Petri net that simulates the P system in question. This is sufficient, because Petri nets in this simple setting are strictly weaker in computational power than Turing machines, see for example [13, 14]. The idea of the proof is similar to that of Theorem 2 in [2], the construction of the Petri net, however, is different.

A *place-transition Petri net* [13] is a quintuple $U = (P, T, F, V, m_0)$ such that P, T are finite sets with $P \cap T = \emptyset$, $P \cup T \neq \emptyset$, the sets of *places* and *transitions*, respectively. The set $F \subseteq (P \times T) \cup (T \times P)$, is a set of "arcs" connecting places and transitions, the *flow relation* of U. The function $V : F \to \mathbb{N}_{>0}$ determines the multiplicity (the *weight*) of the arcs, and $m_0 : P \to \mathbb{N}$ is a function called the initial marking. In general, a *marking* is a function $m : P \to \mathbb{N}$ associating nonnegative integers (the number of *tokens*) to the places of the net. Moreover, for every transition $t \in T$, there is a place $p \in P$ such that $f = (p, t) \in F$ and $V(f) \neq 0$.

Let $x \in P \cup T$. The *pre- and postsets* of x, denoted by ${}^\bullet x$ and $x^\bullet$, respectively, are defined as ${}^\bullet x = \{y \mid (y, x) \in F\}$ and $x^\bullet = \{y \mid (x, y) \in F\}$.

For each transition $t \in T$, we define two markings, $t^-, t^+ : P \to \mathbb{N}$ as follows:

$$t^-(p) = \begin{cases} V(p,t), & \text{if } (p,t) \in F, \\ 0 & \text{otherwise,} \end{cases}$$

$$t^+(p) = \begin{cases} V(t,p), & \text{if } (t,p) \in F, \\ 0 & \text{otherwise.} \end{cases}$$

A transition $t \in T$ is said to be *enabled* if $t^-(p) \leq m(p)$ for all $p \in {}^\bullet t$. Let $\vartriangle t(p) = t^+(p) - t^-(p)$ for $p \in P$, and let us define the *firing of a transition* as follows. A transition $t \in T$ can fire in m (notation: $m \longrightarrow^t$) if t is enabled in m. After the firing of t, the Petri net obtains a new marking $m' : P \to \mathbb{N}$, where $m'(p) = m(p) + \vartriangle t(p)$ for all $p \in P$ (notation: $m \longrightarrow^t m'$).

Petri nets can be considered as computing devices: Starting with the initial marking, going through a series of configuration changes by the firing of a series of transitions, we might obtain a marking where no transitions are enabled. This final marking is the result of the Petri net computation.

Theorem 1. *For any generalized P system with an associated multiset approximation space and inner rules, Π, there is a place-transition Petri net N, such that N generates the same set of numbers as Π in the unsynchronized manner of rule application.*

Proof. Let $\Pi = (O, \mathfrak{B}, w_1^0, w_2^0, \ldots, w_n^0, R, i_o)$ be a generalized P system with an associated multiset approximation space and inner rules, let the underlying set of base sets be $\mathfrak{B} = \{B_i \mid 1 \leq i \leq m\}$, and let $x \in \mathcal{MS}^{<\infty}(O)$ be arbitrary. Then there exists an $h_x \in \mathbb{N}$ such that, for any subset $\{B_1, \ldots, B_s\} \subseteq \mathfrak{B}$, either $x \sqsubseteq \oplus_{h_x} B_1 \sqcup \ldots \sqcup \oplus_{h_x} B_s$ or x cannot be covered by the union of sums of $\{B_1, \ldots, B_s\}$ at all. In fact, if $x = a_1^{j_1} a_2^{j_2} \ldots a_t^{j_t}$, then it is enough to choose $h_x = max\{j_1, \ldots, j_t\}$.

Assume that $r = (x_1, \alpha_1) \ldots (x_k, \alpha_k) \to (y_1, \beta_1) \ldots (y_l, \beta_l) \in R$, and let $h_r = \max\{h_{x_1}, \ldots, h_{x_k}\}$ (which is a positive integer number). Let us denote with $\mathfrak{H}(r)$ the set of all tuples $H = (H_1, \ldots, H_k)$, such that $H_j = \oplus_{h_1^j} B_1^j \sqcup \ldots \sqcup \oplus_{h_{n_j}^j} B_{n_j}^j$ with $h_t^j \leq h_r$ and $x_j \sqsubseteq H_j$ $(1 \leq j \leq k)$. Since $x_i \sqsubseteq \mathsf{l}(u_{\alpha_i})$ if and only if there exists a $H_i \sqsubseteq u_{\alpha_i}$ such that $x_i \sqsubseteq H_i$, in order to check the applicability of r in a configuration $(u_{\alpha_1}, \ldots, u_{\alpha_n})$, it is enough to check whether there exists an $H \in \mathfrak{H}(r)$, such that $H_i \sqsubseteq u_{\alpha_i}$ for every element of H, $1 \leq i \leq k$. We construct the Petri net which makes sure that $x_i \sqsubseteq \mathsf{l}(u_{\alpha_i})$ and simulates the rule application at the same time.

Let us define the Petri net $N = (P, T, F, V, m_0)$ with $P = O \times \{1, \ldots, n\} \cup \{p_{ini}\}$. A place $(a, j) \in P$ represents the number of objects $a \in O$ inside the jth membrane at every step of the computational sequence, so let us set $m_0(p) = w_j^0(a)$ for every place $p = (a, j) \in O \times \{1, \ldots, n\}$, and let also $m_0(p_{ini}) = 1$.

The net N consists of subnets for each pair $(r, H) \in RH = \{(r, H) \mid r \in R, H \in \mathfrak{H}(r)\}$. These subnets are responsible for the simulation of the effect of r together with checking the condition that r is an inner rule. The place p_{ini} makes sure that only one of the subnets can operate at a time, hence the simulation of the rule executions are mutually exclusive.

Let $T = \{t_\delta \mid \delta \in RH\}$ with $\delta = (r, H_1, \ldots, H_k) \in RH$, and let r be denoted as $r = (x_1, \alpha_1) \ldots (x_k, \alpha_k) \to (y_1, \beta_1) \ldots (y_l, \beta_l)$. Then, for $1 \leq j \leq k$,

$$p = (a, \alpha_j) \in {}^\bullet t_\delta \cap t_\delta^\bullet \text{ if and only if } a \in H_j.$$

For $p = (a, \beta_q)$, $1 \leq q \leq l$, we have

$$p = (a, \beta_q) \in {}^\bullet t_\delta \cap t_\delta^\bullet \text{ if and only if } \beta_q = \alpha_j$$

for some $1 \leq j \leq k$ and $a \in H_j$. Otherwise, $p = (a, \beta_q) \in t_\delta^\bullet$. In addition, $p_{ini} \in {}^\bullet t_\delta \cap t_\delta^\bullet$.

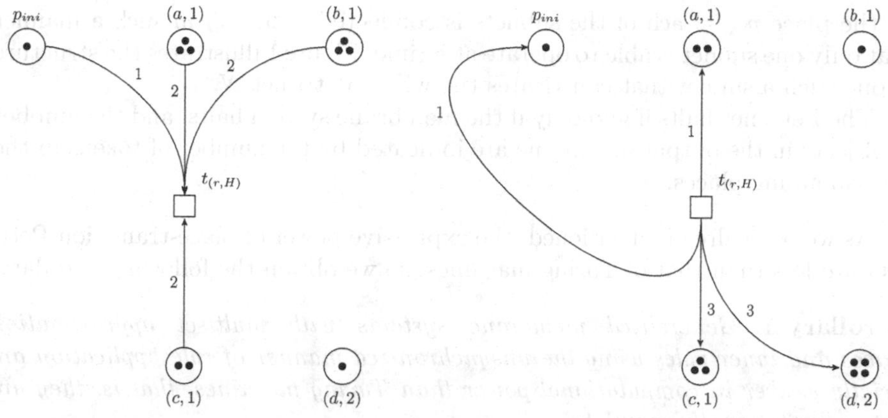

Fig. 1. Assume that $w_1^0 = a^3b^3c^2$, $w_2^0 = d$, $r = (ab^2, 1) \to (c, 1)(d^3, 2)$ and let $B_1 = a^2$, $B_2 = bc$ be base sets. Then, for $H = (H_1) \in \mathfrak{H}(r)$, $H_1 = B_1 \sqcup \oplus_2 B_2 = a^2b^2c^2$ are appropriate. The figure on the left shows the arcs pointing to transition $t_{(r,H)}$ together with their weights, the figure on the right shows the arcs going out from transition $t_{(r,H)}$ together with their weights. (The arcs with zero weight are not indicated explicitly.)

Let $p = (a, \alpha_j)$ with $1 \leq j \leq k$, and let $t_\delta \in T$, $\delta \in RH$, then the weights of the arcs are computed as $V(p, t_\delta) = H_j(a)$, that is, we check whether $H_j(a) \leq u_{\alpha_j}(a)$. Additionally, if $\alpha_j \neq \beta_q$ ($q \in \{1, \ldots, l\}$), then we have $V(t_\delta, p) = H_j(a) - x_{\alpha_j}(a)$, so that the necessary amount of tokens (those which correspond to the objects in $H_j \ominus x_j$) are returned to $p = (a, \alpha_j)$. This way the Petri net transition decreases the number of tokens in p only by $x_j(a)$. When $\alpha_j = \beta_q$ for some $q \in \{1, \ldots, l\}$, then $V(t_\delta, p) = H_j(a) - x_j(a) + y_q(a)$, that is, the righthand side of the rule returns further tokens to u_{α_j}.

For $p = (a, \beta_q)$, $1 \leq q \leq l$, if $\beta_q = \alpha_j$ for some $1 \leq j \leq k$, then the situation is as above. Otherwise, if $\beta_q \neq \alpha_j$ for any $1 \leq j \leq k$, then $V(p, t_\delta) = 0$ and $V(t_\delta, p) = y_{\beta_q}(a)$. Furthermore, $V(p_{ini}, t_\delta) = V(t_\delta, p_{ini}) = 1$.

To summarize the idea of the construction above, the places of the Petri net represent the objects in the different compartments of the P system. For every r, we are able to identify the union of the finite sums of the base sets that must be examined in order to decide whether the multisets $x_1, \ldots, x_k$ appearing on the lefthand side of the rule are in the inner approximation of u_{α_i}, that is, for every x_i we confine ourselves to $(B_1^i, \ldots, B_{k_i}^i) \in \mathfrak{H}(x_i)$, $1 \leq i \leq k$, such that $x_i \sqsubseteq \oplus_{h_1^i} B_1^i \sqcup \ldots \sqcup \oplus_{h_{k_i}^i} B_{k_i}^i$ and $h_t^j \leq h_r$, where $h_r \in \mathbb{N}$ is determined by r. Let $H = (H_1, \ldots, H_k) \in \mathfrak{H}(r)$ be a tuple of such multiset unions. To render r applicable and inner, we have to check whether $H_i \sqsubseteq u_{\alpha_i}$. For each pair $\delta = (r, H)$, where $r = (x_1, \alpha_1) \ldots (x_k, \alpha_k) \to (y_1, \beta_1) \ldots (y_l, \beta_l) \in R$ and $H = (H_1, \ldots, H_k) \in \mathfrak{H}(r)$ is a tuple of elements of $\mathfrak{B}$ with $x_i \sqsubseteq H_i$, we define a subnet consisting of all the places of N and a transition t_δ together with the corresponding arcs. This subnet simulates an application of r while the conditions on H ensure that r is an inner rule. The whole process is controlled by the place p_{ini}. Each of the subnets is connected with p_{ini} in such a manner that only one subnet is able to operate at a time. Figure 1 illustrates the structure of one such a subnet that constitutes the whole Petri net N.

The Petri net halts if and only if the membrane system halts, and the number of objects in the output membrane are indicated by the number of tokens in the corresponding places.

As we have already mentioned, the expressive power of place-transition Petri nets are less than that of Turing machines, so we obtain the following corollary.

Corollary 1. *Generalized membrane systems with multiset approximation spaces and inner rules using the unsynchronized manner of rule application are strictly weaker in computational power than Turing machines, that is, they are not computationally complete.*

Now we continue with the investigation of the case of boundary rules. We show that generalized P systems with boundary rules generate any recursively enumerable set of numbers. We do this by demonstrating how these systems simulate the computations of register machines, a computational model equivalent in power to Turing machines.

A *register machine* is a construct $W = (m, H, l_0, l_h, \mathrm{Inst})$, where m is the number of registers, H is the set of instruction labels, l_0 is the start label, l_h is the halting label, and Inst is the set of instructions. Each label from H labels only one instruction from Inst. There are several types of instructions which can be used. For $l_i, l_j, l_k \in H$ and $r \in \{1, \ldots, m\}$ we have:

- $l_i : (\mathtt{nADD}(r), l_j, l_k)$ - *nondeterministic add*: Add 1 to register r and then go to one of the instructions with labels l_j or l_k, nondeterministically chosen.
- $l_i : (\mathtt{ADD}(r), l_j)$ - *deterministic add*: Add 1 to register r and then go to the instruction with label l_j.
- $l_i : (\mathtt{CHECKSUB}(r), l_j, l_k)$ - *zero check and subtract*: If register r is empty, then go to the instruction with label l_j, if r is non-empty, then subtract one from it and go to the instruction with label l_k.
- $l_h : \mathtt{HALT}$ - *halt*: Stop the machine.

Register machines compute sets of numbers by starting their computation with empty registers and proceeding by applying instructions in the order indicated by the labels, beginning with the instruction l_0. If the machine reaches the halt instruction $l_h : \mathtt{HALT}$, then its work is finished, and the number stored in the first register is said to be the result of the computation. Note that the computed sets of numbers can be infinite, due to the nondeterminism in choosing the continuation of the computation in the case of nondeterministic add instructions, $l_i : (\mathtt{nADD}(r), l_j, l_k)$.

We would like to add here, that register machines can also be defined as deterministic computing devices (without the nondeterministic add instructions). In this case they compute functions of input values placed initially in (some of) the registers. They are able to compute all functions which are Turing computable (see, for example [10]), if they have at least two registers. By providing the machine with the nondeterministic add instruction, as above, we obtain a device which generates sets of numbers starting from a unique initial configuration. Since any recursively enumerable set can be obtained as the range of a Turing computable function on the set of non-negative integers, register machines defined this way are able to generate any recursively enumerable set of numbers.

Theorem 2. *Generalized P systems with associated multiset approximation spaces and boundary rules generate any recursively enumerable set of numbers, even in the unsynchronized manner of rule application.*

Proof. Let L be a recursively enumerable set of numbers, and consider the register machine $W = (m, H, l_0, l_h, \text{Inst})$ generating L. We construct a generalized P system with an associated multiset approximation space and boundary rules, such that it also generates L in the sense that the generated numbers correspond to the multiplicity of a certain object in the output region when the computation halts. Let $\Pi = (O, \mathfrak{B}, w_1, w_2, w_3, R, 2)$ with

$$\begin{aligned} O &= \{l, l' \mid l \in H\} \cup \{a_r, a'_r \mid 1 \leq r \leq m\} \cup \{b_1, b_2, c\}, \\ \mathfrak{B} &= \{a_r a'_r \mid 1 \leq r \leq m\} \cup \{l_h b_1 \mid l_h : \texttt{HALT}\} \cup \{b_2 c, lc \mid l \in H\}, \\ w_1 &= l_0 b_1, \ w_2 = \emptyset, \ w_3 = \emptyset, \\ R &= R_{Add} \cup R_{CheckSub} \cup R_{Ex}, \end{aligned}$$

where

$$\begin{aligned} R_{Add} = \{&(l_i, 1) \to (l_j, 1)(a_r, 2), (l_i, 1) \to (l_k, 1)(a_r, 2) \mid \text{for all} \\ &l_i \colon (\texttt{nADD}(r), l_j, l_k) \in \text{Inst}\} \cup \\ &\{(l_i, 1) \to (l_j, 1)(a_r, 2) \mid \text{for all } l_i : (\texttt{ADD}(r), l_j) \in \text{Inst}\}, \end{aligned}$$

$$\begin{aligned} R_{CheckSub} = \{&(l_i, 1) \to (l'_j, 1)(a'_r, 2), (l'_j, 1)(a'_r, 2) \to (l_j, 1), \\ &(l_i, 1)(a_r, 2) \to (l_k, 1) \mid \text{for all } l_i : (\texttt{CHECKSUB}(r), l_j, l_k) \in \text{Inst}\}, \end{aligned}$$

$$R_{Ex} = \{(b_1, 1) \to (b_2, 3), (b_2, 3) \to (b_1, 1)\}.$$

To see how Π simulates the computations of W, consider its initial configuration $(l_0 b_1, \emptyset, \emptyset)$: it corresponds to the initial configuration of W, as the first region contains l_0, the label of the instruction that has to be executed next, and the number of occurrences of a_r, $1 \leq r \leq m$, in the second region are 0, corresponding to the fact that all registers are initially empty.

Notice that as long as l_h is not present, it is possible to exchange b_1 in the first region with b_2 in the third (and back), since both symbols are in the boundary of the respective regions, so one of the rules of R_{Ex} is always applicable. When l_h

appears in the first region, then after b_1 also appears there, they are "removed" from the boundary of the region (as $l_h b_1 \in \mathfrak{B}$ is a base multiset of the multiset approximation space), and after this happens, no rule of R is applicable. From these considerations we can see that Π reaches a halting configuration only if the label of the halting instruction, l_h appears.

Let us consider the case when the generalized P system Π is in the configuration $(l_i\delta_1 w_{1,in}, ww_{2,in}, \delta_2)$ with $w(a_r) = k_r$, $1 \leq r \leq m$, corresponding to a situation when W is going to execute instruction l_i, and the contents of register r is $k_r \geq 0$, $1 \leq r \leq m$. The symbols δ_1, δ_2 are used to denote either b_1 or b_2, their exact meaning is not important, as they do not interfere with the simulation process until l_h appears. The submultisets $w_{1,in}$ and $w_{2,in}$ denote those elements of the first two regions that are not on the region boundary. By looking at the rules of Π, we might notice that as long as the object l_h is not present, all elements of the first region are on the boundary, thus, we might omit the submultiset $w_{1,in}$ from the above notation, having the configuration $(l_i\delta_1, ww_{2,in}, \delta_2)$. Note also, that $w_{2,in} = (a_r a'_r)^i$ for some $i \geq 0$.

If l_i is the label of an add, or nondeterministic add instruction, then the rule simulating the instruction $l_i : (\mathtt{nADD}(r), l_j, l_k)$ is applicable, yielding the configuration $(l'\delta_1, wa_r w_{2,in}, \delta_2)$ with $l' \in \{l_j, l_k\}$ (or the configuration $(l_j\delta_1, wa_r, \delta_2)$ if $l_i : (\mathtt{ADD}(r), l_j)$ is simulated). In any of these cases, we get a configuration $(l\delta_1, wa_r, \delta_2)$ where l corresponds to the instruction that has to be executed next, while the second region contains one more object a_r, that is, the number stored in register r was incremented, as required by the simulated add instructions.

Suppose now, that Π is in a configuration $(l_i\delta_1, ww_{2,in}, \delta_2)$ and the instruction to be executed is $l_i : (\mathtt{CHECKSUB}(r), l_j, l_k)$. By applying the rules in $R_{CheckSub}$ we might obtain $(l'_j\delta_1, wa'_r w_{2,in}, \delta_2)$. If $w(a_r) = 0$, we get $(l_j\delta_1, ww_{2,in}, \delta_2)$ after the next step, or if $w(a_r) > 0$, then as $a_r a'_r \in \mathfrak{B}$ is a base multiset, one copy of a_r and a'_r is removed from the boundary, so we have $(l'_j\delta_1, w'w'_{2,in}, \delta_2)$ where $a_r w' = w$ and $w'_{2,in} = a_r a'_r w_{2,in}$. In this case l'_j cannot be changed any more, and due to the rules in R_{Ex}, the computation can never reach a halting configuration. On the other hand, if $w(a_r) > 0$, then applying the rule $(l_i, 1)(a_r, 2) \to (l_k, 1)$, we get $(l_k\delta_1, w'w_{2,in}, \delta_2)$ with $w'a_r = w$, thus the checking and subtracting instruction of W is correctly simulated by the system Π.

The simulation is finished when the object l_h appears in the first region. The only rules that are applicable are the rules of R_{Ex}, but when b_1 also appears in the first region, the computation halts, because $l_h b_1$ is a base multiset, so all these objects disappear from the region boundary.

After halting, the result of the computation is the number of a_1 objects in the second region, as they correspond to the contents of the first register (the output register) of the register machine W.

4 Concluding Remarks

We have used multiset approximation spaces to restrict the applicability of multiset evolution rules of generalized P systems. This way we incorporated some additional "dynamics" into the system, as not only the presence or absence of elements, but also the underlying approximation spaces have a role in determining the applicability of the rules.

It turned out that restricting the operation of the rules to the boundaries of compartments increases the computational power of generalized P systems, as they are able to generate any recursively enumerable sets of numbers even in the unsynchronized manner of rule application. We have shown this by demonstrating that they are able to simulate register machines, a computational model equivalent in power to the model of Turing machines. On the other hand, a similar restriction allowing the rules to manipulate only elements of the lower approximation of the compartments of the system does not result in a similar increase of the computational power, as the resulting systems can be simulated by simple place-transition Petri nets, a model which is known to be weaker in computational power than the model of Turing machines.

As a final remark, we would like to add some thoughts on a related model called P systems with anti-matter [1,6]. In P systems with anti-matter, objects have complementary "anti objects", and when they are both present, they annihilate (disappear). In this paper we considered boundary rules which cannot be applied to objects that are not on the boundary: when all the elements of a base multiset are present in a region, they "disappear" from the scope of boundary rules. This effect is similar to the effect of annihilation rules, although not exactly the same. The difference can be seen from a simple example: let two base multisets be $ab, ac \in \mathfrak{B}$. The fact that they form base multisets is not directly modeled by the annihilation rules $ab \to \varepsilon, bc \to \varepsilon$ (as used in the case of P systems with anti-matter), because of the following. If a region contains ab, then these are "invisible" for the boundary rules, but they are not annihilated, as can be seen when an object c enters the region. As bc is also a base multiset, c immediately "disappears" by becoming part of the inner, lower approximation part of the region contents. As we see, the relationship of boundary rules and anti-matter is not as simple as it might look, but it definitely seems to be an interesting topic for further investigations.

Acknowledgments. G. Vaszil was supported by grant K 120558 of the National Research, Development and Innovation Office of Hungary (NKFIH), financed under the K 16 funding scheme. The work is also supported by the EFOP-3.6.1-16-2016-00022 project, co-financed by the European Union and the European Social Fund.

References

1. Alhazov, A., Aman, B., Freund, R.: P systems with anti-matter. In: Gheorghe, M., Rozenberg, G., Salomaa, A., Sosík, P., Zandron, C. (eds.) CMC 2014. LNCS, vol. 8961, pp. 66–85. Springer, Cham (2014). https://doi.org/10.1007/978-3-319-14370-5_5

2. Battyányi, P., Mihálydeák, T., Vaszil, G.: Generalized membrane systems with dynamical structure, Petri nets, and multiset approximation spaces. In: 18th International Conference on Unconventional Computation and Natural Computation. UCNC (2019, Accepted)
3. Bernardini, F., Gheorgue, M., Margenstern, M., Verlan, S.: Networks of cells and Petri nets. In: Díaz-Pernil, D., Graciani, C., Gutiérrez-Naranjo, M.A., Păun, G., Pérez-Hurtado, I., Riscos-Núñez, A. (eds.) Proceedings of the Fifth Brainstorming Week on Membrane Computing, pp. 33–62. Fénix Editora, Sevilla (2007)
4. Csuhaj-Varjú, E., Gheorghe, M., Stannett, M.: P systems controlled by general topologies. In: Durand-Lose, J., Jonoska, N. (eds.) UCNC 2012. LNCS, vol. 7445, pp. 70–81. Springer, Heidelberg (2012). https://doi.org/10.1007/978-3-642-32894-7_8
5. Csuhaj-Varjú, E., Gheorghe, M., Stannett, M., Vaszil, G.: Spatially localised membrane systems. Fundamenta Informaticae **138**(1–2), 193–205 (2015)
6. Leporati, A., Manzoni, L., Mauri, G., Porreca, A.E., Zandron, C.: The counting power of P systems with antimatter. Theoret. Comput. Sci. **701**, 161–173 (2017)
7. Mihálydeák, T., Csajbók, Z.E.: Membranes with boundaries. In: Csuhaj-Varjú, E., Gheorghe, M., Rozenberg, G., Salomaa, A., Vaszil, G. (eds.) CMC 2012. LNCS, vol. 7762, pp. 277–294. Springer, Heidelberg (2013). https://doi.org/10.1007/978-3-642-36751-9_19
8. Mihálydeák, T., Csajbók, Z.E.: On the membrane computations in the presence of membrane boundaries. J. Automata Lang. Comb. **19**(1), 227–238 (2014)
9. Mihálydeák, T., Vaszil, G.: Regulating rule application with membrane boundaries in P systems. In: Rozenberg, G., Salomaa, A., Sempere, J.M., Zandron, C. (eds.) CMC 2015. LNCS, vol. 9504, pp. 304–320. Springer, Cham (2015). https://doi.org/10.1007/978-3-319-28475-0_21
10. Minsky, M.L.: Computation: Finite and Infinite Machines. Prentice-Hall Inc., Upper Saddle River (1967)
11. Pawlak, Z.: Rough sets. Int. J. Comput. Inf. Sci. **11**(5), 341–356 (1982)
12. Pawlak, Z.: Rough Sets: Theoretical Aspects of Reasoning about Data. Kluwer Academic Publishers, Dordrecht (1991)
13. Peterson, J.L.: Petri Net Theory and the Modeling of Systems. Prentice Hall PTR, Upper Saddle River (1981)
14. Popova-Zeugmann, L.: Time and Petri Nets. Springer, Heidelberg (2013). https://doi.org/10.1007/978-3-642-41115-1
15. Păun, G.: Computing with membranes. J. Comput. Syst. Sci. **61**(1), 108–143 (2000)
16. Păun, G.: Membrane Computing: An Introduction. Natural Computing Series, 1st edn. Springer, Heidelberg (2002). https://doi.org/10.1007/978-3-642-56196-2
17. Păun, G., Rozenberg, G., Salomaa, A.: The Oxford Handbook of Membrane Computing. Oxford University Press Inc., New York (2010)

Soft Petri Net

Sibasis Bandyopadhyay[2(✉)], Zbigniew Suraj[1(✉)], and Prasun Kumar Nayak[2]

[1] Department of Computer Science, University of Rzeszów, Rzeszów, Poland
zbigniew.suraj@ur.edu.pl

[2] Department of Mathematics, Midnapore College, Midnapore, India
sibasisbanerjee@rediffmail.com, nayak_prasun@rediffmail.com

Abstract. In this paper, a soft Petri net model has been proposed based on a soft production rule. A soft implication operator has been introduced based on logical and set theoretic operations in a soft set. The truth degree in initial marking is considered as a binary number and boolean operators are used as In, Out_1 and Out_2 operators in the Petri net. Algorithms have been proposed to describe an approximate reasoning process with the soft Petri net. A numerical problem related to the purchase of a beautiful flat by a rational buyer has been discussed to establish relevance of the theory proposed.

Keywords: Soft set · Soft implication · Production rule · Knowledge representation · Approximate reasoning · Petri net

1 Introduction

A soft Petri net is a parameterized graphical representation of a soft production rule to address imprecise situation in an expert system or other decision support system. A soft production rule is used to represent IF-THEN rule, where antecedent or/and consequents are imprecise information, modeled by a soft set. In the recent past, a number of remarkable research with fuzzy production rule and fuzzy Petri net in intelligent system have been executed [11,13,16] with both backward and forward reasoning. A reasoning algorithm was given by Scarpelli et al. [11] related to the construction of a subnet and consequently a high level fuzzy Petri net. Suraj [12–14] proposed a fuzzy Petri net in a different way with input/output operators. Fryc et al. [3] proposed an extended fuzzy Petri net with matrix representation. Suraj and Bandyopadhyay [15] described the fuzzy Petri net with an intuitionistic fuzzy number and intuitionistic fuzzy weights based on a dual structured (N, N') fuzzy Petri net. Bandyopadhyay and Suraj [1] proposed a modified generalized fuzzy Petri (mGFP) with a modified operator binding function δ. There were also some research in the field of a rough Petri net. Peters et al. [7] proposed models of sensors, filters, and sensor fusion using the rough Petri net. Peters et al. [8] used guarded transition in the rough Petri net to simulate conditional computation in a various form of the system. But a membership function in the fuzzy set or rough set is not unique and hence leads to a lot of complications. The reason is the lack of parametrization of the

© Springer Nature Switzerland AG 2019
T. Mihálydeák et al. (Eds.): IJCRS 2019, LNAI 11499, pp. 253–264, 2019.
https://doi.org/10.1007/978-3-030-22815-6_20

imprecision concerned. A soft set can be considered as more generalized than the fuzzy set and rough set with parametrization of the imprecision. The soft Petri net can be a useful parameterized tool in an intelligent system in this situation.

The main objective of the paper is to develop a generalized Petri net model addressing an imprecise situation in an intelligent system. A parameterized model can be useful in this direction. We have used a soft set in a Petri net to serve this purpose. We proposed a soft production rule, i.e., IF-THEN production rule, where antecedent and consequent are the soft set. This production rule is based on different logical and set theoretic operations on the soft set. We have developed the concept of a soft implication operator and also the truth degree of a soft set based on the operations given in [4,5]. The firing rules have been given based on the concept that (In, Out_1, Out_2) operators are respectively the boolean product/boolean sum, boolean product and boolean sum. The truth degree of each soft set in initial position are also represented as a coordinate with elements 0, 1. The certainty factor β and threshold value γ are also coordinate with 0, 1 elements. We proposed algorithms to describe approximate reasoning with the soft Petri net. Ultimately, the proposed theory is established with the help of a numerical example.

The structure of the paper is as follows: In Sect. 2, some preliminaries related to the soft set, logical and set theoretic operations are given. Section 3 proposes the definition of a soft implication operator. A soft production rule and a soft Petri net is also described. In Sect. 4, computational algorithms are proposed to describe the approximate reasoning process. Section 5 makes the proposed theory relevant based on an elaborated numerical example. In Sect. 6, the comparison of the proposed theory is given with the existing literature. In addition, this section provides some conclusions.

2 Preliminaries

We first discuss some basic definitions and results to understand a soft Petri net. A soft set, defined by Molodtsov [5], is based on an initial universe set U and a set of parameters E. $\mathbf{P}(U)$ represents power set of U and $A \subset E$.

Definition 1. *Let $\phi : A \rightarrow \boldsymbol{P}(U)$ be a mapping. Then a soft set over U is defined as an ordered pair (ϕ, A). In other words, a soft set over U can be considered as parameterized family [4] of subsets of the universe U. For any parameter $\delta \in A$, $\phi(\delta)$ represents $\delta-$approximate elements of the soft set (ϕ, A). If $\phi(\epsilon) =$ null set $\forall \epsilon \in A$ then (ϕ, A) is a null soft set. If $\phi(\epsilon) = U$ $\forall \epsilon \in A$ then (ϕ, A) is an absolute soft set.*

Example 1. Let U denote a set of cars under consideration and E - a set of parameters like expensive, comfortable, fashionable, cheap. In this case, a soft set (ϕ, E) represents the "Attractiveness of cars", which somebody may buy. Suppose $U = \{p_1, p_2, p_3, p_4, p_5\}$ represents five cars and their comfort level is given by the set of parameters $E = \{e_1, e_2, e_3, e_4\}$. Parameters e_1, e_2, e_3, e_4 represent expensive, comfortable, fashionable, cheap, respectively. Let us suppose

$$\phi(e_1) = \{p_2, p_4\}, \phi(e_2) = \{p_1\}, \phi(e_3) = \{p_3, p_5\}, \phi(e_4) = \{p_1, p_4, p_5\} \quad (1)$$

which is a parameterized family of subsets of U and describes a collection of approximate representation of an object (Table 1).

Table 1. Tabular representation of soft set from Example 1.

U/E	Expensive	Comfortable	Fashionable	Cheap
p_1	0	1	0	1
p_2	1	0	0	0
p_3	0	0	1	0
p_4	1	0	0	1
p_5	0	0	1	1

Definition 2. *Suppose (ϕ, A) and (Ψ, B) are two soft sets over the same universe U. Then, (ϕ, A) is said to be a soft subset of (ψ, B) if*

1. *$A \subset B$, and*
2. *$\forall \delta \in A$, $\phi(\delta)$ and $\psi(\delta)$ represent the same approximations.*

We denote it as $(\phi, A) \tilde{\subset} (\psi, B)$. If $(\phi, A) \tilde{\subset} (\psi, B)$ and simultaneously $(\psi, B) \tilde{\subset} (\phi, A)$, then we say that a soft set (ϕ, A) is equal to (ψ, B) and write $(\phi, A) = (\psi, B)$.

Definition 3. *$(\phi, A)^c$ is said to be complement of the soft set (ϕ, A) if $(\phi, A)^c = (\phi^c, \rceil A)$, where $\phi^c : \rceil A \rightarrow \boldsymbol{P}(U)$ is a mapping defined as $\phi^c(\alpha) = U - \phi(\rceil \alpha), \forall \alpha \in \rceil A$. $\rceil \alpha$ means not α, i.e., if α represents 'expensive' then $\rceil \alpha$ represents 'not expensive'. $\rceil A$ consists of all such $\rceil \alpha, \forall \alpha \in A$.*

Definition 4. Logical operations on two soft sets: *The logical operations AND and OR on two soft sets (ϕ, A) and (ψ, B) are defined as follows*

1. *(ϕ, A) AND $(\psi, B) = (\phi, A) \wedge (\psi, B) = (\Omega, A \times B)$, where $\Omega(\alpha, \beta) = \phi(\alpha) \cap \psi(\beta), \forall (\alpha, \beta) \in A \times B$*
2. *(ϕ, A) OR $(\psi, B) = (\phi, A) \vee (\psi, B) = (\Gamma, A \times B)$, where $\Gamma(\alpha, \beta) = \phi(\alpha) \cup \psi(\beta), \forall (\alpha, \beta) \in A \times B$*

Proposition 1.

1. *$((\phi, A) \vee (\psi, B))^c = (\phi, A)^c \wedge (\psi, B)^c$*
2. *$((\phi, A) \wedge (\psi, B))^c = (\phi, A)^c \vee (\psi, B)^c$*

Definition 5. Algebraic operations on two soft sets: *The algebraic operations* $\widetilde{\cup}$ *and* $\widetilde{\cup}$ *on two soft sets* (ϕ, A) *and* (ψ, B) *are defined as follows*

1. $(\phi, A) \widetilde{\cup} (\psi, B) = (\Omega, A \cup B)$, *where*
$$\Omega(\alpha) = \begin{cases} \phi(\alpha) \text{ if } \alpha \in A - B, \\ \psi(\alpha) \text{ if } \alpha \in B - A, \\ \phi(\alpha) \cup \psi(\alpha) \text{ if } \alpha \in A \cap B \end{cases}$$
2. $(\phi, A) \widetilde{\cap} (\psi, B) = (O, A \cap B)$, *where* $O(\alpha) = \phi(\alpha)$ *or* $\psi(\alpha)$

Proposition 2.

1. $((\phi, A) \widetilde{\cup} (\psi, B))^c = (\phi, A)^c \widetilde{\cup} (\psi, B)^c$
2. $((\phi, A) \widetilde{\cap} (\psi, B))^c = (\phi, A)^c \widetilde{\cap} (\psi, B)^c$

Proposition 3.

1. $(\phi, A) \widetilde{\cup} (\phi, A) = (\phi, A)$
2. $(\phi, A) \widetilde{\cap} (\phi, A) = (\phi, A)$
3. $(\phi, A) \widetilde{\cup} \Phi = \Phi$, *where* Φ *is the null soft set*
4. $(\phi, A) \widetilde{\cap} \Phi = \Phi$
5. $(\phi, A) \widetilde{\cup} \widetilde{A} = \widetilde{A}$, *where* $\widetilde{A}$ *is the absolute soft set [4]*
6. $(\phi, A) \widetilde{\cap} \widetilde{A} = \widetilde{A}$

3 Soft Implication

In this section, we introduce implication operators on soft sets. The implication on soft set can be described with a statement "If 'Houses are attractive' Then 'Rational buyers will come'". Attractiveness of houses can be described by the terms 'expensive', 'beautiful look', 'modern amenities', 'cheap' etc. On the other hand, 'Rational buyers' also describe houses with the terms 'houses for huge funds', 'houses for low funds', 'selective houses' etc. Let two soft sets (ϕ, A) and (ψ, B) represent 'attractive houses' and 'Rational buyers', respectively. A and B are described over the same universe. Then the implication can be described as

$$\text{if } (\phi, A) \text{ then } (\psi, B) \tag{2}$$
$$\text{or } \ (\phi, A) \rightarrow (\psi, B) \tag{3}$$
$$\text{or}, NOT(\phi, A) \vee (\psi, B) \tag{4}$$

which is equivalent to that in classical logic.

Definition 6. *Implication operator from a soft set* (ϕ, A) *to another soft set* (ψ, B) *denoted as* $(\phi, A) \rightarrow (\psi, B)$ *is defined as* $(\phi^c, \rceil A) \vee (\psi, B) = (\rho, \rceil A \times B), \rho(\alpha, \beta) = \phi^c(\alpha) \widetilde{\cup} \psi(\beta)$.

Definition 7. *Truth degree of a soft set* (ϕ, A) *is defined as* $Tr(\phi, A) = ((0,1)_{A_1}, (0,1)_{A_2}, \cdots, (0,1)_{A_m})$, *where* m *is the cardinality of the set* A *and each* $(0,1)_{A_i}, i = 1, 2, \cdots, m$ *is an nth ordered coordinate,* m *representing the cardinality of set* U.

3.1 Soft Production Rules and Soft Petri Net

The imprecise information in a real world can often be parameterized using the soft set model. We can use a production rule based on soft sets, i.e., a soft production rule (SPR) to process vague or imprecise knowledge. SPRs can be used in an expert system and inference rules can be represented in the form of SPRs. SPRs are considered as IF-THEN rules, where both antecedents and consequents are imprecise terms modeled by soft sets. A SPR is said to be a compound SPR if both antecedents and consequents are associated with connectors AND or OR.

A set of SPRs is given by $\widetilde{R} = \{\widetilde{R_1}, \widetilde{R_2}, \cdots, \widetilde{R_n}\}$, where the ith rule is expressed as:

$$\widetilde{R_i} : \text{IF } (\phi, A) \text{ THEN } (\psi, B)\ (CF = t), \lambda_i \tag{5}$$

where:

- $(\phi, A) = \{(\phi, A_1), (\phi, A_2), \cdots, (\phi, A_k)\}$ represents the antecedent proposition connected by AND or OR connectors;
- $(\psi, B) = \{(\psi, B_1), (\psi, B_2), \cdots, (\psi, B_l)\}$ represents the consequent propositions connected by AND or OR connectors;
- t is the certainty factor of the rule $\widetilde{R_i}, i = 1, 2, \cdots, n$;
- $\lambda_i, i = 1, 2, \cdots, k$ gives the threshold values for each antecedent proposition.

We can describe soft production rules by classifying it as below:

- Type 1: A soft simple rule is expressed as

$$\widetilde{R} : \text{IF } (\phi, A) \text{ THEN } (\psi, B)\ (CF = t), \lambda \tag{6}$$

- Type 2: A soft conjunctive (disjunctive) rule for the antecedent is expressed as $\widetilde{R}$: IF (ϕ, A_1) AND(OR) $(\phi, A_2) \cdots$ AND(OR) (ϕ, A_k) THEN (ψ, B) $(CF = t), \lambda_i, i = 1, 2, \cdots, k$
- Type 3: A soft conjunctive (disjunctive) rule for the consequent is expressed as $\widetilde{R}$: IF (ϕ, A) THEN (ψ, B_1) AND(OR) (ψ, B_2) $\cdots$ AND(OR) (ψ, B_l) $(CF = t), \lambda$

3.2 Soft Petri Net

A soft Petri net (SPN) is described, based on SPRs. The soft Petri net can be used as a tool for graphical representation or mathematical modelling for expert systems and others. The SPN is considered as a parameterized modification of classical Petri nets [9], where the net places assume soft variables with values as nth ordered $(0, 1)$ coordinate representing truth values for different parameters. This truth value is said to be the truth degree of a statement associated with a net place. The transitions can be considered as soft implications where the input places represent premises of an implication and the output places correspond to conclusions of that implication. The threshold values of transitions are given to determine the possibility of a transition firing. The formal definition of an SPN is given as follows:

Definition 8. *An SPN (over a soft set with the universe U) can be defined as a tuple $N = (P, T, S, I, O, \alpha, \beta, \gamma, M_0)$, where:*

1. $P = \{p_1, p_2, \cdots, p_n\}$ *represents a finite set of places,* $n > 0$*;*
2. $T = \{t_1, t_2, \cdots, t_m\}$ *is a finite set of transitions,* $m > 0$*;*
3. $S = \{s_1, s_2, \cdots, s_n\}$ *is a finite set of statements and P,T,S are pairwise disjoint, i.e.,* $P \cap T = S \cap T = P \cap S = \emptyset$ *and* $card(P) = card(S)$*;*
4. $I : T \to \boldsymbol{P}(P)$ *is the input function;*
5. $O : T \to \boldsymbol{P}(P)$ *is the output function;*
6. $\alpha : P \to S$ *is the statement binding function;*
7. $\beta : T \to (0,1)_l$ *is the truth degree function;*
8. $\gamma : T \to (0,1)_l$ *is the threshold function;*
9. $M_0 : P \to (0,1)_l$ *is the initial marking,*

and $\boldsymbol{P}(P)$ represents power set of P, l is the cardinality of set U, $(0,1)_p$ is an p tuple consisting of 0 *and* 1.

Following the convention of the graphical representation, in this discussion, we denote the places by circles and transitions by rectangles. Oriented arcs connecting places with transitions give the function I and the oriented arcs connecting transitions with places express the function O. Logical operators OR and AND are interpreted in SPN as boolean sum and boolean product for binary vectors, and denoted by SUM and PROD, respectively. Based on the theory given so far, we propose a firing rule defined as follows.

3.3 Firing Rule

We suppose that tuple $N = (P, T, S, I, O, \alpha, \beta, \gamma, M_0)$ is a SPN with marking $M : P \to (0,1)_l$. A transition [12] $t \in T$ is considered to be enabled for firing at the marking M, if it satisfies the following conditions

$$In(M(p_{i1}), M(p_{i2}), \cdots, M(p_{in})) \geq \gamma(t) \tag{7}$$

where formula (7) provides the condition for firing transitions in SPN and $I(t) = \{p_{i1}, p_{i2}, \cdots, p_{in}\}$ gives a set of input places corresponding to a transition $t \in T$ and $\beta : T \to (0,1)_l$. In is an input operator and Out_1, Out_2 are represented as output operators corresponding to the transition t. These operators are boolean product/boolean sum, boolean product and boolean sum, respectively. With these assumptions, the mode of firing is proposed as follows.

Let M denotes a marking of N which enables transition t, and M' is the marking derived from M by firing transition t. For each $t \in T$ we have

$$M'(p) = \begin{cases} Out_2(Out_1(In(M(p_{i1}), M(p_{i2}), \cdots, M(p_{in})), \beta(t)), M(p)) \text{ if } p \in O(t), \\ M(p) \text{ otherwise} \end{cases}$$

Now, we will describe a different type of firing rules for different type of Petri nets based on the soft production rules.

– Type 1: A soft simple rule is expressed as (Fig. 1)

$$\widetilde{R} : \text{IF } (\phi, A) \text{ THEN } (\psi, B)\ (CF = t), \lambda \tag{8}$$

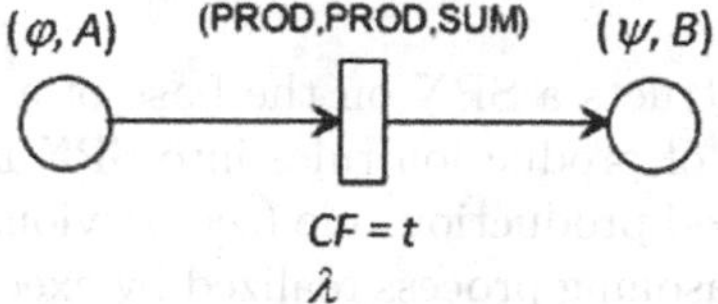

Fig. 1. SPN for type 1 rule.

– Type 2: A soft conjunctive (disjunctive) rule for the antecedent is expressed as $\widetilde{R}$: IF (ϕ, A_1) AND(OR) $(\phi, A_2) \cdots$ AND(OR) (ϕ, A_k) THEN (ψ, B) $(CF = t), \lambda_i, i = 1, 2, \cdots, k$ (Fig. 2)

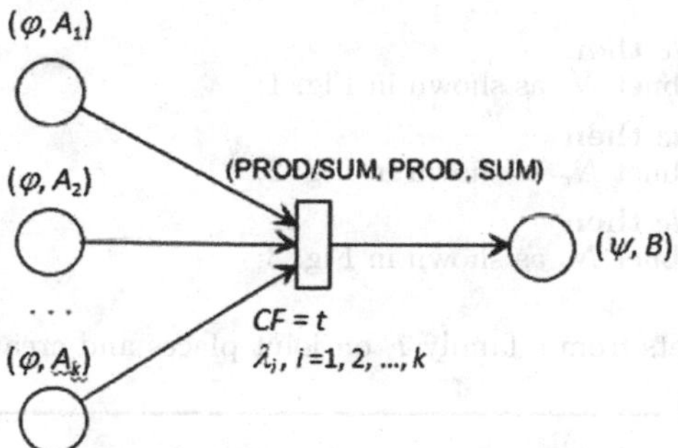

Fig. 2. SPN for type 2 rule.

– Type 3: A soft conjunctive (disjunctive) rule for the consequent is expressed as $\widetilde{R}$: IF (ϕ, A) THEN (ψ, B_1) AND(OR) $(\psi, B_2) \cdots$ AND(OR) (ψ, B_l) $(CF = t), \lambda$ (Fig. 3)

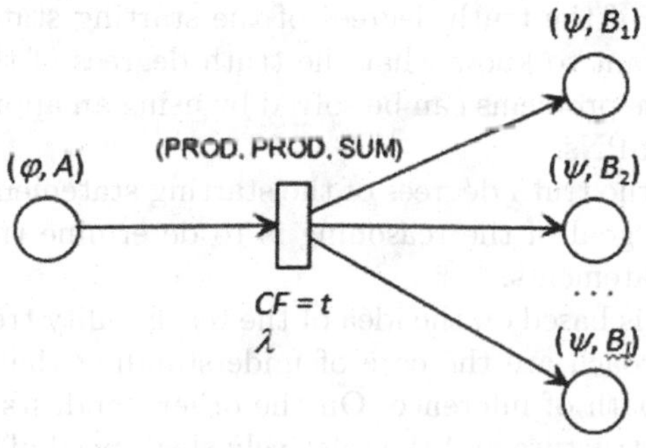

Fig. 3. SPN for type 3 rule.

4 Computational Algorithms

In this section, two algorithms [13] are proposed for the construction of a SPN to express an approximate reasoning process with the production rules given in Sect. 3.3.

The Algorithm 1 constructs a SPN on the base of a given set of production rules; the transformation of production rules into SPN is realized depending on the form of the transformed production rule (see previous section). However, the second one describes a reasoning process realized by execution of SPN representing a given set of production rules.

Algorithm 1. Construction of a SPN with production rules as given in Sect. 3.3.

Input : Finite set R of production rules with parameters.
Output: SPN N.
$F \leftarrow \emptyset$; (* The empty set. *)
for each $r \in R$
if r *is a type 1 rule* **then**
⌊ construct a subnet N_r as shown in Fig. 1;
if r *is a type 2 rule* **then**
⌊ construct a subnet N_r as shown in Fig. 2;
if r *is a type 3 rule* **then**
⌊ construct a subnet N_r as shown in Fig. 3;
$F \leftarrow F \cup \{N_r\}$;
Integrate all subnets from a family F on joint places and create a result net N;
return N;

In order to describe the second algorithm, we need two auxiliary notions. In some situations we may want to determine the antecedence-consequence relationships between two groups of statements: the starting (given) statements $s_{i1}, \ldots, s_{ik}$, and goal (computed) statements $s_{o1}, \ldots, s_{ol}$. In the Petri net representation, the places associated with the first group of statements are called *starting places*, whereas the places associated with the second one are called *goal places*. Furthermore, if the truth degrees of the starting statements $s_{i1}, \ldots, s_{ik}$ are given, we may want to know what the truth degrees of the goal statements $s_{o1}, \ldots, s_{ol}$ are. These problems can be solved by using an approximate reasoning algorithm based on SPNs.

We assume that the truth degrees of the starting statements are given by the domain expert. The goal of the reasoning is to determine the truth degrees of the output (goal) statements.

The Algorithm 2 is based on the idea of the reachability tree [6,10]. The main benefits of this approach are the ease of understanding the algorithm and the ease of finding the path of inference. On the other hand, its weaker side is the more complex data structure and the relatively slow speed of inference (cf. [17]).

Algorithm 2. Approximate reasoning based on SPN.

Input : The initial marking of the starting places with elements of the form (ϕ, A).
Output: The final marking of the goal places with elements of the form (ψ, B).
while *it is not the end of simulation* **do**
    Determine transitions enabled for firing based on firing rule in Section. 3.3
    while *there is a transition enabled for firing* **do**
        Compute a new marking of all places after firing the transition;
        Determine a new transition enabled for firing;
    Read final marking of goal places;
    Reset final marking of all places;

5 Numerical Example

Let us consider that, a 'Rational buyer' wants to buy a 'Beautiful flat'. The rational buyer is inclined to buy a non-expensive, beautiful flat. On other hand, a 'Beautiful flat' should have three or more bedrooms, a top floor, a garden face, satisfactory amenities from the buyer's perspectives. There are three flats $\{H_1, H_2, H_3\}$ to choose from. The buyer first visited flat H_1 having four bedrooms but not on the top floor. Other two flats H_2, H_3 are on top floors but they have only two bedrooms. After satisfying these primary conditions there are two more issues. H_1 and H_3 are expensive but H_2 is non-expensive. Moreover, H_2 and H_3 have a garden face. H_3 has satisfactory amenities but H_2 does not. Now, which house is the most rational according to the buyer's expectations? If (ϕ, A) represents 'Beautiful house' and (ψ, B) represents 'Rational buyer', then we may write the situation as

$$(\phi, A) \to (\psi, B) \tag{9}$$

$\{A_1, A_2, A_3, A_4, A_5\}$ represent parameters 'expensive', 'three or more rooms', 'top floor', 'garden face', 'satisfactory amenities', respectively. We can write

$$A = \bigcup_{i=1}^{5} A_i, (\phi, A_i) \subset (\phi, A), i = 1, 2, \cdots, 5 \tag{10}$$

The situation is represented logically as follows:

1. IF s_2 OR s_3 THEN s_6;
2. IF s_1 AND s_4 AND s_6 THEN s_7;
3. IF s_4 AND s_5 THEN s_8,

where we can represent the variables as follows:

- s_1: = 'The flat is expensive.'
- s_2: = 'The flat has three or more rooms.'
- s_3: = 'The flat is on the top floor.'

- s_4: = 'The flat is garden face.'
- s_5: = 'The flat has satisfactory amenities.'
- s_6: = 'The flat is good for a rational buyer.'
- s_7: = 'The flat is comfortable for a rational buyer.'
- s_8: = 'The flat has a good look.' (Fig. 4)

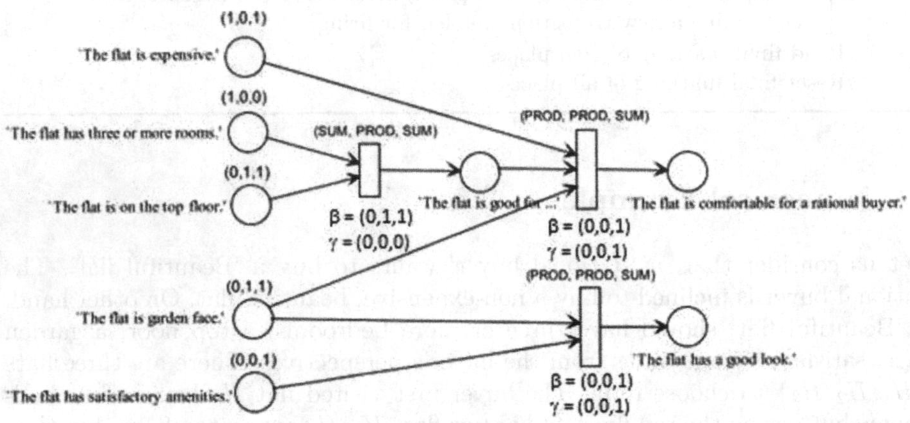

Fig. 4. SPN - before firing.

We see that at place 'The flat is comfortable for a rational buyer.' the truth degree is (0,0,1). It implies that the flat H_3 is acceptable to a rational buyer, which can be described as expensive but it is on the top floor with satisfactory amenities and a garden face. So, (ψ, B) can be described as B ={'expensive', 'top floor', 'garden face', 'satisfactory amenities'} and $\psi(B) = \{H_3\}$ (Fig. 5).

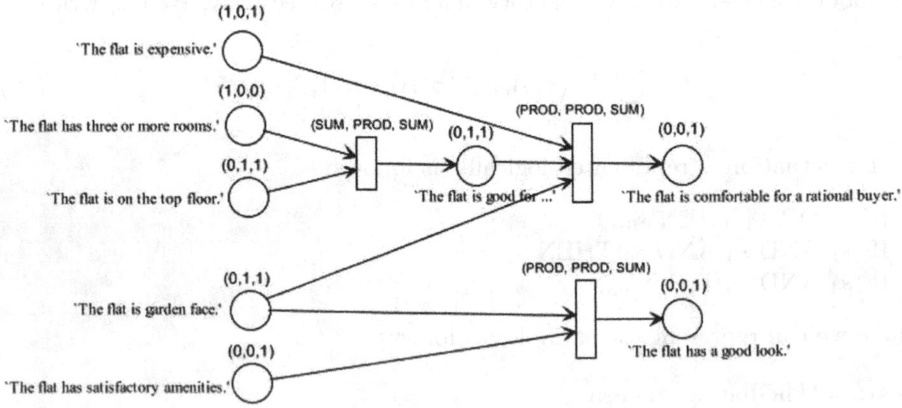

Fig. 5. SPN - after firing.

6 Conclusion

In this paper, a Petri net based on a soft set has been proposed. It holds some advantages compared to others in the literature:

1. This is a parameterized model using the soft sets and hence does not need to configure membership function as in fuzzy Petri net [2].
2. Since SPN involves fewer complex functions, it involves lesser computational complexity.

Here, parameterized knowledge representation have been proposed with the soft Petri net. A soft implication operator and a soft production rule have been proposed to construct the soft Petri net. The In, Out_1, and Out_2 operators are given as boolean operators and the truth degree is represented by binary numbers. This model does not involve any complex operators unlike the fuzzy Petri net [12]. In the fuzzy Petri net, determination of a membership function is sometimes unrealistic and not unique. Moreover, in a backward reasoning an inverted fuzzy implication is not always possible. The parameterized form of the Petri net modeled by the soft set is free from such complications and it presents lesser computational complexity. There is some further scope to use the soft Petri net in the backward reasoning process.

Acknowledgement. This work was partially supported by the Center for Innovation and Transfer of Natural Sciences and Engineering Knowledge at the University of Rzeszów. We would like to thank the anonymous referees for critical remarks and useful suggestions to improve the quality of the paper.

References

1. Bandyopadhyay, S., Suraj, Z., Grochowalski, P.: Modified generalized weighted fuzzy Petri Net in intuitionistic fuzzy environment. In: Flores, V., et al. (eds.) IJCRS 2016. LNCS (LNAI), vol. 9920, pp. 342–351. Springer, Cham (2016). https://doi.org/10.1007/978-3-319-47160-0_31
2. Cardoso, J., Camargo, H. (eds.): Fuzziness in Petri Nets. Physica-Verlag, Berlin (1999)
3. Fryc, B., Pancerz, K., Peters, J.F., Suraj, Z.: On fuzzy reasoning using matrix representation of extended fuzzy Petri Nets. Fundam. Inform. **60**(1–4), 143–157 (2004)
4. Maji, P.K., Biswas, R., Roy, A.R.: Soft set theory. Comput. Math. Appl. **45**, 552–562 (2003)
5. Molodtsov, D.: Soft set theory - first results. Comput. Math. Appl. **37**, 19–31 (1999)
6. Peterson, J.L.: Petri Net Theory and the Modeling of Systems. Prentice-Hall Inc., Englewood Cliffs (1981)
7. Peters, J., Ramanna, S., Borkowski, M., Skowron, A., Suraj, Z.: Sensor, filter, and fusion models with rough Petri Nets. Fundam. Inform. **47**, 307–323 (2001)

8. Peters, J., Skowron, A., Suraj, Z., Ramanna, S.: Guarded transitions in rough Petri Nets. In: Proceedings of 7th European Congress on Intelligent Systems & Soft Computing (EUFIT 1999), Aachen, Germany, 13–16 September 1999, pp. 203–212 (1999)
9. Petri, C.A.: Kommunikation mit Automaten, Schriften des IIM, Nr. 2, Institut für Instrumentelle Mathematik, Bonn 1962. English translation: Technical report RADC-TR-65-377, vol. 1, suppl. 1, Griffiths Air Force Base, New York (1966)
10. Reisig, W.: Petri Nets. Springer Publishing Company, Berlin (1985)
11. Scarpelli, H., Gomide, F., Yager, R.R.: A reasoning algorithm for high-level fuzzy Petri Net. IEEE Trans. Fuzzy Syst. **4**(3), 282–294 (1996)
12. Suraj, Z.: Knowledge representation and reasoning based on generalised fuzzy Petri nets. In: Proceedings of the 12th International Conference on Intelligent Systems Design and Applications (ISDA 2012), Kochi, India, 27–29 November 2012, pp. 101–106. IEEE Press
13. Suraj, Z.: A new class of fuzzy Petri Nets for knowledge representation and reasoning. Fundam. Inform. **128**(1–2), 193–207 (2013)
14. Suraj, Z.: Modified generalised fuzzy Petri Nets for rule-based systems. In: Yao, Y., Hu, Q., Yu, H., Grzymala-Busse, J.W. (eds.) RSFDGrC 2015. LNCS (LNAI), vol. 9437, pp. 196–206. Springer, Cham (2015). https://doi.org/10.1007/978-3-319-25783-9_18
15. Suraj, Z., Bandyopadhyay, S.: Generalized weighted fuzzy Petri Net in intuitionistic fuzzy environment. In: IEEE World Congress on Computational Intelligence 2016, Vancouver, Canada, pp. 2385–2392. IEEE (2016)
16. Suraj, Z., Lasek, A.: Inverted fuzzy implications in backward reasoning. In: Kryszkiewicz, M., Bandyopadhyay, S., Rybinski, H., Pal, S.K. (eds.) PReMI 2015. LNCS, vol. 9124, pp. 354–364. Springer, Cham (2015). https://doi.org/10.1007/978-3-319-19941-2_34
17. Zhou, K.-O., Zain, A.M.: Fuzzy Petri Nets and industrial applications: a review. Artif. Intell. Rev. **45**, 405–446 (2016)

Approximations Induced by Tolerance Relations

Dávid Gégény, Imre Piller(✉), Sándor Radeleczki, and Laura Veres

Institute of Mathematics, University of Miskolc, Miskolc, Hungary
{matgd,matip,matradi,matlaura}@uni-miskolc.hu
http://www.uni-miskolc.hu/~mathint

Abstract. In this paper, we examined by using Formal Concept Analysis methods, the interrelation between the lattices of upper (lower) approximations induced by two tolerance relations $R \subseteq \rho \subseteq U \times U$. These lattices are isomorphic (dually isomorphic) to the concept lattice $\mathcal{L}(U, U, R^c)$, $\mathcal{L}(U, U, \rho^c)$ respectively, where R^c and ρ^c stand for the complements of the corresponding relations. We proved sufficient conditions and we characterized the case when the concept lattice $\mathcal{L}(U, U, \rho^c)$ is a complete sublattice of $\mathcal{L}(U, U, R^c)$. We used the so-called compatibility condition introduced recently and we showed that in the case when ρ is R-compatible and $\mathcal{L}(U, U, \rho^c)$ is a complete sublattice of $\mathcal{L}(U, U, R^c)$, ρ must be an equivalence. Detailed examples for each case were presented.

Keywords: Lower and upper approximation · Rough set · Complete sublattice of a concept lattice · Compatibility condition

1 Introduction

The theory of rough sets was initiated by Pawlak [7]. His idea was that our knowledge about the objects of a universe U is given in the terms of an indistinguishability relation $R \subseteq U \times U$ reflecting the indiscernibility of the objects. Originally, Pawlak assumed that this relation is an equivalence, i.e. a reflexive, symmetric and transitive binary relation, however in the literature numerous studies can be found where approximations are defined by other types of relations (see e.g. [13], [3] or [4]).

For any binary relation $R \subseteq U \times U$ and any element $u \in U$, we denote by $R(u)$ the *R-neighborhood* of u, i.e. $R(u) := \{x \in U \mid (u, x) \in R\}$. Now, for any subset $X \subseteq U$ the *lower approximation* of X is defined as $X_R := \{x \in U \mid R(x) \subseteq U\}$, and the *upper approximation* of X is given by $X^R := \{x \in U \mid R(x) \cap X \neq \emptyset\}$. If R is a reflexive relation then $X_R \subseteq X \subseteq X^R$. The *rough set of X* can be defined as the pair (X_R, X^R), and the *set of all rough sets* is identified with

$$RS(U, R) = \{(X_R, X^R) \mid X \subseteq U\}.$$

The set $RS(U, R)$ may be canonically ordered by the componentwise inclusion:

$$(X_R, X^R) \leq (Y_R, Y^R) \Longleftrightarrow X_R \subseteq Y_R \text{ and } X^R \subseteq Y^R,$$

© Springer Nature Switzerland AG 2019
T. Mihálydeák et al. (Eds.): IJCRS 2019, LNAI 11499, pp. 265–279, 2019.
https://doi.org/10.1007/978-3-030-22815-6_21

resulting in a partially ordered set $\mathbf{RS}(U, R) := (RS(U, R), \leq)$. If R is an equivalence, then $\mathbf{RS}(U, R)$ is a complete distributive lattice (having many other nice properties).

The sets $\wp(U)^R = \{X^R \mid X \subseteq U\}$ and $\wp(U)_R = \{X_R \mid X \subseteq U\}$ in general form dually isomorphic complete lattices $(\wp(U)^R, \subseteq)$ and $(\wp(U)_R, \subseteq)$, called respectively the *lattice of upper approximations* and *the lattice of lower approximations* (see [4]). If R is a *tolerance*, i.e. a reflexive and symmetric relation, then in [4] it is shown that $(\wp(U)_R, \subseteq)$ is isomorphic to the concept lattice $\mathcal{L}(U, U, R^c)$ of the context (U, U, R^c), where $R^c = (U \times U) \setminus R$ is the complement of the relation R. (These notions will be explained in Sect. 2.) Inspired by this observation, we will apply Formal Concept Analysis (FCA) methods to describe the sublattices of the lattices of upper (or lower) approximations. These lattices have an important role in numerous applications of rough set theory (see e.g. [8–11,14]) and comparing different approximations is in the focus of several papers (see e.g. [10], [13] or [6]). In our paper we deduce sufficient conditions which guarantee that for some tolerance relations $R \subseteq \rho \subseteq U \times U$, the lattice $\wp(U)^\rho$ ($\wp(U)_\rho$) is a complete sublattice of $\wp(U)^R$ (of $\wp(U)_R$).

The paper is structured as follows: In Sect. 2, we present some basic properties of the approximation operators $X \to X_R$ and $X \to X^R$, $X \subseteq U$. In case of a tolerance relation R, the interrelation between the concept lattice $\mathcal{L}(U, U, R^c)$ and the lattices $(\wp(U)^R, \subseteq)$ and $(\wp(U)_R, \subseteq)$ is also presented. In Sect. 3, sufficient conditions are proved, and we characterize that when $R \subseteq \rho$, and the concept lattice $\mathcal{L}(U, U, \rho^c)$ is a complete sublattice of $\mathcal{L}(U, U, R^c)$. In Sect. 4, the conditions of our previous results are combined with the so-called "compatibility condition" introduced in [6], and several conclusions are deduced. For instance, we prove that in the case when $R^2 = R \circ R \subseteq \rho$ and $\mathcal{L}(U, U, \rho^c)$ is a complete sublattice of $\mathcal{L}(U, U, R^c)$, ρ must be an equivalence. Our conclusions are summarized in Sect. 5.

2 Preliminaries

The above defined approximations for any $X \subseteq U$ and any $\mathcal{H} \subseteq \mathcal{P}(U)$ have the following properties:

(a) $\left(\bigcup\limits_{X \in \mathcal{H}} X\right)^R = \bigcup\limits_{X \in \mathcal{H}} X^R$ and $\left(\bigcap\limits_{X \in \mathcal{H}} X\right)_R = \bigcap\limits_{X \in \mathcal{H}} X_R$;

(b) $(X^c)^R = (X_R)^c$, $(X^c)_R = \left(X^R\right)^c$.

In view of (a), $X \to X^R$, $X \subseteq U$ is a complete join-homomorphism and $X \to X_R$, $X \subseteq U$ is a complete meet-homomorphism. Thus $\wp(U)_R$ is a *closure system*, being closed under arbitrary intersections and $\wp(U)^R$ is an *interior system*, because it is closed under any union. Therefore, $\wp(U)_R$ and $\wp(U)^R$ are complete lattices with respect to $\subseteq$. If R is a tolerance relation, then these mappings are *adjoint*, i.e. for any $X, Y \subseteq U$ we have: $X^R \subseteq Y \Leftrightarrow X \subseteq Y_R$.

Property (b) implies that the lattices $(\wp(U)_R, \subseteq)$ and $(\wp(U)^R, \subseteq)$ are dually isomorphic via the map $H \colon \wp(U)_R \to \wp(U)^R$, $H \colon X \to X^c$, because we have

$H(X_R) = (X_R)^c = (X^c)^R$. If R is an equivalence, then $\wp(U)_R = \wp(U)^R$ and these approximations form the same Boolean lattice (see e.g. [12]). A detailed analysis of rough approximation operators can be found in [2].

A *formal context* is a triple $\mathcal{K} = (G, M, I)$, where G is a set of *objects, M* is a set of *attributes* and $I \subseteq G \times M$ is a relation, called *incidence relation.* The notations $(g, m) \in I$ and g I m both express that an object g is in relation I with an attribute m, and we read it as "*the object g has the attribute m*". The basics of Formal Concept Analysis (FCA) can be found e.g. in [1]. By defining for all subsets $A \subseteq G$ and $B \subseteq M$

$$A^I = \{m \in M \mid (g, m) \in I, \text{ for all } g \in A\},$$
$$B^I = \{g \in G \mid (g, m) \in I, \text{ for all } m \in B\}$$

we establish a Galois connection between the power-set lattices $(\wp(G), \subseteq)$ and $(\wp(M), \subseteq)$ and the maps $A \to A^{II}, A \subseteq G$ and $B \to B^{II}$, $B \subseteq M$ are closure operators on $\wp(G)$ and $\wp(M)$, respectively.

A small context usually is represented by a cross table (i.e. by a rectangle table) the rows of which are headed by objects and the columns are headed by attributes. A cross in row g and column m means that the object g has the attribute m. The following example is from [5]

Example 1.

Table 1. An example of a formal concept. The objects are geometric forms: 1 = general triangle, 2 = square, 3 = circle, 4 = rectangle, 5 = rhomb.

	Angular (a)	Right angles (r)	Equilateral (e)	Central symmetry (cs)
1	×			
2	×	×	×	×
3				×
4	×	×		×
5	×		×	×

A *formal concept* of the context $\mathcal{K}$ is a pair $(A, B) \in \wp(G) \times \wp(M)$ with $A^I = B$ and $B^I = A$, where the set A is called the *extent* and B is called the *intent* of the concept (A, B). It is easy to check that $(A, B) \in \wp(G) \times \wp(M)$ is a concept if and only if $(A, B) = (A^{II}, A^I) = (B^I, B^{II})$. For instance, $(\{2\}, \{a, r, e, cs\})$, $(\{2, 4\}, \{a, r, cs\})$ or $(\{1, 2, 3, 4, 5\}, \emptyset)$ are some concepts of the formal context from Table 1. The set of all concepts of the context $\mathcal{K}$ is denoted by $\mathcal{L}(\mathcal{K})$. This set is ordered by

$$(A_1, B_1) \leq (A_2, B_2) \Leftrightarrow A_1 \subseteq A_2 \Leftrightarrow B_1 \supseteq B_2,$$

resulting in a complete lattice, called the *concept lattice of the context* $\mathcal{K} = (G, M, I)$, which is denoted by $\mathcal{L}(G, M, I)$. Let $\text{Int}(G, M, I) = \{A^I \mid A \subseteq G\}$

stand for the intents of the context $\mathcal{K}$. Then $\mathcal{L}(G,M,I)$ is isomorphic to $(\mathrm{Int}(G,M,I),\supseteq)$, via the mapping $\varphi(A,B)=B$, $(A,B)\in\mathcal{L}(G,M,I)$.

A relation $J\subseteq I$ is called a *closed subrelation* of the context (G,M,I) if every concept of the context (G,M,J) is also a concept of (G,M,I). In [1] it is proved that this definition is equivalent to the condition that the concept lattice $\mathcal{L}(G,M,J)$ is a complete sublattice of $\mathcal{L}(G,M,I)$.

For a tolerance relation $R\subseteq U\times U$, the relationship between the lattices of approximations and the concept lattice $\mathcal{L}(U,U,R^c)$ was described in [4]. Indeed, let $I=R^c$. Then for any $X\subseteq U$ we have $X^I=\{u\in U\mid xR^cu$, for all $x\in X\}=\{u\in U\mid (x,u)\notin R$, for all $x\in X\}=U\setminus X^R=\left(X^R\right)^c$. Thus

$$X^R=\left(X^I\right)^c \text{ and } X_R=\left(\left(X^c\right)^R\right)^c=\left(X^c\right)^I,$$

according to (b). In [4] it is also proved that $\wp(U)_R$ and $\wp(U)^R$ are complemented lattices such that $(\wp(U)^R,\subseteq)\cong(\wp(U)_R,\supseteq)\cong\mathcal{L}(U,U,R^c)$. In this case, the complement of a concept (A^{II},A^I) is just the concept $\left(A^I,A^{II}\right)$.

3 Complete Sublattices of Approximation Lattices

Now let ρ,R be two tolerance relations such that $R\subseteq\rho\subseteq U\times U$. Consider the formal contexts $\mathcal{K}_R=(U,U,R^c)$ and $\mathcal{K}_\rho=(U,U,\rho^c)$. Since $J:=\rho^c\subseteq R^c:=I$, $\mathcal{K}_\rho$ is a subcontext of $\mathcal{K}_R$. We intend to characterize the case when the lattice $\wp(U)^\rho$ ($\wp(U)_\rho$) is isomorphic (dually isomorphic) to a complete sublattice of $\wp(U)^R$ ($\wp(U)_R$, respectively). The general situation can be formulated as follows:

Lemma 1. *The following conditions are equivalent:*

(1) $(\wp(U)^R,\subseteq)$ *is isomorphic to a complete sublattice of* $(\wp(U)^\rho,\subseteq)$*;*
(2) $(\wp(U)_R,\subseteq)$ *is isomorphic to a complete sublattice of* $(\wp(U)_\rho,\subseteq)$*;*
(3) $\mathcal{L}(U,U,\rho^c)$ *is isomorphic to a complete sublattice of* $\mathcal{L}(U,U,R^c)$*.*

Proof. Since $(\wp(U)^R,\subseteq)\cong(\wp(U)_R,\supseteq)$ and $(\wp(U)^\rho,\subseteq)\cong(\wp(U)_\rho,\supseteq)$ the equivalence of (1) and (2) is obvious. The equivalence of (1) and (3) follows from the facts $(\wp(U)^R,\subseteq)\cong\mathcal{L}(U,U,R^c)$ and $(\wp(U)^\rho,\subseteq)\cong\mathcal{L}(U,U,\rho^c)$. □

Lemma 2. *If* $\mathcal{L}(U,U,\rho^c)$ *is a complete sublattice of* $\mathcal{L}(U,U,R^c)$*, then the following hold:*

(1') $(\wp(U)^\rho,\subseteq)$ *is a complete sublattice of* $(\wp(U)^R,\subseteq)$*,*
(2') $(\wp(U)_\rho,\subseteq)$ *is a complete sublattice of* $(\wp(U)_R,\subseteq)$*.*

Proof. (1'): If our condition holds, then $\mathrm{Int}(U,U,\rho^c)$ is also a complete sublattice of $\mathrm{Int}(U,U,R^c)$. Let $X^\rho\in\wp(U)^\rho$. Then $(X^\rho)^c=X^J\in\mathrm{Int}(U,U,\rho^c)\subseteq\mathrm{Int}(U,U,R^c)$, and hence $(X^\rho)^c=Y^I=\left(Y^R\right)^c$, for some $Y\subseteq U$. Hence $X^\rho=Y^R\in\wp(U)^R$, therefore we get $\wp(U)^\rho\subseteq\wp(U)^R$. Observe that the mapping $X\to X^c$, $X\subseteq\wp(U)^\rho$ is an order-isomorphism between $(\wp(U)^\rho,\subseteq)$ and $(\mathrm{Int}(U,U,\rho^c),\supseteq)$, and also $X\to X^c$, $X\subseteq\wp(U)^R$ is an order-isomorphism between $(\wp(U)^R,\subseteq)$ and $(\mathrm{Int}(U,U,\rho^c),\supseteq)$. Now, from the condition that $\mathrm{Int}(U,U,\rho^c)$ is a complete sublattice of $\mathrm{Int}(U,U,R^c)$, it follows (1'). Assertion (2') is proved similarly. □

Example 2. This example contains two tolerance relations ρ and R, their lower and upper approximations, the contexts and the concept lattices corresponding to R^c and ρ^c on the same universe $U = \{a, b, c, d\}$ (Table 2). Because all the discussed relations are reflexive, loops are not noted in what follows (see Table 3 and Fig. 3).

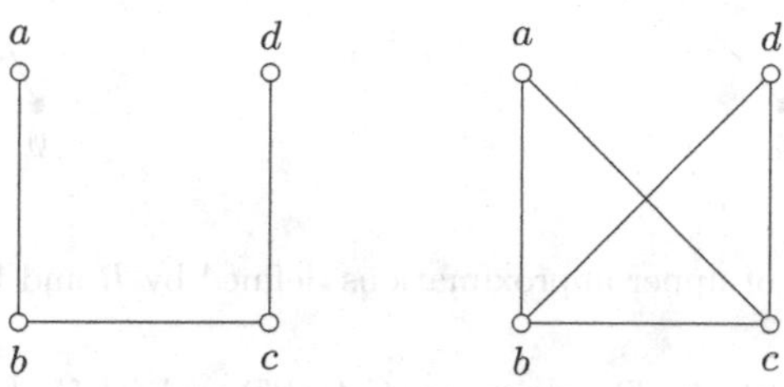

Fig. 1. Relations R and ρ

Table 2. Lower and upper approximations defined by relations R and ρ from Fig. 1.

X	X_R	X^R	X_ρ	X^ρ
$\emptyset$	$\emptyset$	$\emptyset$	$\emptyset$	$\emptyset$
$\{a\}$	$\emptyset$	$\{a, b\}$	$\emptyset$	$\{a, b, c\}$
$\{b\}$	$\emptyset$	$\{a, b, c\}$	$\emptyset$	$\{a, b, c, d\}$
$\{c\}$	$\emptyset$	$\{b, c, d\}$	$\emptyset$	$\{a, b, c, d\}$
$\{d\}$	$\emptyset$	$\{c, d\}$	$\emptyset$	$\{b, c, d\}$
$\{a, b\}$	$\{a\}$	$\{a, b, c\}$	$\emptyset$	$\{a, b, c, d\}$
$\{a, c\}$	$\emptyset$	$\{a, b, c, d\}$	$\emptyset$	$\{a, b, c, d\}$
$\{a, d\}$	$\emptyset$	$\{a, b, c, d\}$	$\emptyset$	$\{a, b, c, d\}$
$\{b, c\}$	$\emptyset$	$\{a, b, c, d\}$	$\emptyset$	$\{a, b, c, d\}$
$\{b, d\}$	$\emptyset$	$\{a, b, c, d\}$	$\emptyset$	$\{a, b, c, d\}$
$\{c, d\}$	$\{d\}$	$\{b, c, d\}$	$\emptyset$	$\{a, b, c, d\}$
$\{a, b, c\}$	$\{a, b\}$	$\{a, b, c, d\}$	$\{a\}$	$\{a, b, c, d\}$
$\{a, b, d\}$	$\{a\}$	$\{a, b, c, d\}$	$\emptyset$	$\{a, b, c, d\}$
$\{a, c, d\}$	$\{d\}$	$\{a, b, c, d\}$	$\emptyset$	$\{a, b, c, d\}$
$\{b, c, d\}$	$\{c, d\}$	$\{a, b, c, d\}$	$\{d\}$	$\{a, b, c, d\}$
$\{a, b, c, d\}$	$\{a, b, c, d\}$	$\{a, b, c, d\}$	$\{a, b, c, d\}$	$\{a, b, c, d\}$

It is visible on (the next) Fig. 2 that the upper approximations defined by ρ form a sublattice of the lattice of upper approximations defined by R. However, the concept lattice of the context (U, U, ρ^c) is not even a subset of the concept lattice of the context (U, U, R^c).

Observe that the conditions (1') and (2') do not imply that $\mathcal{L}(U, U, \rho^c)$ is a complete sublattice of $\mathcal{L}(U, U, R^c)$, they imply only that $\mathcal{L}(U, U, \rho^c)$ is isomorphic to a complete sublattice of $\mathcal{L}(U, U, R^c)$.

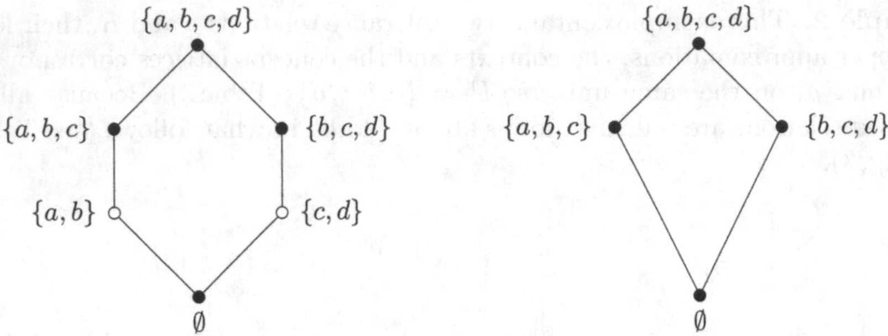

Fig. 2. The lattices of upper approximations defined by R and by ρ, respectively

Table 3. The contexts (U, U, R^c) and (U, U, ρ^c)

R^c	a	b	c	d
a			×	×
b				×
c	×			
d	×	×		

ρ^c	a	b	c	d
a				×
b				
c				
d	×			

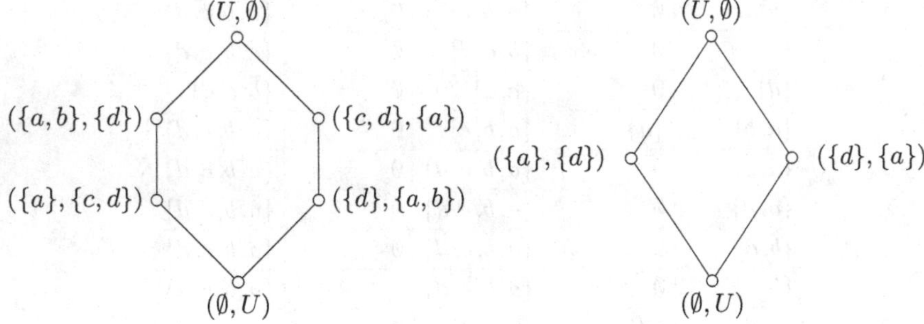

Fig. 3. The Hasse-diagram of the concept lattices $\mathcal{L}(U, U, R^c)$ and $\mathcal{L}(U, U, \rho^c)$

Let (G, M, I) be a formal context and $g \in G$, $m \in M$. The sets $\{g\}^I$ and $\{m\}^I$ will be denoted simply by g^I and m^I. In [1] was proved that a relation $J \subseteq I$ is a closed subrelation of the context (G, M, I), or equivalently, $\mathcal{L}(G, M, J)$ is a complete sublattice of $\mathcal{L}(G, M, I)$ if and only if the following condition holds:

(+) $(g, m) \in I \setminus J$ implies $(h, m) \notin I$ for some $h \in G$ with $g^J \subseteq h^J$ as well as $(g, n) \notin I$ for some $n \in M$ with $m^J \subseteq n^J$.

Let ρ be a tolerance on U. Let us define
$\trianglelefteq (\rho) := \{(x, y) \in U \times U \mid \rho(x) \subseteq \rho(y)\}$ and
$\trianglerighteq (\rho) := \{(x, y) \in U \times U \mid \rho(x) \supseteq \rho(y)\}$.

Clearly, $\trianglelefteq(\rho)$ and $\trianglerighteq(\rho)$ are reflexive and transitive relations, i.e. they are *quasiorders* and $\trianglerighteq(\rho)$ is the inverse relation of $\trianglelefteq(\rho)$. Let the symbol $\circ$ stand for the relational product, in what follows.

Remark 1. Observe that in the case $R \subseteq \rho$ the relations $R\circ \trianglelefteq(\rho) \subseteq \rho$ and $\trianglerighteq(\rho) \circ R \subseteq \rho$ always hold. Indeed, for any $x, y \in U$, $(x,y) \in R\circ \trianglelefteq(\rho)$ means that there exists a $z \in U$ with $(x,z) \in R \subseteq \rho$ and $(z,y) \in \trianglelefteq(\rho)$. Then $x \in \rho(z) \subseteq \rho(y)$ implies $(x,y) \in \rho$, proving $R\circ \trianglelefteq(\rho) \subseteq \rho$. The second inclusion is proved dually.

By using these notions and condition (c) we can formulate:

Theorem 1. *Let ρ, R be tolerance relations satisfying $R \subseteq \rho \subseteq U \times U$. Then the following conditions are equivalent:*

(C) $\mathcal{L}(U,U,\rho^c)$ is a complete sublattice of $\mathcal{L}(U,U,R^c)$;
(D) For any $(a,b) \in \rho \setminus R$ there exist some elements $c,d \in U$ such that $(c,b),(a,d) \in R$ and $\rho(c) \subseteq \rho(a)$, $\rho(d) \subseteq \rho(b)$;
(E) $R\circ \trianglelefteq(\rho) = \rho$;
(E') $\trianglerighteq(\rho) \circ R = \rho$.

Proof. (C) $\Leftrightarrow$ (D). Set $I := R^c$ and $J := \rho^c$. Then $J \subseteq I$ and in view of [1] $\mathcal{L}(U,U,J)$ is a complete sublattice of $\mathcal{L}(U,U,I)$, if and only if condition (+) is satisfied by the incidence relations I and J. We prove that (+) is equivalent to condition (D). Since $G = M = U$ and $R^c \setminus \rho^c = R^c \cap \rho^{cc} = \rho \setminus R$, (by setting $a := g$, $b := m$, $c := h$ and $d := n$) we obtain that for any elements $(a,b) \in \rho \setminus R$ there exist some elements $c,d \in U$ such that $(c,b),(a,d) \in I^c = R$ and $a^J \subseteq c^J$, $b^J \subseteq d^J$. These relations imply $\rho(c) = \{c\}^\rho = \left(c^J\right)^c \subseteq (a^J)^c = \{a\}^\rho = \rho(a)$ and similarly, $\rho(d) \subseteq \rho(b)$ (see Fig. 4).

(D) $\Rightarrow$ (E). Assume that (D) holds. In order to prove (E) it is enough to show $\rho \subseteq R\circ \trianglelefteq(\rho)$. Since $R \subseteq R\circ \trianglelefteq(\rho)$, it is enough to check that $\rho \setminus R \subseteq R\circ \trianglelefteq(\rho)$. Take any $(a,b) \in \rho \setminus R$. Then, in view of (D) there exists an element $d \in U$ with $(a,d) \in R$ and $\rho(d) \subseteq \rho(b)$. Thus $(d,b) \in \trianglelefteq(\rho)$, whence we get $(a,b) \in R\circ \trianglelefteq(\rho)$, proving $\rho \setminus R \subseteq R\circ \trianglelefteq(\rho)$.

(E) $\Leftrightarrow$ (E'). Because we have

$$\rho = R\circ \trianglelefteq(\rho) \Leftrightarrow \rho = \rho^{-1} = \trianglelefteq(\rho)^{-1} \circ R^{-1} = \trianglerighteq(\rho) \circ R.$$

(E') $\Rightarrow$ (D). Suppose that (E') holds. Then (E) is also satisfied, i.e. $R\circ \trianglelefteq(\rho) = \rho = \trianglerighteq(\rho) \circ R$. Now take any $(a,b) \in \rho \setminus R$. Since $(a,b) \in R\circ \trianglelefteq(\rho)$ and $(a,b) \in \trianglerighteq(\rho) \circ R$, there exist some elements $c,d \in U$ with $(a,d),(c,b) \in R$ and $(d,b) \in \trianglelefteq(\rho)$, $(a,c) \in \trianglerighteq(\rho)$. The latter relations imply $\rho(d) \subseteq \rho(b)$ and $\rho(c) \subseteq \rho(a)$. This means that condition (D) is satisfied. $\square$

The next corollary is an immediate consequence of Lemma 2:

Corollary 1. *Let R, ρ be two tolerance relations on U such that $R \subseteq \rho$. If R and ρ satisfy one of the equivalent conditions of Theorem 1, then $(\wp(U)^\rho, \subseteq)$ is a complete sublattice of $\left(\wp(U)^R, \subseteq\right)$ and $(\wp(U)_\rho, \subseteq)$ is a complete sublattice of $(\wp(U)_R, \subseteq)$.*

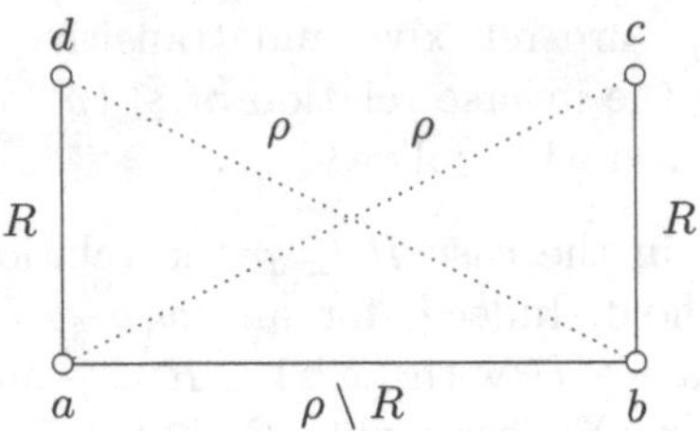

Fig. 4. Condition (D)

Corollary 2. *Let R be a tolerance relation and ρ an equivalence relation on U such that $R \subseteq \rho$. Then the condition (D) is satisfied and $(\wp(U)^{\rho}, \subseteq)$ is a complete sublattice of $(\wp(U)^{R}, \subseteq)$ and of $(\wp(U)_{R}, \subseteq)$.*

Proof. Clearly, we may suppose that $R \neq \rho$. Let $(a, b) \in \rho \setminus R$. Since R is reflexive, we can choose $c := b$ and $d := a$ such that $(c, b), (a, d) \in R$. In this case, since $(c, d) = (b, a) \in \rho$ and ρ is an equivalence relation, we obtain $\rho(c) = \rho(b) = \rho(a) = \rho(d)$. This means that condition (D) is satisfied. Since ρ is an equivalence, we have $\wp(U)^{\rho} = \wp(U)_{\rho}$, and now by using Corollary 1, we obtain that $(\wp(U)^{\rho}, \subseteq)$ is a complete sublattice of $(\wp(U)^{R}, \subseteq)$ and of $(\wp(U)_{R}, \subseteq)$. □

Corollary 3. *Let ρ, T, R be tolerance relations satisfying $R \subseteq T \subseteq \rho \subseteq U \times U$. If $\mathcal{L}(U, U, \rho^c)$ is a complete sublattice of $\mathcal{L}(U, U, R^c)$, then it is also a complete sublattice of $\mathcal{L}(U, U, T^c)$. Accordingly, $(\wp(U)^{\rho}, \subseteq)$ and $(\wp(U)_{\rho}, \subseteq)$ are complete sublattices of $(\wp(U)^{T}, \subseteq)$ and $(\wp(U)_{T}, \subseteq)$, respectively.*

Proof. If the assumptions of this corollary hold, then condition $R\circ \trianglelefteq (\rho) = \rho$ is also satisfied, according to Theorem 1. Now, let $T \subseteq U \times U$ be an arbitrary tolerance relation with $R \subseteq T \subseteq \rho$. We already noted (see Remark 1), that the inclusion $T \subseteq \rho$ implies $T\circ \trianglelefteq (\rho) \subseteq \rho$. On the other hand, we obtain $\rho = R\circ \trianglelefteq (\rho) \subseteq T\circ \trianglelefteq (\rho)$. Thus we get $T\circ \trianglelefteq (\rho) = \rho$. In view of Theorem 1 and Corollary 1, this implies our assertion. □

Example 3. Let R be a tolerance relation and ρ an equivalence relation on the set $U = \{a, b, c, d\}$ in Fig. 5 where $R \subseteq \rho$. Note that in view of Corollary 2, the lattice $\mathcal{L}(U, U, \rho^c)$ is a sublattice of $\mathcal{L}(U, U, R^c)$, and $(\wp(U)^{\rho}, \subseteq)$ is a sublattice of $(\wp(U)^{R}, \subseteq)$ (Tables 4 and 5) (see Figs. 6 and 7).

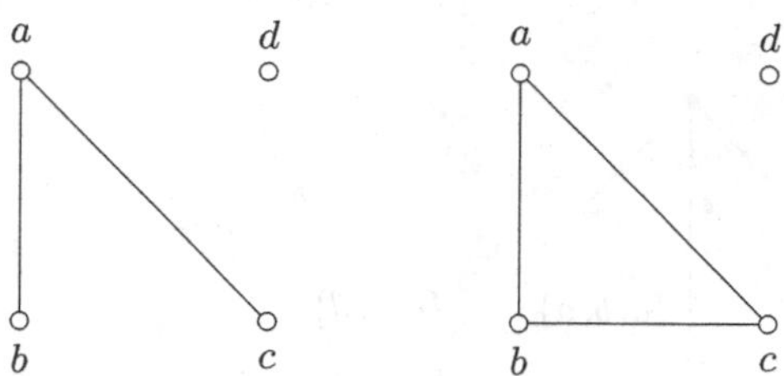

Fig. 5. Relations R and ρ

Table 4. Lower and upper approximations defined by relations R and ρ from Fig. 5.

X	X_R	X^R	X_ρ	X^ρ
$\emptyset$	$\emptyset$	$\emptyset$	$\emptyset$	$\emptyset$
$\{a\}$	$\emptyset$	$\{a,b,c\}$	$\emptyset$	$\{a,b,c\}$
$\{b\}$	$\emptyset$	$\{a,b\}$	$\emptyset$	$\{a,b,c\}$
$\{c\}$	$\emptyset$	$\{a,c\}$	$\emptyset$	$\{a,b,c\}$
$\{d\}$	$\{d\}$	$\{d\}$	$\{d\}$	$\{d\}$
$\{a,b\}$	$\{b\}$	$\{a,b,c\}$	$\emptyset$	$\{a,b,c\}$
$\{a,c\}$	$\{c\}$	$\{a,b,c\}$	$\emptyset$	$\{a,b,c\}$
$\{a,d\}$	$\{d\}$	$\{a,b,c,d\}$	$\{d\}$	$\{a,b,c,d\}$
$\{b,c\}$	$\emptyset$	$\{a,b,c\}$	$\emptyset$	$\{a,b,c\}$
$\{b,d\}$	$\{d\}$	$\{a,b,d\}$	$\{d\}$	$\{a,b,c,d\}$
$\{c,d\}$	$\{d\}$	$\{a,c,d\}$	$\{d\}$	$\{a,b,c,d\}$
$\{a,b,c\}$	$\{a,b,c\}$	$\{a,b,c\}$	$\{a,b,c\}$	$\{a,b,c\}$
$\{a,b,d\}$	$\{b,d\}$	$\{a,b,c,d\}$	$\{d\}$	$\{a,b,c,d\}$
$\{a,c,d\}$	$\{c,d\}$	$\{a,b,c,d\}$	$\{d\}$	$\{a,b,c,d\}$
$\{b,c,d\}$	$\{d\}$	$\{a,b,c,d\}$	$\{d\}$	$\{a,b,c,d\}$
$\{a,b,c,d\}$	$\{a,b,c,d\}$	$\{a,b,c,d\}$	$\{a,b,c,d\}$	$\{a,b,c,d\}$

Table 5. The contexts (U, U, R^c) and (U, U, ρ^c)

R^c	a	b	c	d
a				×
b			×	×
c		×		×
d	×	×	×	

ρ^c	a	b	c	d
a				×
b				×
c				×
d	×	×	×	

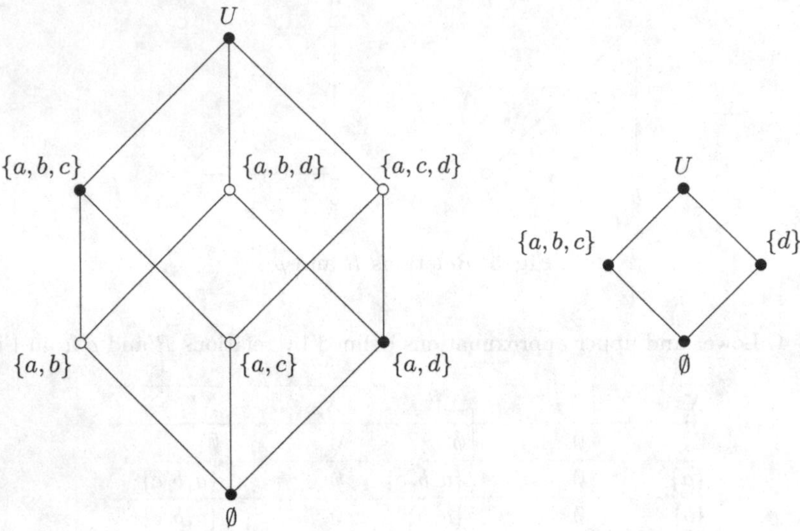

Fig. 6. The lattices of upper approximations defined by R and by ρ respectively

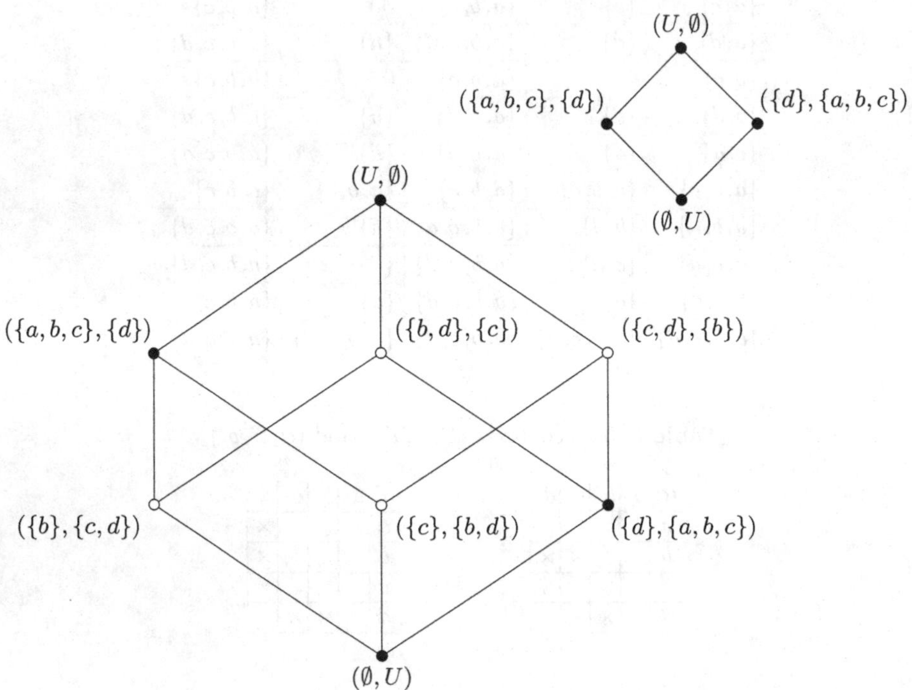

Fig. 7. The concept lattices $\mathcal{L}(U, U, R^c)$ and $\mathcal{L}(U, U, \rho^c)$

4 Compatibility Condition

Let T be a tolerance and E an equivalence relation on a set U. The notion of an E-*compatible tolerance* T was introduced in [6], and means that the condition $E \circ T = T$ holds. The definition was motivated by the fact that in such a case any neighbourhood $T(x)$ is an E-*definable set* (i.e. $T(x)^E = T(x)_E = T(x)$). Then this fact was used to show that the approximation pairs $\{(X_E, X^T) \mid X \subseteq U\}$ form a complete lattice with respect to the componentwise inclusion, generalizing in this way the traditional notion of rough sets.

Let $\rho, R \subseteq U \times U$ be two tolerance relations. Generalizing the above notion we will say that the tolerance ρ is R-*compatible* if $R \circ \rho = \rho$. Clearly, this yields $R \subseteq \rho$ and $R^2 = R \circ R \subseteq \rho$. Since $\rho^{-1} = \rho$ and $(R \circ \rho)^{-1} = \rho^{-1} \circ R^{-1} = \rho \circ R$ we get $R \circ \rho = \rho \Leftrightarrow (R \circ \rho)^{-1} = \rho^{-1} \Leftrightarrow \rho \circ R = \rho$. Hence $R \circ \rho = \rho$ and $\rho \circ R = \rho$ are equivalent conditions.

In [6] we can find several examples of compatible relations. For instance, defining the *kernel* of a tolerance relation ρ as the equivalence $\ker\rho := \{(x, y) \in U \times U \mid \rho(x) = \rho(y)\}$, it is proved that ρ is compatible with an equivalence $R \subseteq U \times U$ iff $R \subseteq \ker\rho$. Obviously, if E and F are equivalence relations on U, then F is E-compatible iff $E \subseteq F$. Another interesting example was given by using a so-called information system.

An *information system* in the sense of Pawlak is a triple $(U, A, \{V\}_{a \in A})$, where U is a set of objects, A is a set of attributes, and V_a is the value set of an $a \in A$. Each attribute is a mapping $a\colon U \to V_a$, and $a(x) \in V_a$ is the value of the attribute a for the object x. For any subset $B \subseteq A$, *the strong indiscernibility relation of* B is defined by

$$\mathrm{Ind}(B) = \{(x, y) \in U \times U \mid a(x) = a(y), \text{ for all } a \in B\}.$$

The *weak indiscernibility relation of* B is given by

$$\mathrm{Wind}(B) = \{(x, y) \in U \times U \mid a(x) = a(y), \text{ for some } a \in B\}.$$

Clearly, $\mathrm{Ind}(B)$ is an equivalence relation and $\mathrm{Wind}(B)$ is a tolerance relation on U. Indeed, $\mathrm{Wind}(B)$ is reflexive by its definition and $(x, y) \in \mathrm{Wind}(B) \Leftrightarrow a(x) = a(y)$ for some $a \in B \Leftrightarrow a(y) = a(x)$ for some $a \in B \Leftrightarrow (y, x) \in \mathrm{Wind}(B)$. Hence the relation $\mathrm{Wind}(B)$ is also symmetric. In [6] is proved that $\mathrm{Ind}(B) \circ \mathrm{Wind}(B) = \mathrm{Wind}(B)$, that is, $\mathrm{Wind}(B)$ is compatible with $\mathrm{Ind}(B)$.

Lemma 3. *Let $C \subseteq B \subseteq A$. Then Ind($B$)$\circ$Wind($C$)= Wind($C$), that is, Wind($C$) is Ind($B$)-compatible.*

Proof. Indeed, the inclusion $\mathrm{Wind}(C) \subseteq \mathrm{Ind}(B) \circ \mathrm{Wind}(C)$, is clear. In order to prove the converse inclusion, take any $(x, y) \in \mathrm{Ind}(B) \circ \mathrm{Wind}(C)$. Then $(x, z) \in \mathrm{Ind}(B)$ and $(z, y) \in \mathrm{Wind}(C)$, for some $z \in U$. As $C \subseteq B$, $(x, z) \in \mathrm{Ind}(B)$ yields $a(x) = a(z)$, for all $a \in C$, and we have also $a'(z) = a'(y)$, for some $a' \in C$, because $(z, y) \in \mathrm{Wind}(C)$. Thus we get $a'(x) = a'(y)$, for some $a' \in C$, i.e. $(x, y) \in \mathrm{Wind}(C)$. This yields also $\mathrm{Ind}(B) \circ \mathrm{Wind}(C) \subseteq \mathrm{Wind}(C)$, completing the proof. □

Let $X \subseteq U$ be arbitrary and let ρ be an R-compatible tolerance. The following relations can be easily proved:

(f) $(X^{\rho})^{R} = X^{\rho \circ R} = X^{\rho} = X^{R \circ \rho} = (X^{R})^{\rho}$;
(g) $(X_{\rho})_{R} = X_{\rho \circ R} = X_{\rho} = X_{R \circ \rho} = (X_{R})_{\rho}$.

Indeed, $X^{\rho \circ R} = X^{\rho} = X^{R \circ \rho}$ is clear. Let us check for instance the equality $(X^{\rho})^{R} = X^{R \circ \rho}$. Now, by definition we have $x \in (X^{\rho})^{R} \Leftrightarrow R(x) \cap X^{\rho} \neq \emptyset \Leftrightarrow$ ($\exists y \in U$ and $\exists z \in X$ with xRy and $y\rho z$) $\Leftrightarrow$ ($(x, z) \in R \circ \rho$, for some $z \in X$) $\Leftrightarrow$ $x \in X^{R \circ \rho}$.

In what follows, we will show that the conditions from Theorem 1 and the compatibility condition are interrelated.

Proposition 1. *Let $\rho, R \subseteq U \times U$ be two tolerance relations satisfying condition (E) and $R \subseteq \rho$. Then $R^2 \subseteq \rho$ if and only if ρ is R-compatible.*

Proof. Let $R \subseteq \rho$. We noted that $R^2 \subseteq \rho$ holds whenever ρ is R-compatible. Suppose that condition (E) also holds. Since $\rho \subseteq R \circ \rho$, to prove the converse inclusion it suffices to show that $R^2 \subseteq \rho$ implies $R \circ (\rho \setminus R) \subseteq \rho$. Take any $(x, b) \in R \circ (\rho \setminus R)$. This means that we may consider any $(a, b) \in \rho \setminus R$ and $x \in U$ with $(x, a) \in R$. Condition (E) and $(a, b) \in \rho$ imply that there exists an element $d \in U$ with $(a, d) \in R$ and $\rho(d) \subseteq \rho(b)$. Then $(x, d) \in R \circ R = R^2 \subseteq \rho$. ρ being symmetric, we get $(d, x) \in \rho$, i.e. $x \in \rho(d) \subseteq \rho(b)$. This implies $(x, b) \in \rho$, proving $R \circ (\rho \setminus R) \subseteq \rho$. □

Corollary 4. *Let R be an equivalence and ρ a tolerance relation satisfying one of the equivalent conditions from Theorem 1, and $R \subseteq \rho$. Then ρ is R-compatible.*

Proof. Since $R^2 = R \subseteq \rho$, by using Proposition 1, we obtain the required assertion. □

Theorem 2. *Let $\rho, R \subseteq U \times U$ be two tolerance relations with $R \subseteq \rho$. Then $\mathcal{L}(U, U, \rho^c)$ is a complete sublattice of $\mathcal{L}(U, U, R^c)$ and ρ is R-compatible if and only if ρ is an equivalence.*

Proof. If ρ is an equivalence, then in view of Corollary 2 condition (D) is satisfied and $R^2 \subseteq \rho^2 = \rho$ implies that ρ is R-compatible, according to Proposition 1. Now, Theorem 1 implies that $\mathcal{L}(U, U, \rho^c)$ is a complete sublattice of $\mathcal{L}(U, U, R^c)$.

Conversely, assume that ρ is R-compatible and that $\mathcal{L}(U, U, \rho^c)$ is a complete sublattice of $\mathcal{L}(U, U, R^c)$. Then, in view of Theorem 1, condition (E') also holds. In order to prove that ρ is an equivalence, let us consider some elements $a, b, e \in U$ such that $(a, b), (b, e) \in \rho$. We will show that $(a, e) \in \rho$, proving in this way the transitivity of ρ, and this means that ρ is an equivalence.

Indeed, in view of condition (E') there exists an element $c \in U$ with $(c, b) \in R \subseteq \rho$ and $(a, c) \in \trianglerighteq(\rho)$, i.e. $\rho(a) \supseteq \rho(c)$. Then $(b, e) \in \rho$ implies $(c, e) \in R \circ \rho = \rho$, i.e. $e \in \rho(c) \subseteq \rho(a)$. This yields $(a, e) \in \rho$, completing our proof. □

Corollary 5. *Let $(U, A, \{V_a\}_{a \in A})$ be an information system, $B \subseteq A$ and $R \subseteq$ Ind(B) a tolerance relation on the set U. Then the equivalence Ind(B) is R-compatible, $(\wp(U)^{Ind(B)}, \subseteq)$ is a complete sublattice of the lattice $(\wp(U)^R, \subseteq)$, and $(\wp(U)_{Ind(B)}, \subseteq)$ is a complete sublattice of $(\wp(U)_R, \subseteq)$. In particular, this holds for any relation $R =$ Ind$(B) \cap$ Wind(C), where $C \subseteq A, C \neq \emptyset$.*

Proof. Since Ind(B) is an equivalence relation, in view of Theorem 2, Ind(B) is R-compatible and the concept lattice $\mathcal{L}(U, U, \text{Ind}(B)^c)$ is a complete sublattice of the concept lattice $\mathcal{L}(U, U, R^c)$. Now by applying Corollary 1, we get our first assertion. Since the meet of two tolerance relations is also a tolerance relation, our second assertion is a simple consequence of this. □

Corollary 6. *Let R be an equivalence and ρ a tolerance relation on a set U such that $R \subseteq \rho$. Then $\mathcal{L}(U, U, \rho^c)$ is a complete sublattice of $\mathcal{L}(U, U, R^c)$ if and only if ρ is an equivalence.*

Proof. In view of Theorem 1, $\mathcal{L}(U, U, \rho^c)$ is a complete sublattice of $\mathcal{L}(U, U, R^c)$ if and only if condition (D) holds. If (D) holds then ρ is R-compatible, according to Corollary 4, and by applying Theorem 2, we obtain that ρ is an equivalence. If ρ is an equivalence, then condition (D) holds, according to Corollary 2. □

5 Conclusions

In this paper, we examined the relationship between the lattices of upper and lower approximations of two tolerance relations $R \subseteq \rho$. We have shown that the approximation lattices corresponding to ρ form complete sublattices of the approximation lattices defined by R, whenever the concept lattice $\mathcal{L}(U, U, \rho^c)$ is a complete sublattice of $\mathcal{L}(U, U, R^c)$. We deduced several conditions (see e.g. (D), (E), (E')) equivalent to this latter condition on concept lattices (i.e. to one being a complete sublattice of the other). We also showed that if the greater relation is an equivalence, then condition (D) automatically holds. However, the converse is not necessarily true. This is shown in Example 4, where condition (D) holds, but neither of the relations is an equivalence. However, if in addition, the tolerance relations satisfy the compatibility condition (i.e. ρ is R-compatible), then the greater relation has to be an equivalence. As a future work we propose to investigate particular types of tolerance relations satisfying the mentioned conditions. For instance, we plan to investigate tolerances induced by an irredundant covering of the universe, which can be considered as a natural generalization of equivalences (see e.g. [5]), or tolerances defined by some particular subsets of the attribute set in an information system.

Example 4. Let R, ρ be two tolerance relations with $R \subseteq \rho$ defined on the universe $U = \{a, b, c, d\}$. This example shows that the concept lattice $\mathcal{L}(U, U, \rho^c)$ is a (complete) sublattice of $\mathcal{L}(U, U, R^c)$, although ρ is not an equivalence relation (Figs. 8, 9 and Table 6).

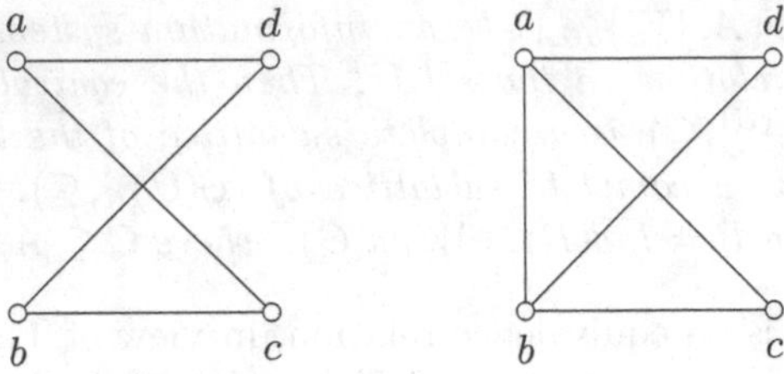

Fig. 8. Tolerances R and ρ satisfy condition (D), but none of them is an equivalence.

Table 6. The contexts (U, U, R^c) and (U, U, ρ^c)

R^c	a	b	c	d
a		×		
b	×			
c				×
d			×	

ρ^c	a	b	c	d
a				
b				
c				×
d			×	

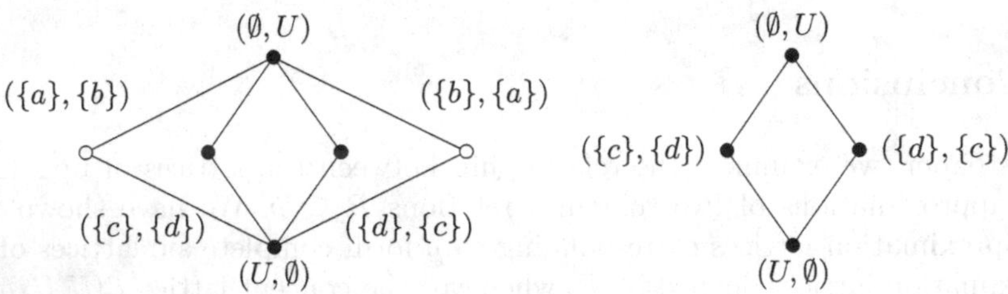

Fig. 9. The Hasse-diagrams of the concept lattices $\mathcal{L}(U, U, R^c)$ and $\mathcal{L}(U, U, \rho^c)$

Acknowledgement. The authors would like to thank the referees for their constructive comments.

References

1. Ganter, B., Wille, R.: Formal Concept Analysis, Mathematical Foundations. Springer, Berlin (1999)
2. Järvinen, J.: Properties of rough approximations. J. Adv. Comput. Intell. Intell. Inform. **9**(5), 502–505 (2005)
3. Järvinen, J., Pagliani, P., Radeleczki, S.: Information completeness in Nelson algebras of rough sets induced by quasiorders. Studia Logica **101**(5), 1073–1092 (2013)
4. Järvinen, J., Radeleczki, S.: Rough sets determined by tolerances. Int. J. Approximate Reasoning **55**(6), 1419–1438 (2014)
5. Järvinen, J., Radeleczki, S.: Irredundant coverings, tolerances, and related algebras. In: Mani, A., Cattaneo, G., Düntsch, I. (eds.) Algebraic Methods in General Rough Sets, pp. 417–458. Birkhäuser, Cham (2019)

6. Järvinen, J., Kovács, L., Radeleczki, S.: Defining rough sets using tolerances compatible with an equivalence (submitted, 2019)
7. Pawlak, Z.: Rough sets. Int J. Comput. Inf. **11**, 341–356 (1982)
8. Pawlak, Z.: Rough set theory and its applications to data analysis. Cybern. Syst. **29**(7), 661–688 (1998)
9. Rissino, S., Lambert-Torres, G.: Rough set theory - fundamental concepts, principals, data extraction, and applications. In: Ponce, J., Karahoca, A. (eds.) Data Mining and Knowledge Discovery in Real Life Applications, pp. 35–58. InTech, Vienna (2009)
10. Pomykala, J.A.: On similarity based approximation of information. Demonstratio Mathematica **XXVII**, 663–671 (1994)
11. Slimani, Th.: Application of rough set theory in data mining, arXiv preprint (2013) arXiv:1311.4121
12. Yang, L., Xu, L.: Algebraic aspects of generalized approximation spaces. Int. J. Approximate Reasoning **51**(1), 151–161 (2009)
13. Yao, Y.Y.: Generalized rough set models. In: Polkowski, L., Skowron, A. (eds.) Rough Sets in Knowledge Discovery, pp. 286–318. Physica-Verlag, Heidelberg (1998)
14. Zakowski, W.: Approximations in the space $(U; \Pi)$. Demonstratio Mathematica **16**, 761–769 (1983)

Three–Way Classification: Ambiguity and Abstention in Machine Learning

Andrea Campagner[1,3], Federico Cabitza[1,2], and Davide Ciucci[1(✉)]

[1] Dipartimento di Informatica, Sistemistica e Comunicazione, University of Milano–Bicocca, Viale Sarca 336, 20126 Milan, Italy
davide.ciucci@unimib.it

[2] IRCCS Istituto Ortopedico Galeazzi, Via Galeazzi 4, 20161 Milan, Italy

[3] Deloitte Italia, Via Tortona 25, Milan, Italy

Abstract. Ambiguity, that is the lack of information to produce a specific classification, is an important issue in decision–making and supervised classification. In case of ambiguity, human–decision makers can resort to abstaining from making precise classifications (especially when error-related costs are high), but this behaviour has been scarcely addressed, and applied, in machine learning algorithms. This contribution grounds on previous works in the areas of three–way decisions, cautious classification and orthopairs, and proposes a set of techniques we developed to address this form of ambiguity, by providing both a general–purpose technique to create three–way algorithms from probabilistic ones, and also more specific techniques which could be applied to popular machine learning frameworks. We also evaluate the proposed idea, by performing a set of experiments where we compare classical classification algorithms with the corresponding three–way generalizations, in order to study the trade–off between classification accuracy and abstention: the results are promising.

Keywords: Machine Learning · Abstention · Three–way decision · Data Mining · Ambiguity · Orthopair · Orthopartition · Uncertainty

1 Introduction

Research in the *Machine Learning* and *Data Mining* fields has recently taken central stage in the Computer Science research community: this interest has been driven by *theoretical* advancements [3,11,17,23], *technological* advancements and, chiefly among all, the promising results in different application areas (driven by the availability of large amounts of data) [13,24,26].

Despite all the attention and recent achievements, a limitation of current Machine Learning methods is the inability to properly deal with uncertainty and biases affecting the training datasets which are fed to learning algorithms as input [15,18,30]. Indeed, as noted in [4] for the healthcare domain, various forms of uncertainties and biases can affect the training data (*missing data*,

© Springer Nature Switzerland AG 2019
T. Mihálydeák et al. (Eds.): IJCRS 2019, LNAI 11499, pp. 280–294, 2019.
https://doi.org/10.1007/978-3-030-22815-6_22

inter–rater disagreement, lack of information, ambiguity, ...) thus hampering the performance and, most relevantly, the *reliability* of the resulting models.

Several uncertainty theories (e.g. *rough-set theory* [27], *fuzzy–set theory* [32], *three–way decisions* [31],...) have been proposed in order to cope with these different forms of uncertainty, also with application to Machine Learning [2,21], but their adoption in mainstream Machine Learning has been lagging, due to different reasons (e.g. those evidenced in [22] for the case of fuzzy sets).

In this work, we will consider a particular type of uncertainty, *lack of information*, also called *ambiguity* in the terminology of [19]. In the decision–making/classification domain, this type of uncertainty occurs when a human (or computational) agent deems the available information insufficient to cast a univocal and reasonable decision.

Whenever possible, the usual strategy that human decision–makers adopt, in order to cope with either ambiguous input or uncertain output, is to reject any pretense of giving a *clear–cut* decision and, instead, *abstain* from expressing a judgment. This approach has the merit of highlighting, in a simple form, which instances are more *uncertain* and, consequently, pointing out which ones would require the acquisition of further information.

While this approach is still little adopted, different authors have tried to address the *abstention* behaviour under a computational perspective: here, we especially mention the work on *cautious classifiers* [16,20] and the work on three–way decisions [31]. In the same direction, in order to develop Machine Learning models with this *abstention* ability, the authors proposed in [5,6] an extended *decision tree–learning* model, based on *orthopairs* [9,10] and three–way decisions.

In this article, we extend this line of research:

- We introduce a general framework for classification with abstention (or three–way classification), based on three–way decisions and orthopartitions, which can be applied to any classification algorithm;
- We define a set of specific strategies which can be used to directly implement three–way classification in the context of popular learning algorithms (e.g. decision trees, random forests, logistic regression);
- We conduct an experimental study, in which we compare different classical learning algorithms with the corresponding three–way ones on various datasets.

More specifically, in Sect. 2, we give a basic introduction to orthopairs and orthopartitions. In Sect. 3 the basic methods are introduced, that is: in Sect. 3.1, we define our approach to convert any classifier into a three–way classification algorithm, both in the binary and multi–class settings, providing also a theoretical–algorithmic analysis of these frameworks; in Sects. 3.2 and 3.3, we describe the strategies to directly implement three–way classification for three popular learning models (i.e., Decision Trees, Random Forests and Convex Learning via Gradient Descent). In Sect. 4.1, we illustrate the setting of the empirical analysis we conducted in order to compare traditional learning algorithms with three–way ones. In Sect. 4.2, we present the results of the conducted experiments, considering the advantages offered by three–way classification algorithms and evaluating the effect of abstention with respect to their performance, supporting our analysis with standard statistical validation techniques. Finally, in Sect. 5 we present our conclusions and outline the set of open problems and issues that we plan to investigate in our future works.

2 Orthopairs and Orthopartitions

Let us recall some basic notions on orthopairs and orthopartitions [6,10].

Let U be a set of objects, an orthopair is a pair $O = \langle P, N \rangle$ of subsets of U such that $P \cap N = \emptyset$. From these two sets we can also define the *boundary* as $Bnd = (P \cup N)^c$. Note that we could take an orthopair as a partially specified set which expresses our (incomplete) knowledge about the assignment of objects in a universe to a certain concept class; in this case, set P represents the positive examples for the concept while N represents the negative ones. We say that a set S is consistent with an orthopair O if it holds that:

$$x \in P \rightarrow x \in S \wedge x \in N \rightarrow x \notin S$$

That is, if we interpret the orthopair O as a partially specified set expressing our degree of knowledge about the belonging (or not) of certain objects to a set, S is coherent with our partial knowledge.

We say that two orthopairs O_1, O_2 are *disjoint* if it holds that:

(Ax D1) $P_1 \cap P_2 = \emptyset$;
(Ax D2) $P_1 \cap Bnd_2 = \emptyset$ and $Bnd_1 \cap P_2 = \emptyset$.

Definition 1. *An* orthopartition *is a set* $\mathcal{O} = \{O_1, ..., O_n\}$ *of orthopairs such that the following axioms hold:*

(Ax O1) $\forall O_i, O_j \in \mathcal{O}$ O_i, O_j *are disjoint;*
(Ax O2) $\bigcup_i (P_i \cup Bnd_i) = U$;
(Ax O3) $\forall x \in U$ $(\exists O_i$ *s.t.* $x \in Bnd_i) \rightarrow (\exists O_j$ *with* $i \neq j$ *s.t.* $x \in Bnd_j)$;
(Ax O4) $|\mathcal{O}| \leq |U|$

It can be observed that an orthopartition represents a *partial classification*, or a *classification with abstentions* (in a multi–class setting): the objects in the boundaries represent those objects whose class assignment is not precisely known (given the available evidence and, hence, the presence of ambiguity).

Definition 2. *A partition* π *is* consistent with an orthopartition $\mathcal{O}$ *iff* $\forall O_i \in \mathcal{O}$, $\exists! S_i \in \pi$ *such that* S_i *is consistent with* O_i. *We denote with* $\Pi_{\mathcal{O}}$ *the set of all partitions consistent with* $\mathcal{O}$: $\Pi_{\mathcal{O}} = \{\pi | \pi$ *is consistent with* $\mathcal{O}\}$.

Viewing an orthopartition as a partial state of knowledge about a multi–class classification (associated with the set $\Pi_{\mathcal{O}}$ which represents all possible consistent complete states of knowledge), we can extend many measures defined on classical partitions to orthopartitions, in particular we will focus on the *entropy* and *accuracy* (the extension of other metrics based on the confusion matrix is analogous). The *logical entropy* [14] of a partition π is defined as:

$$h(\pi) = \frac{dit(\pi)}{|U|^2}$$

where $dit(\pi) = \{(u, u') \in U \times U|\, u, u' \text{ belong to two different blocks of } \pi\}$. We can define three different generalizations of this concept, when applied to orthopartitions:

Definition 3. *Given an orthopartition $\mathcal{O}$, we define the* lower entropy, *the* upper entropy *and the* mean entropy *respectively as:*

$$h_* = min\{h(\pi)|\pi \in \Pi_\mathcal{O}\} \tag{1a}$$

$$h^* = max\{h(\pi)|\pi \in \Pi_\mathcal{O}\} \tag{1b}$$

$$h_A = \frac{1}{|\Pi_\mathcal{O}|} \sum_{\pi \in \Pi_\mathcal{O}} h(\pi) \tag{1c}$$

As shown in [6,7], all three values can be computed in polynomial time. Let π_1, π_2 be two partitions and $f : \pi_1 \mapsto \pi_2$ be a bijection between the blocks of π_1, π_2, the accuracy of π_2 wrt π_1 is defined as:

$$acc_{\pi_1}(\pi_2) = \frac{1}{|U|} \sum_{S_i \in \pi_1} |S_i \cap f(S_i)|$$

Similarly, we can provide three generalizations of the accuracy:

Definition 4. *Given a partition π^*, an orthopartition $\mathcal{O}$, and a bijection f between the respective blocks, we define the* lower accuracy, *the* upper accuracy *and the* mean accuracy *respectively as:*

$$acc_* = min\{acc(\pi)|\pi \in \Pi_\mathcal{O}\} \tag{2a}$$

$$acc^* = max\{acc(\pi)|\pi \in \Pi_\mathcal{O}\} \tag{2b}$$

$$acc_A = \frac{1}{|\Pi_\mathcal{O}|} \sum_{\pi \in \Pi_\mathcal{O}} acc(\pi) \tag{2c}$$

Another interesting measure of accuracy (that we denote as acc_O) is obtained by considering, in the computation of the accuracy value, only the instances which are not in the boundary regions: that is, if $U_r \subseteq U$ is the restriction of U to the objects which are not placed in boundaries for orthopartition $\mathcal{O}$ then:

$$acc_O = \frac{1}{U_r} \sum_{S_i \in \pi_1} |S_i \cap f(S_i)|$$

where S_i and $f(S_i)$ are similarly restricted to U_r.

3 The Methods

In this section, we propose the main method of three-way classification and apply it to different learning strategies.

3.1 Three–Way Classification

Let $Y = \{y_1, ..., y_k\}$ be a set of class labels, $X = \{x^1, ..., x^n\}$ be a set of objects, $C : X \rightarrow Y$ be a function which associates with each object $x^i \in X$ its true classification $y^i_j \in Y$. Let A be a probabilistic classifier, that is, an algorithm which, given an object $x^i \in X$, returns a probability distribution $A(x^i)$ over Y, that is, $A : X \rightarrow \mathcal{P}(Y)$, where $\mathcal{P}(Y)$ is the space of probability distributions over Y. For each $y_j \in Y$, $A(x^i)_j$ represents the probability that algorithm A assigns to the event that y_j is the correct class labeling for object x^i (i.e., the subjective probability that $C(x^i) = y_j$). Typically, in the Machine Learning domain, this *soft* probabilistic classification is then converted into an *hard* one by selecting the $y_j \in Y$ with maximum probability: that is, we define $D(x^i) = argmax_{y_j \in Y} A(x^i)_j$ and we denote with $A(x^i)^*$ the corresponding probability. Note that this classification rule completely hides away the uncertainty of the classifier and, consequently, the ambiguity intrinsic in its input. An approach to let the classifier A fully express its uncertainty, which fully reflects the ambiguity of its input datum, is to let the classifier *abstain* on those instances whose assignment to the classification labels is considered ambiguous.

First, we limit ourselves to a binary classification problem, that is, $Y = \{0, 1\}$. Let ϵ be the cost associated with an erroneous classification, and let τ the cost associated with an abstention. Let $x \in X$ be an object, it is evident and widely known [8,16,31] that, in this context, algorithm A should choose to abstain on x if:

$$\tau < \epsilon * min_{j \in \{0,1\}} A(x^i)_j$$

that is, if choosing to abstain would incur (in the expected value) a lower cost than adopting a clear-cut classification (selected using the standard decision rule). The same decision rule could be given using a probability threshold; it is easy to show that the two formulations are equivalent.

Theorem 1. *Algorithm A should select to abstain iff* $max_{j \in \{0,1\}} A(x^i)_j < 1 - \frac{\tau}{\epsilon}$

Proof. Let $A(x)^* = max_{j \in \{0,1\}} A(x^i)_j$, the rule expressed above is equivalent to $\tau < \epsilon * (1 - A(x)^*) \Rightarrow \frac{\tau}{\epsilon} < 1 - A(x)^* \Rightarrow A(x)^* < 1 - \frac{\tau}{\epsilon}$.

The generalization to the multi–class setting, in which partial decisions could also be expressed, is also feasible and clearly more interesting. Indeed, in [6], a generalization of this classification rule is proposed as follows. Let $Z \subseteq Y$, then in this context we allow the algorithm A to express a decision Z, by which we mean that the algorithm is confident that the true label of x is in Z but it is unsure about its precise identity. Let $A(x)_Z = \sum_{y_j \in Z} A(x)_j$. If, as in the binary classification setting, we adopt a constant abstention cost τ, then the algorithm, with the abstention decision rule, should abstain on instance x if:

$$\tau * A(x)_{Z^*} + \epsilon * A(x)_{Y \setminus Z^*} < \epsilon * (1 - A(x)^*) \tag{3}$$

where $Z^* \subseteq Y$ is the set of labels which minimizes the left hand of the inequality, otherwise it should output the y_j corresponding to $A(x)*$.

Note that, directly translating this definition (as done in [6]) to an algorithm, yields a decision procedure which has complexity *exponential* w.r.t. $|Y|$. However, it is easy to observe that not every $Z \subseteq Y$ should be considered in the above minimization problem. In fact, the above minimization problem can be solved correctly in a *greedy* approach: let $\overset{\wedge}{A(x)} = \langle y_1^*, ..., y_k^* \rangle$ be the result of sorting $A(x)$ in order of decreasing probability. Then the above decision rule can be expressed, without loss of generalization, as:

$$\tau * \sum_{i=1}^{j} \overset{\wedge}{A(x)}_i + \epsilon * \sum_{i=j+1}^{k} \overset{\wedge}{A(x)}_i < \epsilon * (1 - A(x)^*) \tag{4}$$

where j is the index which minimizes the left hand of the inequality.

Theorem 2. *The greedy version of the optimization algorithm is solvable with time complexity $\Theta(n)$ (if $A(x)$ is already sorted).*

Proof. For each j we can pre-compute $\sum_{i=1}^{j} \overset{\wedge}{A(x)}_i$ in constant time (by accumulating the values of the sum over previous js), from this value we can obtain $\sum_{i=j+1}^{k} \overset{\wedge}{A(x)}_i$ in constant time. The result easily follows.

As observed in [6], a constant value of τ has the result that, when the algorithm abstains, Z^* (i.e. the set of labels which minimizes the optimization problem) is always $Z^* = Y$. This problem can be solved in a *regularization* fashion, by penalizing overly uncertain responses from the algorithm. In this case τ is defined as a function $\tau : \{1, ..., |Y|\} \to \mathbb{R}_+$ such that, given $A, B \subseteq Y$, it holds $|A| \leq |B| \to \tau(|A|) \leq \tau(|B|)$.

An interesting aspect to note is that not every value of τ is meaningful in this context, namely the following result holds:

Theorem 3. *Let us consider a* n–class *classification problem. Abstention can be achieved only if $\tau < \epsilon * \frac{n-1}{n}$.*

Proof. Consider the case of constant τ and the formulation given by Eq. 3. Then, we have that the algorithm should decide to abstain iff $\tau < \epsilon * (1 - A(x)^*)$. But $A(x)^* \geq \frac{1}{n}$, thus $\tau < \epsilon * (1 - A(x)^*) \leq \epsilon * (1 - \frac{1}{n})$, from which we obtain the result.

Example 1. Let x be an instance and A a probabilistic algorithm, defined over the label set $Y = \{1, 2, 3, 4\}$ such that $A(x) = [0.3\ 0.3\ 0.2\ 0.2]$, and let $\epsilon = 1$, $\tau = 0.4$. Then, the right hand of Eq. (3) is 0.7, while Z^* can be verified to be, as expected, $Z^* = Y$ with the left hand of inequality (3) assuming value 0.4. If, on the other hand, we do not assume a constant τ but instead adopt $\tau(Z) = 0.4 \cdot \frac{1}{1-\frac{|Z|-2}{|Y|}}$, thus penalizing abstentions over a larger set of alternatives, we have that $Z^* = \{1,2,3\}$ (equivalently, $Z^* = \{1,2,4\}$) and the left hand of the inequality has value 0.63.

3.2 Decision Trees and Random Forests

In [6] an extended Decision Tree model, called Three–Way Decision Tree (TWDT), is proposed. It provides a more tight integration of Decision Trees and Three–Way Classification than the main approach described in this paper. Let $D = \{x_1, ..., x_{|D|}\} \subseteq X$ be a given dataset with a set of features $\{a_1, ..., a_m\}$. We denote by $D_i^a = \{x \in D | v_a(x) = v_i^a\}$ the set of instances that have value v_i^a for feature a. We associate with D_i^a the classification C_i^a, which is obtained by the decision rule described in Sect. 3.1 (note that this class assignment is done locally on the tree nodes, and not only on the final output of the classifier). Since this classification determines an orthopartition $\mathcal{O}_a$, we can then compute the *accuracy* of $\mathcal{O}_a$ w.r.t. D as described in Sect. 2 (selecting among acc_*, acc^*, acc_A) and choose the feature a^* which results in the maximum accuracy value, and then recur (until a termination criterion is met) on the subsets of D determined by feature a^*.

This approach can be easily extended to Random Forests (or other ensemble learning algorithms). Basically, the learning process, as in standard Random Forest learning, first induces a set of n TWDT estimators, which we denote as $T_1, ..., T_n$. Each of these TWDT estimators can be viewed as an orthopartition $\mathcal{O}_i = \{\langle P_{y_1}, N_{y_1}\rangle, ...\langle P_{y_k}, N_{y_k}\rangle\}$ on the set of instances X, which assigns a set of labels $T_i(x) \subseteq Y$ to each instance $x \in X$.

Let $x \in X$ be a new instance to classify, then the ensemble of trees $T_1, ..., T_n$ determines a *basic belief assignment* (BBA) (in the sense of *evidence theory* [28]) $m_x(S) = \frac{|\{T_i | T_i(x) = S\}|}{n}$. This BBA could then be transformed to a probability distribution using the *pignistic transformation* [29] $p(y_j) = \sum_{S \ni y_j} \frac{m(S)}{|S|}$, obtaining a probabilistic classifier to which the decision procedure described in Sect. 3.1 could be applied.

3.3 Convex Learning Approximation

Several ML approaches (e.g. logistic regression, SVMs, multi–layer neural networks, ...) are based on the Gradient Descent algorithm, which is used to iteratively update the parameters of the models by taking in consideration the gradient of a loss function w.r.t. the parameters. A caveat, in order to ensure that the algorithm converges to a global minimum, is that the loss function should be a convex function. It is easy to note that the decision rule described in Sect. 3.1 (which could be seen as a generalized version of the standard *0–1 loss*) does not result in a convex loss function:

Theorem 4. *The loss function determined by the decision procedure described in Sect. 3.1 is not convex.*

Proof. Let $D(x^i) = \begin{cases} Z* & \exists Z^* \text{which solves Eq. 3,} \\ \wedge \\ y^i & otherwise \end{cases}$

Then the loss of algorithm A w.r.t to instance x is $L(x) = \begin{cases} 0 & D(x) = C(x) \\ \tau & C(x) \in D(x) \\ \epsilon & otherwise \end{cases}$

Clearly, $L(x)$ is not convex.

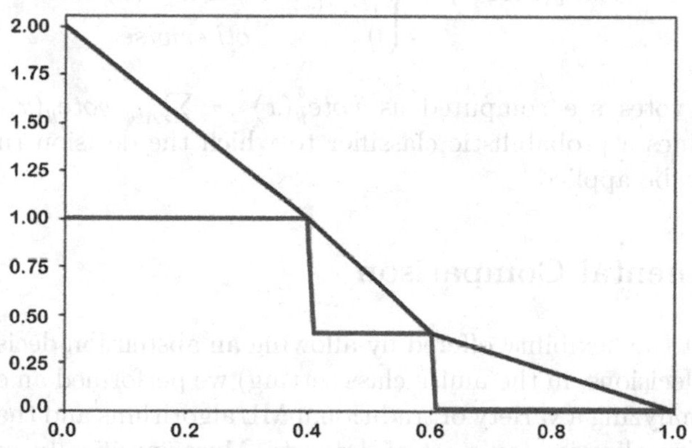

Fig. 1. The depiction of the loss function (in red), and its convex piece–wise linear approximation (in blue), for positive examples. (Color figure online)

We can, however, define a convex approximation of the above described loss function [1]. Consider first a binary classification problem, the loss function described above is depicted in Fig. 1. As shown in Fig. 1 we can, however, define a convex piece–wise linear approximation to the real loss. Consider first a binary classification problem assuming, without loss of generality, that $\epsilon = 1$. For the positive examples (i.e., those $x \in X$ s.t. $C(x) = 1$) we can express an approximation *from above* (so that we never underestimate the error) as:

$$L(w) = max\{0, 1 - w, \frac{(2 * \tau - 1) * w + 3 * \tau - 1 - \tau^2}{2 * \tau - 1}, 2 - \frac{w}{\tau}\} \quad (5)$$

where $w = A(x)_1$ (i.e. the probability that algorithm A assigns object x to the positive class).

Theorem 5. *The loss function described in Eq. 5 is convex.*

Proof. Each of the arguments of the *max* function is linear in w, thus it is convex (every linear function is both convex and concave). Furthermore the point–wise *max* of convex function is convex, from which the statement follows.

The expression for negative examples is equivalent and symmetric. This loss function could be then used to directly train convex learning algorithms and, given a new instance $x \in X$ to classify, we compute $A(x)_1$ and then, we classify x using the decision rule defined in Sect. 3.1.

In order to extend this approach to multi–class classification, we simply adopt a *one–vs–one* learning scheme, in which, for each pair of labels $y_i, y_j \in Y$ we train a classifier $A^{i,j}$ using the convex loss function described above. Then, given a new instance $x \in X$ to classify, we compute for each classifier $A^{i,j}$ its output $D^{i,j}(x)$ and we implement a voting schema:

$$vote_y(x, A^{i,j}) = \begin{cases} 1 & D^{i,j}(x) = y \\ \frac{1}{|D^{i,j}(x)|} & y \in D^{i,j}(x) \\ 0 & otherwise \end{cases}$$

and the final votes are computed as $vote_y(x) = \sum_{A^{i,j}} vote_y(x, A^{i,j})$ which, again, determines a probabilistic classifier to which the decision rule described in Sect. 3.1 can be applied.

4 Experimental Comparison

In order to test the flexibility offered by allowing an abstention decision (or a set of abstention decisions, in the multi–class setting) we performed an experimental comparison, analyzing a variety of traditional ML algorithms and their respective Three–Way generalization, on a set of datasets. More specifically, we considered the following algorithms: *k–nearest neighbors* (KNN), *logistic regression* (LR), *linear discriminant analysis* (LDA), *Naive Bayes* (NB), *SVM*s, *random forest* (RF). For each of these algorithms we also considered the three-way generalization obtained as described in Sect. 3.1 (these algorithms are denoted as TW followed by the acronym of the algorithm as defined previously); in addition, we also considered the three–way decision tree model (in the following denoted as TWDT), described in Sect. 3.2.

4.1 Settings

We compared the algorithms on the following datasets:

- *Iris*: 150 instances, 4 features, 3 classes;
- *Wine*: 178 instance, 13 features, 3 classes;
- *Breast cancer*: 569 instance, 30 features, 2 classes;
- *Digits*: 1797 instances, 64 features, 10 classes;
- *Yeast*: 1484 instances, 8 features, 10 classes;
- *Olivetti faces*: 400 instances, 4096 features, 40 classes;
- *SF12 Mental score* (described in [5]): 462 instances, 10 features, 2 classes;

In order to set the values of τ and ϵ (i.e. the abstention and error costs), we simply selected $\epsilon = 1$ and determined the optimal value of τ using cross-validation. Indeed, for each of the above datasets, we trained the classification algorithms using a 5–fold cross-validation, in order to select the optimal hyper–parameters (which includes τ) of the algorithms (e.g., the tree depth for decision trees). Then, we retrained the algorithms with the best selected hyper–parameters and reported the means and standard deviations of the acc_O accuracy measure (we considered this measure, as motivated in [16], in order to better analyze the trade–off between classification accuracy and abstention). For the three–way classification algorithms, in order to evaluate the trade-off among classification accuracy and coverage (defined as the fraction of objects which are assigned a clear–cut classification), we also measured the *abstention rate*, simply defined as:

$$Abst(A,T) = \sum_{x \in T} \frac{|D(x)|}{|Y|}$$

where A is a three–way classification algorithm, T is a testing set and $D(x) \subseteq Y$, as in Sect. 3.1, is the output labeling of algorithm A on instance x.

In order to more systematically study the trade-off among abstention and classification, for the dataset *Breast cancer* and for algorithms $TWRF$ and $TWSVM$, we also reported the variation with respect to the abstention cost τ of three different metric: accuracy, *true positive rate* (TPR), and *true negative rate* (TNR).

4.2 Results

The results of the experimental comparison are illustrated in Table 1 and, for one specific dataset (Yeast), in Fig. 2.

Table 1. Measured 95% confidence intervals, centered around the mean accuracy, for the considered datasets and algorithms.

Algorithm	Iris	Wine	Breast	Digits	Yeast	Faces	SF12
KNN	0.98 ± 0.03	0.75 ± 0.13	0.93 ± 0.04	0.98 ± 0.03	0.57 ± 0.03	0.90 ± 0.16	0.82 ± 0.02
TWKNN	1.00 ± 0.00	0.99 ± 0.02	0.99 ± 0.01	0.90 ± 0.00	0.67 ± 0.02	0.89 ± 0.01	0.82 ± 0.01
LR	0.95 ± 0.06	0.95 ± 0.05	0.95 ± 0.02	0.93 ± 0.04	0.53 ± 0.03	0.96 ± 0.03	0.73 ± 0.13
TWLR	0.96 ± 0.01	0.98 ± 0.02	0.98 ± 0.01	0.96 ± 0.02	0.78 ± 0.01	0.98 ± 0.01	0.77 ± 0.01
LDA	0.98 ± 0.04	0.98 ± 0.03	0.96 ± 0.03	0.92 ± 0.03	0.59 ± 0.01	0.98 ± 0.01	0.83 ± 0.12
TWLDA	0.98 ± 0.04	0.99 ± 0.00	0.97 ± 0.01	0.94 ± 0.02	0.72 ± 0.07	0.99 ± 0.01	0.83 ± 0.12
NB	0.95 ± 0.04	0.96 ± 0.03	0.94 ± 0.03	0.81 ± 0.06	0.15 ± 0.02	0.82 ± 0.03	0.82 ± 0.07
TWNB	0.97 ± 0.03	0.98 ± 0.03	0.95 ± 0.03	0.83 ± 0.05	0.16 ± 0.02	0.84 ± 0.02	0.86 ± 0.05
SVM	0.98 ± 0.03	0.73 ± 0.09	0.94 ± 0.02	0.97 ± 0.02	0.52 ± 0.03	0.79 ± 0.05	0.74 ± 0.04
TWSVM	0.98 ± 0.01	0.90 ± 0.01	0.96 ± 0.01	1.00 ± 0.00	0.81 ± 0.02	0.87 ± 0.04	0.83 ± 0.06
RF	0.97 ± 0.03	0.98 ± 0.03	0.96 ± 0.02	0.94 ± 0.02	0.58 ± 0.03	0.93 ± 0.02	0.83 ± 0.05
TWRF	0.98 ± 0.01	0.99 ± 0.01	0.99 ± 0.01	0.99 ± 0.01	0.80 ± 0.02	0.98 ± 0.01	0.85 ± 0.03
TWDT	0.97 ± 0.03	0.89 ± 0.07	0.94 ± 0.03	0.83 ± 0.04	0.63 ± 0.02	0.61 ± 0.15	0.84 ± 0.06

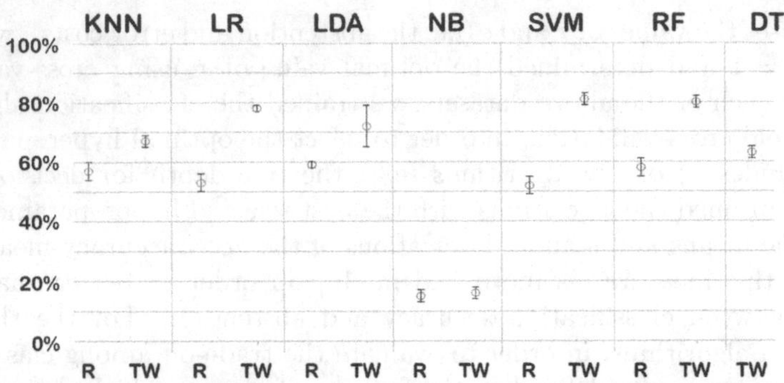

Fig. 2. Measured values of accuracy (and their 95%CIs) for the algorithms under test (regular version, R, on the left) and their three–way version (TW, on the right), on the dataset *Yeast*. Comparing the confidence intervals visually, it is clear that significant differences are observed for 5 model families (namely, KNN, LR, LDA, SVM and RF).

In Table 2, we reported the average ranks of the algorithms (i.e. for each dataset we sorted the algorithms in order of decreasing average accuracy, then we computed the average rank across the datasets).

Table 2. Average ranks of the top 10 performing algorithms.

Alg.	TWRF	TWLDA	TWLR	TWSVM	LDA	RF	TWKNN	KNN	LR	TWNB
Rank	1.75	2.96	3.18	3.57	4.32	4.64	4.64	5.86	6.14	6.78

As can be easily observed from Table 2, in every case the adoption of the possibility of abstention decreases the average rank of the respective algorithm (thus, the algorithm increases its performance). This effect can be explained by noting that the possibility of abstention gives the algorithm the ability to not express a clear–cut decision in those instances which are placed near the decision boundary (i.e. the instances whose class assignment is most uncertain) but, instead, report a list of possible classifications (which, with high confidence, includes the real label). In order to assess if the improvements given by the possibility of abstention were statistically significant, we performed a pair–wise Friedman test [12] for each pair of three–way/classical algorithm, with Li's correction for multiple hypothesis testing [25]: one of the three–way algorithms (TWSVM) was found to be significantly better than the respective classical with a p-value = 0.02, for two others (TWRF, TWLR) there was weak evidence of improvement, albeit with a lower p-value = 0.08 (all other algorithm pairs reported a p–value > 0.1), when considering the standard confidence level of $CL = 95\%$ only the first difference is statistically significant.

In order to investigate the trade–off between classification accuracy and abstention, as mentioned in Sect. 4.1, we measured the abstention rate of the three–way algorithms, as shown in Table 3. It could be easily observed that,

Table 3. Measured abstention rates for the considered datasets and three–way algorithms.

Algorithm	Iris	Wine	Breast	Digits	Yeast	Faces	SF12
TWDT	0.05	0.00	0.08	0.00	0.24	0.08	0.49
TWKNN	0.13	0.58	0.20	0.95	0.11	0.04	0.00
TWLR	0.01	0.19	0.14	0.19	0.27	0.02	0.60
TWLDA	0.00	0.05	0.08	0.30	0.16	0.03	0.00
TWNB	0.07	0.03	0.02	0.05	0.02	0.02	0.34
TWSVM	0.16	0.42	0.05	0.04	0.17	0.11	0.29
TWRF	0.05	0.15	0.13	0.08	0.17	0.05	0.31

in general, the abstention rate is greater than the corresponding increase of accuracy. This effect likely emerges because some of the instances that were classified *correctly* by a classical algorithm, were so *only by chance* (i.e., they were assigned to the correct class label, but with a low confidence level) and, thus, the corresponding three–way algorithm makes this phenomenon apparent (this is particularly evident for the TWKNN, which registered the highest value of abstention rate). An interesting observation is that in the Yeast dataset, the three–way algorithms performed significantly better than the classical ones, with only a moderate increase in abstention rates. It could also be observed that the best performing algorithm (TWRF) was consistently better than the other algorithms in every dataset, although no statistically significant difference (at $CL = 95\%$) could be found with the second ranking algorithm (i.e., TWLDA, $p - value = 0.28$).

Finally, as mentioned in Sect. 4.1, we analyzed the variation of different metrics, that is accuracy, true positive rate (TPR), true negative rate (TNR) and abstention rate, with respect to varying τ on two algorithms: the results are shown in Figs. 3 and 4.

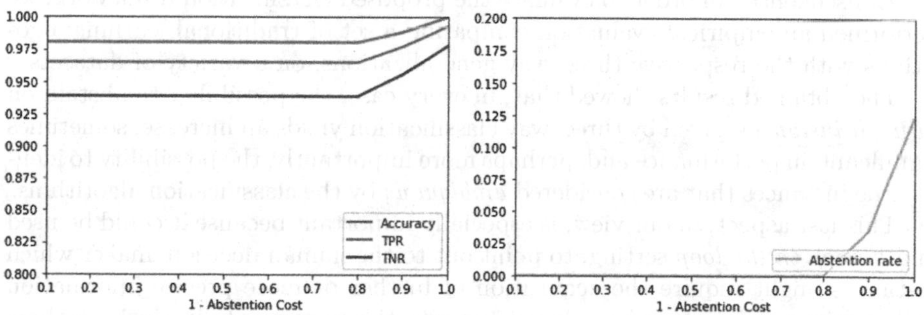

Fig. 3. Variation, w.r.t. abstention cost τ, of different metric for the TWRF algorithm: accuracy, tpr, tnr (left); abstention rate (right).

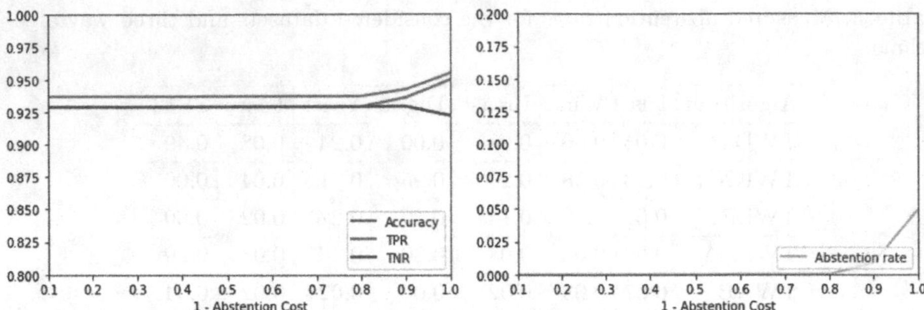

Fig. 4. Variation, w.r.t. abstention cost τ, of different metric for the TWSVM algorithm: accuracy, tpr, tnr (left); abstention rate (right).

As can be easily observed, both the accuracy and the abstention rate increase monotonically with decreasing τ (for both algorithms); furthermore there is a variation in the observed measures only for values $\tau \leq 0.2$ and, even at $\tau = 1$, the observed abstention rates were small; this can be explained, as noted by Theorem 1, as the algorithms assigned great confidence to their predictions.

A final point to note is that the TNR for algorithm TWSVM, shown in Fig. 4, decreases with decreasing abstention cost: this could be related to a deficit in the training dataset, which highlights a possible difficulty in detecting true negative instances.

5 Conclusion

In this work we presented a comprehensive framework to address three–way classification, both in the binary and the multi–class case, by providing a general approach to convert probabilistic classifiers into three–way algorithms. To this aim, we also focused on two techniques to directly embed the possibility of abstention given by this classification approach into three popular learning models. Consequently, in order to evaluate the proposed classification framework, we performed an empirical evaluation comparing a set of traditional learning algorithms with the respective three–way generalizations, on a variety of datasets.

The obtained results showed that, in every case, the possibility to abstain on *difficult instances*, given by three–way classification yields an increase, sometimes significant, in performance and, perhaps more importantly, the possibility to identify the instances that are considered *ambiguous* by the classification algorithms.

This last aspect, in our view, is especially important because it could be used in a *human in the loop* setting, to point out to the human decision–maker which instances might require the acquisition of further or more precise information and require special attention: that is, despite the *uncertainty* intrinsic to these three–way predictions, these could nevertheless be useful to the human decision maker as a way to raise awareness of the weak points and ambiguities affecting the available data.

Given the promising results that we obtained, we plan to continue this line of research considering the following issues and open problems:

- in this paper, we introduced both a general approach to build three–way classifiers and also two more techniques that may be applied to specific learning algorithms. Although we analyzed one such technique (learning of three–way decision trees), we plan to study if directly implementing three–way classification in ensemble tree–based algorithms (e.g. random forests) and convex learning algorithms could be more advantageous than the general post–hoc strategy evaluated in this work;
- in this work, we primarily focused on ambiguity *in the output*, that is, how ambiguity could be managed by allowing three–way, instead of crisp, classifications. However, ambiguity is a multi–faceted problem that could arise also in the input: both in the target attributes (e.g. abstentions are already present in the given gold standard) and the predictor ones (which could present missing or partial values). While we performed some initial works relating to these issues [5,6], we plan to expand this line of research, especially in regard to ambiguity in predictor attributes, in order to build a comprehensive framework for managing ambiguity in machine learning.

References

1. Bartlett, P.L., Wegkamp, M.H.: Classification with a reject option using a hinge loss. J. Mach. Learn. Res. **9**, 1823–1840 (2008)
2. Bello, R., Falcon, R.: Rough sets in machine learning: a review. In: Wang, G., Skowron, A., Yao, Y., Ślęzak, D., Polkowski, L. (eds.) Thriving Rough Sets, pp. 87–118. Springer International Publishing, Cham (2017)
3. Breiman, L.: Random forests. Mach. Learn. **45**(1), 5–32 (2001)
4. Cabitza, F., Ciucci, D., Rasoini, R.: A giant with feet of clay: on the validity of the data that feed machine learning in medicine. In: Cabitza, F., Batini, C., Magni, M. (eds.) Organizing for the Digital World. LNISO, vol. 28, pp. 121–136. Springer, Cham (2019). https://doi.org/10.1007/978-3-319-90503-7_10
5. Campagner, A., Cabitza, F., Ciucci, D.: Exploring medical data classification with three-way decision tree. In: Proceedings of the 12th International Joint Conference on Biomedical Engineering Systems and Technologies (BIOSTEC 2019) - Volume 5: HEALTHINF. pp. 147–158. SCITEPRESS (2019)
6. Campagner, A., Ciucci, D.: Three-way and semi-supervised decision tree learning based on orthopartitions. In: Medina, J., Ojeda Aciego, M., Verdegay, J.L., Pelta, D.A., Cabrera, I.P., Bouchon-Meunier, B., Yager, R.R. (eds.) IPMU 2018. CCIS, vol. 854, pp. 748–759. Springer, Cham (2018). https://doi.org/10.1007/978-3-319-91476-3_61
7. Campagner, A., Ciucci, D.: Orthopartitions and soft clustering. Knowl. Based Syst. (Submitted)
8. Chow, C.: On optimum recognition error and reject tradeoff. IEEE Trans. Inform. Theory **16**, 41–46 (1970)
9. Ciucci, D.: Orthopairs: a simple and widely used way to model uncertainty. Fundamenta Informaticae **108**, 287–304 (2011)

10. Ciucci, D.: Orthopairs and granular computing. Granular Comput. **1**, 159–170 (2016)
11. Cortes, C., Vapnik, V.: Support-vector networks. Mach. Learn. **20**(3), 273–297 (1995)
12. Daniel, W.W.: Applied Nonparametric Statistics. Duxbury Thomson Learning (1990)
13. Deo, R.: Machine learning in medicine. Circulation **132** (2015)
14. Ellerman, D.: An introduction to logical entropy and its relation to Shannon entropy. Int. J. Semant. Comput. **7**(2), 121–145 (2013)
15. Feldman, K., Faust, L., Wu, X., Huang, C., Chawla, N.V.: Beyond volume: the impact of complex healthcare data on the machine learning pipeline. CoRR abs/1706.01513 (2017)
16. Ferri, C., Hernández-Orallo, J.: Cautious classifiers. In: ROC Analysis in Artificial Intelligence, 1st International Workshop, ROCAI-2004, pp. 27–36 (2004)
17. Goodfellow, I., Bengio, Y., Courville, A.: Deep Learning. MIT Press, Cambridge (2016)
18. Hajian, S., Bonchi, F., Castillo, C.: Algorithmic bias: from discrimination discovery to fairness-aware data mining. In: Proceedings of the 22nd ACM SIGKDD International Conference on Knowledge Discovery and Data Mining, pp. 2125–2126, August 2016
19. Han, P.K., Klein, W.M., Arora, N.K.: Varieties of uncertainty in health care: a conceptual taxonomy. Med. Decis. Making **31**(6), 828–838 (2011)
20. Hechtlinger, Y., Póczos, B., Wasserman, L.A.: Cautious deep learning. arXiv/CoRR abs/1805.09460 (2018)
21. Hüllermeier, E.: Fuzzy sets in machine learning and data mining. Appl. Soft Comput. **11**(2), 1493–1505 (2011)
22. Hüllermeier, E.: Does machine learning need fuzzy logic? Fuzzy Sets Syst. **281**, 292–299 (2015). Special Issue Celebrating the 50th Anniversary of Fuzzy Sets
23. Koller, D., Friedman, N.: Probabilistic Graphical Models: Principles and Techniques - Adaptive Computation and Machine Learning. The MIT Press, Cambridge (2009)
24. Kooi, T., et al.: Large scale deep learning for computer aided detection of mammographic lesions. Med. Image Anal. **35**, 303–312 (2017)
25. Li, J.D.: A two-step rejection procedure for testing multiple hypotheses. J. Stat. Plann. Infer. **138**(6), 1521–1527 (2008)
26. Obermeyer, Z., Emanuel, E.J.: Predicting the future - big data, machine learning, and clinical medicine. N. Engl. J. Med. **375**(13), 1216–1219 (2016)
27. Pawlak, Z.: Rough sets. Int. J. Comput. Inform. Sci. **11**(5), 341–356 (1982)
28. Shafer, G.: A Mathematical Theory of Evidence. Princeton University Press, Princeton (1976)
29. Smets, P., Kennes, R.: The transferable belief model. Artif. Intell. **66**(2), 191–234 (1994)
30. Svensson, C., Hübler, R., Figge, M.: Automated classification of circulating tumor cells and the impact of interobsever variability on classifier training and performance. J. Immunol. Res. **2015**, 1–9 (2015)
31. Yao, Y.: An outline of a theory of three-way decisions. In: Yao, J.T., Yang, Y., Słowiński, R., Greco, S., Li, H., Mitra, S., Polkowski, L. (eds.) RSCTC 2012. LNCS (LNAI), vol. 7413, pp. 1–17. Springer, Heidelberg (2012). https://doi.org/10.1007/978-3-642-32115-3_1
32. Zadeh, L.: Fuzzy sets. Inf. Control **8**(3), 338–353 (1965)

Concepts Approximation Through Dialogue with User

Soma Dutta[1(✉)] and Andrzej Skowron[2,3]

[1] University of Warmia and Mazury in Olsztyn, Słoneczna 54, 10-710 Olsztyn, Poland
somadutta9@gmail.com
[2] Systems Research Institute, Polish Academy of Sciences, Newelska 6, 01-447 Warsaw, Poland
skowron@mimuw.edu.pl
[3] Digital Science and Technology Centre, UKSW, Dewajtis 5, 01-815 Warsaw, Poland

Abstract. This paper is an attempt to show how dialogue between a system and a user is important to design a robust system which can learn a user's perspective and revise its knowledge base through interactions. The dialogue is crucial in order to better respond to a given query of a user. The problems, where dialogues can play a role, are discussed from two aspects. One is the aspect in which the system becomes able to learn the perspectives of the user(s) and improve its quality of classifications. The other is the aspect where the system can help a user to get answers to its queries. We have, in particular, considered the problems of (i) learning a user's ontology of concepts, (ii) explaining the system's own classification for a cluster to the user in order to get feedback, and (iii) generating a global description for a cluster, in a user-friendly language, based on a sample of objects available to the system.

Keywords: Rough set · Concept approximation · Dialogue · Classification

1 Introduction

In the present era of information science, looking for relevant information using a system-user interface is a more common practice than enquiring to a next-door neighbour [1]. Different web application systems are increasing in numbers with different information providing facilities. How the system can answer all possible questions of a user related to a domain? Clearly, apart from an initial knowledge base, the system must have a continuous access to this dynamically changing real physical world. So, naturally to build such a system one must include a feature, e.g., continuous learning and developing through interactions with a set of domain experts or users. For this the system requires learning a user's ontology of the domain of concepts. Moreover, a language, through which the user-system interactions can be modeled, is needed. This language, aim of which is to express the behaviour of a community of agents to satisfy their needs, can be treated as a complex dynamic object evolving with time [2–4]. The behaviour of community

© Springer Nature Switzerland AG 2019
T. Mihálydeák et al. (Eds.): IJCRS 2019, LNAI 11499, pp. 295–311, 2019.
https://doi.org/10.1007/978-3-030-22815-6_23

of agents may lead to cooperation, competition or coalition formation. The skills of the community members can improve with time. The continuous learning of new communication forms and strategies causes evolution of the whole communication language. The discussed issues are especially important for constructing intelligent systems, which are able to explain undertaken decisions (see, e.g., home.earthlink.net/~dwaha/research/meetings/ijcai17-xai/).

We shall discuss some exemplary cases concerning the evolution of a communication language among the system and the users. Let there be a database system for a particular domain D. While designing the system the general ontology of the concepts related to D is already embedded to the system by the designer of it. Given a concept from this general ontology, the system is able to classify it with respect to a set of finitely many simple concepts, lying in the lower level of the ontology. These simple concepts can be regarded as the attributes or parameters for classifying the available objects of the system as the positive/negative cases of a certain higher level concepts. In particular, let the domain D be about the papers on *approximation*. There are different branches of approximation, among which some are related by some means and some are non-comparable. While designing the database for the papers on *approximation*, it is expected that this general ontology of D is incorporated in the system. So, one classification problem is sorting the database for the articles on *rough set approximation*. Based on the known domain ontology the system may select a set of keywords such as *rough set, lower approximation, upper approximation* etc. as attributes, and classify the articles as positive and negative instances of *rough set approximation*. If the attributes for classification are changed, the positive and negative instances of the concept vary as well. So, based on the general domain ontology, some concepts are already characterized by the system with respect to the available articles. Now, let us assume that a user is looking for articles related to *rough approximation operators*. If the concept appears new to the system, the challenges for the system are as follows.

1. What is the relationship between the concepts already approximated by the system and the concept, the user is looking for?
2. How to understand the user's ontology of the domain of literature based on the system's own ontology of the domain of literature?
3. What form of communication is needed to provide a well approximation of the user's concept, in the language understandable by the user, with the help of the already existing classifications of the concepts available to the system?

Such problems are typical cases, that one needs to consider while designing an automated intelligent system which can provide answer to a user's query. Let us denote the general ontology of the concepts of a domain, which consists of the concepts of different levels, including the atomic attributes/features, and their interrelations, by the symbol O_G. For simplicity, O_G may be identified with its underlying set of concepts. The system classifies objects based on a subset $\mathcal{A}$ of attributes from O_G. The user's ontology of the same domain of concepts may be different from that of the system. Here, we impose that the system's ontology O_G is somehow embedded in the user's ontology O_U as it is natural that being an expert or interested user of the domain of literature the user is aware of the existing general ontology of the domain. So, $f(O_G) \subseteq O_U$ where f is a mapping which embeds the general ontology O_G into a specific ontology O_U. In particular, f can be a translation of the language of O_G to that of O_U. Now, given $C \in O_U$ either $C \in O_G$ or $C \in O_U \setminus f(O_G)$. For $C \in O_G$, the system must have a classification, and with the help of that classification C can be represented by a set of lower level attributes from O_G. As $f(O_G) \subseteq O_U$, the user can translate the system's classification of the concept C in terms of her ontology. The user may or may not agree with the system's classification. This leads to one direction of exploration. The other possible aspect is when C lies outside the system's ontology of the domain of concepts. Here, the question of learning the user's ontology of the concepts comes; consequently a necessity for the user-system dialogue comes in as well. To explain how dialogues can be integrated in the process of learning, to some extent, we shall depend on [5,6].

This paper attempts to address the challenges of the following aspects of the classification problem. One is to learn a user's ontology of concepts in order to classify a concept given by the user using the database available to the system. The second is to build a communication language between the system and the user so that (i) the system can ask for the user's feedback on a classified concept, (ii) the user can ask for the system's explanation about the method of classification, (iii) and through such interactions the system can learn to improve its quality of classification. The issues discussed in this paper are also relevant to the recently raised problems of *Explainable AI* [7]:

Deep learning approaches, trained on extremely large data sets or using reinforcement learning methods have even exceeded human performance in visual tasks, particularly on playing games [...] *Even in the medical domain there are remarkable results. However, the central problem of such models is that they are regarded as black-box models and even if we understand the underlying mathematical principles of such models they lack an explicit declarative knowledge representation, hence have difficulty in generating the underlying explanatory structures. This calls for systems enabling to make decisions transparent, understandable and explainable.* [...] *medical professionals must have a possibility to understand how and why a machine decision has been made.*

As the possible challenges to the above aspects cannot be all visualized a priori, we focus on a few aspects of such challenges. We will explain the problems and the proposals through some exemplary cases, such as classifying articles from a particular domain of literature, classifying handwritten digits etc. The proposed prescription to the problems is not a complete flow-chart of a step-by-step method; our intention is to see the possibility of using the approach to some fragments of the challenges, and develop the proposal further as future work. In Sect. 2, we present a very brief introduction to the language of dialogue as required for this paper. Sections 3–6 focus on different aspects of the problem mentioned in the last paragraph. Section 7 presents an example.

2 A Language Modeling Dialogues Between User and System

In [5], we developed a notion of dialogue in a dialogue base $(G_1, G_2, \ldots, G_r, R_{int}, R_{ext})$, where by dialogue base we mean a set of databases $G_1, \ldots, G_r$ having an internal relation R_{int} within each database, and an externel relation R_{ext} connecting different databases. The idea was to incorporate a possibility of interaction among databases of different agents. Here, at the simplest case, we assume two databases; one corresponds to the system and the other corresponds to the user. Below we present an example showing the basic forms of dialogues between agents.

Example 1. An outcome of a dialogue is presented just by a sequence of symbols. These symbols can be objects, attributes, values of the attributes, or compound concepts from the ontology of the domain of the database, and some auxilliary symbols. Here we present how the syntax of the language of dialogue would look like.

(i) A sequence of the form $\langle a_1, a_2, \ldots, a_l\rangle$ for any finite l, and attributes $a_1, \ldots, a_l$ is a dialogue. Any concept from the ontology is also regarded as an attribute.
(ii) A sequence of the form $\langle x_1, \ldots, x_k\rangle$ consisting objects of the database is a dialogue.
(iii) A dialogue of the form $\langle x_1, \ldots, x_k \ \S y_1, \ldots, y_q \ \S C\rangle$ consisting of objects $x_1, \ldots, x_k$, $y_1, \ldots, y_q$, and a concept C represents that the first sequence of objects are positive instances of C, and the second sequence of objects are negative instances of C. The symbol § is an auxilliary symbol behaving as a separator.
(iv) A dialogue of the form $\langle x_1, (\,), \ldots, x_k \ \S y_1, \ldots, y_q \ \S C\rangle$ represents dropping some cases from the positive instances. Similarly, one can express dropping some cases from negative instances and/or both.
(v) A dialogue of the form $\langle a \ \S x_1, \ldots, x_k \ \S ?\rangle$ represent a query whether the objects satisfy the attribute a.
(vi) A dialogue of the form $\langle a \ \S x_1, x_2, \ldots, x_k \ \S +, -, \ldots, + \ \S\rangle$ represents that x_1 is a positive instance of a, x_2 is a negative instance of a, and so on.
(vii) A dialogue of the form $\langle x_1, \ldots, x_k \ \S \Box^{?\text{`}}\rangle$ represents a query that which attributes characterize the objects $x_1, \ldots, x_k$.
(viii) A dialogue of the form $\langle \boxtimes \ \S a_1, \ldots, a_l\rangle$ represents that the attributes $a_1, \ldots, a_l$ needs to be modified.
(ix) A dialogue of the form $\langle x_1, \ldots, x_k \ \S C\ ?\rangle$ indicates to verify if $x_1, \ldots, x_k$ satisfies C.
(x) A dialogue of the form $\langle x_1, \ldots, x_k \ \S C \ \Box^{?\text{`}}\rangle$ indicates to explain by which attributes $x_1, \ldots, x_k$ are counted as instances of C.
(xi) A dialogue of the form $\langle a_1, \ldots, a_l \blacktriangleright C\rangle$ represents that the attributes $a_1, \ldots a_l$ define the concept C. Moreover, $C \approx C'$ represents that two concepts C and C' are equivalent. Hence a combined dialogue of the form $\langle a_1, \ldots, a_l \blacktriangleright C \ \S C \approx C'\rangle$ represents an explanation that the attributes defining C can also define C'.

(xii) A dialogue of the form $\langle x_1, \ldots, x_k \blacktriangleright C\rangle$ denotes the objects $x_1, \ldots x_k$ classify C.

(xiii) A dialogue of the form $\langle [x_1, x_2, \ldots, x_{k-1}, x_k]\ \S x_k \blacktriangleright C\ \S a_1, \ldots a_l\rangle$ represents that $x_1, \ldots x_{k-1}, x_k$ belong to a similarity class, where only x_k is classified by C with respect to the attributes $a_1, \ldots, a_l$.

(xiv) Acceptance and rejection notifications for a set of objects classifying a concept C can be presented by $\langle x_1, \ldots, x_k\ \S C\boxtimes\rangle$ and $\langle x_1, \ldots, x_k\ \S C\boxtimes\rangle$ respectively.

3 Learning User's Ontology of Concepts Through Dialogues

Here, the main issue is to learn the user's ontology of a domain through interactions. In this regard, the readers are referred to Fig. 1 and the papers [8,9]. Let C be a concept of $O_U \setminus f(O_G)$, of which the system does not have a classification. The system's task here is to learn the user's concept C, on the basis of its own available objects and already classified concepts, through dialogue. One such possibility is when the user provides a feedback on a sample of objects

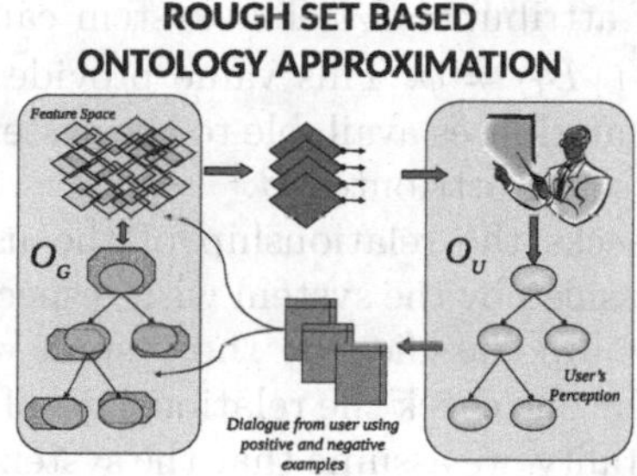

Fig. 1. Learning user's ontology of concepts through dialogue [8,9]

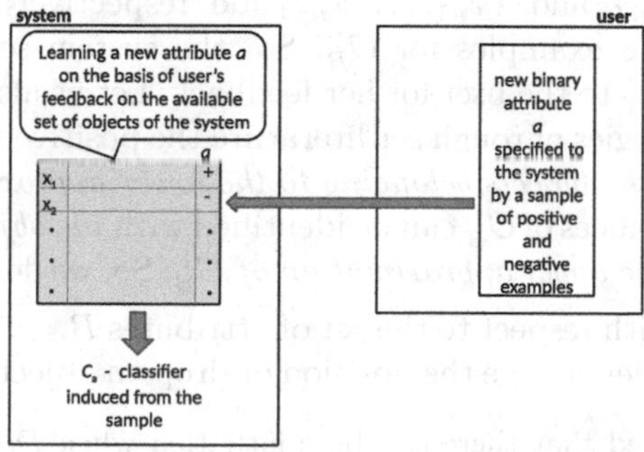

Fig. 2. Learning attributes through user's feedback on classifying objects

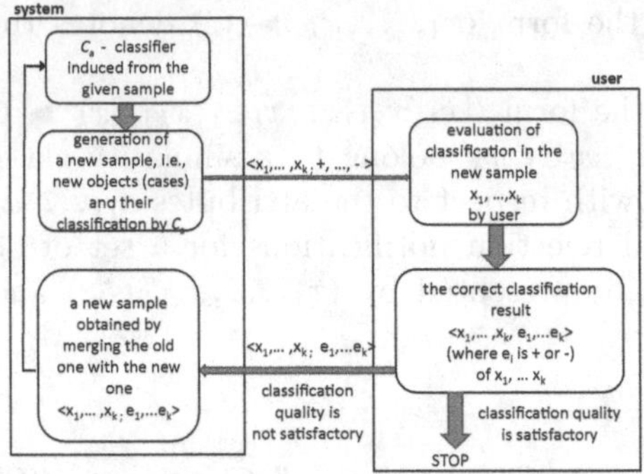

Fig. 3. Learning approximation of concepts through dialogues

with respect to a binary attribute a representing the characteristic function of C (Fig. 2). The system then tries to discover relevant objects to approximate a. The result of approximation may be tested through dialogues with the user (Fig. 3). Another possibility is as follows.

(i) In order to understand $C \in O_U \setminus f(O_G)$, the system first matches the given description of C with its set of attributes $\mathcal{A}$. Let C be described by the user using the set of attributes D_C. The system can check $\frac{|B \cap D_C|}{|D_C|}$ for any $B \subseteq \mathcal{A}$ such that $B \cap D_C \neq \phi$. This value provides a degree of matching between D_C and the attributes available to the system. Based on this value, the system selects a set of attributes B.[1]

(ii) The system then checks the relationship of the user's concept with the concepts that are classified by the system with respect to B. Let $C_B^1, \ldots, C_B^k$ be the concepts that are classified by the system with respect to B. The system now can one by one check the relationships of these concepts with C. Without loss of generality, we assume that the system first starts interaction with the user to check the relationship of C with C_B^1. As $f(C_B^1) \in O_U$, the system's target would be to understand the interrelation between C and $f(C_B^1)$, the embedded image of C_B^1 in O_U.

(iii) Let $\{x_{p1}, \ldots, x_{pi}\}$ and $\{x_{n1}, \ldots, x_{nj}\}$ be respectively the sets of positive and negative examples for C_B^1. So, the system sends $\langle x_{p1}, \ldots, x_{pi} \S x_{n1}, \ldots, x_{nj} \S C_B^1 \rangle$ to the user for her feedback. Let us also note, that following the terminologies of rough set literature the positive instances of C_B^1 can be identified as *the objects belonging to the lower approximation of* C_B^1 and the negative instances of C_B^1 can be identified with *the objects belonging to the complement of the upper approximation of* C_B^1. So, we denote $C_B^{1+} = \underline{C_B^1}$ and $(C_B^{1-})^c = \overline{C_B^1}$, with respect to the set of attributes B.

(iv) The user can either change the position or drop an object from the sequence.

[1] Here, it is to be noted that there can be a situation when D_C does not have non-empty intersection with any available subset of attributes. Such a possibility will also be touched upon in the later part of this discussion (cf. item (c)).

Dropping Instances

(a) If the user drops some objects from the positive instances of C^1_B, then according to the user a subset of the set of positive instances of C^1_B are the positive instances of C. So, there might be a set B', where $B \subseteq B' \subseteq \mathcal{A}$, characterizing the set of positive and negative examples accepted by the user. If the search for such a B' ends in an affirmative result, the system learns a classification of a new concept C with respect to its available sets of attributes and objects.

(b) When the user drops some objects from the negative instances of C^1_B, $\overline{C^1_B} \subseteq \overline{C}$. So, C is considered a more general concept than C^1_B. So, there can be $B'(\subseteq B)$, which classifies C by the pair of sets of objects, recommended by the user.

(c) If the user drops some objects from the positive as well as negative instances, then $\underline{C} \subseteq \underline{C^1_B}$ and $\overline{C^1_B} \subseteq \overline{C}$. So, borderline instances of C is more than that of C^1_B.
In the system there can be a subset B' of attributes such that $B \subseteq B' \subseteq \mathcal{A}$ and it characterizes exactly those positive instances of C^1_B, that are recommended by the user. On the other hand, as $\overline{C^1_B} \subseteq \overline{C}$ there can be some attributes in B by which some possible cases of C cannot be described. In that case, the system needs to learn those attributes of $B'(\supseteq B)$ that are more crucial to understand C. The system initiates a dialogue with the user by sending a sequence $\langle a_1^{B'}, \ldots a_n^{B'} \rangle$ of attributes from B' in order to understand whether the concept C satisfies those properties. In response the user can drop and/or add some new to the list. If all the revised attributes are already available to the system, it starts a new dialogue by sending the set of positive and negative examples, characterized by those attributes, to the user. If some of the attributes are not listed in the database, then the system needs to learn the relationship of that set of attributes, say B'', with the already available ones. So, for each $a \in B''$, the system sends a sequence of the form $\langle a\ \S x_1, \ldots, x_n\ \S? \rangle$ where $\{x_1, \ldots, x_n\}$ is the set of objects present in the system's database. In response, the user returns a sequence where each object x_i is either replaced by $+$ or $-$, or left unchanged. The signs $+$ and $-$ respectively indicate that the respective object is a positive and negative instance of a; if it is left unchanged, that means the user considers it neither as a positive nor as a negative instance of a. Thus, the system learns a classification for each new attribute of B'', and returns to the step (iii).

Changing Position

(d) We suppose that the user changes some positive instances of C^1_B to the negative instances. So, $\underline{C} \subseteq \underline{C^1_B}$ and $(\overline{C^1_B})^c \subseteq (\overline{C})^c$, i.e., $\overline{C} \subseteq \overline{C^1_B}$. This is same as step (a) with one additional constraint that some of the positive instances of C^1_B are counterexamples for C. Here, because $\underline{C} \subseteq \underline{C^1_B}$, there must be some more additional attributes which are not qualifying a few positive instances of C^1_B as the positive instances of C. So, the attributes that characterize the thrown out cases are important. Let $x_{pk_1}, \ldots x_{pk_i}$ be those positive instances of C_B^1 that are considered as negative instances of C. The system takes into account the additional attributes which satisfy $\{x_{pk_1}, \ldots x_{pk_i}\}$ but not the rest of the positive instances of C_B^1. These attributes are sent to the user as a form of a dialogue in order to get feedback from the user. As like case (c), the user can either drop or add some new attributes, and if some of these new attributes are new to the system, system first learns the classifiers for these new attributes.

(e) Let us assume that the user changes some negative instances of C_B^1 to the positive instances. So, $\underline{C_B^1} \subseteq \underline{C}$ and $\overline{C_B^1} \subseteq \overline{C}$. As like above, the system first tries to find the set of attributes that characterizes $\{x_{qk_1}, \ldots, x_{qk_j}\}$, the negative instances from C_B^1 which satisfy C. The process of learning is similar as (c) by asking for the user's feedback on some possible attributes. After that, it returns to the step (iii) with respect to new attributes.

(f) Let us assume that the system changes $x_{pk_1}, \ldots, x_{pk_i}$, a few positive instances of C_B^1 to the negative instances of C and $x_{qk_1}, \ldots, x_{qk_j}$, a few negative instances of C_B^1 to the positive instances of C. Here, three sets of attributes have significant role in designing the required concept C. One is the set, say B', that characterizes $\underline{C} \cap \underline{C_B^1}$. The others are the sets B'' and B''' respectively characterizing $\{x_{qk_1}, \ldots, x_{qk_j}\}$ and $\{x_{pk_1}, \ldots, x_{pk_i}\}$. The sets of attributes B' and B'' need to pass through a close scrutiny as they are satisfying two subsets of $\underline{C}$, and the set B''' of attributes need to be avoided as they satisfy $x_{pk_1}, \ldots, x_{pk_i}$, the cases which are denied to be positive instances of C. So, the system enquires for a feedback on these three sets of attributes. After coming to a consensus with the user, the system repeat the step (iii) with the set of attributes, agreed with the user.

Dropping and Changing

(g) Here the user drops some positive instances from $\underline{C_B^1}$ and changes some negative instances of C_B^1 to the set of positive instances of C. So, here $(\overline{C})^c \subseteq (\overline{C_B^1})^c$, and hence $\overline{C_B^1} \subseteq \overline{C}$. So, first thing is to find a subset B' of attributes such that $B' \subseteq B$ and B' characterizes $\overline{C}$. But as $\underline{C} \subseteq \overline{C}$ the system must find some additional attributes which characterizes exactly $\underline{C}$. One way is to find out the sets of attributes which characterizes respectively $\underline{C} \cap \underline{C_B^1}$ and the subset of negative instances of C_B^1 which are considered as positive instances of C. Learning the set of attributes is same as before by collecting feedback of the user.

(h) Let us consider that the user drop some negative instances of $\overline{C_B^1}$, and change some positive instances of C_B^1 to the negative instances of C. Clearly $\underline{C} \subseteq \underline{C_B^1}$. So, there must be B' such that $B \subseteq B'$ and B' characterizes $\underline{C}$. To find the additional attributes one may look for those attributes which do not satisfy those positive instances of C_B^1 that are considered as negative instances of C.

4 Explanation of System's Classifications Through Dialogues

Here we consider the problem of ontology approximation pertaining to understanding how the system labels a pair of clusters as respectively the positive and negative instances of a concept. Let $C \in O_G$, for which the system has an available classification.

(i) Let us assume that the user enters a query for C, and in response the system provides a sequence of positive and negative instances of C.
(ii) We consider the case when the user is interested to know based on what parameters or reasoning the system chooses the positive and negative instances of C.
(iii) In this regard, the user asks the system to explain its classification for the positive and negative instances of C using the dialogue $\langle x_{p_1}, \ldots, x_{p_k}; x_{n_1}, \ldots, x_{n_i} \ \S\Box^{?'} \rangle$.
(iv) In response the system sends the sequence of attributes $\langle a_1, \ldots, a_l \rangle$, based on which the classification is made. If the user needs more detailed explanation she can ask for next round of explanation by sending a sequence $\langle a_1, \ldots, a_l \blacktriangleright C \ \S\Box^{?'} \rangle$. The explanation may have an indirect explanation of the form $\langle a_1, \ldots a_l \blacktriangleright C' \ \S C' \approx C \rangle$, meaning that the set of parameters $\{a_1, \ldots, a_l\}$ implies C' and C' is similar to C.
(v) One possibility is that the explanation satisfies the user. If not, it may still help the user to guess the differences between the system's way of understanding C and that of her own. This forces the user to be more precise in choosing a set of attributes for describing C. The system then starts a new search based on these new attributes.
(vi) In step (v) if the system learns some new attributes, which were not taken care of by the system before while classifying the positive and negative instances of C, then the system revises its database by introducing new attributes for describing C.

5 Learning How to Label Clusters Through Dialogues

Let the system through dialogues, as discussed above, be able to approximate O_U, the user's ontology (Fig. 1). Now we assume that the system discovers a cluster of objects, obtained by a heuristic method. For instance, the system can notice that some objects with respect to some attributes assume quite similar values, and the system creates a heuristic algorithm for defining the cluster. But the description of such an algorithm may not be very simple to communicate to the user. So, the system needs to learn how to label this cluster with a description that can be understandable by the user.

(i) The system, in this context, sends a sequence of the form $\langle x_{p_1}, \ldots, x_{p_k} \ \S\Box^{?'} \rangle$ to the user asking for the attributes that would describe $x_{p_1}, \ldots, x_{p_k}$.

(ii) The user can send a sequence $\langle C_1, \ldots, C_j \rangle$ of concepts to check if that describe the cluster. As for each concept of O_U, the system already learned a way of approximation, the system checks the positive and the negative instances of each of $C_1, \ldots, C_l$, using the characteristic functions of the approximations of these concepts as attributes, and match with the cluster of objects $x_{p_1}, \ldots, x_{p_k}$.

(iii) If the cluster matches with the lower approximation obtained with respect to each of $C_1, \ldots, C_j$ to an extent greater than or equals to a threshold, the system can label the cluster by the proposed set of properties. If not, the system may simply drop those concepts, say $C_{j1}, \ldots C_{jl}$, for which the clustering mismatches to a greater extent, and label the cluster by the rest of the concepts. But if number of such concepts not matching the cluster is significantly big, the system also can request the user for modification of those attributes by sending a sequence $\langle \boxtimes; C_{j1}, \ldots, C_{jl} \rangle$. Now though for each C_i, $1 \leq i \leq l$, $\{x_{p_1}, \ldots, x_{p_k}\} \cap \underline{\{x_{p_1}, \ldots, x_{p_k}\}}_{C_i}$ has a significant number of elements, each of the lower approximations may not have the same intersection with $\{x_{p_1}, \ldots, x_{p_k}\}$. If $\cap_{i=1}^{l}(\underline{\{x_{p_1}, \ldots, x_{p_k}\}}_{C_i} \cap \{x_{p_1}, \ldots, x_{p_k}\})$ contains significant number of cases, then the resultant set can be considered as the positive instances of the combined concept $C_1 \& \ldots \& C_l$. Otherwise, dropping some of the concepts would help the system to have a more general description of the cluster.

So, when with respect to the available set of attributes all the elements of the cluster become difficult to describe, it needs to be described approximately. In such cases, the descriptions involving generalized quantifiers can help.

6 Searching Relevant Description for a Complex Cluster

In this section we focus on, how through the dialogues the system learns to create a general description for a cluster from the descriptions of the individual objects of the cluster. The description should be understandable by the user. Continuing the example of Sect. 5, we consider a cluster of objects for which the system does not have a simple, straightforward description. The issue is to learn, through dialogues with the user, an invariant description for the whole cluster. We assume that the cluster has a huge number of elements the patterns of which are not easy to formulate with a well-defined description. The system wants to learn a description for the cluster by gathering the user's point of view on a sample of it. To illustrate, let us consider that from a sample $\{x_1, \ldots, x_k\}$ of (images of) objects the system induce a description, such as *a region surrounded by a circular disc*. The system wants to make its description more generalized and of improved quality so that it suits to all objects of the cluster.

(i) In order to get the user's feedback on the given description for the set of objects, the system sends a sequence $\langle x_1, \ldots, x_k \; \S+, -, \cdots + \; \S C \rangle$ to the user where C represents *a region surrounded by a circular disc*, and + and − represent the respective positive and negative instances of C from the sample.

(ii) Observing the individual objects and the description provided by the system, the user suggests a new description viz., *handwritten letter O*. This dialogue formally looks like $\langle x_1, \ldots, x_k \; \S+, -, \ldots, + \; \S C' \rangle$, where C' represents *handwritten letter O*.

(iii) The system then checks if such a description fits to the other objects of the cluster. So, it selects an arbitrary test set of positive instances of C, say $\{x_{p_{k1}}, \ldots, x_{p_{kj}}\}$, from the same cluster, and sends the sequence $\langle x_{p_{k1}}, \ldots, x_{p_{kj}} \; \S C'? \rangle$ to the user for verifying whether according to the user these are positive instances of C'.

(iv) The user either can agree with the same description for this new set of objects from the cluster, or can provide a more general description fitting both the sample and test set of objects. In the latter case, the system again starts a new verification round with the user based on a new test set of objects. In the former case, the user continues the verifications for a few number of times with different test sets of objects. If with a significantly high probability these test sets conform with the description provided by the user, the system accepts the description C' for the whole cluster.

(v) Once a description C' is settled between the system and the user as a relevant description for the whole cluster, the system needs to learn the definition of C' in terms of a set of lower level attributes.
The system thus sends a request for explaining the defining attributes for C' by sending a sequence of the form $\langle x_1, \ldots, x_k \; \S C' \; \square^{?} \rangle$.

(vi) The user replies by sending a set of lower level attributes $\langle a_1, \ldots, a_l \blacktriangleright C' \rangle$ conveying that C' can be defined using $a_1, \ldots, a_l$. For instance, in case of the particular example C' = *handwritten letter O*, the user can suggest a set of attributes such as a_1 = *closed curve*, a_2 = *circular*, a_3 = *oval*, a_4 = *curve with hollow inside* etc.

(vii) Given the recommended attributes, the system now degranulates the sample by considering the values assumed by the individual objects with respect to the proposed attributes. These values can be binary, such as simply + and −, or even a grade rendering the degree of matching the attribute.

(viii) Based on the number of instances of the sample falling into the positive instances of the concepts $a_1, \ldots, a_l$, the system needs to label the whole cluster with the respective properties. So, for each a_i, $1 \leq i \leq l$, the system sends a description of $x_1, \ldots, x_k$ in terms of their binary/graded values in the similar fashion as mentioned in item (i) of this section. The user then may suggest labels of the form 'most of the elements of the sample have the property a_1, 'few elements of the sample have the property a_2', 'many of the elements of the sample satisfy the property a_3' etc.

The question arises that how the system would abstract out properties for the individual objects of the cluster based on the degranulated sample and the descriptions provided by the user involving generalized quantifiers. Through dialogues with the user, the system can introduce some constraints on these properties $a_1, \ldots, a_l$ so that each object from the cluster can be described. This, in turn, helps the system to learn the user's perspectives for the generalized quantifiers, such as 'most', 'few', 'many' etc. For example, if for a property a_i, the user suggests that *most of the elements of the sample have the property* a_i, and the system computes $\frac{|\underline{\{x_1,\ldots,x_k\}}_{a_i} \cap \{x_1,\ldots,x_k\}|}{|\{x_1,\ldots,x_k\}|} \geq \frac{4}{5}$, then the system learns an interpretation of the user for the generalized quantifier 'most'. Additionally, the ratio also helps the system to fix a grade so that it can be claimed that each element of the cluster has property a_i to some specific degree. Moreover, if $\frac{|\underline{\{x_1,\ldots,x_k\}}_{a_j} \cap \{x_1,\ldots,x_k\}|}{|\{x_1,\ldots,x_k\}|} \leq \frac{1}{5}$ for some attribute a_j, the system gets an idea about the relevance of a_j in defining C'.

(a) Let us assume that the system considers an attribute to be relevant if the above mentioned ratio exceeds a prefixed degree, say $\frac{4}{5}$. So the degranulation with respect to a_i deems to be important, and the system looks back to the value of a_i for each element of the sample. Based on the outliers which belong to $\{x_1, \ldots, x_k\} \setminus \underline{\{x_1, \ldots, x_k\}}_{a_i}$, the system fixes a threshold t_i such that the modified property $a_i \geq t_i$ holds for all elements of the sample. As the number of such outliers is comparatively small this process of tuning a threshold t_i is not difficult. In the similar fashion the system can generate modifications of all those concepts for which the concerned ratio exceeds the prefixed grade.

(b) In case of the example of C' = *handwritten letter O* let us assume that a_1, a_2, a_3, a_4 are considered to be relevant following (a). Then the system generates the modified descriptions of the form $a_1 \geq t_1, \ldots, a_4 \geq t_4$ for the sample. With respect to these new attributes, C' now have a straightforward definition fitting to the whole sample.

(c) Through a dialogue $\langle x_1, \ldots, x_k \,\S a_1 \geq t_1, \ldots, a_4 \geq t_4 \rangle$ the system sends this modification to the user. The user may agree with the thresholds, or suggest to drop some positive instances of some of the a_i's. In the latter case, the system again tries to tune the respective t_i. Thus, through interactions with the user the system creates a description for a complicated cluster in the language understandable by the user.

Another relevant aspect of learning a concept is through assimilation of variations of the available objects suggested by the user. This can help the system to induce robust classifiers with respect to deviations of objects. Here we assume, that given a sample set of objects $\{x_1, \ldots, x_k\}$, the user also can suggest some new images to be counted as a positive instance of the concept *handwritten letter O* just by reorienting the objects from the sample itself. By reorientation we mean rotating the image of the object with respect to one of its coordinate in a specific angle, or swapping among some coloured pixels and white pixels within a specific radius of a given point. In general, by allowing different orientations of an object we address the case that if an object falls in a specific category, a similar sets of objects with little variations also can be counted of that category. Moreover, as these variations are imposed by the user, they are likely to reflect the possibilities of deviation of a standard case (such as a typed letter O), caused by real physical noises. In order to present such a recommendation of orientation function in the form of a dialogue, let us introduce the following possibility in the formal set-up.

Definition 1. *Given any object x from the universe of the system, and a parameter $\epsilon > 0$, $Or^{\epsilon}(x)$, called as orientation of x respective to ϵ, is an object falling into a similarity class of x parametrized by ϵ.*

In particular, for images such as handwritten scripts, the notion of $Or^{\epsilon}(x)$ can be interpreted as follows. Let x be an image from the universe of the system. A very small positive number ϵ can be regarded as the degree of angle or radius with respect to a coordinate of the image of x. Then $Or^{\epsilon}(x)$ can be either of the following cases.

- $\ulcorner_{\epsilon}(x)$: object obtained by *rotating x from the upper left corner point to ϵ degree*,
- $(x)_{\epsilon}\urcorner$: object obtained by *rotating x from the upper right corner point to ϵ degree*,
- $\llcorner_{\epsilon}(x)$: object obtained by *rotating x from the lower left corner point to ϵ degree*,
- $(x)_{\epsilon}\lrcorner$: object obtained by *rotating x from the lower right corner point to ϵ degree*,

Similarly one can think about $UL(\maltese, \epsilon)(x)$ to represent an object obtained by *swapping coloured and white pixels in a radius of ϵ with respect to the upper left coordinate of x*, and $UR(\maltese, \epsilon)(x)$ to represent an object obtained by *swapping coloured and white pixels in a radius of ϵ with respect to the upper right coordinate of x*. Similar meanings can be considered for $LL(\maltese, \epsilon)(x)$ and $LR(\maltese, \epsilon)(x)$ where the first letter L stands for *lower*.

We already have discussed the case when the system initiates a dialogue of the form $\langle x_1, \ldots, x_k \, \S +, -, \cdots + \, \S C\rangle$ with the user (cf. (i)). After introducing Or^{ϵ}, a possible reply of the user to the system can be $\langle Or^{\epsilon}(x_1), \ldots, Or^{\epsilon}(x_k) \, \S +, -, \cdots + \, \S C'\rangle$. That is, apart from suggesting a new concept C', the user also can suggest a possible set of similar objects of the sample as instances of the concept C' too. In that case, the system gathers new instances for the concerned cluster. These new instances can be taught to the system in the following way. With respect to a sequence of time points a little variation of the initial object at each turn, as a whole can be regarded as a video. So, an instance of $Or^{\epsilon}(x)$, say $\ulcorner_{\epsilon}(x)$ can be conveyed to the system by a video, which is nothing but a sequence of images consisting x at the initial position and a little variation of x at the consecutive instances. These variations of x are obtained by tilting it from it's leftmost upper corner coordinate to a degree $\beta_1 \leq \epsilon$, and continuing this process finitely many times till reaching the deviation of angle ϵ.

Below is another relevant aspect of learning to be noted. Let us consider the concept *handwritten digit 1.* In the same fashion as for the letter O, the user can teach the system possible variant images of the digit 1 by orienting the available images of 1. So, given a certain instance of 1 by the system, the user creates a sequence of finitely many variations $Or^{\beta_0}(1), Or^{\beta_1}(Or^{\beta_0}(1)), \ldots$, where $0 < \beta_0 < \beta_1 < \ldots < \beta_n \leq \epsilon$; variation is so that two consecutive instances in the sequence are very close to each other. Let us simply denote $1, 1^{\beta_o}, 1^{\beta_1}, \ldots, 1^{\beta_{n-1}}, 1^{\beta_n}$ to be the sequence of variations of one instance of the handwritten digit 1. The variations are such that though $1^{\beta_{n-1}}$ and 1^{β_n} are very close in appearance, the former can be counted as an instance of 1 and the latter is more likely to be counted as an instance of 7. So, the concern is not only to show the possible variations of image 1, but also possible variations of image that are not regarded as 1 or borderline instances of 1. So, we can consider that $1^{\beta_{n-1}}$ and 1^{β_n} belong to the upper approximation of *handwritten digit 1*, and they are the borderline instances of *handwritten digit 1.* In such a situation, the user can make the system learn that with respect to what parameters or attributes these two instances fall into two different categories. Such an information can be conveyed to the system by a dialogue of the form $\langle [1, 1^{\beta_0}, \ldots, 1^{\beta_{n-1}}, 1^{\beta_n}] \; \S \; 1^{\beta_n} \blacktriangleright$ *handwritten digit 7* $\S a_1, \ldots, a_m \rangle$ indicating 1^{β_n} is similar to the handwritten digit 7, and it is distinguished from $1^{\beta_{n-1}}$ with respect to the attributes $a_1, \ldots, a_m$. All these attributes can be stored in the system as an explanation; collecting such explanations for all possible cases provided by the user the system can form arguments for and against to classify a new instance as 1 or 7.

Now, there can be a community of users, like a popular social media, and the system can have separate dialogues with a number of users for understanding and learning the same concept. For example, with respect to different users the system can gather different sets of new images obtained by orienting a given sample set of the digit 1, along with the properties of those classifications. In appearance of a new case, the system can use one or some of these different classifiers to check the belongingness of the case to the concept. This also helps

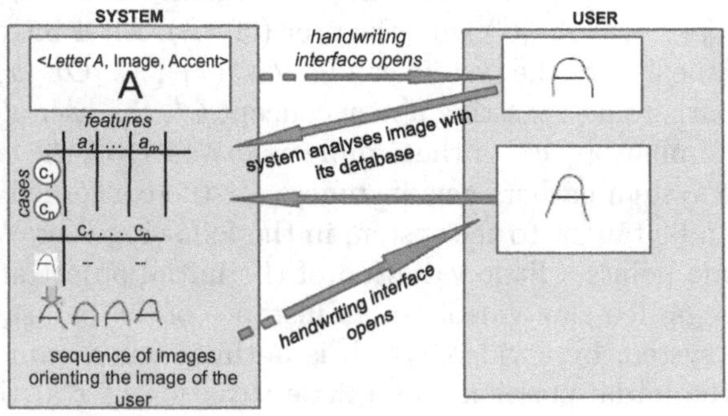

Fig. 4. Language-learning interface

the system to compare different classifiers and resolve the issue of thinning the boundary region by initiating a dialogue with a group of agents.

7 Conclusion: Towards a Practical Application

The idea presented in the above sections can have diverse applications. As discussed in Sect. 1, and from the patterns of examples considered in previous sections, we can expect that this approach can be of use in designing an automated support system with the ability of learning and revising its database based on interactions with a set of agents sitting in the real physical world. One such possible example is given below. The example has two aspects; in one the system learns from the real environment and in the other it provides a support to an apprentice to learn.

Let us think of designing a system which can help a kid to learn basics of a language. To such a system the expectation is to know the letters of the alphabet, a basic list of words and the basic formation rules of sentences. To develop such a system it is fed with a possible set of finitely many images for each letter. These images can be of standard typed form of a letter or various handwritten forms. For each cluster of images there are some properties (i.e., attributes) that every element of the cluster satisfy.

We propose such a system based on dialogues with users. Following the proposed notion of Or^{ϵ} described in Sect. 6, for each letter a sequence of images can be created by a user just by considering different orientations of that very letter. The set of images given by a user is identified as instances of a particular letter with respect to a set of attributes (cf. Sect. 6). The set of images may vary from one user to the other, and so the respective set of attributes too. The system, thus, is enriched with different classifiers for a single letter. Among these different attributes and clusters of images, the system can select a set of attributes and a sample set of images for a letter by stratified sampling.

We assume that the system is embodied with a language-learning interface. When a kid enters to a language-learning interface, the system first provides the sample set of images for each letter through a dialogue. An example can be a sequence $\langle x_1, \ldots, x_k \blacktriangleright \textit{letter A} \rangle$ where $x_1, \ldots, x_k$ are the images for the concept *letter A*. These images have certain attributes in common. One of the attributes of any letter is the *accent* of the letter. After familiarizing the kid with possible images of a letter, the system opens a handwriting-interface, where the kid writes the letter she learnt. This image is analyzed by the system with respect to the available sample and attributes. If it looks closer to a satisfactory degree with the available images, the system includes this image as a new instance of the letter. From this aspect, the system also learns new cases. If the handwritten script does not match to a satisfactory extent, the system takes an attempt to improve the kid's attempt by providing a sequence of orientations, as discussed in Sect. 6, of the script obtained from the kid, and ending at the desired one. Now the writing-interface again gets open to the kid and the kid writes in the interface what she learnt. Following the same process, a few number of time, the kid can learn the proper script for a particular letter (see Fig. 4).

In the similar fashion, at the next level corresponding to each letter there can be a sequence of words starting with that letter. In the simplest case we can consider the words from the names of objects, e.g. fruits, birds, animals, nature, daily-necessary utelsils, colours, shapes etc. Each word is associated to an image, and this association is fed to the system by the designer of the system. At the beginning the system has an object, a name of a real physical object, and as an attribute is has an image. Through dialogues with a group of users, the system at the next round learns a few sets of attributes describing the object. Let, for instance, for the letter A, the system have the word *Apple*. So, in the database of the system there would be a row pertaining to the object *Apple*, and a set of attributes containing an image of apple, as well as some descriptions such as, *fruit*, *colour*, *shape*, *taste*, *living*, *non-living*, *accent* etc. These set of attributes may be fixed by the system by considering stratified sampling among different sets of attributes provided by different users. The sets of attributes for a word must also include the *accent.* Similarly, the system can have a database for a list of basic useful verbs. A possible image to visualize a verb can be an image of an action. Thus, in the database of the system each word can be classified by the concepts, e.g., *subject*, *verb*, *predicate.* Now in the language-learning interface the system sends a sequence consisting of the word *Apple*, followed by the properties *image*, *accent*, *colour*, *shape*, *fruit.* Clicking on each single attribute the kid can enter into a new sub-dialogue box with the image, accent, and spelling of the word. For instance, a click on *colour*, would generate a sequence starting with the word *red*, followed by its *image* through a patch of red, and its *accent.* The writing interface opens automatically after familiarizing the kid with the word *Apple* and its basic properties. In the writing interface the words mis-spelled or differently written by the kid is analyzed by the system, and through dialogues it attempts to improve the writing of the kid, as explained before.

At the next level, the system is taught a few basic typical rules for forming simple sentences. In the form of a dialogue, it can be $\langle(subject), (verb), (predicate)\rangle$. At the lower level, the system already learned words falling into the categories *subject*, *verb* and *predicate.* Here, to be mentioned that teaching the system the syntax of a correct simple sentence is not difficult through interaction; but making the system learn the semantics, and thus forming a meaningful sentence is a challenging task. Without going into the deeper technical aspects, we assume that for each type of sentences, obtained by replacing *subject*, *verb* and *predicate* by respective particular instances, the designer provides a number of sentences with a respective set of images and accent for each. The system then generates new sentences just by following the syntax of the rule. Whether the sentence generated by the system is meaningful that can be verified again by initiating dialogues with the group of users. But this issue would dig into some deeper aspects pertaining to the semantics of a sentence. For instance, let us suppose that the system generates a sentence *Computer eats apple.* It is a syntactically correct sentence. But the semantics of the sentence would depend on how an agent interpret its components. If the user consider the semantics of the sentence with respect to a real physical action, then this sentence does not have meaning in a real physical world. But in a world where a pegasus can exist, *Computer eats apple* may also have meaning. Without going into such debates, we assume that the group of users are agreed to designate a sentence to be meaningful if it corresponds to a real physical action in a real physical world. So, when the system generates a sentence *Computer eats apple*, it is sent through a dialogue to the users. The user sends a rejection notification, and suggests possible replacements for the components of the sentence generated by the system. So, the user sends $\langle (Computer)\ (eats)\ (apple)\ \S\boxtimes \rangle$ notifying that the sentence generated by the system is not accepted. The suggestion for the correct sentence can be sent by $\langle (\ )\ (eats)\ (apple)\ \S\ living \blacktriangleright subject \rangle$ indicating the category *subject* to be replaced by instances satisfying the attribute *living*. This would help the system to learn new sentences which have meaning in the real physical world. The semantics of the sentences can be manipulated by going to the semantics of its individual component already stored in the system. The system can then store this new sentence with its semantics. The learning phase of the kid can follow the similar pattern as above.

Acknowledgments. The research of Andrzej Skowron was partially supported by the NCBiR grant POIR.01. 02.00-00-0184/17-01.

References

1. MacKenzie, I.S.: Human-Computer Interaction: An Empirical Research Perspective. Morgan Kaufmann, Burlington (2013)
2. Loritz, D.: How the Brain Evolved Language. Oxfrod University Press, Oxford (1999)
3. Feldman, J.A.: From Molecule to Metaphor. A Neural Theory of Language. The MIT Press, Cambridge (2006)
4. Ellis, N.C., Larsen-Freeman, D. (eds.): Language as a Complex Adaptive System. Language Learning Research Club, University of Michigan (2009)
5. Dutta, S., Wasilewski, P.: Dialogue in hierarchical learning of concept using prototypes and counterexamples. Fundamenta Informaticae **162**(1), 17–36 (2018). https://doi.org/10.3233/FI-2018-1711
6. Dutta, S., Skowron, A.: Bipolar queries with dialogue: rough set semantics. In: Nguyen, H.S., Ha, Q.-T., Li, T., Przybyła-Kasperek, M. (eds.) IJCRS 2018. LNCS (LNAI), vol. 11103, pp. 229–242. Springer, Cham (2018). https://doi.org/10.1007/978-3-319-99368-3_18
7. Holzinger, A., Biemann, C., Pattichis, C.S., Kell, D.B.: What do we need to build explainable AI systems for the medical domain? pp. 1–28 (2017). https://arxiv.org/abs/1712.09923v1
8. Nguyen, S.H., Bazan, J., Skowron, A., Nguyen, H.S.: Layered learning for concept synthesis. In: Peters, J.F., Skowron, A., Grzymała-Busse, J.W., Kostek, B., Świniarski, R.W., Szczuka, M.S. (eds.) Transactions on Rough Sets I. LNCS, vol. 3100, pp. 187–208. Springer, Heidelberg (2004). https://doi.org/10.1007/978-3-540-27794-1_9
9. Bazan, J.G.: Hierarchical classifiers for complex spatio-temporal concepts. In: Peters, J.F., Skowron, A., Rybiński, H. (eds.) Transactions on Rough Sets IX. LNCS, vol. 5390, pp. 474–750. Springer, Heidelberg (2008). https://doi.org/10.1007/978-3-540-89876-4_26

A Dynamic Dominance-Based Rough Set Approach for Processing Ordered Data

Shaoyong Li and Zhiyong Hong(✉)

Faculty of Intelligent Manufacturing, Wuyi University, Jiangmen 529020, China
meterer@163.com, hongmr@163.com

Abstract. In real-world many information systems are varying over time. How to process efficiently dynamic data is a hot issue in data mining research field. Processing information with preference-ordered attribute domain is an important task in multi-criteria decision making. Since dominance-based rough set approach can process information with preference-ordered attribute domain, it has been applied widely to multi-criteria decision making. In the paper, we introduce a new strategy of updating approximations in DRSA under the cases when a lot of objects being added. For thus, the notation of base boundaries of approximations in DRSA is given to reduce the redundant computation in updating approximations. The feasibility of the approach was validated by a numerical example.

Keywords: Rough set · Dominance relation · Dynamic data mining

1 Introduction

Rough Set Theory (RST) was introduced by Pawlak in 1982 for processing uncertain, inconsistence and incomplete information [1,2]. As a popular tool in data mining research field, it has been widely applied to faults diagnose, image processing, knowledge discovery, intelligent control systems, etc. [3].

To process information with preference-ordered attribute domains, Greco et al. have proposed Dominance-based Rough Set Approach (DRSA) by replacing the indiscernibility relation in RST with a dominance relation [4]. DRSA inherits many advantages of RST in information processing. The great advantage of DRSA is that it can process information with preference-ordered attribute domains. Then DRSA has applied widely in many real applications related to MCDA problems, *e.g.*, rural sustainable development potentialities evaluation [5], airline services evaluation [6–8], group decision [9], multi-criteria web mining [10] and business indicator analysis [11].

Supported by Science and Technology Planning Project on Basic Theory of Jiangmen City (Nos. 2017JC01021, 2018JC01003), Natural Science Foundation of Guangdong Province (No. 2016A030310003), National Science Foundation of China (No. 61603313) and the Fundamental Research Funds for the Central Universities (No. 2682017CX097), China.

© Springer Nature Switzerland AG 2019
T. Mihálydeák et al. (Eds.): IJCRS 2019, LNAI 11499, pp. 312–320, 2019.
https://doi.org/10.1007/978-3-030-22815-6_24

In many real applications, the information systems often evolve over time. New information becomes available while outdated information being sifted out. It raises a problem of how to maintain knowledge in a dynamic information system. In order to solve the problem, many scholars employed the increment update techniques to maintain knowledge, *e.g.*, Blaszczynski et al. discussed the incremental induction of decision rules from dominance-based rough approximations to select the most interesting representatives in the final set of rules [12]. Greco et al. considered incremental induction of decision rules in the context of multiple criteria decision analysis [13]. Li et al. analyzed the variations happened in each step of approximations computation of DRSA when the information system varying and then explored the mechanism of updating approximations of DRSA. They proposed three incremental approaches for updating approximations of DRSA under the object set varying, the attribute set varying and the attributes' values varying, respectively [14–16]. Luo et al. proposed three approaches for updating approximation of set-valued dominance-based rough set approach for three different cases in dynamic data environment, respectively [17–19]. Chen et al. proposed an incremental approach for update approximations of DRSA while attribute values refining or coarsening in incomplete information system [20]. Wang et al. propose an incremental algorithm which can efficiently update approximations of DRSA when objects and attributes increase simultaneously [21]. Luo et al. attempted to apply Dominance-based Rough Set Approach in processing hierarchical attribute value and then provided an approach and corresponding algorithm for updating rough approximations [22].

Following the literature [14], we only investigate the case when some new objects becoming available. In the paper, we introduce a new strategy to maintain the approximations of DRSA. We give a notation of base boundaries of approximations. By the notation, we can reduce the computation in approximations updating when some new objects being added into the information system. The main idea is that new added objects are regarded as a new information system of the same type, then compute approximations of the new information system. By comparison base boundaries of approximations of the new information system and the original one, we reduce redundant computations in approximations updating. A numerical example was employed to illustrate the feasibility of our strategy.

The rest of the paper is organized as follows. In Sect. 2, we review some basic notions of DRSA. Section 3 introduces a new strategy of updating approximations of DRSA. A numerical example is employed to illustrate the feasibility of our strategy in Sect. 4. The paper ends with conclusions and further research work in Sect. 5.

2 Preliminaries

We briefly review some basic notations and concepts of DRSA [4] in the following.

Definition 1. *Let $S = (U, A, V, f)$ be an information system, where*

- *U is a non-empty finite set of objects, which is called the universe;*
- *$A = C \cup \{d\}$, C is a non-empty finite set of condition attributes, and d is the decision attribute;*
- *$V = \bigcup_{a \in A} V_a$ is regarded as the domain of all attributes, where V_a is a domain of attribute a;*
- *$f : U \times A \to V$ is an information function such that $\forall x \in U$, $\exists f(x, a) \in V_a$.*

Definition 2. *Let $P \subseteq C$, then the dominance relation with respect to P on universe U presents as $D_P = \{(x, y) \in U \times U \mid f(x, a) \geq f(y, a), \forall a \in P\}$.*

For $x, y \in U$, x dominates y with respect to the attribute set P, denoted by xD_Py, which means x at least as good as y with respect to each a, $a \in P$.

In DRSA, a dominance relation can partition the universe U into two families of information granules that respectively are two kinds of sets as follows:

- $D_P^+(x) = \{y \in U \mid yD_Px\}$ is the P-dominating set of the object x, which presents the collection of objects that are at least as good as the object x with respect to P;
- $D_P^-(x) = \{y \in U \mid xD_Py\}$ is the P-dominated set of the object x, which presents the collection of objects that are at most as worse as the object x with respect to P.

The decision attribute d can classify the universe U into a family of decision classes, denoted by $\boldsymbol{Cl} = \{Cl_n, n \in T\}$, $T = \{1, \cdots, |V_d|\}$. The concepts characterized are upward and downward unions of decision classes in DRSA. They can be defined respectively as follows:

$$Cl_n^{\geq} = \bigcup_{n' \geq n} Cl_{n'}, \quad Cl_n^{\leq} = \bigcup_{n' \leq n} Cl_{n'}, \quad \forall n, n' \in T.$$

$Cl_n^{\geq}$ is the upward union of decision class Cl_n which means that if $x \in Cl_n^{\geq}$ then x belongs to at least class Cl_n; $Cl_n^{\leq}$ is the downward union of decision class Cl_n which means that if $x \in Cl_n^{\leq}$ then x belongs to at most class Cl_n.

The lower and upper approximations of $Cl_n^{\geq}$ are defined respectively as:

$$\underline{P}(Cl_n^{\geq}) = \{x \in U \mid D_P^+(x) \subseteq Cl_n^{\geq}\}$$
$$\overline{P}(Cl_n^{\geq}) = \{x \in U \mid D_P^-(x) \cap Cl_n^{\geq} \neq \emptyset\}$$

Analogously, The lower and upper approximations of $Cl_n^{\leq}$ are defined respectively as:

$$\underline{P}(Cl_n^{\leq}) = \{x \in U \mid D_P^-(x) \subseteq Cl_n^{\leq}\}$$
$$\overline{P}(Cl_n^{\leq}) = \{x \in U \mid D_P^+(x) \cap Cl_n^{\leq} \neq \emptyset\}$$

For convenience to understand the rest of this paper, we introduce the following definitions.

Definition 3. *The base boundaries of* $\underline{P}(Cl_n^{\geq})$, $\overline{P}(Cl_n^{\geq})$, $\underline{P}(Cl_n^{\leq})$ *and* $\overline{P}(Cl_n^{\leq})$ *are respectively defined as*

- $B_P(\underline{P}(Cl_n^{\geq})) = \langle B_{a_1}(\underline{P}(Cl_n^{\geq})), \ldots, B_{a_i}(\underline{P}(Cl_n^{\geq})), \ldots, B_{a_m}(\underline{P}(Cl_n^{\geq}))\rangle$;
- $B_P(\overline{P}(Cl_n^{\geq})) = \langle B_{a_1}(\overline{P}(Cl_n^{\geq})), \ldots, B_{a_i}(\overline{P}(Cl_n^{\geq})), \ldots, B_{a_m}(\overline{P}(Cl_n^{\geq}))\rangle$;
- $B_P(\underline{P}(Cl_n^{\leq})) = \langle B_{a_1}(\underline{P}(Cl_n^{\leq})), \ldots, B_{a_i}(\underline{P}(Cl_n^{\leq})), \ldots, B_{a_m}(\underline{P}(Cl_n^{\leq}))\rangle$;
- $B_P(\overline{P}(Cl_n^{\leq})) = \langle B_{a_1}(\overline{P}(Cl_n^{\leq})), \ldots, B_{a_i}(\overline{P}(Cl_n^{\leq})), \ldots, B_{a_m}(\overline{P}(Cl_n^{\leq}))\rangle$.

In where

- $B_{a_i}(\underline{P}(Cl_n^{\geq})) = min\{f(x, a_i)|x \in \underline{P}(Cl_n^{\geq})\}$;
- $B_{a_i}(\overline{P}(Cl_n^{\geq})) = min\{f(x, a_i)|x \in \overline{P}(Cl_n^{\geq})\}$;
- $B_{a_i}(\underline{P}(Cl_n^{\leq})) = max\{f(x, a_i)|x \in \underline{P}(Cl_n^{\leq})\}$;
- $B_{a_i}(\overline{P}(Cl_n^{\leq})) = max\{f(x, a_i)|x \in \overline{P}(Cl_n^{\leq})\}$.

Here, $P = \{a_1, \ldots, a_i, \ldots, a_m\}$.

Definition 4. $\forall i, j \in T$ *and* $\forall a \in P$, *the following items hold.*

(1) *If* $B_a(\underline{P}(Cl_i^{\geq})) \geq B_a(\underline{P}(Cl_j^{\geq}))$, *then* $B_P(\underline{P}(Cl_i^{\geq})) \geq B_P(\underline{P}(Cl_j^{\geq}))$;
(2) *If* $B_a(\underline{P}(Cl_i^{\geq})) \leq B_a(\underline{P}(Cl_j^{\geq}))$, *then* $B_P(\underline{P}(Cl_i^{\geq})) \leq B_P(\underline{P}(Cl_j^{\geq}))$;
(3) *If* $B_a(\overline{P}(Cl_i^{\geq})) \geq B_a(\overline{P}(Cl_j^{\geq}))$, *then* $B_P(\overline{P}(Cl_i^{\geq})) \geq B_P(\overline{P}(Cl_j^{\geq}))$;
(4) *If* $B_a(\overline{P}(Cl_i^{\geq})) \leq B_a(\overline{P}(Cl_j^{\geq}))$, *then* $B_P(\overline{P}(Cl_i^{\geq})) \leq B_P(\overline{P}(Cl_j^{\geq}))$;
(5) *If* $B_a(\underline{P}(Cl_i^{\leq})) \geq B_a(\underline{P}(Cl_j^{\leq}))$, *then* $B_P(\underline{P}(Cl_i^{\leq})) \geq B_P(\underline{P}(Cl_j^{\leq}))$;
(6) *If* $B_a(\underline{P}(Cl_i^{\leq})) \leq B_a(\underline{P}(Cl_j^{\leq}))$, *then* $B_P(\underline{P}(Cl_i^{\leq})) \leq B_P(\underline{P}(Cl_j^{\leq}))$;
(7) *If* $B_a(\overline{P}(Cl_i^{\leq})) \geq B_a(\overline{P}(Cl_j^{\leq}))$, *then* $B_P(\overline{P}(Cl_i^{\leq})) \geq B_P(\overline{P}(Cl_j^{\leq}))$;
(8) *If* $B_a(\overline{P}(Cl_i^{\leq})) \leq B_a(\overline{P}(Cl_j^{\leq}))$, *then* $B_P(\overline{P}(Cl_i^{\leq})) \geq B_P(\overline{P}(Cl_j^{\leq}))$.

3 An Strategy for Processing Dynamic Order Data

This strategy focuses on the case when some new objects becoming available. Assume that a dynamic process lasts from time t to $t+1$. Let $\bullet^{(t)}$ and $\bullet^{(t+1)}$ denote the corresponding notations at time t and $t+1$, respectively.

The collection of objects added from time t to $t+1$ may be regarded as the increment of the universe $U^{(t)}$ and denoted by ΔU. ΔCl_n is a decision class with respect to d on the universe ΔU. Accordingly, $\Delta Cl_n^{\geq}$ and $\Delta Cl_n^{\leq}$ are upward and downward unions of the decision class ΔCl_n.

For the base boundaries of approximations on $U^{(t)}$ and ΔU, we present the following proposition.

Proposition 1. *The following items hold.*

1. *If* $B_P(\underline{P}({Cl^{(t)}}_n^{\geq})) \preceq B_P(\underline{P}(\Delta Cl_n^{\geq}))$, *and there is at least an attribute a satisfying* $B_a(\underline{P}({Cl^{(t)}}_n^{\geq})) \geq B_a(\overline{P}(\Delta Cl_{n-1}^{\leq}))$, *then* $\underline{P}({Cl^{(t+1)}}_n^{\geq}) = \underline{P}({Cl^{(t)}}_n^{\geq}) \cup \underline{P}(\Delta Cl_n^{\geq})$;
2. *If* $B_P(\overline{P}({Cl^{(t)}}_n^{\geq})) \preceq B_P(\overline{P}(\Delta Cl_n^{\geq}))$, *then* $\overline{P}({Cl^{(t+1)}}_n^{\geq}) = \overline{P}({Cl^{(t)}}_n^{\geq}) \cup \overline{P}(\Delta Cl_n^{\geq})$;

3. *If* $B_P(\underline{P}(Cl^{(t)}{}^{\leq}_{n})) \succeq B_P(\underline{P}(\Delta Cl^{\leq}_{n}))$, *and there is at least an attribute a satisfying* $B_a(\underline{P}(Cl^{(t)}{}^{\leq}_{n})) \leq B_P(\overline{P}(\Delta Cl^{\geq}_{n+1}))$, *then* $\underline{P}(Cl^{(t+1)}{}^{\leq}_{n}) = \underline{P}(Cl^{(t)}{}^{\leq}_{n}) \cup \underline{P}(\Delta Cl^{\leq}_{n})$;
4. *If* $B_P(\overline{P}(Cl^{(t)}{}^{\leq}_{n})) \succeq B_P(\overline{P}(\Delta Cl^{\leq}_{n}))$, *then* $\overline{P}(Cl^{(t+1)}{}^{\leq}_{n}) = \overline{P}(Cl^{(t)}{}^{\leq}_{n}) \cup \overline{P}(\Delta Cl^{\leq}_{n})$.

For the other cases, we investigate the changes of the dominance relation at first. Let $\Delta_1 D_P$ be the dominance relation from ΔU to $U^{(t)}$ and $\Delta_2 D_P$ be the dominance relation from $U^{(t)}$ to ΔU.

$$\Delta_1 D_P = \{(x,y) \in \Delta U \times U^{(t)} \mid f(x,a) \geq f(y,a), a \in P\} \tag{1}$$
$$\Delta_2 D_P = \{(x,y) \in U^{(t)} \times \Delta U \mid f(x,a) \geq f(y,a), a \in P\} \tag{2}$$

Based on Eqs. (1) and (2), we define the increments of P-dominating and P-dominated sets of $x \in U^{(t)}$ related to ΔU as follows:

$$\Delta_1 D^{+}{}_P(x) = \{y \in \Delta U \mid y \Delta_1 D_P x\} \tag{3}$$
$$\Delta_1 D^{-}{}_P(x) = \{y \in \Delta U \mid x \Delta_1 D_P y\} \tag{4}$$

Analogously, for $x \in \Delta U$, the increments of P-dominating and P-dominated sets related to $U^{(t)}$ are respectively defined as follows:

$$\Delta_2 D^{+}{}_P(x) = \{y \in U^{(t)} \mid y \Delta_2 D_P x\} \tag{5}$$
$$\Delta_2 D^{-}{}_P(x) = \{y \in U^{(t)} \mid x \Delta_2 D_P y\} \tag{6}$$

Proposition 2. *The following items hold.*

$$\underline{P}(Cl^{(t+1)}{}^{\geq}_{n}) = (\underline{P}(Cl^{(t)}{}^{\geq}_{n}) \cap \Delta_1 \underline{P}(Cl^{\geq}_{n})) \cup (\Delta_2 \underline{P}(Cl^{\geq}_{n}) \cap \underline{P}(\Delta Cl^{\geq}_{n})) \tag{7}$$
$$\overline{P}(Cl^{(t+1)}{}^{\geq}_{n}) = \overline{P}(Cl^{(t)}{}^{\geq}_{n}) \cup \Delta_1 \overline{P}(Cl^{\geq}_{n}) \cup \Delta_2 \overline{P}(Cl^{\geq}_{n}) \cup \overline{P}(\Delta Cl^{\geq}_{n}) \tag{8}$$

In where

$$\Delta_1 \underline{P}(Cl^{\geq}_{n}) = \{x \in U^{(t)} \mid \Delta_1 D^{+}{}_P(x) \subseteq \Delta Cl^{\geq}_{n}\} \tag{9}$$
$$\Delta_2 \underline{P}(Cl^{\geq}_{n}) = \{x \in \Delta U \mid \Delta_2 D^{+}{}_P(x) \subseteq Cl^{(t)}{}^{\geq}_{n}\} \tag{10}$$
$$\Delta_1 \overline{P}(Cl^{\geq}_{n}) = \{x \in U^{(t)} \mid \Delta_1 D^{-}{}_P(x) \cap \Delta Cl^{\geq}_{n} \neq \emptyset\} \tag{11}$$
$$\Delta_2 \overline{P}(Cl^{\geq}_{n}) = \{x \in \Delta U \mid \Delta_2 D^{-}{}_P(x) \cap Cl^{(t)}{}^{\geq}_{n} \neq \emptyset\} \tag{12}$$
$$\Delta_1 \underline{P}(Cl^{\leq}_{n}) = \{x \in U^{(t)} \mid \Delta_1 D^{-}{}_P(x) \subseteq \Delta Cl^{\leq}_{n}\} \tag{13}$$
$$\Delta_2 \underline{P}(Cl^{\leq}_{n}) = \{x \in \Delta U \mid \Delta_2 D^{-}{}_P(x) \subseteq Cl^{(t)}{}^{\leq}_{n}\} \tag{14}$$
$$\Delta_1 \overline{P}(Cl^{\leq}_{n}) = \{x \in U^{(t)} \mid \Delta_1 D^{+}{}_P(x) \cap \Delta Cl^{\leq}_{n} \neq \emptyset\} \tag{15}$$
$$\Delta_2 \overline{P}(Cl^{\leq}_{n}) = \{x \in \Delta U \mid \Delta_2 D^{+}{}_P(x) \cap Cl^{(t)}{}^{\leq}_{n} \neq \emptyset\} \tag{16}$$

4 A Numeric Illustration

Example 1. Table 1 presents a information system at time t, and Table 2 presents the collection of new objects added from time t to $t+1$. $C = \{a_1, a_2, a_3\}$ in two tables. Let $P = C$.

Firstly, we list the results related to Table 1 as follows:

- P-dominating and P-dominated sets
  $D_P^-(x_1)^{(t)} = \{x_1, x_2\}$, $D_P^+(x_1)^{(t)} = \{x_1\}$;
  $D_P^-(x_2)^{(t)} = \{x_2\}$, $D_P^+(x_2)^{(t)} = \{x_1, x_2\}$;
  $D_P^-(x_3)^{(t)} = \{x_3\}$, $D_P^+(x_3)^{(t)} = \{x_3\}$;
  $D_P^-(x_4)^{(t)} = \{x_4\}$, $D_P^+(x_4)^{(t)} = \{x_4\}$.
- Upward and downward unions of decision classes
  $Cl^{(t)}{}_1^{\geq} = U^{(t)}, Cl^{(t)}{}_2^{\geq} = \{x_2, x_3\}; Cl^{(t)}{}_1^{\leq} = \{x_1, x_4\}, Cl^{(t)}{}_2^{\leq} = U^{(t)}$.
- Approximations
  $\underline{P}(Cl^{(t)}{}_2^{\geq}) = \{x_3\}, \overline{P}(Cl^{(t)}{}_2^{\geq}) = \{x_1, x_2, x_3\}$
  $\underline{P}(Cl^{(t)}{}_1^{\leq}) = \{x_4\}, \overline{P}(Cl^{(t)}{}_1^{\leq}) = \{x_1, x_2\}$,
  $\underline{P}(Cl^{(t)}{}_2^{\leq}) = \overline{P}(Cl^{(t)}{}_2^{\leq}) = \underline{P}(Cl^{(t)}{}_1^{\geq}) = \overline{P}(Cl^{(t)}{}_1^{\geq}) = U^{(t)}$.

Table 1. An information table at time t.

$U^{(t)}$	a_1	a_2	a_3	d
x_1	2	1	3	1
x_2	2	1	2	2
x_3	3	1	1	2
x_4	2	3	1	1

Table 2. A table of new objects added.

U^+	a_1	a_2	a_3	d
x_5	1	2	3	1
x_6	2	2	1	2
x_7	3	1	2	3

Secondly, we compute the results related to Table 2 as follows:

- P-dominating and P-dominated sets
  $\Delta D_P^-(x_5) = \{x_5\}$, $\Delta D_P^+(x_5) = \{x_5\}$; $\Delta D_P^-(x_6) = \{x_6\}$, $\Delta D_P^+(x_6) = \{x_6\}$;
  $\Delta D_P^-(x_7) = \{x_7\}$, $\Delta D_P^+(x_7) = \{x_7\}$.

- Upward and downward unions of decision classes
  $\Delta Cl_1^{\geq} = \Delta U, \Delta Cl_2^{\geq} = \{x_6, x_7\}, \Delta Cl_3^{\geq} = \{x_7\}; \Delta Cl_1^{\leq} = \{x_5\}, \Delta Cl_2^{\leq} = \{x_5, x_6\}, \Delta Cl_3^{\leq} = \Delta U.$
- Approximations
  $\underline{P}(\Delta Cl_2^{\geq}) = \{x_6, x_7\}, \underline{P}(\Delta Cl_3^{\geq}) = \{x_7\};$
  $\overline{P}(\Delta Cl_2^{\geq}) = \{x_6, x_7\}, \overline{P}(\Delta Cl_3^{\geq}) = \{x_7\};$
  $\underline{P}(\Delta Cl_1^{\leq}) = \{x_5\}, \underline{P}(\Delta Cl_2^{\leq}) = \{x_5, x_6\};$
  $\overline{P}(\Delta Cl_1^{\leq}) = \{x_5\}, \overline{P}(\Delta Cl_2^{\leq}) = \{x_5, x_6\};$
  $\underline{P}(\Delta Cl_3^{\leq}) = \overline{P}(\Delta Cl_3^{\leq}) = \underline{P}(\Delta Cl_1^{\geq}) = \overline{P}(\Delta Cl_1^{\geq}) = \Delta U.$

Thirdly, we compute the base boundaries of approximations and list the results as follows:
$B_P(\underline{P}(Cl^{(t)}{}_2^{\geq})) = \{3,1,1\}$, $B_P(\overline{P}(Cl^{(t)}{}_2^{\geq})) = \{2,1,1\}$;
$B_P(\underline{P}(Cl^{(t)}{}_1^{\leq})) = \{2,3,1\}$, $B_P(\overline{P}(Cl^{(t)}{}_1^{\leq})) = \{2,1,3\}$;
$B_P(\underline{P}(Cl^{(t)}{}_2^{\leq})) = B_P(\overline{P}(Cl^{(t)}{}_2^{\leq})) = \{3,3,3\}$;
$B_P(\underline{P}(Cl^{(t)}{}_1^{\geq})) = B_P(\overline{P}(Cl^{(t)}{}_1^{\geq})) = \{2,1,1\}$;
$B_P(\underline{P}(\Delta Cl_2^{\geq})) = \{2,1,1\}$, $B_P(\underline{P}(\Delta Cl_3^{\geq})) = \{3,1,2\}$,
$B_P(\overline{P}(\Delta Cl_2^{\geq})) = \{2,1,1\}$, $B_P(\overline{P}(\Delta Cl_3^{\geq})) = \{3,1,2\}$,
$B_P(\underline{P}(\Delta Cl_1^{\leq})) = \{1,2,3\}$, $B_P(\underline{P}(\Delta Cl_2^{\leq})) = \{2,2,3\}$,
$B_P(\overline{P}(\Delta Cl_1^{\leq})) = \{1,2,3\}$, $B_P(\overline{P}(\Delta Cl_2^{\leq})) = \{2,2,3\}$,
$B_P(\underline{P}(\Delta Cl_3^{\leq})) = B_P(\overline{P}(\Delta Cl_3^{\leq})) = \{3,2,3\}$;
$B_P(\underline{P}(\Delta Cl_1^{\geq})) = B_P(\overline{P}(\Delta Cl_1^{\geq})) = \{1,1,1\}$.
By Proposition 1, we can obtain some new approximations as follows:
$\overline{P}(Cl^{(t+1)}{}_2^{\geq}) = \{x_1, x_2, x_3, x_6, x_7\}$; $\overline{P}(Cl^{(t+1)}{}_2^{\leq}) = \{x_1, x_2, x_3, x_4, x_5, x_6\}$.
Fourthly, we update other approximations by Propositions 2 as follows:

- Increments of P-dominating and P-dominated sets
  $\Delta_1 D_P^-(x_1) = \emptyset$, $\Delta_1 D_P^+(x_1) = \emptyset$; $\Delta_1 D_P^-(x_2) = \emptyset$, $\Delta_1 D_P^+(x_2) = \{x_7\}$;
  $\Delta_1 D_P^-(x_3) = \emptyset$, $\Delta_1 D_P^+(x_3) = \{x_7\}$; $\Delta_1 D_P^-(x_4) = \{x_6\}$, $\Delta_1 D_P^+(x_4) = \emptyset$;
  $\Delta_2 D_P^-(x_5) = \emptyset$, $\Delta_2 D_P^+(x_5) = \emptyset$; $\Delta_2 D_P^-(x_6) = \emptyset$, $\Delta_2 D_P^+(x_6) = \{x_4\}$;
  $\Delta_2 D_P^-(x_7) = \{x_2, x_3\}$, $\Delta_2 D_P^+(x_7) = \emptyset$.
- Increments of approximations
  $\Delta_1\underline{P}(Cl_2^{\geq}) = U^{(t)}, \Delta_1\underline{P}(Cl_3^{\geq}) = \{x_1, x_2, x_3\}; \Delta_1\overline{P}(Cl_3^{\geq}) = \{x_2, x_3\};$
  $\Delta_1\underline{P}(Cl_1^{\leq}) = \{x_1, x_2, x_3\}, \Delta_1\underline{P}(Cl_2^{\leq}) = U^{(t)}; \Delta_1\overline{P}(Cl_1^{\leq 1})^1 = \emptyset.$
  $\Delta_2\underline{P}(Cl_2^{\geq}) = \{x_5, x_7\}, \Delta_2\underline{P}(Cl_3^{\geq}) = \{x_5, x_7\}; \Delta_2\overline{P}(Cl_3^{\geq}) = \emptyset;$
  $\Delta_2\underline{P}(Cl_1^{\leq}) = \{x_5, x_6\}, \Delta_2\underline{P}(Cl_2^{\leq}) = U^+; \Delta_2\overline{P}(Cl_1^{\leq}) = \{x_6\}.$
- New approximations
  $\underline{P}(Cl^{(t+1)}{}_2^{\geq}) = \{x_3, x_7\}, \underline{P}(Cl^{(t+1)}{}_3^{\geq}) = \{x_7\}; \overline{P}(Cl^{(t+1)}{}_3^{\geq}) = \{x_2, x_3, x_7\}.$
  $\underline{P}(Cl^{(t+1)}{}_1^{\leq}) = \{x_5\}, \underline{P}(Cl^{(t+1)}{}_2^{\leq}) = \{x_1, x_2, x_3, x_4, x_5, x_6\};$
  $\overline{P}(Cl^{(t+1)}{}_2^{\leq}) = \{x_1, x_2, x_3, x_4, x_5, x_6\}.$

5 Conclusions and Future Work

In this paper, we proposed a new strategy of updating approximations of DRSA. In order to reduce the redundant computation in updating approximations, we give the notation of base boundaries of approximation of DRSA. A numerical example is employed to illustrate the feasibility of the strategy. Our future work is to investigate the corresponding parallel strategies for updating approximations of DRSA when the object set varying.

References

1. Pawlak, Z.: Rough sets. Int. J. Comput. Inform. Sci. **11**, 341–356 (1982)
2. Pawlak, Z.: Rough Sets: Theoretical Aspects of Reasoning about Data, System Theory, Knowledge Engineering and Problem Solving, vol. 9. Kluwer Academic Publishers, Dordrecht (1991)
3. Zhang, Q., Xie, Q., Wang, G.: A survey on rough set theory and its applications. CAAI Trans. Intell. Technol. **1**(4), 323–333 (2016)
4. Greco, S., Matarazzo, B., Slowinski, R.: Rough sets theory for multicriteria decision analysis. Eur. J. Oper. Res. **129**, 1–47 (2001)
5. Boggia, A., Rocchi, L., Paolotti, L., Musotti, F., Greco, S.: Assessing rural sustainable development potentialities using a dominance-based rough set approach. J. Environ. Manage. **144**, 160–167 (2014)
6. Liou, J.: A novel decision rules approach for customer relationship management of the airline market. Expert Syst. Appl. **36**, 4374–4381 (2009)
7. Liou, J., Tzeng, G.: A dominance-based rough set approach to customer behavior in the airline market. Inf. Sci. **180**, 2230–2238 (2010a)
8. Liou, J., Yen, L., Tzeng, G.: Using decision rules to achieve mass customization of airline services. Eur. J. Oper. Res. **205**, 680–686 (2010b)
9. Chakhar, S., Ishizaka, A., Labib, A., Saad, I.: Dominance-based rough set approach for group decisions. Eur. J. Oper. Res. **251**(1), 206–224 (2016)
10. do Couto, A.B.G., Gomes, L.F.A.M.: Multi-criteria web mining with DRSA. Procedia Comput. Sci. **91**, 131–140 (2016)
11. Do Couto, A.B.G.: Using a dominance-based rough set approach for analysing business indicators. Procedia Comput. Sci. **55**, 350–359 (2015)
12. Jerzy, B., Slowinski, R.: Incremental induction of decision rules from dominance-base rough approximations. Electron. Notes Theor. Comput. Sci. **82**, 40–45 (2003)
13. Greco, S., Słowiński, R., Stefanowski, J., Żurawski, M.: Incremental versus non-incremental rule induction for multicriteria classification. In: Peters, J.F., Skowron, A., Dubois, D., Grzymała-Busse, J.W., Inuiguchi, M., Polkowski, L. (eds.) Transactions on Rough Sets II. LNCS, vol. 3135, pp. 33–53. Springer, Heidelberg (2004). https://doi.org/10.1007/978-3-540-27778-1_3
14. Li, S., Li, T., Liu, D.: Dynamic maintenance of approximations in dominance-based rough set approach under the variation of the object set. Int. J. Intell. Syst. **28**(8), 729–751 (2013)
15. Li, S., Li, T., Liu, D.: Incremental updating approximations in dominance-based rough sets approach under the variation of the attribute set. Knowl. Based Syst. **40**, 17–26 (2013)
16. Li, S., Li, T.: Incremental update of approximations in dominance-based rough sets approach under the variation of attribute values. Inf. Sci. **294**, 348–361 (2015)

17. Luo, C., Li, T., Chen, H., Liu, D.: Incremental approaches for updating approximations in set-valued ordered information systems. Knowl. Based Syst. **50**, 218–233 (2013)
18. Luo, C., Li, T., Chen, H.: Dynamic maintenance of approximations in set-valued ordered decision systems under the attribute generalization. Inf. Sci. **257**, 210–228 (2014)
19. Luo, C., Li, T., Chen, H., Lu, L.: Fast algorithms for computing rough approximations in set-valued decision systems while updating criteria values. Inf. Sci. **299**, 221–242 (2015)
20. Chen, H., Li, T., Ruan, D.: Maintenance of approximations in incomplete ordered decision systems while attribute values coarsening or refining. Knowl. Based Syst. **31**, 140–161 (2012)
21. Wang, S., Li, T., Luo, C., Fujita, H.: Efficient updating rough approximations with multi-dimensional variation of ordered data. Inf. Sci. **372**, 690–708 (2016)
22. Luo, C., Li, T., Chen, H., Fujita, H., Yi, Z.: Incremental rough set approach for hierarchical multicriteria classification. Inf. Sci. **429**, 72–87 (2018)

CSLI: Cost-Sensitive Collaborative Filtering with Local Information Embedding

Heng-Ru Zhang(✉), Jie Qian, and Fan Min

School of Computer Science, Southwest Petroleum University, Chengdu 610500, China
zhanghrswpu@163.com, swpu_jieqian@163.com, minfanphd@163.com
http://www.fansmale.com

Abstract. Mean absolute error and root mean square error are typically used to evaluate the accuracy of recommender system. However, these evaluation metrics implicitly mean that the cost of different wrong recommendation actions is the same. In this paper, we propose the cost-sensitive collaborative filtering with local information embedding (CSLI) algorithm to handle unequal misclassification costs. First, we employ a clustering algorithm to extract local rating information. Second, we design a collaborative filtering algorithm embedding local rating information to compute the prediction p. Third, we construct a 2×2 cost matrix by considering different misclassification costs. We employ the trichotomy method to obtain the recommendation threshold r_t with the cost matrix. Finally, the recommendation actions are determined based on p and r_t. Combined with the cost matrix, we calculate the average misclassification cost and use it to evaluate the performance of the CSLI algorithm. Experimental results show that the proposed algorithm is lower than the state-of-the-art ones in term of average cost.

Keywords: Collaborative filtering · Cost-sensitive · Local rating information · Misclassification cost

1 Introduction

In recommender systems, mean absolute error (MAE) and root mean square error (RMSE) [18,20] are two classical accuracy metrics. They are used to measure how close the prediction is to the actual rating. However, these evaluation metrics do not take into account the cost of different wrong recommendation actions.

This work is supported in part by the National Natural Science Foundation of China (Grant 41604114), Natural Science Foundation of Sichuan Province (Grant 2019YJ0314), and Scientific Innovation Group for Youths of Sichuan Province (Grant 2019JDTD0017).

© Springer Nature Switzerland AG 2019
T. Mihálydeák et al. (Eds.): IJCRS 2019, LNAI 11499, pp. 321–330, 2019.
https://doi.org/10.1007/978-3-030-22815-6_25

In this paper, we propose the cost-sensitive collaborative filtering with local information embedding (CSLI) algorithm to handle unequal misclassification costs. A misclassification cost [23] is incurred for wrong recommender actions, e.g., recommending items to users who dislike them or not recommending items to users who would like them. In fact, misclassification cost exists widely in different data mining applications [14,25]. It is a major issue in cost-sensitive learning [1,5,16,17,31].

First, we employ M-distance [30] to construct user and item clusters. Users in the same cluster have similar item preferences, and items in the same cluster have similar popularity. Compared with other general purpose clustering algorithms, the M-distance based algorithm is very efficient. The time complexity is only $O(mn)$, where m is the number of users, and n is the number of items.

Second, we design a collaborative filtering (CF) algorithm embedding local rating information [15] to obtain the prediction p. The local information correspond to the rating of a cluster of users to a cluster of items. There are three methods of local rating embedding. One is that local-user and global-item rating information acts as input, another is global-user and local-item rating information, and the third is local-user and local-item rating information.

Third, we construct a 2×2 cost matrix by considering different misclassification cost. The cost is incurred for the wrong recommendation actions [22,23]. The system will set different costs for different application scenarios [7]. Based on the misclassification cost, we employ trichotomy method [23] to compute the recommendation threshold r_t.

Finally, the recommendation actions are determined based on p and r_t. If $p > r_t$, the item is recommended to the user. Otherwise, the item is not recommended to the user. There are two wrong actions. One is that users do not like the recommended items. The other is that the system does not recommend items liked by users. Misclassification cost is incurred to the two actions. The average cost is computed based on misclassification one [22].

Experiments on the well-known MovieLens data set (http://www.movielens.org/). The results show that our proposed algorithm is lower than the state-of-the-art ones in term of the average cost metric.

2 Related Works

This section first reviews the definition of the rating system. Second, we outline the popular CF algorithms. Finally, we introduce the cost-sensitive learning.

2.1 Rating System

We revisit the definition of a rating matrix R [22]. Let $U = \{u_1, u_2, \ldots, u_m\}$ be the user set and $T = \{t_1, t_2, \ldots, t_n\}$ be the item set. The rating matrix is defined as

$$R : U \times T \rightarrow R_L, \tag{1}$$

where $R_L = \{r_l, \ldots, r_h\}$, r_l is the lowest rating level, and r_h is the highest rating level. $r_{i,j} = R(u_i, t_j)$ represents the rating for u_i to t_j. Naturally, we have $r_{i,j} \in [r_l, r_h]$.

Table 1 shows an example of a rating matrix R, where $n = 6$, $m = 6$, $r_l = 1$, and $r_h = 5$. "–" indicates that the user has not rated the corresponding item.

Table 1. The rating matrix (R)

UID\TID	t_1	t_2	t_3	t_4	t_5	t_6
u_1	–	3	–	2	1	5
u_2	3	1	4	3	–	1
u_3	2	5	2	3	1	–
u_4	5	2	–	4	3	1
u_5	4	–	5	–	5	1
u_6	–	2	3	5	3	2

2.2 Collaborative Filtering

The core idea of CF [13,18,24,29] is to use the preferences of users with similar interests and common experiences. User-based and item-based CF are widely used in recommender system. They make recommendations by calculating the similarities between users or items. User-based approaches [29] make recommendations by calculating the similarities between users. As the number of users increases, user-based approaches will increase the computation time. Item-based approaches [18] can handle this issue due to the stability of item number.

The k-nearest-neighbor (kNN) [11,17] algorithm is one of the most fundamental CF recommendation ones. One key factor in the performance of kNN algorithm is the definition of the distance metric. Popular metrics include the cosine distance [28], Eucliden distance [4], and Pearson distance [2].

Slope one [13] is an item-based CF algorithm based on linear regression. It employs Global-User and Global-Item ($GUGI$) information to make prediction (see $GUGI$ of Table 2). There are four steps to predict the rating of u_u to t_j: (1) Obtain a set of users who rate t_i and t_j; (2) Compute the rating deviation between t_i and t_j; (3) Obtain a set of items rated by u_u; and (4) Compute the prediction for u_u to t_j.

2.3 Cost-Sensitive Learning

Most research in cost-sensitive learning [12,26,31] is an extension of machine learning approaches through the consideration of different costs. Hunt et al. [10] proposed misclassification cost and test cost. Yao et al. [21] considered misclassification cost and delay cost. Turney [19] proposed a taxonomy of costs, including test cost, misclassification cost, delay cost, teacher cost, etc.

Our work only considers misclassification cost. It is that incurred as a result of classification errors. AdaCost [8] used the misclassification cost to update the training distribution of successive boosting rounds. An optimal Bayesian classification rule has been constructed to minimize the expected misclassification cost for various cost functions [3]. Zhang et al. constructed random-forest-based [22] and regression-based [23] recommender systems to minimize classification costs. In real applications, how to evaluate misclassification cost is a very difficult problem. In some cases, relative values are assigned to represent misclassification cost [27].

3 Algorithms

This section first employs clustering algorithm to extract the local information. The new algorithm is designed through embedding the local information [15] in the popular CF algorithm. Then, we compute the average cost to evaluate the performance of the new algorithm.

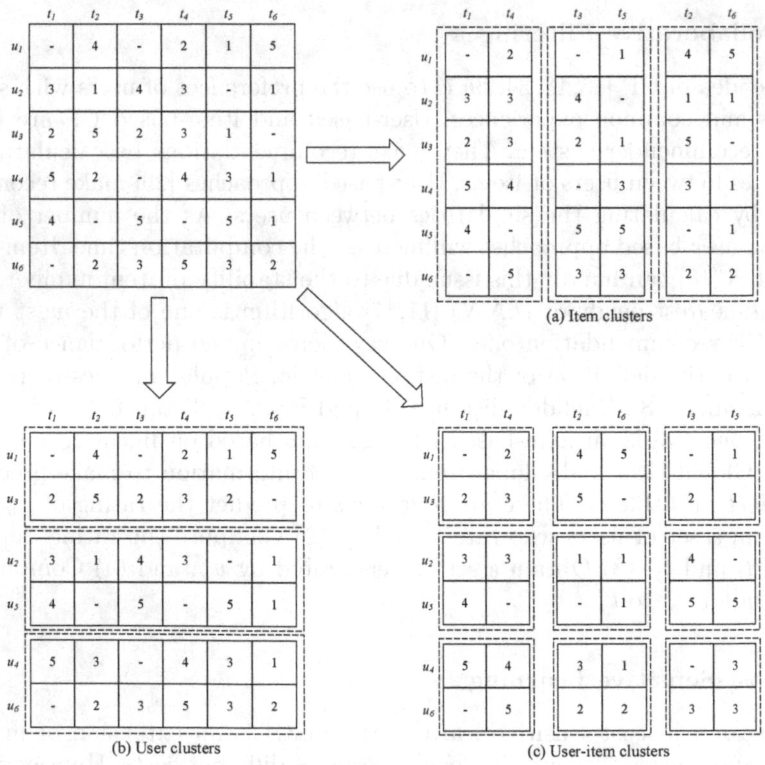

Fig. 1. Local information extraction

3.1 Collaborative Filtering for Local Information Embedding

CF computes a predicted rating for items to users based on their history ratings [6,9]. Existing CF algorithms usually use global ratings as input. This will lead to under-fitting problem. To handle this issue, we employ clustering algorithms to extract local information, and then embed it to popular CF algorithms.

Figure 1 depicts an example for local rating extraction. User cluster represents that a group of users have similar preferences. Let U^i be i-th user cluster. We have $U = U^1 \cup U^2 \cdots \cup U^g$ and $U^1 \cap U^2 \cdots \cap U^g = \emptyset$, where $i \in [1, g]$. u_1 and u_3 purchased the same items t_2 and t_4, and their ratings for the same item are very close. Therefore, they are combined in a user cluster U^1.

Item cluster represents a group of items with similar popularity. Let T^j be j-th item cluster. We have $T = T^1 \cup T^2 \cdots \cup T^h$ and $T^1 \cap T^2 \cdots \cap T^h = \emptyset$, where $j \in [1, h]$. t_1 and t_4 are purchased by the users u_2, u_3 and u_4. For the same user, the ratings for the two items are very close. Therefore, t_1 and t_4 are combined in a item cluster T^1.

The user-item clusters include local user and item information. Local user and item information are obtained based on global user and item information, respectively. It describes user preference or item popularity in a more granular way. $\{u_1, u_3\}$ and $\{t_1, t_4\}$ construct a user-item cluster.

There are three local information embedding methods. *LUGI* of Table 2 depicts Local-User and Global-Item algorithm that embeds local user information. *GULI* of Table 2 depicts Global-User and Local-Item algorithm that embeds local item information. *LULI* of Table 2 depicts Local-User and Local-Item algorithm that embeds local user and item information.

3.2 Average Cost Computation

Table 3 lists the cost matrix for correct and incorrect recommendation actions, where L represents the set of users liked by users, $\widetilde{L}$ represents the set of users disliked by users. The costs of incorrect recommendation actions are λ_{12} and λ_{21}. λ_{12} indicates that the items liked by users are not recommended. λ_{21} indicates that the items disliked by users are recommended. In general, there are no costs when the recommendation actions are correct [7]. Mathematically, it should always be the case that $\lambda_{11} = 0$ and $\lambda_{22} = 0$.

Similarly, we use two notations (RN, NY) to indicate the classification numbers of incorrect recommendation actions. RN is the number of users who dislike the recommended items, which will result in a cost corresponding to λ_{12}. NY is the number of users who like the no recommended items, which will result in a cost corresponding to λ_{21}.

Algorithm 1 depicts three steps for average-cost computation.

Step 1. We use a CF algorithm embedding local rating information to predict the rating p. This step corresponds to line 3 of the algorithm.

Step 2. We employ the trichotomy method [23] to obtain the recommendation threshold r_t. This step corresponds to lines 5–17 of the algorithm.

Step 3. We calculate the average cost. This step corresponds to line 18 of the algorithm.

Table 2. The prediction process for different slope one algorithms

Approach	Step	Formula
GUGI	1	$S^{\cdot,\cdot}_{i,j} = \{u_k \mid r_{k,i} > 0, r_{k,j} > 0, u_k \in U\}$
	2	$dev^{\cdot,\cdot}_{i,j} = \frac{\sum_{u_k \in S^{\cdot,\cdot}_{i,j}} (r_{k,j} - r_{k,i})}{\lvert S^{\cdot,\cdot}_{i,j}\rvert}$
	3	$N^{\cdot,\cdot}_{u,j} = \{t_i \mid r_{u,i} > 0, r_{u,j} > 0, t_i \in T\}$
	4	$p^{\cdot,\cdot}_{u,j} = \frac{\sum_{t_i \in N^{\cdot,\cdot}_{u,j}} (r_{u,i} + dev^{\cdot,\cdot}_{i,j})}{\lvert N^{\cdot,\cdot}_{u,j}\rvert}$
LUGI	1	$S^{g,\cdot}_{i,j} = \{u_k \mid r_{k,i} > 0, r_{k,j} > 0, u_k \in U^g\}$
	2	$dev^{g,\cdot}_{i,j} = \frac{\sum_{u_k \in S^{g,\cdot}_{i,j}} (r_{k,j} - r_{k,i})}{\lvert S^{g,\cdot}_{i,j}\rvert}$
	3	$N^{g,\cdot}_{u,j} = \{t_i \mid r_{u,i} > 0, r_{u,j} > 0, t_i \in T\}$
	4	$p^{g,\cdot}_{u,j} = \frac{\sum_{t_i \in N^{g,\cdot}_{u,j}} (r_{u,i} + dev^{g,\cdot}_{i,j})}{\lvert N^{g,\cdot}_{u,j}\rvert}$
	Example	$p^{1,\cdot}_{1,1} = \frac{(2-5)+4+(2-3)+2+(2-2)+1}{3} = 1$
GULI	1	$S^{\cdot,h}_{i,j} = \{u_k \mid r_{k,i} > 0, r_{k,j} > 0, u_k \in U\}$
	2	$dev^{\cdot,h}_{i,j} = \frac{\sum_{u_k \in S^{\cdot,h}_{i,j}} (r_{k,j} - r_{k,i})}{\lvert S^{\cdot,h}_{i,j}\rvert}$
	3	$N^{\cdot,h}_{u,j} = \{t_i \mid r_{u,i} > 0, r_{u,j} > 0, t_i \in T^h\}$
	4	$p^{\cdot,h}_{u,j} = \frac{\sum_{t_i \in N^{\cdot,h}_{u,j}} (r_{u,i} + dev_{i,j})}{\lvert N^{\cdot,h}_{u,j}\rvert}$
	Example	$p^{\cdot,1}_{1,1} = \frac{(3-3)+(2-3)+(5-4)}{3} + 2 = 2$
LULI	1	$S^{g,h}_{i,j} = \{u_k \mid r_{k,i} > 0, r_{k,j} > 0, u_k \in U^g\}$
	2	$dev^{g,h}_{i,j} = \frac{\sum_{u_k \in S^{g,h}_{i,j}} (r_{k,j} - r_{k,i})}{\lvert S^{g,h}_{i,j}\rvert}$
	3	$N^{g,h}_{u,j} = \{t_i \mid r_{u,i} > 0, r_{u,j} > 0, t_i \in T^h\}$
	4	$p^{g,h}_{u,j} = \frac{\sum_{t_i \in N^{g,h}_{u,j}} (r_{u,i} + dev_{i,j})}{\lvert N^{g,h}_{u,j}\rvert}$
	Example	$p^{1,1}_{1,1} = \frac{2-3}{1} + 2 = 1$

Table 3. Misclassification cost matrix

Actions	L	$\tilde{L}$
Recommend	λ_{11}	λ_{12}
Not recommend	λ_{21}	λ_{22}

Algorithm 1. Average-cost computation

Input: Rating matrix (R), like threshold (r_s)
Output: average cost (A_c)
Method: costComputation

R is divided into the training set (T_r) and the testing set (T_e);
//Step 1. Predict T_e based on T_r
$P = \{p_{i,j} | i \in [1,m], j \in [1,n]\}$ is predicted based on Table 2;
//Step 2. Employ the trichotomy method to obtain the recommendation threshold (r_t)
$left = r_l, right = r_h$;
while $|left - right| > \varepsilon$ **do**
  $md = \frac{left+right}{2}$;
  $mmd = \frac{md+right}{2}$;
  $ls(R,P,md,r_s) = \frac{\lambda_{12} \times RN(R,P,md,r_s) + \lambda_{21} \times NY(R,P,md,r_s)}{|T_e|}$;
  $ls(R,P,mmd,r_s) = \frac{\lambda_{12} \times RN(R,P,mmd,r_s) + \lambda_{21} \times NY(R,P,mmd,r_s)}{|T_e|}$;
  if $ls(R,P,md,r_s) < ls(R,P,mmd,r_s)$ **then**
    $right = mmd$;
  else
    $left = md$;
  end if
end while
$r_t = left$;
//Step 3. Compute the lowest average cost
$A_c = ls(R,P,r_t,r_s)$;
return A_c;

4 Experiments

In this section, we conduct several experiments on the well-known MovieLens data set. The original set is repeated 10 times to divide into the training set and the testing set with different random partitionings (i.e., 10× cross-validation). We choose 80% of the original set as the training set and the rest as the testing set.

We set the total misclassification cost to 135 and change $\frac{\lambda_{10}}{\lambda_{01}}$ from 0.125 to 8. To evaluate the performance of local information embedding, we conducted three basic methods and three integrating ones. They compare with the original slope one algorithm $GUGI$ and kNN. The three basic algorithms are $LULI$, $LUGI$ and $GULI$. $LULGI$ represents the integration of $LULI$ and $LUGI$. $LGULI$ represents the integration of $LULI$ and $GULI$. $LGULGI$ represents the integration of $LUGI$ and $GULI$.

Based on Algorithm 1, we obtain the recommendation threshold r_t listed in Table 4. The recommendation threshold decreases with the increase of $\frac{\lambda_{21}}{\lambda_{12}}$. When $\frac{\lambda_{21}}{\lambda_{12}} = 0.125$, the recommendation threshold has been close to the highest rating level. When $\frac{\lambda_{21}}{\lambda_{12}} = 8$, the recommendation threshold has been close to the lowest rating level. Therefore, the range of $\frac{\lambda_{21}}{\lambda_{12}}$ we set is appropriate.

Table 4. r_t for different ratio of $\frac{\lambda_{21}}{\lambda_{12}}$

$\frac{\lambda_{21}}{\lambda_{12}}$	kNN	$GUGI$	$LULI$	$LUGI$	$GULI$	$LULGI$	$LGULI$	$LGULGI$
0.125	4.6	4.5	4.6	4.5	4.6	4.5	4.6	4.5
0.25	4.2	4.1	4.2	4.1	4.2	4.2	4.2	4.1
0.5	3.9	3.8	3.7	3.8	3.7	3.8	3.7	3.8
1	3.6	3.4	3.4	3.4	3.4	3.4	3.4	3.4
2	2.6	3.1	2.9	3.1	3.0	3.1	3.1	3.1
4	2.2	2.7	2.6	2.7	2.6	2.6	2.6	2.6
8	1.0	2.3	2.3	2.1	2.0	2.1	1.8	2.3

Table 5 lists the average costs of the eight algorithms under different setting of misclassification costs. Compared to kNN, our proposed algorithms have lower average costs. When $\frac{\lambda_{21}}{\lambda_{12}} = 0.125$ or 4 or 8, $LGULGI$ algorithm obtains the lowest average cost. When $\frac{\lambda_{21}}{\lambda_{12}} = 0.25$, $LGULI$ algorithm obtains the lowest average cost. When $\frac{\lambda_{21}}{\lambda_{12}} = 1$, $GULI$ algorithm obtains the lowest average cost. In summary, our proposed algorithm performs better than the counterparts on five out of seven ratio settings. The other two are close to the lowest average cost.

Table 5. Average cost for different ratio of $\frac{\lambda_{21}}{\lambda_{12}}$

$\frac{\lambda_{21}}{\lambda_{12}}$	kNN	$GUGI$	$LULI$	$LUGI$	$GULI$	$LULGI$	$LGULI$	$LGULGI$
0.125	8.30	8.16	8.23	8.15	8.19	8.14	8.19	**8.13**
0.25	14.20	13.58	13.58	13.55	13.52	13.55	**13.50**	13.52
0.5	20.60	**17.88**	18.16	17.97	18.06	18.04	18.05	17.95
1	22.50	18.98	19.10	19.11	**18.91**	18.98	19.03	18.93
2	18.40	**15.88**	16.18	15.99	16.13	15.95	16.16	15.91
4	11.90	10.90	10.95	11.01	10.90	10.91	10.87	**10.85**
8	6.70	6.44	6.48	**6.43**	6.47	6.47	6.48	**6.43**

5 Conclusions

In this paper, we have proposed the CSLI algorithm to handle unequal misclassification costs. M-distance algorithm is employed to extract local rating information. The local information is embedded into the slope one algorithm. According to our experiments on Movielens data set, the performance of our proposed algorithm is mostly lower than kNN and the traditional slope one algorithm in terms of average cost. In the future, we will consider more costs, such as promotion one, and test one.

References

1. Abe, N., Zadrozny, B., Langford, J.: An iterative method for multi-class cost-sensitive learning. In: Proceedings of the Tenth ACM SIGKDD International Conference on Knowledge Discovery and Data Mining, pp. 3–11. ACM (2004)
2. Benesty, J., Chen, J., Huang, Y., Cohen, I.: Pearson correlation coefficient. In: Cohen, I., Huang, Y., Chen, Jacob, J., Benesty, J. (eds.) Noise Reduction in Speech Processing. STSP, vol. 2, pp. 1–4. Springer, Heidelberg (2009). https://doi.org/10.1007/978-3-642-00296-0_5
3. Boutros, N., Nasrallah, H., Leighty, R., Torello, M., Tueting, P., Olson, S.: Auditory evoked potentials, clinical vs. research applications. Psychiatry Res. **69**(2–3), 183–195 (1997)
4. Danielsson, P.E.: Euclidean distance mapping. Comput. Graph. Image Process. **14**(3), 227–248 (1980)
5. Domingos, P.: MetaCost: a general method for making classifiers cost-sensitive. In: Proceedings of the Fifth ACM SIGKDD International Conference on Knowledge Discovery and Data Mining, pp. 155–164. ACM (1999)
6. Ekstrand, M.D., Riedl, J.T., Konstan, J.A., et al.: Collaborative filtering recommender systems. Found. Trends® Hum.-Comput. Interact. **4**(2), 81–173 (2011)
7. Elkan, C.: The foundations of cost-sensitive learning. In: International Joint Conference on Artificial Intelligence, vol. 17, pp. 973–978. Lawrence Erlbaum Associates Ltd. (2001)
8. Fan, W., Stolfo, S.J., Zhang, J.X., Chan, P.K.: AdaCost: misclassification cost-sensitive boosting. In: ICML, pp. 97–105 (1999)
9. Hu, Y.F., Koren, Y., Volinsky, C.: Collaborative filtering for implicit feedback datasets. In: Eighth IEEE International Conference on Data Mining, ICDM 2008, pp. 263–272. IEEE (2008)
10. Hunt, E.B., Marin, J., Stone, P.J.: Experiments in induction (1966)
11. Jadhav, S.D., Channe, H.: Comparative study of *knn*, Naive Bayes and decisiontree classification techniques. Int. J. Sci. Res. **5**(1), 1842 (2016)
12. Jia, X.Y., Liao, W.H., Tang, Z.M., Shang, L.: Minimum cost attribute reduction in decision-theoretic rough set models. Inf. Sci. **219**, 151–167 (2013)
13. Lemire, D., Maclachlan, A.: Slope one predictors for online rating-based collaborative filtering. In: Proceedings of the 2005 SIAM International Conference on Data Mining, pp. 471–475. SIAM (2005)
14. Li, T.R., Ruan, D., Geert, W., Song, J., Xu, Y.: A rough sets based characteristic relation approach for dynamic attribute generalization in data mining. Knowl.-Based Syst. **20**(5), 485–494 (2007)
15. Ma, Y.Y., Zhang, H.R., Xu, Y.Y., Min, F., Gao, L.: Three-way recommendation integrating global and local information. J. Eng. **2018**(16), 1397–1401 (2018)
16. Min, F., He, H.P., Qian, Y.H., Zhu, W.: Test cost-sensitive attribute reduction. Inf. Sci. **181**(22), 4928–4942 (2011)
17. Min, F., Liu, F.L., Wen, L.Y., Zhang, Z.H.: Tri-partition cost-sensitive active learning through kNN. Soft Comput. **23**, 1–16 (2017)
18. Sarwar, B., Karypis, G., Konstan, J., Riedl, J.: Item-based collaborative filtering recommendation algorithms. In: Proceedings of the 10th International Conference on World Wide Web, pp. 285–295. ACM (2001)
19. Turney, P.D.: Cost-sensitive classification: empirical evaluation of a hybrid genetic decision tree induction algorithm. J. Artif. Intell. Res. **2**, 369–409 (1994)

20. Willmott, C.J., Matsuura, K.: Advantages of the mean absolute error (MAE) over the root mean square error (RMSE) in assessing average model performance. Clim. Res. **30**(1), 79–82 (2005)
21. Yao, Y.: Three-way decisions with probabilistic rough sets. Inf. Sci. **180**(3), 341–353 (2010)
22. Zhang, H.R., Min, F.: Three-way recommender systems based on random forests. Knowl.-Based Syst. **91**, 275–286 (2016)
23. Zhang, H.R., Min, F., Shi, B.: Regression-based three-way recommendation. Inf. Sci. **378**, 444–461 (2017)
24. Zhang, H.R., Min, F., Zhang, Z.H., Wang, S.: Efficient collaborative filtering recommendations with multi-channel feature vectors. Int. J. Mach. Learn. Cybern. **10**, 1–8 (2018)
25. Zhang, J.B., Li, T.R., Ruan, D., Liu, D.: Neighborhood rough sets for dynamic data mining. Int. J. Intell. Syst. **27**(4), 317–342 (2012)
26. Zhang, L.M., et al.: A forward modeling method based on electromagnetic theory to measure the parameters of hydraulic fracture. Fuel **251**, 466–473 (2019)
27. Zhang, S.C.: Cost-sensitive classification with respect to waiting cost. Knowl.-Based Syst. **23**(5), 369–378 (2010)
28. Zhang, Y., Callan, J., Minka, T.: Novelty and redundancy detection in adaptive filtering. In: Proceedings of the 25th Annual International ACM SIGIR Conference on Research and Development in Information Retrieval, pp. 81–88. ACM (2002)
29. Zhao, Z.D., Shang, M.S.: User-based collaborative-filtering recommendation algorithms on hadoop. In: Third International Conference on Knowledge Discovery and Data Mining, WKDD 2010, pp. 478–481. IEEE (2010)
30. Zheng, M., Min, F., Zhang, H.R., Chen, W.B.: Fast recommendations with the m-distance. IEEE Access **4**, 1464–1468 (2016)
31. Zhou, Z.H., Liu, X.Y.: Training cost-sensitive neural networks with methods addressing the class imbalance problem. IEEE Trans. Knowl. Data Eng. **18**(1), 63–77 (2006)

Attribute Reduction Based on Optimistic Multi-granulation Information Systems

Binli Ou[1] and Tian Yang[2,3](✉)

[1] College of Logistics and Transportation, Central South University of Forestry and Technology, Changsha 410004, Hunan, People's Republic of China
oubinli2010@126.com

[2] Hunan Provincial Key Laboratory of Intelligent Computing and Language Information Processing, Hunan Normal University, Changsha 410081, Hunan, People's Republic of China
math_yangtian@126.com

[3] College of System Engineering, National University of Defense Technology, Changsha 410073, Hunan, People's Republic of China

Abstract. Attribute reduction is the most important and widely applied part in rough sets. Multi-granulation rough set model is a significant generalization of classical rough sets, which includes pessimistic and optimistic multi-granulation models. Several attribute reduction algorithms based on multi-granulation models arc designed in literatures, but all of them are based on pessimistic models, while the attribute reduction based on optimistic models has not been developed. Thus, in this paper, we propose an attribute reduction approach, named related family, for the first and the second optimistic multi-granulation covering rough set models, which is the basis of attribute reduction of all optimistic multi-granulation rough set models and decrease the time complexity of attribute reduction.

Keywords: Optimistic multi-granulation · Covering rough set · Related family · Attribute reduction

1 Introduction

Since Pawlak proposed rough set theory in 1982 [7], it has been a powerful tool to deal with uncertainty and incompleteness. Compared with fuzzy sets, evidence theory and probability theory, rough set is a data driving method, since it can be used to process data without any prior knowledge. Attribute reduction is the most important research topic in rough set theory, it has been widely used in data compression, pattern recognition, management systems and so on.

Covering rough set was first proposed by Zakowski [14], which extended the strict theoretical base–equivalence relation to covering. Because of the complexity of covering relation, there are much more approximation operators of covering rough sets than those of Pawlak rough sets. Yao et al. [13] classify covering

© Springer Nature Switzerland AG 2019
T. Mihálydeák et al. (Eds.): IJCRS 2019, LNAI 11499, pp. 331–340, 2019.
https://doi.org/10.1007/978-3-030-22815-6_26

rough sets into three categories: element based, granule based and subsystem based approximation operators. The properties of some approximation operators resulted in that the attribute reduction methods based on Pawlak's rough set are not suitable for all types of covering set approximation operators. In reference [12], yang demonstrated that the discernibility matrix is not suitable for the third type of covering set approximations, and proposed the related family to replace discernibility matrices, which provided a fast attribute reduction algorithm based on covering rough sets.

Qian et al. [8,10] proposed a multi-granulation rough set model where the set approximation operators are defined by using multi equivalence relations on the universe, and developed two different multi-granulation rough sets (MGRS) called optimistic and pessimistic, respectively. Then on the basis of multi-granulation rough set model, Qian and Liang [9] extended MGRS to incomplete information systems with respect to multiple tolerance relations. Lin [4] proposed covering based multi-granulation rough sets to solve problems in multi-source information systems with a covering environment. Liu and Pedrycz [6] proposed multi-granulation fuzzy rough sets in the covering approximation space. Similar with covering rough sets, the classical attribute reduction based on discernibility matrix, dependency degree, information entropy and evidence theory can not be applied to all types of multi-granulation rough sets. Scholars proposed several elegant attribute reduction algorithms for pessimistic multi-granulation models based on discernibility matrix, dependency degree, information entropy or evidence theory [2,3,5]. However, as the other important part of multi-granulation rough sets, optimistic multi-granulation models have never been studied from the view of attribute reduction. As a result, we focus on attribute reduction of optimistic multi-granulation rough set models in this paper. The related family is defined based on optimistic multi-granulation rough set models as the first step. Then we design the attribute reduction procedure based on related family and Boolean operation. The complexity of proposed method is lower than existing algorithms, and it can calculate all attribute reducts.

The structure of this paper is as follows. Section 2 introduces the background knowledge of Pawlak's rough set, covering rough set and approximation operators of multi-granulation rough set models. Section 3 introduces related family attribute reduction methods for the first and the second types of multi-granulation optimistic covering rough set. Finally, conclusion and problems needed to be further studied are presented in Sect. 4.

2 Background Knowledge

2.1 Basic Notions of Pawlak's Rough Set

Let $\langle U, AT\rangle$ be an information system, where $AT = \{R_1, R_2, ..., R_m\}$ is a finite non-empty set of attributes on $U = \{x_1, x_2, ..., x_n\}$. For any $x_i \in U$, we use $a(x_i)$ to denote the value of x_i under the attribute $R_i (R_i \in AT)$. Given $R \subseteq AT$, an indiscernibility relation $ind(R)$ can be defined as: $ind(R) = \{(x, y) \in U \times U | a(x) = a(y), a \in R\}$.

Based the indiscernibility relation $ind(R)$, we can define the lower and upper approximation of X on U as $\underline{R}(X) = \{x \in U | [x]_R \subseteq X\}$, $\overline{R}(X) = \{x \in U | [x]_R \cap X \neq \varnothing\}$, respectively, where $[x]_R = \{y \in U | (x, y) \in ind(R)\}$ is an equivalence class containing x.

2.2 Covering Rough Sets

Suppose $\mathcal{C}$ is a family of nonempty subsets of U and $\cup\mathcal{C} = U$, we call $\mathcal{C}$ a covering and $\langle U, \mathcal{C} \rangle$ a covering approximation space. Obviously, a partition on U is definitely a covering on U, thus partition is a special case of covering.

Yao et al. [13] classify covering rough sets into three categories: element based, granule based and subsystem based approximation operators. We mainly investigate granule based approximation operators in this paper.

Definition 1 *[1] (Minimal description). Let $\mathcal{C}$ be a covering on U, $Md_{\mathcal{C}}(x) = \{K \in \mathcal{C} | x \in K \wedge (\forall S \in \mathcal{C} \wedge x \in S \wedge S \subseteq K) \Rightarrow K = S\}$ is the minimal description of x. When there is no confusion, we omit $\mathcal{C}$ from the subscript.*

Definition 2 *[16] (Neighborhood). Let $\mathcal{C}$ be a covering on U, For $x \in U$, the neighborhood of x is defined as $N_{\mathcal{C}}(x) = \cap\{K \in \mathcal{C} | x \in K\}$. When there is no confusion, we omit $\mathcal{C}$ from the subscript.*

2.3 Multi-granulation Rough Set

Qian et al. [8] propose a generalized rough set model called multi-granulation rough set model. According to two different approximation strategies, Qian developed two different multigranulation rough sets (MGRS) including optimistic and pessimistic models. Then, Lin et al. [4] introduce multi-granulation notions into covering based rough sets by the idea of MGRS. In this paper, we mainly investigate optimistic multi-granulation covering rough sets.

The Optimistic Multi-granulation Rough Set

Definition 3 *[11]. Let I be an information system in which $A_1, A_2, ..., A_m \subseteq AT$; then $\forall X \subseteq U$, the optimistic multi-granulation lower and upper approximations are denoted by $\underline{\sum_{i=1}^{m} A_i}^O(X)$ and $\overline{\sum_{i=1}^{m} A_i}^O(X)$, respectively.*

$\underline{\sum_{i=1}^{m} A_i}^O(X) = \{x \in U | [x]_{A_1} \subseteq X \vee [x]_{A_2} \subseteq X \vee ... \vee [x]_{A_m} \subseteq X\}$;

$\overline{\sum_{i=1}^{m} A_i}^O(X) = \sim (\underline{\sum_{i=1}^{m} A_i}^O(\sim X))$

where $[x]_{A_i} (1 \leq i \leq m)$ is the equivalence class of x in terms of set of attributes A_i, and $\sim X$ is the complement of X.

The First Type of Optimistic CMGRS

Definition 4 *[4]. Let $\langle U, \Omega\rangle$ be a multi-granulation covering information system, $\Omega = \{\mathcal{C}_1, \mathcal{C}_2, ..., \mathcal{C}_n\}$ a family of coverings of U with $\mathcal{C}_i = \{K_{i1}, K_{i2}, ..., K_{im_i}\}$, and $X \subseteq U$. An optimistic lower approximation and an optimistic upper approximation of X with respect to Ω are denoted by $\underline{\sum_{i=1}^n \mathcal{C}_i}^{O_1}(X)$ and $\overline{\sum_{i=1}^n \mathcal{C}_i}^{O_1}(X)$, respectively.*

$$\underline{\sum_{i=1}^n \mathcal{C}_i}^{O_1}(X) = \cup\{K_{ij} \in \mathcal{C}_i | \vee (K_{ij} \subseteq X), i \in \{1, 2, ..., n\}; j = 1, 2, ..., m_i\}$$

$$\overline{\sum_{i=1}^n \mathcal{C}_i}^{O_1}(X) = \sim \underline{\sum_{i=1}^n \mathcal{C}_i}^{O_1}(\sim X)$$

The Second Type of Optimistic CMGRS

Definition 5 *[15]. Let $\langle U, \Omega\rangle$ be a multi-granulation covering information system, $\Omega = \{\mathcal{C}_1, \mathcal{C}_2, ..., \mathcal{C}_n\}$ a family of coverings of U with $\mathcal{C}_i = \{K_{i1}, K_{i2}, ..., K_{im_i}\}$, and $X \subseteq U$. Let $\mathcal{N}_i = \{N_i(x_1), N_i(x_2), ..., N_i(x_{|U|})\}$, where $N_i(x_j) = \cap\{K_{it} \in \mathcal{C}_i | x_j \in K_{it}\}$, $i = 1, 2, ..., n, j = 1, 2, ..., |U|, t = 1, 2, ..., m_i$. An optimistic lower approximation and an optimistic upper approximation of X with respect to Ω are denoted by $\underline{\sum_{i=1}^n \mathcal{C}_i}^{O_2}(X)$ and $\overline{\sum_{i=1}^n \mathcal{C}_i}^{O_2}(X)$, respectively.*

$\underline{\sum_{i=1}^n \mathcal{C}_i}^{O_2}(X) = \cup\{N_i(x_j) \in \mathcal{N}_i | \vee (N_i(x_j) \subseteq X), i \in \{1, 2, ..., n\}; j = 1, 2, ..., |U|\}$

$$\overline{\sum_{i=1}^n \mathcal{C}_i}^{O_2}(X) = \sim \underline{\sum_{i=1}^n \mathcal{C}_i}^{O_2}(\sim X)$$

Example 1. Let $\langle U, \Omega\rangle$ be a multi-granulation covering information system, where $\Omega = \{\mathcal{C}_1, \mathcal{C}_2\}$, where $U = \{x_1, x_2, x_3, x_4, x_5, x_6, x_7\}$, $\mathcal{C}_1 = \{\{x_1, x_2, x_3, x_4\}, \{x_1, x_3, x_6\}, \{x_4, x_6, x_7\}, \{x_5\}\}$, $\mathcal{C}_2 = \{\{x_1, x_3\}, \{x_1, x_2, x_3, x_7\}, \{x_2, x_3, x_4, x_6\}, \{x_4, x_5, x_7\}\}$.

For $X = \{x_1, x_3, x_4, x_6\}$, we have
$\underline{\mathcal{C}_1 + \mathcal{C}_2}^{O_1}(X) = \{x_1, x_3, x_6\}$,
$\overline{\mathcal{C}_1 + \mathcal{C}_2}^{O_1}(X) = \{x_1, x_2, x_3, x_4, x_6, x_7\}$.
For $X = \{x_1, x_3, x_4, x_6\}$, we have
$\mathcal{N}_1 = \{\{x_1, x_3\}, \{x_1, x_2, x_3, x_4\}, \{x_4\}, \{x_5\}, \{x_6\}, \{x_4, x_6, x_7\}\}$,
$\mathcal{N}_2 = \{\{x_1, x_3\}, \{x_2, x_3\}, \{x_3\}, \{x_4\}, \{x_4, x_5, x_7\}, \{x_2, x_3, x_4, x_6\}, \{x_7\}\}$,
$\underline{\mathcal{C}_1 + \mathcal{C}_2}^{O_2}(X) = \{x_1, x_3, x_4, x_6\}$,
$\overline{\mathcal{C}_1 + \mathcal{C}_2}^{O_2}(X) = \{x_1, x_2, x_3, x_4, x_6\}$.

It can be seen that the lower and upper approximations of the first type of optimistic CMGRS are derived from the original covering set, and the lower and upper approximations of the second type of optimistic CMGRS are derived from the covering set, which is formed by the neighbourhood induced by the original covering set $\mathcal{C}_i$.

3 Attribute Reduction

Suppose $\langle U, \Omega\rangle$ is a multi-granulation covering information system, $\Omega = \{\mathcal{C}_1, \mathcal{C}_2, ..., \mathcal{C}_n\}$ be a family of coverings on U. The goal of reduction of multi-granulation covering information system is to find the minimal granulation set $\mathbb{G}$ of Ω such that the approximations of any $X \subseteq U$ are invariant.

In the above three pairs of approximation operators, we can see that the lower approximation operator and upper approximation operator are dual operators, and the upper approximation operator is defined by the lower approximation operator. In other words, as long as the lower approximation operator is invariant, the upper approximation operator will also be invariant. Therefore, we only need to investigate lower approximation operators in this reduction method.

Additionally, if any $C_i \in \Omega, i = \{1, 2, ..., n\}$ is a partition on the universe U, the lower approximation operator in the first and second types of optimistic CMGRS will degenerate into the MGRS. Namely, MGRS is a special case of the first and second types of optimistic CMGRS. Therefore, any one of the following two methods can be used to reduce granulation in MGRS.

3.1 Reduction Based the First Type of Optimistic CMGRS

In the first type of optimistic CMGRS, there is reducible elements in approximation space formed by lower approximation operator. In Reference [17], it is proved that the same lower approximation can be generated in the approximation space after deleting and reducible elements. And $\cup\{Md(x)|x \in U\}$ is the minimal covering set which can generate the same lower approximation.

Definition 6 *(type-1 space). Suppose $\langle U, \Omega\rangle$ is a multi-granulation covering information system, $\Omega = \{C_1, C_2, ..., C_n\}$ be a family of coverings on U, and $C_i = \{K_{i1}, K_{i2}, ..., K_{im_i}\}$. Since $\cup\Omega$ is also a covering on U, we define $\mathcal{T}^1(\Omega) = \cup\{Md_{\cup\Omega}(x)|x \in U\}$ as the first type of multi-granulation space, denoted $\mathcal{T}^1$-space.*

$\cup\Omega = \{K_{ij}|K_{ij} \in C_i, i = 1, 2, ..., n; j = 1, 2, ..., m_i\}$. We define attribute reduction of the first type of optimistic CMGRS based on the first type of multi-granulation space, since the upper and the lower approximations of the first type of optimistic CMGRS are invariant as long as the first type of multi-granulation space is unchanged.

Definition 7. *Let $\langle U, \Omega\rangle$ be a multi-granulation covering information system, $\Omega = \{C_1, C_2, ..., C_n\}$ be a family of coverings on U. C_i is dispensable in Ω if $\mathcal{T}^1(\Omega) = \mathcal{T}^1(\Omega - \{C_i\})$. Otherwise, C_i is indispensable in Ω. If $\mathbb{G} \subseteq \Omega$, $\mathcal{T}^1(\Omega) = \mathcal{T}^1(\mathbb{G})$ and every attribute in $\mathbb{G}$ is indispensable, then $\mathbb{G}$ is called a reduct of Ω. The collection of all reducts of Ω is denoted by $RED(\Omega)$. The collection of all indispensable granulations in Ω is called the core of Ω, denoted as $CORE(\Omega)$.*

Yang et al. [12] proposed an attribute reduction method, called related family to compute the reducts of the third types of covering rough sets. We define the related set and the related family based on $\mathcal{T}^1$-space for multi-granulation covering rough sets.

Definition 8 *(Related family on type-1 space). Let $\langle U, \Omega\rangle$ be a multi-granulation covering information system, $\Omega = \{\mathcal{C}_1, \mathcal{C}_2, ..., \mathcal{C}_n\}$ be a family of coverings on U, and $\mathcal{T}^1(\Omega) = \{K_1, K_2, ...K_m\}$ be the $\mathcal{T}^1$-space of Ω. For any $K_i \in \mathcal{T}^1(\Omega)$, we define $r^1(K_i) = \{\mathcal{C}_t | K_i \in \mathcal{C}_t \in \Omega\}$ is the related set of K_i, and $R^1(U, \Omega) = \{r^1(K_i) | i = 1, 2, ..., m\}$ is the related family of (U, Ω).*

Proposition 1. $CORE(\Omega) = \{\mathcal{C} \in \Omega \mid \exists r^1(K_i) \in \mathcal{T}^1(\Omega) s.t. r^1(K_i) = \{\mathcal{C}\}\}$.

Proof. Suppose $\mathcal{C} \in CORE(\Omega)$, then $\mathcal{T}^1(\Omega) \neq \mathcal{T}^1(\Omega - \{\mathcal{C}\})$. There must be $K_i \in \mathcal{T}^1(\Omega) s.t. K_i \notin \mathcal{T}^1(\Omega - \{\mathcal{C}\})$. Since K_i is join irreducible in $\cup\Omega$, it is evident that $K_i \in \mathcal{C}$ and $K_i \notin \cup(\Omega - \{\mathcal{C}\})$. In another word, $\mathcal{C}$ is the only attribute containing K_i, thus $r^1(K_i) = \{\mathcal{C}\}$.

Suppose there is $r^1(K_i)$ such that $r^1(K_i) = \{\mathcal{C}\}$, it is evident that $\mathcal{C} \in CORE(\Omega)$.

Proposition 2. *Let $\Omega' \subseteq \Omega$, then $\mathcal{T}^1(\Omega') = \mathcal{T}^1(\Omega)$ if and only if $\Omega' \cap r^1(K_i) \neq \varnothing$, for any $r^1(K_i), i = 1, 2, ..., m$.*

Proof. Suppose $\mathcal{T}^1(\Omega') = \mathcal{T}^1(\Omega)$, then for any $K_i \in \mathcal{T}^1(\Omega)$, $K_i \in \mathcal{T}^1(\Omega')$. That means there is $\mathcal{C} \in \Omega'$ such that $K_i \in \mathcal{C}$. It is evident that $\mathcal{C} \in \Omega'$ and $\mathcal{C} \in r^1(K_i)$, thus $\Omega' \cap r^1(K_i) \neq \varnothing$.

Suppose $\Omega' \cap r^1(K_i) \neq \varnothing$, for any $r^1(K_i), i = 1, 2, ..., m$. Then for any $K_i \in \mathcal{T}^1(\Omega)$, since $\Omega' \cap r^1(K_i) \neq \varnothing$, suppose $\mathcal{C} \in (\Omega' \cap r^1(K_i))$. It is obvious that $K_i \in \mathcal{C} \in \Omega'$, then $K_i \in \mathcal{T}^1(\Omega')$. Thus $\mathcal{T}^1(\Omega') = \mathcal{T}^1(\Omega)$.

Proposition 3. *Let $\Omega' \subseteq \Omega$, then Ω' is a reduct of Ω, if and only if it is a minimal subset satisfying $\Omega' \cap r^1(K_i) \neq \varnothing$, for any $r^1(K_i) \neq \varnothing, i = 1, 2, ..., m$.*

Definition 9. *Let $\langle U, \Omega\rangle$ be a multi-granulation covering information system, $\Omega = \{\mathcal{C}_1, \mathcal{C}_2, ..., \mathcal{C}_n\}$ be a family of coverings on U, $R^1(U, \Omega) = \{r^1(K_i) | i = 1, 2, ..., m\}$ be the related family. A related function $f^1(U, \Omega)$ is a Boolean function of n boolean variables $\overline{\mathcal{C}_1}, \overline{\mathcal{C}_2}, ..., \overline{\mathcal{C}_n}$ corresponding to the coverings $\mathcal{C}_1, \mathcal{C}_2, ..., \mathcal{C}_n$, respectively, which is defined as: $f^1(U, \Omega)(\overline{\mathcal{C}_1}, \overline{\mathcal{C}_2}, ..., \overline{\mathcal{C}_n}) = \wedge\{\vee(r^1(K_i) | r^1(K_i) \in R^1(U, \Omega)\}$.*

Theorem 1. *Let $\langle U, \Omega\rangle$ be a multi-granulation covering information system, $f^1(U, \Omega)$ be a related function. If $g(U, \Omega) = (\wedge\Omega_1) \vee ... \vee (\wedge\Omega_l)$ is a reduced disjunctive derived from $f^1(U, \Omega)$ via the laws of multiplication and absorption. Namely, for any $\Omega_k \subseteq \Omega, k = 1, 2, ..., l$, there is no repeated element in Ω_k. Then $RED(\Omega) = \{\Omega_1, \Omega_2, ..., \Omega_l\}$.*

Proof. For any $r^1(K_i) \in R^1(U, \Omega)$, $\wedge\Omega_k \leq \vee r^1(K_i)$ is established for every $k = 1, 2, ..., l$, so $\Omega_k \cap r^1(K_i) \neq \varnothing$. Let $\Omega'_k = \Omega_k - \{\mathcal{C}\}$ for any $\mathcal{C} \in \Omega_k$, then $g(U, \Omega) \lneqq \vee_{t=1}^{k-1}(\wedge\Omega_t) \vee (\wedge\Omega'_k) \vee (\vee_{t=k+1}^{l}(\wedge\Omega_t))$. If for every $r^1(K_i) \in R^1(U, \Omega)$, we have $\Omega'_k \cap r^1(K_i) \neq \varnothing$, then $\wedge\Omega'_k \leq \vee r^1(K_i)$ for every $r^1(K_i) \in R^1(U, \Omega)$. Namely, $g(U, \Omega) \geq \vee_{t=1}^{k-1}(\wedge\Omega_t) \vee (\wedge\Omega'_k) \vee (\vee_{t=k+1}^{l}(\wedge\Omega_t))$, which is a contradiction. It means there is $r_{i0}^1 \in R^1(U, \Omega)$ such that $\Omega'_k \cap r_{i0}^1 = \varnothing$. Therefore, Ω_k is a reduct of Ω.

For any $X \in RED(\Omega)$, $X \cap r^1(K_i) \neq \varnothing$ is established for every $r^1(K_i) \in R^1(U,\Omega)$, so $f^1(U,\Omega) \wedge (\wedge X) = \wedge(\vee r^1(K_i)) \wedge (\wedge X) = \wedge X$, which means $\wedge X \leq f^1(U,\Omega) = g(U,\Omega)$. Suppose $\Omega_k - X \neq \varnothing$ for every $k = 1,2,...,l$. Then there is $\mathcal{C}_k \in \Omega_k - X$ for every k. Through rewriting $g(U,\Omega) = (\vee_{k=1}^{l}\mathcal{C}_k) \wedge \Phi$, $\wedge X \leq \vee_{k=1}^{l}\mathcal{C}_k$. Thus, there is $\mathcal{C}_{k_0}$ such that $\wedge X \leq \mathcal{C}_{k_0}$, that is to say, $\mathcal{C}_{k_0} \in X$, which is a contradiction. So $\Omega_{k_0} \subseteq X$ for some k_0, since both X and Ω_{k_0} are reducts, it is clearly that $X = \Omega_{k_0}$. Therefore, $RED(\Omega) = \{\Omega_1, \Omega_2, ..., \Omega_l\}$.

From the above theorem, we can compute all attribute reducts of Ω with related functions.

Example 2. Let $\langle U,\Omega\rangle$ be a multi-granulation covering information system, where $\Omega = \{\mathcal{C}_1,\mathcal{C}_2,\mathcal{C}_3\}$, where $U = \{x_1,x_2,x_3,x_4\}$, $\mathcal{C}_1 = \{\{x_1\},\{x_2,x_3\},\{x_3,x_4\}\}$, $\mathcal{C}_2 = \{\{x_1,x_2,x_4\},\{x_1,x_2,x_3\},\{x_3,x_4\}\}$, $\mathcal{C}_3 = \{\{x_1,x_2,x_3\},\{x_2,x_4\},\{x_1,x_4\}\}$.

We can see that

$\cup\Omega = \{\{x_1\},\{x_2,x_3\},\{x_3,x_4\},\{x_1,x_2,x_4\},\{x_1,x_2,x_3\},\{x_2,x_4\},\{x_1,x_4\}\}$.

It is easy to calculate that

$Md_{\cup\Omega}(x_1) = \{\{x_1\}\}$,

$Md_{\cup\Omega}(x_2) = \{\{x_2,x_3\},\{x_2,x_4\}\}$,

$Md_{\cup\Omega}(x_3) = \{\{x_2,x_3\},\{x_3,x_4\}\}$,

$Md_{\cup\Omega}(x_4) = \{\{x_3,x_4\},\{x_2,x_4\},\{x_1,x_4\}\}$.

Obviously,

$\cup\{Md_{\cup\Omega}(x)|x \in U\} = \{\{x_1\},\{x_1,x_4\},\{x_2,x_3\},\{x_2,x_4\},\{x_3,x_4\}\}$

$r^1(\{x_1\}) = \{\mathcal{C}_1\}$,

$r^1(\{x_1,x_4\}) = \{\mathcal{C}_3\}$,

$r^1(\{x_2,x_3\}) = \{\mathcal{C}_1\}$,

$r^1(\{x_2,x_4\}) = \{\mathcal{C}_3\}$,

$r^1(\{x_3,x_4\}) = \{\mathcal{C}_1,\mathcal{C}_2\}$.

$f^1(U,\Omega)(\overline{\mathcal{C}_1},\overline{\mathcal{C}_2},\overline{\mathcal{C}_3}) = \wedge\{\vee(r^1(K_i)|r^1(K_i) \in R^1(U,\Omega)\} = \mathcal{C}_1 \wedge \mathcal{C}_3 \wedge (\mathcal{C}_1 \vee \mathcal{C}_2) = \mathcal{C}_1 \wedge \mathcal{C}_3$.

That is to say, $\{\mathcal{C}_1,\mathcal{C}_3\}$ is a reduct of these granular structures in the the first type of optimistic CMGRS.

3.2 Reduction Based the Second Type of Optimistic CMGRS

Suppose $\langle U,\Omega\rangle$ is a multi-granulation covering information system, $\Omega = \{\mathcal{C}_1,\mathcal{C}_2,...,\mathcal{C}_m\}$ be a family of coverings on U, and $\mathcal{N}_i = \{N_i(x_1),N_i(x_2),...N_i(x_{|U|})\}$ is the set of all neighborhoods induced by $\mathcal{C}_i$. It is clear that $\cup\mathcal{N}_i = \{N_i(x) \in \mathcal{N}_i|x \in U, i = 1,2,...,m\}$ is still a covering of U. Obviously, it is the minimal covering set which can generate the same lower approximation in the the second type of optimistic CMGRS.

Definition 10 *(type-2 space). Suppose $\langle U,\Omega\rangle$ is a multi-granulation covering information system, $\Omega = \{\mathcal{C}_1,\mathcal{C}_2,...,\mathcal{C}_m\}$ be a family of coverings on U, and $\mathcal{N}_i = \{N_i(x_1),N_i(x_2),...N_i(x_{|U|})\}$. We define $\mathcal{T}^2(\Omega) = \cup_{i=1}^{m}\mathcal{N}_i$ as the second type of multi-granulation space, denoted $\mathcal{T}^2$-space.*

Definition 11. *Let $\langle U, \Omega \rangle$ be a multi-granulation covering information system, $\Omega = \{\mathcal{C}_1, \mathcal{C}_2, ..., \mathcal{C}_m\}$ be a family of coverings on U, and $\mathcal{N}_i = \{N_i(x_1), N_i(x_2), ...N_i(x_{|U|})\}$. $\mathcal{C}_i$ is said to be dispensable granulation in Ω if $\mathcal{T}^2(\Omega) = \mathcal{T}^2(\Omega - \{\mathcal{C}_i\})$. Otherwise, $\mathcal{C}_i$ is indispensable granulation in Ω. For every $\mathbb{G} \subseteq \Omega$, if $\mathcal{T}^2(\Omega) = \mathcal{T}^2(\mathbb{G})$ and every granulation in $\mathbb{G}$ is indispensable, $\mathbb{G}$ is called a reduct of Ω, the collection of all reducts of Ω is denoted by $RED(\Omega)$. The collection of all indispensable granulations in Ω is called the core of Ω, denoted as $CORE(\Omega)$.*

Definition 12 *(Related family on type-2 space). Let $\langle U, \Omega \rangle$ be a multi-granulation covering information system, $\Omega = \{\mathcal{C}_1, \mathcal{C}_2, ..., \mathcal{C}_m\}$ a family of coverings of U, and $\mathcal{N}_i = \{N_i(x_1), N_i(x_2), ...N_i(x_{|U|})\}$. For any $N_i(x_j) \in \mathcal{T}^2(\Omega)$, we define $r^2(N_i(x_j)) = \{\mathcal{C}_i \in \Omega | N_i(x_j) \in \mathcal{T}^2(\Omega)\}$ is the related set of $N_i(x_j)$, and $R^2(U, \Omega) = \{r^2(N_i(x_j)) | N_i(x_j) \in \mathcal{T}^2(\Omega)\}$ is the related family of (U, Ω).*

Proposition 4. $CORE(\Omega) = \{\mathcal{C} \in \Omega \mid \exists r^2(N_i(x_j)) \in \mathcal{T}^2(\Omega) s.t. r^2(N_i(x_j)) = \{\mathcal{C}\}\}$.

Proof. Suppose $\mathcal{C} \in CORE(\Omega)$, then $\mathcal{T}^2(\Omega) \neq \mathcal{T}^2(\Omega - \{\mathcal{C}\})$. Let $\mathcal{T}^2(\Omega) = \{N_i(x_j), i = 1, 2, ..., m; j = 1, 2, ..., |U|\}$ and $\mathcal{T}^2(\Omega - \{\mathcal{C}\}) = \{N_i'(x_j), i = 1, 2, ..., m; j = 1, 2, ..., |U|\}$. There must be $1 \leq i \leq m$ and $1 \leq i \leq |U|$ such that $N_i(x_j) \neq N_i'(x_j)$, then $N_i(x_j) \subset N_i'(x_j)$. Suppose $y \in (N_i'(x_j) - N_i(x_j))$, then there must be $K \in \mathcal{C}$ such that $y \in K$ and $x_i \notin K$ and for any $K' \in \cup(\Omega - \{\mathcal{C}\})$, if $y \in K'$ then $x_j \in K'$. Thus $r^2(N_i(y)) = \{\mathcal{C}\}$.

Suppose there is $r^2(N_i(x_j))$ such that $r^2(N_i(x_j)) = \{\mathcal{C}\}$, thus $N_i(x_j)$ will be changed if we delete $\mathcal{C}$ from Ω. It is evident that $\mathcal{C} \in CORE(\Omega)$.

Proposition 5. *Let $\Omega' \subseteq \Omega$, then $\mathcal{T}^2(\Omega') = \mathcal{T}^2(\Omega)$ if and only if $\Omega' \cap r^2(N_i(x_j)) \neq \varnothing$, for any $r^2(N_i(x_j)), i = 1, 2, ..., m$.*

Proof. Suppose $\mathcal{T}^2(\Omega') = \mathcal{T}^2(\Omega)$, then for any $N_i(x_j) \in \mathcal{T}^2(\Omega)$, $N_i(x_j)$ can be induced by Ω'. That means $\mathcal{C}_i \in \Omega'$. Based on the definition of related set, we get that $\mathcal{C}_i \in r^2(N_i(x_j))$. It is obvious that $\mathcal{C}_i \in (r^2(N_i(x_j)) \cap \Omega')$. Then $\Omega' \cap r^2(N_i(x_j)) \neq \varnothing$, for any $r^2(N_i(x_j)), i = 1, 2, ..., m$.

Suppose $\Omega' \cap r^2(N_i(x_j)) \neq \varnothing$, for any $r^2(N_i(x_j)), i = 1, 2, ..., m$. Then for any $N_i(x_j) \in \mathcal{T}^2(\Omega)$, since $\Omega' \cap r^2(N_i(x_j)) \neq \varnothing$, suppose $\mathcal{C} \in (\Omega' \cap r^2(N_i(x_j)))$. It is obvious that $N_i(x_j)$ can be induced by $\mathcal{C}$, then $\mathcal{T}^2(\Omega)$ can be induced by Ω'. Thus $\mathcal{T}^2(\Omega') = \mathcal{T}^2(\Omega)$.

Proposition 6. *Let $\Omega' \subseteq \Omega$, then Ω' is a reduct of Ω, if and only if it is a minimal subset satisfying $\Omega' \cap r^2(N_i(x_j)) \neq \varnothing$, for any $r^2(N_i(x_j)) \neq \varnothing, i = 1, 2, ..., m$.*

Definition 13. *Let $\langle U, \Omega \rangle$ be a multi-granulation covering information system, $\Omega = \{\mathcal{C}_1, \mathcal{C}_2, ..., \mathcal{C}_m\}$ be a family of coverings on U, $R^2(U, \Omega) = \{r^2(N_i(x_j)) | N_i(x_j) \in \mathcal{T}^2(\Omega)\}$. A related function $f^2(U, \Omega)$ is a Boolean function of m boolean variables $\overline{\mathcal{C}_1}, \overline{\mathcal{C}_2}, ..., \overline{\mathcal{C}_m}$ corresponding to the coverings $\mathcal{C}_1, \mathcal{C}_2, ..., \mathcal{C}_m$, respectively, which is defined as: $f^2(U, \Omega)(\overline{\mathcal{C}_1}, \overline{\mathcal{C}_2}, ..., \overline{\mathcal{C}_m}) = \wedge\{\vee(r^2(N_i(x_j)) | r^2(N_i(x_j)) \in R^2(U, \Omega)\}$.*

Theorem 2. *Let $\langle U, \Omega \rangle$ be a multi-granulation covering information system, $f^2(U, \Omega)$ be a related function. If $g(U, \Omega) = (\wedge \Omega_1) \vee ... \vee (\wedge \Omega_l)$ is a reduced disjunctive derived from $f^2(U, \Omega)$ via the laws of multiplication and absorption. Namely, for any $\Omega_k \subseteq \Omega$, $k = 1, 2, ..., l$, there is no repeated element in Ω_k. Then $RED(\Omega) = \{\Omega_1, \Omega_2, ..., \Omega_l\}$.*

Proof. The proof is similar to that of Theorem 1.

From the above theorem, we can compute all attribute reducts of Ω with related functions.

Example 3. Let $\langle U, \Omega \rangle$ be a multi-granulation covering information system, and $\Omega = \{\mathcal{C}_1, \mathcal{C}_2, \mathcal{C}_3\}$, where $U = \{x_1, x_2, x_3, x_4\}, \mathcal{C}_1 = \{\{x_1, x_2, x_3\}, \{x_2, x_3, x_4\}, \{x_1, x_3, x_4\}\}$, $\mathcal{C}_2 = \{\{x_1, x_2, x_3\}, \{x_1, x_2, x_4\}, \{x_2, x_4\}\}, \mathcal{C}_3 = \{\{x_1, x_2\}, \{x_3, x_4\}, \{x_1, x_2, x_3\}\}$.

It is easy to calculate that

$\mathcal{N}_1 = \{\{x_1, x_3\}, \{x_2, x_3\}, \{x_3\}, \{x_3, x_4\}\}$,

$\mathcal{N}_2 = \{\{x_1, x_2\}, \{x_2\}, \{x_1, x_2, x_3\}, \{x_2, x_4\}\}$,

$\mathcal{N}_3 = \{\{x_1, x_2\}, \{x_3\}, \{x_3, x_4\}\}$.

Obviously,

$\cup_{i=1}^{3} \mathcal{N}_i = \{\{x_1, x_2\}, \{x_1, x_3\}, \{x_2\}, \{x_2, x_3\}, \{x_1, x_2, x_3\}, \{x_2, x_4\}, \{x_3\}, \{x_3, x_4\}\}$

$r^2(\{x_1, x_2\}) = \{\mathcal{C}_2, \mathcal{C}_3\}$,

$r^2(\{x_1, x_3\}) = \{\mathcal{C}_1\}$,

$r^2(\{x_2\}) = \{\mathcal{C}_2\}$,

$r^2(\{x_2, x_3\}) = \{\mathcal{C}_1\}$,

$r^2(\{x_1, x_2, x_3\}) = \{\mathcal{C}_2\}$,

$r^2(\{x_2, x_4\}) = \{\mathcal{C}_2\}$,

$r^2(\{x_3\}) = \{\mathcal{C}_1, \mathcal{C}_3\}$,

$r^2(\{x_3, x_4\}) = \{\mathcal{C}_1, \mathcal{C}_3\}$.

$f^2(U, \Omega)(\overline{\mathcal{C}_1}, \overline{\mathcal{C}_2}, \overline{\mathcal{C}_3}) = \wedge\{\vee(r^2(N_i(x_j))|r^2(N_i(x_j)) \in R^2(U, \Omega)\} = (\mathcal{C}_2 \vee \mathcal{C}_3) \wedge \mathcal{C}_1 \wedge \mathcal{C}_2 \wedge (\mathcal{C}_1 \vee \mathcal{C}_3) = \mathcal{C}_1 \wedge \mathcal{C}_2$.

That is to say, $\{\mathcal{C}_1, \mathcal{C}_2\}$ is the attribute reduct of the second type of optimistic CMGRS.

4 Conclusions

We propose two approximation spaces for two types of optimistic multi-granulation rough set models, the first type of multi-granulation space and the second type of multi-granulation space, respectively. Based on the two approximation spaces, we defined related family to calculate all attribute reducts of two types of optimistic multi-granulation rough set models, which enriches the theoretical basis of attribute reduction. In the future, we will continue to investigate other multi-granulation approximation operators, and try to design efficient and effective attribute reduction algorithms.

Acknowledgments. This work is supported by the National Natural Science Foundation of China (Grant No. 11201490), special financial grant from the Postdoctoral Science Foundation NO. 2017T100795, Natural Science Foundation of Hunan Province (Grant No. 2017JJ2408).

References

1. Bonikowski, Z., Bryniarski, E., Wybraniec-Skardowska, U.: Extensions and intentions in the rough set theory. Inf. Sci. **107**(1–4), 149–167 (1998)
2. Jing, Y., Li, T., Fujita, H., Yu, Z., Wang, B.: An incremental attribute reduction approach based on knowledge granularity with a multi-granulation view. Inf. Sci. **411**, 23–38 (2017)
3. Liang, J., Feng, W., Dang, C., Qian, Y.: An efficient rough feature selection algorithm with a multi-granulation view. Int. J. Approximate Reasoning **53**(6), 912–926 (2012)
4. Lin, G., Liang, J., Qian, Y.: Multigranulation rough sets: from partition to covering. Inf. Sci. **241**, 101–118 (2013)
5. Lin, Y., Li, J., Lin, P., Lin, G., Chen, J.: Feature selection via neighborhood multi-granulation fusion. Knowl.-Based Syst. **67**, 162–168 (2014)
6. Liu, C., Pedrycz, W.: Covering-based multi-granulation fuzzy rough sets. J. Intell. Fuzzy Syst. **30**(1), 303–318 (2016)
7. Pawlak, Z.: Rough sets. Int. J. Comput. Inf. Sci. **11**(5), 341–356 (1982)
8. Qian, Y., Liang, J.: Rough set method based on multi-granulations. In: 2006 5th IEEE International Conference on Cognitive Informatics, vol. 1, pp. 297–304. IEEE (2006)
9. Qian, Y., Liang, J., Dang, C.: Incomplete multigranulation rough set. IEEE Trans. Syst. Man Cybern. Part A Syst. Hum. **40**(2), 420–431 (2010)
10. Qian, Y., Liang, J., Yao, Y., Dang, C.: MGRS: a multi-granulation rough set. Inf. Sci. **180**(6), 949–970 (2010)
11. Qian, Y., Zhang, H., Sang, Y., Liang, J.: Multigranulation decision-theoretic rough sets. Int. J. Approximate Reasoning **55**(1), 225–237 (2014)
12. Yang, T., Li, Q., Zhou, B.: Related family: a new method for attribute reduction of covering information systems. Inf. Sci. **228**, 175–191 (2013)
13. Yao, Y., Yao, B.: Covering based rough set approximations. Inf. Sci. **200**, 91–107 (2012)
14. Zakowski, W.: Approximations in the space (u, π). Demonstratio mathematica **16**(3), 761–770 (1983)
15. Zhang, X., Kong, Q.: On four types of multi-covering rough sets. Fundamenta Informaticae **147**(4), 457–476 (2016)
16. Zhu, W.: Relationship between generalized rough sets based on binary relation and covering. Inf. Sci. **179**(3), 210–225 (2009)
17. Zhu, W., Wang, F.Y.: Reduction and axiomization of covering generalized rough sets. Inf. Sci. **152**, 217–230 (2003)

Constructing the Optimal Approximation Sets of Rough Sets in Multi-granularity Spaces

Qinghua Zhang[1,2(✉)], Fan Zhao[1,2], Xu Yubin[2], and Jie Yang[1]

[1] Chongqing Key Laboratory of Computational Intelligence, Chongqing University of Posts and Telecommunications, Chongqing 400065, China
zhangqh@cqupt.edu.cn

[2] School of Science, Chongqing University of Posts and Telecommunications, Chongqing 400065, China

Abstract. Rough set theory is an important tool to solve the uncertain problems. How to use the existing knowledge granules to approximately describe an uncertain target concept X has been a key issue. However, current research on theories and methods is still not comprehensive enough. $R_{0.5}(X)$, a kind of approximation sets of an uncertain concept, was proposed and analyzed in detail in our previous research work. However, whether $R_{0.5}(X)$ is the optimal approximation set of an uncertain concept X is still unable to determine. As a result, in this paper, based on the approximation of an uncertain concept, the existence of the optimal approximation set is explored. Then an optimal approximation set $R_{Best}(X)$ is proposed and discussed. At first, the definition of $R_{Best}(X)$ is defined. Then several comparative analysis between $R_{Best}(X)$ and other approximation sets is carried out. Next, operation properties of $R_{Best}(X)$ are presented and proved respectively. Finally, with changing knowledge granularity spaces, the change rules of the similarity between an uncertain set X and its $R_{Best}(X)$ are revealed.

Keywords: Rough sets · Uncertain concept · Similarity · Knowledge granularity · Granular computing

1 Introduction

Recently, computer technology and automatic control technology have rapidly developed, and research on the uncertain information system has attracted more and more researchers' attention [18,22]. Fuzzy set theory, rough set theory and quotient space theory are three basic granular computing models which have been successfully applied to process uncertain information. As a simple computing model, rough set theory [4,13] is an important method for handling uncertain problems as well as probability theory, fuzzy set theory and evidence theory. In the view of Pawlak's rough sets, people usually research how to acquire decision rules from the upper approximation set $\overline{R}(X)$ and the lower approximation

© Springer Nature Switzerland AG 2019
T. Mihálydeák et al. (Eds.): IJCRS 2019, LNAI 11499, pp. 341–355, 2019.
https://doi.org/10.1007/978-3-030-22815-6_27

set $\underline{R}(X)$. Furthermore, many extended rough set models are proposed to deal with the real-life uncertain information, such as variable precision rough set model [26], probability rough set model [11], game-theoretic rough sets [2] and so on [10,12,25]. Pawlak and Skworn analyzed and summarized these extended models referred to [11]. Mi analyzed the variable precision rough set model and discussed how to use this model to obtain attribute reduction [8]. Yao and Ziarko et al., combining probability and inclusion degree, proposed probability rough set model and obtained many related results [17,19,20]. However, these methods mainly focus on constructing the extended approximation operators of traditional rough set model. There is little research on how to use the existing knowledge granules in knowledge base to construct an approximation set of X. Could we construct an approximation set which is more approximate to X than $\overline{R}(X)$ or $\underline{R}(X)$? And does an optimal approximation set exist? The first problem is solved in our previous work [21], and the second problem is our main motivations in this paper.

Based on above assumptions, the related models and results on the approximation set of an uncertain set were proposed in our other paper referred to [21,23]. In these papers, the basic idea is translating rough sets into fuzzy sets according to the different membership degree of elements in boundary region and constructing an approximation set of an uncertain concept by using cut-set of the fuzzy set with some thresholds. With this construction method, the approximation sets with the existing knowledge granules can be obtained directly. In the literature [21], a general approximation set was constructed and it had many good properties. Experimental results show that $R_{0.5}(X)$ is a better model dealing with uncertain information systems. Better classification results could be obtained with $R_{0.5}(X)$. The amount of correct classification objects increases and amount of uncertain classification objects reduces.However, that $R_{0.5}(X)$ is the optimal approximation set of an uncertain set X is still unable to determine, and the related concepts and results on the optimal approximation were not presented in [21]. It is difficult to search for the optimal approximation set of an uncertain set directly. Based on the research referred to [21], through minimizing similarity between the target concept and its approximation sets, the optimal approximation set $R_{Best}(X)$ is defined, and an algorithm for constructing the optimal approximation set $R_{Best}(X)$ is proposed in this paper. And several comparative analysis between $R_{Best}(X)$ and other approximation sets is carried out In addition, the operations properties of $R_{Best}(X)$ is analyzed. Finally we discuss the change rules of the similarity degree between X and its $R_{Best}(X)$ in different knowledge granularity levels.

The rest of this paper is organized as follows. In Sect. 2, the related basic concepts and preliminary knowledge are reviewed. The $R_{Best}(X)$ of an uncertain set X in rough approximation spaces is proposed in Sect. 3. Besides, several comparative analysis between $R_{Best}(X)$ and other approximation sets and many operation rules related the $R_{Best}(X)$ are given in Sect. 3. The change rules of the similarity degree between X and its $R_{Best}(X)$ in the different knowledge granularity levels are discussed in Sect. 4. Finally, the conclusions are drawn in Sect. 5.

2 Preliminaries

In order to better present the context of this paper, many preliminary concepts, definitions and results related to rough set and uncertainty measurement are reviewed as follows.

Definition 1 (Information table of knowledge system [9,14]**).** *A knowledge system can be described as* $S = \langle U, A, V, f\rangle$. *U is the domain.* $A = C \cup D$ *is the set of all attributes. Subset C is the set of conditional attributes, and D is the set of decision attributes.* $V = \cup_{r\in A} V_r$ *is the set of attribute values.* V_r *describes the range of attribute values r where* $r \in A$. $f : U \times A \to V$ *is a function which describes attribute values of object x in U.*

Definition 2 (Indiscernibility Relation [9,14]**).** *For any attribute set* $R \subseteq A$*, an indiscernibility relation is defined as*

$$IND(R) = \{(x, y)|(x, y) \in U^2 \wedge \forall_{b\in R}(b(x) = b(y))\}.$$

Definition 3 (Upper Approximation Set and Lower Approximation Set [9,14]**).** *Let* $S = \langle U, A, V, f\rangle$ *be a knowledge System, for any* $X \subseteq U$ *and* $R \subseteq A$*, the upper approximation set* $\overline{R}(X)$ *and the lower approximation set* $\underline{R}(X)$ *of X are defined as follows,*

$$\overline{R}(X) = \cup\{Y_i|Y_i \in U/IND(R) \wedge Y_i \cap X \neq \emptyset\},$$
$$\underline{R}(X) = \cup\{Y_i|Y_i \in U/IND(R) \wedge Y_i \subseteq X\},$$

where $U/IND(R) = \{X|X \subseteq U \wedge \forall_{x\in X, y\in Y, b\in R}\ (b(x) = b(y))\}$ *is a partition of equivalence relation R on U. The upper approximation set and the lower approximation set of X on R can be defined in another form as follows,*

$$\overline{R}(X) = \{x|x \in U \wedge [x]_R \cap X \neq \emptyset\},$$
$$\underline{R}(X) = \{x|x \in U \wedge [x]_R \subseteq X\},$$

where $[x]_R \in U/IND(R)$*, and* $[x]_R$ *is an equivalence class of x on relation R.* $\underline{R}(X)$ *is a set of objects which certainly belong to U according to knowledge R;* $\overline{R}(X)$ *is a set of objects which possibly belong to U according to knowledge R.* $BND_R(X) = \overline{R}(X) - \underline{R}(X)$ *is called as* ***Boundary region*** *of the target concept X on attribute set R.* $POS_R(X) = \underline{R}(X)$ *is called as* ***Positive region*** *of target concept X on attribute set R.* $NEG_R(X) = U - \overline{R}(X)$ *is called as* ***Negative region*** *of target concept X on attribute set R.* $BND_R(X)$ *is a set of objects which just possibly belong to target concept X.*

Let U be a finite domain. Let $X \subseteq U$*,* $x \in U$*, and the membership degree of x belong to set X is defined as*

$$\mu_X^R(x) = \frac{|X \cap [x]_R|}{|[x]_R|},$$

obviously, $0 \leq \mu_X^R(x) \leq 1$.

Definition 4 (λ-Approximation Sets of X [21]). *Let X be a subset (the target concept) of U, let*

$$R_{\lambda}(X) = \{x | x \in U \wedge \mu_X^R(x) \geq \lambda\} \ \ (1 \geq \lambda > 0),$$

then $R_{\lambda}(X)$ is called as λ-approximation sets of X. Let

$$\underset{\cdot}{R}_{\lambda}(X) = \{x | x \in U \wedge \mu_X^R(x) > \lambda\} \ \ (1 \geq \lambda > 0),$$

then $\underset{\cdot}{R}_{\lambda}(X)$ is called as λ-strong approximation sets of X.

So when $\lambda = 0.5$, $R_{0.5}(X)$ is called as 0.5-approximation sets of X and $\underset{\cdot}{R}_{0.5}(X)$ is called as 0.5-strong approximation sets of X.

Definition 5 [14]**.** *Let $U = \{x_1, x_2, \cdots, x_n\}$ be a non-empty finite set, $P' = \{P'_1, P'_2, \cdots, P'_l\}$ and $P'' = \{P''_1, P''_2, \cdots, P''_m\}$ be two partition spaces on U. If $\forall_{P'_i \in P'}(\exists_{P''_j \in P''}(P'_i \subseteq P''_j))$, then P' is finer than P'', denoted by $P' \preceq P''$.*

Definition 6 [14]**.** *Let $U = \{x_1, x_2, \cdots, x_n\}$ be a non-empty finite set, $P' = \{P'_1, P'_2, \cdots P'_l\}$ and $P'' = \{P''_1, P''_2, \cdots P''_m\}$ be two partition spaces on U. If $P' \preceq P''$, and $\exists_{P'_i \in P'}(\exists_{P''_j \in P''}(P'_i \subset P''_j))(P'_i \subset P''_j))$, then P' is strictly finer than P'', denoted by $P' \prec P''$.*

Definition 7 (Similarity Degree [21]**).** *Let A and B be two subsets of U, the mapping: $S : U \times U \to [0, 1]$. $S(A, B)$ is called as* ***similarity degree*** *between A and B, if and only if $S(A, B)$ satisfy the following properties:*

(1) For any $A, B \subseteq U$, $0 \leq S(A, B) \leq 1$ (Boundedness),

(2) For any $A, B \subseteq U$, $S(A, B) = S(B, A)$ (Symmetry),

(3) For any $A \subseteq U$, $S(A, A) = 1$; $S(A, B) = 0$ if and only if $A \cap B = \emptyset$.

Any formula that satisfy above (1), (2) and (3) is a similarity degree formula between two sets. In similarity measurement of rough sets, because of its universality and effectiveness, most experts and scholars have adopted a similarity degree formula in reference [7] as

$$S(A, B) = \frac{|A \cap B|}{|A \cup B|},$$

where $|\cdot|$ represents cardinality of elements in finite subset. Obviously, this formula satisfies above (1), (2) and (3).

3 Optimal Approximation Set of Rough Set

In our previous works, we find that $R_{0.5}(X)$, as an approximation set of X, has many excellent properties. However, whether $R_{0.5}(X)$ is the optimal approximation set of X when $0 \leq \lambda < 0.5$? Let us analyze according to the following example.

Table 1. Decision information table

	x_1	x_2	x_3	x_4	x_5	x_6	x_7	x_8	x_9
a	1	1	1	1	1	1	1	1	1
b	1	1	0	1	1	0	1	1	0
c	1	0	0	1	0	0	1	0	0
d	1	1	1	0	0	0	0	0	0

Example 1. In a decision information table (Table 1), let $U = \{x_1, x_2, \ldots, x_9\}$, the condition attribute set $C = \{a, b, c\}$ and the decision attribute set $D = \{d\}$.

According to rough set theory, the following partitions can be obtained easily,

$$IND(C) = \{\{x_1, x_4, x_7\}, \{x_2, x_5, x_8\}, \{x_3, x_6, x_9\}\},$$
$$IND(D) = \{\{x_1, x_2, x_3\}, \{x_4, x_5, x_6, x_7, x_8, x_9\}\}.$$

Let $X_1 = \{x_1, x_2, x_3\}$, $X_2 = \{x_4, x_5, x_6, x_7, x_8, x_9\}$. According to the definition $\mu_X^R(x) = \frac{|[x]_R \cap X|}{|[x]_R|}$, a fuzzy set can be obtained as

$$F_{X_1}(U) = \left\{\frac{1/3}{x_1}, \frac{1/3}{x_2}, \frac{1/3}{x_3}, \frac{1/3}{x_4}, \frac{1/3}{x_5}, \frac{1/3}{x_6}, \frac{1/3}{x_7}, \frac{1/3}{x_8}, \frac{1/3}{x_9}\right\}.$$

Then

$$R_{0.5}(X_1) = \emptyset, \quad R_{0.3}(X_1) = U.$$

While

$$S(X_1, R_{0.5}(X_1)) = \frac{|X_1 \cap R_{0.5}(X_1)|}{|X_1 \cup R_{0.5}(X_1)|} = \frac{0}{3} = 0, S(X_1, R_{0.3}(X_1)) = \frac{1}{3}.$$

So, $S(X_1, R_{0.3}(X_1)) > S(X_1, R_{0.5}(X_1))$.

In the same way, the following fuzzy set can be obtained,

$$F_{X_2}(U) = \left\{\frac{2/3}{x_1}, \frac{2/3}{x_2}, \frac{2/3}{x_3}, \frac{2/3}{x_4}, \frac{2/3}{x_5}, \frac{2/3}{x_6}, \frac{2/3}{x_7}, \frac{2/3}{x_8}, \frac{2/3}{x_9}\right\}.$$

Then we have $R_{0.5}(X_2) = U$, $R_{0.3}(X_2) = U$. Here we have $S(X_2, R_{0.3}(X_2)) = S(X_2, R_{0.5}(X_2))$.

So, we find that the approximation set $R_{0.5}(X_1)$ is not the optimal approximation set of X_1, on the contrary, $R_{0.5}(X_2)$ is the optimal approximation set of X_2. Therefore, the approximation set $R_{0.5}(X)$ may not be the optimal approximation set of X and an optimal approximation set of X in rough approximation spaces must exist. Thus, based on the membership degree function $\mu_X^R(x)$, the optimal approximation set of X in rough approximation space is proposed through minimizing similarity between the target concept and its approximation sets.

Definition 8. *(Optimal approximation set) Let X be a subset (target concept) of U, $R_\lambda(X)$ be a $\lambda-$ approximation set of X, and $S_{Best} = \max_{0<\lambda\leq 1} \{S(X, R_\lambda(X))\}$. For any $\lambda(0 < \lambda \leq 1)$, if $S(X, R_\lambda(X)) = S_{Best}$, then the $R_\lambda(X)$ is called the optimal approximation set of X, denoted by $R_{Best}(X)$. Namely, $S(X, R_{Best}(X)) = S_{Best}$.*

According to Definition 8, we know if $R_\lambda(X)$ is the $R_{Best}(X)$, λ must be in the interval $(0, 0.5]$. In order to more clearly show the Definition 8, an example with a decision information table (Table 2) is presented.

Example 2. Let domain $U = \{x_1, x_2, \ldots, x_{15}\}$, the condition attribute set $C = \{a, b, c\}$ and the decision attribute set $D = \{d\}$.

Table 2. Decision information table

	x_1	x_2	x_3	x_4	x_5	x_6	x_7	x_8	x_9	x_{10}	x_{11}	x_{12}	x_{13}	x_{14}	x_{15}
a	1	1	1	1	2	2	1	1	1	2	1	2	1	2	2
b	2	2	0	0	1	2	2	0	0	1	2	2	0	1	2
c	0	0	2	2	1	1	0	2	2	1	0	1	2	1	1
d	1	1	1	1	1	2	2	2	2	2	3	3	3	3	3

According to rough set theory, the following partition is obtained easily,

$$U/IND(D) = \{\{x_1, x_2, x_3, x_4, x_5\}, \{x_6, x_7, x_8, x_9, x_{10}\}, \{x_{11}, x_{12}, x_{13}, x_{14}, x_{15}\}\}.$$

Here, three decision concepts induced by decision attribute set are generated, and they are $X_1 = \{x_1, x_2, x_3, x_4, x_5\}$, $X_2 = \{x_6, x_7, x_8, x_9, x_{10}\}$ and $X_3 = \{x_{11}, x_{12}, x_{13}, x_{14}, x_{15}\}$ respectively. For the decision concept X_1, computing $U/IND(C)$,

$$U/IND(C) = \{\{x_1, x_2, x_7, x_{11}\}, \{x_3, x_4, x_8, x_9, x_{13}\}, \{x_5, x_{10}, x_{14}\}, \{x_6, x_{12}, x_{15}\}\}.$$

Then let $Y_1 = \{x_1, x_2, x_7, x_{11}\}$, $Y_2 = \{x_3, x_4, x_8, x_9, x_{13}\}$, $Y_3 = \{x_5, x_{10}, x_{14}\}$, $Y_4 = \{x_6, x_{12}, x_{15}\}$. Computing the membership degree $\mu(x)$ of x $(x \in U)$, where $\mu(x) = \frac{|Y_i \cap X|}{|Y_i|}$, that is to say, every object in equivalence class Y_i has same membership degree. For X_1, the membership degrees are shown as follows,
(1) For equivalence class Y_1, the membership degree is1/2;
(2) For equivalence class Y_2, the membership degree is 2/5;
(3) For equivalence class Y_3, the membership degree is 1/3;
(4) For equivalence class Y_4, the membership degree is 0.

Computing $R_{0.5}(X_1)$ and $R_{\mu_i}(X_1)$ and sorting the μ_i, then we can get

$$R_{1/3}(X_1) = \{x_1, x_2, x_3, x_4, x_5, x_7, x_8, x_9, x_{10}, x_{11}, x_{13}, x_{14}\};$$
$$R_{2/5}(X_1) = \{x_1, x_2, x_3, x_4, x_7, x_8, x_9, x_{11}, x_{13}\}; R_{0.5}(X_1) = \{x_1, x_2, x_7, x_{11}\}.$$

Further, we can obtain:

$$S\left(X_1, R_{1/3}\left(X_1\right)\right) = \frac{5}{12}, S\left(X_1, R_{2/5}\left(X_1\right)\right) = \frac{2}{5}, S\left(X_1, R_{0.5}\left(X_1\right)\right) = \frac{2}{7}.$$

We find that $S_{1/3}(X_1)$ is maximum value, that is to say, the approximation set $R_{1/3}(X_1)$, is the optimal approximation set of X_1. Then we can obtain $R_{Best}(X_1)$ as follows,

$$R_{Best}(X_1) = R_{1/3}(X_1) = \{x_1, x_2, x_3, x_4, x_5, x_7, x_8, x_9, x_{10}, x_{11}, x_{13}, x_{14}\}.$$

Similarly,

$$R_{Best}(X_2) = \{x_3, x_4, x_5, x_6, x_8, x_9, x_{10}, x_{12}, x_{13}, x_{14}, x_{15}\};$$
$$R_{Best}(X_3) = \{x_5, x_6, x_{10}, x_{12}, x_{14}, x_{15}\}.$$

The purpose of selecting $R_{Best}(X)$ is to characterize a target concept and further acquire rules. So compared with $\overline{R}(X)$, $\underline{R}(X)$ and $R_{0.5}(X)$, what advantages does the $R_{Best}(X)$ have? Then, relative analysis is shown as follows:
(Continue)Example 3. In Table 2, we can get three decision concepts $X_1 = \{x_1, x_2, x_3, x_4, x_5\}$, $X_2 = \{x_6, x_7, x_8, x_9, x_{10}\}$ and $X_3 = \{x_{11}, x_{12}, x_{13}, x_{14}, x_{15}\}$. With these decision concepts, we can obtain,

$$\underline{R}(X_1) = \emptyset, \underline{R}(X_2) = \emptyset, \underline{R}(X_3) = \emptyset;$$
$$\overline{R}(X_1) = \{x_1, x_2, x_3, x_4, x_5, x_7, x_8, x_9, x_{10}, x_{11}, x_{13}, x_{14}\},$$
$$\overline{R}(X_2) = \{x_1, x_2, x_3, x_4, x_5, x_6, x_7, x_8, x_9, x_{10}, x_{11}, x_{12}, x_{13}, x_{14}, x_{15}\},$$
$$\overline{R}(X_3) = \{x_1, x_2, x_3, x_4, x_5, x_6, x_7, x_8, x_9, x_{10}, x_{11}, x_{12}, x_{13}, x_{14}, x_{15}\};$$
$$R_{0.5}(X_1) = \{x_1, x_2, x_7, x_{11}\}, R_{0.5}(X_2) = \emptyset, R_{0.5}(X_3) = \emptyset;$$
$$R_{Best}(X_1) = \{x_1, x_2, x_3, x_4, x_5, x_7, x_8, x_9, x_{10}, x_{11}, x_{13}, x_{14}\},$$
$$R_{Best}(X_2) = \{x_3, x_4, x_5, x_6, x_8, x_9, x_{10}, x_{12}, x_{13}, x_{14}, x_{15}\},$$
$$R_{Best}(X_3) = \{x_5, x_6, x_{10}, x_{12}, x_{14}, x_{15}\}.$$

From the decision information Table 2 we can acquire many decision rules based on $\underline{R}(X)$ are shown as follows,

(1) For X_1, the corresponding approximation set is $\emptyset$;
(2) For X_2, the corresponding approximation set is $\emptyset$;
(3) For X_3, the corresponding approximation set is $\emptyset$.

We can acquire many decision rules based on $\overline{R}(X)$ are shown as follows,

(1) For X_1, the decision rule is $(a = 1 \wedge b = 2 \wedge c = 0) \vee (a = 1 \wedge b = 0 \wedge c = 2) \vee (a = 2 \wedge b = 1 \wedge c = 1) \rightarrow d = 1$;

(2) For X_2, the decision rule is $(a = 2 \wedge b = 2 \wedge c = 1) \vee (a = 1 \wedge b = 2 \wedge c = 0) \vee (a = 1 \wedge b = 0 \wedge c = 2) \vee (a = 2 \wedge b = 1 \wedge c = 1) \rightarrow d = 2$;

(3) For X_3, the decision rule is $(a = 1 \wedge b = 2 \wedge c = 0) \vee (a = 2 \wedge b = 2 \wedge c = 1) \vee (a = 1 \wedge b = 0 \wedge c = 2) \vee (a = 2 \wedge b = 1 \wedge c = 1) \rightarrow d = 3$.

We can acquire many decision rules based on $R_{0.5}(X)$ are shown as follows,

(1) For X_1, the decision rule is $(a = 1 \wedge b = 2 \wedge c = 0) \to d = 1$;
(2) For X_2, the corresponding approximate set is $\emptyset$;
(3) For X_3, the corresponding approximate set is $\emptyset$.
We can acquire many decision rules based on $R_{Best}(X)$ are shown as follows,
(1) For X_1, the decision rule is $(a = 1 \wedge b = 2 \wedge c = 0) \vee (a = 1 \wedge b = 0 \wedge c = 2) \vee (a = 2 \wedge b = 2 \wedge c = 1) \vee (a = 2 \wedge b = 1 \wedge c = 1) \to d = 1$;
(2) For X_2, the decision rule is $(a = 2 \wedge b = 1 \wedge c = 1) \vee (a = 1 \wedge b = 0 \wedge c = 2) \vee (a = 2 \wedge b = 2 \wedge c = 1) \to d = 2$;
(3) For X_3, the decision rule is $(a = 2 \wedge b = 1 \wedge c = 1) \vee (a = 2 \wedge b = 2 \wedge c = 1) \to d = 3$.

Table 3. Comparative analysis

		Supporting amount	Wrong amount	Unrecognized amount
$\underline{R}(X)$	X_1	0	0	5
	X_2	0	0	5
	X_3	0	0	5
$\overline{R}(X)$	X_1	5	7	0
	X_2	5	10	0
	X_3	5	10	0
$R_{0.5}(X)$	X_1	2	2	3
	X_2	0	0	5
	X_3	0	0	5
$R_{Best}(X)$	X_1	5	7	0
	X_2	4	7	1
	X_3	3	3	2

A comparative analysis Table 3 is constructed according to the above decision rules. From these above rules acquired from $\underline{R}(X)$, $\overline{R}(X)$, $R_{0.5}(X)$ and $R_{Best}(X)$ and Table 3, the qualitative and quantitative comparisons could be made. It obvious that many objects can not determine decision classification if the decision rules are acquired based on $R_{0.5}(X)$ and $\underline{R}(X)$, and this is not an expected result in actual decision problems. Though each object belongs to a certain decision classification if the decision rules are acquired based on $\overline{R}(X)$ the objects in the boundary cannot be assigned to a certain decision classification, and the error classifications will be produced. Compared with $R_{0.5}(X)$ and $\underline{R}(X)$, the amount of correct classification objects increases, and amount of uncertain classification objects reduces if the decision rules are acquired based on $R_{Best}(X)$. Compared with $\overline{R}(X)$, although the amount of correct classification objects weakly declines, the amount of error classification objects also reduces if the decision rules are acquired based on $R_{Best}(X)$.

According to above comparison results, we find that the decision rules based on $R_{Best}(X)$ have more powerful classification ability than the decision rules based on $\underline{R}(X)$, $\overline{R}(X)$ and $R_{0.5}(X)$. The $R_{Best}(X)$ provide a novel perspective for approximate characterization of a target concept in multi-granularity spaces. Furthermore, it would be an effective method that could be suitable to real-life knowledge discovery from the uncertain information systems.

4 Operation Properties of $R_{Best}(X)$

It is well known that $\underline{R}(X)$ and $\underline{R}(X)$ have many important operation properties as literature [8,24]. Now, we will prove that the optimal approximation set $R_{Best}(X)$ has many similar operation properties with the upper approximation set and lower approximation set also. For convenience, let $R_{Best}(X) = R_k(X)$, $(0 < k \leq 0.5)$, U be a finite domain, and X, Y be two subsets on U, we have

(1) $R_k(\sim X) =\sim \underset{\cdot}{R}_{1-k}(X)$, $\underset{\cdot}{R}_k(\sim X) =\sim R_{1-k}(X)$;
(2) if $X \subseteq Y$, then $R_k(X) \subseteq R_k(Y)$, $\underset{\cdot}{R}_k(X) \subseteq \underset{\cdot}{R}_k(Y)$;
(3) $R_k(X \cap Y) \subseteq R_k(X) \cap R_k(Y)$, $\underset{\cdot}{R}_k(X \cap Y) \subseteq \underset{\cdot}{R}_k(X) \cap \underset{\cdot}{R}_k(Y)$;
(4) $R_k(X \cup Y) \supseteq R_k(X) \cup R_k(Y)$, $\underset{\cdot}{R}_k(X \cup Y) \supseteq \underset{\cdot}{R}_k(X) \cup \underset{\cdot}{R}_k(Y)$.

Proof. (1) Because

$$R_k(\sim X) = \left\{x|\mu_X^R(x) \geq k\right\} = \left\{x|\frac{|[x]_R \cap (\sim X)|}{|[x]_R|} \geq k\right\} = \left\{x|\frac{|[x]_R \cap (U - X)|}{|[x]_R|} \geq k\right\}$$
$$= \left\{x|1 - \frac{|[x]_R \cap X|}{|[x]_R|} \geq k\right\} = \left\{x|\frac{|[x]_R \cap X|}{|[x]_R|} \leq 1 - k\right\} =\sim \underset{\cdot}{R}_{1-k}(X),$$

Similarly, $\underset{\cdot}{R}_k(\sim X) =\sim R_{1-k}(X)$ holds. Hence, the proposition (1) is proved.

(2) $\forall x \in R_k(X)$, we have $[x]_R$ satisfying $\frac{|[x]_R \cap X|}{|[x]_R|} \geq k$; because $X \subseteq Y$, we have $|[x]_R \cap X| \leq |[x]_R \cap Y|$, then $\frac{|[x]_R \cap X|}{|[x]_R|} \leq \frac{|[x]_R \cap Y|}{|[x]_R|}$, then $\frac{|[x]_R \cap Y|}{|[x]_R|} \geq k$. So we can get $x \in R_k(Y)$, and then $R_k(X) \subseteq R_k(Y)$. Similarly, $\forall x \in \underset{\cdot}{R}_k(X)$, we have $[x]_R$ satisfying $\frac{|[x]_R \cap X|}{|[x]_R|} > k$; because $X \subseteq Y$, we have $|[x]_R \cap X| \leq |[x]_R \cap Y|$, then $\frac{|[x]_R \cap X|}{|[x]_R|} \leq \frac{|[x]_R \cap Y|}{|[x]_R|}$, and we can get $\frac{|[x]_R \cap Y|}{|[x]_R|} > k$. Therefore, we have $x \in \underset{\cdot}{R}_k(Y)$, and then $\underset{\cdot}{R}_k(X) \subseteq \underset{\cdot}{R}_k(Y)$. So, the proposition (2) is proved successfully.

(3) Because $X \cap Y \subseteq X$ and $X \cap Y \subseteq Y$, according to proposition (2) we have $R_k(X \cap Y) \subseteq R_k(X)$ and $R_k(X \cap Y) \subseteq R_k(Y)$, and we have $R_k(X \cap Y) \subseteq R_k(X) \cap R_k(Y)$. Similarly, we can get $\underset{\cdot}{R}_k(X \cap Y) \subseteq \underset{\cdot}{R}_k(X) \cap \underset{\cdot}{\mathrm{R}}\mathrm{k}(Y)$. So the proposition (3) holds.

(4) Because $X \subseteq X \cup Y$ and $Y \subseteq X \cup Y$, according to proposition (2), $R_k(X) \subseteq R_k(X \cup Y)$ and $R_k(Y) \subseteq R_k(X \cup Y)$, so $R_k(X \cup Y) \supseteq R_k(X) \cup R_k(Y)$ is held. Similarly because $X \subseteq X \cup Y$ and $Y \subseteq X \cup Y$, both $\underset{\cdot}{R}_k(X) \subseteq \underset{\cdot}{R}_k(X \cup Y)$ and $\underset{\cdot}{R}_k(Y) \subseteq \underset{\cdot}{R}_k(X \cup Y)$ are held, so we have $\underset{\cdot}{R}_k(X \cup Y) \supseteq \underset{\cdot}{R}_k(X) \cup \underset{\cdot}{R}_k(Y)$. Then the proposition (4) is proved successfully.

5 Change Rules of $R_{Best}(X)$ with Changing Knowledge Granularity

Currently, in different knowledge granularity levels, change rules of rough set uncertainty is one of important issues to measure the uncertainty of knowledge

[1,3,5,6,15,16]. Therefore, in different knowledge granularity levels, change rules of $S(X, R_{Best}(X))$ are also focus on our attention. Next, we will discuss the change rules in detail. Firstly, suppose a, b, c, d, e and f be all real number, and some basic results and lemmas are reviewed in order to discuss the relevant theorems easily.

Lemma 1. *[21] If $0 < a < b$ and $0 < c < d$, then $a/b < (a+d)/(b+c)$.*

Lemma 2. *[21] If $f/e = (b+d)/(a+c)$ and $b/a < f/e$, then $d/c > f/e$.*

Lemma 3. *[19] Let $0 < c < a$, $0 < d < b$. If $a/b \geq c/d$, then $a/b \leq (a-c)/(b-d)$. On the contrary, if $a/b \leq c/d$, then $a/b \geq (a-c)/(b-d)$.*

Next, we would discuss the relationship between $S(X,\ R_{Best}(X))$ and $S(X, R'_{Best}(X))$ in different knowledge granularity levels. Let $R_{Best}(X) = R_\lambda(X)$ then $(0 < \lambda \leq 0.5)$. And let $[x_{i_1}]_R, [x_{i_2}]_R, \ldots, [x_{i_k}]_R$ be the equivalence classes induced by an equivalence relation R on U, and $[x_{i_1}]_{R'}, [x_{i_2}]_{R'}, \ldots, [x_{i_k}]_{R'}$ be the equivalence classes induced by another equivalence relation R' on U. If the partition U/R' is finer than U/R namely $U/R' \preceq U/R$, for any $x \in U$, there is $[x]_{R'} \subseteq [x]_R$. For convenience, let $R_\lambda(X) = \underline{R}(X) \cup [x_{i_1}]_R \cup [x_{i_2}]_R \cup \ldots \cup [x_{i_k}]_R$, and $BND_R(X) = [x_{i_1}]_R \cup [x_{i_2}]_R \cup \ldots \cup [x_{i_m}]_R$. The equivalence classes divided into finer equivalence classes (sub-granules) may be in $NEG_R(X)$, $POS_R(X)$ or $BND_R(X)$. When the equivalence classes divided into finer equivalence classes are in $NEG_R(X)$ or $POS_R(X)$, $S(X, R_\lambda(X)) = S(X, R'_\lambda(X))$ is held. Next we will focus on $S(X, R'_\lambda(X))$ when the equivalence classes divided into finer equivalence classes are in $BND_R(X)$. For simplicity, suppose there is only one granule subdivided to two disjoint granules and others remain unchanged. That is to say, suppose $[x_{i_t}]_R = [x^1_{i_t}]_{R'} \cup [x^2_{i_t}]_{R'}$. This situation will be discussed as follows.

Theorem 1. *If $\lambda = 0.5$, $1 \leq t \leq k$, that is $[x_{i_t}]_R \subset R_\lambda(X)$:*
(1) If $[x^1_{i_t}]_{R'} \subset R'_\lambda(X)$, $[x^2_{i_t}]_{R'} \subset R'_\lambda(X)$, then $S(X, R_\lambda(X)) = S(X, R'_\lambda(X))$;
(2) If $[x^1_{i_t}]_{R'} \subset R'_\lambda(X)$, $[x^2_{i_t}]_{R'} \not\subset R'_\lambda(X)$, then $S(X, R_\lambda(X)) \leq S(X, R'_\lambda(X))$.

Proof. $\forall x \in R_{0.5}(X)$, we have $\mu^R_X(x) = \frac{|[x]_R \cap X|}{[x]_R} \geq 0.5$. Then we can obtain,

$$R_{0.5}(X) = \left\{x|\mu^R_X(x) \geq 0.5\right\} = \left\{x|\mu^R_X(x) = 1\right\} \cup \left\{x|0.5 \leq \mu^R_X(x) < 1\right\}.$$

Obviously, $\left\{x|\mu^R_X(x) = 1\right\} = \underline{R}(X)$, and then Let $\left\{x|0.5 \leq \mu^R_X(x) < 1\right\} = [x_{i_1}]_R \cup [x_{i_2}]_R \cup \ldots \cup [x_{i_k}]_R$, we have

$$X \cap R_{0.5}(X) = X \cap (\underline{R}(X) \cup [x_{i_1}]_R \cup [x_{i_2}]_R \cup \ldots \cup [x_{i_k}]_R).$$

Since the intersection sets between any two elements in $\underline{R}(X)$, $[x_{i_1}]_R, [x_{i_2}]_R, \ldots,$ $[x_{i_k}]_R$ is empty set, we have

$$\begin{aligned}|X \cap R_{0.5}(X)| =& |X \cap \underline{R}(X)| + |X \cap [x_{i_1}]_R| + |X \cap [x_{i_2}]_R| + \ldots + |X \cap [x_{i_n}]_R| \\ =& |\underline{R}(X)| + |X \cap [x_{i_1}]_R| + |X \cap [x_{i_2}]_R| + |\ldots| + |X \cap [x_{i_n}]_R|.\end{aligned}$$

Since $X \cup R_\lambda(X) = X \cup ([x_{i_1}]_R - X) \cup ([x_{i_2}]_R - X) \cup \ldots \cup ([x_{i_k}]_R - X)$ and the intersection between any two elements in X, $([x_{i_1}]_R - X), ([x_{i_2}]_R - X), \ldots,$ $([x_{i_k}]_R - X)$ is empty set, we have

$$|X \cup R_{0.5}(X)| = |X| + |([x_{i_1}]_R - X)| + |([x_{i_2}]_R - X)| + |\ldots| + |([x_{i_k}]_R - X)|.$$

Therefore,

$$S(X, R_{0.5}(X)) = \frac{|R(X)| + |X \cap [x_{i_1}]_R| + |X \cap [x_{i_2}]_R| + \ldots + |X \cap [x_{i_k}]_R|}{|X| + |[x_{i_1}]_R - X| + |[x_{i_2}]_R - X| + \ldots + |[x_{i_k}]_R - X|}.$$

While $[x_{i_t}^1]_{R'} \subset R'_\lambda(X)$ and $[x_{i_t}^2]_{R'} \subset R'_\lambda(X)$, we can get

$$\begin{aligned} &S(X, R'_{0.5}(X)) \\ &= \frac{|R'(X)| + |X \cap [x_{i_1}]_{R'}| + \ldots + |X \cap [x_{i_t}^1]_{R'}| + |X \cap [x_{i_t}^2]_{R'}| + \ldots + |X \cap [x_{i_k}]_{R'}|}{|X| + |[x_{i_1}]_{R'} - X| + \ldots + |[x_{i_t}^1]_{R'} - X| + |[x_{i_t}^2]_{R'} - X| + \ldots + |[x_{i_k}]_{R'} - X|} \\ &= \frac{|R(X)| + |X \cap [x_{i_1}]_R| + |X \cap [x_{i_2}]_R| + \ldots + |X \cap [x_{i_k}]_R|}{|X| + |[x_{i_1}]_R - X| + |[x_{i_2}]_R - X| + \ldots + |[x_{i_k}]_R - X|} \\ &= S(X, R_\lambda(X)). \end{aligned}$$

So the part (1) is proved successfully.

For the part (2) when $[x_{i_t}^2]_{R'} \not\subset R'_\lambda(X)$, we can have the equality as follows,

$$S(X, R'_{0.5}(X)) = \frac{|\underline{R}'(X)| + |X \cap [x_{i_1}]_{R'}| + \ldots + |X \cap [x_{i_t}^1]_{R'}| + \ldots + |X \cap [x_{i_k}]_{R'}|}{|X| + |[x_{i_1}]_{R'} - X| + \ldots + |[x_{i_t}^1]_{R'} - X| + \ldots + |[x_{i_k}]_{R'} - X|}.$$

With Lemma 2, we have the inequality $\frac{|X \cap [x_{i_t}^1]_{R'}|}{|[x_{i_t}^1]_{R'} - X|} > \frac{|X \cap [x_{i_t}]_{R'}|}{|[x_{i_t}]_{R'} - X|}$, let $\frac{|X \cap [x_{i_t}^1]_{R'}|}{|[x_{i_t}^1]_{R'} - X|} = \frac{|X \cap [x_{i_t}]_{R'}| + p}{|[x_{i_t}]_{R'} - X| + q}$, then $\frac{p}{q} > \frac{|X \cap [x_{i_t}]_{R'}|}{|[x_{i_t}]_{R'} - X|}$, then we can get

$$\begin{aligned} &S(X, R'_\lambda(X)) \\ &= \frac{|\underline{R}'(X)| + |X \cap [x_{i_1}]_{R'}| + |X \cap [x_{i_2}]_{R'}| + \ldots + |X \cap [x_{i_t}^1]_{R'}| + \ldots + |X \cap [x_{i_k}]_{R'}|}{|X| + |[x_{i_1}]_{R'} - X| + |[x_{i_2}]_{R'} - X| + \ldots + |[x_{i_t}^1]_{R'} - X| + \ldots + |[x_{i_k}]_{R'} - X|} \\ &= \frac{|\underline{R}(X)| + |X \cap [x_{i_1}]_R| + |X \cap [x_{i_2}]_R| + \ldots + |X \cap [x_{i_k}]_R| + p}{|X| + |[x_{i_1}]_R - X| + |[x_{i_2}]_R - X| + \ldots + |[x_{i_k}]_R - X| + q}. \end{aligned}$$

We know

$$S(X, R_{0.5}(X)) = \frac{|\underline{R}(X)| + |X \cap [x_{i_1}]_R| + |X \cap [x_{i_2}]_R| + \ldots + |X \cap [x_{i_k}]_R|}{|X| + |[x_{i_1}]_R - X| + |[x_{i_2}]_R - X| + \ldots + |[x_{i_k}]_R - X|},$$

according to Definition 6 we have $0 \le S(X, R_{0.5}(X)) \le 1$, then we can easily get the inequality as follow,

$$\begin{aligned} &|\underline{R}(X)| + |X \cap [x_{i_1}]_R| + |X \cap [x_{i_2}]_R| + \ldots + |X \cap [x_{i_k}]_R| \\ &\le |X| + |[x_{i_1}]_R - X| + |[x_{i_2}]_R - X| + \ldots + |[x_{i_k}]_R - X|. \end{aligned}$$

Since $\frac{p}{q} > \frac{|X \cap [x_{i_t}]_{R'}|}{|[x_{i_t}]_{R'} - X|}$ we have $p > q$. With Lemma 1,

$$\frac{|\underline{R}(X)| + |X \cap [x_{i_1}]_R| + \ldots + |X \cap [x_{i_k}]_R| + p}{|X| + |[x_{i_1}]_R - X| + \ldots + |[x_{i_k}]_R - X| + q} \geq \frac{|\underline{R}(X)| + |X \cap [x_{i_1}]_R| + \ldots + |X \cap [x_{i_k}]_R|}{|X| + |[x_{i_1}]_R - X| + \ldots + |[x_{i_k}]_R - X|},$$

namely $S(X, R_{0.5}(X)) \leq S(X, R'_{0.5}(X))$. Hence the part (2) is proved successfully.

Theorem 1 show that when $\lambda = 0.5$, no matter the equivalence classes divided into many sub-granules are in $NEG_R(X)$, $POS_R(X)$ or $BND_R(X)$, similarity degree $S(X, R'_\lambda(X))$ is not smaller than $S(X, R_\lambda(X))$.

Theorem 2. *If $0 < \lambda < 0.5$, $k < t \leq m$, that is $[x_{i_t}]_R \not\subset R_\lambda(X)$:*
(1) If $[x^1_{i_t}]_{R'} \subset R'_\lambda(X)$, $[x^2_{i_t}]_{R'} \not\subset R'_\lambda(X)$ and $S(X, R_\lambda(X)) \leq |X \cap [x^1_{i_t}]_{R'}|/|[x^1_{i_t}]_{R'} - X|$, then $S(X, R_\lambda(X)) \leq S(X, R'_\lambda(X))$;
(2) If $[x^1_{i_t}]_{R'} \not\subset R'_\lambda(X)$, $[x^2_{i_t}]_{R'} \not\subset R'_\lambda(X)$, then $S(X, R_\lambda(X)) = S(X, R'_\lambda(X))$.

Proof. According to the Theorem 1 we can get the equality as follow,

$$S(X, R_\lambda(X)) = \frac{|\underline{R}(X)| + |X \cap [x_{i_1}]_R| + |X \cap [x_{i_2}]_R| + \ldots + |X \cap [x_{i_k}]_R|}{|X| + |[x_{i_1}]_R - X| + |[x_{i_2}]_R - X| + \ldots + |[x_{i_k}]_R - X|}.$$

Since $[x^1_{i_t}]_{R'} \subset R'_\lambda(X)$ and $[x^2_{i_t}]_{R'} \not\subset R'_\lambda(X)$ we can obtain the following equality,

$$\begin{aligned} S(X, R'_\lambda(X)) &= \frac{|X \cap R'_\lambda(X)|}{|X \cup R'_\lambda(X)|} \\ &= \frac{|\underline{R}'(X)| + |X \cap [x_{i_1}]_{R'}| + |X \cap [x_{i_2}]_{R'}| + \ldots + |X \cap [x_{i_k}]_{R'}| + |X \cap [x^1_{i_t}]_{R'}|}{|X| + |[x_{i_1}]_{R'} - X| + |[x_{i_2}]_{R'} - X| + \ldots + |[x_{i_k}]_{R'} - X| + |[x^1_{i_t}]_{R'} - X|} \\ &= \frac{|\underline{R}(X)| + |X \cap [x_{i_1}]_R| + |X \cap [x_{i_2}]_R| + \ldots + |X \cap [x_{i_k}]_R| + |X \cap [x^1_{i_t}]_{R'}|}{|X| + |[x_{i_1}]_R - X| + |[x_{i_2}]_R - X| + \ldots + |[x_{i_k}]_R - X| + |[x^1_{i_t}]_{R'} - X|}. \end{aligned}$$

Let $S(X, R_\lambda(X)) = k$, $|\underline{R}(X)| + |X \cap [x_{i_1}]_R| + \ldots + |X \cap [x_{i_k}]_R| = k_1$ and $|X| + |[x_{i_1}]_R - X| + \ldots + |[x_{i_k}]_R - X| = k_2$, we can get the equality as follow,

$$S(X, R'_\lambda(X)) = \frac{k_1 + |X \cap [x^1_{i_t}]_{R'}|}{k_2 + |[x^1_{i_t}]_{R'} - X|}.$$

For $S(X, R_\lambda(X)) = k = \frac{k_1}{k_2} \leq \frac{|X \cap [x^1_{i_t}]_{R'}|}{|[x^1_{i_t}]_{R'} - X|}$, and according to Lemma 2 we can easily obtain the following inequality,

$$S(X, R'_\lambda(X)) = \frac{k_1 + |X \cap [x^1_{i_t}]_{R'}|}{k_2 + |[x^1_{i_t}]_{R'} - X|} \geq \frac{k_1 + k_1}{k_2 + k_2} = S(X, R_\lambda(X)).$$

Therefore the part (1) is proved.

For the part (2), since $[x_{i_t}^1]_{R'} \not\subset R'_\lambda(X)$ and $[x_{i_t}^2]_{R'} \not\subset R'_\lambda(X)$, we can easily get the equality,

$$\begin{aligned}S(X, R'_\lambda(X)) &= \frac{|X \cap R'_\lambda(X)|}{|X \cup R'_\lambda(X)|}\\ &= \frac{|\underline{R}'(X)| + |X \cap [x_{i_1}]_{R'}| + |X \cap [x_{i_2}]_{R'}| + \ldots + |X \cap [x_{i_k}]_{R'}|}{|X| + |[x_{i_1}]_{R'} - X| + |[x_{i_2}]_{R'} - X| + \ldots + |[x_{i_k}]_{R'} - X|}\\ &= \frac{|\underline{R}(X)| + |X \cap [x_{i_1}]_{R}| + |X \cap [x_{i_2}]_{R}| + \ldots + |X \cap [x_{i_k}]_{R}|}{|X| + |[x_{i_1}]_{R} - X| + |[x_{i_2}]_{R} - X| + \ldots + |[x_{i_k}]_{R} - X|}\\ &= S(X, R_\lambda(X)).\end{aligned}$$

Therefore the part (2) is proved completely.

Note: since $[x_{i_t}]_R \not\subset R_\lambda(X)$, according to the Lemma 2, the inclusion relations $[x_{i_t}^1]_{R'} \subset R'_\lambda(X)$ and $[x_{i_t}^2]_{R'} \subset R'_\lambda(X)$ are not existed.

Theorem 3. *If $0 < \lambda < 0.5$, $k < t \leq m$, that is $[x_{i_t}]_R \subset R_\lambda(X)$:*

(1) If $[x_{i_t}^1]_{R'} \subset \underline{R}'_\lambda(X)$, $[x_{i_t}^2]_{R'} \subset \underline{R}'_\lambda(X)$, then $S(X, R_\lambda(X)) = S(X, R_\lambda(X))$.

(2) If $[x_{i_t}^1]_{R'} \subset R'_\lambda(X)$, $[x_{i_t}^2]_{R'} \not\subset R'_\lambda(X)$, as well as $S(X, R_\lambda(X)) \geq |X \cap [x_{i_t}^2]_{R'}|/|[x_{i_t}^2]_{R'} - X|$, then $S(X, R_\lambda(X)) \leq S(X, R'_\lambda(X))$.

Proof. For the part (1), since $[x_{i_t}^1]_{R'} \subset R'_\lambda(X)$, and $[x_{i_t}^2]_{R'} \subset R'_\lambda(X)$, we can easily get the equality as follow,

$$\begin{aligned}S(X, R'_\lambda(X)) &= \frac{|X \cap R'_\lambda(X)|}{|X \cup R'_\lambda(X)|}\\ &= \frac{|\underline{R}'(X)| + |X \cap [x_{i_1}]_{R'}| + |X \cap [x_{i_2}]_{R'}| + \ldots + |X \cap [x_{i_k}]_{R'}|}{|X| + |[x_{i_1}]_{R'} - X| + |[x_{i_2}]_{R'} - X| + \ldots + |[x_{i_k}]_{R'} - X|}\\ &= \frac{|\underline{R}(X)| + |X \cap [x_{i_1}]_{R}| + |X \cap [x_{i_2}]_{R}| + \ldots + |X \cap [x_{i_k}]_{R}|}{|X| + |[x_{i_1}]_{R} - X| + |[x_{i_2}]_{R} - X| + \ldots + |[x_{i_k}]_{R} - X|} = S(X, R_\lambda(X)).\end{aligned}$$

Hence the part (1) is proved.

For the part (2), since $[x_{i_t}^2]_{R'} \not\subset R'_\lambda(X)$, we can easily get the equality as follow,

$$\begin{aligned}S(X, R'_\lambda(X)) &= \frac{|X \cap R'_\lambda(X)|}{|X \cup R'_\lambda(X)|}\\ &= \frac{|\underline{R}'(X)| + |X \cap [x_{i_1}]_{R'}| + |X \cap [x_{i_2}]_{R'}| + \ldots + |X \cap [x_{i_t}^1]_{R'}| + \ldots + |X \cap [x_{i_k}]_{R'}|}{|X| + |[x_{i_1}]_{R'} - X| + |[x_{i_2}]_{R'} - X| + \ldots + |[x_{i_t}^1]_{R'} - X| + \ldots + |[x_{i_k}]_{R'} - X|}\\ &= \frac{|\underline{R}'(X)| + |X \cap [x_{i_1}]_{R'}| + \ldots + |X \cap [x_{i_t}]_{R'}| + \ldots + |X \cap [x_{i_k}]_{R'}| - |X \cap [x_{i_t}^2]_{R'}|}{|X| + |[x_{i_1}]_{R'} - X| + \ldots + |[x_{i_t}]_{R'} - X| + \ldots + |[x_{i_k}]_{R'} - X| - |[x_{i_t}^2]_{R'} - X|}\\ &= \frac{|X \cap R_\lambda(X)| - |X \cap [x_{i_t}^2]_{R'}|}{|X \cup R_\lambda(X)| - |[x_{i_t}^2]_{R'} - X|}.\end{aligned}$$

Since $S(X, R_\lambda(X)) \geq |X \cap [x_{i_t}^2]_{R'}|/|[x_{i_t}^2]_{R'} - X|$, then according to Lemma 3, we can get $S(X, R_\lambda(X)) \leq S(X, R'_\lambda(X))$. So the part (2) is proved perfectly.

Note: Since $[x_{i_t}]_R \subset R_\lambda(X)$, according to the Lemma 2, the inclusion relations $[x_{i_t}^1]_{R'} \not\subset R'_\lambda(X)$ and $[x_{i_t}^2]_{R'} \not\subset R'_\lambda(X)$ are not held either.

Theorems 2 and 3 show that when $0 < \lambda < 0.5$ and the equivalence classes are subdivided into many finer equivalence classes (sub-granules) by R', the similarity degree between $R'_\lambda(X)$ and X is not generally lower than the similarity degree between $R_\lambda(X)$ and X.

6 Conclusions

Since rough set theory was proposed in 1982, it has developed more than 30 years. Many scholars have made some improvements for the traditional models and obtained many extended rough set models which overcome some shortcomings of the traditional models. Combining with the fuzzy set theory, we have constructed an approximation set of an uncertain set X with the cut-set and proposed a general approximation model $R_{0.5}(X)$, but the optimal approximation set of X still is not established. In order to solve this problem, in this paper the optimal approximation set through minimizing similarity between the uncertain concept and its approximation sets is defined. Then comparative analysis between $R_{Best}(X)$ and other approximation sets is given. Next, the operation properties of $R_{Best}(X)$ are presented and proved successfully. Finally, change rules of the similarity degree between X and $R_{Best}(X)$ in different knowledge granularity levels are discussed in detail. These research presents a computational method for establishing or searching an optimal approximation set of X from the perspective of similarity. We hope these works can expand the range of rough set theory model to deal with uncertain problems in the real world and promote the development of uncertainty artificial intelligence.

Acknowledgments. This work was supported in part by the National Natural Science Foundation of China under Grant 61876201, in part by the National Key Research and Development Program of China under Grant 2017YFC0 804002, in part by the Chongqing Postgraduate Scientific Research and Innovation Project under Grant CYS18244.

References

1. Guan, L.H., Wang, G.Y., Yu, H.: Incremental algorithm of Pawlak reduction based on attribute order. J. Southwest Jiaotong Univ. **46**(3), 461–468 (2011)
2. Herbert, J.P., Yao, J.T.: Game-theoretic rough sets. Fundamenta Informaticae **108**(3–4), 267–286 (2011)
3. Hu, F., Wang, G.Y.: Quick algorithm for certain rule acquisition based on divide and conquer method. Pattern Recognit. Artif. Intell. **23**(3), 349–356 (2010)
4. Li, T.Y., et al.: Rough sets and knowledge technology. Fundamenta Informaticae **7414**(2–3), I–II (2008)
5. Liang, J.Y., Qian, Y.H.: Granulation monotonicity of entropy measure in information systems. J. Shanxi Univ. **30**(2), 156–162 (2007)

6. Liang, J.Y., Shi, Z.Z.: The information entropy, rough entropy and knowledge granulation in rough set theory. Int. J. Uncertainty Fuzziness Knowl. Based Syst. **12**(1), 37–46 (2011)
7. Liu, X.C.: Entropy, distance measure and similarity measure of fuzzy sets and their relations. Elsevier North-Holland Inc. (1992)
8. Mi, J.S., Wu, W.Z., Zhang, W.X.: Approaches to knowledge reduction based on variable precision rough set model. Inf. Sci. **159**(3–4), 255–272 (2004)
9. Pawlak, Z.: Rough sets. Int. J. Comput. Inf. Sci. **11**(5), 341–365 (1982)
10. Pawlak, Z., Skowron, A.: Rough sets: some extensions. Inf. Sci. **177**(1), 28–40 (2007)
11. Pawlak, Z., Wong, S.K.M., Ziarko, W.: Rough sets: probabilistic versus deterministic approach*. Int. J. Man-Mach. Stud. **29**(1), 81–95 (1988)
12. Qian, Y.H., Zhang, H., Sang, Y.L., Liang, J.Y.: Multigranulation decision-theoretic rough sets. Int. J. Approximate Reasoning **55**(1), 225–237 (2014)
13. Tiwari, S.P., Srivastava, A.K.: Fuzzy rough sets, fuzzy preorders and fuzzy topologies. Elsevier North-Holland Inc. (2013)
14. Wang, G.Y.: Rough set theory and knowledge discovery. China Xi'an Jiaotong University Press (2007)
15. Wang, G.Y., Zhang, Q.H.: Uncertainty of rough sets in different knowledge granularities: uncertainty of rough sets in different knowledge granularities. Chin. J. Comput. **31**(9), 1588–1598 (2008)
16. Yao, J.T., Vasilakos, A.V., Pedrycz, W.: Granular computing: perspectives and challenges. IEEE Trans. Cybern. **43**(6), 1977–1989 (2013)
17. Yao, Y.Y.: Two semantic issues in a probabilistic rough set model. Fundamenta Informaticae **108**(3), 249–265 (2011)
18. Yao, Y.Y.: Rough-set concept analysis: interpreting RS-definable concepts based on ideas from formal concept analysis. Inf. Sci. **346**, 442–462 (2016)
19. Yao, Y.Y., Li, X.N.: Comparison of rough-set and interval-set models for uncertain reasoning. Fundamenta Informaticae **27**(2), 289–298 (1996)
20. Yao, Y.Y., She, Y.H.: Rough set models in multigranulation spaces. Inf. Sci. **327**, 40 (2016)
21. Zhang, Q.H., Wang, G.Y., Xiao, Y.: Approximation sets of rough sets. J. Softw. **23**(7), 1745–1759 (2012)
22. Zhang, Q.H., Xie, Q., Wang, G.Y.: A survey on rough set theory and its applications. CAAI Trans. Intell. Technol. **1**(4), 323–333 (2016)
23. Zhang, Q.H., Zhang, P., Wang, G.Y.: Research on approximation set of rough set based on fuzzy similarity. J. Intell. Fuzzy Syst. **32**(3), 2549–2562 (2017)
24. Zhang, W.X., Wu, W.Z.: An introduction and a survey for the studies of rough set theory. Fuzzy Syst. Math. **14**(4), 1–12 (2000)
25. Zhang, X., Mo, Z.: Variable precision rough sets. Pattern Recognit. Artif. Intell. **17**(2), 151–155 (2004)
26. Ziarko, W.: Variable precision rough set model. J. Comput. Syst. Sci. **46**(1), 39–59 (1993)

Discovering Flow Graphs from Data Tables Using the Classification and Prediction Software System (CLAPSS)

Krzysztof Pancerz[1(✉)], Arkadiusz Lewicki[2], and Jaromir Sarzyński[1]

[1] Faculty of Mathematics and Natural Sciences, University of Rzeszów, Pigonia Street 1, 35-310 Rzeszów, Poland
kpancerz@ur.edu.pl

[2] University of Information Technology and Management in Rzeszów, Sucharskiego Street 2, 35-225 Rzeszów, Poland
alewicki@wsiz.rzeszow.pl

Abstract. In the paper, theoretical background, as well as practical implementation, of discovering flow graphs (both fuzzy and rough set) from data tables are presented. We assume that data tables represent information/decision systems in the Pawlak's sense. The implementation was made in a software tool called the Classification and Prediction Software System (CLAPSS). CLAPSS is a tool developed in the Java technology for solving different classification and prediction problems using, among others, some specialized approaches based mainly on fuzzy sets and rough sets. In general, those specialized approaches implemented in CLAPSS are not available in other tools.

Keywords: Flow graphs · Rough sets · Fuzzy sets · Software tool

1 Introduction

Information flow distribution is the kind of knowledge that can be helpful in solving different problems appearing in data analysis, especially, if we deal with temporal data, i.e., results of observations or measurements are ordered in time, (cf. [5,10]). In the literature, different approaches based on flow graphs were proposed. The fundamental one, called flow networks, was proposed by Ford and Fulkerson [3]. In the paper, we are interested in two other approaches introduced in the area of data mining, namely fuzzy flow graphs proposed by Mieszkowicz-Rolka and Rolka [6] and flow graphs (called here, rough set flow graphs) proposed by Pawlak [12]. These approaches were implemented in a software tool called Classification and Prediction Software System (CLAPSS) [8,9]. It is a tool developed over the last few years for solving different classification and prediction problems using, among others, some specialized approaches based mainly on fuzzy sets and rough sets. Last time, the tool was supplemented with the possibility of discovering fuzzy flow graphs and rough set flow graphs from data

© Springer Nature Switzerland AG 2019
T. Mihálydeák et al. (Eds.): IJCRS 2019, LNAI 11499, pp. 356–368, 2019.
https://doi.org/10.1007/978-3-030-22815-6_28

tables representing information/decision systems in the Pawlak's sense. On the entry side, the tool accepts popular text formats of data tables used in other data mining and machine learning tools (WEKA [4], RSES [1]) and the XML format used in ROSETTA [7]. The output in the form of a flow graph can be exported to the popular DOT format used in the Graphviz tool [2].

Discovering fuzzy flow graphs is preceded by the fuzzification process of attribute values. The fuzzification process can be made in CLAPSS in one of the two ways, graphical and scripting. The details are described in Sect. 3. It is worth noting that the graphical way is very useful for the users. Discovering rough set flow graphs can be preceded by the discretization process. The discretization process can be made in CLAPSS in a scripting way.

There has been a lack of a general purpose software tool in which discovering fuzzy as well as rough set flow graphs is implemented. The implementation made in CLAPSS bridges this gap. The paper describes this new possibility, added to CLAPSS lately.

In the remaining part of the paper, theoretical background as well as practical implementation of discovering flow graphs (both fuzzy and rough set) from data tables are presented.

2 Theoretical Background

2.1 Information and Decision Systems

In the presented approaches, information systems are understood as Pawlak's knowledge representation systems.

Formally, an information system IS is a quadruple

$$IS = (U, A, \{V_a\}_{a \in A}, f_{inf}),$$

where:

- U is the nonempty, finite set of objects,
- A is the nonempty, finite set of attributes,
- $\{V_a\}_{a \in A}$ is the family of nonempty sets of attribute values,
- $f_{inf} : A \times U \to \bigcup_{a \in A} V_a$ is the information function such that $f_{inf}(a, u) \in V_a$ for each $a \in A$ and $u \in U$.

In many applications (e.g. supervised learning), two classes of attributes (the so-called condition attributes and decision attributes) are distinguished in the set of attributes of an information system. The information system, with two classes of attributes distinguished, is called a decision system. Formally, the decision system DS is a tuple

$$DS = (U, C, D, \{V_a\}_{a \in C \cup D}, f_{inf}, f_{dec}),$$

where:

- U is the nonempty, finite set of objects,
- C is the nonempty, finite set of condition attributes,
- D is the nonempty, finite set of decision attributes,
- $\{V_a\}_{a \in C \cup D}$ is the family of nonempty sets of condition and decision attribute values,
- $f_{inf} : C \times U \rightarrow \bigcup_{c \in C} V_c$ is the information function such that $f_{inf}(c, u) \in V_c$ for each $c \in C$ and $u \in U$.
- $f_{dec} : D \times U \rightarrow \bigcup_{d \in D} V_d$ is the decision function such that $f_{dec}(d, u) \in V_d$ for each $d \in D$ and $u \in U$.

We often consider a decision system with one decision attribute, i.e., $D = \{d\}$.

Further, only information systems will be considered since a decision system can be treated as a special case of an information system.

There are two key types of values: numerical and symbolic. Numerical values are expressed by numbers (e.g., real numbers, integers, prime numbers, etc.). Symbolic values usually describe qualitative concepts. Let $\mathbb{R}$ be a set of real numbers. A numerical attribute is an attribute whose set of values is a nonempty subset of $\mathbb{R}$. A symbolic attribute is an attribute whose set of values includes symbolic values only.

2.2 Fuzzification

Fuzzification is the process that transforms the real value variables into linguistic variables whose domains contain linguistic values which can be described by fuzzy sets (their membership functions).

Now, we are interested in information systems with numerical attributes only. Let $IS = (U, A, \{V_a\}_{a \in A}, f_{inf})$ be an information system such that $V_a \subseteq \mathbb{R}$ for each $a \in A$. For each attribute $a \in A$, we can define a linguistic variable λ_a. With each linguistic variable λ_a, a set L^{λ_a} of linguistic values is associated:

$$L^{\lambda_a} = \{l_1^a, l_2^a, \ldots, l_{k_a}^a\}.$$

Each linguistic value l_i^a, where $i = 1, 2, \ldots, k_a$, is described by a membership function $\mu_{l_i^a} : \mathbb{R} \rightarrow [0, 1]$. Many types of membership functions can be used to describe linguistic values. The following types of membership functions have been implemented in CLAPSS:

- triangular shaped membership function,
- trapezoidal shaped membership function,
- Gaussian shaped membership function,
- generalized bell shaped membership function,
- S shaped membership function,
- π shaped membership function,
- sigmoidal shaped membership function,
- fuzzy singleton membership function,
- sinusoidal shaped membership function,
- Z shaped membership function.

2.3 Fuzzified Information Systems

We are interested in information systems with numerical attributes only. Let:

- $IS = (U, A, \{V_a\}_{a \in A}, f_{inf})$ be an information system with $U = \{u_1, u_2, \ldots, u_n\}$ and $A = \{a_1, a_2, \ldots, a_m\}$, such that $V_a \subseteq \mathbb{R}$ for each $a \in A$,
- $\{L^{\lambda_a}\}_{a \in A}$ be the family of sets of linguistic values associated with linguistic variables from the family $\{\lambda_a\}_{a \in A}$ defined for attributes from A, where $L^{\lambda_a} = \{l_1^a, l_2^a, \ldots, l_{k_a}^a\}$ for each $a \in A$.

A fuzzified information system $\mathcal{F}(IS)$ corresponding to IS, is a quadruple

$$\mathcal{F}(IS) = (U^{\mathcal{F}}, \Phi, \{V_\phi\}_{\phi \in \Phi}, f_{inf}^{\mathcal{F}}),$$

where:

- $U^{\mathcal{F}}$ is the nonempty, finite set of objects such that each $u^* \in U^{\mathcal{F}}$ corresponds exactly to one $u \in U$,
- $\Phi = \Phi_{a_1} \cup \Phi_{a_2} \cup \cdots \cup \Phi_{a_m}$ is the nonempty, finite set of fuzzified attributes, such that
  - $\Phi_{a_1} = \{a_1^{l_1^{a_1}}, a_1^{l_2^{a_1}}, \ldots, a_1^{l_{k_{a_1}}^{a_1}}\}$,
  - $\Phi_{a_2} = \{a_2^{l_1^{a_2}}, a_2^{l_2^{a_2}}, \ldots, a_2^{l_{k_{a_2}}^{a_2}}\}$,
  - $\ldots$,
  - $\Phi_{a_m} = \{a_m^{l_1^{a_m}}, a_m^{l_2^{a_m}}, \ldots, a_m^{l_{k_{a_m}}^{a_m}}\}$,
- $\{V_\phi\}_{\phi \in \Phi}$ is the family of sets of fuzzified attribute values and $V_\phi = [0, 1]$ for each $\phi \in \Phi$,
- $f_{inf}^{\mathcal{F}} : \Phi \times U^{\mathcal{F}} \rightarrow \bigcup_{\phi \in \Phi} V_\phi$ is the information function such that
  - $f_{inf}^{\mathcal{F}}(a^{l_i^a}, u^*) \in V_\phi$ for each $a^{l_i^a} \in \Phi$ and $u^* \in U^{\mathcal{F}}$,
  - $f_{inf}^{\mathcal{F}}(a^{l_i^a}, u^*) = \mu_{l_i^a}(f_{inf}(a, u))$, where $\mu_{l_i^a}$ is a membership function describing l_i^a and $u^* \in U^{\mathcal{F}}$ corresponds to $u \in U$,

  for each $a \in A$ and $i = 1, 2, \ldots, k_a$.

If some attributes of an information system are symbolic (this situation is common for decision attributes), then we can use the so-called binary fuzzification for them.

If for a given $a \in A$, the value set of a is a finite set $V_a = \{v_1, v_2, \ldots, v_{k_a}\}$ of symbolic values, then

- $\Phi_a = \{a^{v_1}, a^{v_2}, \ldots, a^{v_{k_a}}\}$,
- $f_{inf}^{\mathcal{F}}(a^{v_i}, u^*) = \begin{cases} 1, & f_{inf}(a, u) = v_i, \\ 0, & f_{inf}(a, u) \neq v_i, \end{cases}$

where $u^* \in U^{\mathcal{F}}$ corresponds to $u \in U$ and $i = 1, 2, \ldots, k_a$. One can see that in this case we use, in fact, a fuzzy singleton membership function.

2.4 Fuzzy Flow Graphs

Fuzzy flow graphs were proposed by Mieszkowicz-Rolka and Rolka (see [6]) to allow representation of information/decision tables with fuzzy attributes. In our software tool, we have adopted the following formal definition of a fuzzy flow graph. Let $\mathcal{F}(IS) = (U^{\mathcal{F}}, \Phi, \{V_\phi\}_{\phi\in\Phi}, f^{\mathcal{F}}_{inf})$ be a fuzzified information system corresponding to an information system $IS = (U, A, \{V_a\}_{a\in A}, f_{inf})$ with $U = \{u_1, u_2, \ldots, u_n\}$ and $A = \{a_1, a_2, \ldots, a_m\}$. A fuzzy flow graph corresponding to $\mathcal{F}(IS)$ is a tuple

$$\mathcal{FFG}(\mathcal{F}(IS)) = (N, B, cer),$$

where:

- $N = N_{a_1} \cup N_{a_2} \cup \cdots \cup N_{a_m}$ is the set of nodes such that for each $a \in \{a_1, a_2, \ldots, a_m\}$: $N_a = \{\widehat{a}^{l^a_1}, \widehat{a}^{l^a_2}, \ldots, \widehat{a}^{l^a_{k_a}}\}$,
- $B \subseteq N \times N$ is a set of labelled directed branches such that for any $(\widehat{\phi}^x, \widehat{\phi}^y) \in B$, $\widehat{\phi}^x \in N_{a_{i-1}}$ and $\widehat{\phi}^y \in N_{a_i}$ and $i \in \{2, 3, \ldots, m\}$,
- $cer : B \to [0, 1]$ is a certainty function labelling branches such that:

$$cer(\widehat{a}_j^{l^{a_j}_x}, \widehat{a}_k^{l^{a_k}_y}) = \frac{\sum\limits_{u^*\in U^{\mathcal{F}}} f_{inf}(a_j^{l^{a_j}_x}, u^*) f_{inf}(a_k^{l^{a_k}_y}, u^*)}{card(U)}$$

  for any $(\widehat{a}_j^{l^{a_j}_x}, \widehat{a}_k^{l^{a_k}_y}) \in B$.

One can see that we can distinguish particular layers in the set N of nodes of $\mathcal{FFG}(\mathcal{F}(IS))$. The layer N_a, where $a \in \{a_1, a_2, \ldots, a_m\}$, corresponds exactly to one attribute $a \in A$. Each node in the layer N_a corresponds exactly to one linguistic value from the set L^{λ_a} of linguistic values assigned to a linguistic variable λ_a defined for the attribute a. It is worth noting that, in the numerator of the fraction defining the value of the certainty function, the so-called fuzzy cardinality (power) is calculated.

2.5 Rough Set Flow Graphs

Rough set flow graphs were defined by Pawlak (see [12]) as a tool for reasoning from data. In our software tool, we have adopted the following formal definition of a rough set flow graph. Let $IS = (U, A, \{V_a\}_{a\in A}, f_{inf})$ be an information system with $U = \{u_1, u_2, \ldots, u_n\}$ and $A = \{a_1, a_2, \ldots, a_m\}$, such that $V_a = \{v_a^1, v_a^2, \ldots, v_a^{k_a}\}$ for each $a \in A$. A rough set flow graph corresponding to IS is a tuple

$$\mathcal{RSFG}(IS) = (N, B, cer, str, cov),$$

where:

- $N = N_{a_1} \cup N_{a_2} \cup \cdots \cup N_{a_m}$ is the set of nodes such that for each $a \in \{a_1, a_2, \ldots, a_m\}$: $N_a = \{\widehat{a}^{v_a^1}, \widehat{a}^{v_a^2}, \ldots, \widehat{a}^{v_a^{k_a}}\}$,
- $B \subseteq N \times N$ is a set of multi-labelled directed branches such that for any $(n^x, n^y) \in B$, $n^x \in N_{a_{i-1}}$ and $n^y \in N_{a_i}$ and $i \in \{2, 3, \ldots, m\}$,

- $cer : B \to [0,1]$ is a certainty function labelling branches such that:

$$cer(\widehat{a}_{i-1}^{v_{a-1}^{x}}, \widehat{a}_{i}^{v_{a}^{y}}) = \frac{card(\{u \in U : f_{inf}(a_{i-1}, u) = v_{a_{i-1}}^{x} \wedge f_{inf}(a_i, u) = v_{a_i}^{y}\})}{card(\{u \in U : a_{i-1}(u) = v_{a_{i-1}}^{x}\})},$$

for any $(\widehat{a}_{i-1}^{v_{a-1}^{x}}, \widehat{a}_{i}^{v_{a}^{y}}) \in B$,
- $str : B \to [0,1]$ is a strength function labelling branches such that:

$$str(\widehat{a}_{i-1}^{v_{a-1}^{x}}, \widehat{a}_{i}^{v_{a}^{y}}) = \frac{card(\{u \in U : f_{inf}(a_{i-1}, u) = v_{a_{i-1}}^{x} \wedge f_{inf}(a_i, u) = v_{a_i}^{y}\})}{card(U)},$$

for any $(\widehat{a}_{i-1}^{v_{a-1}^{x}}, \widehat{a}_{i}^{v_{a}^{y}}) \in B$,
- $cov : B \to [0,1]$ is a covering function labelling branches such that:

$$cov(\widehat{a}_{i-1}^{v_{a-1}^{x}}, \widehat{a}_{i}^{v_{a}^{y}}) = \frac{card(\{u \in U : f_{inf}(a_{i-1}, u) = v_{a_{i-1}}^{x} \wedge f_{inf}(a_i, u) = v_{a_i}^{y}\})}{card(\{u \in U : a_i(u) = v_{a_i}^{y}\})},$$

for any $(\widehat{a}_{i-1}^{v_{a-1}^{x}}, \widehat{a}_{i}^{v_{a}^{y}}) \in B$.

One can see that we can distinguish particular layers in the set N of nodes of $\mathcal{RSFG}(IS)$. The layer N_a, where $a \in \{a_1, a_2, \ldots, a_m\}$, corresponds exactly to one attribute $a \in A$. Each node in the layer N_a corresponds exactly to one value from the set V_a of values of a.

3 CLAPSS Implementation

CLAPSS (Classification and Prediction Software System) is a software tool developed for solving different classification and prediction problems using, among others, some specialized approaches based mainly on fuzzy sets and rough sets. The tool is developed in the Java technology. Selected functionalities of the earlier versions of CLAPSS were described in [8] and [9]. The main features of CLAPSS are the following:

- *Portability.* Thanks to the Java technology, the application works on various software and hardware platforms.
- *User-friendly interface* (see Fig. 1).
- *Modularity.* CLAPSS implementation takes into consideration modularity.

In this section, we present a new possibility added to CLAPSS lately. It is the possibility to discover fuzzy flow graphs and rough set flow graphs form data tables representing information/decision systems in the Pawlak's sense. The general scheme of this functionality of CLAPSS is shown in Fig. 2. The tool uses its own text format to enter the input data. Moreover, the tool accepts popular text formats of data tables used in other data mining and machine learning tools (WEKA [4], RSES [1]) and the XML format used in ROSETTA [7]. The output flow graph can be exported to the popular DOT format used in the Graphviz tool [2].

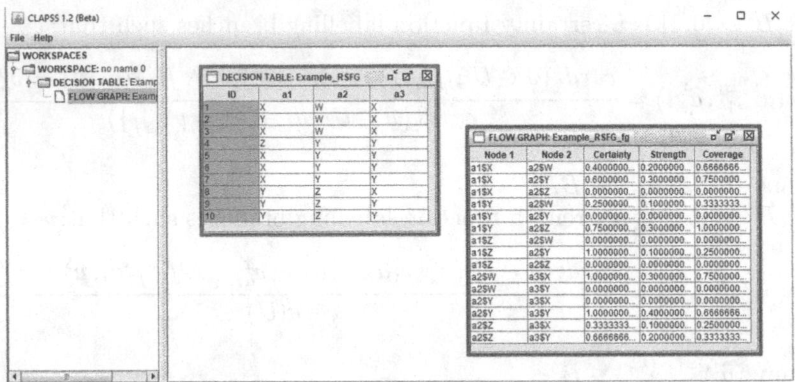

Fig. 1. A user-friendly interface of CLAPSS.

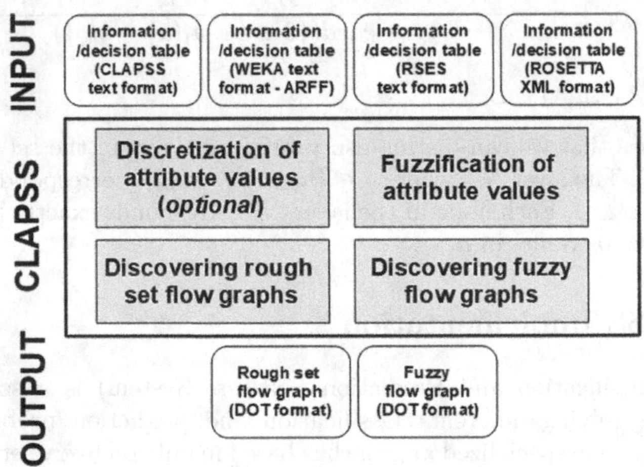

Fig. 2. A general scheme of the possibility to discover fuzzy flow graphs and rough set flow graphs form data tables in CLAPSS.

Rough set flow graphs require symbolic or discrete numerical attribute values in the input data tables. Therefore, in case of numerical attribute values, discovering rough set flow graphs in CLAPSS can be preceded by the discretization process that can be made in a scripting way using a special language. Discovering fuzzy flow graphs is preceded by the required fuzzification process of attribute values. In CLAPSS, two ways of the fuzzification process were implemented, namely the graphical way and the scripting one. Particularly, the tool aiding the fuzzification process in a graphical way is important and useful. The main window of this tool is shown in Fig. 4.

In the remaining part of this section, we present two simple examples clarifying the new possibility added to CLAPSS.

Example 1. Let us consider a simple information system $IS^1 = (U^1, A^1, \{V_a\}_{a \in A^1}, f^1_{inf})$, where $U^1 = \{u_1, u_2, \ldots, u_{10}\}$, $A^1 = \{a_1, a_2, a_3\}$, $V_{a_1} = \{X, Y, Z\}$, $V_{a_2} = \{W, Y, Z\}$, and $V_{a_3} = \{X, Y\}$. This information system is shown in Table 1 as a data table. This table was included in a text file using the CLAPSS format:

```
TABLE Example_RSFG
a1 a2 a3
string string string
condition condition condition
X W X
Y W X
X W X
Z Y Y
X Y Y
X Y Y
X Y Y
Y Z X
Y Z Y
Y Z Y
```

Table 1. The information system IS^1.

U^1 A^1	a_1	a_2	a_3
u_1	X	W	X
u_2	Y	W	X
u_3	X	W	X
u_4	Z	Y	Y
u_5	X	Y	Y
u_6	X	Y	Y
u_7	X	Y	Y
u_8	Y	Z	X
u_9	Y	Z	Y
u_{10}	Y	Z	Y

The rough set flow graph corresponding to the information system IS^1, generated in CLAPSS was exported to the DOT format:

```
digraph FuzzyFlowGraph {
label="RSFG" fontsize=20
labelloc=top
N0 [label=<<TABLE><TR><TD>a1</TD></TR><TR><TD>X</TD></TR></TABLE>> ]
N1 [label=<<TABLE><TR><TD>a1</TD></TR><TR><TD>Y</TD></TR></TABLE>> ]
N2 [label=<<TABLE><TR><TD>a1</TD></TR><TR><TD>Z</TD></TR></TABLE>> ]
N3 [label=<<TABLE><TR><TD>a2</TD></TR><TR><TD>W</TD></TR></TABLE>> ]
N4 [label=<<TABLE><TR><TD>a2</TD></TR><TR><TD>Y</TD></TR></TABLE>> ]
N5 [label=<<TABLE><TR><TD>a2</TD></TR><TR><TD>Z</TD></TR></TABLE>> ]
N6 [label=<<TABLE><TR><TD>a3</TD></TR><TR><TD>X</TD></TR></TABLE>> ]
N7 [label=<<TABLE><TR><TD>a3</TD></TR><TR><TD>Y</TD></TR></TABLE>> ]
N0->N3 [label="cer=0.4000 str=0.2000 cov=0.6667" ]
N0->N4 [label="cer=0.6000 str=0.3000 cov=0.7500" ]
N0->N5 [label="cer=0.0000 str=0.0000 cov=0.0000" ]
```

```
N1->N3 [label="cer=0.2500 str=0.1000 cov=0.3333" ]
N1->N4 [label="cer=0.0000 str=0.0000 cov=0.0000" ]
N1->N5 [label="cer=0.7500 str=0.3000 cov=1.0000" ]
N2->N3 [label="cer=0.0000 str=0.0000 cov=0.0000" ]
N2->N4 [label="cer=1.0000 str=0.1000 cov=0.2500" ]
N2->N5 [label="cer=0.0000 str=0.0000 cov=0.0000" ]
N3->N6 [label="cer=1.0000 str=0.3000 cov=0.7500" ]
N3->N7 [label="cer=0.0000 str=0.0000 cov=0.0000" ]
N4->N6 [label="cer=0.0000 str=0.0000 cov=0.0000" ]
N4->N7 [label="cer=1.0000 str=0.4000 cov=0.6667" ]
N5->N6 [label="cer=0.3333 str=0.1000 cov=0.2500" ]
N5->N7 [label="cer=0.6667 str=0.2000 cov=0.3333" ]
}
```

Visualization of this graph in the Graphviz tool is shown in Fig. 3.

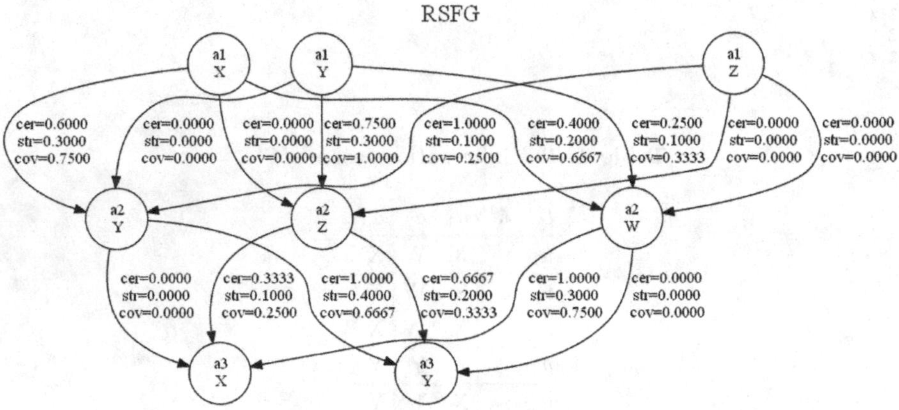

Fig. 3. The rough set flow graph corresponding to the information system IS^1.

Example 2. Let us consider a simple information system $IS^2 = (U^2, A^2, \{V_a\}_{a \in A^2}, f^2_{inf})$, where $U^1 = \{u_1, u_2, \ldots, u_{10}\}$, $A^2 = \{a_1, a_2, a_3\}$. This information system is shown in Table 2 as a data table. For each attribute $a \in A^2$, we defined a linguistic variable λ_a with a set L^{λ_a} of linguistic values $L^{\lambda_a} = \{low, medium, high\}$. The following membership functions were used to fuzzify values of attributes from A^2.

1. For the linguistic value *low*:

$$\mu_{low}(x) = \begin{cases} 1 & \text{if } x \geq 0 \text{ and } x \leq 1.0, \\ 1 - \frac{x-1.0}{3.0-1.0} & \text{if } x > 1.0 \text{ and } x \leq 3.0, \\ 0 & \text{otherwise.} \end{cases}$$

2. For the linguistic value *medium*:

$$\mu_{medium}(x) = \begin{cases} \frac{x-1.0}{2.5-1.0} & \text{if } x \geq 1.0 \text{ and } x \leq 2.5, \\ 1 - \frac{x-2.5}{4.0-2.5} & \text{if } x > 2.5 \text{ and } x \leq 4.0, \\ 0 & \text{otherwise.} \end{cases}$$

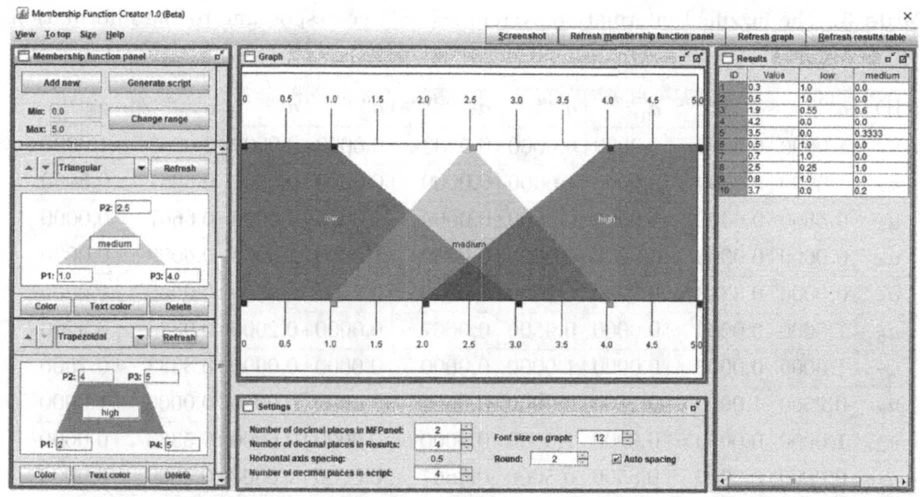

Fig. 4. The main window of the tool aiding the fuzzification process in a graphical way.

3. For the linguistic value *high*:

$$\mu_{high}(x) = \begin{cases} \frac{x-2.0}{4.0-2.0} & \text{if } x \geq 2.0 \text{ and } x \leq 4.0, \\ 1 & \text{if } x > 4.0 \text{ and } x \leq 5.0, \\ 0 & \text{otherwise.} \end{cases}$$

In fact, for *low* and *high*, we used trapezoidal shaped membership functions whereas for *medium* we used a triangular shaped membership function.

Table 2. The information system IS^2.

U^2 A^2	a_1	a_2	a_3
u_1	0.3	3.2	4.9
u_2	0.5	0.9	3.0
u_3	1.9	4.5	2.0
u_4	4.2	4.3	0.8
u_5	3.5	2.5	0.4
u_6	0.5	1.1	2.6
u_7	0.7	0.7	3.5
u_8	2.5	3.0	4.9
u_9	0.8	0.8	1.8
u_{10}	3.7	2.0	0.1

Table 3. The fuzzified information system $\mathcal{F}(IS^2)$ corresponding to the information system IS^2.

ID	a_1^{low}	a_1^{medium}	a_1^{high}	a_2^{low}	a_2^{medium}	a_2^{high}	a_3^{low}	a_3^{medium}	a_3^{high}
u_1	1.0000	0.0000	0.0000	0.0000	0.5333	0.6000	0.0000	0.0000	1.0000
u_2	1.0000	0.0000	0.0000	1.0000	0.0000	0.0000	0.0000	0.6667	0.5000
u_3	0.5500	0.6000	0.0000	0.0000	0.0000	1.0000	0.5000	0.6667	0.0000
u_4	0.0000	0.0000	1.0000	0.0000	0.0000	1.0000	1.0000	0.0000	0.0000
u_5	0.0000	0.3333	0.7500	0.2500	1.0000	0.2500	1.0000	0.0000	0.0000
u_6	1.0000	0.0000	0.0000	0.9500	0.0667	0.0000	0.2000	0.9333	0.3000
u_7	1.0000	0.0000	0.0000	1.0000	0.0000	0.0000	0.0000	0.3333	0.7500
u_8	0.2500	1.0000	0.2500	0.0000	0.6667	0.5000	0.0000	0.0000	1.0000
u_9	1.0000	0.0000	0.0000	1.0000	0.0000	0.0000	0.6000	0.5333	0.0000
u_{10}	0.0000	0.2000	0.8500	0.5000	0.6667	0.0000	1.0000	0.0000	0.0000

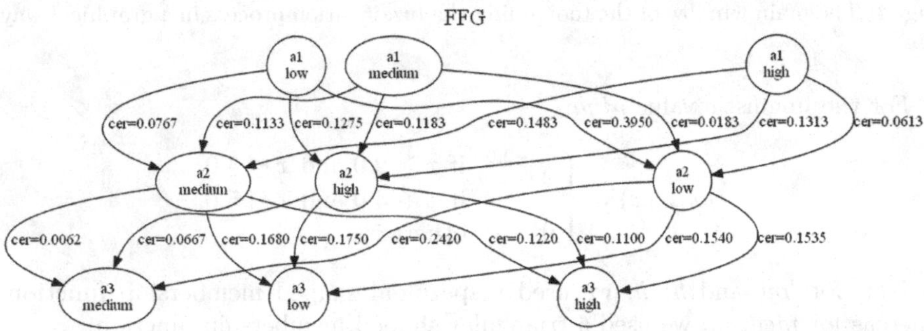

Fig. 5. The fuzzy flow graph corresponding to the information system IS^2.

In case of a scripting fuzzification process, the user must put the script in a special language implemented in CLAPSS. In this script, linguistic values and membership functions (their shapes and parameters) associated with them are determined. The script for the fuzzification process of the attribute values of the information system shown in Table 2 has the form:

```
ATTR[0]->fuzzification(lingvalues={low=(trapezoidal,0.0000,0.0000,1.0000,3.0000),
medium=(triangular,1.0000,2.5000,4.0000),high=(trapezoidal,2.0000,4.0000,5.0000,5.0000)});
ATTR[1]->fuzzification(lingvalues={low=(trapezoidal,0.0000,0.0000,1.0000,3.0000),
medium=(triangular,1.0000,2.5000,4.0000),high=(trapezoidal,2.0000,4.0000,5.0000,5.0000)});
ATTR[2]->fuzzification(lingvalues={low=(trapezoidal,0.0000,0.0000,1.0000,3.0000),
medium=(triangular,1.0000,2.5000,4.0000),high=(trapezoidal,2.0000,4.0000,5.0000,5.0000)});
```

In case of a graphical fuzzification process, the user determines linguistic values and membership functions (their shapes and parameters) associated with them in a special window shown in Fig. 4. On the left-hand side, the membership function panel can be seen. In the centre of the window, one can see the panel with defined membership functions for the whole range of a given attribute. In this panel, the membership functions created earlier can be manually modified (e.g., characteristic points or slopes can be moved). On the right-hand side, the panel with calculated values of the fuzzified attribute can be seen. These values are automatically updated if some changes in membership functions are made by the user. The fuzzified information system $\mathcal{F}(IS^2)$ corresponding to the information system IS^2 is shown in Table 3 as a data table. The fuzzy flow graph corresponding to the information system IS^2 visualized using the Graphviz tool is shown in Fig. 5.

4 Conclusions

In the paper, we have presented a new possibility added to the Classification and Prediction Software System (CLAPSS). This possibility enables the users to discover flow graphs (both fuzzy and rough set) from data tables. These flow graphs can be used further as the spaces to extract the useful knowledge (in a form of rules or episodes) hidden in the analysed data. One of the main directions in further developing of CLAPSS is to implement the approaches in which the semantics of data (for example, expressed by ontologies) is taken into consideration in the processes of discovering flow graphs (cf. [11]).

Acknowledgments. This work was partially supported by the Center for Innovation and Transfer of Natural Sciences and Engineering Knowledge at the University of Rzeszów.

References

1. Bazan, J.G., Szczuka, M.: The rough set exploration system. In: Peters, J.F., Skowron, A. (eds.) Transactions on Rough Sets III. LNCS, vol. 3400, pp. 37–56. Springer, Heidelberg (2005). https://doi.org/10.1007/11427834_2
2. Ellson, J., Gansner, E.R., Koutsofios, E., North, S.C., Woodhull, G.: Graphviz and dynagraph - static and dynamic graph drawing tools. In: Jünger, M., Mutzel, P. (eds.) Graph Drawing Software. MATHVISUAL, pp. 127–148. Springer, Heidelberg (2004). https://doi.org/10.1007/978-3-642-18638-7_6
3. Ford, L.R., Fulkerson, D.: Flows in Networks. Princeton University Press, Princeton (1962)
4. Hall, M., Frank, E., Holmes, G., Pfahringer, B., Reutemann, P., Witten, I.H.: The WEKA data mining software: an update. ACM SIGKDD Explor. Newslett. **11**(1), 10–18 (2009)
5. Kostek, B., Czyzewski, A.: Processing of musical metadata employing Pawlak's flow graphs. In: Peters, J.F., Skowron, A., Grzymała-Busse, J.W., Kostek, B., Świniarski, R.W., Szczuka, M.S. (eds.) Transactions on Rough Sets I. LNCS, vol. 3100, pp. 279–298. Springer, Heidelberg (2004). https://doi.org/10.1007/978-3-540-27794-1_13
6. Mieszkowicz-Rolka, A., Rolka, L.: Flow graphs and decision tables with fuzzy attributes. In: Rutkowski, L., Tadeusiewicz, R., Zadeh, L.A., Żurada, J.M. (eds.) ICAISC 2006. LNCS (LNAI), vol. 4029, pp. 268–277. Springer, Heidelberg (2006). https://doi.org/10.1007/11785231_29

7. Øhrn, A., Komorowski, J., Skowron, A., Synak, P.: The ROSETTA software system. In: Polkowski, L., Skowron, A. (eds.) Rough Sets in Knowledge Discovery 2, Studies in Fuzziness and Soft Computing, vol. 19, pp. 572–576. Physica-Verlag, Heidelberg (1998)
8. Pancerz, K.: On selected functionality of the classification and prediction software system (CLAPSS). In: Proceedings of the International Conference on Information and Digital Technologies (IDT 2015), Zilina, Slovakia, pp. 278–285 (2015)
9. Pancerz, K., Grochowalski, P., Paja, W.: On selected data preprocessing procedures with the classification and prediction software system (CLAPSS). In: Proceedings of the International Conference on Information and Digital Technologies (IDT 2016), Rzeszow, Poland, pp. 219–226 (2016)
10. Pancerz, K., Lewicki, A., Tadeusiewicz, R., Warchoł, J.: Ant-based clustering in delta episode information systems based on temporal rough set flow graphs. Fundamenta Informaticae **128**(1–2), 143–158 (2013)
11. Pancerz, K.: Paradigmatic and syntagmatic relations in information systems over ontological graphs. Fundamenta Informaticae **148**(1–2), 229–242 (2016)
12. Pawlak, Z.: Flow graphs and data mining. In: Peters, J.F., Skowron, A. (eds.) Transactions on Rough Sets III. LNCS, vol. 3400, pp. 1–36. Springer, Heidelberg (2005). https://doi.org/10.1007/11427834_1

Methods to Edit Multi-label Training Sets Using Rough Sets Theory

Marilyn Bello[1,2(✉)], Gonzalo Nápoles[2], Koen Vanhoof[2], and Rafael Bello[1]

[1] Computer Science Department, Universidad Central de Las Villas, Santa Clara, Cuba
mbgarcia@uclv.cu

[2] Faculty of Business Economics, Hasselt University, Hasselt, Belgium

Abstract. In multi-label classification problems, instances can be associated with several decision classes (labels) simultaneously. One of the most successful algorithms to deal with this kind of problem is the ML-kNN method, which is lazy learner adapted to the multi-label scenario. All the computational models that realize inferences from examples have the common problem of the selection of those examples that should be included into the training set to increase the algorithm's efficiency. This problem in known as *training sets edition*. Despite the extensive work in multi-label classification, there is a lack of methods for editing multi-label training sets. In this research, we propose three reduction techniques for editing multi-label training sets that rely on the Rough Set Theory. The simulations show that these methods reduce the number of examples in the training sets without affecting the overall performance, while in some case the performance is even improved.

Keywords: Multi-label classification · Rough Set Theory · Granular Computing · Machine learning · Edit training set

1 Introduction

In multi-label classification (MLC), each example in the training set belongs to several classes from a set of predefined labels [15,29]. MLC continues to receive attention within the machine learning community because of the wide variety of real-world problems that can be modeled in that context. In text categorization, an electronic document can be referred to sport topics as to politics and society. In semantic scene classification, an image may contain multiple objects. In video annotation, a film can be annotated with several labels or tags. In bioinformatics, each protein may be labeled with multiple functional labels such as metabolism, energy and cellular biogenesis [9,17,25,26].

ML-kNN [38] is an adaptation of the k-NN algorithm [11] to the multi-label scenario. Given a set of n training examples, upon receiving a new instance to be predicted, the k-NN classifier identifies k nearest examples of that instance and then it assigns a set of labels to that instance. The probability of associating

© Springer Nature Switzerland AG 2019
T. Mihálydeák et al. (Eds.): IJCRS 2019, LNAI 11499, pp. 369–380, 2019.
https://doi.org/10.1007/978-3-030-22815-6_29

the instance with a certain decision class is determined based on the number of neighbors that contain the target label.

Some aspects that have a pivotal relevance on k-NN's performance are the reduction of the classification error and the reduction of the computational cost. The k-NN method is very sensitive to incorrectly labeled examples close to the decision boundary as such instances are liable to create a region around them where new examples will also be misclassified [1,5]. Moreover, in large data collections, searching for the nearest neighbor can be quite a time-consuming task. A major problem of instance-based learners is that classification time increases as more examples are added to training set.

All reasoning models that perform the inference process from examples have the problem of selecting the examples that should be included in the training set to increase the efficiency. This problem is known as the training sets edition [4,32]. In the literature, several papers [4,13,14,30] have been proposed to cope with this problem in the context of single-label learning. These techniques are usually based on the reduction or edition of training instances [5] with the goal of reducing the learning matrix. Overall, it decreases the algorithm's workload even when it might yield a little less precision [33].

Moreover, edition techniques can eliminate instances that induce an incorrect classification, even though it is certain that they produce elimination of examples, their fundamental objective is to obtain a training sample of better quality to have a better precision with the system [16].

Despite the extensive work in multi-label classification [21], there is a lack of methods for editing multi-label datasets. Existing methods reported in the literature mainly focus on selecting prototypes [6,18,19]. This fact became a driving-force to study this problem in the multi-label learning context. By doing so, we rely on Rough Set Theory (RST) [22] which is a mathematical theory for data analysis and reasoning [1,20]. The advantage of RST include (1) it only uses the original data and does not need any external information, (2) no assumption about the data is necessary, and (3) it is useful to analyze both qualitative and quantitative attributes [34] in a straightforward manner.

Being more explicit, this paper presents three training edition methods for MLC problems with the goal of increasing algorithms' efficacy without significantly harming their efficacy. Those methods rely on the lower and upper approximations as computed in RST to determine a suitable granularity degree in the training set. The first method builds a training set as the union of the lower approximations attached with each decision class. The second method additionally includes objects that are in the boundary region, which have been relabeled by taking into account the membership degree to each decision class. The third method is similar to the second one, but it omits the connection among decisions classes, so that labels are treated independently.

The rest of the paper is organized as follows. Section 2 presents the theoretical background on rough sets, while Sect. 3 introduces the edition methods for MLC datasets. Section 4 is dedicated to evaluating the performance of the ML-kNN algorithm on synthetic datasets that have been improved with the proposed algorithms. Finally, in Sect. 5 we formalize relevant concluding remarks and future research directions to be explored.

2 Rough Set Theory

RST is a methodology proposed in the early 1980's for handling uncertainty that is manifested in the form of inconsistent data [2,22]. It uses two main components: an information system and an inseparability relationship. The former is defined as $IS = (U, A)$, where U is a non-empty finite set of objects, and A is a non-empty finite set of attributes that describe each object. A particular case are the decision systems where $DS = (U, A \cup \{d\})$, whereas $d \notin A$ is the decision attribute. The inseparability relation allows granulating the universe of discourse using the principles behind rough sets [23,36].

According to [37] the information granulation involves partitioning objects into granules, with a granule being a clump of objects which are drawn together by indistinguishability, similarity or functionality. Any subset $X \subseteq U$ can be approximated by two crisp sets [3]: the lower and the upper approximation. They are defined as $B_*X = \{x \in U : [x]_B \subseteq X\}$ and $B^*X = \{x \in U : [x]_B \cap X \neq \emptyset\}$ respectively, where $[x]_B$ denotes the set of inseparable objects associated to x using an indiscernibility relation defined by $B \subseteq A$.

The objects in B_*X are categorically members of X, whereas the objects in B^*X are possible members of the subset X. This model does not consider any tolerance of errors: if two inseparable objects belong to different classes then the decision system will be inconsistent. However, the definition of indiscernibility as an equivalence relation is excessively strict. It means that two inseparable objects could incorrectly be labeled as separable.

This problems can be alleviated in some extent by extending the concept of inseparability relation [27] and replacing the equivalence relation with a weaker binary relation. Equation (1) shows an indiscernibility relation, where $0 \leq \delta(x, y) \leq 1$ is a similarity function. This weak binary relation states that objects x and y are inseparable as long as their similarity degree $\delta(x, y)$ exceeds a similarity threshold $0 \leq \xi_1 \leq 1$. This relation actually defines a similarity class $\overline{R}(x) = \{y \in U : yRx\}$ that replaces the equivalence class.

$$R_1 : xRy \Longleftrightarrow \delta(x, y) \geq \xi \tag{1}$$

The similarity function could be formulated in a variety of ways, for example, $\varphi(x, y) = 1 - \delta(x, y)$ with $\delta(x, y)$ being the distance between objects x and y. In reference [31] the authors studied the properties of several distance functions which allow comparing heterogeneous instances, i.e., objects comprising both numerical and nominal attributes. In this paper, we have adopted the Heterogeneous Euclidean-Overlap Metric (HEOM) defined in Eq. (2), which computes the normalized Euclidean distance between numerical attributes and an overlap metric for nominal attributes,

$$\delta(x,y) = \sqrt{\frac{\sum_{j=1}^{M} \omega_j \sigma_j(x,y)}{\sum_{j=1}^{M} \omega_j}} \tag{2}$$

such that

$$\sigma_j(x,y) = \begin{cases} 0 & \text{if } A_j \text{ is nominal } \wedge x(j) = y(j) \\ 1 & \text{if } A_j \text{ is nominal } \wedge x(j) \neq y(j) \\ (x(j) - y(j))^2 & \text{if } A_j \text{ is numerical} \end{cases} \tag{3}$$

where A represents the set of features describing the problem, $0 \leq \omega_j \leq 1$ is the relative relevance of the jth attribute, $x(j)$ and $y(j)$ denote the values of the jth attribute associated to objects x and y respectively.

Equations (4) and (5) formalize how to compute lower and upper approximations respectively, based on the elements described above,

$$B_*X = \{x \in U : \overline{R}(x) \subseteq X\} \tag{4}$$

$$B^*X = \bigcup_{x \in X} \overline{R}(x). \tag{5}$$

The lower and upper approximations allow computing three well-defined regions of set X. The *positive region* $POS(X) = B_*X$ includes those objects that are certainly contained in X, the *negative region* $NEG(X) = \mathcal{U} - B^*X$ denotes those objects that are certainly not related to X, while the *boundary region* $BND(X) = B_*X - B^*X$ captures the objects whose membership to X is uncertain, i.e., they might be members of X [35].

3 Edition Methods for Multi-label Training Sets

In this section, we propose three methods for editing multi-label training sets which employ the upper and lower approximation concepts as defined in RST. It offers a pattern-oriented model the deal with uncertainty in the form of *inconsistency*, as often happen in the presence of noise. In fact, the lower approximation eliminates the cases having a noisy behavior.

As mentioned, in MLC scenarios an instance may be associated with multiple labels. Let $mlDS = (U, A \cup L)$ be a multi-label decision system, where the set U is a non-empty finite set of objects, A is a non-empty finite set of attributes that describe each observation, and $L = \{L_1, L_2, \ldots, L_k\}$ is a non-empty finite set of labels such that the label domain is $L_i = \{0, 1\}$.

Aiming at extending the reduction technique proposed in [5] for the multi-label case, we must define what is considered to be a decision class in the MLC context. In this paper, the following variants are contemplated:

- Each combination C_i of labels represents a decision value. For example, let $L = \{L_1, L_2, L_3\}$ denote the set of labels, a combination of labels could be "101", pointing out that the object belongs to the labels L_1 and L_3, then "101" defines a decision class, so that all objects associated with labels L_1 and L_3 belong to that decision class.
- Each label (L_i) is considered a decision value, so that all the objects associated that label belong to this decision class. According with this definition, in the example above there are three decision classes.

3.1 Each Label Combination Is a Decision Value

The basic idea behind the *first edition method* is summarized as follows. First, we detect all possible combinations of labels (i.e., decision classes according with the first definition) that have been observed in the dataset. Afterwards, we compute the lower approximations and construct the news dataset as the union of those information granules as formalized in the following equation,

$$mlTS = B_*(X_1) \cap B_*(X_2) \cap \ldots \cap B_*(X_k) \tag{6}$$

where $B_*(X_i)$ is the lower approximation associated with the i-th decision class and k is the number of label combinations. This is equivalent to saying that the training set will be the positive region of the decision system since objects that are incorrectly labeled or near to the decision boundary will be eliminated from the dataset [10]. Algorithm 1 formalized this procedure.

Algorithm 1. Edit1mlTS

1: Form the sets $X_i \subseteq U$, where X_i denotes the ith decision class that contains all objects associated with the ith combination C_i.
2: For each set X_i, calculate its lower approximation $B_*(X_i)$.
3: Construct the edited training set as the union of all the sets $B_*(X_i)$ as defined in Equation (6).

In the *second edition method*, we use the lower approximation and the boundary region of each label combination to create the edited training set. Besides the objects that belong to the lowers approximation, this method detects suspicious objects by computing the boundary regions while changing the decision class of some of them by using a likelihood measure.

If an object y belongs to the ith boundary region, then there are inseparable objects to x which are associated with different label combinations. Therefore y is considered suspicious. The relabeling method uses the membership function in Eq. (7) as a likelihood measure,

$$\mu_{X_i}(x) = \frac{|\ X_i \cap \overline{R}(x)\ |}{|\ \overline{R}(x)\ |} \tag{7}$$

where X_i comprises all objects associated with the ith label combination, and $\overline{R}(x)$ is the similarity class for object x. Hence, for each object y in the boundary regions, we calculate the membership degree to each combination (as defined in Eq. (7) so that we can assign to x the combination that reaches the highest membership degree. Algorithm 2 displays this method.

3.2 Each Label Is Considered a Decision Value

As mentioned, each label combination observed in the dataset can be considered as a decision class in the MLC problem. The *third edition method* takes this idea into account, as described in the Algorithm 3.

Algorithm 2. Edit2mlTS

1: Form the sets $X_i \subseteq U$, where X_i denotes the ith decision class that contains all objects associated with the ith combination (C_i).
2: For each set X_i do
  Calculate their lower approximation $B_*(X_i)$ and upper approximation $B^*(X_i)$
  $mlST = mlST \cup B_*(X_i)$
  $T_i = BND(X_i) = B^*(X_i) - B_*(X_i)$
3: $T = \bigcup_{i=1}^{k} T_i$, where k represents the number of label combinations.
4: For each object $x \in T$, calculate the membership degree to each X_i, and re-labelling the x object with the C_i combination associated with the ith decision class in which the highest membership degree is reached.
5: $mlST = mlST \cup T'$

Algorithm 3. Edit3mlTS

1: Form the sets $X_i \subseteq U$, where X_i is the ith decision class that contains all the objects that have the ith label (L_i).
2: For each set X_i do
  Calculate their lower approximation $B_*(X_i)$ and upper approximation $B^*(X_i)$
  $mlST = mlST \cup B_*(X_i)$
  $T_i = BND(X_i) = B^*(X_i) - B_*(X_i)$
3: $T = \bigcup_{i=1}^{k} T_i$, where k denotes the number of labels.
4: For each object $x \in T$, calculate the membership degree to each X_i, and re-labelling the x object with the L_i label when the membership grade is greater than a β threshold.
5: $mlST = mlST \cup T'$

As a first step, we compute the lower approximations with respect to each existing label in the MLC dataset. Afterwards, we consider the elements that are on the boundary, that is, those objects in which there is suspicion about their membership to the concept denoted that label. These objects are relabeled according with their degree of membership to a label. In other words, suspicious objects will be linked with the labels reporting a membership degree greater than a threshold β. Therefore, the edited training set will composed of relabeled objects and objects belonging to the lower approximations.

4 Experimental Results

In this section, we explore the global performance of our edition methods when coupled with the ML-kNN classification algorithm. To do this, we used *Hamming Loss* (HL) metric which is probably the most used performance metric in MLC scenarios [15,24]. This metric is defined in Eq. (8),

$$HammingLoss = \frac{1}{n}\frac{1}{k}\sum_{i=1}^{n}|Y_i \Delta Z_i| \tag{8}$$

where Δ operator returns the symmetric difference between Y_i (the real label set of the ith instance) and Z_i (the predicted one). Observe that, since the mistakes counter is divided by the number of labels (k), this metric will result in different assessments for the same amount of errors when used with MLDs having a label set with a different cardinality.

We leaned upon 12 multi-label datasets corresponding with three application areas in which multi-label data is often observed: text categorization, multimedia classification and bioinformatics. Such datasets were taken from the MULAN [28] and RUMDR [7] repositories.

Table 1 summarizes the number of instances, attributes, and labels for each dataset. The number of distinct labelsets, calculated as the number of distinct combinations of labels found in the dataset is also given. The *TCS* metric [8] in Eq. (9) is adopted as a theoretical complexity indicator. The higher the value, the more complex the MLC dataset.

$$TCS = log(attributes \times labels \times distinct) \tag{9}$$

Remark that TCS values are logarithmic, thus a difference of only one unit implies one order of magnitude lower or higher.

Figure 1 displays the reduction percent in the original datasets applying the proposed editions methods to the original training sets. In this experiment, the similarity threshold ξ used in Eq. (1) ranges from 0.9 to 0.99, while the parameter β in Algorithm 3 goes from 0.5 to 0.75.

From the results in Fig. 1 we can conclude that the proposed edition methods always obtain in the most of cases multi-label datasets with a less number of examples than the original. In the case of *bibtex*, *corel5k*, *enron*, *medical*, *scene*, *slashdot* and *yeast* the first edition method reported higher reduction rates when compared with the other two variants.

To determine whether the reduction is statistically significant, we computed the Friedman two-way analysis of variances by ranks [12]. The test suggests rejecting the null hypothesis H_0 (p-value $= 3.8867\text{E}{-7} < 0.05$) for a confidence interval of 95%, so that there are significant differences in the number of instances after applying the proposed edition methods.

Table 2 shows the HL values achieved by the ML-kNN method for the original multi-label data sets without the preprocessing step, and the results obtained after applying the three edition methods proposed in the paper.

Table 1. Characterization of the MCL datasets used in our study.

Dataset	Domain	Instances	Attributes	Labels	Labelsets	TCS
bibtex	text	7395	1836	159	2856	20.541
birds	audio	645	260	19	133	13.395
cal500	music	502	68	174	502	15.597
corel5k	images	5000	499	374	3175	20.2
emotions	music	593	72	6	27	9.364
enron	text	1702	1001	53	753	17.503
flags	images	194	19	7	54	8.879
genbase	biology	662	1186	27	32	13.84
medical	text	978	1449	45	94	15.629
scene	images	2407	294	6	15	10.183
slashdot	text	3785	1079	22	156	15.125
yeast	biology	2417	103	14	198	12.562

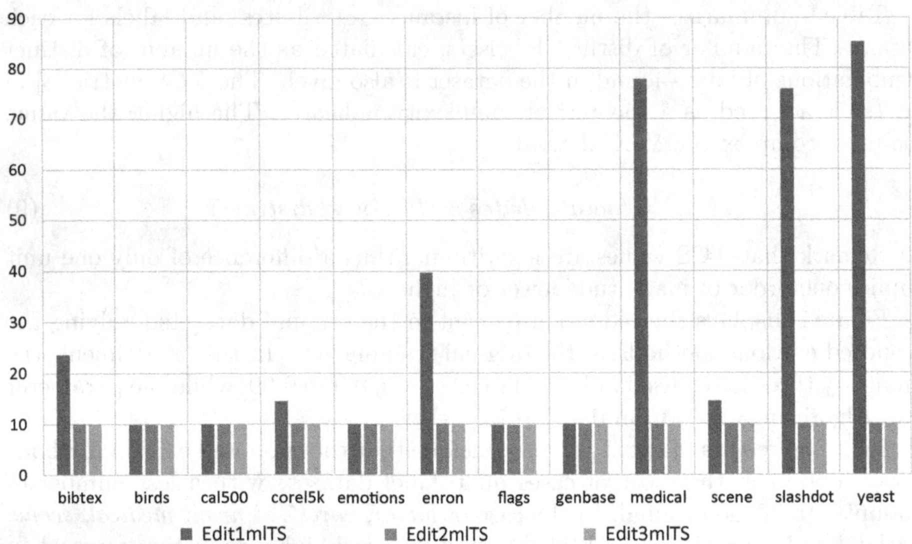

Fig. 1. Reduction percent achieved by each edition method.

For each datasets, we have estimated the HL value by using a 10-fold cross validation scheme. For each fold, this procedure splits the whole training set into two data pieces, namely, the training set and the test set. It should be highlighted that, while the training set is preprocessed with the editing methods, the test set is never modified so that it only serves to compute the HL associated with the current fold.

Similarly, Table 2 shows that the reduced datasets lead to similar HL values. Actually, in some cases the ML-kNN method using the edited datasets reports better results (e.g., such as *bibtex*, *corel5k*, *flags*, *genbase* and *slashdot*) than those obtained with the original datasets. This might be a result of addressing the uncertainty that comes in the form of inconsistent patterns. For example, no classification algorithm will properly recognize two inseparable objects with different decision classes when using equivalence classes to derive the rough granules.

Table 2. HL values achieved by the ML-kNN method.

	OriginalmlTS	Edit1mlTS	Edit2mlTS	Edit3mlTS
bibtex	0.01364	0.01341	0.01363	0.01364
birds	0.04725	0.04725	0.04725	0.04725
cal500	0.13881	0.13843	0.13881	0.13883
corel5k	0.00936	0.00934	0.00937	0.00936
emotions	0.19512	0.19512	0.19512	0.19512
enron	0.05235	0.0514	0.05251	0.05278
flags	0.25357	0.25132	0.25132	0.25357
genbase	0.0048	0.00475	0.00475	0.0048
medical	0.01511	0.02308	0.01561	0.01516
scene	0.08621	0.08434	0.08621	0.08628
slashdot	0.05169	0.04986	0.05188	0.05169
yeast	0.1933	0.22805	0.1933	0.1933

In order to examine the existence of statistically differences in performance, we computed the Friedman two-way analysis of variances by ranks. For this experiment, the test suggests accepting the null hypothesis H_0 (p-value $= 0.142 > 0.05$) for a confidence interval of 95%. Therefore, we can conclude that the proposed methods reduce the number of examples in the training sets, while preserving the efficacy of the ML-kNN method.

5 Concluding Remarks

The efficacy and efficiency of the Machine Learning models depend on the quantity and quality of data. In the case of lazy learners, such as the k-NN method, the number of instances is relevant in the efficiency since these algorithms go over all examples each time a new problem is presented. Moreover, these models are sensitive to incorrectly classifier instances. One alternative to deal with these issues is to edit the training set as a pre-processing step with the intention either to reduce the number of instances or improve data quality.

This paper proposed three methods based on Granular Computing to editing training sets in MLC environments. These methods rely on the concepts of lower and upper approximations while introducing different approaches to define the semantics of decision classes. The experiments using several multi-label training sets have shown that the methods allow to significantly reduce the number of instances in the training set without affects the performance of the classification. Actually, in some cases we observed an increase the discriminatory power of the ML-kNN algorithm when operating on edited datasets.

References

1. Barandela, R., Cortés, N., Palacios, A.: The nearest neighbor rule and the reduction of the training sample size. In: Proceedings 9th Symposium on Pattern Recognition and Image Analysis, vol. 1, pp. 103–108 (2001)
2. Bello, R., Falcón, R., Pedrycz, W.: Granular Computing: At the Junction of Rough Sets and Fuzzy Sets. Studies in Fuzziness and Soft Computing, vol. 224. Springer, Heidelberg (2007). https://doi.org/10.1007/978-3-540-76973-6
3. Bello, R., Verdegay, J.L.: Rough sets in the soft computing environment. Inf. Sci. **212**, 1–14 (2012)
4. Brighton, H., Mellish, C.: Advances in instance selection for instance-based learning algorithms. Data Min. Knowl. Discov. **6**(2), 153–172 (2002)
5. Caballero, Y., Bello, R., Salgado, Y., Garcia, M.M.: A method to edit training set based on rough sets. Int. J. Comput. Intell. Res. **3**(3), 219–229 (2007)
6. Calvo-Zaragoza, J., Valero-Mas, J.J., Rico-Juan, J.R.: Improving kNN multi-label classification in prototype selection scenarios using class proposals. Pattern Recognit. **48**(5), 1608–1622 (2015)
7. Charte, F., Charte, D., Rivera, A., del Jesus, M.J., Herrera, F.: R ultimate multilabel dataset repository. In: Martínez-Álvarez, F., Troncoso, A., Quintián, H., Corchado, E. (eds.) HAIS 2016. LNCS (LNAI), vol. 9648, pp. 487–499. Springer, Cham (2016). https://doi.org/10.1007/978-3-319-32034-2_41
8. Charte, F., Rivera, A., del Jesus, M.J., Herrera, F.: On the impact of dataset complexity and sampling strategy in multilabel classifiers performance. In: Martínez-Álvarez, F., Troncoso, A., Quintián, H., Corchado, E. (eds.) HAIS 2016. LNCS (LNAI), vol. 9648, pp. 500–511. Springer, Cham (2016). https://doi.org/10.1007/978-3-319-32034-2_42
9. Chen, X.j., Zhan, Y.z., Ke, J., Chen, X.b.: Complex video event detection via pairwise fusion of trajectory and multi-label hypergraphs. Multimed. Tools Appl. **75**(22), 15079–15100 (2016)
10. Cortijo, J.: Techniques of approximation II: non parametric approximation. Ph.D. thesis, Thesis. Department of Computer Science and Artificial Intelligence, Universidad de Granada, Spain (2001)
11. Dasarathy, B.V.: Nearest neighbor ({NN}) norms:{NN} pattern classification techniques (1991)
12. Friedman, M.: The use of ranks to avoid the assumption of normality implicit in the analysis of variance. J. Am. Stat. Assoc. **32**(200), 675–701 (1937)
13. Garcia, S., Derrac, J., Cano, J., Herrera, F.: Prototype selection for nearest neighbor classification: taxonomy and empirical study. IEEE Trans. Pattern Anal. Mach. Intell. **34**(3), 417–435 (2012)

14. Guan, D., Yuan, W., Lee, Y.K., Lee, S.: Nearest neighbor editing aided by unlabeled data. Inf. Sci. **179**(13), 2273–2282 (2009)
15. Herrera, F., Charte, F., Rivera, A.J., Del Jesus, M.J.: Multilabel classification. In: Multilabel Classification, pp. 17–31. Springer (2016). https://doi.org/10.1007/978-3-319-41111-8_2
16. Jiang, Y., Zhou, Z.-H.: Editing training data for $k$nn classifiers with neural network ensemble. In: Yin, F.-L., Wang, J., Guo, C. (eds.) ISNN 2004. LNCS, vol. 3173, pp. 356–361. Springer, Heidelberg (2004). https://doi.org/10.1007/978-3-540-28647-9_60
17. Jin, B., Muller, B., Zhai, C., Lu, X.: Multi-label literature classification based on the gene ontology graph. BMC Bioinform. **9**(1), 525 (2008)
18. Kanj, S., Abdallah, F., Denœux, T.: Purifying training data to improve performance of multi-label classification algorithms. In: 2012 15th International Conference on Information Fusion (FUSION), pp. 1784–1791. IEEE (2012)
19. Kanj, S., Abdallah, F., Denœux, T., Tout, K.: Editing training data for multi-label classification with the k-nearest neighbor rule. Pattern Anal. Appl. **19**(1), 145–161 (2016)
20. Komorowski, J., Pawlal, Z., Polkowski, L., Skowron, A.: B6. A rough set perspective on data and knowledge. In: Klösgen, W., Zytkow, J.M. (eds.) The Handbook of Data Mining and Knowledge Discovery. Oxford University Press, Oxford (1999)
21. Madjarov, G., Kocev, D., Gjorgjevikj, D., Džeroski, S.: An extensive experimental comparison of methods for multi-label learning. Pattern Recognit. **45**(9), 3084–3104 (2012)
22. Pawlak, Z.: Rough sets. Int. J. Comput. Inf. Sci. **11**(5), 341–356 (1982)
23. Pedrycz, W., Skowron, A., Kreinovich, V.: Handbook of Granular Computing. Wiley, New York (2008)
24. Pereira, R.B., Plastino, A., Zadrozny, B., Merschmann, L.H.: Correlation analysis of performance measures for multi-label classification. Inf. Process. Manage. **54**(3), 359–369 (2018)
25. Qi, G.J., Hua, X.S., Rui, Y., Tang, J., Mei, T., Zhang, H.J.: Correlative multi-label video annotation. In: Proceedings of the 15th ACM International Conference on Multimedia, pp. 17–26. ACM (2007)
26. Sebastiani, F.: Machine learning in automated text categorization. ACM Comput. Surv. (CSUR) **34**(1), 1–47 (2002)
27. Slowinski, R., Vanderpooten, D.: A generalized definition of rough approximations based on similarity. IEEE Trans. Knowl. Data Eng. **12**(2), 331–336 (2000)
28. Tsoumakas, G., Xioufis, E., Vilcek, J., Vlahavas, I.: MULAN multi-label dataset repository (2014). http://mulan.sourceforge.net/datasets.html
29. Tsoumakas, G., Katakis, I., Vlahavas, I.: Mining multi-label data. In: Maimon, O., Rokach, L. (eds.) Data Mining and Knowledge Discovery Handbook, pp. 667–685. Springer, Boston (2000). https://doi.org/10.1007/978-0-387-09823-4_34
30. Van Hulse, J., Khoshgoftaar, T.: Knowledge discovery from imbalanced and noisy data. Data Knowl. Eng. **68**(12), 1513–1542 (2009)
31. Wilson, D.R., Martinez, T.R.: Improved heterogeneous distance functions. J. Artif. Intell. Res. **6**, 1–34 (1997)
32. Wilson, D.L.: Asymptotic properties of nearest neighbor rules using edited data. IEEE Trans. Syst. Man Cybern. **3**, 408–421 (1972)
33. Wilson, R., Martinez, T.R.: Reduction techniques for exemplar-based learning algorithms. Machine Learning. Computer Science Department, Brigham Young University, USA (1998)

34. Xu, Z., Liang, J., Dang, C., Chin, K.S.: Inclusion degree: a perspective on measures for rough set data analysis. Inf. Sci. **141**(3–4), 227–236 (2002)
35. Yao, Y.: Three-way decisions with probabilistic rough sets. Inf. Sci. **180**(3), 341–353 (2010)
36. Yao, Y.: Information granulation and rough set approximation. Int. J. Intell. Syst. **16**(1), 87–104 (2001)
37. Zadeh, L.A.: Key roles of information granulation and fuzzy logic in human reasoning, concept formulation and computing with words. In: Proceedings of IEEE 5th International Fuzzy Systems, vol. 1, p. 1. IEEE (1996)
38. Zhang, M.L., Zhou, Z.H.: Ml-KNN: a lazy learning approach to multi-label learning. Pattern Recognit. **40**(7), 2038–2048 (2007)

Areas of Application

The Impact of Rough Set Conferences

JingTao Yao(✉)

Department of Computer Science, University of Regina,
Regina, SK S4S 0A2, Canada
jtyao@cs.uregina.ca

Abstract. Thousands of research papers on rough set theory and its applications have been published since Pawlak's seminal work Rough Sets in 1982. A large number of rough sets papers were published in proceedings of conferences such as the International Conference on Rough Sets and Current Trends in Computing (RSCTC); International Conference on Rough Sets, Fuzzy Sets, Data Mining, and Granular Computing (RSFDGrC); International Conference on Rough Sets and Knowledge Technology (RSKT); International Conference on Rough Sets and Intelligent Systems Paradigms (RSEISP); and more recently, merged International Joint Conference on Rough Sets (IJCRS). The aim of this paper is to analyse the influences and impact of rough set conference papers using Scientometrics approaches based on the information obtained from Web of Science. The results suggest that rough set conference papers have an impact in research which is similar to other computer science domains.

1 Introduction

Conference papers are often viewed as important as journal articles in computer science. In many cases, conference papers, especially those appearing in highly rated conference proceedings, are considered as top research outcomes. A recent study confirms that our particular discipline "values conferences as a publication venue more highly than any other academic field of study" [24]. It also shows that top computer science conferences papers have received higher average citations than journal or book chapters. This could be true for rough set conference papers but has yet to be confirmed. As the first step of the confirmation, we would like to study the impact and influence of rough set conference papers in this article by using scientometrics approaches.

Thousands of research papers on rough set theory and its applications appeared in journals, books, and conference proceedings since the birth of rough set theory in 1982 [44]. Scientometrics can be viewed as a science of science which studies the measuring and analysis of scientific literature. It has been used for identifying research areas, trends, relationships, development and future directions [20]. Yao and Zhang [31] classified rough set research into three categories, namely, content based approach that focuses on the contents of rough set research, method based approach that focuses on the constructive and algebraic (axiomatic) methods of rough sets, and scientometrics approach that focuses on

© Springer Nature Switzerland AG 2019
T. Mihálydeák et al. (Eds.): IJCRS 2019, LNAI 11499, pp. 383–394, 2019.
https://doi.org/10.1007/978-3-030-22815-6_30

quantitatively analyzing the contents and citations of rough set publications. This paper focuses on scientometrics study of rough sets and can be viewed as a followup of previous work published in 2013 [31] and 2017 [29]. We would like to understand rough set research contribution, development, and impact especially in flagship conferences in this article.

2 Methodology and Search Setting

The study is based on search of Institute for Scientific Information's Web of Science (https://webofknowledge.com) database including its subsection Conference Proceedings Citation Index. We used "rough sets" or "rough set" or "rough computing" or "rough computation" in the Topic field in Web of Science as the definition of rough set papers in our previous studies [29,31]. By using the same query, we have 12,066 rough set papers. The h-index is 130 and Highly Cited in Field is 72. Web of Science defines highly cited papers as papers that "received enough citations to place [them] in the top 1% of the academic field based on a highly cited threshold for the field and publication year". Due to the fact that more than 10, 000 papers appeared in the result, Web of Science analysis in terms of citation report cannot be obtained. In the current study, we use the same keywords in the Conference field which result in 2,341 items in an early March 2019 search. In fact, when we changed the keywords to "rough sets" or "rough set" the same results were generated as no one uses "rough computing" or "rough computation" in conference titles. Further examining Source Titles, we found that there are some non rough set conferences and journals that published special issues of rough set conferences.

We only consider flagship rough set conferences including International Conference on Rough Sets and Current Trends in Computing (RSCTC); International Conference on Rough Sets, Fuzzy Sets, Data Mining, and Granular Computing (RSFDGRC); International Conference on Rough Sets and Knowledge Technology (RSKT); Rough Sets and Intelligent Systems Paradigms (RSEISP); and newly merged International Joint Conference on Rough Sets (IJCRS) in this research. Out of a total of 2,341 papers, there are 2,044 papers from the above mentioned 5 flagship conferences. Some statistical results are:

- Number of papers: 2,044
- H-index: 34
- Average citations per item: 4.34
- Sum of times cited: 8,870
- Sum of times cited without self citations: 7,827
- Citing articles: 5,849
- Citing articles without self citations: 5,265

We list some of the results we obtained in June 2013 [31],

Number of papers: 7,088; H-index: 80; Average citations per item: 5.90; Sum of Times Cited: 41, 844; and June 2016 [29], Number of papers: 9,570; H-index: 106; Average citations per item: 8.02; Sum of Times Cited: 76,733 for comparison.

3 Papers in Rough Set Flagship Conferences

We show some statistics and analysis of research papers published in rough set flagship conference proceedings in this section. Table 1 shows the number of papers and shares in percentage of the total papers published each year. It is noted that all rough set conference proceedings were published as Lecture Notes in Computer Science or Lecture Notes in Artificial Intelligence series by Springer. The earliest conference included in Web of Science is RSFDGrC 1999. However, there are early conference proceedings such as the first RSCTC in 1998, RSCTC 2000, and the first International Workshop on Rough Sets and Knowledge Discovery in 1993, all published by Springer, which are not included in the database.

Table 1. Number of publications per year

Year	1999	2002	2003	2004	2005	2006	2007	2008
No of papers	68	83	127	64	152	213	225	152
% of total	3.33%	4.06%	6.21%	3.13%	7.44%	10.42%	11.01%	7.44%
Year	2009	2010	2011	2013	2014	2015	2016	2017
No of papers	151	182	151	81	158	90	53	94
% of total	7.39%	8.90%	7.39%	3.96%	7.73%	4.40%	2.59%	4.60%

Table 2 lists the most prolific authors with the number of papers they published and the share of total publication. The top 5 prolific authors are Yao YY, Wang GY, Slowinski R, Skowron A and Tsumoto S. By comparing with our 2013 (Slowinski R, Skowron A, Yao YY, Wang GY, and Peters JF) and 2017 (Slowinski R, Yao YY, Skowron A, Wang GY, and Zhu W) results, we found that the top 4 of the top 5 are the same except position shifts. The fifth one is different from year to year.

Table 2. The most prolific authors

Rank	Author names	Papers	Share in %	Rank	Author name	Papers	Share in %
1	Yao YY	51	2.50%	11	Peters JF	24	1.17%
2	Wang GY	49	2.40%	12	Nakata M	23	1.13%
3	Slowinski R	46	2.25%	13	Sakai H	23	1.13%
4	Skowron A	43	2.10%	14	Lingras P	22	1.08%
5	Tsumoto S	37	1.81%	15	Suraj Z	22	1.08%
6	Slezak D	35	1.71%	16	Zhu W	22	1.08%
7	Greco S	33	1.61%	17	Li TR	21	1.03%
8	Wu WZ	31	1.52%	18	Moshkov M	21	1.03%
9	Miao DQ	30	1.47%	19	Yao JT	21	1.03%
10	Grzymala-Busse JW	26	1.27%	20	Hirano S	19	0.93%

Table 3. No of papers by organizations

Rank	Organization	Papers	Share in %
1	University of Warsaw	121	5.92%
2	University of Regina	109	5.33%
3	Polish Academy of Sciences	78	3.82%
4	Southwest Jiaotong University	70	3.42%
5	Chongqing University of Posts Telecommunications	59	2.89%
6	Poznan University of Technology	56	2.75%
7	Polsko Japonska Akademia Technik Komputerowych	54	2.64%
7	Warsaw University of Technology	54	2.64%
9	Chinese Academy of Sciences	50	2.45%
10	Shimane University	43	2.10%
11	Zhejiang Ocean University	42	2.05%
12	University of Rzeszow	38	1.86%
13	Tongji University	37	1.81%
14	Xi'an Jiaotong University	36	1.76%
15	University of Catania	34	1.66%
16	University of Silesia	29	1.41%
17	Nanjing University	28	1.37%
18	University of Manitoba	27	1.32%
19	Anhui University	26	1.27%
19	University of Kansas	26	1.27%
21	Kyushu Institute of Technology	24	1.17%
22	Tsinghua University	24	1.17%
23	University of Information Technology Management Rzeszow	23	1.13%
24	Harbin Institute of Technology	22	1.08%
24	University of Electronic Science Technology of China	22	1.08%
26	Beijing Jiaotong University	21	1.03%
26	California State University System	21	1.03%
26	Indian Statistical Institute Kolkata	21	1.03%
26	Infobright Inc.	21	1.03%
26	Josai International University	21	1.03%
31	Jilin University	20	0.98%
31	Minnan Normal University	20	0.98%
31	Osaka University	20	0.98%

Table 3 shows the top 30 organizations each published at least 21 or 1% of papers in rough set conference proceedings. The similar results for countries are shown in Table 4.

There are 5,354 articles citing the 2,044 rough set conference papers. There are 4,770 articles remaining after removing self citations. The citing articles are published in journals, conference proceedings, and some featured books. The

Table 4. No of papers from countries

Rank	Country	Papers	Share in %	Rank	Country	Papers	Share in %
1	China	749	36.64%	9	Spain	47	2.30%
2	Poland	452	22.11%	10	Taiwan	43	2.10%
3	Canada	232	11.35%	11	Australia	38	1.86%
4	Japan	169	8.27%	12	Russia	23	1.13%
5	USA	152	7.44%	13	Malaysia	22	1.08%
6	India	99	4.84%	14	England	20	0.98%
7	Italy	62	3.03%	15	Germany	20	0.98%
8	South Korea	56	2.74%	16	Sweden	19	0.93%

total Source Titles number is 2,077. Table 5 shows top 10 journals, which are all SCI indexed journals, with articles citing rough set conference papers. Please be noted that we exclude self citation in this statistics. The Articles column shows the number of articles in that journal citing rough set conference papers, the Share in % shows the percentage of the articles in the journal amongst the total number of citing articles. It is noted 6 out of the 10 top journals are Q1 journals, i.e. top 25% of journals based on impact factor (IF) distribution. Similarly, Q2 denotes the middle-high position (between top 50% and top 25%), Q3 middle-low position (between top 75% to top 50%), and Q4 the lowest position (bottom 25% of the IF distribution). The data was retrieved from SCImago Journal & Country Rank (https://www.scimagojr.com/). There are at least 1,050 journal articles that cited these rough set conference papers based on Table 5's numbers. This is about 22% of total citing articles. Further, consider Q1 journals, the total number is at least 715 which is about 15% of total citing articles.

Table 5. Top 10 journals citing rough set conference papers

Rank	Journal name (IF quartile)	Articles	Share in %
1	Information Sciences (Q1)	254	5.33%
2	Fundamenta Informaticae (Q3)	190	3.98%
3	Knowledge Based Systems (Q1)	136	2.85%
4	Int. Journal of Approximate Reasoning (Q1)	134	2.81%
5	Expert Systems with Applications (Q1)	79	1.66%
6	Applied Soft Computing (Q1)	66	1.38%
7	Journal of Intelligent Fuzzy Systems (Q2)	57	1.20%
8	Int. Journal of Machine Learning & Cybernetics (Q2)	52	1.09%
9	Neurocomputing (Q1)	46	0.96%
10	Soft Computing (Q2)	36	0.76%

4 Most Influential Rough Set Conference Papers

We analyzed the most influential rough set papers and authors in the previous study [29,31]. Similarly, we analyze the most influential papers published in rough set conference proceedings. As shown above, the h-index of these conference papers is 34. Table 6 shows the top 38 papers with at least 34 citations.

Table 7 shows number of papers published each year. Tables 8 and 9 list organizations and countries the authors of these papers are from respectively. We can observe that these influential papers are more or less equally distributed from 1999–2010. There is one paper in 2013, two papers in year 2014, and no papers in 2015 onward due to the short time of publication. University of Regina with 11 papers topped the organization list. Poland with 12 papers topped the country list. There are 7 authors who published at least 2 influential papers.

From Table 6, we also find that these top cited papers attracted at least 2 citations per year. The top 5 in terms of citation per year are 11.38 [32], 8.91 [34], 7.00 [14], 7.00 [5], and 6.86 [35]. It is interesting to know that the top two most cited papers by Yao YY are also top 2 cited per year.

We classify the top 38 influential papers listed in Table 6 into three groups, rough set theory (RS theory), rough set applications (RS app) and other theory and applications (Others). Although we keep the same number of groups, we combined rough set theory and hybrid groups as classified in previous research. The Others group is a new class. It is interesting to know that the similarity of the clustering result identified by Wei, Miao and Li [27] for rough set research are classified into four groups, dimensionality reduction, hybrid, three-way view decision, and others (Table 10).

There are 15 papers in RS theory, 9 papers in RS app, and 14 papers in others. The theory papers includes decision-theoretic rough sets [32], three-way decisions [34,35], granular computing [12,33], generalize rough sets [5,22,38], covering based rough sets [17,18], game-theoretic rough sets [7], Bayesian rough sets [37], probabilistic rough sets [43], reduction of rough sets [36], and a hybrid model of fuzzy sets and rough sets [3]. The application paper includes association rules [15], dealing with missing values [4], a system with various kind of rough set approaches [1], classification rules [30], reduction algorithm [8], incomplete data [6], Web-based support systems [28], dominance rough sets [19], and information measures [16]. Other papers include classifications [14], fuzzy sets [11,13], soft sets [23], decision trees [41], support vector machines [10,42], neural networks [21,26], GIS [9], mutual information [2], concept lattice [40], and quotient space [39].

Three-way decision is a newly emerged research originally extended from rough set theory.

Please be noted that more than one-third of top papers are in Others group. This means that not only are our flagship conferences a good venue for rough set research but they are also a good venue for other research.

Table 6. Top 38 cited papers contributed to H-index

Rank	Authors	Year	Cites	Avg	Category
1	Yao YY [32]	2007	148	11.38	RS theory
2	Yao YY [34]	2009	98	8.91	RS theory
3	Stefanowski J & Tsoukias A [22]	1999	96	4.54	RS theory
4	Nguyen HS & Slezak D [15]	1999	95	4.52	RS app
5	Yao YY [38]	2003	80	4.71	RS theory
6	Qin KY, Gao Y & Pei Z [17]	2007	77	5.92	RS theory
7	Napierala K, Stefanowski J & Wilk S [14]	2010	70	7.00	Other
8	Greco S, Matarazzo B & Slowinski R [4]	1999	68	3.24	RS app
9	Bazan JG, Szczuka MS & Wroblewski J [1]	2002	67	3.72	RS app
10	Cornelis C, De Cock M & Radzikowska AM [3]	2007	65	5.00	RS theory
11	Yao JT & Yao YY [30]	2002	58	3.22	RS app
12	Lin, TY [12]	2003	57	3.35	RS theory
13	Yao YY [33]	2007	56	4.31	RS theory
14	Hu F, Wang GY, Huang H & Wu Y[8]	2005	52	3.47	RS app
14	Grzymala-Busse JW [6]	2004	52	3.25	RS app
16	Li D, Deogun J, Spaulding W & Shuart B[11]	2004	49	3.06	Other
17	Yao YY [35]	2013	48	6.86	RS theory
17	Sun Q-M, Zhang Z-L & Liu J [23]	2008	48	4.00	Other
17	Zhou ZH & Tang W [41]	2003	47	2.76	Other
20	Yao JT & Herbert JP [28]	2007	44	3.38	RS app
21	Samanta P & Chakraborty MK [18]	2009	42	3.82	RS theory
21	Lai KK, Yu L, Zhou LG & Wang SY [10]	2006	42	3.00	Other
23	Stateczny A & Wlodarczyk-Sielicka M[21]	2014	40	6.67	Other
23	Herbert JP & YaoJT [7]	2008	40	3.33	RS theory
23	Greco S, Matarazzo B & Slowinski R[5]	2005	40	7.00	RS theory
26	Maji PK [13]	2009	39	3.55	Other
27	Yao YY & Zhou B [37]	2010	38	3.80	RS theory
28	Ziarko W [43]	2005	37	2.47	RS theory
28	Slowinski R, Greco S & Matarazzo B[19]	2002	37	2.06	RS app
30	Janowski A, Nowak A, Przyborski M & Szulwic J[9]	2014	36	6.00	Other
31	Walters-Williams J & Li Y [25]	2009	36	3.27	Other
32	Blaszczynski J, Deckert M, Stefanowski J & Wilk S[2]	2010	35	3.50	Other
32	Zhang WX, Wei L & Qi JJ [40]	2005	35	2.33	Other
34	Qian YH & Liang JY [16]	2006	34	2.43	RS app
34	Yao YY, Zhao Y & Wang J [36]	2006	34	2.43	RS theory
34	Wang, SJ [26]	2003	34	2.00	Other
34	Zhang, L and Zhang, B [39]	2003	34	2.00	Other
34	Zhu ML, Wang Y, Chen SF & Liu XD [42]	2003	34	2.00	Other

Table 7. Number of publications per year for top 38 papers

Year	1999	2002	2003	2004	2005	2006	2007	2008	2009	2010	2013	2014
No of papers	3	3	6	2	4	3	5	2	4	3	1	2
% of total	7.90	7.90	15.79	5.26	10.53	7.90	13.16	5.26	10.53	7.90	2.63	5.26

Table 8. Number of top papers by organizations

Organization	Papers	Share in %
University of Regina	11	28.95%
Poznan University Technology	6	15.79%
Chinese Academy Sciences	3	7.90%
University Catania	3	7.90%
Nanjing University	2	5.26%
Polish Academy Sciences	2	5.26%
Warsaw University	2	5.26%

Table 9. No of top papers from countries

Country	Papers	Share in %	Country	Papers	Share in %
Poland	12	31.56%	Italy	3	7.90%
Canada	11	28.95%	USA	3	7.90%
China	11	28.95%	India	2	5.26 %

Table 10. Authors who contributed top cited papers

Papers	Author names	Share in %
8	Yao YY	21.05%
3	Greco S	7.90%
3	Matarazzo B	7.90%
3	Slowinski R	7.90%
3	Yao JT	7.90%
2	Herbert JP	5.26%
2	Wilk S	5.26%

5 Concluding Remarks

We presented our analysis of productivity and impact of rough set conference papers based on search results on Web of Science database. There are a total of 2,044 papers published from 1999 to 2017 and these papers received 8,870 citations. The average citation per paper is 4.34. The top 5 most prolific authors are Yao YY, Wang GY, Slowinski R, Skowron A and Tsumoto S. The h-index of these papers is 34 and 38 papers received at least 34 citations. Yao YY, Greco S, Matarazzo B, Slowinski R, Yao JT, Herbert JP, and Wilk S contributed at least 2 influential top 38 papers. In addition, journals are also paying attention to papers published in rough set conferences. More than 20% of citations are from 10 journals which cite the most rough set conference papers. Amongst the 10 journals, 6 are Q1 journals and they contributed at least 15% of citation counts. The fact of most cited papers, average citation per paper, citing journals show that rough set research presented in conference proceedings have certain impact on research. It is recommend that rough set researchers should continue to contribute to rough set flagship conferences to get your research result known.

References

1. Bazan, J.G., Szczuka, M.S., Wróblewski, J.: A new version of rough set exploration system. In: Alpigini, J.J., Peters, J.F., Skowron, A., Zhong, N. (eds.) RSCTC 2002. LNCS (LNAI), vol. 2475, pp. 397–404. Springer, Heidelberg (2002). https://doi.org/10.1007/3-540-45813-1_52
2. Błaszczyński, J., Deckert, M., Stefanowski, J., Wilk, S.: Integrating selective pre-processing of imbalanced data with Ivotes ensemble. In: Szczuka, M., Kryszkiewicz, M., Ramanna, S., Jensen, R., Hu, Q. (eds.) RSCTC 2010. LNCS (LNAI), vol. 6086, pp. 148–157. Springer, Heidelberg (2010). https://doi.org/10.1007/978-3-642-13529-3_17
3. Cornelis, C., De Cock, M., Radzikowska, A.M.: Vaguely quantified rough sets. In: An, A., Stefanowski, J., Ramanna, S., Butz, C.J., Pedrycz, W., Wang, G. (eds.) RSFDGrC 2007. LNCS (LNAI), vol. 4482, pp. 87–94. Springer, Heidelberg (2007). https://doi.org/10.1007/978-3-540-72530-5_10
4. Greco, S., Matarazzo, B., Slowinski, R.: Handling missing values in rough set analysis of multi-attribute and multi-criteria decision problems. In: Zhong, N., Skowron, A., Ohsuga, S. (eds.) RSFDGrC 1999. LNCS (LNAI), vol. 1711, pp. 146–157. Springer, Heidelberg (1999). https://doi.org/10.1007/978-3-540-48061-7_19
5. Greco, S., Matarazzo, B., Słowiński, R.: Rough membership and Bayesian confirmation measures for parameterized rough sets. In: Ślęzak, D., Wang, G., Szczuka, M., Düntsch, I., Yao, Y. (eds.) RSFDGrC 2005. LNCS (LNAI), vol. 3641, pp. 314–324. Springer, Heidelberg (2005). https://doi.org/10.1007/11548669_33
6. Grzymała-Busse, J.W.: Characteristic relations for incomplete data: a generalization of the indiscernibility relation. In: Tsumoto, S., Słowiński, R., Komorowski, J., Grzymała-Busse, J.W. (eds.) RSCTC 2004. LNCS (LNAI), vol. 3066, pp. 244–253. Springer, Heidelberg (2004). https://doi.org/10.1007/978-3-540-25929-9_29
7. Herbert, J.P., Yao, J.T.: Game-theoretic risk analysis in decision-theoretic rough sets. In: Wang, G., Li, T., Grzymala-Busse, J.W., Miao, D., Skowron, A., Yao, Y. (eds.) RSKT 2008. LNCS (LNAI), vol. 5009, pp. 132–139. Springer, Heidelberg (2008). https://doi.org/10.1007/978-3-540-79721-0_22

8. Hu, F., Wang, G.Y., Huang, H., Wu, Y.: Incremental attribute reduction based on elementary sets. In: Ślęzak, D., Wang, G., Szczuka, M., Düntsch, I., Yao, Y. (eds.) RSFDGrC 2005. LNCS (LNAI), vol. 3641, pp. 185–193. Springer, Heidelberg (2005). https://doi.org/10.1007/11548669_20
9. Janowski, A., Nowak, A., Przyborski, M., Szulwic, J.: Mobile indicators in GIS and GPS positioning accuracy in cities. In: Kryszkiewicz, M., Cornelis, C., Ciucci, D., Medina-Moreno, J., Motoda, H., Raś, Z.W. (eds.) RSEISP 2014. LNCS (LNAI), vol. 8537, pp. 309–318. Springer, Cham (2014). https://doi.org/10.1007/978-3-319-08729-0_31
10. Lai, K.K., Yu, L., Zhou, L., Wang, S.: Credit risk evaluation with least square support vector machine. In: Wang, G.-Y., Peters, J.F., Skowron, A., Yao, Y. (eds.) RSKT 2006. LNCS (LNAI), vol. 4062, pp. 490–495. Springer, Heidelberg (2006). https://doi.org/10.1007/11795131_71
11. Li, D., Deogun, J., Spaulding, W., Shuart, B.: Towards missing data imputation: a study of fuzzy k-means clustering method. In: Tsumoto, S., Słowiński, R., Komorowski, J., Grzymała-Busse, J.W. (eds.) RSCTC 2004. LNCS (LNAI), vol. 3066, pp. 573–579. Springer, Heidelberg (2004). https://doi.org/10.1007/978-3-540-25929-9_70
12. Lin, T.Y.: Granular computing. In: Wang, G., Liu, Q., Yao, Y., Skowron, A. (eds.) RSFDGrC 2003. LNCS (LNAI), vol. 2639, pp. 16–24. Springer, Heidelberg (2003). https://doi.org/10.1007/3-540-39205-X_3
13. Maji, P.K.: More on intuitionistic fuzzy soft sets. In: Sakai, H., Chakraborty, M.K., Hassanien, A.E., Ślęzak, D., Zhu, W. (eds.) RSFDGrC 2009. LNCS (LNAI), vol. 5908, pp. 231–240. Springer, Heidelberg (2009). https://doi.org/10.1007/978-3-642-10646-0_28
14. Napierała, K., Stefanowski, J., Wilk, S.: Learning from imbalanced data in presence of noisy and borderline examples. In: Szczuka, M., Kryszkiewicz, M., Ramanna, S., Jensen, R., Hu, Q. (eds.) RSCTC 2010. LNCS (LNAI), vol. 6086, pp. 158–167. Springer, Heidelberg (2010). https://doi.org/10.1007/978-3-642-13529-3_18
15. Nguyen, H.S., Ślęzak, D.: Approximate reducts and association rules. In: Zhong, N., Skowron, A., Ohsuga, S. (eds.) RSFDGrC 1999. LNCS (LNAI), vol. 1711, pp. 137–145. Springer, Heidelberg (1999). https://doi.org/10.1007/978-3-540-48061-7_18
16. Qian, Y.H., Liang, J.: Combination entropy and combination granulation in incomplete information system. In: Wang, G.-Y., Peters, J.F., Skowron, A., Yao, Y. (eds.) RSKT 2006. LNCS (LNAI), vol. 4062, pp. 184–190. Springer, Heidelberg (2006). https://doi.org/10.1007/11795131_27
17. Qin, K.Y., Gao, Y., Pei, Z.: On covering rough sets. In: Yao, J.T., Lingras, P., Wu, W.-Z., Szczuka, M., Cercone, N.J., Ślęzak, D. (eds.) RSKT 2007. LNCS (LNAI), vol. 4481, pp. 34–41. Springer, Heidelberg (2007). https://doi.org/10.1007/978-3-540-72458-2_4
18. Samanta, P., Chakraborty, M.K.: Covering based approaches to rough sets and implication lattices. In: Sakai, H., Chakraborty, M.K., Hassanien, A.E., Ślęzak, D., Zhu, W. (eds.) RSFDGrC 2009. LNCS (LNAI), vol. 5908, pp. 127–134. Springer, Heidelberg (2009). https://doi.org/10.1007/978-3-642-10646-0_15
19. Słowiński, R., Greco, S., Matarazzo, B.: Rough set analysis of preference-ordered data. In: Alpigini, J.J., Peters, J.F., Skowron, A., Zhong, N. (eds.) RSCTC 2002. LNCS (LNAI), vol. 2475, pp. 44–59. Springer, Heidelberg (2002). https://doi.org/10.1007/3-540-45813-1_6
20. Small, H.: Tracking and predicting growth areas in science. Scientometrics **68**(3), 595–610 (2006)

21. Stateczny, A., Wlodarczyk-Sielicka, M.: Self-organizing artificial neural networks into hydrographic big data reduction process. In: Kryszkiewicz, M., Cornelis, C., Ciucci, D., Medina-Moreno, J., Motoda, H., Raś, Z.W. (eds.) RSEISP 2014. LNCS (LNAI), vol. 8537, pp. 335–342. Springer, Cham (2014). https://doi.org/10.1007/978-3-319-08729-0_34
22. Stefanowski, J., Tsoukiàs, A.: On the extension of rough sets under incomplete information. In: Zhong, N., Skowron, A., Ohsuga, S. (eds.) RSFDGrC 1999. LNCS (LNAI), vol. 1711, pp. 73–81. Springer, Heidelberg (1999). https://doi.org/10.1007/978-3-540-48061-7_11
23. Sun, Q.-M., Zhang, Z.-L., Liu, J.: Soft sets and soft modules. In: Wang, G., Li, T., Grzymala-Busse, J.W., Miao, D., Skowron, A., Yao, Y. (eds.) RSKT 2008. LNCS (LNAI), vol. 5009, pp. 403–409. Springer, Heidelberg (2008). https://doi.org/10.1007/978-3-540-79721-0_56
24. Vrettas, G., Sanderson, M.: Conferences versus journals in computer science. J. Assoc. Inf. Sci. Technol. **6**(12), 2674–2684 (2015)
25. Walters-Williams, J., Li, Y.: Estimation of mutual information: a survey. In: Wen, P., Li, Y., Polkowski, L., Yao, Y., Tsumoto, S., Wang, G. (eds.) RSKT 2009. LNCS (LNAI), vol. 5589, pp. 389–396. Springer, Heidelberg (2009). https://doi.org/10.1007/978-3-642-02962-2_49
26. Wang, S.: A new development on ANN in China — biomimetic pattern recognition and multi weight vector neurons. In: Wang, G., Liu, Q., Yao, Y., Skowron, A. (eds.) RSFDGrC 2003. LNCS (LNAI), vol. 2639, pp. 35–43. Springer, Heidelberg (2003). https://doi.org/10.1007/3-540-39205-X_5
27. Wei, W., Miao, D., Li, Y.: A bibliometric profile of research on rough sets. In: Mihálydeák, T., et al. (eds.) IJCRS 2019. LNAI, vol. 11499, pp. 534–548. Springer, Cham (2019)
28. Yao, J.T., Herbert, J.P.: Web-based support systems with rough set analysis. In: Kryszkiewicz, M., Peters, J.F., Rybinski, H., Skowron, A. (eds.) RSEISP 2007. LNCS (LNAI), vol. 4585, pp. 360–370. Springer, Heidelberg (2007). https://doi.org/10.1007/978-3-540-73451-2_38
29. Yao, J.T., Onasanya, A.: Recent development of rough computing: a scientometrics view. In: Wang, G., Skowron, A., Yao, Y., Ślęzak, D., Polkowski, L. (eds.) Thriving Rough Sets. SCI, vol. 708, pp. 21–45. Springer, Cham (2017). https://doi.org/10.1007/978-3-319-54966-8_3
30. Yao, J.T., Yao, Y.Y.: Induction of classification rules by granular computing. In: Alpigini, J.J., Peters, J.F., Skowron, A., Zhong, N. (eds.) RSCTC 2002. LNCS (LNAI), vol. 2475, pp. 331–338. Springer, Heidelberg (2002). https://doi.org/10.1007/3-540-45813-1_43
31. Yao, J.T., Zhang, Y.: A scientometrics study of rough sets in three decades. In: Lingras, P., Wolski, M., Cornelis, C., Mitra, S., Wasilewski, P. (eds.) RSKT 2013. LNCS (LNAI), vol. 8171, pp. 28–40. Springer, Heidelberg (2013). https://doi.org/10.1007/978-3-642-41299-8_4
32. Yao, Y.Y.: Decision-theoretic rough set models. In: Yao, J.T., Lingras, P., Wu, W.-Z., Szczuka, M., Cercone, N.J., Ślęzak, D. (eds.) RSKT 2007. LNCS (LNAI), vol. 4481, pp. 1–12. Springer, Heidelberg (2007). https://doi.org/10.1007/978-3-540-72458-2_1
33. Yao, Y.Y.: The art of granular computing. In: Kryszkiewicz, M., Peters, J.F., Rybinski, H., Skowron, A. (eds.) RSEISP 2007. LNCS (LNAI), vol. 4585, pp. 101–112. Springer, Heidelberg (2007). https://doi.org/10.1007/978-3-540-73451-2_12

34. Yao, Y.Y.: Three-way decision: an interpretation of rules in rough set theory. In: Wen, P., Li, Y.Y., Polkowski, L., Yao, Y.Y., Tsumoto, S., Wang, G. (eds.) RSKT 2009. LNCS (LNAI), vol. 5589, pp. 642–649. Springer, Heidelberg (2009). https://doi.org/10.1007/978-3-642-02962-2_81
35. Yao, Y.Y.: Granular computing and sequential three-way decisions. In: Lingras, P., Wolski, M., Cornelis, C., Mitra, S., Wasilewski, P. (eds.) RSKT 2013. LNCS (LNAI), vol. 8171, pp. 16–27. Springer, Heidelberg (2013). https://doi.org/10.1007/978-3-642-41299-8_3
36. Yao, Y.Y., Zhao, Y., Wang, J.: On reduct construction algorithms. In: Wang, G.-Y., Peters, J.F., Skowron, A., Yao, Y.Y. (eds.) RSKT 2006. LNCS (LNAI), vol. 4062, pp. 297–304. Springer, Heidelberg (2006). https://doi.org/10.1007/11795131_43
37. Yao, Y.Y., Zhou, B.: Naive Bayesian rough sets. In: Yu, J., Greco, S., Lingras, P., Wang, G., Skowron, A. (eds.) RSKT 2010. LNCS (LNAI), vol. 6401, pp. 719–726. Springer, Heidelberg (2010). https://doi.org/10.1007/978-3-642-16248-0_97
38. Yao, Y.Y.: On generalizing rough set theory. In: Wang, G., Liu, Q., Yao, Y., Skowron, A. (eds.) RSFDGrC 2003. LNCS (LNAI), vol. 2639, pp. 44–51. Springer, Heidelberg (2003). https://doi.org/10.1007/3-540-39205-X_6
39. Zhang, L., Zhang, B.: The quotient space theory of problem solving. In: Wang, G., Liu, Q., Yao, Y., Skowron, A. (eds.) RSFDGrC 2003. LNCS (LNAI), vol. 2639, pp. 11–15. Springer, Heidelberg (2003). https://doi.org/10.1007/3-540-39205-X_2
40. Zhang, W.-X., Wei, L., Qi, J.-J.: Attribute reduction in concept lattice based on discernibility matrix. In: Ślęzak, D., Yao, J.T., Peters, J.F., Ziarko, W., Hu, X. (eds.) RSFDGrC 2005. LNCS (LNAI), vol. 3642, pp. 157–165. Springer, Heidelberg (2005). https://doi.org/10.1007/11548706_17
41. Zhou, Z.-H., Tang, W.: Selective ensemble of decision trees. In: Wang, G., Liu, Q., Yao, Y., Skowron, A. (eds.) RSFDGrC 2003. LNCS (LNAI), vol. 2639, pp. 476–483. Springer, Heidelberg (2003). https://doi.org/10.1007/3-540-39205-X_81
42. Zhu, M., Wang, Y., Chen, S., Liu, X.: Sphere-structured support vector machines for multi-class pattern recognition. In: Wang, G., Liu, Q., Yao, Y., Skowron, A. (eds.) RSFDGrC 2003. LNCS (LNAI), vol. 2639, pp. 589–593. Springer, Heidelberg (2003). https://doi.org/10.1007/3-540-39205-X_95
43. Ziarko, W.: Probabilistic rough sets. In: Ślęzak, D., Wang, G., Szczuka, M., Düntsch, I., Yao, Y. (eds.) RSFDGrC 2005. LNCS (LNAI), vol. 3641, pp. 283–293. Springer, Heidelberg (2005). https://doi.org/10.1007/11548669_30
44. Pawlak, Z.: Rough sets. Int. J. Parallel Prog. **11**(5), 341–356 (1982)

Multivariate Ovulation Window Detection at OvuFriend

Joanna Fedorowicz[1], Łukasz Sosnowski[2], Dominik Ślęzak[3(✉)], Iwona Szymusik[4], Wojciech Chaber[1], Łukasz Miłobędzki[1], Tomasz Penza[1], Jadwiga Sosnowska[1], Katarzyna Wójcicka[1], and Karol Zaleski[1]

[1] OvuFriend Sp. z o.o., Warsaw, Poland
{joanna.fedorowicz,wojciech.chaber,lukasz.milobedzki,tomasz.penza, jadwiga.sosnowska,katarzyna.wojcicka,karol.zaleski}@ovufriend.pl

[2] Systems Research Institute, Polish Academy of Sciences, Warsaw, Poland
lukasz.sosnowski@ibspan.waw.pl

[3] Institute of Informatics, University of Warsaw, Warsaw, Poland
slezak@mimuw.edu.pl

[4] Department of Obstetrics and Gynecology, Medical University of Warsaw, Warsaw, Poland
iwona.szymusik@gmail.com

Abstract. We present new results related to retrospective detection of ovulation days basing on information entered by the users of one of online platforms available in the market. Comparing to our previous studies, we improve the accuracy of algorithms which are based on evaluation and synthesis of multivariate data sources. Results are reported for 224 menstrual cycles which were labeled by medical experts. In the experiments, we pay special attention to the aspect of uncertainty associated with the tagging process.

Keywords: Ovulation window detection · e-Health advisory systems · Multivariate time series · Uncertain data tagging

1 Introduction

The problem of infertility is increasing all over the world. Female partner in every fifth couple actively trying to have children is not able to get pregnant within the first year of attempts. As a result, women are ready to share within various online platforms the detailed data about their family status, age, weight, physical symptoms associated with menstruation, results of ovulation and pregnancy tests, temperature measurements, mood, medications, etc. – everything that lets them compare themselves with similar cases of other couples and that

Co-financed by the EU Smart Growth Operational Programme 2014–2020 under the project "Development of New World Scale Solutions in the Field of Machine Learning Supporting Family Planning and Overcoming the Infertility Problem", POIR.01.01.01-00-0831/17-00.

© Springer Nature Switzerland AG 2019
T. Mihálydeák et al. (Eds.): IJCRS 2019, LNAI 11499, pp. 395–408, 2019.
https://doi.org/10.1007/978-3-030-22815-6_31

might be useful for them to support further attempts, e.g., by means of more accurate detection of ovulation windows.

In [1], we showed how to construct an ovulation day detection layer within an online decision support platform – delivered by a digital health company OvuFriend – which utilizes intelligent data analysis to support women in building their families. The platform is designed to help the users to get pregnant faster by basing on insights derived from information which they enter. There are a great number of documented cases of women who got pregnant while being the OvuFriend's users.

In our studies up to now, we concentrated on retrospective detection, i.e., the ability to automatically tag the already-finished menstrual cycles. We relied on our own multivariate similarity-based methods summarized in [2], so we could leverage information about the past menstrual cycles (of the given user and/or users with similar profiles) while analyzing the new ones. Besides investigating standard sequences of day-to-day temperature jumps, we proposed the detection process basing on more diversified information, whereby time series instances of body measurements that are difficult for the users are combined with other data sources which are more convenient for women to enter via the OvuFriend's questionnaires and interfaces.

In this paper, we explain our approach to aggregating information from local, single-scope similarity-based indicators, especially from the perspective of evaluating information sources which can vary with respect to their reliability. Comparing to our previous research in this area, we test our method against a smaller data set derived from the OvuFriend's database, but now more completely (and iteratively) annotated by medical experts. In particular, in our experiments we pay attention to the influence of the experts' confidence levels (which they mark by themselves while assigning ovulation time windows to specific menstrual cycles) on the accuracy of each of single indicators, as well as the whole proposed multivariate detection schema.

The paper is organized as follows. In Sect. 2, we refer to the literature. In Sect. 3, we recall the OvuFriend's platform and its database backend. In Sect. 4, we outline previously unpublished aspects of our detection method. In Sect. 5, we discuss how to cooperate with medical experts in order to enrich the training data. In Sect. 6, we report our new experimental results. Section 7 concludes our work.

2 Related Work

There are many approaches to menstrual cycle modeling [3,4]. There are also a number of self-awareness-supporting applications for controlling ovulation [5,6]. Still, there is a need to develop new methods which could cope incompleteness and uncertainty of the data gathered by such applications. One possible solution is to let the users register their wearable devices which can measure parameters of stress, sleep, etc. [7,8]. The corresponding signals can be a source of information beyond the data entered manually. Although precision of such signals is often questionable, one can work on models which learn from both manually- and machine-generated data sources.

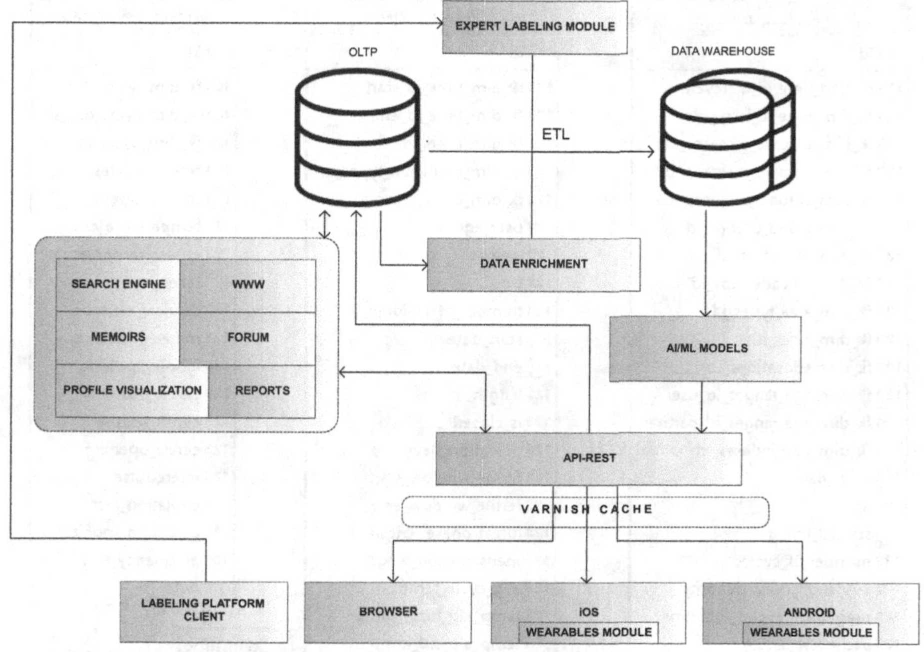

Fig. 1. The OvuFriend's system architecture [1].

In the considered area, a special attention should be paid to time series analysis [9,10], including techniques developed for the purpose of dealing with incomplete and imprecise multidimensional data sequences. Moreover, given the specifics of our approach, various aspects of similarity-based time series processing are worth taking into account as well [11,12]. Last but not least, it is important to refer to interactive methods which let machine learning algorithms accommodate feedback from subject matter experts [13] and to remember that the outcomes of detection models can lead the experts toward changing their opinions about particular examples [14].

3 The System and the Data

Figure 1 illustrates the OvuFriend's system architecture. It composes of the user modules (reports, memoires, etc.), the browser-based GUI, as well as the OLTP layer including menstrual cycle and ovulation day detection AI/ML (where ML stands for machine learning) algorithms (data enrichment). The OLTP layer outputs are stored together with the original data (data warehouse). The AI/ML algorithms can be used for both prognostic (day-by-day) and retrospective (after a given cycle has finished) ovulation day detection. That latter mode – the topic of next sections – is useful while discovering cycle anomalies and irregularities, leading toward higher-level advisory reports indicating, e.g., a risk of suffering from various diseases [15,16].

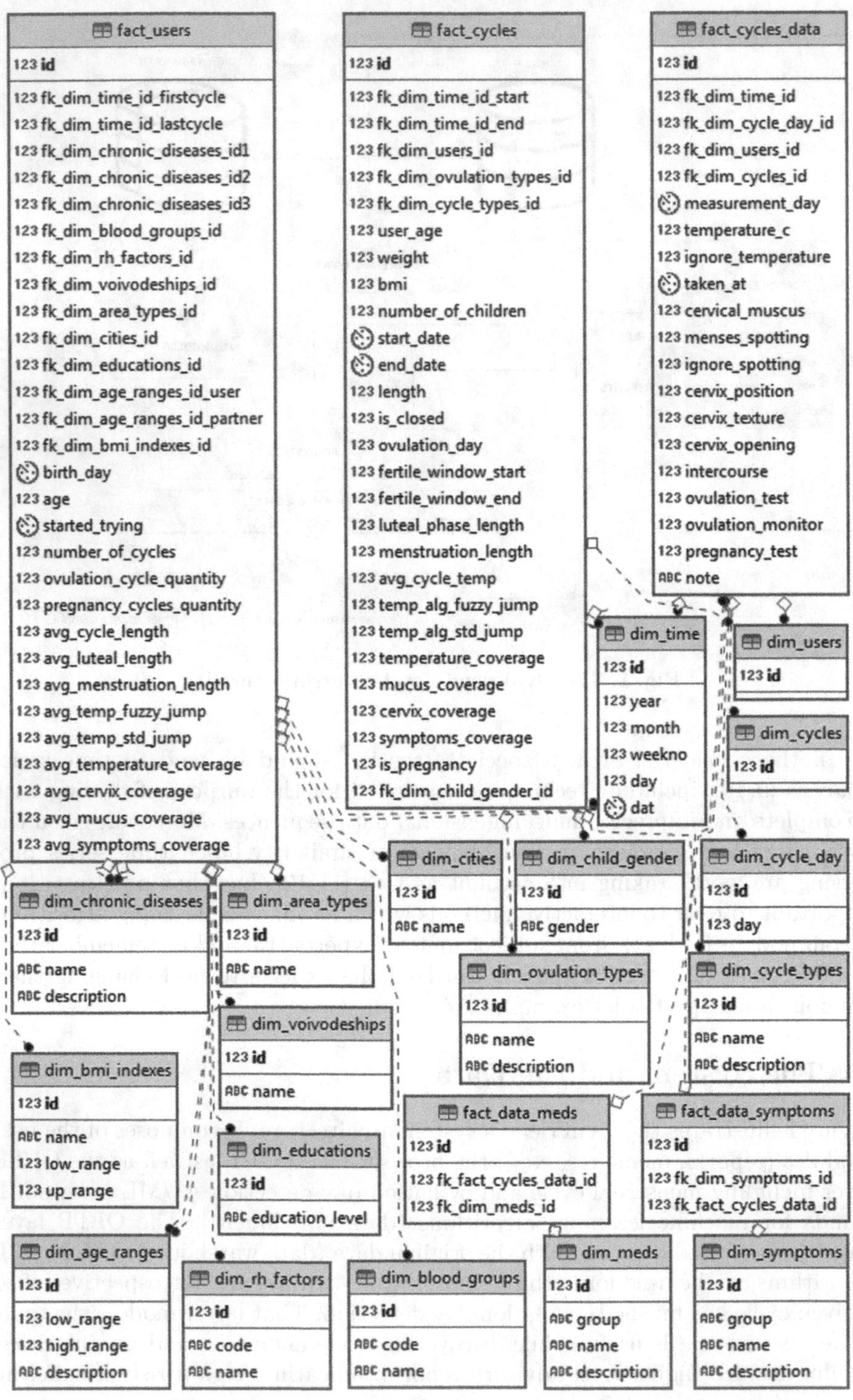

Fig. 2. The current OvuFriend's data warehouse schema (revised comparing to [1]).

The OvuFriend's GUI displays such components as ovulation day detector (with a checkpoint whether any ovulation days have been already identified), pregnancy detector (including information about observations which make pregnancy more likely), temperature trend (based on manual inputs and interpolations), etc. It contains also a dialog window to enter the new data, which will be combined in the future with acquisition of wearable device signals and a forum post analysis. Surely, the quality of information entered by the users is in relation with the quality of the outcomes of AI/ML algorithms. One of solutions with this respect is to make the dialog more iterative, whereby the user is requested to answer just a few adaptively generated questions every time and the underlying detection models adjust their work to partial answers.

Figure 2 shows the current OvuFriend's data warehouse schema. From the data model perspective, it resembles some of electronic health record system architectures [17], although it is devoted mainly to manual inputs and user feedback. By the end of 2018, it included information about over 4,000,000 single day measurements and observations (table `fact_cycles_data`). Given such sizes, the system can still perform efficiently on standard PostgreSQL, especially when taking into account the ability of query sharding [18]. However, as both the volumes and the complexity of queries are expected to grow over time, other scalable solutions should be considered, particularly those of them which are compatible with PostgreSQL [19].

Columns such as `ovulation_day` represent information that can be derived using the method outlined in Sect. 4. The algorithm is simple enough to be executed as a part of the enriched ETL process. Although it refers to the previous cycles of the given user (and/or some statistics characterizing similar users), the data schema is optimized well enough to run the corresponding scripts in real time (and to recalculate aggregations such as `avg_cycle_length` in table `fact_users` as well). In the future, the OvuFriend's data warehouse will also contain fully integrated feedback from medical experts acquired using the expert labeling module (through the labeling platform client) visible in Fig. 1. As for now, such feedback is registered remotely for experimental purposes. We will go back to this aspect in Sect. 6.

4 Ovulation Window Detection

Overall, our retrospective detection approach is similar to the one presented in [1]. For the given menstrual cycle of the given user, the algorithm first applies local detectors corresponding to particular types of registered information and then combines them using a simple procedure. Such single-scope detectors are quite standard, as they refer to temperature, mucus and cervix measurements (if available), as well as to additional aspects of manually entered information reported in Table 1. As for the temperature, the corresponding component of the detection system is now significantly improved using selected tools of fuzzy data analysis and modeling [9].

The combination procedure requires additional explanation which was not provided in our previous papers. Each of single detectors/indicators delivers a

Table 1. Examples of additional fertility indicators.

Name	Description
Ovulation test	It works with a series of ovulatory tests (based on urine, saliva, etc.). The case of positive test preceded by negative test means that the LH hormone begins to rise. It marks one of the next three days as the ovulation start, while the remaining days are labeled with appropriately computed fuzzy weights
Simple stats	Ovulation window is determined from basic statistics of the user's cycles (shortest, longest, etc.). Ovulation start is estimated as an average length of the user's luteal phase in the past
User history	Cycles of the given user are featurized and, for the considered cycle, the most similar historical cases are chosen using similarity-based techniques introduced in [2]. Ovulation days are estimated using best-matching cycles and defuzzyfication techniques reported in [22]
Profile history	Analogous detector which takes into account historical cycles of other users belonging to the same profile cluster
Symptoms	Detector based on automatically derived rules which correlate ovulation days with over 70 kinds of symptoms occurring in table `fact_cycles_data_symptoms`
Ovulation monitor	Indicator based on the monitoring device measurements which are entered manually by the user via the OvuFriend's GUI. In the future, the scope of such measurements will be extended and their acquisition will be automatized
USG monitor	Feedback from examination by a medical doctor. Such types of feedback are not yet represented in GUI but we are able to extract them partially from text comments
Moon cycle	A detector which takes into account the user's biorhythms determined on the basis of her date and time of birth

fuzzy vector of its beliefs regarding ovulation days spanned over the given menstrual cycle. The resulting matrix is then aggregated into a unified vector which assigns the overall belief to each of days. One can use a number of aggregation mechanisms utilizing, e.g., OWA operators [20] or adapting the ideas of weak classifier combination developed in machine learning [21]. The choice of particular method may depend on interpretation of output vectors produced by single indicators (e.g.: fuzzy relation or probability distribution) and it should take into account the aspect of day-by-day detection.

Currently, as illustrated by Fig. 3, the algorithm annotates the most reliable day of ovulation as a day with the maximum weighted total score of single-indicator beliefs. However, the following aspects are worth additional attention:

1. Although our aggregation method does not seem to take into account any temporal aspects, they are implicitly expressed by "fuzzified" outputs of

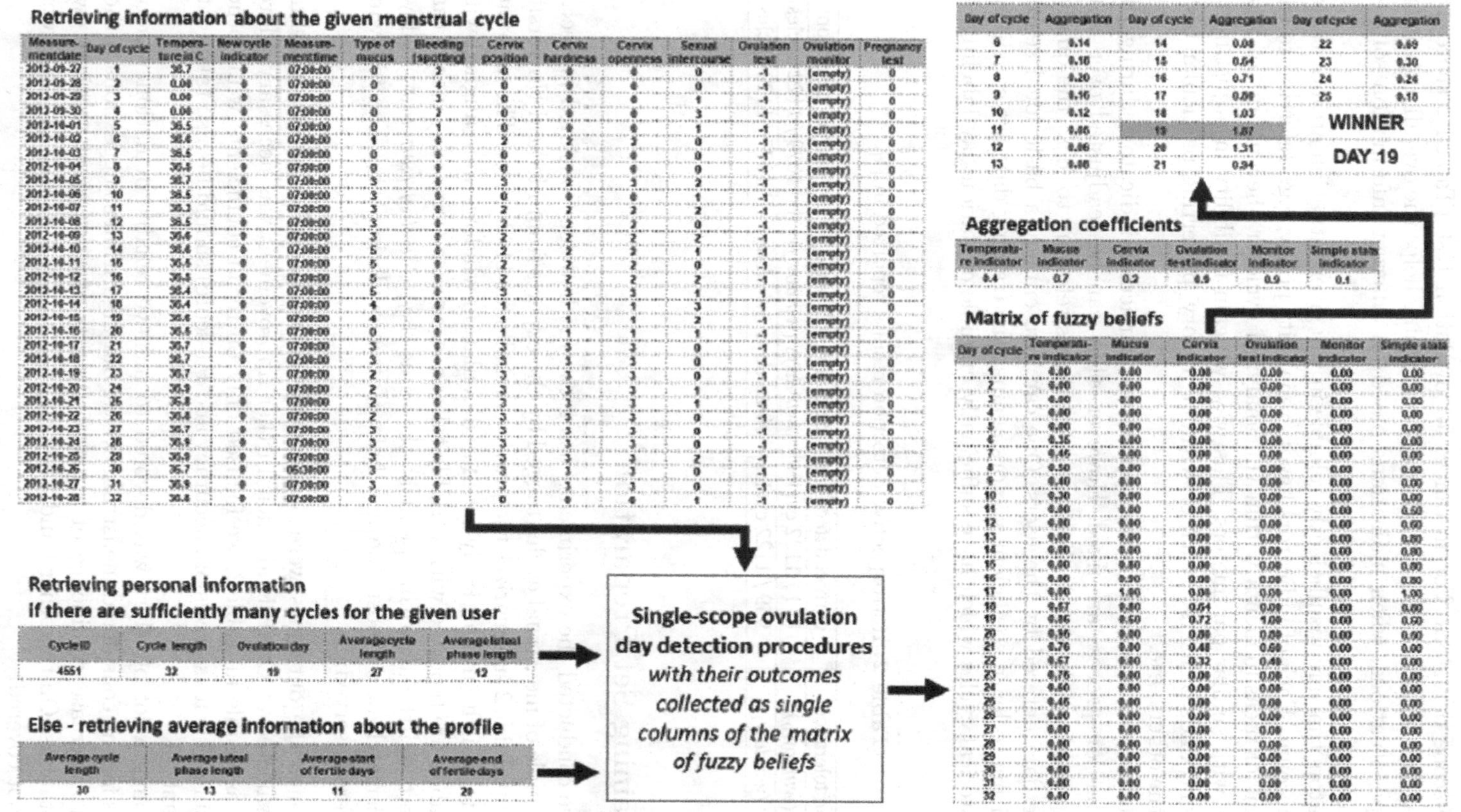

Fig. 3. Simple aggregation process implemented in the current detection schema.

single-scope detectors. For instance, for ovulation test (Table 1), there are several days with non-zero beliefs. Such smoothing helps the overall algorithm to cope situations, when indications based on different information sources are shifted in time.

2. Indicators that are insufficiently reliable in the given situation are not taken into account. Each of indicators has its reliability policy, e.g., referring to the amount of measurements within the analyzed cycle or the amount of past cycles with similar parameters. This way, we gain flexibility in adjusting to different kinds of users – with or without history, preferring particular tests over the others, etc.
3. For new users, the coefficients applied for particular indicators during the combination process are fixed. However, they can be readjusted over time basing on the indicators' efficiency assessments referring to, e.g., the quality of the data entered by the users. One can also employ a basic local search algorithm to learn coefficient vectors against the training data set consisting of historical cycles which have sufficiently credible ovulation day labelings.

Table 2. Retrospective detection results reported in [1].

detection tolerance	correct detections	detection tolerance	correct detections
precise (window$_0$)	811 / 1122 cycles	*two-days (window$_2$)*	1029 / 1122 cycles
one-day (window$_1$)	959 / 1122 cycles	*three-days (window$_3$)*	1077 / 1122 cycles

5 Training Set Preparation

In [1], we conducted the experiments using the methodology outlined in Sect. 4 over a sample of menstrual cycles tagged by the experts from Medical University of Warsaw. Table 2 summarizes the previous results for 1122 cycles. The amounts of cases for which automatically estimated ovulation windows were aligned with expert labels are reported with variable tolerance: *precise (window$_0$)* means that the expert and the algorithm point at the same day, *one-day (window$_1$)* allows for a mismatch of at most one day before/after ovulation, etc. The following issues were observed:

- Although the outcomes were better than in the case of original algorithms implemented at OvuFriend, their further improvement was necessary. Moreover, as visible in Table 2, we focused only on recall coefficients while other measures (such as precision, accuracy and F1-score) should be also taken into account.
- Although a lot of effort was devoted to achieve reliable ovulation window labels for the considered menstrual cycles, our algorithm's suggestions turned out to decrease confidence of some of the experts in regard to their own annotations. On the other hand, they claimed that some types of cycles were still not covered.

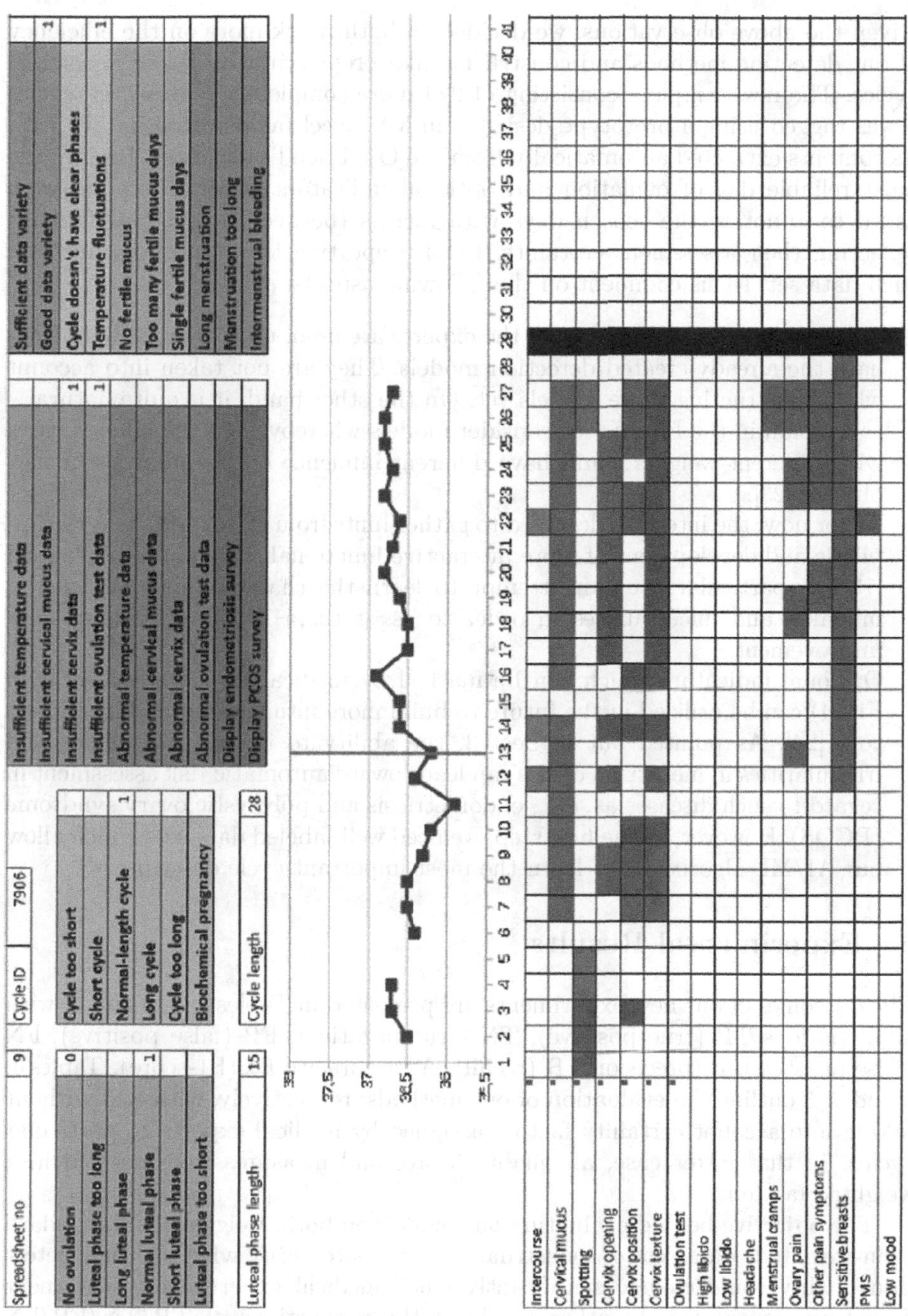

Fig. 4. Exemplary feedback. The 13th cycle's day is assessed as the most reliable day of ovulation. It is also indicated that, in the expert's opinion, this cycle has normal luteal phase and length, insufficient cervix and ovulation test data, good data variety and temperature fluctuations.

Given the above observations, we decided to both, work more on the efficiency of our detection methods and construct a more representative set of menstrual cycles. The new sample – consisting of 224 more completely represented cycles – was tagged using a prototype designed in MS Excel (as reported in [1]), with its contents extracted automatically from the OvuFriend's database. Besides the most reliable day of ovulation and optional indicators, medical experts were asked to annotate the chosen days with factors (between 0 and 1, default 1) reflecting their assessment certainty. Before reporting the results obtained on such data set, let us comment on the following aspects:

- Certainty factors acquired from the experts are used, up to now, only to evaluate the already-created detection models. They are not taken into account while constructing those models yet. On the other hand, it is quite natural – as we plan in the future – to consider models whereby, e.g., the labeled cases with different weights could have different influence on the analysis of new objects [23].
- As for now, the interface designed to gather hints from the experts is very simplistic and development of more interactive functionalities should be planned [13]. In particular, we can attempt to learn the characteristics of experts' mistakes and uncertainties in order to assist them in the process of data improvement.
- Optional indicators which can be marked by medical experts (as visible in Fig. 4) can be utilized in the future to build more intuitive hierarchical detectors [24]. As pointed out in Sect. 3, the ability to identify anomalies and irregularities in menstrual cycles can lead toward automatic risk assessment in regard to such diseases as, e.g., endometriosis and polycystic ovary syndrome (PCOS). However, as the first step, we need well-labeled data sets which allow our AI/ML algorithms to learn the most important cycle descriptors.

6 Experimental Results

The outcomes of our new experiments are presented in Tables 3, 4, 5 and 6, with abbreviations TP (true positive), TN (true negative), FP (false positive), FN (false negative), P (precision), R (recall), A (accuracy), F1 (F1-score). Tables 4, 6 and 3, 5 outline the evaluation of our methods, respectively, with and without taking into account certainty factors assigned by medical experts to particular cycles. In that latter case, all quality scores and measures were derived in a weighted fashion.

The objective behind evaluating our models on both, weighted and standard (non-weighted) data sets of menstrual cycles was to verify whether their detection quality could change significantly when medical expert's self assessment scores are considered. In other words, if the proportions of TP/TN/FP/FN amounts were too different from each other for both scenarios, then it would mean that our overall algorithm may not cope relatively more difficult cases (where the experts' certainty is lower). However, the respective P/R/A/F1 scores turn out to be quite comparable to each other.

Table 3. Comparison of ovulation day detection methods using the new data (standard scenario).

Tolerance	Previous algorithm [1]								Revised algorithm							
$window_1$	TP	146	TN	13	P	0.94	R	0.73	TP	180	TN	16	P	0.96	R	0.90
	FP	10	FN	55	A	0.71	F1	0.82	FP	7	FN	21	A	0.88	F1	0.93
$window_2$	TP	176	TN	13	P	0.95	R	0.88	TP	191	TN	16	P	0.96	R	0.95
	FP	10	FN	25	A	0.84	F1	0.91	FP	7	FN	10	A	0.92	F1	0.96

Table 4. Comparison of ovulation day detection methods using the new data (weighted scenario).

Tolerance	Previous algorithm [1]								Revised algorithm							
$window_1$	TP	112.15	TN	12.00	P	0.92	R	0.74	TP	138.00	TN	14.75	P	0.95	R	0.90
	FP	9.50	FN	40.35	A	0.71	F1	0.82	FP	6.75	FN	14.50	A	0.88	F1	0.93
$window_2$	TP	134.15	TN	12.00	P	0.93	R	0.88	TP	146.05	TN	14.75	P	0.96	R	0.96
	FP	9.50	FN	18.35	A	0.84	F1	0.91	FP	6.75	FN	6.45	A	0.92	F1	0.96

When looking at Tables 3, 4, we can conclude that improvements comparing to our previously reported research are truly substantial. Although we report the cases of $window_1$ and $window_2$ tolerance, the same tendency would be visible for $window_0$ and $window_3$ too. As a side note, let us mention that – by definition – the TN/FP amounts are invariant with respect to the length of detection tolerance windows. The difference between $window_1$- and $window_2$-specific outcomes can be seen on TP/FN.

Tables 5, 6 illustrate the detection quality coefficients which would be obtained without applying the combination procedure on top of single-scope outputs. The total TP+TN+FP+FN amounts of analyzed cycles are different for Temperature (174 cases), Mucus (199), Ovulation test (201) and Cervix (74). The same would happen also for other examples of indicators in Table 1. These differences occur because of the reliability thresholds implemented for particular indicators (item 2 in Sect. 4).

Let us emphasize that all reported experimental results were obtained for fixed aggregation coefficients (see Fig. 3), without any learning/readjusting procedure implemented yet. Therefore, one may expect that the quality of our multivariate detection model can be further increased. We are now in the process of introducing two parameter learning mechanisms. The first one (compare with item 3 in Sect. 4) relies on supervised learning over the data set consisting of the whole history of those of the OvuFriend's users for whom the aforementioned sample contained a menstrual cycle annotated correctly at $window_0$ level. The second mechanism is designed to tune the considered coefficients by basing on

Table 5. Ovulation day detection using each of revised indicators separately (standard scenario).

Tolerance	Temperature				Mucus				Ovulation test				Cervix			
$window_1$	TP	96	P	0.83	TP	146	P	0.90	TP	165	P	0.95	TP	40	P	0.82
	FP	20	A	0.56	FP	17	A	0.77	FP	9	A	0.87	FP	9	A	0.58
	TN	2	R	0.63	TN	6	R	0.83	TN	9	R	0.90	TN	3	R	0.65
	FN	56	F1	0.72	FN	30	F1	0.86	FN	18	F1	0.92	FN	22	F1	0.72
$window_2$	TP	121	P	0.86	TP	158	P	0.90	TP	176	P	0.95	TP	47	P	0.84
	FP	20	A	0.71	FP	17	A	0.82	FP	9	A	0.92	FP	9	A	0.68
	TN	2	R	0.80	TN	6	R	0.90	TN	9	R	0.96	TN	3	R	0.76
	FN	31	F1	0.83	FN	18	F1	0.90	FN	7	F1	0.96	FN	15	F1	0.80

Table 6. Ovulation day detection using each of revised indicators separately (weighted scenario).

Tolerance	Temperature				Mucus				Ovulation test				Cervix			
$window_1$	TP	75.55	P	0.80	TP	114.20	P	0.88	TP	125.85	P	0.93	TP	29.60	P	0.77
	FP	18.75	A	0.56	FP	16.00	A	0.78	FP	8.75	A	0.86	FP	8.75	A	0.57
	TN	1.75	R	0.65	TN	5.50	R	0.86	TN	8.50	R	0.91	TN	2.75	R	0.66
	FN	41.20	F1	0.72	FN	18.50	F1	0.87	FN	12.50	F1	0.92	FN	15.30	F1	0.71
$window_2$	TP	94.30	P	0.83	TP	122.05	P	0.88	TP	133.85	P	0.93	TP	35.25	P	0.80
	FP	18.75	A	0.70	FP	16.00	A	0.83	FP	8.75	A	0.91	FP	8.75	A	0.67
	TN	1.75	R	0.81	TN	5.50	R	0.92	TN	8.75	R	0.97	TN	2.75	R	0.79
	FN	22.45	F1	0.82	FN	10.65	F1	0.90	FN	4.5	F1	0.95	FN	9.65	F1	0.79

information about applicability of particular indicators, as well as their ability to point at particular days with high (close to 1) confidence. In the future, such mechanisms can be extended by taking into account the already-discussed certainty factors of the learning cases too.

7 Conclusions

We presented the new research on ovulation window detection methods embedded into the OvuFriend's online platform which is designed to assist women in increasing their chances of getting pregnant. Comparing to our previous studies, we improved the accuracy of retrospective detection. Our method is based on dynamic evaluation and synthesis of multivariate data sources acquired from the platform users. From the perspective of quality assessment, it required to build a representative sample of menstrual cycles (including also the access to information about the past cycles of the corresponding users, if available) which were carefully tagged by medical experts.

It is important to note that approaches introduced in this paper are just a part of the currently running R&D project related to the OvuFriend's system architecture. Herein, among the others, we can point out the tasks of optimizing the user dialog interfaces (so the data sources utilized as inputs to the implemented machine learning algorithms are of better quality), introducing new machine learning algorithms (so they provide higher-accuracy results and become more robust with respect to partially uncertain information), making a better usage of the available – and potentially available – information sources (so the machine learning algorithms can be based on richer data), providing the users with relevant examples of other cases within the same behavioral/medical profiles, as well as assuring that the platform is scalable, it can serve many users simultaneously and the data analytics outputs are provided to them without any delays.

One of our specific challenges refers to – as already mentioned above – partially unreliable information expressed in this paper by means of uncertainty factors assigned to particular cases by medical experts. In the future, we intend to utilize these factors at the phase of learning/tuning the coefficients of our detection models as well. On the other hand, one needs to remember that there are also many other aspects of uncertainty related to the data stored by OvuFriend, e.g., with respect to observations entered manually by the platform's users. Any progress in the area of investigation of such kinds of data can be of great importance in many real-world applications.

References

1. Sosnowski, Ł., et al.: Similarity-based detection of fertile days at OvuFriend. In: Proceedings of IEEE BigData 2018, pp. 2663–2668 (2018)
2. Sosnowski, Ł.: Compound objects comparators in application to similarity detection and object recognition. Trans. Rough Sets **21**, 169–300 (2019)
3. Scherwitzl, E.B., Hirschberg, A.L., Scherwitzl, R.: Identification and prediction of the fertile window using NaturalCycles. Eur. J. Contracept. Reprod. Health Care **20**(5), 403–408 (2015)
4. Su, H.W., Yi, Y.C., Wei, T.Y., Chang, T.C., Cheng, C.M.: Detection of ovulation, a review of currently available methods. Bioeng. Transl. Med. **2**(3), 238–246 (2017)
5. Sohda, S., Suzuki, K., Igari, I.: Relationship between the menstrual cycle and timing of ovulation revealed by new protocols: analysis of data from a self-tracking health app. J. Med. Internet Res. **19**(11), e391 (2017)
6. Koch, M., et al.: Improving usability and pregnancy rates of a fertility monitor by an additional mobile application: results of a retrospective efficacy study of Daysy and DaysyView app. Reprod. Health **15**, 37 (2018)
7. Shilaih, M., Goodale, B.M., Falco, L., Kübler, F., De Clerck, V., Leeners, B.: Modern fertility awareness methods: wrist wearables capture the changes of temperature associated with the menstrual cycle. Biosci. Rep. **38**(6) (2018). https://www.ncbi.nlm.nih.gov/pmc/articles/PMC6265623/
8. Kutt, K., Nalepa, G.J., Giżycka, B., Jemioło, P., Adamczyk, M.: BandReader - a mobile application for data acquisition from wearable devices in affective computing experiments. In: Proceedings of HSI 2018, pp. 42–48 (2018)

9. Afanasieva, T., Yarushkina, N., Toneryan, M., Zavarzin, D., Sapunkov, A., Sibirev, I.: Time series forecasting using fuzzy techniques. In: Proceedings of IFSA-EUSFLAT 2015, pp. 1068–1075 (2015)
10. Fukaya, K., Kawamori, A., Osada, Y., Kitazawa, M., Ishiguro, M.: The forecasting of menstruation based on a state-space modeling of basal body temperature time series. Stat. Med. **36**(21), 3361–3379 (2017)
11. Iglesias, F., Kastner, W.: Analysis of similarity measures in times series clustering for the discovery of building energy patterns. Energies **6**(2), 579–597 (2013)
12. Tsumoto, S., Iwata, H., Hirano, S., Tsumoto, Y.: Similarity-based behavior and process mining of medical practices. Future Gener. Comput. Syst. **33**, 21–31 (2014)
13. Yimam, S.M., Biemann, C., Majnaric, L., Šabanović, Š., Holzinger, A.: Interactive and iterative annotation for biomedical entity recognition. In: Guo, Y., Friston, K., Aldo, F., Hill, S., Peng, H. (eds.) BIH 2015. LNCS (LNAI), vol. 9250, pp. 347–357. Springer, Cham (2015). https://doi.org/10.1007/978-3-319-23344-4_34
14. Hu, B., Jiang, X., de Souza, E.N., Pelot, R., Matwin, S.: Identifying fishing activities from AIS data with conditional random fields. In: Proceedings of FedCSIS 2016, pp. 47–52 (2016)
15. Dunselman, G.A.J., et al.: ESHRE guideline: management of women with endometriosis. Hum. Reprod. **29**(3), 400–412 (2014)
16. Meena, D.K., Manimekalai, D.M., Rethinavalli, S.: Implementing neural fuzzy rough set and artificial neural network for predicting PCOS. Int. J. Recent Innov. Trends Comput. Commun. **3**(12), 6722–6727 (2015)
17. Fox, F., Aggarwal, V.R., Whelton, H., Johnson, O.: A data quality framework for process mining of electronic health record data. In: Proceedings of ICHI 2018, pp. 12–21 (2018)
18. Szczuka, M.S., Sosnowski, Ł., Krasuski, A., Kreński, K.: Using domain knowledge in initial stages of KDD: optimization of compound object processing. Fundamenta Informaticae **129**(4), 341–364 (2014)
19. Ślęzak, D., Glick, R., Betliński, P., Synak, P.: A new approximate query engine based on intelligent capture and fast transformations of granulated data summaries. J. Intell. Inf. Syst. **50**(2), 385–414 (2018)
20. Emrouznejad, A., Marra, M.: Ordered weighted averaging operators 1988–2014: a citation-based literature survey. Int. J. Intell. Syst. **29**(11), 994–1014 (2014)
21. Kuncheva, L.I., Diez, J.J.R.: A weighted voting framework for classifiers ensembles. Knowl. Inf. Syst. **38**(2), 259–275 (2014)
22. Sosnowski, Ł., Szczuka, M.: Defuzzyfication in interpretation of comparator networks. In: Medina, J., et al. (eds.) IPMU 2018, Part II. CCIS, vol. 854, pp. 467–479. Springer, Cham (2018). https://doi.org/10.1007/978-3-319-91476-3_39
23. Widz, S.: Introducing NRough framework. In: Polkowski, L., et al. (eds.) IJCRS 2017, Part I. LNCS (LNAI), vol. 10313, pp. 669–689. Springer, Cham (2017). https://doi.org/10.1007/978-3-319-60837-2_53
24. Skowron, A., Wang, H., Wojna, A., Bazan, J.: A hierarchical approach to multimodal classification. In: Ślęzak, D., Yao, J.T., Peters, J.F., Ziarko, W., Hu, X. (eds.) RSFDGrC 2005, Part II. LNCS (LNAI), vol. 3642, pp. 119–127. Springer, Heidelberg (2005). https://doi.org/10.1007/11548706_13

Incremental Sequential Three-Way Decision Using a Deep Stacked Autoencoder

Hong Chen[1], Huaxiong Li[1](✉), Bing Huang[2], Xiuyi Jia[3], and Xianzhong Zhou[1]

[1] Department of Control and Systems Engineering, School of Management and Engineering, Nanjing University, Nanjing 210093, China
huaxiongli@nju.edu.cn

[2] School of Technology, Nanjing Audit University, Nanjing 211815, China

[3] School of Computer Science and Engineering, Nanjing University of Science and Technology, Nanjing 210094, China

Abstract. Most traditional face recognition classifiers attempt to minimize recognition error rate rather than misclassification costs, which is unreasonable in many real world applications. On the other hand, many facial images are usually unlabeled, and the label process may result in high costs. Considering imbalanced misclassification costs and the hardship of gathering sufficient labeled images, an incremental sequential three-way decisions (3WD) model for cost-sensitive face recognition is proposed, in which a deep stacked autoencoder (DSAE) is used to extract an efficient deep feature set. The model takes full account of the costs of obtaining labeled data in real life. In addition, the model incorporates the boundary decision into the process of making decision, leading to a delayed decision with insufficient labeled images, which simulates the decision-making process from a small amount to a large amount of data. In summary, the model aims to select an optimal decision step so as to gain the desirable recognition results with the least amount of data. This strategy is applied to two facial image databases, which validate the effectiveness of the proposed methods.

Keywords: Face recognition · Cost-sensitive · Incremental learning · Sequential three-way decisions

1 Introduction

In many traditional face recognition systems, the recognition accuracy is the evaluation indicator of most previous methods, which is to minimize the recognition error rate [24]. However, this assumption is not reasonable because it neglects the unbalanced misclassification costs in real-world scenarios. For many real-world applications, the losses caused by different types of errors are different [23]. Thus, a face recognition system with superior performance should

© Springer Nature Switzerland AG 2019
T. Mihálydeák et al. (Eds.): IJCRS 2019, LNAI 11499, pp. 409–423, 2019.
https://doi.org/10.1007/978-3-030-22815-6_32

take the misclassification costs into account rather than simply minimizing the classification error rate.

In addition to the cost of misclassification errors mentioned above, cost of computation and cost of acquiring cases are also available in real-world applications. Turney [11] made a detailed induction of the cost of concept learning. Typically the performance of machine learning algorithm is positively correlated with the number of data in a certain range. However, using sufficient cases in the real world is either impossible or expensive. The main reasons lie in two aspects. First of all, the cost of acquiring cases, especially labeled ones, should not be ignored. Furthermore, computing resources are limited. Thus, it makes sense to consider the cost of computation. In the case that only a few facial images are available, a delayed decision is better than an immediate decision, since the cost of making a wrong decision is much higher than the cost of making a delayed decision. It is likely to make wrong decisions when labeled images are scarce. However, it is no longer an advantage to make a delayed decision after gathering more labeled images, which would be more costly. In this case, the samples that were previously delayed might be explicitly determined. Such a decision method denotes a three-way decision strategy [16].

In a traditional two-way decisions model, only two options of acceptance and rejection are considered [22]. In the application of two-way decisions model, people also have to make decisions under the condition of incomplete information, which may lead to irreparable consequences. As a generalization of the traditional two-way decisions theory, three-way decisions model adds a third alternative to acceptance and rejection, that is non-commitment, namely indecision or delayed decision [17]. When information is not enough to accept or reject, delayed decision is more in line with the way people deal with practical problems. Thus, researchers apply three-way decisions in numerous fields, including, clustering analysis [21], email spam filtering [3], face recognition [5], etc.

Three-way decisions can be divided into two categories: a single one-step three-way decision and a sequential three-way decision. The latter is a multi-step decision-making and the former can be regarded as a step of the latter [20]. Delayed decision is the key of sequential three-way decisions. Sequential three-way decisions use a sequential strategy, which can use granular structure [14]. When the information used to make decisions is not sufficient, people have rough granular feature of objects, so it is difficult to make a precise decision. Therefore, the decision can be postponed, that is to say, the boundary decision is adopted. In this case, the decision cost is the least, which conforms to the minimum risk Bayes decision theory. The boundary decision indicates that the information available for decision making is insufficient at present. In order to make the correct decision, we need to collect more information needed for the making decision. After collecting information fully, the object is transformed from rough granule to precise granule, which makes it possible to make precise decisions. These multi-step decision-making processes from rough granule to precise granule establish multi-granulation three-way decisions. A sequential, multi-step three-way decision-making strategy is constituted, which describe the dynamic progressive decision-making process used in many practical problems. It is worth mentioning that accurate conditional probability is necessary to make the correct decision. Thus, DSAE is used to extract the deep features so that the accurate conditional probability can be obtained.

In order to transform decision objects from rough granule to precise granule, we need to collect enough information. There are several ways to gather more information: more decision attributes and features [5], or increasing the number of samples [8]. Increasing the number of labeled samples is what we are researching. To the best of our knowledge, if enough labeled samples can be collected to train the classifier, a classifier with superior generalization performance can be obtained. Collecting images is cheap while labeling them is very high cost or even impossible in real applications [4]. Generally speaking, the more training samples, the better the effect of the classifier. But when the number of samples reaches a certain number, the classifier effect is improved little or no longer with the increase of samples. In other words, the problem is thought in threes [19]: insufficient samples leads to poor performance, sufficient samples results in required performance and minimal cost while over-sufficient samples cause better performance and high cost. Thus, we attempt to achieve a required level of accuracy with a minimal cost under the consideration of the cost of obtaining samples.

The remainder of this paper is organized as follows. In Sect. 2, we briefly introduce some related work on cost sensitivity and three-way decision. In Sect. 3, we propose a dynamic cost-sensitive incremental sequential three-way decision model. In Sect. 4, we present the implementation of incremental sequential learning. In Sect. 5, we introduce a popular deep neural network to obtain the conditional probability. In Sect. 6, the experimental results and analysis are shown and verified. In the last section, we make a conclusion of this paper.

2 Related Work

Three-way decision strategy was proposed by Yao in [15]. Three-way decision was further extended into sequential three-way decision in consideration of the update and supplement of information [20]. Many researches related to three-way decisions mainly studied the extension researches of rough sets. There are several typical types of three-way decisions, such as three-way decisions based on decision-theoretic rough sets (DTRS), fuzzy sets, interval-valued fuzzy rough sets (IVFRS), interval sets, random sets, probability rough sets and so on. These were discussed in detail by Hu [2]. Some preferred to extend static three-way decision to the dynamic version. Yao [18] attempted to interpret sequential three-way decisions based on multiple levels of granularity. Li et al. proposed a cost-sensitive sequential three-way decision model. This model is dynamic and based on DTRS. In recent months, Yao [19] put forward a TAO model, which integrated three-way decision and granular computing and proposed thinking in threes.

In addition, the applications of three-way decisions in many fields and disciplines were widely studied. Zhang et al. used random forests to build recommender systems which applied three-way decisions. Li et al. [6] focused on the application of cost-sensitive sequential three-way decisions in image recognition. Luo et al. [8] put forward some propositions on incremental three-way decisions. In order to incrementally updating three-way probabilistic regions, Yang et al. [13] proposed a unified dynamic framework of decision-theoretic rough sets. Yu et al. [21] considered that the data we collected are incremental in real world and combined incremental clustering method with three-way decision theory. It is obvious that three-way decisions can be applied in many realistic decisions problems. Despite many studies on incremental three-way decisions, only a few are applied to face recognition.

3 Incremental Sequential Three-Way Decision

3.1 Sequential Three-Way Decision Model

As mentioned in Sect. 1, it is difficult to have enough information at the beginning in practical applications. Relatively speaking, it is much easier to get a small piece of information. In this case, for these subjects that are difficult to categorize correctly, delayed decisions are better choices, which allows us to make a further precise decision after gathering more information. The process continues with the increase of information until satisfactory results are achieved or more information is difficult to obtain. This process forms a sequential three-way decision from rough granularity to precise granularity, which is similar to that of the way human make decisions. In this section, we will make a problem formulation and exhibit the decision cost and process of sequential three-way decisions.

When training a cost sensitive face recognition classifier, many face images form a training set $X = \{X_1, X_2, \ldots, X_M\}$, where M is the number of images. Each image $X_i \in R^{p\times q}$ is an $p\times q$ image matrix. The class label of X_i is denoted as $l_i \in \{1, 2, \ldots, n_G, n_G+1, \ldots, n\}$, where the former n_G labels are gallery subjects and the last n_I ($n_I = n - n_G$) labels are impostor subjects. For convenience, P denotes gallery subjects label and the label of impostor subjects is represented as N.

For a cost-sensitive sequential three-way decision face recognition problem, the decision choices involve three options: positive decision (gallery subject), negative decision (impostor subject) and delayed decision (boundary), which are described as $D = \{a_P, a_N, a_B\}$. These three options lead to six results and the cost is determined by the recognition results. These six costs are displayed and explained below:

1. λ_{PP}: a gallery subject is correctly classified as a gallery subject.
2. λ_{NN}: an impostor subject is correctly classified as an impostor subject.
3. λ_{PN}: an impostor subject is wrongly classified as a gallery subject.
4. λ_{NP}: a gallery subject is wrongly classified as an impostor subject.
5. λ_{BP}: a gallery subject is classified to the boundary for further decision.
6. λ_{BN}: an impostor subject is classified to the boundary for further decision.

These six costs form a matrix $(\lambda_{ij})_{2\times3}$, where $i \in \{P, B, N\}$, $j \in \{P, N\}$. It is obvious that the classification costs of different decisions should not be equal. Generally speaking, the cost of making a right decision is lower than that of boundary decision and error classification. In addition, the cost of delayed decision is lower than that of wrong decision, otherwise there is no need for delayed decision. So $\lambda_{PP} \leqslant \lambda_{BP} \leqslant \lambda_{NP}$, $\lambda_{NN} \leqslant \lambda_{BN} \leqslant \lambda_{PN}$. For a cost-sensitive problem, false rejection is easier to accept than false acceptance, namely, $\lambda_{NP} < \lambda_{PN}$. To facilitate the discussion, this paper ignores the costs of correct classification, that is, $\lambda_{PP} = 0$, $\lambda_{NN} = 0$.

Definition 1. *Let X be an image data set. X^i represents the data set obtained by acquiring images i times. $X^1 \subsetneqq X^2 \subsetneqq \cdots \subsetneqq X^t \subseteq X$, where t is number of increases in data. $X^T = \{X^1, X^2, \ldots, X^t\}$ is called as a incremental update data set on X. Assume $D = \{a_P, a_N, a_B\}$ is a decision set. $Cost(d|X^l)$ denotes the cost of deciding X^l as d, then we can obtain the following series, which is called sequential three-way decisions:*

$$SD = (SD_1, \ldots, SD_l, \ldots, SD_n) = (\phi^*(X^1), \ldots, \phi^*(X^l), \ldots, \phi^*(X^t)). \quad (1)$$

where $\phi^*(X^l) = \text{argmin}_{d\in D}\, Cost(d|X^l)$.

As mentioned above, $\phi^*(X^l) = \text{argmin}_{d\in D}\, Cost(d|X^l)$. So we can transform the solution of SD into minimizing $Cost(d|X^l)$. Based on Bayesian decision procedure, the decision costs($cost(a_i|X^l), i = P, B, N$) are computed as follows:

$$\begin{aligned} cost(a_P|X^l) &= \lambda_{PP}\Pr(P|X^l) + \lambda_{PN}\Pr(N|X^l), \\ cost(a_N|X^l) &= \lambda_{NN}\Pr(N|X^l) + \lambda_{NP}\Pr(P|X^l), \\ cost(a_B|X^l) &= \lambda_{BP}\Pr(P|X^l) + \lambda_{BN}\Pr(N|X^l). \end{aligned} \quad (2)$$

Then, select the minimum cost from the three costs computed according to (2). We can get the optimal decision for the l-th step, i.e. SD_l, which is formulated as follows [7]:

$$SD_l = \phi^*(X^l) = \arg\min_{i\in P,N,B} Cost(a_i|X^l) \quad (3)$$

As mentioned in Sect. 1, besides decision cost, the training cost is not negligible in practical applications. The optimal decision step SD_l is closely related to the training cost. The calculation of the total cost will be introduced later.

3.2 Cost Calculation for Three-Way Decision

Generally, the overall cost consists of misclassification costs $Cost_M$ and training costs. In this paper, the time cost $Cost_T$ and the data cost $Cost_D$ constitute the training cost. These two costs should be an increasing function of the decision step. Time cost is the time spent training classifiers and the time cost is given by the program. Data cost refers to the cost of tagging pictures. It is worth mentioning that we assume that the cost of labeling each image is the same and the initial training set is also counted into the number of labeled images D. To sum up, the total cost consists of misclassification cost, time cost and data cost, so the total cost is a function for pooling together these three costs [18], that is:

$$Cost = F(Cost_M, Cost_T, Cost_D) \tag{4}$$

We take a linear combination to calculate the cost as (5).

$$Cost = \omega_M \times Cost_M + \omega_T \times Cost_T + \omega_D \times Cost_D \tag{5}$$

The cost weight ω_M, ω_T and ω_D can be changed according to the actual situation.

Equation (5) shows that if we acquire a low decision cost with a large number of labeled images, the total cost will also be high. Labeling much fewer images with little increase in decision costs can lead to a lower total cost, which requires a superior sampling strategy. How to obtain the optimal incremental data set X^i will be investigated in the next section.

4 Incremental Sampling for Sequential 3WD

In the information age where the Internet industry is highly developed, it is no longer as difficult to obtain data as before, but it is still costly to label data manually. How to select the most informative samples from a large number of samples so as to train a satisfactory model with as few labeled data as possible and finally obtain satisfactory results at a relatively low marking cost is a hot research topic in recent years. It must be noted that expert marking is expensive and slow. When the labeled data is insufficient, three-way decisions is a better choice. The process of incremental sequential three-way decision is as in Algorithm 1. It can be seen that the sampling strategy is the core of the incremental sequential three-way decision. The quality of the sampling strategy directly determines the information of the training set and then influences the effect of the classifier. Among all the sampling strategies, the uncertainty sampling strategy is the most commonly used sampling strategy, because it is simple, effective and universal. Firstly, it does not depend on the base classifier, whether in the probabilistic model or in the non-probabilistic model such as decision tree, it has a good effect. Secondly, its computational complexity is smaller than the version space reduction and expected error reduction strategy. The simplest uncertainty sampling strategy is the minimum confidence strategy. For the multiclass problem, the strategy selects the example with the least posterior probability in the most probable class, that is, $\arg\min_{x \in X^u}(\arg\min_{j \in l} P(y = j \mid x))$.

The disadvantage of the minimum confidence strategy is that it only considers the most probable class, while ignoring the probability of the other categories. Although the final classification results are determined by the maximum posterior probability, it is not appropriate to consider only this probability when selecting examples in multiclass classification problems. In order to overcome this shortcoming, marginal sampling strategy for multiclass classification problems is proposed. In the marginal sampling strategy, in addition to the maximum

Algorithm 1. The process of incremental sequential three-way decision

Input: Initial training set X^1; The unlabeled image pool X_u; The number of decision steps s; The number of incremental images per step n

Output: The total cost of two-way decisions C_{2WD} and three-way decisions C_{3WD};

for $1 \leq i \leq s$ **do**
    Train a classifier Γ_i with training set X_i
    Use Γ_i to calculate the total cost C^i_{2WD} and C^i_{3WD}
    Extract n samples from X_u to form I^i
    Let experts label I^i
    $X^{i+1} \leftarrow X^i \cup I^i$, $X_u \leftarrow X_u \setminus I^i$
end for
$C_{2WD} = \{C^1_{2WD}, C^2_{2WD}, \ldots, C^s_{2WD}\}$
$C_{3WD} = \{C^1_{3WD}, C^2_{3WD}, \ldots, C^s_{3WD}\}$
return C_{2WD} and C_{3WD}

probability, the information of the second largest probability category is used rather than discarded directly. It calculates the margin of each sample, that is, the posterior probability difference between the most probable category and the second possible category, and selects the example x^* with the smallest margin according to Eq. 6.

$$x^* = \arg\min_{x \in X^u}(P(y = j_1 \mid x) - P(y = j_2 \mid x)) \tag{6}$$

where $j_1 = \arg\min_{j \in l} P(y = j \mid x)$ is the most probable class and $j_2 = \arg\min_{j \in l \setminus j_1} P(y = j \mid x)$ represents the second most probable class. Obviously, the example with high margin is easily distinguished by the classifier, such an example can be regarded as a small amount of information, marking the example may lead to a waste of resources. Those samples with very small margin are liable to be misclassified by the classifier, and obtaining the labels of these examples will greatly improve the performance of the classifier.

Such a reasonable assumption can be made that the most probable category is considered as the final category and the second most probable category is regarded as the most easily misclassified category in decision making. Based on this assumption, the cost sensitive margin can be calculated. The strategy of selecting the most informative examples with cost-sensitive margin is called cost-sensitive minimum margin strategy (CSMM), which makes the minimum margin strategy cost sensitive. Compared with simple minimum margin strategy, the cost-sensitive minimum margin strategy can be targeted to select such impostor $\tilde{x}^*$ that are easily misclassified as galleries and these galleries that are easily misclassified as impostors.

$$\tilde{x}^* = \arg\min_{x \in X^u}(P(y = j_1 \mid x) - P(y = j_2 \mid x))W(j_1, j_2) \tag{7}$$

where $W(j_1, j_2)$ is the weight of misclassifying j_1 to j_2.

In addition to incremental image set, the optimal decision SD_l is also closely related to conditional probability $\Pr(P|X^l)$, as (2) and (3) shows. Compared with traditional approaches, the probability using deep neural network is more accurate. The calculation method of conditional probability will be explained in detail in the next section.

5 DSAE Extract Features for 3WD

To estimate the conditional probability mentioned in the previous section, DSAE is used to extract features. Deep neural network has been a common and effective classifier since it was proposed because it has more powerful ability than other classifiers in image recognition. It is generally believed that the development of deep learning began in 2006. Hinton et al. [1] proposed a method to build multi-layer neural networks on unsupervised data and used greedy layerwise approach to extract features, which formed a deep stacked autoencoder.

The autoencoder, an unsupervised deep neural network, whose purpose is to reconstruct the input data and to approximate an identity mapping function [10] (obviously, input size = output size), namely, $\hat{x} \approx x$, where $\hat{x}$ represents the real output of neural networks and x represents input data. If the number of hidden layers is limited to less than the input size, the autoencoder is forced to learn the compressed representation of the input data which makes it useful to extract features from unlabeled data, and then classify or label them [9]. Encoder phase and decoder phase are the two phases of the autoencoder's execution process, respectively. A mapping from the input layer X to the hidden layer Y is established in encoder phase, represented as $H(x) = Y = f(W_1X + b_1)$, where W_1 and b_1 are kernel vector and bias, respectively. f depicts the sigmoid function. The decoding phase is the process of reconstruction, in which the hidden layer Y to the output layer $\hat{X}$ is mapped as $\hat{X} = f(W_2X + b_2)$, where W_2 is used to represent kernel vector and b_1 is the bias.

Let $\{x^1, x^2, \ldots, x^m\}$ represent the data set, the mean squared reconstruction error (MSRE) representing the dispersion of the input and the actual output is calculated as $J_{cost} = \frac{1}{2}\|\hat{X} - X\|^2$. To reduce the impact of overfitting problems, a weight decay term is added to the loss function in order to diminish the magnitude of the weight, mathematical representation as $J_{weight} = W_1^2 + W_2^2$. If a sparsity constraint on the hidden units was imposed on autoencoder, it constructs a sparse autoencoder. The sparsity constraint forces most neural to be inactive, that is, the output value is close to zero, which helps the neural network to remove the feature of useless information. Taking an average on training set, then the average activation of hidden units j is denoted as $\hat{\rho}_j = \frac{1}{m}\sum_{i=1}^{m} y_j^{(i)}$, where m is the number of training images.

Let $\{y_1, y_2, \ldots, y_n\}$ denotes the hidden layer Y, in order to implement the above sparsity constraint, an additional penalty function based on Kullback-Leibler (KL) divergence is defined as $J_{sparse} = \sum_{j=1}^{n} KL(\rho\|\hat{\rho}_j)$, where ρ is a sparse parameter, whose value is close to zero. And the formula to calculate Kullback-Leibler (KL) divergence is represented as in 8:

$$\sum_{j=1}^{n} KL(\rho\|\hat{\rho}_j) = \sum_{j=1}^{n}(\rho \log \frac{\rho}{\hat{\rho}_j} + (1-\rho)\log \frac{1-\rho}{1-\hat{\rho}_j}) \tag{8}$$

The ultimate optimization objective of an sparse autoencoder is presented as follows:

$$J(W,b) = J_{cost} + \frac{\lambda}{2} J_{weight} + \beta J_{sparse} \tag{9}$$

where λ is the weight decay parameter and β decides the importance of the sparse penalty function. Then the back-propagation algorithm is applied to decrease the loss function, which will make the output as perfect as possible to reconstruct the input, and finally get the weights W_1, W_2 and biases b_1, b_2.

According to the above method, we can train a deep stacked autoencoder (DSAE) by greedy layerwise approach. The DSAE is a deep neural network with multiple layers of sparse autoencoders. In DSAE, the output of k-th layer is the input of $k+1$ layer. Assuming that DSAE has $(2n-1)$ hidden layer, the n layer is its deepest hidden layer. The activation value of the neuron in this layer is a higher order representation of the input value and contains the information that we are most interested in. Labels of training data are unnecessary when training a DSAE, that is, it is an unsupervised learning scheme. A classifier such as softmax is necessary when the features from the DSAE is used for classification tasks. Here we use the softmax classifier to obtain conditional probability. Usually the decoder is discarded and the output of the n layer is fed as the input feature of the classifier [12]. In order to obtain better results, after training the classifier, the parameters of all layers are adjusted simultaneously. Finally, the output layer of softmax is the conditional probability needed to make the three-way decision.

6 Experiments

In this section, experiments based on two popular face databases Extended YaleB and PIE are designed to verify the superiority of incremental sequential three-way decisions and the effectiveness of the proposed incremental selection method in the case of insufficient data. We compare the performance of sequential three-way decisions with two-way decisions, and make a comparison of the filter incremental method proposed and simple incremental method.

6.1 Face Databases

The Extended YaleB database contains 38 subjects, with a total of 2414 frontal images collected under various strictly controlled illumination conditions. We take all five subsets of frontal face images and randomly obtain part of them as training set and test set. The PIE database consists of over 40000 face images of 68 subjects,

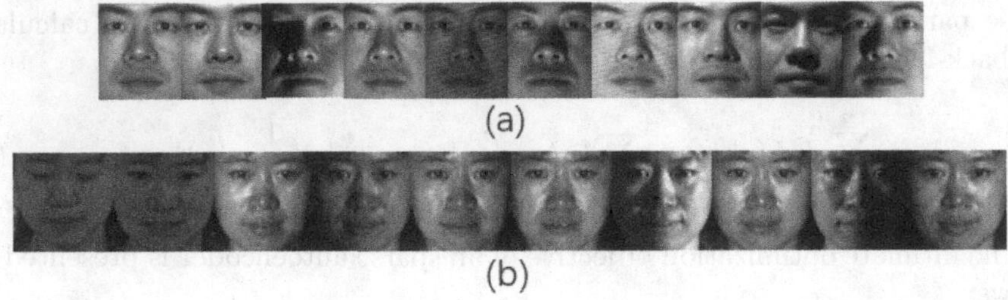

Fig. 1. Some samples of EYaleB (a) and PIE (b) face databases used in experiments

including 13 poses, 43 illumination conditions and 4 emotions. We selected all the images under various light conditions from a subset, that contains 24 images per subject. All images are cropped and resized to 50×50 pixels. The partial images of the two databases are shown in Fig. 1. Since we mainly study cost-sensitive sequential three-way decision, we would better set different parameters for gallery subjects and impostor subjects. Moreover, to avoid the effects of class imbalance, we make an assumption that the number of images in each class is the same. For each gallery and impostor subject, T_r images are included in training set to gain incremental data sets, while T_e images are used to calculate the accuracy and misclassification. The initial training set is extracted from the training set, and the others are used as unlabeled image pools. The number of gallery and impostor subjects selected from each database is represented by G and I, respectively. The number of additional labeled samples per step is represented by N_i. The cost matrix $(\lambda_{ij})_{2\times 3}$ used to calculate misclassification cost and threshold α and β is given by experience. Since it is the proportion rather than the value of the cost matrix that affects the experimental results, we set the minimum value λ_{BP} as 1 while ignoring the cost of the correct classification. All parameter are presented in Table 1. Because the size of the training set and the test set are different between the two databases, the weights in Eq. (5) should also be different in our experiments. For simplicity, ω_M is set to 1 for two databases. ω_T and ω_D are set to 4 and 0.7 for EYaleB while 10 and 2.5 for PIE, respectively, to ensure that the data cost is more important than the time cost.

Table 1. Experimental parameter settings

Database	G	I	T_r	T_e	N_i	$\lambda_{PN} : \lambda_{NP} : \lambda_{BN} : \lambda_{BP}$
EYaleB	20	15	24	24	25	$12 : 3 : 2 : 1$
PIE	20	15	8	16	8	$12 : 3 : 2 : 1$

In order to meet the practical situation and avoid the experimental errors caused by randomness, the sequence of images obtained is fixed in each experiment. In this case we can accurately compare the experimental results of simple incremental method and active incremental method. In each experiment, the training set, the set to be marked and the test set of random increment method and active increment method should be identical, so that the validity of the experimental results can be proved. Thus, for each database, 10 different orderly fixed data sets are acquired in the random way. One data set is used in each experiment, and the average value of 10 experiments is regarded as the result. In addition, all experiments were carried out on a computer equipped with GTX1050Ti.

6.2 Costs and Errors of Sequential Incremental Learning

In order to verify the superiority of the proposed sequential three-way decision over the sequential two-way decision in incremental learning, we compare their decision cost and decision errors in this section. All experimental parameters are set strictly according to the data in Table 1. We mainly compare and show the following items:

1. **Decision cost**: the cost of misclassifying the image into wrong categories, that is, the sum of all the six costs mentioned in Sect. 3.
2. **Total cost**: the sum of decision cost and training cost.
3. **Error rate**: the ratio of the number of images misclassified to the number of test sets. It should be pointed out that the boundary regions of three decisions are considered to be classified errors in order to ensure fairness when compared with two decisions.
4. **Three regions**: the changes of three regions with the decision steps in the three-way decision.

All four items of these experimental results on EYaleB and PIE are shown in Fig. 2. As shown in Fig. 2, the variation trend of decision cost and error rate of sequential two-way decision and sequential three-way decision are the same. In most cases the decision cost and error rate decrease with the increase of decision steps. This is because in sequential incremental learning, as the decision steps increase, more images are labeled and added to the training set, which increases the available information for decision. More available information can help the classifier to make more correct decision and finally reduce the decision cost and error rate. In addition, we should also see that when the decision cost and error rate are low, there also are scenarios where the decision steps increase while the decision cost and error rate increase. The existence of anomalies is also normal because generally only the overall trend satisfies the consistency of decision.

Comparing the decision cost of sequential three-way decision with sequential two-way decision, we find that the former is lower than the latter, especially at the beginning, which is the benefit of boundary decision. As we discussed earlier, when the available information is not sufficient, if the classifier has to make a decision, it will easily lead to errors, so the boundary decision is the better choice. However, the advantages of boundary decision are no longer apparent when available information is not scarce, which can be seen from the cost-sensitive minimum margin strategy in Fig. 2. Cost-sensitive minimum margin strategy

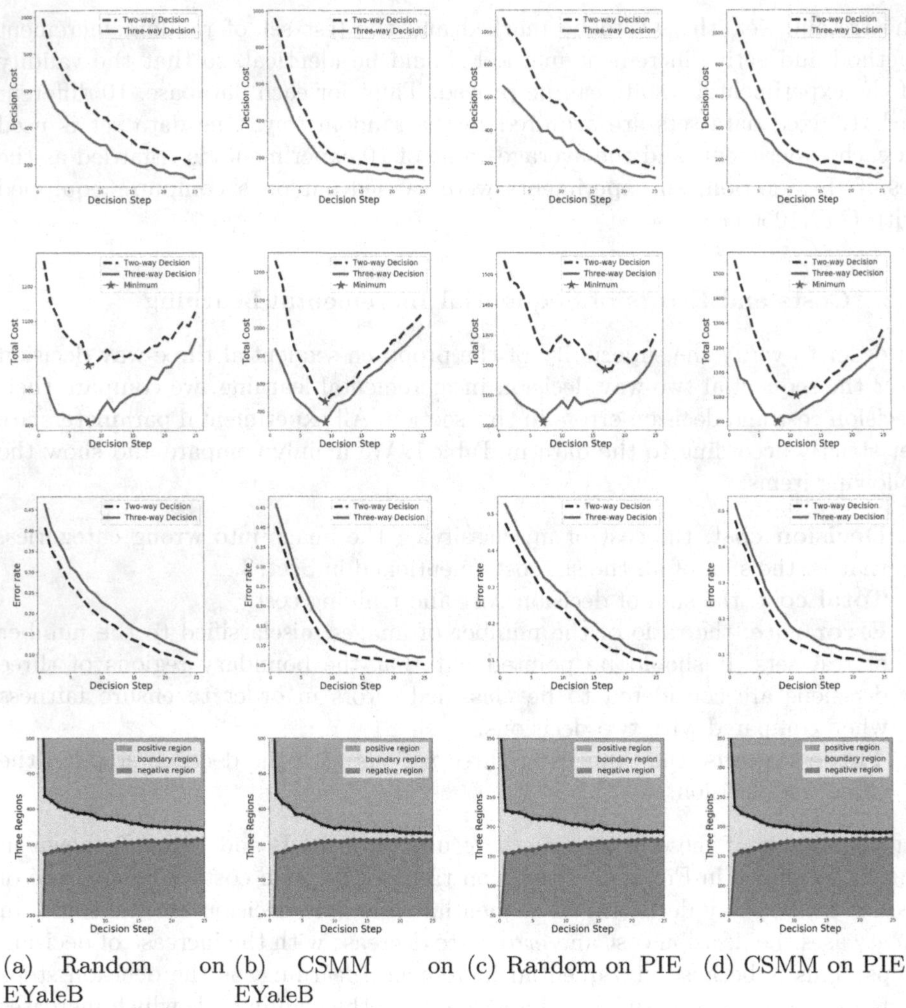

(a) Random on EYaleB (b) CSMM on EYaleB (c) Random on PIE (d) CSMM on PIE

Fig. 2. Comparison of sequential two-way decision and sequential three-way decision on two databases

can obtain information quickly, so in the later decision stage, the superiority of three-way decision is whittled away. In addition, the total cost, which is the sum of decision cost and training cost, is also shown in Fig. 2. The increase of training data not only brings more information, but also increases the time cost and data cost. Although more information brings lower decision cost, the total cost decreases at first and then increases because of the existence of training cost. As we emphasize in Fig. 2, the total cost is minimized at some middle decision step. The results show that the decision cost and total cost of sequential three-way decision are lower than that of sequential two-way decision, which

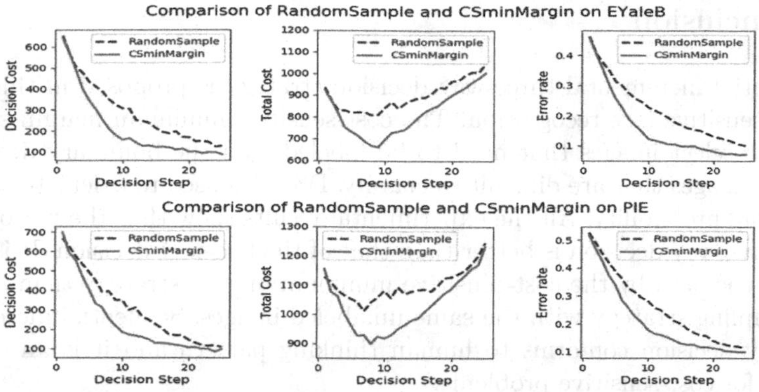

Fig. 3. Comparison of random sampling strategy and cost-sensitive minimum margin strategy on two databases.

verifies the superiority of three-way decision in dealing with the cost-sensitive face recognition problem.

Besides decision cost and total cost, we compare the error rates between sequential two-way decision and sequential three-way decision. The error rate of sequential three-way decision is higher than that of sequential two-way decision, which is the result of boundary decision. The uncertain face image is classified into the boundary region in three-way decision, which is considered to be a misclassification but they may be classified correctly in two decisions. Boundary decision not only brings lower decision cost, but also leads to higher error rate, which shows that three-way decision are not suitable for cost-insensitive face recognition.

The changes of the three regions are also shown and marked in Fig. 2. The increase of information is accompanied by the increase of decision steps, which makes the classifier more effective, so the range of the boundary region has been reduced. With enough information, the boundary region disappears completely and all samples are classified precisely.

A comparison between random sampling strategy and cost sensitive minimum margin strategy is made, as shown in Fig. 3. In the initial stage, due to the small number of labeled samples, the cost-sensitive minimum margin strategy has no obvious advantage, but as the number of labeled samples increases, the cost and error rate of the cost-sensitive minimum marginal strategy are significantly lower than that of the random sampling strategy. The reason is that the cost-sensitive minimum margin strategy selects the most informative sample and takes into account the misclassification cost of gallery and impostor. The random sampling strategy also gets a good effect with enough labeled data, and the performance of the two sampling strategies tends to be the same.

7 Conclusion

A sequential incremental three-way decision strategy is proposed in this paper for cost-sensitive face recognition. The cost-sensitive minimum margin strategy is used to select images that need to be labeled, and the boundary decision is made for images that are difficult to classify. DSAE is used to calculate accurate conditional probability. All the experimental results show that the performance of the three-way decision is better than that of the two-way decision. It is better to select the data by the cost-sensitive minimum margin strategy than the random sampling strategy with the same unlabeled images. Sequential incremental three-way decision conforms to human thinking pattern and it is an effective strategy for cost-sensitive problems.

Acknowledgments. This work was supported by the Natural Science Foundation of China (Nos. 71671086, 61773208, 71732003, and 61876157), the National Key Research and Development Program of China (No. 2016 YFD0702100), and the Fundamental Research Funds for the Central Universities (No. 14380037).

References

1. Hinton, G.E., Salakhutdinov, R.: Reducing the dimensionality of data with neural networks. Science **313**(5786), 504–507 (2006)
2. Hu, B.Q.: Three-way decisions space and three-way decisions. Inf. Sci. **281**, 21–52 (2014)
3. Jia, X., Zheng, K., Li, W., Liu, T., Shang, L.: Three-way decisions solution to filter spam email: an empirical study. In: Yao, J.T., et al. (eds.) RSCTC 2012. LNCS (LNAI), vol. 7413, pp. 287–296. Springer, Heidelberg (2012). https://doi.org/10.1007/978-3-642-32115-3_34
4. Li, H., Wang, M., Zhou, X., Zhao, J.: An interval set model for learning rules from incomplete information table. Int. J. Approximate Reasoning **53**(1), 24–37 (2012)
5. Li, H., Zhang, L., Huang, B., Zhou, X.: Sequential three-way decision and granulation for cost-sensitive face recognition. Knowl.-Based Syst. **91**, 241–251 (2016)
6. Li, H., Zhang, L., Zhou, X., Huang, B.: Cost-sensitive sequential three-way decision modeling using a deep neural network. Int. J. Approximate Reasoning **85**, 68–78 (2017)
7. Li, H., Zhou, X., Huang, B., Liu, D.: Cost-sensitive three-way decision: a sequential strategy. In: Lingras, P., Wolski, M., Cornelis, C., Mitra, S., Wasilewski, P. (eds.) RSKT 2013. LNCS (LNAI), vol. 8171, pp. 325–337. Springer, Heidelberg (2013). https://doi.org/10.1007/978-3-642-41299-8_31
8. Luo, C., Li, T.: Incremental three-way decisions with incomplete information. In: Cornelis, C., Kryszkiewicz, M., Ślęzak, D., Ruiz, E.M., Bello, R., Shang, L. (eds.) RSCTC 2014. LNCS (LNAI), vol. 8536, pp. 128–135. Springer, Cham (2014). https://doi.org/10.1007/978-3-319-08644-6_13
9. Minar, M.R., Naher, J.: Recent advances in deep learning: an overview. arXiv: Learning (2018)
10. Raza, M., Sharif, M., Yasmin, M., Khan, M.A., Saba, T., Fernandes, S.L.: Appearance based pedestrians gender recognition by employing stacked auto encoders in deep learning. Future Gener. Comput. Syst. **88**, 28–39 (2018)

11. Turney, P.D.: Types of cost in inductive concept learning. arXiv preprint cs/0212034 (2002)
12. Xu, J., et al.: Stacked sparse autoencoder (SSAE) for nuclei detection on breast cancer histopathology images. IEEE Trans. Med. Imaging **35**(1), 119–130 (2016)
13. Yang, X., Li, T., Liu, D., Chen, H., Luo, C.: A unified framework of dynamic three-way probabilistic rough sets. Inf. Sci. **420**, 126–147 (2017)
14. Yao, J.T., Vasilakos, A.V., Pedrycz, W.: Granular computing: perspectives and challenges. IEEE Trans. Cybern. **43**(6), 1977–1989 (2013)
15. Yao, Y.: Three-way decision: an interpretation of rules in rough set theory. In: Wen, P., Li, Y., Polkowski, L., Yao, Y., Tsumoto, S., Wang, G. (eds.) RSKT 2009. LNCS (LNAI), vol. 5589, pp. 642–649. Springer, Heidelberg (2009). https://doi.org/10.1007/978-3-642-02962-2_81
16. Yao, Y.: Three-way decisions with probabilistic rough sets. Inf. Sci. **180**(3), 341–353 (2010)
17. Yao, Y.: An outline of a theory of three-way decisions. In: Yao, J.T., et al. (eds.) RSCTC 2012. LNCS (LNAI), vol. 7413, pp. 1–17. Springer, Heidelberg (2012). https://doi.org/10.1007/978-3-642-32115-3_1
18. Yao, Y.: Granular computing and sequential three-way decisions. In: Lingras, P., Wolski, M., Cornelis, C., Mitra, S., Wasilewski, P. (eds.) RSKT 2013. LNCS (LNAI), vol. 8171, pp. 16–27. Springer, Heidelberg (2013). https://doi.org/10.1007/978-3-642-41299-8_3
19. Yao, Y.: Three-way decision and granular computing. Int. J. Approximate Reasoning **103**, 107–123 (2018)
20. Yao, Y., Deng, X.: Sequential three-way decisions with probabilistic rough sets. In: IEEE ICCI* CC, pp. 120–125 (2011)
21. Yu, H., Zhang, C., Wang, G.: A tree-based incremental overlapping clustering method using the three-way decision theory. Knowl.-Based Syst. **91**, 189–203 (2016)
22. Zhang, L., Li, H., Zhou, X., Huang, B., Shang, L.: Cost-sensitive sequential three-way decision for face recognition. In: Kryszkiewicz, M., Cornelis, C., Ciucci, D., Medina-Moreno, J., Motoda, H., Raś, Z.W. (eds.) RSEISP 2014. LNCS (LNAI), vol. 8537, pp. 375–383. Springer, Cham (2014). https://doi.org/10.1007/978-3-319-08729-0_39
23. Zhang, Y., Zhou, Z.H.: Cost-sensitive face recognition. IEEE Trans. Pattern Anal. Mach. Intell. **32**(10), 1758–1769 (2010)
24. Zhou, Z.H., Liu, X.Y.: Training cost-sensitive neural networks with methods addressing the class imbalance problem. IEEE Trans. Knowl. Data Eng. **18**(1), 63–77 (2006)

Three-Way Decision Collaborative Recommendation Algorithm Based on User Reputation

Fulan Qian[1(✉)], Qianqian Min[1], Shu Zhao[1], Jie Chen[1], Xiangyang Wang[2], and Yanping Zhang[1]

[1] School of Computer Science and Technology, Anhui University, Hefei 230601, Anhui, People's Republic of China
qianfulan@hotmail.com

[2] Anhui Electrical Engineering Professional Technique College, Hefei 230051, Anhui, People's Republic of China

Abstract. Collaborative filtering algorithm is a widely used personalized recommendation technology in e-commerce system. However, due to data sparsity, obtained information is insufficient, so that recommendation accuracy is insufficient. By analyzing user rating data to establish user reputation system, and taking full advantage of user reputation to supplement information contribute to improve recommendation accuracy. In this paper, we use three-way decision to make delayed recommendation and propose an algorithm called Three-way Decision Collaborative Recommendation Algorithm Based on User Reputation (TWDA). Firstly, based on Beta distribution, we introduce three-way decision to the process of calculating user reputation, and we use boundary region parameter to reasonably assign ratings in boundary region into positive or negative region. Then, we combine user reputation with matrix factorization model of collaborative filtering recommendation field. Experimental results on two classic data sets show that TWDA improves recommendation accuracy compared with existing recommendation algorithms.

Keywords: Recommendation system · Collaborative filtering · User reputation · Three-way decision

1 Introduction

Collaborative filtering algorithm is a widely used recommendation technology in personalized recommendation system. It has received widespread attention from scholars and made great progress. However, with the continuous expansion

Supported by Natural Science Foundation of China (No. 61702003, No. 61673020, No. 61876001, and No. 61602003), Natural Science Foundation of Anhui Province (1808085MF175).

© Springer Nature Switzerland AG 2019
T. Mihálydeák et al. (Eds.): IJCRS 2019, LNAI 11499, pp. 424–438, 2019.
https://doi.org/10.1007/978-3-030-22815-6_33

of scale, sparsity of user-item rating matrix seriously affects the effect of recommender algorithm. Due to data sparsity, obtained information is lacking, so that recommendation accuracy is insufficient. By analyzing user rating data to establish user reputation system, and taking full advantage of user reputation and three-way decision to supplement information and make delayed recommendation respectively, both contribute to improve recommendation accuracy.

Three-way decision may be viewed as an extension of rough set theory, based on the same philosophy but goes beyond [1]. It is a decision-making model based on human cognition and an extension of two-way decision theory. Three-way decision is to segment a finite non-empty universe of objects or observations U into three-pairwise disjoint regions (positive region, negative region and boundary region) based on a set of criteria or conditions [2]. Ideas of dividing and processing the universe with three regions have been widely used in many fields [3–6], such as medicine [7], social networks [6], recommender system [8–11]. The key of the three-way decision theory is that introducing boundary region (delayed decision) to the two-way decision except positive region and negative region, and the three-way decision gives boundary region the semantic of delayed decision, so there is a problem of further characterization of the boundary region. Three-way decision first selects the most important decision attributes to classify the universe roughly, then select an attribute to classify the boundary after obtaining the boundary region, and so on until the results are satisfactory.

The so-called three-way recommendation introduces delay recommendation strategy based on the two-way recommendation. Zhang et al. [8] proposed a three-way recommendation system based on regression, which aims to minimize the average cost by adjusting the thresholds of different behaviors. Zhang et al. [9] proposed an algorithm that integrates three-way decision and random forests to construct a recommendation system. In addition to recommendation and non-recommendation, the third option is to consider the teacher cost of delay decision. Xu et al. [10] proposed a three-way decision method for recommendation system. In addition to recommended items and unrecommended items, the model adds a set of items that may be recommended to users. Huang et al. [11] proposed a new three-way recommended system considering variable cost, where the variable cost is a function of project popularity. In summary, existing research on three-way recommendation mainly focuses on single rating granulation, it is completely based on the framework of granular computing, so recommended accuracy is not sufficient.

In this paper, we take advantage of user reputation to supplement information, and use three-way decision to make delayed recommendation contribute to improve recommendation accuracy, therefore, we propose a three-way decision collaborative recommendation algorithm based on user reputation (TWDA), by giving each user a corresponding reputation coefficient. For the less-reputed users, their influence is suppressed by the reputation coefficient, and the impact on accuracy of natural factors of real users with lower reputation in the system can be corrected. By introducing delay decision, three-way decision can further improve the recommendation accuracy through correction of boundary region.

Experimental results show that TWDA can improve recommendation accuracy to a certain extent compared with existing recommendation models.

The following sections of this paper are organized as follows: Sect. 2 briefly introduces the matrix factorization model in collaborative filtering recommendation algorithm; Sect. 3 describes three-way decision collaborative recommendation algorithm based on user reputation in detail; Sect. 4 presents the experimental results and analysis; Sect. 5 is the conclusion of the full paper.

2 Related Concepts

The matrix factorization method is used most commonly in the current recommendation system, which has achieved outstanding results in the Netflix Prize recommendation system competition. Taking the user-item rating matrix as an example, matrix factorization predicts missing values in the rating matrix and then recommends them to the user in some way based on the predicted values. Common matrix factorization methods are basic matrix factorization (basic MF), regularized matrix factorization (Regularized MF), probabilistic matrix factorization (PMF) [12], etc., where Regularized MF is known to be one of the most successful methods for rating prediction outperforming other methods like Pearson-correlation based KNN or co-clustering [13–16]. So we use the regularized matrix factorization (Regularized MF) model in this paper.

Basic MF is the most basic factorization method. The high-dimensional rating matrix R is decomposed into two low-dimensional user matrices U and project matrices V. Through continuous iterative training, the product of U and V is closer to the real matrices. Regularized MF is an optimization of Basic MF, which solves the over-fitting problem caused by MF. It adds a normalization factor $\lambda(\|U_i\|^2 + \|V_i\|^2)$ based on the loss function, then considering the whole as a loss function as shown in Eq. (1),

$$L = \sum_{(i,\gamma)\in train} (r_{i\gamma} - \sum_{f=1}^{F} U_{if} V_{if})^2 + \lambda(\|U_i\|^2 + \|V_i\|^2) \tag{1}$$

When solving U and V, we still use gradient descent method to minimize the loss function, and the iteration formula becomes Eqs. (2) and (3):

$$U_{if} = U_{if} + \alpha(E_{i\gamma} V_{\gamma f} - \lambda U_{if}) \tag{2}$$

$$V_{\gamma f} = V_{\gamma f} + \alpha(E_{i\gamma} U_{if} - \lambda V_{\gamma f}) \tag{3}$$

where,

$$E_{i\gamma} = r_{i\gamma} - \sum_{f=1}^{F} U_{if} V_{if} \tag{4}$$

3 Collaborative Recommendation Algorithm Based on User Reputation and Three-Way Decision

The rating system can be described by a weighted bipartite network, which consists of users denoted by set U and objects denoted by set O. We use the Latin and Greek letters to represent the users and objects. The rating $r_{i\gamma}$ given by user i to object γ is the weight of the link in the bipartite network and all the ratings could be described as a rating matrix A. The degree of user i is denoted as k_i. Moreover, the reputation of user i and the quality of object γ are denoted as R_i and Q_γ [17].

3.1 User Reputation Calculation Based on the Three-Way Decision

Beta distribution can be used to represent the posterior probability of a binary event. The general Beta distribution can be represented by the gamma function Γ as shown in Eq. (5):

$$f(p|\alpha,\beta) = \frac{\Gamma(\alpha+\beta)}{\Gamma(\alpha)\Gamma(\beta)} p^{\alpha-1}(1-p)^{\beta-1}, where \quad 0 \le p \le 1, \alpha > 0, \beta > 0 \quad (5)$$

And the probability expectation value of the Beta distribution is given by Eq. (6) [18]:

$$E(p) = \alpha/(\alpha+\beta) \quad (6)$$

Based on Beta distribution, we evaluate user's reputation based on the probability expectation value and distribution that user i will provide a fair rating. Firstly, considering user's personalization that different users tend to have different rating standards. Some people tend to give high ratings, while others tend to give low ratings. Therefore, we use a normalized method to transform a rating to the extent of fanciness as shown in Eq. (7):

$$r'_{i\gamma} = \begin{cases} \frac{2(r_{i\gamma}-r_i^{min})}{(r_i^{max}-r_i^{min})} - 1, & r_i^{max} \ne r_i^{min} \\ 0, & r_i^{max} = r_i^{min} \end{cases} \quad (7)$$

where r_i^{max} and r_i^{min} denote the maximum and minimum rating user i gives. In this way, all ratings given by a specific user will be mapped to $[-1, 1]$. Specifically, for users who always give the same ratings, their ratings will be normalized to 0. The normalized rating matrix is denoted by A', where each element represents the ratings' extent of fanciness. The positive and negative value of the element could be interpreted as the positive and negative opinion given by users. P_γ and N_γ denote the number of users who have positive and negative attitude toward object γ, respectively. When the normalized rating $r'_{i\gamma} > 0$, we define a rating as fair rating if $P_\gamma > N_\gamma$, otherwise as unfair rating if $P_\gamma < N_\gamma$. Else when the normalized rating $r'_{i\gamma} < 0$, we define a rating as fair rating if $P_\gamma < N_\gamma$, otherwise as unfair rating if $P_\gamma > N_\gamma$.

However, there are a large number of indistinguishable ratings except distinguishable fair and unfair ratings in recommendation system. And the indistinguishable ratings including two situations. One is $P_\gamma = N_\gamma$, for the ratings given by users to object γ, half of the users have a positive attitude and half of the users have a negative attitude. The other is $r'_{i\gamma} = 0$, that is when the normalized rating is 0. In both situations, we are unable to determine the fairness of ratings. Therefore, based on the idea of three-way decision, we define fair ratings as positive region (POS), unfair ratings as negative region (NEG), and indistinguishable ratings that need to be judged twice as boundary region (BND). Thus, the fairness of all user ratings in the system can be described in Table 1.

Table 1. Fairness of all user ratings.

$r'_{i\gamma} > 0$	$P_\gamma > N_\gamma$	fair	POS
	$P_\gamma < N_\gamma$	unfair	NEG
	$P_\gamma = N_\gamma$	***secondary decision***	***BND***
$r'_{i\gamma} < 0$	$P_\gamma > N_\gamma$	unfair	NEG
	$P_\gamma < N_\gamma$	fair	POS
	$P_\gamma = N_\gamma$	***secondary decision***	***BND***
$r'_{i\gamma} = 0$	-	***secondary decision***	***BND***

Regarding a large number of indistinguishable ratings existing in the boundary region, a second determination is made by defining the rating difference degree $|r_{i\gamma} - Q_\gamma|$. The rating is considered to be fair when the rating difference degree is not greater than threshold value α, otherwise, the rating is considered to be unfair. Therefore, the fairness of ratings in the boundary region can be described in Table 2.

Table 2. Fairness of ratings in the boundary region

$P_\gamma = N_\gamma$	$\lvert r_{i\gamma} - Q_\gamma\rvert \le \alpha$	fair	POS
	$\lvert r_{i\gamma} - Q_\gamma\rvert > \alpha$	unfair	NEG
$r'_{i\gamma} = 0$	$\lvert r_{i\gamma} - Q_\gamma\rvert \le \alpha$	fair	POS
	$\lvert r_{i\gamma} - Q_\gamma\rvert > \alpha$	unfair	NEG

In summary, we can get a fairness matrix for all user ratings in the system. And all the symbols describing the fairness of the ratings are collectively shown in Table 3. If all the ratings in the recommendation system can be considered as two types, namely, fair ratings and unfair ratings. We use the Bayesian analysis to model the user reputation. Bayesian analysis adopts a binary event to measure each of users ratings: Fair rating (denoted by 1) or unfair rating (denoted by 0). Therefore, we first use distinguishable rating data of user i as training set to obtain Beta prior distribution as shown in Eq. (8):

$$\begin{aligned} \alpha &= r_i^0 + 1, r_i^0 \geq 0 \\ \beta &= s_i^0 + 1, s_i^0 \geq 0 \end{aligned} \tag{8}$$

thus, the probability expectation value of Beta distribution, which is defined as user reputation, is given by Eq. (9):

$$R_i = E(p_i) = (r_i^0 + 1)/(r_i^0 + s_i^0 + 2) \tag{9}$$

where r_i^0 is the number of distinguishable fair ratings and s_i^0 is the number of distinguishable unfair ratings.

Table 3. Symbol description

r^0	# of fair ratings in non-boundary regions
s^0	# of unfair ratings in non-boundary regions
r^1	# of fair ratings in boundary region
s^1	# of unfair ratings in boundary region
$r(r^0 + r^1)$	# of fair ratings in all user ratings
$s(s^0 + s^1)$	# of unfair ratings across all user ratings

Then, after making a secondary decision on the boundary region, we use all the rating data of user i as training set to obtain a new Beta prior distribution as shown in Eq. (10):

$$\begin{aligned} \alpha &= r_i^0 + 1 + r_i^1, r_i^0 \geq 0, r_i^1 \geq 0 \\ \beta &= s_i^0 + 1 + s_i^1, s_i^0 \geq 0, s_i^1 \geq 0 \end{aligned} \tag{10}$$

and its expectation is shown in Eq. (11):

$$E(p_i) = (r_i^0 + 1 + r_i^1)/(r_i^0 + r_i^1 + s_i^0 + s_i^1 + 2) \tag{11}$$

consequently, we can get the reputation of user i as shown in Eq. (12):

$$\begin{aligned} R_i = E(p_i) &= (r_i^0 + 1 + r_i^1)/(r_i^0 + r_i^1 + s_i^0 + s_i^1 + 2) \\ &= \frac{r_i + 1}{k_i + 2} \end{aligned} \tag{12}$$

where $k_i = r_i^0 + r_i^1 + s_i^0 + s_i^1$, which is all ratings that user i gave, this formula indicates that the more the percentage of fair ratings user i gives, the larger reputation he/she will have.

In this paper, we still use the method in enhanced iterative algorithm with reputation redistribution (short for IARR2) [19] to calculate the quality of objects. For the IARR2 method, the quality of an object is not only determined by the received weighted average rating, but also relied on the maximum reputation of the users who rate it, thus the quality of object γ could be expressed as Eq. (13):

$$Q_\gamma = \max_{i \in U_\gamma}\{R_i\} \frac{\sum\limits_{i \in U_\gamma} R_i r_{i\gamma}}{\sum\limits_{i \in U_\gamma} R_i} \tag{13}$$

3.2 Matrix Factorization Model Based on User Reputation

Introducing reputation coefficient into the Regularized MF model to obtain a reputation-based regularization matrix factorization model, which can be shown as Eq. (14):

$$L = \sum_{(i,\gamma) \in train} R_i (r_{i\gamma} - \sum_{f=1}^{F} U_{if} V_{if})^2 + \lambda(\|U_i\|^2 + \|V_i\|^2) \tag{14}$$

In the process of optimizing loss function, when reputation value is higher, forcing the predicted value and the user's real rating to get closer, users with high reputation value has a greater influence on recommendation result. Conversely, it will weaken impact of users with low reputation value on recommendation result. Regarding natural noise users who are not rigorous in actual system, their ratings are less correlated with weighted average reputation of the object, that is, reputation of not rigorous users is lower, which weakens influence of natural noise on recommendation results, thus achieve the purpose of improving recommendation quality.

We adopt a stochastic gradient descent algorithm to optimize above model. The final U and V, according to the formula $\hat{r}_{i\gamma} = \sum\limits_{f=1}^{F} U_{if} V_{\gamma f}$, can be used to obtain corresponding prediction rating.

The flow description and schematic illustration of TWDA algorithm are shown in Algorithm 1 and Fig. 1.

Algorithm 1. Three-way Decision Collaborative Recommendation Algorithm Based on User Reputation(TWDA)

Input: rating matrix A_{M*N} (M: the number of users in the rating system, N: the number of items in the rating system)

Output: predicted rating $\widehat{r}_{i\gamma}$, user reputation R_i

1: Normalizing rating matrix A by formula (7);

2: Obtain fairness of all user ratings by Table 1;

3: Get user reputation by formula (8);

4: Introducing reputation coefficient into the Regularized MF model and optimizing it by formula (14);

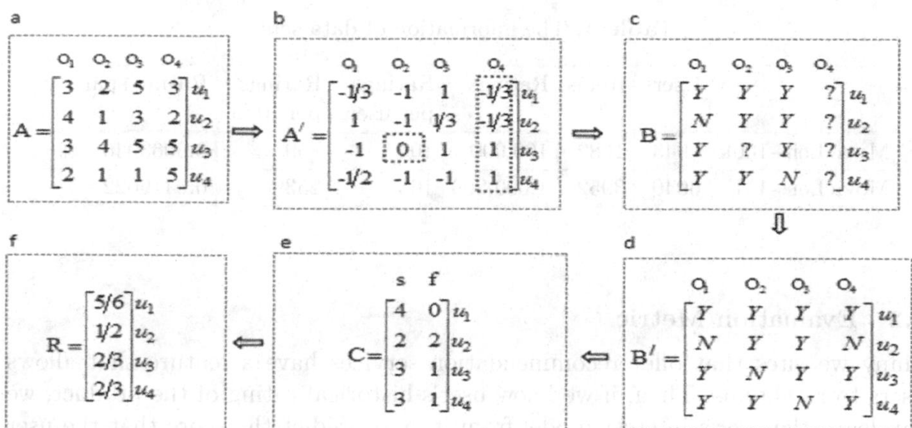

Fig. 1. A schematic illustration of TWDA algorithm. The black arrow shows the steps of the procedure. (a) The corresponding rating matrix, A. The row and column correspond to users and objects, respectively. (b) The normalized rating matrix, A'. Take U_2 as an example, $r_{21} = 4$, $r'_{21} = 2 * (4 - 1)/(4 - 1) - 1 = 1$. (c) Whether each rating is fair or not could be represented as the matrix B. Take O_1 as an example, since $r'_{11} < 0, r'_{21} > 0, r'_{31} < 0, r'_{41} < 0$, the ratings given by U_1, U_3 and U_4 to O_1 are regarded as fair ratings (denoted by Y) and the rating given by U_2 is defined as an unfair rating (denoted by N). (d) Fairness matrix for all ratings, B'. Since there are indistinguishable ratings in matrix B, it is necessary to make a second decision on these ratings. (e) The number of fair and unfair ratings, say s and f, given by each of users, could be denoted as matrix C. Take U_4 as an example, U_4 gives 3 fair ratings and 1 unfair rating, so $s_4 = 3$, $f_4 = 1$. (f) The reputation matrix, R. $R_4 = (1 + s_4)/(2 + s_4 + f_4) = 2/3$.

4 Experimental Analysis

In order to verify the effectiveness of TWDA algorithm, this section will further analyze the role of user reputation coefficient in the recommendation process through experiments, that is, its impact on the recommendation accuracy.

4.1 Data Description

In this paper, we employe two experimental datasets which are well-known and widely used in recommender systems, MovieLens-100k and MovieLens-1M[1]. Specifically, MovieLens-100K dataset contains nearly 100,000 rating records of 943 users on 1,682 movies, and MovieLens-1M dataset contains 1,000,209 rating records from 6,040 users for 3,952 movies, both of two data sets come from the MovieLens website. In addition, each user has rated at least 20 movies, the integer rating scale from 1 to 5, representing the degree of preference from low to high, 1 point means very dissatisfied, and 5 points means very satisfied. Table 4 is a description of basic statistical characteristics of the two datasets.

Table 4. The information of data sets

	Users	Items	Ratings	Ratings per user	Ratings per item	Rating sparsity
MovieLens-100k	943	1682	100,000	106	59	0.0063046
MovieLens-1M	6040	3952	1,000,209	165	253	0.0419022

4.2 Evaluation Metric

Many websites that offer recommendation services have a feature that allows users to rate items. Then, if we know user's historical rating of the product, we can learn the user's interest model from it, and predict the score that the user will rate when he sees an unrated item in the future. The behavior of predicting a user's rating of an object is called rating prediction.

The prediction accuracy of rating prediction is generally calculated by Root Mean Square Error (RMSE) and Mean Absolute Error (MAE). We adopt MAE, which is widely used in many fields, including recommendation systems, to measure the average deviation between true rating and predicted rating. Regarding a user u and item i in test set, MAE is defined as shown in Eq. (15):

$$MAE = \frac{\sum\limits_{u,i \in T} |r_{ui} - \widehat{r}_{ui}|}{|T|} \tag{15}$$

where r_{ui} is actual rating of item i given by user u, $\widehat{r}_{ui}$ is prediction rating given by the recommendation algorithm and $|T|$ is the number of testing data samples.

Obviously, the lower value of MAE indicates better performance of the model.

[1] https://grouplens.org/datasets/movielens/.

4.3 Selection of Boundary Region Parameter

Three-way decision model acquires decision rules according to positive region, negative region, and boundary region. When dealing with ambiguous and incomplete data, they can give non-commitment rules, which can reduce false decisions and improve decision accuracy. The existence of boundary region is temporary. As the information is increased, the knowledge of the objects in boundary region is more refined, the final division must be clear, that is, the positive region or the negative region. Regarding to ratings in boundary region, rating difference degree $|r_{i\gamma} - Q_{\gamma}|$ is used to represent absolute value of difference between the rating and quality of the object, the rating is considered fair when difference degree is less than threshold α, otherwise, the rating is considered unfair, thereby achieving a secondary division of boundary region.

In order to evaluate the influence of threshold α on recommender system, it is necessary to clarify the range of α. For that reason, we conduct experiments to get distribution of rating differences, and the proportion of training set is set to 80%. From Fig. 2(a) and (b), we can see the distribution of rating differences on two datasets.

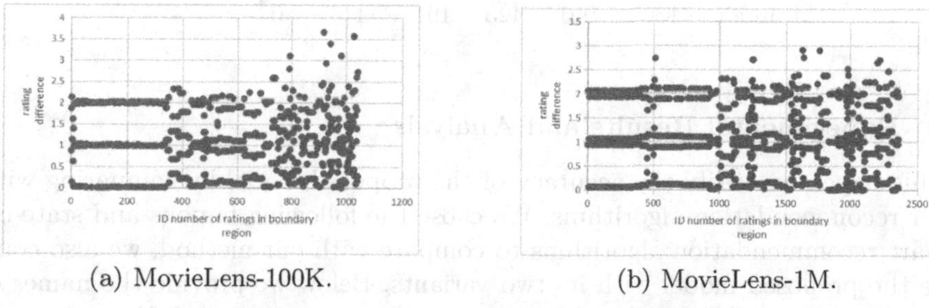

(a) MovieLens-100K. (b) MovieLens-1M.

Fig. 2. Distribution of rating differences on two datasets

Then, we compare the accuracy of our method on two datasets under different α. From Fig. 3(a), it can be seen that our method obtains optimal value when $\alpha = 1$ and eventually tend to be stable on MovieLens-100k dataset. In addition, Fig. 3(b) shows that our method has best performance when $\alpha = 0.5$ and eventually tend to be stable on MovieLens-1M dataset. (The abscissa is rating difference). Furthermore, we change the proportion of training set. The training ratio is set to 50%, 60%, 70%, and 80%, it also shows that our method gets best performance when training ratio is set to 80%.

Table 5 shows the number of fair and unfair ratings under different α on two datasets, and the proportion of training set is set to 80%. It can be seen that more and more ratings are divided into positive region, but our method only obtains optimal value when $\alpha = 1$ on MovieLens-100k dataset, and when $\alpha = 0.5$ on MovieLens-1M dataset. That is, the performance of our method didn't increase as the number of fair rating increases. Thus, following experiments set α to 1 on MovieLens-100k dataset, and set α to 0.5 on MovieLens-1M dataset.

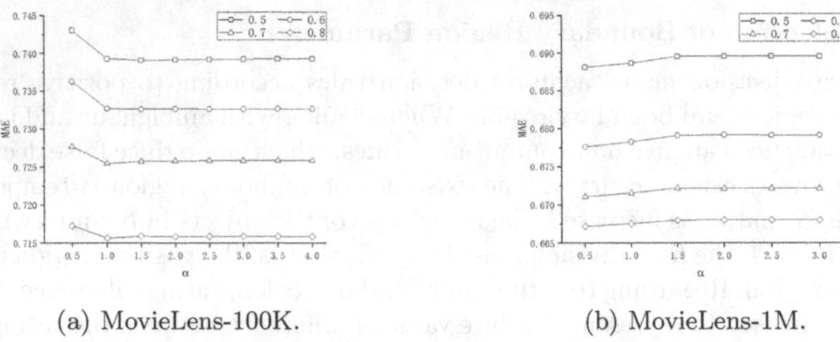

(a) MovieLens-100K. (b) MovieLens-1M.

Fig. 3. Performance on two datasets under different α

Table 5. Two types of ratings in boundary region on two dataset

α	0.5	1	1.5	2	2.5	3	3.5	4
100k-fair	316	556	795	930	1028	1037	1040	1043
100k-unfair	727	487	248	113	15	6	3	0
1M-fair	857	1370	1936	2170	2350	2361	-	-
1M-unfair	1504	991	425	191	11	0	-	-

4.4 Experimental Results and Analysis

In this paper, we verify the accuracy of the proposed model by comparing with other recommendation algorithms. We chose the following famous and state-of-the-art recommendation algorithms to compare with our method, we also compare the proposed model with its two variants. Below we provide the names of algorithms that will be used in following experiments, and its brief introduction.

- **PMF**: Probabilistic matrix factorization model, which is a widely used matrix factorization model.
- SR^{imp}: Exploiting users implicit social relationships for recommendation.
- **PRMF**: Learning users dependencies without prior information.
- **RegSVD**: A rating prediction algorithm based on SVD.
- **LOD-MF**: A novel recommender model based on matrix factorization and semantic similarity measure.
- **FRAIPA**: A fast recommendation algorithm based on self-adaptation and multi-thresholding.
- **TWA**: An original algorithm without processing boundary region.
- **TWA-POS**: A variant of the proposed model which adopts two-way decision method and divided all ratings in the boundary region into positive region.
- **TWA-NEG**: A variant of the proposed model which adopts two-way decision method and divided all ratings in the boundary region into negative region.
- **TWDA**: Our proposed model which adopts three-way decision method to process the boundary region and divided all ratings in boundary region into positive region or negative region reasonably.

After extensive experiments, we found that TWA-NEG and TWA always have the same result, because calculation of user reputation is only related to the ratings in positive region, dividing all the ratings into negative region is equivalent to no improvement. Therefore, following figures only give experimental results of TWA.

Impact of Latent Factors. In order to examine our method in depth, we compare our method with other variant algorithms under different dimension of latent factor (F), ranging from 10 to 100. Similar to previous sections, we randomly choose 80% of original data as training set and remaining 20% as test set. In addition, the number of iterations is set to 100, learning rate is set to 0.02 and the parameter of regularization term is set to 0.03 on both two datasets.

Figure 4(a) shows that with increase of F, performance of all methods keep improving and eventually tend to be stable on MovieLens-100K, and we can see that TWDA outperforms other variant algorithms. As for performance on MovieLens-1M, it is unstable with the change of F. Figure 4(b) shows MAE of TWDA maintains a gradual decline when $F < 50$. We can see that MAE of TWDA is obviously lower than other algorithms, though thc improvement becomes smaller and smaller when $F > 50$. According to above analysis, we can conclude that the performance of TWDA becomes better and gradually reaches a stable state with increase of F, but obviously, the computational complexity of matrix factorization is proportional to F. Therefore, we should consider the balance between accuracy and efficiency according to actual situation.

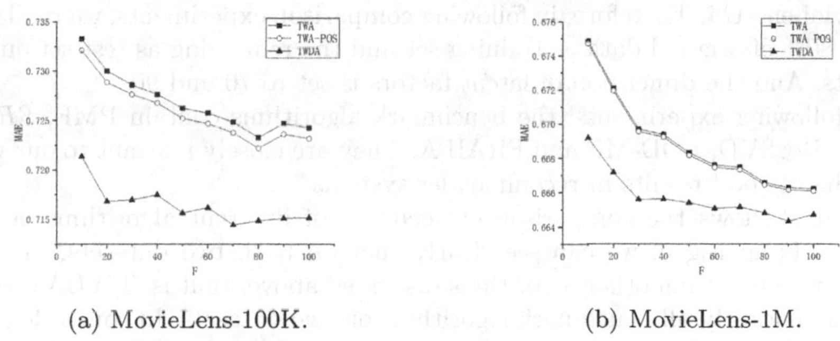

(a) MovieLens-100K. (b) MovieLens-1M.

Fig. 4. Performance on two datasets under different F

Impact of Sparsity. Sparsity is one of the most important factors that affect the performance of recommender system. To further evaluate our method, we change the proportion of training set. The training ratio is set to 50%, 60%, 70% and 80%. The dimension F of latent factor is set to 70 on MovieLens-100K dataset and 90 on MovieLens-1M dataset.

As we have expected, the sparsity of dataset greatly affects the performance of recommendation algorithm. Figure 5(a) shows the results on MovieLens-100K dataset, we can see that training ratio from 50% to 80%, MAE of TWDA is always maintained at a relatively low level, and TWDA performs substantially well over all other variant algorithms. From Fig. 5(b), we can see clearly that on Movielens-1M dataset, our method outperforms all methods discussed here under different data sparsity, even though the improvement of performance becomes relatively low. That is because of the influence of latent factor. In conclusion, our method can make less prediction error on extremely sparse datasets.

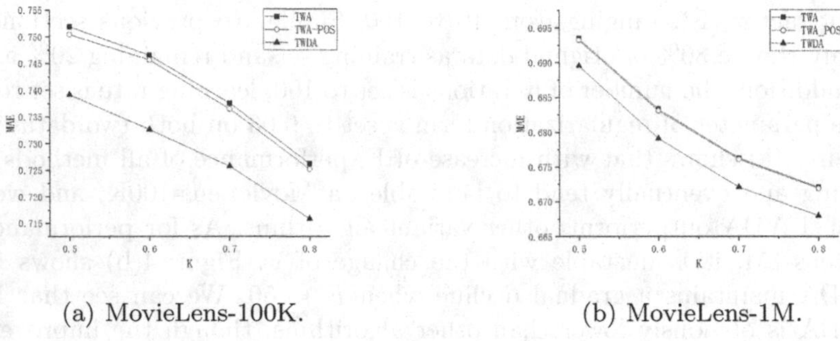

(a) MovieLens-100K. (b) MovieLens-1M.

Fig. 5. Performance on two datasets under different sparsity

Above experimental results demonstrate that TWDA obtains less prediction error when latent factor F = 70 and training ratio K = 0.8 on MovieLens-100K, and has better performance when latent factor F = 90 and training ratio K = 0.8 on MovieLens-1M. Therefore, in following comparison experiments, we randomly choose 80% of original data as training set and the remaining as test set on two datasets. And the dimension of latent factors is set to 70 and 90.

In following experiments, the benchmark algorithms contain PMF, SR^{imp}, PRMF, RegSVD, LOD-MF and FRAIPA. They are closely relevant to our work and achieve good results in recommender systems.

Figure 6 shows the comparison of accuracy of different algorithms on two datasets. From Fig. 6, we can see clearly that on both two datasets, MAE of TWDA is better than other algorithms discussed above, that is, TWDA outperforms significantly all benchmark algorithms on two datasets. In conclusion, our method that combined with rating fairness has better performance on accuracy.

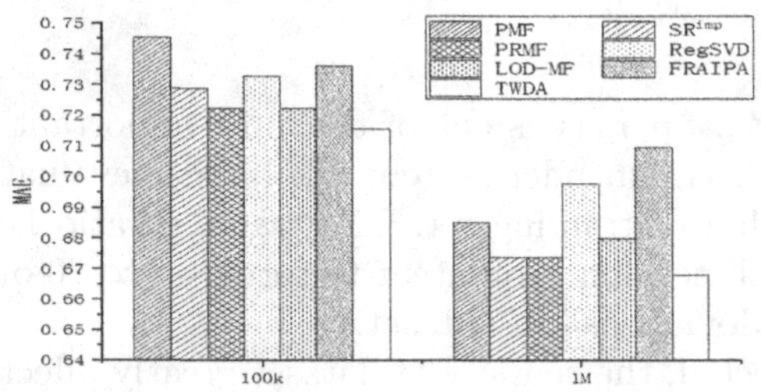

Fig. 6. Comparison of MAE on two datasets

5 Conclusion

In this paper, we propose TWDA algorithm to make recommendation. Firstly, based on Beta distribution, we introduce three-way decision to the process of calculating user reputation, and using boundary region parameter α to reasonably assign ratings in boundary region into positive or negative region. Then user reputation is calculated by the number of ratings in the positive region. Finally, we combine user reputation with matrix factorization model, through user reputation coefficient limits the role of some less rigorous users in the recommendation process, that is, filtering out natural noise to improve accuracy. Experimental results on two classic data sets show that TWDA algorithm improves recommendation accuracy compared with existing recommendation algorithms.

References

1. Yao, Y.: Granular computing and sequential three-way decisions. In: Lingras, P., Wolski, M., Cornelis, C., Mitra, S., Wasilcwski, P. (eds.) RSKT 2013. LNCS (LNAI), vol. 8171, pp. 16–27. Springer, Heidelberg (2013). https://doi.org/10.1007/978-3-642-41299-8_3
2. Peters, J.F., Ramanna, S.: Proximal three-way decisions: theory and applications in social networks. Knowl.-Based Syst. **91**, 4–15 (2016)
3. Gao, C., Yao, Y.: Actionable strategies in three-way decisions. Knowl.-Based Syst. **133**(1), 141–155 (2017)
4. Zhang, Y., Yao, J.T.: Gini objective functions for three-way classifications. Int. J. Approximate Reasoning **81**, 103–114 (2016)
5. Zhang, X., Miao, D.: Three-way attribute reducts. Int. J. Approximate Reasoning **88**, 401–434 (2017)
6. Cabitza, F., Ciucci, D., Locoro, A.: Exploiting collective knowledge with three-way decision theory: cases from the questionnaire-based research. Int. J. Approximate Reasoning **83**, 356–370 (2016)
7. Iserson, K.V., Moskop, J.C.: Triage in medicine, part I: concept, history, and types. Ann. Emerg. Med. **49**(3), 275–281 (2007)
8. Zhang, H.R., Min, F., Shi, B.: Regression-based three-way recommendation. Inf. Sci. **378**, 444–461 (2016)
9. Zhang, H.R., Min, F.: Three-way recommender systems based on random forests. Knowl.-Based Syst. **91**, 275–286 (2016)
10. Xu, Y.-Y., Zhang, H.-R., Min, F.: A three-way recommender system for popularity-based costs. In: Polkowski, L., et al. (eds.) IJCRS 2017. LNCS (LNAI), vol. 10314, pp. 278–289. Springer, Cham (2017). https://doi.org/10.1007/978-3-319-60840-2_20
11. Huang, J., Wang, J., Yao, Y., Zhong, N.: Cost-sensitive three-way recommendations by learning pair-wise preferences. Int. J. Approximate Reasoning **86**, 28–40 (2017)
12. Salakhutdinov, R., Mnih, A.: Probabilistic matrix factorization. In: International Conference on Neural Information Processing Systems, pp. 1257–1264 (2007)
13. Rendle, S., SchmidtThieme, L.: Online-updating regularized kernel matrix factorization models for large-scale recommender systems. In: ACM Conference on Recommender Systems 2008, pp. 251–258 (2008)

14. Takacs, G., Pilaszy, I., Nemeth, B., Tikk, D.: On the gravity recommendation system. In: KDD Cup Workshop at the ACM SIGKDD International Conference on Knowledge Discovery and Data Mining, pp. 303–313 (2007)
15. Bell, R.M., Koren, Y.: Scalable collaborative filtering with jointly derived neighborhood interpolation weights. In: IEEE International Conference on Data Mining, pp. 43–52 (2008)
16. George, T., Merugu, S.: A scalable collaborative filtering framework based on co-clustering. In: IEEE International Conference on Data Mining, pp. 625–628 (2005)
17. Liu, X.L., et al.: Identifying online user reputation of user-object bipartite networks. Phys. A **467**, 508–516 (2017)
18. Jøsang, A., Ismail, R.: The beta reputation system. In: Bled Conference on Electronic Commerce (2002)
19. Liao, H., Zeng, A., Xiao, R., Ren, Z.M., Chen, D.B., Zhang, Y.C.: Ranking reputation and quality in online rating systems. PLoS ONE **9**(5), e97146 (2014)

Multi-graded Hybrid MRDM Model for Assisting Financial Performance Evaluation Decisions: A Preliminary Work

Kao-Yi Shen[1(✉)], Hiroshi Sakai[2], and Gwo-Hshiung Tzeng[3]

[1] Department of Banking and Finance, Chinese Culture University, Taipei, Taiwan
atrategy@gmail.com

[2] Mathematical Sciences Section, Department of Basic Science, Faculty of Engineering, Kyushu Institute of Technology, Kitakyushu, Japan
sakai@mns.kyutech.ac.jp

[3] Graduate Institute of Urban Planning, College of Public Affairs, National Taipei University, New Taipei City, Taiwan
ghtzeng@gm.ntpu.edu.tw

Abstract. This study proposes a novel multiple rule-base decision-making (MRDM) model to transform the current bipolar model into a multi-graded one based on the theoretical foundation of rough set approximations. In the existing bipolar model, the decision class (DC) comprises only three classes: positive, others, and negative ones, and the induced positive or negative rules by the dominance-based rough set approach (DRSA) or variable-consistency dominance-based rough set approach (VC-DRSA) are constrained by the dominance relationship. In certain scenarios or applications, the decision attribute of a bipolar model might need to be transformed into multi-graded DCs to meet practices; examples are the commonly observed Likert 5-point scale questionnaire adopted in a marketing survey. In other words, by eliciting a decision maker's (DM's) preferential judgements on the preferred degree of each DC, the newly proposed model may be more flexible to reflect the DM's preferences or knowledge on modeling an application in a more delicate manner. To reach this goal, the present study proposes a novel MRDM model with multi-graded preferential degree of each DC. Furthermore, the performance of each alternative's score on each rule can be assessed by the crisp (i.e., binary) or fuzzy set technique (FST) and aggregated by a linear or nonlinear operator. This study provides an exemplary case by evaluating the performance of a group of financial holding companies in Taiwan by using the binary assessment and the simple additive weight (SAW) aggregator. The obtained ranking by evaluating their financial data in 2016 is consistent with their actual financial performance in 2017, which suggests the validity of the proposed model.

Keywords: Multiple criteria decision-making (MCDM) · Multiple rule-based decision-making (MRDM) · Dominance-based decision approach (DRSA) · Bipolar decision model · Fuzzy set theory (FST) · Rough set theory (RST)

© Springer Nature Switzerland AG 2019
T. Mihálydeák et al. (Eds.): IJCRS 2019, LNAI 11499, pp. 439–453, 2019.
https://doi.org/10.1007/978-3-030-22815-6_34

1 Introduction

The decision rule approach [1] has gained increasing interests in the multiple criteria decision-making (MCDM) research in recent years, also termed as the multiple-rule-based decision-making (MRDM) [2, 3] approach. Though there are various theoretical foundations of the decision rule approach (e.g., the decision tree [4]), in this study, we focus on the methodology yielded (or extended) from the rough set theory (RST) [5]. Around the early 2000s, the eminent Laboratory of Intelligent Decision Support Systems (IDSS) research group proposed the dominance-based rough set approach (DRSA) [6, 7] and the subsequent variable-consistency dominance-based rough set approach (VC-DRSA) [8, 9] to consider the dominance relationship among the condition and decision attributes, to replace the indiscernibility of the classical RST. Ever since, the advantages of the decision rule approach have been acknowledged in MCDM research [1]; examples are the insights brought by the DRSA approximations, in the form of "***if*…*then*…**" decision rules.

At the early stage, this approach was regarded as a classification tool rather than a decision model. Later on, Greco et al. [10] discussed how to apply the RST methodology for sorting problems in the presence of multiple criteria. It is worthwhile to mention that both the DRSA and VC-DRSA algorithms depend on the dominance relationship among the condition and decision attributes to conduct approximations. Though most criteria (attributes) exhibit a preference-order characteristic, such as the higher the better (or the lower the better), still some criteria might have one or more than one superior range over their full spectrum. Take the debt ratio of a company—regarding investment—for example. The lower debt ratio often implies lower financial risk; however, low debt ratio also intimates losing the opportunity to leverage external capital from financial markets to gaining higher profitability for shareholders. Some other ratios, such as the cash and profitability ratios should be jointly considered to judging if a low debt ratio is superior to a higher one.

To resolve the kind of issues, Greco et al. [10] proposed a framework to cover three types of relationship in the context of MCDM: (1) indiscernibility (from the classical RST), (2) dominance relation, and (3) similarity. In this regard, the aforementioned issue of lacking preference-order of an attribute (or termed criterion in MCDM) can be modeled by considering more than one of the abovementioned relationships in a decision model.

The framework proposed by Greco et al. [10] paved a theoretic foundation to applying DRSA/VC-DRSA approximations to resolve the sorting problem while considering multiple heterogeneous criteria. Furthermore, to model the preference of a decision maker (DM), Greco et al. [6] introduced a pairwise comparison table (PCT) to elicit a DM's preference from a partial set of reference alternatives (points) that the DM has confidence in. Once a decision model induced multiple rules to serve as a decision model, the DM may calculate the net flow score (NFS) of a new alternative (object), yielded from the four-valued logic outranking [11]. It can be termed as the reference-point-based approach. This innovative design bridged the conventional outranking theory with the RST to resolve the ranking problems in MCDM.

Though the combination of the PCT and NFS by the DRSA/VC-DRSA approximations has served as an innovative approach to indicate the preference of a DM, several limitations remain. First, this approach mainly depends on a DM's confidential judgements to form an initial PCT, which is more suitable to deal with the subjective opinions or preferences of an individual. However, in many MCDM industrial applications (e.g., engineering and finance), the research goal is to explore the implicit/hidden patterns or knowledge from a data set (historical records) and domain expert's judgements. To obtain reliable results, this kind of applications often involve multiple experts. And the final outcome is usually calculated by averaging the opinions from those experts. A DM's preference thus plays a marginal role in handling this kind of problem.

Second, the NFS comprises four outranking situations: (1) true, (2) false, (3) contradictory, and (4) unknown ones. A PCT yielded from one DM may include multiple contradictory situations; the increased number of experts might deteriorate the consistency of the obtained result, which devastates the credibility of the obtained NFS for ranking new alternatives (objects). Therefore, the reference-point-based approach has been rarely applied for the data-centric industrial applications.

To bridge the gap, Shen and Tzeng [12, 13] proposed a bipolar model based on the DRSA/VC-DRSA approximations, to deal with the data-centric problems with the minimal requirements from DMs. To form a bipolar model, a DM merely has to classify the decision class (DC) of an alternative as the Positive (preferred) or the Negative (unwanted) or the Neutral/Others (unknown) one. The bipolar approach structures an information system (*IS*), similar to the classical DRSA one, by defining a 4-tuple *IS* as: $IS = \langle U, A^{\odot}, V, f \rangle$. In which, U is a finite set of n alternatives or objects (i.e., $U = (o_1, \ldots, o_j, \ldots, o_n)$), and $A^{\odot}$ comprises two types of attributes (considering a time factor), the condition attributes C^t and the decision one D^{t+1} with a predefined time-lag (this setting is revised from the original DRSA to induce causal rules for decision-aids). In the DRSA, the condition and decision attributes are two disjoint sets (i.e., $C \cap D = \varnothing$ and $C \cup D = A$). Similarly, in a bipolar model, D^{t+1} (usually a singleton $\{d^{t+1}\}$) can be treated as the consequence (in the subsequent t + 1 period) of an alternative that a DM can categorize it without hesitation (also $C^t \cap D^{t+1} = \varnothing$ and $C^t \cup D^{t+1} = A^{\odot}$).

The associated condition attributes C^t of an alternative are organized in an information table with the corresponding values at time period t; the consequence as a DC at the period t + 1. The approximations may yield two groups of rules: the positive and the negative ones, which are regarded as the *new criteria* of a bipolar model. The bipolar model is devised to unveil the cause-effect logical consequences under a predefined time lag, which has been applied on modeling the yearly or quarterly financial performance (FP) of companies in previous research [12].

While applying a bipolar model to assess a group of new alternatives (objects), an alternative would be ranked higher if it is more similar to the positive rules and dissimilar to the negatives ones. The details of the bipolar model will be provided in the next section. Thus, the bipolar model can be regarded as a hybrid decision model.

It requires a DM (or DMs) to assign the values of a group of alternatives (objects) at the beginning, and it also needs a DM (or DMs) to assess a group of new alternatives regarding the degree of its similarity with the positive rules and the dissimilarity to the negatives ones, for supporting the ranking decisions.

Though the bipolar model has contributed to devising a logical approach to rank a group of alternatives with the combination of DMs' inputs and the rough approximations, several limitations remain. For instance, the idea of a bipolar model might not be ideal for all practical scenarios. Once a business practice is designed to classify a decision attribute in five or more DCs (e.g., the well-known Likert 5-point scale questionnaire [14], from 1 (very unsatisfied) to 5 (very satisfied)), a more delicate classification of the decision attribute would be needed. To amend this potential weakness, the present study proposes a flexible hybrid MRDM model to address this issue.

The new approach also attempts to resolve the limitation of the bipolar model that assumes the equal importance of the positive and the negative aspects while setting a threshold to cover the supporting objects. In other words, if a DM thought that being similar to the Positive rules is much more important than dissimilar to the Negative group (e.g., two- or threefold) while ranking new alternatives, the bipolar model lacks a mechanism to make the adjustments. In this study, we propose a weighting mechanism by adjusting the relative importance of each DC associated rules, similar to the concept of the analytical hierarchy process (AHP) [15]. The new multi-graded MRDM approach could be more flexible to fit business practices on various applications.

Overall, this study focuses on devising a hybrid MRDM model that can explore the hidden knowledge of a complicated data-centric problem with two sources of inputs: (1) a historical data set and (2) a group of domain experts. This hybrid approach transforms the decision attribute of the bipolar model into a multi-graded one considering the weighting of each DC. Such a design aims to meet business practices and specific requirements or preferences of DMs, which offers more flexibility to DMs based on the encountered scenarios. A numerical case will illustrate the associated calculations in Sect. 4.

2 Multi-graded Hybrid MRDM Model

This section begins with the essentials of the DRSA and the extended bipolar model, which are the predecessors of the proposed approach of this work. In addition, the conceptual framework that fuses the inputs from a group of data set and DMs is illustrated in Fig. 1.

2.1 Briefing of Dominance-Based Rough Set Approach (DRSA)

As mentioned in the previous Section, the DRSA extended the classical RST by adopting the dominance relationship to replace the indiscernibility one while making approximations. In a typical DRSA model, we can define an *IS* as: $IS = \langle U, A^{\odot}, V, f \rangle$, where U and $A^{\odot}$ (i.e., $A^{\odot} = \{a_1, a_2, \ldots, a_m, a_{m+1}\}$, which indicates that there are m condition attributes and a decision attribute) are mentioned in the previous section.

Also, V is the value domain of $A^{\odot}$; more specifically, V_{a_i} denotes the value domain of a_i. $V = \cup_{a_i \in A} V_{a_i}$ and $i = 1, 2, \ldots, m+1$. In this *IS*, f denotes a total function, such that $f(o_j, a_i) \in V_{a_i}$ for $j = 1, \ldots, n$ and $i = 1, 2, \ldots, m+1$. Generally, $D^{t+1} = \{d\} = \{a_{m+1}\}$, which can be assigned as the last attribute at time period $t+1$ of an information table. For any object (alternative) in U (i.e., $o_j \in U$), o_j can only be assigned to a specific class regarding an attribute a_i. To express the idea of a class in DRSA formally, let $\boldsymbol{Cl} = \{Cl_\alpha,\ \alpha \in \{1, \ldots, \mathrm{N}\}\}$ and thus for each $o \in U$ belong to only one $Cl_\alpha \in \boldsymbol{Cl}$ on each attribute (for any $a \in A$).

Assume that there are n objects in U; in an *IS* that considers the dominance relationship, Cl_α can be defined as a preference-ordered classification. And two unions of sets can be formed as: (1) an upward union $Cl_{\overline{\alpha}}^{\succeq} = \cup_{\beta \succeq \alpha} Cl_\beta$ and (2) a downward union $Cl_{\overline{\alpha}}^{\preceq} = \cup_{\beta \preceq \alpha} Cl_\beta$ of *Cls*. Start from here, we merely use the upward union to explain the followings. The downward union can be reasoned by analogy. Therefore, to describe the dominance relationship between any two objects o_p and o_q in U, if o_p dominates o_q on a partial set of C (i.e., $P \subseteq C$), it can be denoted as $o_p P_{Dom} o_q$ (i.e., $o_p \succeq_{P_{Dom}} o_q$). In this regard, two sets termed as P-dominating ($D^{\uparrow}_{P_{Dom}}(\bullet)$) and P-dominated ($D^{\downarrow}_{P_{Dom}}(\bullet)$) sets are defined in Eqs. (1)–(2):

$$D^{\uparrow}_{P_{Dom}}(o_p) = \{o_p \in U : o_p P_{Dom} o_q\} \tag{1}$$

$$D^{\downarrow}_{P_{Dom}}(o_p) = \{o_p \in U : o_q P_{Dom} o_p\}. \tag{2}$$

In the next, the DRSA further defines the P-upper and the P-lower approximations of an upward union $Cl_{\overline{\alpha}}^{\succeq}$ as: (1) $\overline{P}\left(Cl_{\overline{\alpha}}^{\succeq}\right) = \cup_{o_p \in Cl_{\overline{\alpha}}^{\succeq}} D^{\uparrow}_{P_{Dom}}(o_p)$ and (2) $\underline{P}\left(Cl_{\overline{\alpha}}^{\succeq}\right) = \left\{o_p \in U : D^{\uparrow}_{P_{Dom}}(o_p) \subseteq Cl_{\overline{\alpha}}^{\succeq}\right\}$, where $Cl_\alpha \in \boldsymbol{Cl}$. In the P-lower approximation, $\overline{P}\left(Cl_{\overline{\alpha}}^{\succeq}\right)$ also can be defined as $\overline{P}\left(Cl_{\overline{\alpha}}^{\succeq}\right) = \left\{o_p \in U : D^{\downarrow}_{P_{Dom}}(o_p) \cap Cl_{\overline{\alpha}}^{\succeq} \neq \varnothing\right\}$. The differences between the P-upper and the P-lower approximations is the P-boundary set ($B_P(\bullet)$), shown in Eq. (3) (the P-boundary set of a downward union is similar as Eq. (3)). The issues relate to measuring the approximation quality and the concept of REDUCT are not necessary to the understanding of the DRSA approximations, which can be found in the previous research [1, 6, 7].

$$B_P\left(Cl_{\overline{\alpha}}^{\succeq}\right) = \overline{P}\left(Cl_{\overline{\alpha}}^{\succeq}\right) - P\left(Cl_{\overline{\alpha}}^{\succeq}\right) \tag{3}$$

With the lower and upper approximations of the upward and/or downward unions of DCs, the DRSA can induce a group of decision rules, which is the cornerstone of the MRDM approach. Although Greco et al. [1] defined five types of decision rules, only two types of certain rules (i.e., the $D^{t+1}_{\succeq}$ and the $D^{t+1}_{\preceq}$ decision rules) are mainly applied for decision aids. Take the $D^{t+1}_{\succeq}$ certain rule regarding a class Cl_α for instance, which provides descriptions of the objects/alternatives belonging to the $Cl_{\overline{\alpha}}^{\succeq}$ union with certainty. In other words, if an object satisfies the ${}^{ith}D^{t+1}_{\succeq}$ certain rule that only has three antecedents regarding a_1, a_2, and a_3 (i.e., ***if*** "$o_\beta \succeq_{a_1} o_\alpha$" and "$o_\beta \succeq_{a_3} o_\alpha$" and "$o_\beta \succeq_{a_5} o_\alpha$", ***then*** "$o_\beta \in Cl_\alpha^{\succeq}$"), then any $o_i \in U$ that satisfies the ${}^{ith}D_{\succeq}^{t+1}$ certain rule should be categorized at least as good as Cl_α.

2.2 Bipolar MRDM Model

The bipolar model was inspired by a well-known MCDM method: the TOPSIS (Technique for Order Preference by Similarity to Ideal Solution) [19]. And the bipolar model is based on an idea to handle the ranking problem: to be more similar (closer) to the positive rules and dissimilar to the negative ones. It begins with dividing the decision attribute in DRSA (D^{t+1}) into three disjoint *Cls*: Positive (Cl_{POS}), Others (Cl_{OTR}), and Negative (Cl_{NEG}). While organizing an existing data with known records (objects), DMs may base on their knowledge/experience or preferences to classify the objects on hands as Cl_{POS} and Cl_{NEG} at t + 1 time period. Therefore, the two group of rules associated with $Cl_{POS}^{\succeq}$ and $Cl_{NEG}^{\preceq}$ with certainty can be transformed into the new criteria of a bipolar decision model.

Since the DRSA approximations may generate multiple rules, a researcher (or a DM) has to set up a threshold Ψ $(0 < \Psi \leq 100\%)$ regarding the percentage of how many objects (instances or alternatives) should be covered in each group (the Positive/Negative group). Therefore, the bipolar model proposes a mechanism to select the rules that should be kept in each group. Let's assume that DMs categorize ε and ϕ objects as Cl_{POS} and Cl_{NEG} of a data set respectively, where $\varepsilon + \phi \leq n$ (n is the total number of the objects of this data set). Also, $| \bullet |$ indicates cardinality in here; thus, $|o_i^{POS}|$ and $|o_j^{NEG}|$ denotes the number of objects while $1 \leq i \leq \varepsilon \leq n$ and $1 \leq j \leq \phi \leq n$ in Eqs. (4)–(5):

$$\frac{\left| \cup_{i \in \{1,\ldots,\varepsilon_{\min}\}} o_i^{POS} \right|}{\varepsilon} \geq \Psi \tag{4}$$

$$\frac{\left| \cup_{j \in \{1,\ldots,\phi_{\min}\}} o_j^{NEG} \right|}{\phi} \geq \Psi. \tag{5}$$

In here, $\varepsilon_{\min}$ and $\phi_{\min}$ denote the minimum numbers that may satisfy Eq. (4) and Eq. (5), respectively. Also, if a Positive (or Negative) object satisfied all the requirements (antecedents) of a rule at time period t and its DC belonged to $Cl_{POS}^{\succeq}$ (or $Cl_{NEG}^{\preceq}$) at time period t + 1, this object can be defined as a support of this rule. The higher number of supports of a rule indicates its higher importance (or influence), vice versa.

In each group of rules (i.e., the positive and negative ones), the certain positive (associated with $Cl_{POS}^{\succeq}$) and the negative (associated with $Cl_{NEG}^{\preceq}$) rules should be listed in sequence ($\text{Rule}_{i \in \{1,\ldots,s\}}^{POS}$ or $\text{Rule}_{j \in \{1,\ldots,t\}}^{NEG}$) according to their supporting numbers, from high to low (i.e., $SUPP_{\text{Rule}_1^{POS}} > SUPP_{\text{Rule}_s^{POS}}$ and $SUPP_{\text{Rule}_1^{NEG}} > SUPP_{\text{Rule}_t^{NEG}}$). Each group of rules should keep the minimal number of certain rules by cross-reference the positive group sequence with Eq. (4) or the negative group sequence with Eq. (5).

In this regard, the raw weight of the *i*-th strong positive rule is $\left|o_{ith}^{POS}\right|/\varepsilon$ and the raw weight of the *j*-th strong negative rule is $\left|o_{jth}^{NEG}\right|/\phi$. After calculating the raw weight of each rule of the newly formed bipolar model, all the raw weights have to be normalized to sum up to one for a bipolar model. Recently, the stability issue of a bipolar was discussed [13], and it can enhance to deal with non-deterministic or semi-nondeterministic condition attributes by adopting an extended approach.

2.3 Multi-graded MRDM Model

A multi-graded MRDM model, just like the bipolar decision model, focuses on dealing with those data-centric problems that also require the knowledge or preference from DMs. The bipolar model begins with dividing a DRSA decision attribute into three disjoint *Cls*. Similarly, the **First Step** of a multi-graded MRDM model begins with defining the number of *Cls* for a decision attribute (D^{t+1}) in a DRSA *IS*. For instance, let's assume that there are five disjoint *Cls* of a decision attribute (i.e., $D^{t+1} = \left\{d^{(4)}, d^{(3)}, d^{(2)}, d^{(1)}, d^{\otimes}\right\}$, where $d^{(4)} \cap d^{(3)} \cap d^{(2)} \cap d^{(1)} \cap d^{\otimes} = \varnothing$ and $d^{(4)} \succ d^{(3)} \succ d^{(2)} \succ d^{(1)} \succ\succ d^{\otimes}$).

While applying a multi-graded MRDM model, the lowest-ranked $d^{\otimes}$ denotes the unwanted *Cl*, and $d^{(1)}$ is the one (a granule or an interval) that a DM can accept with the minimal requirement. Take the case of investment as an example, $d^{\otimes}$ might suggest net losses, and $d^{(1)}$ could be within the minimal range of profitability (e.g., $0\% \leq d^{(1)} \leq 1.5\%$) that a DM feels acceptable. Intuitively, while ranking a new object (alternative), an object that is more similar to a rule associated with $Cl_{d^{(4)}}^{\succeq}$ should be ranked higher comparing with another object that is more similar to a rule associated with $Cl_{d^{(1)}}^{\succeq}$. One thing should be noticed in here. Unlike the bipolar model that considers both the upper and the downward unions, the multi-graded MRDM model only involves the upper unions associated decision rules by excluding the rules associated with $Cl_{d^{\otimes}}^{\preceq}$. Though we merely use five *Cls* as an illustration, it can be reduced to three or extended to more than five *Cls* to fit a specific ranking problem.

In the **Second Step**, a DM needs to set a threshold, similar to the mechanism in the bipolar model, to select the minimal associated rules for each involved *Cls*. Suppose that a threshold is assigned as Ω ($0 < \Omega \leq 100\%$). Take the objects covered by the rules associated with $Cl_{d^{(4)}}^{\succeq}$ for instance. Assuming that there are φ objects categorized in $d^{(4)}$, which leads to the following Eq. (6):

$$\frac{\left|\cup_{j\in\{1,\ldots,\varphi_{\min}\}} o_k^{(4)}\right|}{\varphi} \geq \Omega, \text{ for } k = 1,\ldots,\varphi. \quad (6)$$

Similar to the rule selection mechanism in the bipolar decision model, all the certain rules should be ranked from high to low supports in a sequence. And the rules that satisfy Eq. (6) should be reserved in this multi-graded MRDM model. This rule-selection mechanism applies for the other DCs' associated rules, and all the decision rules associated with $Cl^{\succeq}_{d^{(i)}}$ (for $i = 1, 2, .., 4$ in here) can be regarded as the (i)-th group of rule. The raw support weight of each rule in the (i)-th group can be named as $SUPP^{p\text{-th}}_{\mathrm{R}^{d^{(i)}}}$ for the p-th rule (assume that there are $\omega^{(i)}$ decision rules in this group of rules and $1 \le p \le \omega^{(i)}$). Then the initial support weight of the p-th decision rule can be calculated as: $SUPP^{p\text{-th}}_{\mathrm{R}^{d^{(i)}}} \Big/ \sum_{p=1}^{\omega^{(i)}} \left(SUPP^{p\text{-th}}_{\mathrm{R}^{d^{(i)}}} \right)$; the sum of all the initial support weight of each rule in the (i)-th group should be 100%. Unlike the bipolar decision model, the multi-graded MRDM one further relaxes the equal weighting assumption of each new criterion, transformed from the decision rules, by allowing a DM(s) to denote his/her preferences toward each DC.

Thus, the **Third Step** adopts the concept of the AHP method by soliciting a DM's preferential opinions by a questionnaire. A DM should denote his/her preference regarding the relative importance of each DC over the other ones, and we follow the relative importance scale proposed by Saaty [15] in here (i.e., from "9 : 1" to "1 : 9"). Though there several approaches in the AHP to calculate the relative importance of each criterion, this study adopts the row geometric mean method (RGMM) [16] to simplify the calculation procedures.

Let $\boldsymbol{D} = \left(d_{ij}\right)_{m \times m}$, where $m = 4$ for $\left\{d^{(4)},\ d^{(3)},\ d^{(2)},\ d^{(1)}\right\}$ and $1 \le i, j \le 4$ in here. $\boldsymbol{D}$ is a preferential judgement matrix, where d_{ij} denotes the relative importance of $d^{(i)}$ to $d^{(j)}$ and $d_{11} = d_{22} = d_{33} = d_{44} = 1$. In RGMM, Aguaron and Moreno-Jiménez [17] suggested that the geometric means of the rows of matrix $\boldsymbol{D}$ can be applied as the relative weight for each DC in Eq. (7):

$$w_{d^{(i)}} = \frac{\sqrt[1/4]{\prod_{j=1}^{4} d_{ij}}}{\sum_{i=1}^{4} \left(\sqrt[1/4]{\prod_{j=1}^{4} d_{ij}} \right)}. \tag{7}$$

The relative importance of each DC can be applied to adjust the initial support weight of each rule for a specific $Cl^{\succeq}_{d^{(i)}}$ associated group. Thus, the **Fourth Step** calculates the final supporting weight of each rule by multiplying $w_{d^{(i)}}$ (from Eq. (7) with all the rules' initial support weights in the (i)-th group. This step normalizes all the supporting weights of the rules in a multi-graded MRDM model.

In the **Fifth Step**, while applying a multi-graded MRDM model to evaluate a group of new alternatives with the same information structure, a DM needs to decide if each antecedent of each rule was satisfied or not to assign the assessment score of each alternative on each rule. There are at least four combinations to forming the final score of a new alternative at this step. On one side, a DM may choose a binary assessment (Yes or No) to decide if a new alternative satisfied an antecedent of a rule, also termed as the crisp binary assessment. Or, the assessment can be extended into a verbal assessment with the fuzzy judgement in various techniques. On the other side, the final aggregation of all the scores from each rule can be either linear (e.g., simple additive way; SAW) or nonlinear (e.g., the fuzzy integral [18]). To simplifie the illustration for this preliminary work, we chose the combination of the binary assessment and the SAW aggregation in the following section.

3 An Illustrative Case

In here, we illustrate how to apply the multi-graded MRDM model to rank the financial performance of a group of financial holding companies by using their historical financial data.

3.1 Data

Since 2011, there are 13 financial holding companies in the public listed stock market of Taiwan. Because those financial holding companies require certain distinct measures to monitor their financial soundness and sustainability, the Taiwan Economic Journal (TEJ) database [19] offers a group of indicators that comprises general financial ratios and three specific ones (e.g., Bank of International Settlement (BIS) capital adequacy ratio) for this industry.

Among this group of indicators, there are 60 ratios from six dimensions. To simplify the modeling, we interviewed with experts to remove certain redundant indicators. The research flows and the associated experimental settings are shown in Fig. 1.

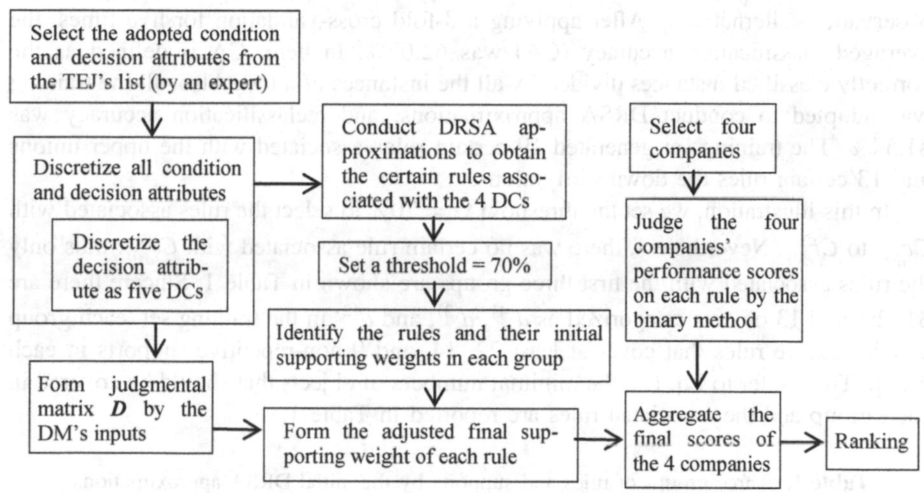

Fig. 1. Research flow of the hybrid multi-graded MRDM model in this case.

To simplify the modeling, we conducted several rounds of interviews with two domain experts, and the intersection of their selected indicators (ratios) were adopted as the 22 attributes in this case (i.e., 21 condition attributes and one decision attribute (return on equity, *ROE*)). To illustrate the proposed multi-graded MRDM model, only one domain expert (a retired vice president of a financial institution and works as a senior consultant for a think tank) was invited for the following experiments, including the assessments for the testing set.

In this case, we used the companies' yearly data to form two sets: a training set (from 2011 to 2015) and a testing one (the condition attributes in 2016 and the decision attribute in 2017). Therefore, the timeframe is structured based on the annual data, all retrieved from the TEJ database [19].

3.2 Forming a Multi-graded MRDM Model

After collecting the raw data (i.e., the 24 indicators or ratios) of the 13 financial holding companies from 2011 to 2017, all the 23 condition attributes were discretized into three values: High (*H*), Middle (*M*), and Low (*L*), based on the percentile method (e.g., the top 33.33% were categorized as *H*) in each year. The 13 financial holding companies' decision attribute (ROE^{t+1}) was discretized into five DCs: $\{d^{(4)}, d^{(3)}, d^{(2)}, d^{(1)}, d^{\otimes}\}$, where $d^{\otimes}$ denotes unwanted net losses in the time period $t + 1$ (i.e., $d^{(\otimes)} < 0\%$). The other four DCs were defined as: (1) $d^{(4)} \geq 9\%$, (2) $9\% > d^{(3)} \geq 6\%$, (3) $6\% > d^{(2)} \geq 3\%$, and (4) $3\% > d^{(1)} \geq 0\%$, based on the historical data and the expert's suggestion. The following experiments only adopted the four DCs. The settings are in line with $d^{(4)} \succ d^{(3)} \succ d^{(2)} \succ d^{(1)} \succ\succ d^{\otimes}$.

In the next, we conducted the DRSA approximations for the training set (65 observations/alternatives). After applying a 3-fold cross-validation for five times, the averaged classification accuracy (CA) was 62.07%. In here, CA is defined as the correctly classified instances divided by all the instances of a test. Also, all the training was adopted to conduct DRSA approximations, and reclassification accuracy was 81.54%. The training set generated 19 certain rules associated with the upper unions and 13 certain rules the downward unions.

In this illustration, we set the threshold $\Omega = 70\%$ to select the rules associated with $Cl^{\succeq}_{d^{(4)}}$ to $Cl^{\succeq}_{d^{(1)}}$. Nevertheless, there was no certain rule associated with $Cl^{\succeq}_{d^{(1)}}$; thus only the rules associated with the first three groups are shown in Table 1. Sincere there are 31, 20, and 13 objects categorized as $d^{(4)}$, $d^{(3)}$, and $d^{(2)}$ in the training set, each group should reserve rules that cover at least 22, 14, and 9 non-repetitive supports in each group. Thus, refer to Eq. (6), the minimal numbers of objects that should be covered in each group and the associated rules are reported in Table 1.

Table 1. Three groups of rules and supports by the initial DRSA approximations.

	Associated certain rules	Supports	Supporting instances
$Cl^{\succeq}_{d^{(4)}}$	R1	12	2, 7, 15, 24, 29, 31, 41, 42, 44, 46, 54, 57
	R2	10	2, 5, 15, 16, 24, 29, 33, 42, 46, 54
	R3	10	2, 7, 15, 20, 24, 28, 41, 46, 54, 63
	R4	10	5, 18, 23, 24, 27, 37, 44, 50, 57, 64
$Cl^{\succeq}_{d^{(3)}}$	R16	23	2, 3, 7, 12, 15, 16, 20, 22, 24, 28, 29, 33, 41, 42, 44, 46, 54, 55, 57, 59, 60, 63, 65
$Cl^{\succeq}_{d^{(2)}}$	R17	46	2, 3, 4, 6, 7, 9, 10, 11, 12, 14, 15, 16, 17, 19, 20, 22, 23, 24, 27, 28, 29, 30, 32, 33, 35, 36, 37, 41, 42, 43, 44, 45, 46, 48, 49, 50, 54, 55, 57, 58, 59, 60, 61, 62, 63, 65

Though there is no certain rules associated with $Cl^{\succeq}_{d^{(1)}}$, the present study still requested the expert to denote his opinions (i.e., only one DM in this case) to forming the judgmental matrix $\boldsymbol{D}$ (Table 2). Also, by referring to Eq. (7), the relative importance of each DC, the initial support weight and the final supporting weight of each rule are reported in Table 3. The details (including antecedents and consequences) of the involved rules are shown in Table 4.

Table 2. Judgmental matrix $\boldsymbol{D}$.

	$d^{(4)}$	$d^{(3)}$	$d^{(2)}$	$d^{(1)}$
$d^{(4)}$	1	3	5	7
$d^{(3)}$	1/3	1	3	5
$d^{(2)}$	1/5	1/3	1	3
$d^{(1)}$	1/7	1/5	1/3	1

Table 3. Final supporting weight of each rule.

	Relative importance of each DC	Rules	Initial supporting weights	Adjusted supporting weights	Normalized final weights
$d^{(4)}$	56.46%	R1	26.09%*	14.73%	16.43%
		R2	21.74%	12.27%	13.70%
		R3	21.74%	12.27%	13.70%
		R4	21.74%	12.27%	13.70%
$d^{(3)}$	26.31%	R16	100.00%	26.31%	29.36%
$d^{(2)}$	11.76%	R17	100.00%	11.76%	13.21%
$d^{(1)}$	5.47%	–	–	–	–

*Note: The total number of supports of R1, R2, R3, and R4 are 44 (12 + 10 + 10 + 10 = 44). Refer to Table 2, the initial support weight of R1 is 26.09% (12/44 = 26.09%).

Table 4. The involved rules of this multi-graded MRDM model.

Rules	Antecedents	Consequences
R1	$(PerExpense \leq L)$ & $(Worth\_U \geq H)$	$\succeq d^{(4)}$
R2	$(PerExpense \leq L)$ & $(CASHshare \geq H)$ & $(Worth\_U \geq M)$	$\succeq d^{(4)}$
R3	$(PerExpense \leq M)$ & $(ROA \geq H)$ & $(Worth\_U \geq H)$	$\succeq d^{(4)}$
R4	$(UsuEPS \geq M)$ & $(CASHshare \geq M)$ & $(Issu\_Deposit \leq L)$	$\succeq d^{(4)}$
R16	$(PerExpense \leq M)$ & $(DivAuth \geq M)$ & $(UsuEPS \geq M)$	$\succeq d^{(3)}$
R17	$(DivAuth \geq M)$	$\succeq d^{(2)}$

Until here, we finished the four steps mentioned in Subsect. 2.3. In the next step, four financial holding companies' financial figures in 2017 would be provided to the expert for performance assessments. One thing needs to be noted in here, all the antecedents of a rule were assumed to be equal in this model.

3.3 Ranking and Discussions

The four selected financial holding companies are: (1) Fubon Financial Holdings (code: 2881), (2) China Development Financial (code: 2883), (3) Yuanta Financial Holdings (code: 2885), and (4) SinoPac Holdings (code: 2890). As mentioned in Subsect. 2.3, we adopted the binary approach (Yes/No) for the expert to decide whether the four companies' associated financial indicators in 2016 could satisfy the antecedents (requirements) in Table 4.

For brevity, only the condition attributes appeared in Table 4 are briefly described in Appendix A. And the four companies' performance scores on each rule and their final performance scores, aggregated by the SAW method, are reported in Table 5.

Table 5. The four companies' final scores and ranking.

Rules Normalized final supporting weights	R1 16.43%	R2 13.70%	R3 13.70%	R4 13.70%	R16 29.36%	R17 13.21%	Final performance (Ranking)
Fubon (2881)	1.00	0.66*	1.00	0.66	1.00	1.00	90.78% (1)
China_Dev (2883)	0.00	0.00	0.00	0.33	0.00	0.00	4.52% (4)
Yuanta (2885)	0.00	0.33	0.33	1.00	0.66	1.00	55.33% (2)
SinoPac (2890)	0.00	0.33	0.00	0.66	0.33	1.00	36.46% (3)

*Note: Fubon (2881) satisfied two antecedents ("$PerExpense \leq L$" & "$Worth\_U \geq M$") of the three antecedents on R2; therefore, its performance score on R2 is 0.66 ($0.33 + 0 + 0.33 = 0.66$).

The actual *ROE* figures of the four companies in 2017 are 11.07% (*Fubon*), 4.69% (*China_Dev*), 7.10% (*Yuanta*), and 6.57% (*SinoPac*), which is fully consistent with the model's ranking: $Fubon \succ Yuanta \succ SinoPac \succ China\_Dev$. This finding suggests the ranking capability of the hybrid multi-graded MRDM model by adopting the binary assessment and the SAW aggregation method.

Although this preliminary test revealed consistent result, which lacks sufficient sensitivity analysis and robust checks. Regarding the sensitivity analysis, several thresholds can be applied to induce different sets of rules to forming different models. Also, in an extended experimental design, the training and the testing can be extended to several combinations to ensure the robustness of the model. The two extended analytics are the limitations of the present work, which is still at the preliminary stage.

4 Concluding Remarks

To conclude, the present study is still at the preliminary stage, which proposes a hybrid multi-graded MRDM model. And a group of financial holding companies was used as an example to illustrate this new model with an affirmative result. The meaning of "hybrid" in this work denotes the collaborations between the expert (or a DM) and the DRSA approximated decision rules.

At the beginning, a DM may define the number of DCs to forming a suitable spectrum for the confronted problem. This step enables a DM(s) to express his/her preferences or emphases toward the disjointed DCs (granules) of the full spectrum of decision space. This design is the pivotal point to bridge a DM's preferences/knowledge to the outcomes induced by the rough machine learning in the next stage.

We presume that a complicated logical reasoning problem, no matter based on the indiscernibility, dominance or similarity relations, is very difficult for human beings to conclude objective and consistent logics (i.e., rules). However, with the supports of DRSA or some other soft computing techniques, the obstacle can be resolved. Therefore, how to devise a reasonable mechanism that may leverage the strength of human beings and machine learning techniques should be a promising field in decision-making.

Still, there are many limitations of the proposed model, and there are several directions to enhance it in the future. Examples are incorporating the FST-based verbal assessment techniques and adopting non-linear aggregators to forming the final score for each alternative. Or, different discretization approaches can be applied using the unsupervised machine learning methods. We intend to exchange opinions and learn from the other researchers' feedbacks by this preliminary work.

Acknowledgement. The authors appreciate the funding supports from the two grants of the Ministry of Technology and Science (MOST) of Taiwan: MOST-105-2410-H-034-019-MY2 and MOST-107-2410-H-034-018-MY2.

Appendix A

See Table 6.

Table 6. Brief description of the symbols used in Table 4.

Symbols	Descriptions
PerExpense	The average cost or expense for manpower
Worth_U	Net return rate-Net income-exc dispo
CASHshare	Cash flow per share
WorthProf	Net after-tax return ratio

(*continued*)

Table 6. (*continued*)

Symbols	Descriptions
ROA	Return on total asset
UsuEPS	Earnings per share in the recent four quarters
Issu_Deposit	Deposit-loan ratio
DivAuth	Interest receivable divided by loan

References

1. Greco, S., Matarazzo, B., Słowiński, R.: Decision rule approach. In: Greco, S., Ehrgott, M., Figueira, J. (eds.) Multiple Criteria Decision Analysis. International Series in Operations Research & Management Science, vol. 233, pp. 497–552. Springer, New York (2016). https://doi.org/10.1007/978-1-4939-3094-4_13
2. Tzeng, G.H., Shen, K.Y.: New Concepts and Trends of Hybrid Multiple Criteria Decision Making. CRC Press (Taylor & Francis Group), New York (2017)
3. Shen, K.Y., Zavadskas, E.K., Tzeng, G.H.: Updated discussions on 'Hybrid multiple criteria decision-making methods: a review of applications for sustainability issues'. Econ. Res.-Ekonomska Istraživanja **31**(1), 1437–1452 (2018)
4. Safavian, S.R., Landgrebe, D.: A survey of decision tree classifier methodology. IEEE Trans. Syst. Man Cybern. **21**(3), 660–674 (1991)
5. Pawlak, Z.: Rough sets. Int. J. Comput. Inform. Sci. **11**(5), 341–356 (1982)
6. Greco, S., Matarazzo, B., Slowinski, R.: Rough approximation of a preference relation by dominance relations. Eur. J. Oper. Res. **117**(1), 63–83 (1999)
7. Greco, S., Matarazzo, B., Slowinski, R.: Rough sets theory for multicriteria decision analysis. Eur. J. Oper. Res. **129**(1), 1–47 (2001)
8. Greco, S., Matarazzo, B., Slowinski, R., Stefanowski, J.: Variable consistency model of dominance-based rough sets approach. In: Ziarko, W., Yao, Y. (eds.) RSCTC 2000. LNCS, vol. 2005, pp. 170–181. Springer, Heidelberg (2001). https://doi.org/10.1007/3-540-45554-X_20
9. Błaszczyński, J., Greco, S., Słowiński, R., Szelg, M.: Monotonic variable consistency rough set approaches. Int. J. Approximate Reasoning **50**(7), 979–999 (2009)
10. Greco, S., Matarazzo, B., Slowinski, R.: Rough sets methodology for sorting problems in presence of multiple attributes and criteria. Eur. J. Oper. Res. **138**(2), 247–259 (2002)
11. Greco, S., Matarazzo, B., Slowinski, R., Tsoukiàs, A.: Exploitation of a rough approximation of the outranking relation in multicriteria choice and ranking. In: Stewart, T.J., van den Honert, R.C. (eds.) Trends in Multicriteria Decision Making. LNE, vol. 465, pp. 45–60. Springer, Heidelberg (1998). https://doi.org/10.1007/978-3-642-45772-2_4
12. Shen, K.Y., Tzeng, G.H.: Contextual improvement planning by fuzzy-rough machine learning: a novel bipolar approach for business analytics. Int. J. Fuzzy Syst. **18**(6), 940–955 (2016)
13. Shen, K.Y., Sakai, H., Tzeng, G.H.: Comparing two novel hybrid MRDM approaches to consumer credit scoring under uncertainty and fuzzy judgments. Int. J. Fuzzy Syst. **21**(1), 194–212 (2019)

14. Albaum, G.: The Likert scale revisited. J. Res. Market Res. Soc. **39**(2), 1–21 (1997)
15. Saaty, T.L.: Decision making—the analytic hierarchy and network processes (AHP/ANP). J. Syst. Sci. Syst. Eng. **13**(1), 1–35 (2004)
16. Dong, Y., Zhang, G., Hong, W.C., Xu, Y.: Consensus models for AHP group decision making under row geometric mean prioritization method. Decis. Support Syst. **49**(3), 281–289 (2010)
17. Aguaron, J., Moreno-Jiménez, J.M.: The geometric consistency index: approximated thresholds. Eur. J. Oper. Res. **147**(1), 137–145 (2003)
18. Grabisch, M.: The application of fuzzy integrals in multicriteria decision making. Eur. J. Oper. Res. **89**(3), 445–456 (1996)
19. TEJ Homepage. http://www.finasia.biz/ensite/. Accessed 15 Oct 2018

3D Face Recognition Based on Hybrid Data

Xinxin Li[1] and Xun Gong[2(✉)]

[1] School of Computer and Software, Jincheng College of Sichuan University, Chengdu 611731, China
[2] School of Information Science and Technology, Southwest Jiaotong University, Chengdu 611756, China
xgong@swjtu.edu.cn

Abstract. Unlike 2D face recognition (FR), the problem of insufficient training data is a major difficulty in 3D face recognition. Traditional Convolutional neural networks (CNNs) can not comprehensively learn all proper filters for FR applications. We embed a handcrafted feature map into our CNN framework—A hybrid data representation is proposed for 3D face. Furthermore, we use a Squeeze-Excitation block to learn the weights of data channels from training face datasets. To overcome the bias of training model based on a small 3D dataset, transfer learning is applied by fine-turning pre-training models, which is trained based on a large 2D face datasets. Tests show that, under challenge conditions such as expression and occlusion, our method outperforms other state-of-the-art methods and can run in real-time.

Keywords: 3D face · Face recognition · Deep learning · Local binary pattern

1 Introduction

The human face is the most important biometric feature due to its accessibility and non-intrusiveness nature. Face recognition (FR) has been an active research topic for many years. FR has a number of applications in a broad fields of surveillance, security, entertainment, etc. Even already applied in many commercial applications, FR is still a challenging problem under many uncontrolled scenarios, the facial appearance and surface of a person can be vary greatly due to changes in pose, illumination, make-up, expression and occlusions.

Extracting effective face feature is essential to facial recognition. Convolutional Neural Networks (CNNs) are effective in feature extraction, especially perform well on images. A big amount of prior research has focused on designing novel architectures of convolutional layers or loss functions, or on strengthening the representational power of CNNs to find a more effective learning mechanism.

Supported by The National Natural Science Foundation of China (61876158), Sichuan Science and Technology Program (2019YFS0432).

© Springer Nature Switzerland AG 2019
T. Mihálydeák et al. (Eds.): IJCRS 2019, LNAI 11499, pp. 454–464, 2019.
https://doi.org/10.1007/978-3-030-22815-6_35

As the first CNNs application in FR task, DeepFace [1] outperformed human in face recognition for the first time in unconstrained scenarios. More recently, trained on million-scale databases (Ms-celeb-1M [2]), Cao et al. [3] achieved the state-of-the-art recognition results on the challenging data sets IJB-A [4] and IJB-B [5].

Nowadays, with the rapid development of 3D acquisition technologies, 3D face recognition (FR) has drawn growing attention due to its potential capability to overcome the inherent disadvantage of its 2D counterpart. However, current 3D face modeling and recognition algorithms are still suffering from one main challenge:

- Limitation of 3D face data. The 3D face datasets are so limited, most of current existing 3D datasets are less than 1000 identities [6]. It's very difficult to overcome pose, illumination and data degeneration problem when training CNNs with limited 3D face data. Some challenges can be seen in Fig. 1, where we can see 2 models, one is a high-resolution 3D face while the other is low-resolution. Hair occlusion and pose variations can impose dramatic negative effects on 3D models.

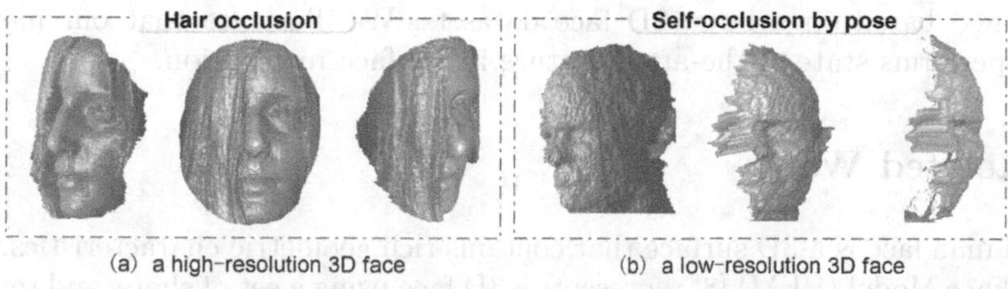

Fig. 1. Illustration of major challenges in building feature representation for 3D faces: missing parts, occlusions and data degeneration.

As lacking large number of 3D faces, CNNs can not comprehensively learn all proper filters for FR tasks. Therefore we propose to add a handcrafted feature map into our CNN framework. As shown in Fig. 2, we have a hybrid data which contains a depth image and a LBP feature image. Furthermore, in order to learn a powerful 3D feature model, we choose to use transfer learning which is trained on a huge 2D dataset previously. And then we fine-tune the model on a 3D dataset.

The contribution of each channel in hybrid data is unknown, in order to allocate an appropriate weights for each channel, we propose to use a "Squeeze and Excitation" (SE) block [7] to learn the importance of each layer during the training stage. SE is used as channels' weights adjusting mechanism for any hybrid data D. These weights are taken to fuse the channels of D to generate the output of the SE block, which is then fed to the subsequent convolutional neural network.

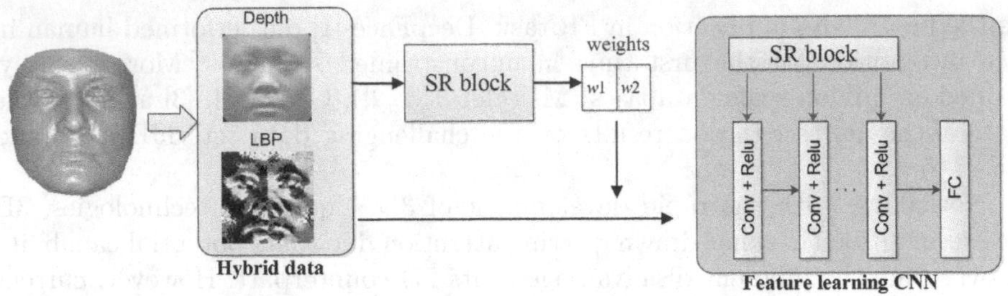

Fig. 2. Flowchart of the proposed hybrid data 3D face feature CNN

Our main contributions can be summarized as follows:

1. A hybrid data for 3D facial shape is proposed. It comprises of 2 types of data that can represent different aspects of a 3D data.
2. In order to modulate channels' contribution by allocate a suitable weight for each layer, SE block is used to learn the importance of each channel at the training stage.
3. Transfer learning is applied by fine-turning a pre-training model, which is trained based on a large 2D face datasets. We illustrate that our method outperforms state-of-the-art literature in 3D face recognition.

2 Related Work

The human face is a 3D surface that contains rich geometric characteristics. The Basel Face Model (BFM) [8] represents a 3D face using a set of shape and texture parameters, under various lighting conditions and various poses. These parameters can be used to identify different people. Iterative Closest Point (ICP) [9] is another commonly used holistic method for 3D face matching. Generally, holistic methods of those type are sensitive to variance in head poses, resolutions and illuminations.

Local feature models human face more robustly against expressions and occlusions [10]. Mehryar et al. [11] use the nasal region to develop an expression-insensitive 3D face representation, which was quite robust in recognition. Lei et al. [12] propose a local facial descriptor. They set the descriptor by calculating four types of geometric features in a keypoint area. Then they use a two-phase classification framework with extracted local descriptors for recognition. The main advantage of local feature-based methods is robustness in the presence of changes in facial expression and occlusion.

Data augmentation is always used for creating a big data set for CNNs methods. Kim et al. [13] propose a 3D face augmentation technique which synthesizes a number of different facial expressions from a single 3D face scan. And they conduct transfer learning from a CNN trained on 2D face images to alleviate the data insufficiency problem. Li et al. [14] proposed an efficient 2D+3D facial expression recognition (FER) deep fusion convolutional neural network (DF-CNN), which fully utilizes the geometric information of the 3D model to improve the recognition rate. To overcome the limited available RGB-D data for deep learning, Lee et al. [15] first train their network using color and grayscale images from a 2D face dataset, and later fine-tune them using depth images. Their result is robust against variations in head rotation and environmental illumination.

3 Hybrid Data and Weights Learning

To make full use of 2D data with sophisticated network architectures, using multiple projected 2D views of a 3D face is a natural and widely adopted strategy in 3D face analysis. For example, Li et al. [14] used six 2D facial attribute maps to represent a textured 3D face scan. All the six maps are fed into a CNN and contribute equally to subsequent feature learning and fusion. However, when all these maps are fed into the network and treated independently, there is no way to learn the mutual information among those different views. Our idea is not only merge different types of information into an integrated representation, but also design a mechanism to explore the mutual information.

3.1 Hybrid Data

Our hybrid facial shape data is built from 2 types of information: (a) depth map, (b) handcrafted feature map, as illustrated in Fig. 2.

Geometric Map. A simple yet effective way to use 3D geometric information is to convert it to a depth image. Extensive experiments have proved that depth image is effective for FR once it has been created properly [16]. Otherwise, the depth image will introduce errors in feature extraction. We map the fitted face plane to the 2D plane, and translate the face so that the nose tip is projected to the plane center. The final image is scaled to a specified resolution, i.e., 0.5 mm per pixel. This depth map keeps the real scale of the human face. In general, the position of nose tip is easy to obtain by finding the vertex with the largest Z value. Note that this method sometimes may fail to find the actual nose tip due to the burrs, which can be easily wiped off by a filter. So it is still the most significant geometrical feature in 3D space that is widely used in 3D FR area.

Handcrafted Feature Map. With the geometric images created in the previous step, we also introduce a handcrafted feature map. Theoretically, similar features may be trainable through deep neural networks directly from images. However, this is the case only when large volume of training data is available. With the limited number of subjects and scans, automatically extracting all effective feature maps are difficult. Therefore we choose adopt LBP [17] as our handcrafted filters, which is proved a high effective feature extractor.

3.2 Weights Learning by Squeeze and Excitation Block

Unlike common strategies adopted in existing literature [13,14,20] which separates different channels of a 3D face into independent images. Our goal is to build an integrated scheme with 2 different channels coupled together (see Fig. 2). To effectively learn/extract the coupling relationship between channels, we introduce an SE block scheme to compute the importance weight for each channel from training data.

Obviously, different channel of our hybrid data encodes distinctive information of a 3D face. They usually play different roles in FR task. Instead of predetermining the importance of each channel by heuristics, we hope to train the network to find weights for face recognition scenarios. However, conventional CNN takes all input channels equally. In other words, given a convolutional operator $\mathbf{F}_{cov}$ on the input data layer, we have

$$\mathbf{Y} = \mathbf{F}_{cov}(\mathbf{D}), \mathbf{D} \in \Re^{H \times W \times C}, \mathbf{Y} \in \Re^{H' \times W' \times C'} \tag{1}$$

let

$$\mathbf{y}_j = \mathbf{f}_j * \mathbf{D} = \sum_{i=1}^{C} f_j^i * \mathbf{d}^i, 1 \leq j \leq C' \tag{2}$$

where $*$ denotes the convolutional operation, $\mathbf{f}_j = \left[f_j^i, f_j^i, ..., f_j^C\right]$ is the convolution kernels. $\mathbf{D} = \left[\mathbf{d}^1, \mathbf{d}^2, ..., \mathbf{d}^C\right]$ is the input hybrid data with C channels. f_j^i is a 2D spatial kernel representing a single channel that acts on the corresponding channel of $\mathbf{D}$. As we can see, just as fully discussed in CNNs related work [21], convolutional kernel $\mathbf{f}_j$ has local effects on the inputs across all channels. As a consequence, the channel relationships modeled by convolution are inherently local. In order to weigh the importance of each input channel, we need to consider the relationship of each channel in a global manner. Inspired by SENet [7], we use *squeeze* and *excitation* steps before data fed into CNNs.

Squeeze. In order to assign weights for channels in a reasonable and global way, like SENet, we use a global average pooling to generate channel-wise statistics. Formally, a statistic $\mathbf{z} \in \Re^C$ is generated by condensing $\mathbf{D}$ throughout its spatial dimensions ($H \times W$), such that the c-th element of $\mathbf{z}$ is calculated by:

$$z_c = \mathbf{F}_{sq}(\mathbf{d}^m) = \frac{1}{H \times W} \sum_{h=1}^{H} \sum_{w=1}^{W} d_c(i, j) \tag{3}$$

Excitation. To capture channel-wise dependencies, after the *squeeze* operation, *excitation* operation is applied, which can flexibly learning a nonlinear interaction between channels. A gating mechanism with sigmoid activation is used:

$$\mathbf{s} = \sigma(\mathbf{W}_2 * \delta(\mathbf{W}_1 * \mathbf{z})) \tag{4}$$

where δ refers to the ReLU function, σ is sigmoid activation. $\mathbf{W}_1$ and $\mathbf{W}_2$ are 1×1 point-wise convolution filters used for dimension reduction and dimension increasing respectively. The final output of the block is obtained by rescaling each channel of hybrid data:

$$\mathbf{S}(\mathbf{d}_m) = s_m \cdot \mathbf{d}_m \tag{5}$$

This is a channel-wise multiplication between the scalar s_m and the data channel $\mathbf{d}_m \in \Re^{H \times W}$. The *excitation* operator assigns each channel of the input with a specific weight, which is learned automatically based on the channel data itself. Undoubtedly, this procedure helps to boost feature discriminability in the feature extraction CNNs.

3.3 Network Architecture

We choose to use a 36-layer Resnet CNN [22], which performs well in 2D face recognition, as our backbone for feature extraction. As shown in Table 1, we can see that SE block is applied after the data layer and at the end of every convolutional unit. A 2D dataset CASIAWebFace [23] is used in pre-trained model. Then we transfer the model to 3D by fine-tuning on our hybrid data of 3D models.

Table 1. CNN architecture. Conv1.x, Conv2.x and Conv3.x denote convolution units that contain multiple convolution layers and residual units. The last column denotes the position of the SE block, where 'Y' means SE is applied after the corresponding layer, 'N' means there is no SE block after the layer.

Layer	20-layer CNN	SE block
Data		Y
Conv1. x	[3 * 3, 64] * 1, S2	N
	[3 * 3, 64] * 2	N
	[3 * 3, 64] * 2	Y
Conv2. x	[3 * 3, 128] * 1, S2	N
	[3 * 3, 128] * 4	N
	[3 * 3, 128] * 4	Y
Conv3. x	[3 * 3, 256] * 1, S2	N
	[3 * 3, 256] * 8	N
	[3 * 3, 256] * 8	Y
Conv4. x	[3 * 3, 512] * 1, S2	N
	[3 * 3, 512] * 2	N
	[3 * 3, 512] * 2	Y
FC1	512	
Slice&Eltwise(Max)	256	

4 Experiments

In this section, we carry out extensive experiments to test the robustness of our methods in 3D FR application. 2 high-resolution 3D face datasets are tested, i.e, Bosphorus [24] and FRGCv2 [25].

All 3D faces are normalized to upright pose as a preprocessing step [16]. Below are the details of datasets and experiments.

- Bosphorus: contains 4652 facial scans belonging to 105 individuals (60 men and 45 women aged between 25 and 35). These scans have been recorded under different poses, expressions and external occlusions, which therefore provide a challenging and large benchmark for the evaluation of 3D face recognition algorithms. These occlusions include: (1) occlusion of the mouth with hand, (2) glasses, (3) occlusion of the face with hair, and (4) occlusion of the left eye and forehead regions by hands.
- FRGCv2: includes 4007 3D facial scans. The validation dataset contains 2410 facial scans with neutral expression, and 1597 facial scans with various facial expressions including disgust, happiness, sadness, surprise, and anger. Based on test protocol [26], we select the first neutral scan of each individual to form the gallery set (466 in total). Then, three experimental subsets are constructed: (1) "neutral vs. neutral" (N-N) experiment (1944 probes); (2) "neutral vs. non-neutral" (N-NN) experiment (1,597 probes); (3) "neutral vs. all" (N-A) experiment (3541 probes).

4.1 FR Evaluation on FRGCv2 Dataset

A accuracy comparison between our approach and the state-of-the-art approaches is given in Table 2. We follow the FRGCv2 protocols of face recognition and include only one scan of each individual (466 in total) in the gallery. Our approach gets an overall Rank-1 IR of 98.88% and a VR of 98.80% (0.1% FAR) in the N-A experiment. In the N-NN experiment, our approach achieves a Rank-1 FRR of 97.23% and a VR of 97.22% (0.1% FAR), respectively. A Rank-1 FRR of 99.80% and a VR of 99.79% (0.1% FAR) are achieved respectively for the easier case of N-N experiment. However, since all of the facial scans in the

Table 2. Comparison with the state-of-the-art on the FRGCv2 dataset. (a) 0.1% FAR VRs for the "neutral vs. neutral", "neutral vs. nonneutral" and "neutral vs. all" experiments and (b) the rank-1 IDs for the "neutral vs. all" experiment. (%)

Approaches	Neutral		Non-neutral		All	
	I-Rate	V-Rate	I-Rate	V-Rate	I-Rate	V-Rate
Gilani et al. [27]	**99.90**	**99.90**	96.90	96.60	98.50	98.70
Li et al. [28]					96.30	
Guo et al. [29]	99.90	97.00		97.18		**99.01**
Mian et al. [30]	99.40	99.90	92.10	96.60	96.10	98.60
Drira et al. [31]	99.20		96.80		97.70	97.10
Mehryar et al. [11]	98.45		98.50		97.90	93.50
Ours	99.80	99.79	**97.23**	**97.22**	**98.88**	98.80

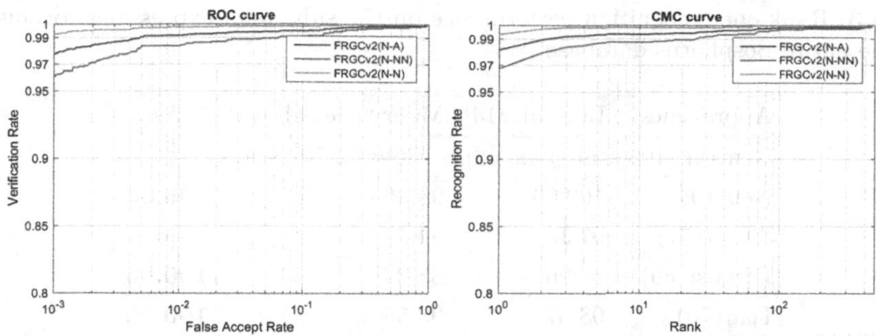

Fig. 3. Face recognition results on the FRGCv2 dataset. CMC results for face identification. ROC results for face verification.

FRGCv2 dataset are in nearly frontal views with very high quality compared to other 3D face datasets (e.g., the Bosphorus), it is not difficult to achieve a good recognition performance by most of the existing 3D FR approaches. Nevertheless, our approach achieves a very competitive performance (a rank-1 FRR of 98.88% and a 0.1% FAR VR of 98.80%) in the N-A experiment. Our results are comparable to the best results reported in the literature (a 0.1% FAR VR of 99.01%). ROC and CMC charts are shown in Fig. 3.

4.2 FR Evaluation on Bosphorus Dataset

For experiments on Bosphorus dataset, we follow the test protocols used in [28], where the first neutral scan of each subject is used to construct the gallery, and the remaining scans or their subsets are used as probes. Our approach gets an overall Rank-1 FRR of 99.64% and VR of 99.03% (0.1% FAR) in the Bos(all) experiment. In the Bos(exp) experiment, our approach achieves a Rank-1 FRR of 99.66% and VR of 99.34% (0.1% FAR), respectively. A Rank-1 FRR of 99.63% and a VR of 96.3% (0.1% FAR) are achieved respectively for the difficult case of Bos(occ) experiment.

Comparative results are given in Table 3. Our proposed technique significantly outperforms the state-of-the-art in pose invariant face recognition. And at the same time it performs well in expression and occlusion cases. Although occlusion is commonly a hard problem for traditional 3D FR methods, most of existing methods cannot achieve good results on the Bos(occ) tests, however, our method can get a 99.68% accuracy in this subset.

Table 3. Rank-one recognition performance on the subsets of expressions, occlusions and the entire Bosphorus database (%)

Approaches	Li et al. [14]	Mehryar et al. [15]	Ours
Neutral (105 scans) vs. expressions (2,797 scans)			
Neu(194)	100.00	98.96	**100.00**
Anger(71)	97.18	94.12	**98.01**
Disgust(69)	86.96	88.24	**100.00**
Fear(70)	98.57	98.55	**100.00**
Happy(106)	98.11	98.08	**100.00**
Sadness(66)	100.00	96.92	**100.00**
Surprise(71)	98.59	100.00	**100.00**
LFAU(1,509)	98.84		**99.80**
CAU(169)	100.00		**100.00**
UFAU(432)	100.00		
All	98.82		**99.86**
Neutral (105 scans) vs. Occlusions (381 scans)			
Eye(105)	100.00		**100.00**
Glass(104)	100.00		**100.00**
Hair(67)	95.52		**96.51**
Mouth(105)	100.00		**100.00**
All(381)	99.21		**99.42**
Neutral (105 scans) vs. all scans (3178 scans)			
All	96.56	95.35	**99.68**

5 Conclusions

Insufficient training data is a major difficulty in learning 3D facial features. We propose to use a hybrid data including a handcrafted feature map as 3D face representation. What's more, we propose to use SE block to learn the importance of channels from training data. SE is a light-weight block that can easily apply to current feature extraction networks without bring any computational complexities. Extensive experiments have conducted on 2 challenging high-resolution 3D face datasets. Tests show that, under challenge conditions such as expression and occlusion, our method outperforms other state-of-the-art methods and can run in real-time.

References

1. Taigman, Y., Yang, M., Ranzato, M., Wolf, L.: DeepFace: closing the gap to human-level performance in face verification. In: IEEE Conference on Computer Vision and Pattern Recognition 2014, pp. 1701–1708 (2014)
2. Guo, Y., Zhang, L., Hu, Y., He, X., Gao, J.: MS-Celeb-1M: a dataset and benchmark for large-scale face recognition. In: Leibe, B., Matas, J., Sebe, N., Welling, M. (eds.) ECCV 2016. LNCS, vol. 9907, pp. 87–102. Springer, Cham (2016). https://doi.org/10.1007/978-3-319-46487-9_6
3. Cao, Q., Shen, L., Xie, W., Parkhi, O.M., Zisserman, A.: VGGFace2: a dataset for recognising faces across pose and age. In: 2017 13th IEEE International Conference on Automatic Face and Gesture Recognition (FG 2018), pp. 67–74 (2017)
4. Klare, B.F., et al.: Pushing the frontiers of unconstrained face detection and recognition: IARPA Janus benchmark A, pp. 1931–1939 (2015)
5. Whitelam, C., et al.: IARPA Janus benchmark-B face dataset, July 2017
6. Faltemier, T.C., Bowyer, K.W., Flynn, P.J.: Using a multi-instance enrollment representation to improve 3D face recognition, pp. 1–6 (2007)
7. Hu, J., Shen, L., Albanie, S., Sun, G., Wu, E.: Squeeze-and-excitation networks. IEEE Conf. Comput. Vis. Pattern Recognit. **2018**, 1–14 (2018)
8. Paysan, P., Knothe, R., Amberg, B., Romdhani, S., Vetter, T.: A 3D face model for pose and illumination invariant face recognition. In: IEEE International Conference on Advanced Video and Signal Based Surveillance. IEEE Computer Society, Washington, DC, USA, pp. 296–301 (2009)
9. Besl, P.J., McKay, N.D.: A method for registration of 3-D shapes. IEEE Trans. Pattern Anal. Mach. Intell. **14**, 239–256 (1992). https://doi.org/10.1109/34.121791
10. Soltanpour, S., Boufama, B., Wu, Q.M.J.: A survey of local feature methods for 3D face recognition. Pattern Recognit. **72**, 391–406 (2017). https://doi.org/10.1016/j.patcog.2017.08.003
11. Emambakhsh, M., Evans, A.: Nasal patches and curves for expression-robust 3D face recognition. IEEE Trans. Pattern Anal. Mach. Intell. **39**(5), 995–1007 (2017). https://doi.org/10.1109/TPAMI.2016.2565473
12. Lei, Y., Guo, Y., Hayat, M., Bennamoun, M., Zhou, X.: A two-phase weighted collaborative representation for 3D partial face recognition with single sample. Pattern Recognit. **52**(C), 218–237 (2016). https://doi.org/10.1016/j.patcog.2015.09.035
13. Kim, D., Hernandez, M., Choi, J., Medioni, G.: Deep 3D face identification. In: 2017 IEEE International Joint Conference on Biometrics (IJCB), pp. 133–142 (2017)
14. Li, H., Sun, J., Xu, Z., Chen, L.: Multimodal 2D+3D facial expression recognition with deep fusion convolutional neural network. IEEE Trans. Multimed. **19**(12), 2816–2831 (2017). https://doi.org/10.1109/TMM.2017.2713408
15. Lee, Y., Chen, J., Tseng, C.W., Lai, S.-H.: Accurate and robust face recognition from RGB-D approach. In: Wilson, R.C., Hancock, E.R., Smith, W.A.P. (eds.) Proceedings of the British Machine Vision Conference (BMVC). BMVA Press, pp. 1–14, September 2016. https://doi.org/10.5244/C.30.123
16. Gong, X., Luo, J., Fu, Z.: Normalization for unconstrained pose-invariant 3D face recognition. In: Sun, Z., Shan, S., Yang, G., Zhou, J., Wang, Y., Yin, Y.L. (eds.) CCBR 2013. LNCS, vol. 8232, pp. 1–8. Springer, Cham (2013). https://doi.org/10.1007/978-3-319-02961-0_1
17. Ahonen, T., Hadid, A., Pietikäinen, M.: Face recognition with local binary patterns. In: Pajdla, T., Matas, J. (eds.) ECCV 2004. LNCS, vol. 3021, pp. 469–481. Springer, Heidelberg (2004). https://doi.org/10.1007/978-3-540-24670-1_36

18. Dalal, N., Triggs, B.: Histograms of oriented gradients for human detection. In: CVPR 2005, pp. 886–893. IEEE Computer Society, Washington, DC, USA (2005)
19. Yang, M., Zhang, L.: Gabor feature based sparse representation for face recognition with gabor occlusion dictionary. In: Daniilidis, K., Maragos, P., Paragios, N. (eds.) ECCV 2010. LNCS, vol. 6316, pp. 448–461. Springer, Heidelberg (2010). https://doi.org/10.1007/978-3-642-15567-3_33
20. Sang, G., Li, J., Zhao, Q.: Pose-invariant face recognition via RGB-D images. Comput. Intell. Neurosci. **2016**, 1–9 (2016). https://doi.org/10.1155/2016/3563758. Identifier: 3563758
21. Krizhevsky, A., Sutskever, I., Hinton, G.E.: ImageNet classification with deep convolutional neural networks. In: Pereira, F., Burges, C.J.C., Bottou, L., Weinberger, K.Q. (eds.) The 25th International Conference on Neural Information Processing Systems. Curran Associates Inc., pp. 1097–1105 (2012)
22. Liu, W., Wen, Y., Yu, Z., Li, M., Raj, B., Song, L.: SphereFace: deep hypersphere embedding for face recognition. In: The IEEE Conference on Computer Vision and Pattern Recognition (2017)
23. Yi, D., Lei, Z., Liao, S., Li, S.Z.: Learning Face Representation from Scratch. eprint arXiv:1411.7923 (2014)
24. Savran, A., et al.: Bosphorus database for 3D face analysis. In: Schouten, B., Juul, N.C., Drygajlo, A., Tistarelli, M. (eds.) BioID 2008. LNCS, vol. 5372, pp. 47–56. Springer, Heidelberg (2008). https://doi.org/10.1007/978-3-540-89991-4_6
25. Phillips, P.J., et al.: Overview of the face recognition grand challenge. In: CVPR 2005. IEEE Computer Society, San Diego, CA, USA, pp. 947–954, 1069015 (2005)
26. Mian, A., Bennamoun, M., Owens, R.: An efficient multimodal 2D–3D hybrid approach to automatic face recognition. IEEE Trans. Pattern Anal. Mach. Intell. **29**(11), 1927–1943 (2007). https://doi.org/10.1109/TPAMI.2007.1105
27. Gilani, S.Z., Mian, A., Eastwood, P.: Deep, dense and accurate 3D face correspondence for generating population specific deformable models. Pattern Recognit. **69**(C), 238–250 (2017). https://doi.org/10.1016/j.patcog.2017.04.013
28. Li, H., Huang, D., Morvan, J.-M., Wang, Y., Chen, L.: Towards 3D face recognition in the real: a registration-free approach using fine-grained matching of 3D keypoint descriptors. Int. J. Comput. Vis. **113**, 1–15 (2014). https://doi.org/10.1007/s11263-014-0785-6
29. Guo, Y., Lei, Y., Liu, L., Wang, Y., Bennamoun, M., Sohel, F.: Ei3D: expression-invariant 3D face recognition based on feature and shape matching. Pattern Recognition Letters **83**, 403–412 (2016). https://doi.org/10.1016/j.patrec.2016.04.003. Efficient Shape Representation, Matching, Ranking, and its Applications
30. Mian, A.S., Bennamoun, M.: Keypoint detection and local feature matching for textured 3D face recognition. Int. J. Comput. Vis. **79**(1), 1–12 (2008). https://doi.org/10.1007/s11263-007-0085-5
31. Drira, H., Amor, B.B., Srivastava, A., Daoudi, M., Slama, R.: 3D face recognition under expressions, occlusions, and pose variations. IEEE Trans. Pattern Anal. Mach. Intell. **35**(9), 2270–2283 (2013). https://doi.org/10.1109/TPAMI.2013.48

Rough Sets and Local Texture Features for Diagnosis of Connective Tissue Disorders

Debamita Kumar(✉) and Pradipta Maji

Biomedical Imaging and Bioinformatics Lab, Machine Intelligence Unit, Indian Statistical Institute, Kolkata, India
{debamita_r,pmaji}@isical.ac.in

Abstract. The standard method for diagnosis of connective tissue disorders is based on the automatic classification of antinuclear autoantibodies by analyzing indirect immunofluorescence images of human epithelial type 2 (HEp-2) cells. In this regard, the paper presents a new method to select relevant texture features for HEp-2 cell staining pattern recognition. The proposed method is developed by judiciously integrating the theory of rough sets and the merits of local texture descriptors. While hypercuboid equivalence partition matrix of rough sets helps to select important texture descriptors for HEp-2 cell classification, the maximum relevance-maximum significance criterion of feature selection facilitates identification of significant and relevant features under important descriptors. Finally, support vector machine with different kernels as well as extreme learning machine are used to recognize one of the known staining patterns present in HEp-2 cell images. The effectiveness of the proposed method, along with a comparison with related approaches, is demonstrated on publicly available MIVIA HEp-2 cell image database.

Keywords: Connective tissue disorders · HEp-2 cell classification · Local texture descriptor · Feature selection · Rough sets

1 Introduction

A group of disorders, which has connective tissues of the body as a target of pathology, is referred to as connective tissue disease (CTD). Connective tissue, being a biological tissue with an extensive extracellular matrix, supports, binds together, and protects organs. It forms a matrix for the body and is composed of two major structural protein molecules - collagen and elastin. In patients with CTD, collagen and elastin become injured by inflammation. Many CTDs feature abnormal immune system activity with inflammation in tissues, as a result of an immune system that is directed against one's own body tissues. The classic CTDs include systemic lupus erythematosus, Sjögren's syndrome, Scleroderma, and rheumatoid arthritis. These types of CTDs are mainly characterized by the presence of antinuclear autoantibodies (ANAs) in the blood of patients.

© Springer Nature Switzerland AG 2019
T. Mihálydeák et al. (Eds.): IJCRS 2019, LNAI 11499, pp. 465–479, 2019.
https://doi.org/10.1007/978-3-030-22815-6_36

The ANAs are a specific class of autoantibodies that have the capability of binding and destroying certain structures within the nucleus of the cells [1]. Indirect immunofluorescence (IIF) is a technique that is becoming increasingly important for the diagnosis of several autoimmune diseases. The correct interpretation of the IIF-ANA results is important and must always be correlated with the patient's symptoms and signs. In case of ANA tests, the most used substrate is the human epithelial type 2 (HEp-2) cells. HEp-2 cells allow the recognition of more than 30 different nuclear and cytoplasmic patterns, which are given by upwards of 100 different autoantibodies [2].

The evaluation of a patient's serum is made by looking for specific fluorescent staining patterns in the HEp-2 cells of IIF images. The manual analysis of IIF images usually consists of the classification of the staining pattern for each slide where the patient serum is dispensed, diluted and incubated to react with the HEp-2 cells. During this task, the doctor has to recognize staining patterns, each corresponding to a different autoimmune disease. This task is really challenging because of the high number of classes that can be recognized. There can be more than 30 different nucleolar, nucleoplasmic and cytoplasmic staining patterns [2]. However, usually the following six ANA patterns are considered: Centromere, Nucleolar, Homogeneous, Fine Speckled, Coarse Speckled and Cytoplasmic. In recent years, there has been a growing interest towards the realization of Computer-Aided Diagnosis (CAD) systems for the analysis of IIF images. Results produced by these techniques can be used to support the scientists' subjective analysis, leading to test results being more reliable and consistent across laboratories [3–5]. Image processing and pattern recognition play an important role for the CAD system development to diagnose CTDs. The primary objective is to identify one of the known staining patterns present in the HEp-2 cell images.

Texture is generally considered to distinguish different HEp-2 patterns, as it carries much information about the surface of the HEp-2 cells. The inherent textures in different HEp-2 cell images are quite different from each other. The HEp-2 cell pattern images have also unpredictably ambiguous texture. This difficulty exists in both inter-class and intra-class examples. An important aspect of texture is also scale. To characterize a cell image, several local texture descriptors, namely, local binary pattern (LBP) [6], rotation-invariant uniform LBP [7], completed LBP [8], co-occurrence of adjacent LBPs [9], rotation-invariant co-occurrence of adjacent LBPs [10], are used in [11], while the concept of gradient-oriented co-occurrence of LBP is introduced in [12]. One of the main problems in HEp-2 cell classification is uncertainty. Some of the sources of this uncertainty include incompleteness and vagueness in HEp-2 cell staining pattern class definition. The theory of rough sets [13] is an effective paradigm to deal with uncertainty, vagueness, and incompleteness. It provides a mathematical framework to capture uncertainties associated with the data. While rough set has been applied successfully for feature selection of discrete valued data [13–15], rough hypercuboid approach [16,17] is found to be suitable for numerical data sets.

In this background, the paper presents a new method for staining pattern recognition of HEp-2 cell IIF images. It helps in the diagnosis of CTDs through automatic identification of ANAs. Integrating judiciously the merits of local texture features and the theory of rough sets, the proposed method classifies staining pattern present in HEp-2 cell images. Given an HEp-2 cell image data set, the proposed method first identifies a set of relevant local texture descriptors for a pair of staining pattern classes, and then selects an important feature set corresponding to each relevant descriptor. The theory of rough hypercuboid approach is used to compute the relevance of a descriptor. It helps to find important features, characterizing the HEp-2 cell images as well as HEp-2 staining pattern classes. The feature set for multiple classes is formed from all the important features selected under several relevant local descriptors for all pairs of classes. The performance of the proposed method is evaluated by using support vector machine with different kernels and extreme learning machine. The effectiveness of the proposed method, along with a comparison with related approaches, is demonstrated on MIVIA HEp-2 cell image databases.

2 Preliminaries

This section presents a brief description of some local texture descriptors, along with the theory of rough hypercuboid.

2.1 Rough Hypercuboid

A hyperrectangle, in geometry, is the generalization of a rectangle for higher dimensions. An m-dimensional hyperrectangle or hypercuboid is defined as the Cartesian product of orthogonal intervals, in the m-dimensional Euclidean space represented by m features of objects or samples [17]. An m-dimensional hypercuboid with m attributes as its dimensions is defined as the Cartesian product of m orthogonal intervals. It encloses a region in the m-dimensional space, where each dimension corresponds to a certain attribute. The value domain of

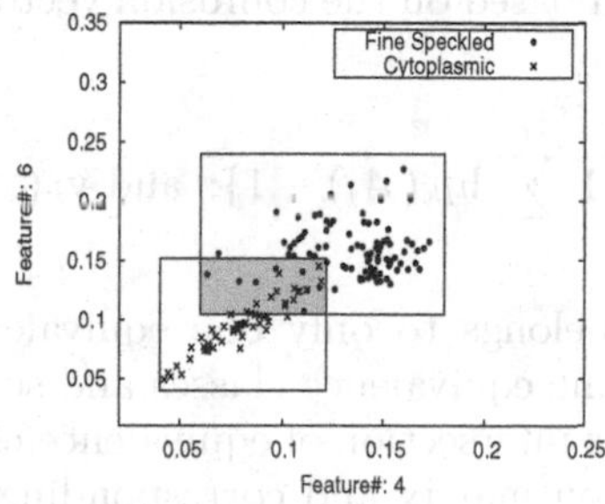

Fig. 1. Examples of rough hypercuboids in two dimension.

each dimension is the value range or interval that corresponds to a particular class. For all hypercuboids, any two objects that belong to a same class hypercuboid are said to be indiscernible with respect to that particular class [16,17]. In real data analysis, uncertainty arises due to overlapping class boundaries, marked by shaded region of Fig. 1 for HEp-2 cell data as example. Hence, every two class hypercuboids may intersect with each other. The intersection of two hypercuboids forms an implicit hypercuboid, which encompasses the misclassified samples or objects those belong to more than one classes. The degree of dependency of the decision attribute set or class label on the condition attribute set depends on the cardinality of the implicit hypercuboids. The degree of dependency increases with the decrease in cardinality [16,17].

Let $\mathbb{U} = \{O_1, \cdots, O_k, \cdots, O_n\}$ be the finite set of n objects or samples, and $\mathbb{C} = \{\mathcal{A}_1, \cdots, \mathcal{A}_j, \cdots, \mathcal{A}_m\}$ and $\mathbb{D}$ are the condition and decision attribute sets in $\mathbb{U}$, respectively. If $\mathbb{U}/\mathbb{D} = \{\beta_1, \cdots, \beta_i, \cdots, \beta_c\}$ denotes c equivalence classes or information granules of $\mathbb{U}$ generated by the equivalence relation induced from the decision attribute set $\mathbb{D}$, then c equivalence classes of $\mathbb{U}$ can also be generated by the equivalence relation induced from each condition attribute $\mathcal{A}_j \in \mathbb{C}$. If $\mathbb{U}/\mathcal{A}_j = \{\delta_1, \cdots, \delta_i, \cdots, \delta_c\}$ denotes c equivalence classes or information granules of $\mathbb{U}$ induced by the condition attribute $\mathcal{A}_j$ and n is the number of objects in $\mathbb{U}$, then c-partitions of $\mathbb{U}$ are the sets of (cn) values $\{\mathrm{h}_{ik}(\mathcal{A}_j)\}$ that can be conveniently arrayed as a $(c \times n)$ matrix $\mathbb{H}(\mathcal{A}_j) = [\mathrm{h}_{ik}(\mathcal{A}_j)]$, termed as hypercuboid equivalence partition matrix of the condition attribute $\mathcal{A}_j$ [16], where $\mathrm{h}_{ik}(\mathcal{A}_j) = 1$ if $O_k(\mathcal{A}_j) \in [\mathrm{L}_i, \mathrm{U}_i]$ and 0 otherwise.

A $c \times n$ hypercuboid equivalence partition matrix $\mathbb{H}(\mathcal{A}_j)$ represents the c-hypercuboid equivalence partitions of the universe generated by an equivalence relation induced by $\mathcal{A}_j$. Each row of the matrix $\mathbb{H}(\mathcal{A}_j)$ is a hypercuboid equivalence partition or class. Here $\mathrm{h}_{ik}(\mathcal{A}_j) \in \{0, 1\}$ represents the membership of object O_k in the ith equivalence partition or class β_i. The interval $[\mathrm{L}_i, \mathrm{U}_i]$ is the value range of condition attribute $\mathcal{A}_j$ with respect to class β_i. The value of each object O_k with class label β_i falls within interval $[\mathrm{L}_i, \mathrm{U}_i]$. The intersection between every two intervals may form the implicit hypercuboids (marked by shaded region of Fig. 1 for HEp-2 cell data as example). Using the concept of hypercuboid equivalence partition matrix, the misclassified objects of implicit hypercuboids are identified based on the confusion vector $\mathbb{V}(\mathcal{A}_j) = [\mathrm{v}_k(\mathcal{A}_j)]$ [16], where

$$\mathrm{v}_k(\mathcal{A}_j) = \min\{1, \sum_{i=1}^{c} \mathrm{h}_{ik}(\mathcal{A}_j) - 1\}; \text{ and } \mathrm{v}_k(\mathcal{A}_j) \in \{0, 1\}. \tag{1}$$

If $\mathrm{v}_k(\mathcal{A}_j) = 0$, then O_k belongs to only one equivalence partition; otherwise O_k belongs to more than one equivalence classes and so falls within the implicit hypercuboid, formed at the intersection of equivalence classes. Hence, the hypercuboid equivalence partition matrix and corresponding confusion vector of the condition attribute $\mathcal{A}_j$ can be used to define the lower and upper approximations of the ith class β_i of the decision attribute set $\mathbb{D}$. Let $\beta_i \subseteq \mathbb{U}$. β_i can be approximated using only the information contained within $\mathcal{A}_j$ by constructing the A-lower and A-upper approximations of β_i [16]:

$$\underline{A}(\beta_i) = \{O_k |\ \mathrm{h}_{ik}(\mathcal{A}_j) = 1 \text{ and } \mathrm{v}_k(\mathcal{A}_j) = 0\}; \tag{2}$$

$$\overline{A}(\beta_i) = \{O_k |\ \mathrm{h}_{ik}(\mathcal{A}_j) = 1\}; \tag{3}$$

where equivalence relation A is induced from attribute $\mathcal{A}_j$. Based on the definitions of lower and upper approximations, the cardinality of positive region of decision attribute set $\mathbb{D}$ can be defined as:

$$|POS_A(\mathbb{D})| = \left| \bigcup_{\beta_i \in \mathbb{U}/\mathbb{D}} \underline{A}(\beta_i) \right| = \sum_{i=1}^{c} \sum_{k=1}^{n} \mathrm{h}_{ik}(\mathcal{A}_j)[1 - \mathrm{v}_k(\mathcal{A}_j)]. \tag{4}$$

Hence, the dependency between condition attribute $\mathcal{A}_j$ and decision attribute $\mathbb{D}$ is defined as follows:

$$\gamma_{\mathcal{A}_j}(\mathbb{D}) = \frac{1}{n} \sum_{i=1}^{c} \sum_{k=1}^{n} \mathrm{h}_{ik}(\mathcal{A}_j)[1 - \mathrm{v}_k(\mathcal{A}_j)] = 1 - \frac{1}{n} \sum_{k=1}^{n} \mathrm{v}_k(\mathcal{A}_j), \tag{5}$$

where $0 \leq \gamma_{\mathcal{A}_j}(\mathbb{D}) \leq 1$. If $\gamma_{\mathcal{A}_j}(\mathbb{D}) = 1$, $\mathbb{D}$ depends totally on $\mathcal{A}_j$, if $0 < \gamma_{\mathcal{A}_j}(\mathbb{D}) < 1$, $\mathbb{D}$ depends partially on $\mathcal{A}_j$, and if $\gamma_{\mathcal{A}_j}(\mathbb{D}) = 0$, then $\mathbb{D}$ does not depend on $\mathcal{A}_j$.

2.2 Local Texture Descriptors

Four local descriptors, namely, local binary pattern (LBP) [6], rotation invariant LBP ($\mathrm{LBP}^{\mathrm{ri}}$) [7], rotation invariant uniform LBP ($\mathrm{LBP}^{\mathrm{riu2}}$) [7], and co-occurrence of adjacent LBPs (CoALBP) [9], which are considered in the current study, are presented next.

Local Binary Pattern (LBP): LBP, proposed by Ojala et al. [18], is an operator that describes texture within a small region around a pixel as a binary pattern. A local neighborhood is thresholded by the gray value of the center pixel and a binary bit pattern is generated, which is then converted to a corresponding decimal number to assign unique label to the local textural element. From a gray scale image, LBP values are computed as follow:

$$\mathrm{LBP}_{N,R} = \sum_{i=0}^{N-1} \mathrm{S}(g_i - g_c)2^i; \ \mathrm{S}(x) = \begin{cases} 1, x \geq 0 \\ 0, x < 0 \end{cases} \tag{6}$$

where N is the number of neighboring pixels considered and R is the radius of the neighborhood. g_c is the center pixel intensity and g_i denotes the intensity of ith neighboring pixel whose coordinates are given by $(-R\sin(2\pi i/N)$, $R\cos(2\pi i/N))$. So, the N bit binary pattern in LBP characterizes the micro-patterns of the image, formed by the intensity variation of a pixel along with it's immediate neighbors. Computation of LBP on an example 3×3 neighborhood is

original

6	5	2
7	6	1
9	3	7

thresholded

1	0	0
1		0
1	0	1

weights

4	2	1
8		128
16	32	64

$\text{LBP}_{8,1} = (01011100)_2 = 92$

$\text{LBP}^{\text{ri}}_{8,1} = (00010111)_2 = 23$

$\text{LBP}^{\text{riu2}}_{8,1} = 8 + 1 = 9$

Fig. 2. Computation of LBP, LBP^{ri} and LBP^{riu2} from a 3×3 neighborhood.

illustrated in Fig. 2, where the 8 bit binary pattern and the corresponding decimal value is presented. Histogram of LBP values can be used for further analysis of the textural properties of the image. Instead of using original intensity values, LBP uses relative intensities of pixels for computation which makes it robust to monotonic gray scale transformation. Again, histogram of the micro-patterns is considered, which implies locations of the patterns are not preserved, and so, LBP is invariant to image translation as well. The distribution of the 2^N LBP values thus characterizes the texture of an image. The LBP operator can be extended for multiscale analysis by simply varying the parameters (N, R).

Rotation Invariant LBP ($\mathbf{LBP^{ri}}$, $\mathbf{LBP^{riu2}}$): The operator $\text{LBP}_{N,R}$ is not rotationally invariant. When the image is rotated, pattern obtained from the neighborhood of a center pixel will also be rotated. Rotation of a pattern will automatically lead to a different $LBP_{N,R}$ value. This is not applicable for patterns consisting of only 0s or 1s, as they are inherently rotation invariant. In order to remove the effect of rotation from the patterns, LBP^{ri} is introduced by Ojala et al. in [7]. Robustness to rotation is achieved by grouping all the binary patterns together that are basically rotated versions of the same pattern. An unique identifier is assigned to each of the rotationally invariant patterns as:

$$\text{LBP}^{\text{ri}}_{N,R} = \min\{ROR(\text{LBP}_{N,R}, i) \mid i = 0, 1, .., N-1\} \tag{7}$$

where $ROR(x, i)$ represents i times circular bit-wise right shift operation on the N bit number x. In the example, presented in Fig. 2, the LBP^{ri} and LBP^{riu2} values are turn out to be 23 and 9, respectively. Histogram of $\text{LBP}^{\text{ri}}_{N,R}$ quantifies the occurrence statistics of each rotation invariant patterns corresponding to certain micro-structures in the image, and hence, can be interpreted as feature vector for the image. Considering only the unique rotation invariant patterns, LBP^{ri} leads to a significant reduction in the feature vector dimensionality. For example, only 36 unique patterns can occur in case of 8 neighborhood operation. The LBP^{ri} losses directional information, which can be crucial for certain applications. However, it has proven to be efficient than LBP for the analysis of homogeneous textures.

Ojala et al. [7] observed that not all local patterns are able to model the characteristics of textures in the images. Certain binary patterns, known as uniform patterns, represent fundamental micro-structures of local image textures. The circular representation of these patterns exhibits limited number of discontinuities. So, they appear uniform in nature and hence, they are termed as 'uniform' patterns. These patterns constitute a major portion of textural patterns present in the images. The most frequently occurring uniform patterns correspond to primitive micro-structures, such as spots, edges, and corners. A uniformity measure $U(pattern)$ is introduced, which corresponds to the number of bitwise transition from 0 to 1 and 1 to 0 in the circular representation of the pattern. For example, 11111111 has uniformity value 0, 11110000 has uniformity value 2, whereas 11001100 has uniformity value 4. Patterns with uniformity value at most 2 are referred to as uniform patterns. The rotation invariant uniform operator is defined as:

$$\text{LBP}_{N,R}^{\text{riu2}} = \begin{cases} \sum_{i=0}^{N-1} \text{S}(g_i - g_c) & \text{if} \quad U(\text{LBP}_{N,R}) \leq 2 \\ N+1 & \text{otherwise;} \end{cases} \tag{8}$$

$$U(\text{LBP}_{N,R}) = \mid \text{S}(g_{N-1} - g_c) - \text{S}(g_0 - g_c) \mid + \sum_{i=1}^{N-1} \mid \text{S}(g_i - g_c) - \text{S}(g_{i-1} - g_c) \mid .$$

According to (8), exactly $N+1$ uniform patterns can occur in a circularly symmetric neighborhood of N pixels. Each of such uniform patterns is assigned an unique quantifier, corresponding to the number of 1s present in the pattern (0 to N). Rest of the non-uniform patterns are accumulated under a miscellaneous label $N+1$, which leads to a significant amount of suppression of noise like patterns, and reduction in the dimensionality of texture descriptor as well.

Co-occurrence of Adjacent LBP (CoALBP): Conventional LBP histogram describes textural properties of an image for a small neighborhood, where each LBP value, corresponding to certain micro-structure, is accumulated into a single bin. Thus, spatial relation information among the LBP values is lost. The concept of co-occurrence is introduced in [9] to take into account the spatial relationship among the LBP values. Co-occurrence among LBP values measures how often a LBP value has co-occurred with another LBP value spatially. So, incorporation of co-occurrence among LBPs extracts global structures from the image along with the micro-structures. CoALBP at $\mathbf{r}$ is defined as:

$$\text{CoALBP}(\mathbf{r}) = (\text{LBP}(\mathbf{r}), \text{LBP}(\mathbf{r} + \Delta\mathbf{r})) \tag{9}$$

where $\Delta\mathbf{r} = (r\cos\theta, r\sin\theta)$ denotes the displacement vector between a pair of LBP values. The magnitude of $\mathbf{r}$ signifies the distance between the LBP pair, whereas θ defines the angle LBP pair makes with the positive horizontal axis. Generally, co-occurrence of LBP pair along horizontal, left diagonal, vertical and right diagonal directions ($\theta = 0, \pi/4, \pi/2, 3\pi/4$) are considered. Since LBP offers 2^N ($= N_P$) number of possible patterns, CoALBP computes four $N_P \times N_P$ auto-correlation matrices of spatial co-occurrence of adjacent LBPs at 4 possible directions. The matrices are vectorized and feature vector of dimension $4N_P^2$ is obtained. Surely, CoALBP produces feature vector of dimensionality significantly greater than conventional LBP, but it complements the descriptive ability of features representing the intricate structures of the image.

3 Proposed Algorithm

The proposed method assumes that a particular descriptor at a given scale may be effective in differentiating a specific pair of staining pattern classes, but may not be able to capture the intrinsic properties of other pairs of pattern classes. So, the proposed method selects important features from the effective local descriptors for each pair of classes, and then forms the final feature set for multiple staining pattern classes.

Let us consider a set of n training HEp-2 cell images, denoted by $\mathbb{U} = \{O_1, \cdots, O_k, \cdots, O_n\}$, where each image $O_k \in \Re^m$. Each O_k is represented by a set of m features $\mathbb{C} = \{\mathcal{A}_1, \cdots, \mathcal{A}_j, \cdots, \mathcal{A}_m\}$. So, a set $L_k = \{L_{k1}, \cdots, L_{kj}, \cdots, L_{km}\}$, consisting of m feature values, represents an image O_k, where $L_{kj} = O_k(\mathcal{A}_j)$ is the value of the j-th feature $\mathcal{A}_j$ of the k-th image O_k. In the current study, four local descriptors, namely, LBP [18], LBP^{ri} [7], LBP^{riu2} [7], and CoALBP [9], are considered. Hence, L_k represents the normalized histogram of the image O_k, obtained from either of the four local descriptors. Let, L_k be sorted in descending order and represented by $\tilde{L}_k = \{\tilde{L}_{k1}, \cdots, \tilde{L}_{kj}, \cdots, \tilde{L}_{km}\}$, and the corresponding feature index of $\tilde{L}_k$ is maintained in $J_k = \{J_{k1}, \cdots, J_{kj}, \cdots, J_{km}\}$. Let us also assume that the images of $\mathbb{U}$ are classified into one of the known c staining pattern classes $\mathbb{U}/\mathbb{D} = \{\beta_1, \cdots, \beta_i, \cdots, \beta_c\}$, where $\mathbb{D}$ represents the set of sample categories.

Primarily, the important characteristics of an image O_k can be represented in terms of the feature values L_{kj}'s of the normalized histogram L_k. The current study is based on the assumption that all the feature values of L_k do not contribute uniformly in describing the properties of the image O_k. Indeed, significant characteristics of O_k can be efficiently demonstrated by only a subset of features of $\mathbb{C}$, which is defined as the important set of features of O_k and denoted by I_k, where $I_k \subseteq \mathbb{C}$. However, different images have their own characteristics, which can be described with different sets of important features.

In order to identify the important features of each sample image O_k, cumulative sum of first q features of corresponding $\tilde{L}_k$ is computed and denoted as $\mathcal{E}(O_k, q)$. Clearly, $\mathcal{E}(O_k, q) \in [0, 1]$ and $\mathcal{E}(O_k, m) = 1, \forall O_k \in \mathbb{U}$. Here, $\mathcal{E}$ is termed as the energy function. It represents the fraction of total energy, contained in the sorted normalized histogram $\tilde{L}_k$, preserved by the first q features of O_k. So, the important information about the sample O_k can be expressed in terms of the energy of O_k obtained from $\tilde{L}_k$. For each sample O_k, the number of important features d_k required to preserve a given fraction of energy $\mathcal{E}_0$ is computed from the sorted histogram $\tilde{L}_k$. The average number of important features $\overline{d}$ is computed from the d_ks corresponding to the entire set of samples $\mathbb{U}$, while the set of important features I_k, corresponding to the sample O_k, is defined as

$$I_k = \{\mathcal{A}_j \mid J_{kq} = j \text{ and } q \leq \overline{d}\}. \tag{10}$$

Hence, the set $I_k \subseteq \mathbb{C}$ contains only the important features of O_k, which can sufficiently represent the significant characteristics of the image O_k.

Now, it is expected that the samples belonging to the same class will have similar sets of important features, while the samples belonging to different classes will have different sets of important features. So, the probability of occurrence $Pr(\mathcal{A}_j|\beta_i)$ of a feature $\mathcal{A}_j$ in the important sets of samples of a particular class β_i is computed. In order to discard the noisy features, the feature set $\mathcal{C}(\beta_i)$ corresponding to the class β_i is formed, based on the value of a threshold ϵ, as follows:

$$\mathcal{C}(\beta_i) = \{\mathcal{A}_j \mid Pr(\mathcal{A}_j|\beta_i) \geq \epsilon\}. \tag{11}$$

A feature $\mathcal{A}_j$ will only be present in the set $\mathcal{C}(\beta_i)$ if it is important in most of the samples of β_i as well as bears significant amount of information regarding the characteristics of class β_i.

Let, $\mathcal{C}(\{\beta_i, \beta_r\})$ denote the feature set corresponding to the pair of classes $\{\beta_i, \beta_r\}$. It contains the features that represent significant characteristics, common to the particular pair of classes β_i and β_r, and is defined as follows:

$$\mathcal{C}(\{\beta_i, \beta_r\}) = \{\mathcal{A}_j \mid Pr(\mathcal{A}_j|\beta_i) \geq \epsilon \text{ and } Pr(\mathcal{A}_j|\beta_r) \geq \epsilon\}. \tag{12}$$

In the proposed method, a modality refers to a particular local texture descriptor considered under a specific scale. Considering a set of t number of modalities $\mathcal{M} = \{\mathcal{M}_1, \cdots, \mathcal{M}_p, \cdots, \mathcal{M}_t\}$, the proposed method computes the relevance $\Gamma_p(\{\beta_i, \beta_r\})$ of the feature set $\mathcal{C}_p(\{\beta_i, \beta_r\})$, corresponding to the pair of classes $\{\beta_i, \beta_r\}$ under the p-th modality $\mathcal{M}_p$. The concept of hypercuboid equivalence partition matrix of rough hypercuboid approach [16] is used to compute the relevance. The relevance $\Gamma_p(\{\beta_i, \beta_r\})$ of the feature set $\mathcal{C}_p(\{\beta_i, \beta_r\})$ represents the relevance of the p-th modality $\mathcal{M}_p$, with respect to the class-pair $\{\beta_i, \beta_r\}$, and is computed based on (5) as follows:

$$\Gamma_p(\{\beta_i, \beta_r\}) = 1 - \frac{1}{n_{ir}} \sum_{k=1}^{n_{ir}} v_k(\mathcal{C}_p(\{\beta_i, \beta_r\})) \tag{13}$$

where n_{ir} is the number of samples belonging to class-pair $\{\beta_i, \beta_r\}$ and

$$\mathbb{V}(\mathcal{C}_p) = [v_1(\mathcal{C}_p), \cdots, v_k(\mathcal{C}_p), \cdots, v_n(\mathcal{C}_p)] \tag{14}$$

is termed as the confusion vector for the feature set $\mathcal{C}_p(\{\beta_i, \beta_r\})$, which can be computed from the corresponding hypercuboid equivalence partition matrix $\mathbb{H}(\mathcal{C}_p)$. So, the relevance $\Gamma_p(\{\beta_i, \beta_r\}) \in [0, 1]$. If $\Gamma = 1$, $\{\beta_i, \beta_r\}$ depends totally on $\mathcal{C}_p$, if $0 < \Gamma < 1$, $\{\beta_i, \beta_r\}$ depends partially on $\mathcal{C}_p$, and if $\Gamma = 0$, then $\{\beta_i, \beta_r\}$ does not depend on $\mathcal{C}_p$.

After computing the relevance of the feature set $\mathcal{C}_p(\{\beta_i, \beta_r\})$ corresponding to t modalities for each pair of classes $\{\beta_i, \beta_r\}$, $\tilde{t}$ most relevant feature sets $\{\mathcal{C}_p(\{\beta_i, \beta_r\})\}$ are chosen, and the set $\tilde{\mathcal{C}}_{ir} = \{\mathcal{C}_p(\{\beta_i, \beta_r\})\}$ is formed for the pair of classes β_i and β_r. The final feature set is obtained as the union of the feature set $\tilde{\mathcal{C}}_{ir}$ over all possible pairs of classes. The set $\mathcal{C}_p(\{\beta_i, \beta_r\})$ corresponding to the class-pair $\{\beta_i, \beta_r\}$, obtained in the proposed method, may contain redundant information that has no significant contribution in differentiating the samples of class β_i from that of β_r. So, during the computation of relevance $\Gamma_p(\{\beta_i, \beta_r\})$ of the feature set $\mathcal{C}_p(\{\beta_i, \beta_r\})$, irrelevant and insignificant features are removed from the set, based on maximum relevance-maximum significance criterion of feature selection reported in [15].

4 Performance Analysis

The performance of the proposed method is studied extensively and the corresponding results are reported in this section, along with a comparison with related methods.

4.1 Algorithms Compared and Data Set Used

In order to validate the proficiency of the proposed method in recognizing the HEp-2 cell staining pattern classes, the performance of the proposed method is extensively compared with various local texture descriptors, namely, local binary pattern (LBP) [18], rotation invariant LBP (LBP^{ri}) [7], rotation invariant uniform LBP (LBP^{riu2}) [7] and co-occurrence among adjacent LBPs (CoALBP) [9], computed at different scales, such as scale 1 ($\mathcal{S}_1$), scale 2 ($\mathcal{S}_2$), scale 3 ($\mathcal{S}_3$), scale 4 ($\mathcal{S}_4$), concatenation of $\mathcal{S}_1$, $\mathcal{S}_2$ and $\mathcal{S}_3$ ($\mathcal{S}_{123}$), concatenation of $\mathcal{S}_1$, $\mathcal{S}_2$ and $\mathcal{S}_4$ ($\mathcal{S}_{124}$). The descriptors, used in the current study, are chosen at arbitrary and hence, the proposed descriptor selection method is equally compatible with any other sets of descriptors. For CoALBP, 4 neighboring pixels are considered to capture the micro-pattern around a center pixel, while 8 neighboring pixels are considered for the rest of the descriptors. The performance of the proposed method is also studied with reference to the existing multimodal data integration methods. Four different classifiers, namely, support vector machine (SVM) [19] with polynomial (SVM_P), radial basis function (SVM_R) and linear (SVM_L) kernels, and extreme learning machine (ELM) [20] are used to evaluate the performance of different approaches.

In this section, a brief description of the data set, which is used for validation of the proposed method, is provided. This data set is the ICPR 2012 HEp-2 cell classification contest data set, termed as MIVIA image database [21]. The data set contains 1455 cells from 28 images among which four images belong to cytoplasmic, fine speckled and nucleolar staining patterns each, five images belong to coarse speckled and homogeneous each, and six centromere images are present. The images have 24 bits color depth with uncompressed 1388×1038 pixels resolution. The data set contains 721 and 734 Hep-2 cell images in the training set and test set, respectively. The robustness of the proposed method as well as existing approaches is studied through evaluation on this data set. Table 1 indicates the number of training and testing cells with respect to different HEp-2 patterns of this data set, which are used to validate the performance of the proposed algorithm as well as existing methods.

4.2 Optimum Values of Parameters

In the proposed method, energy $\mathcal{E}$ of an image represents the important information specific to that image, whereas threshold ϵ is defined to identify the features that significantly contribute in characterizing a class of HEp-2 cell images. The feature set $\mathcal{C}(\beta_i)$ corresponding to the class β_i is formed based on the values of $\mathcal{E}$ and ϵ. So, the performance of the proposed method depends on the values of

Table 1. Description of data set used

Data sets	Number of cells					
	Centromere	Homogen.	Nucleolar	Coarse Speckl.	Fine Speckl.	Cytoplasmic
Training	208	150	102	109	94	58
Test	149	180	139	101	114	51

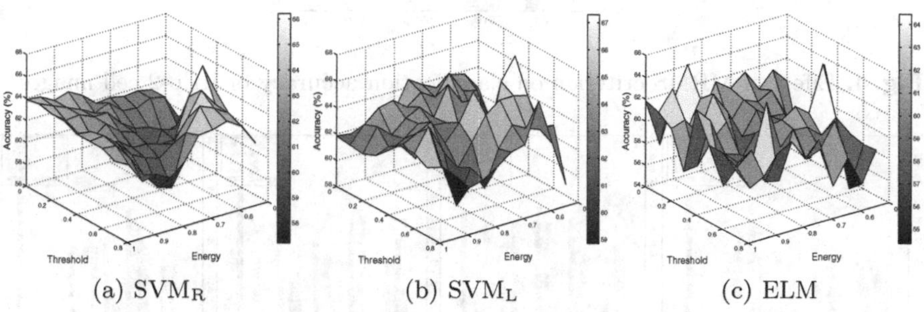

(a) SVM_R (b) SVM_L (c) ELM

Fig. 3. Variation of classification accuracy with respect to energy $\mathcal{E}$ and threshold ϵ.

$\mathcal{E}$ and ϵ to a great extent. In order to obtain the optimum values of $\mathcal{E}$ and ϵ, extensive experiment is carried out on MIVIA data set by varying the value of $\mathcal{E}$ from 0.50 to 1.00 with an interval of 0.05 and ϵ is varied from 0.00 to 0.70 with an interval of 0.10. Figure 3(a-c) depicts the variation of classification accuracy, obtained using three classifiers - SVM_R, SVM_L, and ELM, on HEp-2 cell images of test data set, with respect to $\mathcal{E}$ and ϵ. Similar results are found for SVM_P also. From the results reported in Fig. 3, it can be observed that the proposed method achieves significantly better accuracy at $\mathcal{E} = 0.55$ and $\epsilon = 0.40$, irrespective of classifiers used. So, the optimum values of $\mathcal{E}$ and ϵ are chosen to be 0.55 and 0.40 in the current study for the MIVIA data set. It ensures compact representation of the feature set $\mathcal{C}(\beta_i)$, corresponding to the class β_i, by considering only the features which preserve 55% of total energy by at least 40% samples of β_i.

4.3 Importance of Max Relevance-Max Significance Criterion

In the proposed method, maximum relevance-maximum significance (MRMS) criterion [15] is employed to discard the insignificant and irrelevant features from the feature set, corresponding to a pair of classes. Using the MRMS criterion, a significantly reduced feature set can be obtained, while the relevance value of the set remains unaltered. To establish the effectiveness of MRMS criterion in the proposed method, Fig. 4 compares the classification accuracy obtained using MRMS criterion with the accuracy without applying MRMS. It can be noticed from Fig. 4 that discarding irrelevant and insignificant features not only reduces the cardinality of the final feature set from 546 to 277, but also increases the classification accuracy obtained using different classifiers.

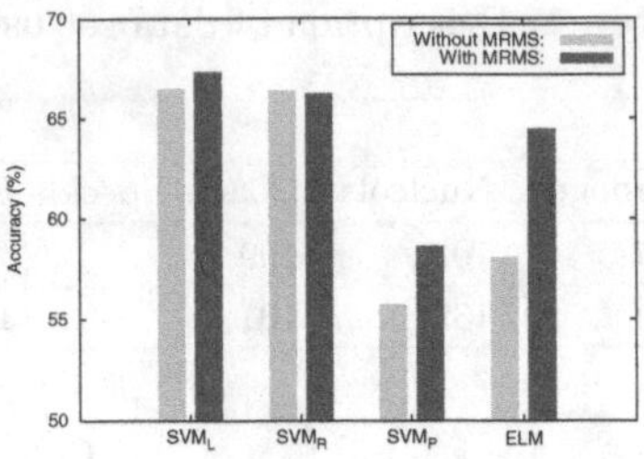

Fig. 4. Effect of MRMS criterion on classification accuracy of HEp-2 cell images.

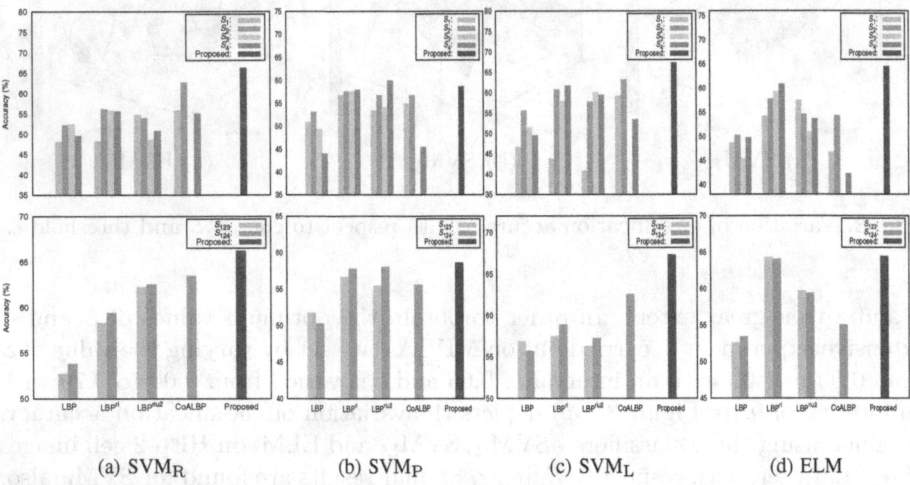

(a) SVM_R (b) SVM_P (c) SVM_L (d) ELM

Fig. 5. Comparative performance analysis of proposed method with different local texture descriptors at various scales (top: single scale; bottom: multiple scales).

4.4 Comparison with Existing Approaches

A particular descriptor, computed at a specific scale, is generally used to classify different staining pattern classes of HEp-2 cell images. However, the proposed method first identifies relevant modalities for each pair of classes, and then final feature set is formed from the significant features of selected modalities. To validate the importance of class-pair specific modalities over uniform modalities, the proposed method is compared with the four local descriptors LBP, LBP^{ri}, LBP^{riu2} and CoALBP, each computed at different single scales and concatenated scales as well. Figure 5(a-d) exhibits the comparative performance analysis between proposed method and existing approaches on the HEp-2 cell images of MIVIA data set for different classifiers. Top row of Fig. 5 presents the performance of existing approaches at single scales, while the bottom row describes the same for concatenated scales. From the results presented in Fig. 5, it can be seen that the proposed method attains better classification accuracy, irrespective of descriptors, scales and classifiers, except for LBP^{riu2} at $\mathcal{S}_4$ for SVM_P.

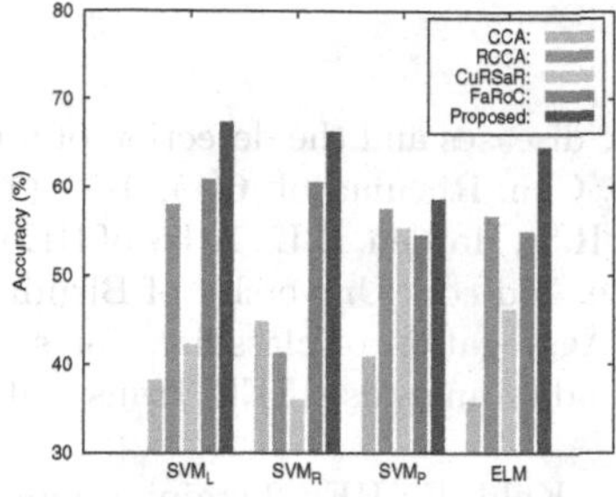

Fig. 6. Classification accuracy of proposed and different data integration methods.

Finally, the performance of the proposed method is studied with reference to different statistical data integration methods, such as canonical correlation analysis (CCA) [22], regularized CCA (RCCA) [23], CuRSaR [24], and FaRoC [25]. In these methods, information from different modalities is combined to categorize the HEp-2 cell staining patterns into one of the known classes. Figure 6 reports the performance of proposed method along with that of four existing data integration methods, on HEp-2 cell images of MIVIA database for four different classifiers. It is evident from Fig. 6 that the proposed method achieves highest classification accuracy with respect to all the data integration methods for each of the classifiers.

5 Conclusion

The paper presents a new method to select relevant textural features of important modalities for diagnosis of connective tissue disorders. The theory of rough sets and the merits of local textural descriptors have been integrated judiciously to develop the proposed method. While hypercuboid equivalence partition matrix helps to select class-pair specific important modalities, the maximum relevance-maximum significance criterion of feature selection facilitates identification of significant and relevant features under important modalities. Finally, support vector machine and extreme learning machine are used to recognize one of the known staining patterns present in IIF images. The effectiveness of the proposed method, along with a comparison with related approaches, has been demonstrated on publicly available MIVIA HEp-2 cell image database.

Acknowledgment. This publication is an outcome of the R&D work undertaken in the project under the Visvesvaraya PhD Scheme of Ministry of Electronics and Information Technology, Government of India, being implemented by Digital India Corporation. The authors would like to thank Ankita Mandal of Indian Statistical Institute, Kolkata for her valuable experimental support.

References

1. Walravens, M.: Systemic diseases and the detection of nuclear and anticytoplasmic antibodies. A Hist. Rev. Clin. Rheumatol. **6**(1), 9–17 (1987)
2. Bradwell, A.R., Hushes, R.S., Harden, E.L.: Atlas of HEp-2 Patterns: & Laboratory Techniques. Binding Site. 2nd edn. University of Birmingham (2003)
3. Soda, P., Iannello, G.: Aggregation of classifiers for staining pattern recognition in antinuclear autoantibodies analysis. IEEE Trans. Inf. Technol. Biomed. **13**(3), 322–329 (2009)
4. Strandmark, P., Ulen, J., Kahl, F.: HEp-2 staining pattern classification. In: Proceedings of the 21st International Conference on Pattern Recognition, pp. 33–36 (2012)
5. Wiliem, A., Wong, Y., Sanderson, C., Hobson, P., Chen, S., Lovell, B.C.: Classification of human epithelial type 2 cell indirect immunofluoresence images via codebook based descriptors. In: Proceedings of the IEEE Workshop on Applications of Computer Vision, pp. 95–102 (2013)
6. Ojala, T., Pietikainen, M., Harwood, D.: Performance evaluation of texture measures with classification based on Kullback discrimination of distributions. In: Proceedings of the 12th IAPR International Conference on Pattern Recognition, Conference A: Computer Vision & Image Processing, pp. 582–585 (1994)
7. Ojala, T., Pietikainen, M., Maenpaa, T.: Multiresolution gray-scale and rotation invariant texture classification with local binary patterns. IEEE Trans. Pattern Anal. Mach. Intell. **24**(7), 971–987 (2002)
8. Guo, Z., Zhang, L., Zhang, D.: A completed modeling of local binary pattern operator for texture classification. IEEE Trans. Image Process. **19**(6), 1657–1663 (2010)
9. Nosaka, R., Ohkawa, Y., Fukui, K.: Feature extraction based on co-occurrence of adjacent local binary patterns. In: Ho, Y.-S. (ed.) PSIVT 2011. LNCS, vol. 7088, pp. 82–91. Springer, Heidelberg (2011). https://doi.org/10.1007/978-3-642-25346-1_8
10. Nosaka, R., Fukui, K.: HEp-2 cell classification using rotation invariant co-occurrence among local binary patterns. Pattern Recognit. **47**(7), 2428–2436 (2014)
11. Cataldo, S.D., Bottino, A., Islam, I., Vieira, T.F., Ficarra, E.: Subclass discriminant analysis of morphological and textural features for HEp-2 staining pattern classification. Pattern Recognit. **47**(7), 2389–2399 (2014)
12. Theodorakopoulos, I., Kastaniotis, D., Economou, G., Fotopoulos, S.: HEp-2 cells classification via sparse representation of textural features fused into dissimilarity space. Pattern Recognit. **47**(7), 2367–2378 (2014)
13. Pawlak, Z.: Rough Sets: Theoretical Aspects of Reasoning about Data. Kluwer Academic Publishers, Dordrecht (1991). ISBN 0792314727
14. Maji, P., Pal, S.K.: Rough-Fuzzy Pattern Recognition: Applications in Bioinformatics and Medical Imaging. Wiley/IEEE Computer Society Press, Hoboken/New Jersey (2012)
15. Maji, P., Paul, S.: Rough set based maximum relevance-maximum significance criterion and gene selection from microarray data. Int. J. Approx. Reason. **52**(3), 408–426 (2011)
16. Maji, P.: A rough hypercuboid approach for feature selection in approximation spaces. IEEE Trans. Knowl. Data Eng. **26**(1), 16–29 (2014)
17. Wei, J.M., Wang, S.Q., Yuan, X.J.: Ensemble rough hypercuboid approach for classifying cancers. IEEE Trans. Knowl. Data Eng. **22**(3), 381–391 (2010)

18. Ojala, T., Pietikäinen, M., Harwood, D.: A comparative study of texture measures with classification based on feature distributions. Pattern Recognit. **29**(1), 51–59 (1996)
19. Vapnik, V.: The Nature of Statistical Learning Theory. Springer, New York (1995). https://doi.org/10.1007/978-1-4757-2440-0
20. Huang, G.B., Zhu, Q.Y., Siew, C.K.: Extreme learning machine: theory and applications. Neurocomputing **70**(1), 489–501 (2006)
21. Foggia, P., Percannella, G., Soda, P., Vento, M.: Benchmarking HEp-2 cells classification methods. IEEE Trans. Med. Imaging **32**(10), 1878–1889 (2013)
22. Hotelling, H.: Relations between two sets of variates. Biometrika **28**(3/4), 321–377 (1936)
23. Vinod, H.D.: Canonical ridge and econometrics of joint production. J. Econ. **4**(2), 147–166 (1976)
24. Maji, P., Mandal, A.: Multimodal omics data integration using max relevance-max significance criterion. IEEE Trans. Biomed. Eng. **64**(8), 1841–1851 (2017)
25. Mandal, A., Maji, P.: FaRoC: fast and robust supervised canonical correlation analysis for multimodal omics data. IEEE Trans. Cybern. **48**(4), 1229–1241 (2018)

Developing Artwork Pricing Models for Online Art Sales Using Text Analytics

Laurel Powell[1(✉)], Anna Gelich[1,2], and Zbigniew W. Ras[1,3]

[1] College of Computing and Informatics, University of North Carolina at Charlotte, Charlotte, NC 28223, USA
{lpowel28,agelich,ras}@uncc.edu
[2] New Media Arts Department, Polish-Japanese Academy of Information Technology, 02-008 Warsaw, Poland
[3] Computer Science Department, Polish-Japanese Academy of Information Technology, 02-008 Warsaw, Poland

Abstract. This work explores utilizing a combination of features, built with text analytics, and other features to predict prices of works of art. Basic metrics, such as the length of the text descriptions and the presence of the artist's social media links are considered as attributes for predicting the price of art. This work also utilizes the Paragraph2Vec algorithm combined with clustering as a method of classifying artworks for price.

Keywords: Data analytics · Art market · Social media

1 Introduction

How is the ideal price of a work of art determined? This question is relevant for art collectors, art investors, artists, and art dealers. The art market is highly uncertain, and has proven to be a challenge to explain using traditional economic theories [10]. The value of a work of art comes from its aesthetic quality and the reputation of the artist that created it rather than simply the cost of the materials or the number of hours it took to create [10]. In art auctions, aspects such as the format and the relationship between bidders can all influence the results [14]. In today's primary art market, many artists sell their works themselves on Internet platforms such as SaatchiArt or Artfinder [6,7]. How should an artist on one of these platforms without a previous sales record decide how to price their work?

Online sales are a critical part of today's art market and are growing in importance. According to [5], online driven sales make up 29% of total gallery sales and those sales are valued at 4.22 billion USD. However, some collectors prefer being able to examine works in person and object to losing the communal aspects of buying art [16]. Another obstacle to the growth of the online market is buyer confidence. In the 2018 Hiscox trade report on the art market it is stated that:

© Springer Nature Switzerland AG 2019
T. Mihálydeák et al. (Eds.): IJCRS 2019, LNAI 11499, pp. 480–494, 2019.
https://doi.org/10.1007/978-3-030-22815-6_37

> Although existing collectors are used to secrecy and non-transparency when it comes to pricing, this is an aspect which clearly doesn't sit comfortably with new buyers. In this year's survey, 90% of new buyers and 92% of small spenders said that price transparency was a key consideration when buying art online [5].

Buyers making online purchases from markets such as Saatchi or Artfinder have limited information about the work. Without doing external research on the artist on other sites, their information is limited to what the artist, or artist representative, posted on the site. Usually, this is restricted to one or more photographs of the work, a description of the work and its characteristics, and biographical information about the artist. An artist's reputation is vital for establishing the quality of that artist's work and reducing buyer uncertainty [10]. However, when this information is limited or the buyer is not an expert on judging an artist's achievements, how can they feel certain in a price?

This work proposes a method of predicting the prices of works of art using artist provided information combined with text analytics. This work will discuss using methods such as the word count and presence or absence of provided social media links on the prices of the given works. Next, this work proposes using the Paragraph2Vec algorithm [11,21], which is sometimes referred to as Doc2Vec or the Distributed Memory Model of Paragraph Vectors (PV-DM), to create vector representations of the provided text which can then be clustered. These methods are tested on a dataset collected from an art sales site and the results will be discussed in this paper.

2 Recommender Systems

In this section, we will provide a short overview of existing recommendation approaches and some examples of their application domains [17,25].

Collaborative filtering is a method of making automatic predictions (filtering) about the interests of a user by collecting preferences or taste information from many users (collaborating). The underlying assumption of the collaborative filtering approach is that if a person A has the same opinion as a person B on an issue, A is more likely to have B's opinion on a different issue than that of a randomly chosen person [20].

Content-based filtering compares the content of already consumed items with new items that can potentially be recommended to the user, i.e., to find items that are similar to those already rated positively by the user. This approach was proposed by Pazzani and Billsus [31].

Knowledge-based recommender systems have been introduced by Felfernig [17]. They are based on explicit knowledge, rules or constraints about the item assortment, user preferences, and recommendation criteria (i.e., which item should be recommended in which context). For instance, system presented in [24] belongs to that group. It is based on the knowledge extracted from HCUP datasets.

Group recommender systems, presented in [13], are based on the idea that recommendations are not determined for a single user but the whole group should be satisfied with the given recommendation (e.g., a family's decision regarding a smart home solution). Recommendations in this context are often determined on the basis of group decision heuristics. For example, least misery is a heuristic that prefers recommendations with the property that the misery of all group members is minimized. In contrast, most pleasure tries to maximize the pleasure of individual group members.

Hybrid recommendation is based on the idea of combining basic recommendation approaches in such a way that one helps to compensate the weaknesses of the other. For example, when combining content-based filtering with collaborative recommendation, content-based recommendation helps to recommend unrated items. If a user has already consumed some items (e.g., purchased them), the content description of a new item can be compared with the descriptions of items already purchased by the user. System CLIRS (presented in [31,33]) belongs to that group.

Recommender systems are utilized in a variety of areas including healthcare [24], business [33,36], tourism [40], social life [28], healthy food [37], and the art market [30].

Existing recommender systems in the art market domain do not use data analytics but human experts to evaluate fine art pieces and make recommendations. For instance, MutualArt (https://www.mutualart.com/artappraisal) has the world's most comprehensive database of past sale results but the number of features describing these sales is quite small. Its advisors are assigning price tags to new pieces of art by comparing them with similar pieces in the MutualArt database. The charge for a single service (one piece of art) is $49 and the waiting time to get a recommendation is 72 h.

FINDARTINFO (http://www.findartinfo.com/english.html) is a similar but free art appraisal service which contains information about 438,003 artists and 3,775,762 art prices. With this art appraisal tool, an artist can value his/her fine art by comparing it with recent auction prices of similar pieces. There are also websites providing free art appraisal hints. For instance, wikiHow (https://www.wikihow.com/Value-Your-Art) helps to value artworks. There are professional art appraisers available, but they charge $300+ for a single service and the waiting time to get price recommendations is still relatively long.

Our knowledge-based recommender system, called ArtIST, will be based on big data analytics. Knowing the artist's name, appraisal of the piece of art will be done by its personalized module built from the data describing similar artists and similar art pieces including information about their sales. To evaluate an art piece using ArtIST, the user needs to submit the same information about it as is required by existing art appraisal tools/websites. It may appear that we will be in a competition with art consulting/appraising companies and professional art appraisers, but our recommendations will be more reliable, more accurate, less expensive, and delivered in real time.

3 Methods

3.1 Dataset

This work exclusively models prices for the primary art market. Artworks are sold online by their original creators or by a representative working for them. These creators may or may not have gallery representation, but the art used in this work is posted on a third party site. Predicting prices in the secondary market would involve adding a number of other features, such as the provenance of the artwork, its previous owners, and its past sales value, which would go considerably beyond the scope of this work.

In this work, all artworks were extracted from Artfinder.com [6] using a webscraper. To ensure consistence across records, only one sales site was used. Artfinder was an ideal choice for this analysis because of its diverse collection of artists and artworks. Many countries, artistic styles, mediums and artistic subjects are represented on their site. The dataset collected includes artists from over 60 countries and their works represent a variety of styles and mediums. These mediums include photography, printmaking, drawing, collage, and sculpture in addition to painting and mixed media. These works were scraped from the Artfinder website using a combination of Python, Apache Selenium [2] to handle the Javascript, and Beautiful Soup [1]. A dataset of approximately 160,000 works representing over 2000 artists were collected for use in this work. On Artfinder, a page for a single artwork provides a number of features for use in classification. All pages include at least one photograph of the work, as well as information such as the medium used, the size of the work, and if the work is signed. A number of the features collected were provided by the artists themselves and are used for tagging in the site's search engine. For example, features such as the artist's country and the medium used are searchable. Other tags, such as the subject of the work or it style can also be used to refine the search process. The initial list of features used takes cues from the features developed in [30]. Another important piece of information provided by the artists is the artwork's text description which will be discussed in detail in another section.

Artists also have the option to provide information about themselves. This information is listed in "About" pages, which have optional sections for the artist's biography, education, past and future events, awards received, their current country and links to other locations where they can be found online. In addition to this information, a number of artists have visible reviews from past customers. However, these are extremely limited. The only visible reviews have come from previous customers and unless the reviewers directly stated it in their comment, it is not always obvious what work they are complimenting or criticizing. On Artfinder, artists are rated on a 5-star scale and comments may or may not be provided. However, a significant number of artists do not have any reviews visible on their profiles. In the dataset collected, approximately 50% of the artists have one or more reviews. Additionally, reviews are something which an artist will accumulate over a span of time on a site. This would make a model that relies excessively on comments far less useful for artists that have spent less time on Artfinder.

Prices in the dataset range from a maximum of 1,000,000 USD to 12.97 USD. However, significantly more lower priced works appear in the dataset. 85.66% of the works are 1000 USD or less. Only 0.26% of the artworks are valued at greater than 10,000 USD. According to [18], the majority of online art sales are valued at 1,500 USD or less. The exact price distribution is represented in Fig. 1. This figure omits works greater than 10,000 USD for reasons of scale. For the purposes of analysis, the price dimension was reduced to discrete intervals. These intervals are as follows: (0–105), (105–205), (205–405), (405–605), (605–810), (810–1030), (1030–1445), (1445–1825), (1825–2455), (2455–3855), (3855–5000), (5000–10,000), (>10000). All prices are in USD, however, a number were converted from their original currency. The intervals were selected by searching for areas where very few works were priced and splitting the price feature in these gaps. However, this led to an excessive number of very small intervals. When examining the price distribution in detail, a significant number of works cluster around the 50 and 100 values which creates many gaps. Therefore, sets of intervals were merged to form the discretization described above.

The features listed below were used as the base set of attributes. All attributes are discrete. Their effects, alone and paired with other features, is discussed in detail in the Sect. 4.

- artistID - A unique identifier for an artist
- artistCountry - The artist's current country of residence as listed on their profile
- percent_five_stars - Percentage of five star reviews out of the total reviews.
- percent_four_stars - Percentage of four star reviews out of the total reviews.
- percent_three_stars - Percentage of three star reviews out of the total reviews.
- percent_two_stars - Percentage of two star reviews out of the total reviews.
- percent_one_star - Percentage of one star reviews out of the total reviews.
- medium - Artist provided medium of the artwork.
- style - Artist provided style of the artwork.
- subject - Artist provided subject of the artwork.
- authentication - Artist provided method of authenticating the work.
- artwork_width - The width of the artwork.
- artwork_height - The height of the artwork.

3.2 Product Descriptions

On Artfinder, as with many other sales sites, artists write descriptions of their products. Product descriptions can be used to capture customer interest, but they are more factual in nature than advertising [38]. In the Shotfarm Product Information Report, they found that:

> Ninety-five percent of those surveyed say product information is important when making a purchase decision, with nearly four in five indicating that it is very important [4].

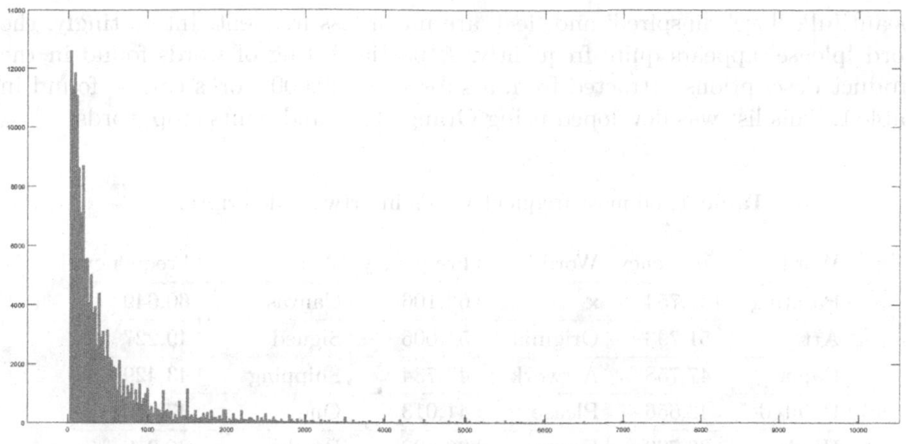

Fig. 1. Price distribution

However, there is little certainty about what is considered a complete or informative description.

For example, the ideal length of a description and its impact on sales is an open question. In [34], the researchers determined that between 40 to 55 words should be focused on product description for eBay sellers. They also found that the use of words denoting uncertainty, such as "probably" or jargon harmed sales [34]. However, that length for descriptions is far from universally agreed upon. In [29], retailers are advised to keep product descriptions between 350 and 400 words.

In [34], a number of other features were determined to have an influence on buyer behavior and the price that a piece sold for in an online auction. Rawlins et al. determined that readability is crucial when encouraging buyers, and also that attributes such as the length of the description and the seller's use of slang also had a bearing on the resulting price [34]. The impact of the length of the description on the price is explored in detail later in this work.

Using text analytics as a method of predicting consumer behavior has received attention from the research community. In [19], the direction of movement of the price of stocks was predicted using text analytics, with a particular focus on modeling sentiment. Another work [23], used text analytics of news articles to predict shifts in price in oil. In [3], the use of word clouds and counting the frequency of words is used as method of determining the sentiment of speeches given in the financial sector to predict future market behavior.

Product descriptions for the artworks in this dataset are largely factual in nature, with many being concerned primarily with listing attributes such as the medium, size, subject and other similar features. The descriptions are heavily dominated by words describing basic facts. Factual terms such as 'canvas', 'x' (which is used to indicate the dimensions of the work), 'original', 'signed', and 'shipping' are heavily emphasized. More emotionally evocative words, such as 'beautiful', 'love', 'inspired' and 'feel' are much less frequent. Interestingly, the word 'please' appears quite frequently. A partial listing of words found in the product descriptions extracted from a subset of 150,000 works can be found in Table 1. This list was developed using Orange [15], and omits stop words.

Table 1. 60 most frequent words in artwork descriptions

Word	Frequency	Word	Frequency	Word	Frequency
Painting	131,754	x	62,106	Canvas	60,649
Art	51,793	Original	51,006	Signed	49,227
Paper	47,758	Artwork	47,734	Shipping	43,429
Painted	42,656	Please	41,013	One	37,370
Work	36,729	Frame	36,102	Ready	36,018
Acrylic	35,046	Oil	34,225	Hang	32,638
Print	31,577	Certificate	31,007	Artist	30,693
Paintings	29,333	cm	28,998	Authenticity	28,674
Back	28,550	Size	27,259	Quality	25,583
Abstract	23,422	Piece	23,252	Colors	21,963
Paint	21,786	Made	20,372	Using	20,242
Series	19,963	Front	19,666	White	18,917
Also	18,811	5	18,758	May	18,395
1	18,112	Edition	18,097	Different	17,875
Like	17,644	Time	17,588	Created	17,392
Contact	17,366	Image	17,281	Note	16,592
Free	16,531	Days	16,319	3	16,156
Materials	15,875	Hands	15,154	Stretched	14,532
Shipped	14,484	Printed	14,457	Life	14,400
Works	14,079	High	13,617	Artworks	13,608

3.3 Social Media

A number of artists cultivate a social media presence to improve their sales. In the Hiscox online art trade review, they found that Instagram is the preferred social media platform for the art market [5]. Instagram is popular because of its tight focus on visuals, so artist's work can be appreciated without sharing space with other types of content [8]. A number of artists also have their own websites to display their work.

On the Artfinder platform, the artist's "About" pages have a section for social media links. In the collected dataset of approximately 2,000 artists, 85% have Facebook accounts listed. 42% have Twitter accounts provided on their profiles and 54% have Instagram links.

The section on the artist's about pages for social media links were checked for the presence or absence of the words, upper and lower case, "Facebook", "Twitter" and "Instagram". This was used as a Boolean attribute in the classifier.

3.4 Clustering Text Using Doc2Vec

In this work, text describing the artists and their works were converted into vectors and those vectors were clustered. The algorithm used for the construction of these features was Doc2Vec, a Gensim implementation of Paragraph Vector [21].

Paragraph Vector, and its parent Word2Vec, have received considerable attention from the research community. Word2Vec combined with decision trees was used in [39]. In [22], Doc2Vec was used to classify product descriptions into categories. This methodology has also been applied to other tasks such as finding item similarities [9], and in determining the similarity of pieces of text [12].

Word2Vec represents words as vectors [11,27]. It makes use of the Skip-Gram model which was created in [26]. This model uses a neural network to predict a word given other words in a sequence of words [11,26,27]. "The training objective of the Skip-gram model is to find word representations that are useful for predicting the surrounding words in a sentence or a document." [27] This model was then used in [27] to create Word2Vec. It can be used to find words that are used in similar contexts and is more efficient than other methods developed at the time [27]. Another interesting feature is that vectors can be combined mathematically with simple vector addition to find words near the sum of two terms [27].

Paragraph Vector, which sometimes referred to as the "Distributed Memory Model of Paragraph Vectors (PV-DM)", is an adaptation of the original Word2Vec [11,21]. Paragraph Vector expands the model so that vector representations can be created for sentences or lengthy documents [21]. Each paragraph is represented as a unique vector in a matrix which is concatenated with the vectors for each word in that matrix [21]. The identifier for the paragraph "remembers" the subject of the document but otherwise functions as another word [21].

In this work, to create the vectors, the Gensim [35] implementation of Paragraph Vector, which is termed Doc2Vec, was trained on text provided by the artist. The vectors were calculated for the text of the artist's biographies, the text of the artist provided description of the artwork, the title of the artwork, the artist provided description of their education, the artist provided description of awards received, and the artist provided description of events they have held. The biography, awards section, education section and events section were found on the artist's about page. Gensim includes pre-built packages for basic text preprocessing which were used on the text before it was used in the creation of the model. Then, the resulting model was given the original text for each piece of training text. This created a set of 100 term vectors. They were then clustered using K-Means, implemented with Python's Sci-Kit Learn Library [32].

4 Results

The following experimental results were obtained on a randomly selected subset of 150,000 works using Orange [15], and tested using 10 fold cross validation.

4.1 Base Results

Table 2 gives the results of testing the base set of features with k-Nearest Neighbors, Support Vector Machines and Random Forest classifiers. These results were obtained on a subset of 150,000 artworks. These attributes were all listed in Sect. 3.1. The k value used was 5 and 100 trees were used in the Random Forest classifier. As can be readily seen here, Random Forest had the best results, and so it was used for testing the extensions of the base set of features.

Table 2. Results with base features

Method	AUC	CA	F1	Precision	Recall
kNN	0.884	0.625	0.621	0.619	0.625
SVM	0.608	0.169	0.17	0.221	0.169
Random Forest	0.938	0.667	0.663	0.662	0.667

4.2 Word Counts

The number of words used in the biography and in the description of artworks do have an impact on the price of a work of art. The results found when testing this impact can be found in Table 3. As can be seen from the table, the word count of the description has a more notable impact on the accuracy of the classifier than the word count of either the title or the artist's biography. However, the word count of the biography combined with the word count of the description has a slight increase in the accuracy of the classifier. Adding the word count of the title to these two features does not have a notable impact on the accuracy of the classifier.

4.3 Social Media Presence

As can be seen in Table 4, the presence or absence of social media does not have a significant impact on the accuracy of the classifier.

4.4 Document Vector Based Clusters

In Tables 5 and 6 the results of testing the quality of the classifier extracted from the dataset with base features enlarged by the document vector based clusters are shown. In Table 5, each feature is placed in one of 10 clusters. Extending the feature set with the awards cluster, biography cluster, education cluster, or events cluster in some combination has the highest positive impact.

Table 3. Results with word counts

	AUC	CA	F1	Precision	Recall
Base Features	0.938	0.667	0.664	0.663	0.667
Base Features and Biography Word Count (BWC)	0.939	0.67	0.667	0.665	0.67
Base Features and Description Word Count (DWC)	0.94	0.674	0.67	0.669	0.674
Base Features and Title Word Count (TWC)	0.938	0.668	0.664	0.663	0.668
Base Features, BWC, and DWC	0.941	0.677	0.673	0.672	0.676
Base Features, BWC, and TWC	0.939	0.671	0.668	0.666	0.671
Base Features, DWC, and TWC	0.94	0.674	0.67	0.669	0.674
Base Features, BWC, DWC and TWC	0.941	0.677	0.673	0.672	0.677

Table 4. Results with social media

	AUC	CA	F1	Precision	Recall
Base Features	0.938	0.667	0.664	0.663	0.667
Base Features and Facebook (FB)	0.938	0.667	0.663	0.662	0.667
Base Features and Twitter (TWT)	0.938	0.668	0.665	0.664	0.668
Base Features and Instagram (INST)	0.938	0.669	0.665	0.664	0.669
Base Features, FB and TWT	0.939	0.669	0.666	0.665	0.669
Base Features, FB and INST	0.939	0.669	0.665	0.664	0.669
Base Features, TWT and INST	0.939	0.669	0.666	0.665	0.669
Base Features, FB, TWT and INST	0.9339	0.67	0.667	0.665	0.67

Table 5. Results with 10 clusters

	AUC	CA	F1	Precision	Recall
Base Features (BF)	0.938	0.667	0.664	0.663	0.667
BF and Awards Cluster (Cl)	0.938	0.669	0.665	0.664	0.669
BF and Biography (Bio) Cl	0.939	0.671	0.668	0.667	0.671
BF and Description (Desc) Cl	0.936	0.662	0.658	0.657	0.662
BF and Education (Edu) Cl	0.939	0.67	0.667	0.666	0.67
BF and Events Cl	0.939	0.669	0.666	0.665	0.669
BF and Title Cl	0.934	0.656	0.652	0.651	0.656
Base Features, Awards, Bio, Desc, Edu, Events, Title Clusters	0.939	0.667	0.663	0.662	0.667

Table 6. Results with 25 clusters

	AUC	CA	F1	Precision	Recall
Base Features (BF)	0.938	0.667	0.664	0.663	0.667
BF and Awards Cluster (Cl)	0.939	0.669	0.666	0.665	0.669
BF and Biography (Bio) Cl	0.939	0.671	0.668	0.667	0.671
BF and Description (Desc) Cl	0.936	0.664	0.66	0.659	0.664
BF and Education (Edu) Cl	0.939	0.6	0.667	0.665	0.67
BF and Events Cl	0.939	0.67	0.667	0.665	0.67
BF and Title Cl	0.935	0.659	0.655	0.654	0.659
Base Features, Awards, Bio, Desc, Edu, Events, Title Clusters	0.941	0.672	0.668	0.667	0.672

4.5 Combined Features

In Table 7, the results of extending the dataset with different set of features combined are shown. The social media features combined with the word count features have a positive impact on the accuracy of the classifier. The text cluster features when combined with the social media and word count features have a noticeable positive impact on the accuracy of the classifier. In keeping with the results when testing the cluster features alone, the largest gain appeared when the awards, biography, education or events clusters were used.

5 Discussion

How artists present their work to the public can be used as a predictor of the price of that work. The length of the artist's descriptions and biography, both tied to how a potential customer will perceive their work, is a predictive feature. Interestingly, whether an artist links to social media in the profile has no apparent impact on the price of their work when it is not combined with other features. However, their language and word choice can be used to classify artworks by price.

All artworks analyzed in this work were selected from a single source. A relevant topic to be explored in later research is the impact of the sales platform on the price. This could function as a method for artists to improve their profits by placing their works at more opportune sales locations. Do customers have different expectations about how an artist presents their work depending on the sales platform? Do auction based sites have different results? This could be used to help artists better tailor their messages to their platforms audiences.

Why the number of words used in the description of an artwork or the biography of an artist impacts the price is unclear. The length of the artist's biography may be a measure of that artist's experience and accomplishments. An excessively short description of a product may not give a potential customer sufficient information to decide to make a purchase. As it does have an impact, this may

Table 7. Results with features combined

	AUC	CA	F1	Precision	Recall
Base Features (BF)	0.938	0.667	0.664	0.663	0.667
BF, FB, TWT, INST, BWC, DWC, TWC	0.942	0.681	0.677	0.676	0.681
BF, FB, TWT, INST, BWC, DWC, TWC, Awards Cl (10 Clusters)	0.943	0.682	0.678	0.677	0.682
BF, FB, TWT, INST, BWC, DWC, TWC, Bio Cl (10 Clusters)	0.943	0.683	0.679	0.678	0.683
BF, FB, TWT, INST, BWC, DWC, TWC, Desc Cl (10 Clusters)	0.941	0.677	0.673	0.672	0.677
BF, FB, TWT, INST, BWC, DWC, TWC, Edu Cl (10 Clusters)	0.943	0.682	0.679	0.678	0.682
BF, FB, TWT, INST, BWC, DWC, TWC, Events Cl (10 Clusters)	0.943	0.682	0.678	0.677	0.682
BF, FB, TWT, INST, BWC, DWC, TWC, Title Cl (10 Clusters)	0.941	0.674	0.67	0.669	0.674
BF, FB, TWT, INST, BWC, DWC, TWC, Awards, Bio, Desc, Edu, Events, Title Clusters (10 Clusters)	0.943	0.678	0.674	0.674	0.678
BF, FB, TWT, INST, BWC, DWC, TWC, Awards Cl (25 Clusters)	0.943	0.683	0.679	0.678	0.683
BF, FB, TWT, INST, BWC, DWC, TWC, Bio Cl (25 Clusters)	0.943	0.683	0.6679	0.670	0.683
BF, FB, TWT, INST, BWC, DWC, TWC, Desc Cl (25 Clusters)	0.941	0.677	0.672	0.672	0.677
BF, FB, TWT, INST, BWC, DWC, TWC, Edu Cl (25 Clusters)	0.942	0.683	0.68	0.679	0.683
BF, FB, TWT, INST, BWC, DWC, TWC, Events Cl (25 Clusters)	0.942	0.683	0.679	0.678	0.683
BF, FB, TWT, INST, BWC, DWC, TWC, Title Cl (25 Clusters)	0.941	0.674	0.67	0.669	0.674
BF, FB, TWT, INST, BWC, DWC, TWC, Awards, Bio, Desc, Edu, Events, Title Clusters (25 Clusters)	0.944	0.68	0.676	0.675	0.68

prove useful when making online sales or talking about a work on social media. Determining the ideal length, and if that changes depending on the medium, platform, or subject is a topic for further study.

The artist's social media presence had little to no impact on the accuracy of the classifier. It could be argued that the use of social media is a measure of the artist skill in marketing or their level of experience as an artist. Therefore, these features may be too strongly associated with other features, such as the artist's identifier, to be of any use. It is also possible that the presence of a social media page is not a relevant feature, but their popularity, measured through follower counts or a similar metric would be.

Another potential topic for later research is determining the optimum number of clusters for the text features. In this work, 10 clusters and 25 clusters were used. The number of clusters did not significantly impact the accuracy of the classifier. However, the potential of this method is still largely unexplored. The highest positive impact on the accuracy of the classifier comes from the awards cluster, biography cluster, education cluster, or events cluster in some combination. Each of these features comes from the artist's about page, so there are only approximately 2,000 points being clustered. In contrast, the features with the lowest positive impact are from the clusters made from the artwork description or artwork title. These are unique to each artwork, so over 160,000 points were clustered. A greater number of clusters for these features or a more guided approach may improve the results.

Text analytics can be used as a price predictor in artworks. This work explored the possibility of using vector representation of posted information about artworks and artists to form clusters that can be used as predictors for the prices of artworks in a commercial setting.

Acknowledgement. This research is supported by the National Science Foundation under grant IIP 1749105. Any opinions, findings, and conclusions or recommendations expressed in this material are those of the authors and do not necessarily reflect the views of the National Science Foundation.

References

1. Beautiful Soup. https://www.crummy.com/software/BeautifulSoup/
2. Selenium. https://www.seleniumhq.org/
3. New directions in sentiment analysis: charting words. In: Sentiment Indicators, pp. 227–250. Wiley, October 2015. https://doi.org/10.1002/9781119204398.ch12
4. 2015/2016 The Shotfarm Product Information Report. Technical report (2016)
5. The Hiscox Online Art Trade Report 2018. Technical report, ArtTactic (2018). https://arttactic.com/product/hiscox-online-art-trade-report-2018/
6. Artfinder.com (2019). https://www.artfinder.com/
7. Saatchiart.com (2019). https://www.saatchiart.com/
8. Bamberger, A.: How Artists Use Instagram to Present and Sell Their Art. https://www.artbusiness.com/artists-how-to-use-post-sell-art-on-instagram.html
9. Barkan, O., Koenigstein, N.: Item2Vec: neural item embedding for collaborative filtering (2016). arXiv:1603.04259v3
10. Beckert, J., Rössel, J.: The price of art: uncertainty and reputation in the art field. Eur. Soc. **15**(2), 178–195 (2013)
11. Beysolow II, T.: Topic modeling and word embeddings. In: Applied Natural Language Processing with Python: Implementing Machine Learning and Deep Learning Algorithms for Natural Language Processing, pp. 77–119. Apress, Berkeley (2018)
12. Dai, A.M., Olah, C., Le, Q.V.: Document embedding with paragraph vectors (2015). arXiv:1507.07998v1
13. Dara, S., Chowdary, C.R., Kumar, C.: A survey on group recommender systems. J. Intell. Inf. Syst. (2019). https://doi.org/10.1007/s10844-018-0542-3

14. Dass, M., Reddy, S.K., Iacobucci, D.: A network bidder behavior model in online auctions: a case of fine art auctions. J. Retail. **90**(4), 445–462 (2014)
15. Demšar, J., et al.: Orange: data mining toolbox in Python. J. Mach. Learn. Res. **14**, 2349–2353 (2013)
16. Evans, D.: The current and future influence of online art sales on the art market. Ph.D. thesis (2015)
17. Felfernig, A., et al.: An overview of recommender systems in the Internet of Things. J. Intell. Inf. Syst. **52**(2), 285–309 (2019)
18. Fischer, M.S.: Online Art Sales Gathers Steam Among Buyers. ThinkAdvisor, April 2015
19. de Fortuny, E.J., Smedt, T.D., Martens, D., Daelemans, W.: Evaluating and understanding text-based stock price prediction models. Inf. Process. Manag. **50**(2), 426–441 (2014)
20. Guo, L., Liang, J., Zhu, Y., Luo, Y., Sun, L., Zheng, X.: Collaborative filtering recommendation based on trust and emotion. J. Intell. Inf. Syst. (2018). https://doi.org/10.1007/s10844-018-0517-4
21. Le, Q.V., Mikolov, T.: Distributed representations of sentences and documents (2014). arXiv:1405.4053v2
22. Lee, H., Yoon, Y.: Engineering doc2vec for automatic classification of product descriptions on O2O applications. Electron. Commer. Res. **18**(3), 433–456 (2018)
23. Li, J., Xu, Z., Yu, L., Tang, L.: Forecasting oil price trends with sentiment of online news articles. Procedia Comput. Sci. **91**, 1081–1087 (2016)
24. Mardini, M.T., Raś, Z.W.: Extraction of actionable knowledge to reduce hospital readmissions through patients personalization. Inf. Sci. **485**, 1–17 (2019)
25. Mendoza, M., Torres, N.: Evaluating content novelty in recommender systems. J. Intell. Inf. Syst. (2019). https://doi.org/10.1007/s10844-019-00548-x
26. Mikolov, T., Chen, K., Corrado, G., Dean, J.: Efficient estimation of word representations in vector space (2013). arXiv:1301.3781v3
27. Mikolov, T., Sutskever, I., Chen, K., Corrado, G., Dean, J.: Distributed representations of words and phrases and their compositionality (2013). arXiv:1310.4546v1
28. Nobahari, V., Jalali, M., Seyyed Mahdavi, S.J.: ISoTrustSeq: a social recommender system based on implicit interest, trust and sequential behaviors of users using matrix factorization. J. Intell. Inf. Syst. **52**(2), 239–268 (2019)
29. Parish, S.: Product Description Word Counts: Why Length Matters. https://content26.com/blog/product-description-word-counts-length-matters-2/
30. Pawlowski, C., Gelich, A., Raś, Z.W.: Can we build recommender system for artwork evaluation? In: Bembenik, R., Skonieczny, Ł., Protaziuk, G., Kryszkiewicz, M., Rybinski, H. (eds.) Intelligent Methods and Big Data in Industrial Applications. SBD, vol. 40, pp. 41–52. Springer, Cham (2019). https://doi.org/10.1007/978-3-319-77604-0_4
31. Pazzani, M., Billsus, D.: Learning and revising user profiles: the identification of interesting web sites. Mach. Learn. **27**(3), 313–331 (1997)
32. Pedregosa, F., et al.: Scikit-learn: machine learning in Python. J. Mach. Learn. Res. **12**, 2825–2830 (2011)
33. Ras, Z.W., Tarnowska, K.A., Kuang, J., Daniel, L., Fowler, D.: User friendly NPS-based recommender system for driving business revenue. In: Polkowski, L., et al. (eds.) IJCRS 2017. LNCS (LNAI), vol. 10313, pp. 34–48. Springer, Cham (2017). https://doi.org/10.1007/978-3-319-60837-2_4
34. Rawlins, C., Johnson, P.: Selling on eBay: persuasive communication advice based on analysis of auction item descriptions. J. Strat. E-Commer. **5**(1&2), 75–81 (2007)

35. Řehůřek, R., Sojka, P.: Software framework for topic modelling with large corpora. In: Proceedings of the LREC 2010 Workshop on New Challenges for NLP Frameworks, pp. 45–50. ELRA, Valletta, Malta, May 2010
36. Tarnowska, K., Ras, Z.W., Daniel, L.: Recommender System for Improving Customer Loyalty. SBD, vol. 55. Springer, Cham (2020). https://doi.org/10.1007/978-3-030-13438-9
37. Trang Tran, T.N., Atas, M., Felfernig, A., Stettinger, M.: An overview of recommender systems in the healthy food domain. J. Intell. Inf. Syst. **50**(3), 501–526 (2018)
38. Tseng, M.Y.: Describing creative products in an intercultural context: toward a pragmatic and empirical account. J. Pragmat. **80**, 52–69 (2015)
39. Zharmagambetov, A.S., Pak, A.A.: Sentiment analysis of a document using deep learning approach and decision trees. In: 2015 Twelve International Conference on Electronics Computer and Computation (ICECCO). IEEE, September 2015. https://doi.org/10.1109/icecco.2015.7416902
40. Zheng, X., Luo, Y., Sun, L., Zhang, J., Chen, F.: A tourism destination recommender system using users' sentiment and temporal dynamics. J. Intell. Inf. Syst. **51**(3), 557–578 (2018)

Hardware Implementation on Field Programmable Gate Array of Two-Stage Algorithm for Rough Set Reduct Generation

Tomasz Grzes(✉) and Maciej Kopczynski

Faculty of Computer Science, Bialystok University of Technology, Bialystok, Poland
{t.grzes,m.kopczynski}@pb.edu.pl
http://www.wi.pb.edu.pl

Abstract. The rough sets theory developed by Prof. Z. Pawlak is one of the tools used in intelligent systems for data analysis and processing. In modern systems, the amount of the collected data is increasing quickly, so the computation speed becomes the critical factor. One of the solutions to this problem is data reduction. Removing the redundancy in the rough sets can be achieved with the reduct. Most of the algorithms of generating the reduct are only software implementations, therefore having many limitations coming from using the fixed word length, as well as consuming time for fetching and processing of the instruction and data. These limitations make the software-based implementations relatively slow. Unlike a software, the hardware systems can process the data faster than software. In this paper, the hardware implementation of the two-stage greedy algorithm to find the one reduct is presented. The first stage of the algorithm is calculating the core using the discernibility matrix, and the second is enriching the core with the attributes that are necessary to build the reduct. The presented algorithm was implemented in Field Programmable Gate Array (FPGA) as a digital device consisting of blocks that process the data in a single step. For the research purpose, the algorithm was also implemented in C language and run on a PC. The times of execution of the reduct calculation in hardware and software were considered. Obtained results show an increase in the speed of data processing.

Keywords: Data reduction · Digital systems design · Field Programmable Gate Array (FPGA) · Reduct · Rough set

1 Introduction

The rough sets theory developed in the eighties of the twentieth century by Prof. Z. Pawlak is one of the tools that can be used in the intelligent systems for data analysis and processing. Banking, medicine, image recognition and security are among the possible fields of utilization. In all these fields the amount of

© Springer Nature Switzerland AG 2019
T. Mihálydeák et al. (Eds.): IJCRS 2019, LNAI 11499, pp. 495–506, 2019.
https://doi.org/10.1007/978-3-030-22815-6_38

the collected data is increasing quickly, but with the increase of the data, the computation speed become the critical factor.

Data reduction is one of the solutions to this problem. Removing the redundancy in the rough sets can be achieved with the reduct, which is the subset of the decision attributes that provides the discernibility of the objects. For the given decision table there can be more than one reduct, but for the reduction purposes one reduct is sufficient.

A lot of algorithms of generating the reduct were developed, but most of them are only software implementations, therefore have many limitations. Microprocessor uses the fixed word length, consumes a lot of time for either fetching as well as processing of the instruction and data, consequently the software based implementations are relatively slow. Hardware systems don't have these limitations and can process the data faster than a software, what was shown in previous authors' papers. Connecting all hardware implementations into single system will allow to build a solution, that is capable of processing large collections of data in significantly shorter time comparing to pure software system. Of course, there is also a disadvantage of hardware approach: created system is not flexible in terms of processing datasets with different structure. Each time structure changes, system has to modified to different data characteristic. But as long, as data have the same structure, what is the case for most practical use cases, hardware system will perform much better than it's software version.

Some hardware implementations of the specific rough set methods exists at the moment. Moreover only few of them were implemented in a hardware devices, a bulk of the ideas were only described theoretically and left unimplemented in the real devices. The idea of a sample processor generating the decision rules from the decision tables was described in [5]. In [3] the design, simulation, implementation and experiment of the rough set processor was described. Authors of paper [7] have presented the design for generating a reduct from the binary discernibility matrix.

Foregoing authors' research results focused on a subject of the hardware implementations of the rough sets methods can be found in the previously published papers. Simple solution for the hardware supported reduct calculation was described in [1] and a core generation using the FPGA based solution was presented in [2]. In [6] the core computation algorithm was optimized for using with the large datasets. Solution described in this paper is using the discernibility matrix for the two-staged reduct generation. First stage of this algorithm is a calculating the core, and the second is enriching the core with the attributes necessary to generate the reduct.

The paper is organized as follows. In Sect. 2 some information about the notion of the discernibility matrix, core and reduct are provided. Also the pseudocodes of the algorithms are presented as well as the dataset used during the research is described. The Sect. 3 focuses on a description of the hardware solution, while the Sect. 4 is devoted to the experimental results.

2 Introductory Information

In this section one can find the definitions of a discernibility matrix, a core and a reduct, as well as the algorithms for generating all of the mentioned above structures and sets.

2.1 The Notion of Discernibility Matrix, Core and Reduct in the Rough Set Theory

Some of the condition attributes in the decision table may be superfluous (redundant in other words). Removing any of the redundant attribute should not lead to worsen the classification and thus preserve the discernibility of the objects. There can be a few condition attributes that cannot be removed without affecting the classification power of all condition attributes. The set of all indispensable condition attributes is called the core.

One can also observe that the core is the intersection of all reducts – each element of the core belongs to every reduct. Thus, in a sense, the core is the most important subset of condition attributes. In order to compute the core we can use discernibility matrix.

The notion of a discernibility matrix was introduced by Prof. A. Skowron and first described in [4]. Both the rows and columns of the discernibility matrix are labeled by the objects. An entry of the discernibility matrix is the set that consists of all condition attributes on which the corresponding two objects have distinct values. If an entry consists of only one attribute, the unique attribute must be a member of core. A much more detailed description of the concept of the core can be found, for example, in the book [8].

We can compute a reduct using the discernibility matrix by the following observation: If a condition attribute is more frequent in the discernibility matrix, then the more important this attribute might be.

2.2 Pseudocode for Generating the Discernibility Matrix – DMgen

Below one can find pseudocode for generating discernibility matrix. The discernibility matrix generated by this pseudocode is used by the algorithms described in the following subsections.

INPUT: decision table $(U, A \cup \{d\})$
OUTPUT: discernibility matrix DM

```
for x ∈ U do
  for y ∈ U do
    DM(x, y) ← ∅
    if d(x) ≠ d(y) then
      for a ∈ A do
        if a(x) ≠ a(y) then
          DM(x, y) ← DM(x, y) ∪ {a}
        end if
```

```
 9:         end for
10:      end if
11:   end for
12: end for
```

The discernibility matrix entry $DM(x, y)$ is generated in lines 3–10. First in line the entry $DM(x, y)$ is cleared (ie. loaded with empty set). Then the decision attributes $d(x)$ and $d(y)$ are compared and if they are not equal the discernibility matrix entry generation is performed. Loop in lines 5–9 is used for finding the differences between the conditional attributes $a(x)$ and $a(y)$ for every attribute a. In line 6 the values are compared and if they are not equal the attribute a is added to the entry $DM(x, y)$. All calculations are repeated for every entry of the decision table (lines 1 and 2).

2.3 Pseudocode for Generating a Core Using the Discernibility Matrix – COREgen

Pseudocode for algorithm of calculating the core using discernibility matrix can be found below. This is the first stage of the algorithm for the reduct generation.
INPUT: discernibility matrix DM
OUTPUT: core $C \subseteq A$

```
1: C ← ∅
2: for x ∈ U do
3:    for y ∈ U do
4:       if |DM(x,y)| = 1 and DM(x,y) ∉ C then
5:          C ← C ∪ DM(x,y)
6:       end if
7:    end for
8: end for
```

This algorithm is using a singleton for generating the core. Singleton is a cell from the discernibility matrix consisted of the only one attribute. Singleton detection is done in line 4 of the algorithm. Lines 2 and 3 iterates all of the discernibility matrix entries. After finding the singleton the attribute is added to the core in line 5.

2.4 Pseudocode for Generating a Reduct Using the Discernibility Matrix – REDgen

Below is the pseudocode for calculating reduct using discernibility matrix. This is the second stage of the algorithm for the reduct generation.
INPUT: discernibility matrix DM, core C
OUTPUT: reduct R

```
1: R ← C
2: for x ∈ U do
3:    for y ∈ U do
4:       for a ∈ C do
```

```
5:          if a ∈ DM(x, y) then
6:              DM(x, y) ← ∅
7:          end if
8:      end for
9:    end for
10: end for
11: while DM ≠ ∅ do
12:   for a ∈ A do
13:      counts(a) ← 0
14:   end for
15:   for x ∈ U do
16:      for y ∈ U do
17:          for a ∈ A do
18:              if a ∈ DM(x, y) then
19:                  counts(a) ← counts(a) + 1
20:              end if
21:          end for
22:      end for
23:   end for
24:   redAttr ← {b ∈ A \ R :
          counts(b) = max{counts(a) : a ∈ A \ R}
25:   R ← R ∪ {redAttr}
26:   for x ∈ U do
27:      for y ∈ U do
28:          if redAttr ∈ DM(x, y) then
29:              DM(x, y) ← ∅
30:          end if
31:      end for
32:   end for
33: end while
```

In the first step in the line 1 the variable storing the reduct R is initialized with the value of the core C obtained in the first stage. Then all of the entries having any attribute from the core are removed from the discernibility matrix (lines 2–10). Line 11 contains main loop which continues an execution while the discernibility matrix DM has any attribute within its cells. In lines 12 to 14 all cells of *counts* vector for counting the occurrences of each condition attribute in discernibility matrix is set to 0. Two loops in lines 15 and 16 iterates through a whole discernibility matrix. Loop in line 17 processes each condition attribute. If given attribute exists in DM cell (line 18), then the value of the *counts* vector for this attribute is incremented by 1. Line 24 chooses attribute with maximum number of occurrences. This attribute, stored in *redAttr* variable, is added to the final reduct R in line 25. Two loops in lines 26 and 27 iterates over the discernibility matrix again. Condition in line 28 checks if attribute stored in *redAttr* exists in DM cell corresponding to x and y objects. If so, then whole cell in the discernibility matrix is cleared (contains no attributes).

2.5 Data for Experimental Results

In this paper, we conduct experimental studies using data about children with insulin-dependent diabetes mellitus (type 1). Insulin-dependent diabetes mellitus is a chronic disease of the body's metabolism characterized by an inability to produce enough insulin to process carbohydrates, fat, and protein efficiently. Treatment requires injections of insulin. Twelve condition attributes, which include the results of physical and laboratory examinations and one decision attribute (microalbuminuria) describe the database used in our experiments.

The data collection so far consists of 107 cases. Out of twelve condition attributes eight attributes describe the results of physical examinations, one attribute describes insulin therapy type and three attributes describe the results of laboratory examinations. The former eight attributes include sex, the age at which the disease was diagnosed and other diabetological findings. The latter three attributes include the criteria of the metabolic balance, hypercholesterolemia and hypertriglyceridemia. The decision attribute describes the presence or absence of microalbuminuria. All this information is collected during treatment of diabetes mellitus.

The database is shown at the end of the paper [9]. A detailed analysis of the above data (only with the use of software systems) is in chapter 6 of the book [8].

3 Hardware Implementation

This section describes the architecture of the system for the reduct generation using two-stage algorithm.

All of the blocks included in this system were designed to perform the actions described is the pseudocodes DMgen, COREgen and REDgen.

System architecture was presented as block diagram in Fig. 1. It consists of the listed below blocks:

- **Decision Table DT** – memory for storing the values from the decision table; this block must be capable to store all the objects.
- **Comparators** – block consisting of the comparators used for comparing the objects from the decision table and generating the entries of the discernibility matrix.
- **Comparator Block CB** – single comparator used to compare the decision table entries.
- **Discernibility matrix** – memory for storing the discernibility matrix entries.
- **Core generator** – block for generating the core from the discernibility matrix.
- **Reduct generator** – block for finding the attribute which is a candidate for the reduct.
- **Control logic** – block for controlling the system.
- **AMR/RED** (Attribute Mask Register/REDuct register) – register used for masking the attributes in following steps of the algorithm and for storing the result.

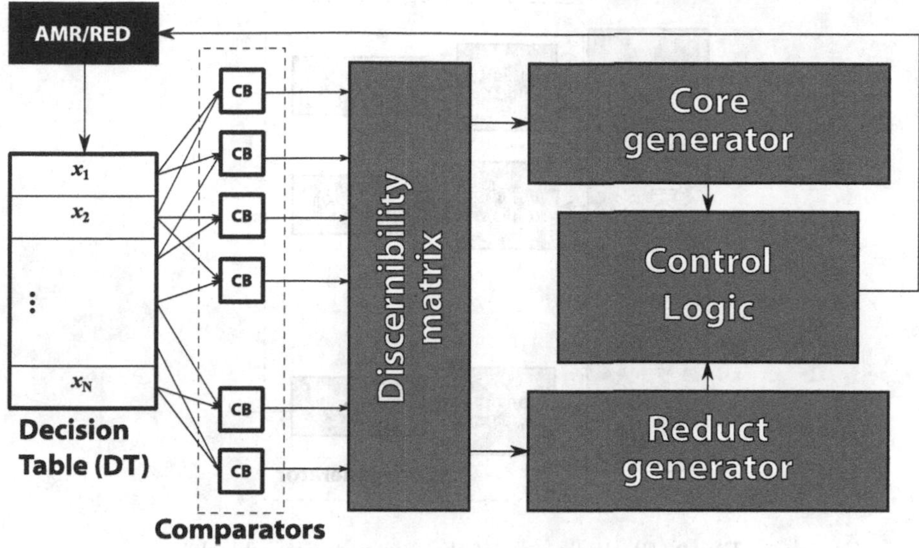

Fig. 1. Block diagram of the two-stage reduct generation system

At the beginning DT is filled with the values from the decision table. Then the values of the decision table are passed to the set of the comparators (CB), which are related to the condition in line 7 of the *DMgen* pseudocode. Because there are as many CB blocks as the entries in the discernibility matrix, the loops in lines 4 and 5 are not needed.

Moreover, AMR/RED register is used to masking the attributes in the discernibility matrix and therefore there is no need for the loops in lines 2–10 and 26–32 of the REDgen pseudocode. The only loop implemented in hardware was line 11 from REDgen pseudocode in the Control logic block.

Discernibility matrix entries are passed to two blocks: Core generator and Reduct generator. These blocks form the two-stage system for the reduct generation and are described below in details.

Architecture of the core generator block was presented as block diagram in Fig. 2. It consists of the listed below blocks:

- **Singleton detector** – block for detecting if the input value is a singleton, ie. consists of the only one "1".
- **OR-cascade** – block for generating the core by concatenating the singletons from all singleton detectors.

The singleton detector function is described in line 4 of the COREgen pseudocode. OR-cascade behaves in a way described in the line 5 of the COREgen pseudocode. Because there are as many mentioned above blocks as the entries in the discernibility matrix, the loops in lines 2 and 3 are not needed.

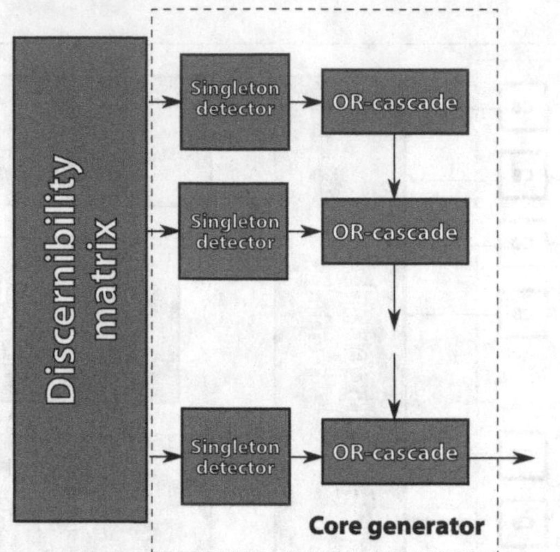

Fig. 2. Block diagram of the core generation block

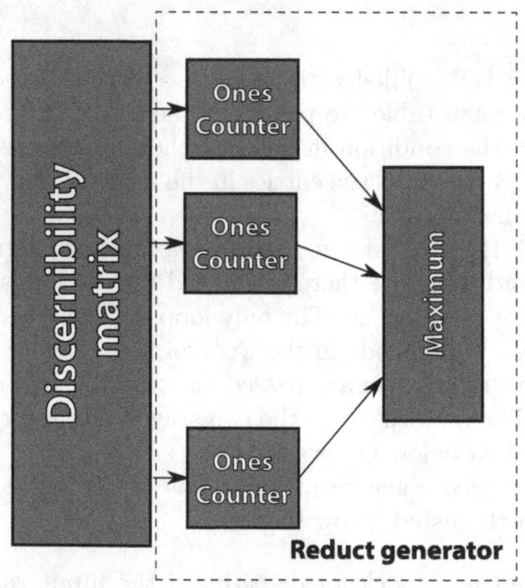

Fig. 3. Block diagram of the reduct generation block

Architecture of the core generator block was presented as block diagram in Fig. 3. It consists of the listed below blocks:

- **Ones counter** – block for counting the number of "ones" in the input word.
- **Maximum** – block for selecting the maximum from the input words.

The ones counter is a block performing all the activities described in lines 15–23 of the REDgen pseudocode. Maximum is a block equivalent to the line 24 of the REDgen pseudocode. All these blocks are repeated for all the conditional attributes and therefore there is no need to create the blocks for controlling the loops in lines 15–23.

4 Experimental Results

For the research purpose algorithms described in Sects. 2.2, 2.3 and 2.4 were implemented in C language. The main reason for choosing such language was deterministic program execution time, huge flexibility in the software creation, easiness of low-level communication implementation and the future plans of moving control program to the microprocessor independent from PC. The role of the microprocessor would be controlling operation of rough sets hardware implementation modules. Microcontroller, due to the limited memory and computational resources in comparison to the PC, should not use additional runtime environments required by e.g. Java.

The results of the software implementation were obtained using a PC equipped with an 8 GB of RAM and 2-core Intel Core i5-4210U with maximum 2.7 GHz in Turbo mode clock speed running Windows 10 operational system. The source code of application was compiled using the GNU GCC 4.9.2 compiler. Given times are averaged for 10 000 runs of each algorithm with the same data set.

The hardware implementation was written in VHDL. Quartus II 13.1 was used for creation, compilation, synthesis and verifying simulation of the hardware implementation. Synthesized hardware blocks were downloaded and run on TeraSIC DE-3 equipped with Stratix III EP3SL150F1152C2N FPGA chip. FPGA clock running at 50 MHz (64 times slower than PCs clock) for the sequential parts of the project was derived from development board oscillator. Timing results were obtained using Tektronix TDS3052B (500 MHz bandwidth, 5 GS/s) oscilloscope.

All calculations were performed using diabetes dataset. Full decision table has 107 objects and 13 attributes (12 condition and 1 decision). Presented results have been obtained using full set and smaller subsets (in terms of smaller number of objects).

Table 1 presents the results of the time elapsed for software and hardware reduct calculation. Column “Objects” provides the number of the objects used as input of the algorithm. Column “Software - t_S” presents the times of the execution of the software implementation of the algorithm while column “Hardware - t_H” presents the times of the execution of the hardware implementation. Last column “$\frac{t_S}{t_H}$” shows the quotient of the execution times.

Obtained results show a huge increase in the speed of data processing. Hardware module execution times compared to the software implementation is at least a 100 times shorter for reduct calculation what is shown in Table 1 in column $\frac{t_S}{t_H}$ and is increasing with larger data sets.

Because the Control logic block is a sequential circuit, one must remember to take the clock speed difference between FPGA and PC under consideration. FPGA in our configuration uses 50 MHz clock and the PCs clock runs at 2.7 GHz (54 times faster). Despite of this, the FPGA is still significantly faster.

Table 1. Comparison of execution time for calculating reduct using hardware and software version.

Objects —	Software - t_S [μs]	Hardware - t_H [μs]	$\frac{t_S}{t_H}$ —
15	343	1.88	182.45
30	1360	4.46	304.93
45	3 170	6.91	458.76
60	5 560	10.20	545.10
90	12 900	17.32	744.80
107	19 700	26.71	737.55

Table 2 presents the FPGA structure utilization in Logical Elements (LEs) basis for hardware implementation of the full system.

Table 2. FPGA structure utilization in LEs for full hardware system.

Objects	LEs
15	2 196
30	8 661
45	19 400
60	34 415
90	77 271
107	109 146

Big FPGA resources consumption is caused by the fact, that the complete discernibility matrix is generated. Consequently for all entries are other blocks generated (e.g. comparator block, singleton detector). In general, the combinational solutions consume more resources than the sequential types, but these are few orders of magnitude faster. For practical solutions sequential units are preferred than combinational because of their resources saving implementation.

For the most real datasets it is impossible to create a single hardware structure capacious enough to store the entire data. Besides the real datasets are so differential, so there is no universal structure of the processing unit.

5 Conclusion

Proposed hardware implementation of the two-stage reduct generation algorithm shows a big acceleration comparing to the software. The hardware implementation is very competitive to high computing power PCs, especially in real time solutions. The factor of acceleration is greater than a 180 and is increasing with the size of dataset. But when taking into account the difference in clock speeds between a PC and a FPGA this acceleration is factor rises to over 9000. Major disadvantage of proposed solution is a resources consumption and hence need an additional software processing such as the data decomposition. Hardware module must be also recompiled for different structures of datasets. Of course the biggest advantage of FPGA is possibility of its reprogramming, but it needs a supplementary software processing unit.

Further research can include using the decomposition algorithms for the big datasets, multiplying the hardware modules processing subsequent parts of dataset, as well as create other hardware blocks for supporting different parts of the algorithm.

Acknowledgment. This work was supported by grant S/WI/1/2018 from Bialystok University of Technology and funded with resources for research by the Ministry of Science and Higher Education in Poland.

References

1. Kopczyński, M., Grześ, T., Stepaniuk, J.: FPGA in rough-granular computing: reduct generation. In: WI 2014: The 2014 IEEE/WCI/ACM International Joint Conferences on Web Intelligence, Warsaw, vol. 2, pp. 364–370. IEEE Computer Society (2014)
2. Kopczynski, M., Grzes, T., Stepaniuk, J.: Generating core in rough set theory: design and implementation on FPGA. In: Kryszkiewicz, M., Cornelis, C., Ciucci, D., Medina-Moreno, J., Motoda, H., Raś, Z.W. (eds.) RSEISP 2014. LNCS (LNAI), vol. 8537, pp. 209–216. Springer, Cham (2014). https://doi.org/10.1007/978-3-319-08729-0_20
3. Kanasugi, A., Matsumoto, M.: Design and implementation of rough rules generation from logical rules on FPGA board. In: Kryszkiewicz, M., Peters, J.F., Rybinski, H., Skowron, A. (eds.) RSEISP 2007. LNCS (LNAI), vol. 4585, pp. 594–602. Springer, Heidelberg (2007). https://doi.org/10.1007/978-3-540-73451-2_62
4. Skowron, A., Rauszer, C.: The discernibility matrices and functions in information systems. In: Slowinski, R. (ed.) Intelligent Decision Support. Handbook of Applications and Advances of the Rough Sets Theory, pp. 331–362. Kluwer, Dordrecht (1992)
5. Pawlak, Z.: Elementary rough set granules: toward a rough set processor. In: Pal, S.K., Polkowski, L., Skowron, A. (eds.) Rough-Neural Computing. Cognitive Technologies, pp. 5–13. Springer, Heidelberg (2004). https://doi.org/10.1007/978-3-642-18859-6_1
6. Kopczynski, M., Grzes, T., Stepaniuk, J.: Computation of cores in big datasets: an FPGA approach. In: Ciucci, D., Wang, G., Mitra, S., Wu, W.-Z. (eds.) RSKT 2015. LNCS (LNAI), vol. 9436, pp. 153–163. Springer, Cham (2015). https://doi.org/10.1007/978-3-319-25754-9_14
7. Tiwari, K.S., Kothari, A.G., Keskar, A.G.: Reduct generation from binary discernibility matrix: an hardware approach. Int. J. Future Comput. Commun. **1**(3), 270–272 (2012)
8. Stepaniuk, J.: Rough – Granular Computing in Knowledge Discovery and Data Mining. SCI, vol. 152. Springer, Heidelberg (2008). https://doi.org/10.1007/978-3-540-70801-8

9. Stepaniuk, J.: Knowledge discovery by application of rough set models. In: Polkowski, L., Tsumoto, S., Lin, T.Y. (eds.) Rough Set Methods and Applications. New Developments in Knowledge Discovery in Information Systems. Studies in Fuzziness and Soft Computing, vol. 56, pp. 137–233. Physica-Verlag, Heidelberg (2000). https://doi.org/10.1007/978-3-7908-1840-6_5

A Multi-Granularity Representation Learning Framework for User Identification Across Social Networks

Shun Fu, Guoyin Wang(✉), Shuyin Xia, and Li Liu

Chongqing Key Laboratory of Computational Intelligence,
Chongqing University of Posts and Telecommunications, Chongqing, China
fushun@cigit.ac.cn, wanggy@ieee.org, {shuyinxia,liliu}@cqupt.edu.cn

Abstract. Predicting latent user identity across social networks has many application scenarios and it can be demonstrated by using the similarity of network structure caused by the similarity of friend relationships. Former related research works consider the structural similarities only while not sufficiently modeling the higher order structural properties. Moreover, the very limited supervisory anchor pairs, which are crucial for the task of user identification across social networks, are not utilized effectively. Based on the idea of multi-granularity cognitive computing and for partly solving the problem of multiple granularity representation of data proposed in DGCC (Data-driven granular cognitive computing) [23], this paper proposes a high-performance framework called multi-granularity representation learning (MGRL) framework for user identification across social networks which facilitates a well-designed heuristic mechanism to weight the edges on which a guided sampling strategy is conducted for vertex sequence generation. This enhances the model's capability of capturing the higher-order structural proximity. By integrating two aspects of structural properties, the multi-granularity structural features are preserved well. Experiments on real life social networks demonstrate that the MGRL significantly outperforms other state-of-the-art methods on the task of identifying latent corresponding users across social networks.

Keywords: Granular computing · Network representation learning · Social network analysis · User identity alignment

1 Introduction

With the rapid development of social networks, a person usually registers on several websites for diverse usages, such as Twitter, Flickers, Facebook, and LinkedIn [19,24]. It is a large group of people who choose not to share account information (also called user identity) on multiple social networks due to some considerations such as privacy. This poses the problem of identifying the corresponding accounts that belong to one natural person across different social networks (also called *user identity linkage across social networks*). Research about

© Springer Nature Switzerland AG 2019
T. Mihálydeák et al. (Eds.): IJCRS 2019, LNAI 11499, pp. 507–521, 2019.
https://doi.org/10.1007/978-3-030-22815-6_39

this problem can support network inference tasks such as cross-domain recommendation [5], link prediction [27], and network fusion [26].

The name or users' profile can be used for identity prediction across social networks, but these information cannot be always reliable due to privacy problems. Alternatively, as one person's circle of friends is highly individual and it's almost impossible for two users share the same circle of friends, the circle of friends has strong distinction in one's identity. Moreover, the connections on social networks can be quite directly acquired and the relationships among users can be represented by graph structures. Comparing with other kinds of information such as user profile and the user generated content in social networks, the structure-based approach has the strongest universality and it is compatible with the similarities extracted from other kinds of information. A typical way of merging the structural features with the features extracted by other information is simply concatenate the feature vectors [28].

Existing approaches that leverage network structure for identity prediction across social networks can be categorized into two classes: *Feature-based* and *Embedding-based*. In the feature-based methods, the macro structural features are used to find the certain structural similarity between vertices across networks. They address it as a problem of network alignment [8]. This kind of methods suffers NP-Hard combinatorial optimization problem and the scope of application of this method is very limited [1]. The feature-based methods utilize micro structural features such as in/out degree [10], the number of involved triangles [9] and common neighbors [3] to calculate the match degree between two vertices across network. These feature-based models are not robust against any slight disturbance or noise.

Network representation learning (NRL), which is also called network embedding, maps the structural properties of each vertex on network into a low-dimensional vector and its effectiveness for structural property preserving is well-demonstrated on many network inference tasks in last several years [6,7]. Figure 1 shows the basic ideas of NRL based identity alignment across social networks. The user "001" in source network and user "001" in target network are two accounts but they belong to one person. This is called one anchor pair. If the anchor pair is given beforehand, it's called *supervisory anchor pair* (SAP). The relationship (graph structure) of each user in the original network space can be represented in a latent vector space and the users' similarities can be calculated by their structural representation vectors (embeddings). In the NRL based approaches, the PALE [13] predicts anchor links by using network embedding of observed links as the supervised information. It suffers the difference of scales and initialization of parameters of two networks. The human-designed objective function is deficient on unifying the semantics of embeddings for two networks due to the structural complexity and non-linear properties in networks [22]. Inspired by the 1st-order proximity which means the weight of linkage and 2nd-order proximity which means the number of friends shared by two vertices [21], works such as IONE [12] and MAH [20] considers the second order proximity and regard source network G^s and target network G^t as an entire network and map it to a hyper-graph to learn latent network features. The latent semantic of network structure is extracted by them and models can be stable when the local statistic parameters such as vertex degree, number of neighbors slightly changed. However they are deficient on modeling the higher-order structural proximity. For instance, in Fig. 1, the unknown user "007" and user "011" has the higher-order structural proximity with user "001" and user "003" respectively, but they are far away from the known SAPs. The problem of how to represent the network structural features into low-dimensional vectors is still open. Designing a novel framework which utilizes the higher-order structural proximity and take advantage of the supervisory information to discover the hidden anchor links across social networks is still of great significance.

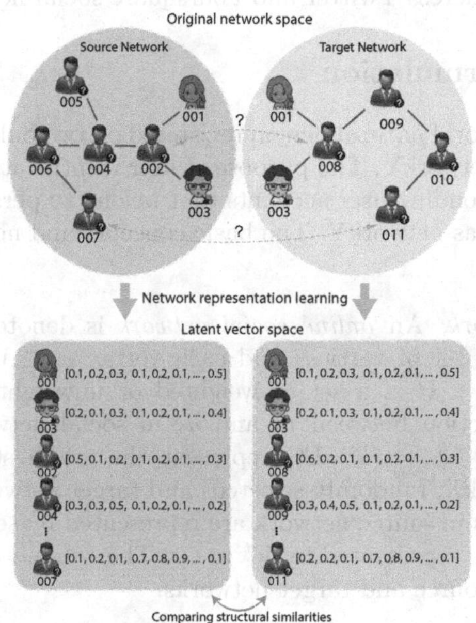

Fig. 1. NRL based identity alignment across social networks

In order to utilize the structural information to improve the performance of aligning user identities across social networks, we consider the multi-layer characteristics which essentially exist in networks when looking into network properties [2,17,18]. Inspired by the idea of multi-granularity cognitive computing, this paper proposes a novel multi-granularity representation learning (MGRL) framework for user identification across social networks. Specifically, (1) On the local structural feature learning, the higher order structural properties are captured through the establishment of the corpus obtained by random walk. (2) For modeling the anchor pairs oriented structural features, we train a heuristic weight assigner based on the SAPs, and the edge to which the higher weight is assigned will guide the random walker in the direction more likely to touch the potential anchors. Through modeling the network features on two granularities, this representation framework not only satisfies adequately capturing the local structural features, but also emphasizes the anchor pairs oriented features across the source and target networks. In this way, the model enables that the corresponding user accounts that belongs to one natural person share more context across networks to get closer vector representations in the latent low-dimensional vector space.

To evaluate the quality of the proposed MGRL, we conduct experiments on the task of aligning user identities across two real-world social networks: Twitter and Foursquare. Comparing with the baselines, under the unify metrics, the proposed MGRL model achieves higher precision in the task of corresponding user pair prediction across Twitter and Foursquare social networks.

2 Problem Formulation

Given a single real-world natural person registered on two online social networks, e.g. networkX and networkY. The purpose of user identification across networks is to find the corresponding user accounts that belong to persons who registered in networkX as well as networkY. The basic concepts and notations are defined as below:

Online Social Network: An *online social network* is denoted as a graph $G = (V, E)$ where V is a set of vertices and each vertex $v_i \in V$ represents a user account u_i. $E \subseteq V \times V$ is a set of weighted or unweighted edges, $e_{ij} \in E$ represent the connection between u_i and u_j in social network. The notation $G^s = (V^s, E^s)$ and $G^t = (V^t, E^t)$ represent the graph structure for source network (e.g. networkX, randomly selected) and target network (e.g. networkY) respectively. Vertices in source network are represented by set of V^s while those in target network are represented by V^t in G^t. The U^s and U^t denotes the sets of user accounts in source and target networks.

Adjacency Matrix A: $A \in \mathbb{R}^{|V| \times |V|}$. For an unweighted graph, $A_{ij} = 1$ if and only if there exists an edge from v_i to v_j and $A_{ij} = 0$ otherwise. For a weighted graph, A_{ij} is a real number w_{ij} called the weight of the edge e_{ij}. We only discuss the weight from 0 to 1 since the other range of weight can be normalized into $[0, 1]$. Note the w_{ij} also denotes the weight of e_{ij} in this paper. For a graph with adjacency matrix A, the *degree matrix D* of that graph can be defined as:

$$D_{ij} = \begin{cases} \sum_p A_{ip}, & \text{if } i = j \\ 0, & \text{if } i \neq j \end{cases} \tag{1}$$

Transition Probabilities Matrix T: For a random walker on a graph, the *transition matrix* determines the probability of the walker transits from one vertex to another. The *1-step probability transition matrix T* is defined as:

$$T = D^{-1}A \tag{2}$$

Where T_{ij} is the probability of a transition from v_i to vertex v_j within one step, and the weight w_{ij} is proportional to the transition probability T_{ij}: $w_{ij} \propto T_{ij}$.

Cross-net Identification: Given two *online social networks* G^s and G^t, the task of *Cross-net identification* is to predict whether a pair of vertices v_i and v_j chosen from V^s and V^t respectively belong to a same real natural person, i.e., $f : V^s \times V^t \mapsto \{0,1\}$ such that,

$$f(v_i, v_j) = \begin{cases} 1, & \text{if } u_i \text{ and } u_j \text{ belong to same person,} \\ 0, & \text{otherwise.} \end{cases} \quad (3)$$

where $f(\cdot)$ is the prediction function that our model wants to learn. u_i is the user account of vertex v_i and u_j is the user account of vertex v_j.

Network Representation Learning: The *network representation learning* can be viewed as the extraction for structural features of network components. *Network representation learning* is to learn a mapping $\Phi : v_i \mapsto \boldsymbol{z}_i$ where the low-dimensional vector $\boldsymbol{z}_i \in \mathbb{R}^d$ is the embedding for vertex v_i, and the integer $d \ll |V|$ is the dimensionality of vector $\boldsymbol{z}_i$.

Considering that fully aligned networks hardly exist in the real world, in this paper, we also adopt the assumption of partially aligned social platforms as proposed. Table 1 summarizes the notations in this paper.

3 Multi-granularity Representation Learning for User Identification Across Social Networks

User Identity Alignment Across Social Networks. The *User identity alignment across social networks* is searching for a projection functions Φ by minimizing the following objective function:

$$\min \sum_{u_i \in U^s, u_j \in U^t} D(\Phi(u_i), \Phi(u_j)) \quad (4)$$

where u_i and u_j belong to one person, while $D(\cdot)$ is a distance measuring function, such as cosine distance.

3.1 Parameter Sharing by Zippering SAPs

In the embedding-base methods, it's intuitive that the optimal embeddings can be obtained by minimizing the difference between the embedding of G^s and G^t with the help of supervisory anchor pairs (SAPs, some known corresponding user pairs beforehand). This method cannot completely bypass the disunity of the embeddings in different latent feature space due to the elusiveness of latent feature space. However, MGRL conducts this uniform in the process of generating embeddings, and this is regarded as the hard constrain for our model. Figure 2 shows how the MGRL zippers the G^s and G^t together by SAPs. Each

Table 1. Table of notations

Notation	Description
G^s, G^t	Graph of source network and target network
G^m	Graph of the zippered network
G^m_w	Weighted Graph of the zippered network
V^s, V^t	Sets of vertices in G^s and G^t
U^s, U^t	Sets of user accounts in source and target networks
E^s, E^t	Sets of edges in G^s and G^t
V^m, E^m	Set of vertices and edges in the zippered network
E_p	Set of edges that connect supervisory anchors
E_n	Set of edges that do not connect supervisory anchors
u_i	User account for vertex v_i in graph G
v_i	Vertex of user u_i in G^s
v_i'	Vertex of user u_i in G^t, (v_i, v_i') is a corresponding vertex pair
u^s, u^t	User account from graph G^s and G^t
z_i	Low-dimensional representing vector for vertex v_i on un-weighted graph G^m
z_i^m	Low-dimensional representing vector for vertex v_i on weighted graph G^m_w
z_{ij}	The low-dimensional representation vector for edge e_{ij}
$\Phi(\cdot)$	Mapping from one vertex into its low-dimensional vectorial representation
$\Omega(\cdot)$	Mapping from G^s and G^t into G^m, the graph of zippered network
$\sigma(\cdot)$	Sigmoid function
P^{acr}	The set of supervisory anchor pairs
p_k	The supervisory anchor pair, $p_k \in P^{acr}$
(v_k, v_k')	The k-th supervisory anchor pair, merged into v_k in zippering operation
$\mathcal{Y}$	Set of labels for edge properties
w_{ij}	Weigh for edge e_{ij} given by the heuristic weight assigning model
$r^{(ij)}$	The feature representation for edge e_{ij}
T_{ij}	The random walker's transition probability from v_i to v_j, $T_{ij} \propto w_{ij}$

SAP consists of one user in the G^s and one user in the G^t. After zippering, each SAP is treated as one vertex in the network G^m. Such that, the G^s and G^t share same process of iteration of parameters. Although this operation may lead to slight changes in structural characteristics, multi-step random walk concerns more about the relationships in the entire sampled vertex sequence. The experiment result indicates that this operation is beneficial to the improvement of model performance.

The zippering operation can be viewed as a mapping of two graphs into one graph. We note it as $\Omega : G^s, G^t \mapsto G^m$. Assume we have a n-size supervisory anchor pair set: $P^{acr} = \{p_1, \ldots, p_n\}$ and $p_k = \{(v_k, v_k') | v_k \in V_a^s, v_k' \in V_a^t\}$ where

$V_a^s \subset V^s$ is subset of vertices in G^s and $V_a^t \subset V^t$ is subset of vertices in G^t. The vertices v_k and v_k' belongs to one natural person's identity: $u_k \in U^{acr} = (U^s \cap U^t)$ where U^{acr} is all the corresponding user accounts contains known and unknown corresponding account pairs.

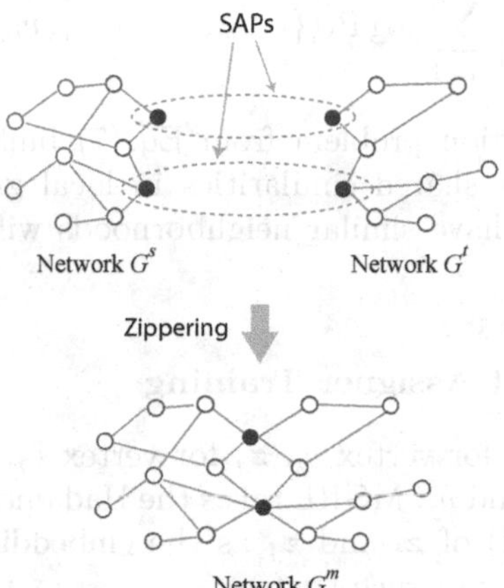

Fig. 2. Zipper the SAPs for parameter sharing

3.2 Higher-Order Proximity Sampling and Embedding

In order to obtain the low-dimensional vectors preserving the structural properties of vertices in graph, MGRL borrows the methods of DeepWalk [16]. Walkers sample the connecting information of vertices such that the sequences of vertices, namely corpora, are generated under the transition probability of T_{ij}. The low-dimensional vector is the result of optimization using model of word2vec [14,15]. The sequence of vertices generated by random walk can be denoted as: $v_1, v_2, \ldots, v_n$ under the transition probability of $T_{12}, T_{23}, \ldots, T_{(n-1)n}$.

In the DeepWalk model, the likelihood of observing vertex v_i given all the previous vertices visited so far in the random walk:

$$\Pr\left(v_i | (v_1, v_2, \ldots, v_{i-1})\right) \tag{5}$$

When we define the mapping function $\Phi : v_i \mapsto \boldsymbol{z}_i$, the problem is to estimate the likelyhood:

$$\Pr\left(v_i | (\Phi(v_1), \Phi(v_2), \ldots, \Phi(v_{i-1}))\right) = \Pr\left(v_i | (\boldsymbol{z}_1, \boldsymbol{z}_2, \ldots, \boldsymbol{z}_{i-1})\right) \tag{6}$$

Refer to the recent relaxation in language modeling [14,15], the model is required to maximize the probability of any word appearing in the context without the knowledge of its offset from the given word. This yields the optimization problem for network representation:

$$\arg\min_{\Phi} \quad -\frac{1}{|V^m|}\sum_{i=1}^{|V^m|} \log \Pr(\{v_{i-w}, \cdots, v_{i-1}, v_{i+1}, \ldots, v_{i+w}\}|v_i) \qquad (7)$$

Solving the optimization problem from Eq. (7) builds representations, e.g. vectors that capture the shared similarities in local graph structure between vertices. Vertices which have similar neighborhoods will acquire similar representations.

3.3 Heuristic Weight Assigner Training

Given the embedding $\boldsymbol{z}_i$ for vertex v_i, $\boldsymbol{z}_j$ for vertex v_j, and the edge $e_{ij} \in E^m$ that connects vertex v_i and v_j. MGRL takes the Hadamand product (also known as the Schur product [4]) of $\boldsymbol{z}_i$ and $\boldsymbol{z}_j$ as the embedding of edge e_{ij}, and the embedding is denoted as $\boldsymbol{z}_{ij}$, such that:

$$\boldsymbol{z}_{ij} = \boldsymbol{z}_i \circ \boldsymbol{z}_j = [a_{i1}b_{j1}, \ldots, a_{id}b_{jd}] \qquad (8)$$

where $a_{ik} \in \mathbb{R}^d$ is the k-th dimension of vector $\boldsymbol{z}_i = [a_{i1}, \ldots, a_{id}]$ and $b_k \in \mathbb{R}^d$ is the k-th dimension of vector $\boldsymbol{z}_j = [b_{j1}, \ldots, b_{jd}]$. The Hadamard product of $\boldsymbol{z}_i$ and $\boldsymbol{z}_i$ is denoted as: $\boldsymbol{z}_i \circ \boldsymbol{z}_j$.

For the edge $e_{ij} \in E^m$ connecting vertex v_i and v_j, the weight w_{ij} can be given by a trained weigh assigning function $\mathcal{W}(\boldsymbol{z}_{ij}) \mapsto w_{ij}$ where w_{ij} is the weight for edge e_{ij}.

Particularly, we have a set of training examples $D = [\boldsymbol{z}_{ij}, \mathcal{Y}_{ij}]$, where $i, j \in [1, \ldots, |V^s + V^t|]$ and $i \neq j$. The value of $\mathcal{Y}_{ij}$ is defined below:

$$\mathcal{Y}_{ij} = \begin{cases} 1, & \text{if } v_i \text{ or } v_j \text{ belongs to known anchor set,} \\ 0, & \text{otherwise.} \end{cases} \qquad (9)$$

The $\boldsymbol{\Theta} \in \mathbb{R}^d$ notes the parameter vector for model $\mathcal{W}$, d is the dimensionality of embedding vector $\boldsymbol{z}_i$. We define the weight of edge e_{ij} is the conditional probability models of the form:

$$w_{ij} = \Pr(\mathcal{Y}_{ij} = 1|\boldsymbol{\Theta}, \boldsymbol{z}_{ij}) = \mathcal{W}(\boldsymbol{\Theta}^T \boldsymbol{z}_{ij}) = \mathcal{W}(\sum_{e_{ij} \in E^m} \boldsymbol{\theta}_{ij}(\boldsymbol{z}_i \circ \boldsymbol{z}_j)) \qquad (10)$$

In what follows we use the logistic link function:

$$\mathcal{W}(r) = \frac{\exp(r)}{1 + \exp(r)} \qquad (11)$$

$$r^{(ij)} = \sum_{e_{ij} \in E^m} \boldsymbol{\theta}_{ij}(\boldsymbol{z}_i \circ \boldsymbol{z}_j) + b \tag{12}$$

thereby producing a logistic regression model. The r_{ij} is the feature representation for edge e_{ij} weighted by parameter vector $\boldsymbol{\theta}_{ij}$ and added bias b.

The cost function:

$$J = -\frac{1}{m}[\sum_{k=1}^{m} \mathcal{Y}^{(k)} \log \mathcal{W}(r^{(k)}) + (1 - \mathcal{Y}^{(k)}) \log(1 - \mathcal{W}(r^{(k)}))] \tag{13}$$

The parameter $\boldsymbol{\Theta}$ can be updated by:

$$\Theta := \Theta + \Delta\Theta,\ \Delta\Theta = -\eta \nabla J(\Theta) \tag{14}$$

And the derivative of J can be calculated by:

$$\frac{\partial J}{\partial \Theta} = \frac{1}{m} \cdot x^{(k)} \cdot [\sum_{k=1}^{m} \mathcal{W}(r^{(k)}) - \mathcal{Y}^{(k)}] \tag{15}$$

where the k-th sample $x_k = \boldsymbol{z}_i \circ \boldsymbol{z}_j$, and its label $\mathcal{Y}_k = \mathcal{Y}_{ij}$. The m is number of samples sampled from zippered graph G^m. Finally, use the update equation below to get the parameter vector $\boldsymbol{\Theta}$:

$$\Theta := \Theta + \eta \sum_{k=1}^{m} (\mathcal{Y}^{(k)} - \mathcal{W}(r^{(k)}))x^{(k)} \tag{16}$$

Where η is the step of iteration. The algorithms such as stochastic gradient descent (SGD) or mini batch SGD [11] can be used when large number of edges are sampled.

3.4 Multi-granularity Cooperative Representation

The weighted network G_w^m can be obtained by assigning weight on the zippered graph G^m using the heuristic weight assigner trained in Sect. 3.3. Perform weighted random walk on G_w^m can obtain the corpus C_w. Initialize the representation of each vertex of G_w^m and update the representation with the model of Skip-Gram with the similar steps of Sect. 3.2. The representation of v_i on G_w^m is denoted as $\boldsymbol{z}_i^m$.

MGRL concatenates the embeddings obtained on two granular layers, i.e. the embeddings from the unweighted zippered network and the embeddings from the weighted network. The concatenating operation can be denoted as:

$$\boldsymbol{z}_i := [\boldsymbol{z}_i, \boldsymbol{z}_i^m] \tag{17}$$

Through this way, the structural semantic on two networks will be preserved in the concatenated embeddings.

4 Experiments

In this section, some preliminary experiments are conducted to compare proposed MGRL with existing baselines on two realistic datasets.

4.1 Dataset

Twitter - Foursquare dataset. The real-world social network datasets collected from Foursquare and Twitter [25]. The ground truth of anchors are provided in Foursquare profiles. Twitter dataset consists of 5,120 users and 164,919 connections while Foursquare dataset consists of 5,313 users and 76,972 connections. There are 1,609 user pairs are known as corresponding user pairs.

4.2 Experimental Settings

In the step of embedding the zippered network, we set the walk-length parameter as 80 for sampling longer sequence of corpus and the higher-order structural proximity can be contained in corpus. The window size for word2vec model in DeepWalk is set to 4, the number of walks is set to 10 by default. When calculating the distance between two vectors, we use the cosine distance in general way.

4.3 Comparative Methods

We compare the proposed MGRL model with several state-of-the-art methods, which are summarized as follows:

- **PALE-DeepWalk** [13]: This method performs reconciliation in an embedding-matching framework in which the representation method is DeepWalk [16]. We implemented this method with the matching of MLP (multilayer perceptron).
- **IONE** [12]: This method considers the following/followee relations are approximated in the latent space with an explicit constraint to ensure that latent feature vectors are equal. We evaluate this method by using the source code they offer.
- **MAG** [20]: It computes the user-to-user pairwise weight and builds a social graph for each network. The identification is conducted by manifold alignment.
- **CRW** [25]: A method called collective random walk with restart that is essentially a collective link fusion across partially aligned probabilistic networks.
- **Mego2Vec-struct** [28]: This method pre-aligns the potential corresponding identity pairs by the similarity of user names and uses structural and attribute information to learn a neural network classifier. In order to equally compare the performance of identifying purely using structural information, we only use the structural embedding model of Mego2Vec framework. This is noted as *Mego2Vec-struct* in this paper.

4.4 Evaluation

We have the test set $V^s_{test} \subset V^s$ consists of hidden anchors $v_i \in V_{test}$, where $i = 1, \ldots, m$ and $m = |V^s_{test}|$. MGRL generates the embeddings $\boldsymbol{z}_i$ for v_i, and $\boldsymbol{z}_j$ for $v_j \in V^t$, where V^t is the set of vertices on target network G^t. For each $v_i \in V_{test}$, calculate the cosine distance between $\boldsymbol{z}_i$ and $\boldsymbol{z}_j$. Note the number of v_j is $|V^t|$, such that we got $|V^t|$ distances. Rank the $|V^t|$ distances, we pick out the v_js that are in top-K distances as the set of candidates H. If the corresponding vertex $v'_i \in V^t$ of v_i is in the set of candidates H, we mark v_i as 'hit', otherwise, v_i is 'not hit'. Do the distance calculating for every v_i in V_{test} and summarize the number of 'hit's. The number of 'hit's described above is denoted as *hits_number* in the equation below:

$$hits\_number = \sum_{\substack{v_i \in V_{test} \\ v_j \in V^t}} f(v_i, v_j)$$

where

$$f(v_i, v_j) = \begin{cases} 1, & \text{if } u^s_i \text{ and } u^t_j \text{ belong to one person} \\ 0, & \text{otherwise} \end{cases}$$

The u^s_i is the user account for vertex v_i in source network while the u^t_j is the user account for vertex v_j in target network. Thus, the precision at K-size candidate set can be calculated as:

$$precision@K = \frac{hits\_number@K}{|V^s_{test}|}$$

To be consistent, we use same metrics of IONE [12]. The value of K determines the size of candidate set. Larger K makes the model have higher probability to find the corresponding vertex of $v_i \in V_{test}$ while the bigger size of candidate means the less accurate the result is.

4.5 Experiment Result

The overall performance is evaluated by precision defined in Sect. 4.4 and Fig. 3 shows the precision got from the MGRL model and the comparative methods. We can find the proposed MGRL out-performs the baselines given different experimental conditions. In Fig. 3(a), K, the size of candidate set, is set to 30. When the ratio of training ground truth is 10%, the proposed MGRL algorithm can get about 50% promotion over the best performance among the base line methods. With the increase of the ratio of training ground truth, the promotion is decreasing and finally the proposed MGRL has about 7% promosion over IONE [12]. As we discussed in the Sect. 1, comparing with the IONE or IONE like methods, the proposed MGRL considers more about capturing the higher-order structural proximity by building the corpus of vertices via random walking. It's worth mentioning that with the increase of the ratio of training ground truth (training ratio), the ratio of higher-order anchor pairs that need to be explored is decreasing. The experiment result matches the analysis. On the other hand, Fig. 3(b) shows the changing of precision with different size of candidate, i.e. the value of K, under the condition that training ratio (the ratio of ground truth that used for training) is set to 90%. The MGRL also out-performs with the baselines with respect to precision. We can see that when the K is at 5 to 10, the proposed MGRL has the most significant promotion over the baselines. The smaller K at which the model achieves the significant promotion, the higher performance is the model for user identification across social networks since the smaller K means the model has higher accuracy for identifying the corresponding user account in target network.

The detailed results of precision with respect to training ratio and K are shown in Tables 2 and 3.

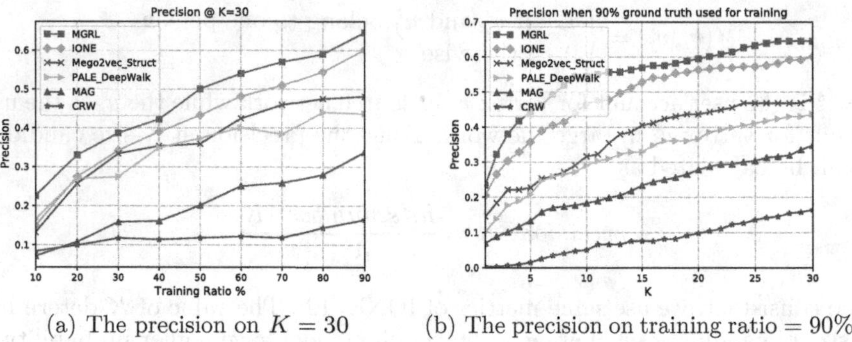

(a) The precision on $K = 30$ (b) The precision on training ratio = 90%

Fig. 3. Detailed performance comparison on Twitter-Foursquare Dataset

Table 2. The precision for different ratio of ground truth used for training

	Precision at different training ratio ($K = 30$)								
	10%	20%	30%	40%	50%	60%	70%	80%	90%
CRW	0.0843	0.0992	0.1172	0.1137	0.1179	0.1216	0.1181	0.1437	0.1646
MAG	0.0718	0.1079	0.1584	0.1609	0.2013	0.2503	0.2584	0.2798	0.3381
PALE-DeepWalk	0.1659	0.2698	0.2739	0.3492	0.3739	0.4009	0.3840	0.4404	0.4367
Mego2Vec-struct	0.1324	0.2572	0.3337	0.3534	0.3591	0.4242	0.4592	0.4752	0.4810
IONE	0.1492	0.2746	0.3419	0.3903	0.4328	0.4812	0.5094	0.5443	0.6012
MGRL	**0.2255**	**0.3302**	**0.4073**	**0.4223**	**0.5006**	**0.5397**	**0.5696**	**0.5902**	**0.6456**

Table 3. The precision for different value of K

	Precision at different K (Training ratio = 90%)							
	1	5	9	13	17	21	25	30
CRW	0.0031	0.0158	0.0474	0.0664	0.0822	0.1044	0.1360	0.1646
MAG	0.0696	0.1329	0.1772	0.2056	0.2500	0.2879	0.3037	0.3381
PALE-DeepWalk	0.1012	0.2088	0.2721	0.3227	0.3607	0.3860	0.4113	0.4367
Mego2Vec-struct	0.1455	0.2278	0.2911	0.3797	0.4241	0.4430	0.4684	0.4810
IONE	0.2056	0.3575	0.4398	0.4968	0.5411	0.5664	0.5727	0.6012
MGRL	**0.2341**	**0.4430**	**0.5443**	**0.5570**	**0.5759**	**0.6013**	**0.6329**	**0.6456**

5 Conclusion and Future Work

In this paper, we rethink how to employ the structural features for hidden corresponding user identity alignment. With the contribution of known anchors' structural features, the proposed MGRL generate a heuristic weigh assigner which emphasize the edge that contains structural feature of supervisory anchors. This heuristic weighting mechanism makes the corpus of vertices contain context with more SAPs. And finally, the distance of hidden corresponding users especially the higher-order latent anchor pairs will be close after model's iteration. In the future, we plan to explore more discriminating structural features and the multi-granularity models for effectively extracting them.

Acknowledgement. This work is supported in part by the National Key Research and Development Program of China (grant no. 2016QY01W0200), National Natural Science Foundation of China (grant no. 61772096, 61572091 and 61806031), Graduate Research and Innovation Project Plan of Chongqing Municipal Education Commission (grant no. CYB18166) and Doctor Training Program of Chongqing University of Posts and Telecommunications (grant no. BYJS201809).

References

1. Bayati, M., Gerritsen, M., Gleich, D.F., Saberi, A., Wang, Y.: Algorithms for large, sparse network alignment problems. In: Ninth IEEE International Conference on Data Mining, ICDM 2009, pp. 705–710. IEEE (2009)
2. Clauset, A., Moore, C., Newman, M.E.: Hierarchical structure and the prediction of missing links in networks. Nature **453**(7191), 98 (2008)
3. Cui, Y., Pei, J., Tang, G., Luk, W.S., Jiang, D., Hua, M.: Finding email correspondents in online social networks. World Wide Web **16**(2), 195–218 (2013)
4. Davis, C.: The norm of the schur product operation. Numer. Math. **4**(1), 343–344 (1962)
5. Dong, Y., et al.: Link prediction and recommendation across heterogeneous social networks. In: 2012 IEEE 12th International Conference on Data Mining, pp. 181–190. IEEE (2012)
6. Goyal, P., Ferrara, E.: Graph embedding techniques, applications, and performance: a survey. Knowl. Based Syst. **151**, 78–94 (2018)
7. Hamilton, W.L., Ying, Z., Leskovec, J.: Representation learning on graphs: methods and applications. IEEE Data Eng. Bull. **40**, 52–74 (2017)

8. Kollias, G., Mohammadi, S., Grama, A.: Network similarity decomposition (NSD): a fast and scalable approach to network alignment. IEEE Trans. Knowl. Data Eng. **24**(12), 2232–2243 (2012)
9. Kong, X., Zhang, J., Yu, P.S.: Inferring anchor links across multiple heterogeneous social networks. In: Proceedings of the 22nd ACM International Conference on Information & Knowledge Management, pp. 179–188. ACM (2013)
10. Korula, N., Lattanzi, S.: An efficient reconciliation algorithm for social networks. Proc. VLDB Endowment **7**(5), 377–388 (2014)
11. Li, M., Zhang, T., Chen, Y., Smola, A.J.: Efficient mini-batch training for stochastic optimization. In: Proceedings of the 20th ACM SIGKDD International Conference on Knowledge Discovery and Data Mining, pp. 661–670. ACM (2014)
12. Liu, L., Cheung, W.K., Li, X., Liao, L.: Aligning users across social networks using network embedding. In: IJCAI, pp. 1774–1780 (2016)
13. Man, T., Shen, H., Liu, S., Jin, X., Cheng, X.: Predict anchor links across social networks via an embedding approach. In: IJCAI, vol. 16, pp. 1823–1829 (2016)
14. Mikolov, T., Chen, K., Corrado, G., Dean, J.: Efficient estimation of word representations in vector space. arXiv preprint arXiv:1301.3781 (2013)
15. Mikolov, T., Sutskever, I., Chen, K., Corrado, G.S., Dean, J.: Distributed representations of words and phrases and their compositionality. In: Advances in Neural Information Processing Systems, pp. 3111–3119 (2013)
16. Perozzi, B., Al-Rfou, R., Skiena, S.: DeepWalk: online learning of social representations. In: Proceedings of the 20th ACM SIGKDD International Conference on Knowledge Discovery and Data Mining, pp. 701–710. ACM (2014)
17. Ravasz, E., Somera, A.L., Mongru, D.A., Oltvai, Z.N., Barabási, A.L.: Hierarchical organization of modularity in metabolic networks. Science **297**(5586), 1551–1555 (2002)
18. Sales-Pardo, M., Guimerà, R., Moreira, A.A., Amaral, L.A.N.: Extracting the hierarchical organization of complex systems. Proc. Natl. Acad Sci. U S A **104**(39), 15224–15229 (2007)
19. Sun, Y., Han, J., Yan, X., Yu, P.S.: Mining knowledge from interconnected data: a heterogeneous information network analysis approach. Proc. VLDB Endowment **5**(12), 2022–2023 (2012)
20. Tan, S., Guan, Z., Cai, D., Qin, X., Bu, J., Chen, C.: Mapping users across networks by manifold alignment on hypergraph. In: AAAI, vol. 14, pp. 159–165 (2014)
21. Tang, J., Qu, M., Wang, M., Zhang, M., Yan, J., Mei, Q.: LINE: large-scale information network embedding. In: Proceedings of the 24th International Conference on World Wide Web, pp. 1067–1077. International World Wide Web Conferences Steering Committee (2015)
22. Wang, D., Cui, P., Zhu, W.: Structural deep network embedding. In: Proceedings of the 22nd ACM SIGKDD International Conference on Knowledge Discovery and Data Mining, pp. 1225–1234. ACM (2016)
23. Wang, G.: Data-driven granular cognitive computing. In: Polkowski, L., et al. (eds.) IJCRS 2017. LNCS (LNAI), vol. 10313, pp. 13–24. Springer, Cham (2017). https://doi.org/10.1007/978-3-319-60837-2_2
24. Zafarani, R., Liu, H.: Connecting users across social media sites: a behavioral-modeling approach. In: Proceedings of the 19th ACM SIGKDD International Conference on Knowledge Discovery and Data Mining, pp. 41–49. ACM (2013)
25. Zhang, J., Philip, S.Y.: Integrated anchor and social link predictions across social networks. In: IJCAI, pp. 2125–2132 (2015)

26. Zhang, J., Yu, P.S.: PCT: partial co-alignment of social networks. In: Proceedings of the 25th International Conference on World Wide Web, pp. 749–759. International World Wide Web Conferences Steering Committee (2016)
27. Zhang, J., Yu, P.S., Zhou, Z.H.: Meta-path based multi-network collective link prediction. In: Proceedings of the 20th ACM SIGKDD International Conference on Knowledge Discovery and Data Mining, pp. 1286–1295. ACM (2014)
28. Zhang, J., et al.: Mego2vec: embedding matched ego networks for user alignment across social networks. In: Proceedings of the 27th ACM International Conference on Information and Knowledge Management, pp. 327–336. ACM (2018)

A Robust Long-Term Pedestrian Tracking-by-Detection Algorithm Based on Three-Way Decision

Ziye Wang[1], Duoqian Miao[1,2](✉), Cairong Zhao[1,2], Sheng Luo[1], and Zhihua Wei[1,2]

[1] Department of Computer Science and Technology, Tongji University, Shanghai 201804, China
{yeziwang, zhaocairong, zhihuawei}@tongji.edu.cn, miaoduoqian@163.com, tjluosheng@gmail.com
[2] Key Laboratory of Embedded System and Service Computing, Ministry of Education, Tongji University, Shanghai, China

Abstract. Pedestrian Detection Technology has become a hot research topic in target detection field recent years. But how to track the pedestrian target accurately in real time is still a challenge problem. Recently deep learning has got the extensive research and application in both target tracking and target detection. However, the tracking effect based on deep learning needs to be improved in the motion blur and occlusion cases. In this paper, we propose a new model that combines the target tracking and target detection and introduce the idea of granular computing to realize high-precision long-term robust pedestrian tracking. In this model, we use a pre-trained tracking model to track the specified object and use the three-way decision theory to judge the color histogram feature and correct the results by the detector. Compared with the separated tracker, our model invokes the target detector to detect the current frame when the tracking result is wrong and the detection result which is the most similar to the target is selected as the tracking result. Experimental results show that our model can significantly improve the tracking accuracy especially in the complex situations, compared with the separated tracker and the detector.

Keywords: Long-term tracking · Tracking by detection · Color histogram · Granular computing · Three-way decision

1 Introduction

Conventional target tracking methods cannot handle and adapt to complex tracking changes, and its robustness and accuracy remain to be improved. It is because that these classical algorithms do not rely on the prior knowledge, but use the method of probability density, manual setting or other methods to detect the moving target directly from the image, and finally locate the interest moving target in each frame of a video.

The first author is a student.

© Springer Nature Switzerland AG 2019
T. Mihálydeák et al. (Eds.): IJCRS 2019, LNAI 11499, pp. 522–533, 2019.
https://doi.org/10.1007/978-3-030-22815-6_40

On the contrary, the methods based on deep learning [1] could learn the valid features from big data automatically, which would cost years for conventional methods. That is the reason why deep learning methods are superior to the conventional methods [2] which use the hand-designed features such as HOG [3] or CN in the representation of the feature.

Deep learning has long been used in various fields of computer vision, such as image classification, target detection, semantic segmentation and so on. Until recently, with the great development of big data and the continuous improvement of computing power, deep learning began to be applied in the fields of target tracking and target detection.

In 2013, Wang and Yeung [4] proposed the use of the stacked denoising autoencoder (SADE) to extract the target features from a large number of data by unsupervised pre-training, and then use the particle filter to track online. This is the first tracking algorithm that applies depth models to single target tracking tasks. In 2014, Wang and Yeung proposed SO-DLT [5], which is a successful application of large-scale CNN in target tracking. Long and Shelhamer proposed the FCNT [6], its characteristic is expressing the difference and connection in the target attributes by exploring the CNN features of different layers and using tiny convolution neutral network to make them sparse. These measures can effectively prevent the drift of the tracker and have better robustness to the deformation of the target itself. Nam and Han proposed pre-training CNN by tracking videos directly to obtain general target expression ability and using an innovative multi-domain training method [7]. It wined over other contestants in the VOT-2015 Challenge, it is also the first time that there is alternate training in tracking. In 2016, Held put forward a deep vision tracking algorithm named Generic Object Tracking Using Regression Networks. GOTURN uses offline learning to learn through a large number of video and picture samples [8], so that the network can learn the appearance models and motion models of objects, it is the first time that the target tracking algorithm using depth learning achieves 100 FPS.

At present, the mainstream target detection algorithm is mainly based on the deep learning model. It can be divided into R-CNN algorithm based on region proposal [15], such as Fast R-CNN [16], Faster R-CNN [17], and end to end algorithm, such as YOLO [18], SSD [19]. The Faster R-CNN algorithm is widely used for the moment, which can be regarded as the combination of Region Proposal Network (RPN) [17] and fast R-CNN and uses the idea of shared convolutional layer to reduce the computational burden of proposal generation.

Before GOTURN was proposed, most of the tracking methods based on deep learning cannot meet the real-time requirement [12–15], it is the first model which makes the real time tracking with deep learning is possible. However, the tracking effect of GOTURN algorithm still needs to be improved in the cases of blurred target and target occlusion, and the loss of target often occur. Therefore, we hope to establish a tracking model to achieve impressive success in robust tracking.

Pedestrian is a non-rigid target, the complexity of its scenes and shape changing, view changing makes pedestrian tracking has been a difficult challenge in the field of computer vision research. The tracker could use the information from previous frames to lock the target efficiently but do not work well in complex situations. On the contrary, a well pre-trained target detector could detect the pedestrians easily under the condition of motion blur or partial occlusion. Therefore, we consider integrating the advantages of tracking and detection and propose a new model which combines detection and tracking to achieve long term robust tracking for pedestrians using the idea of tracking-by-detection.

The pedestrian tracking model built by us includes 3 layers, the first layer is a target tracker used GOTURN algorithm and the third layer is a target detector used Faster R-CNN. The most important second layer is to judge the tracking result and decide whether to call the detector according to the judge result.

The way to judge the result is to calculate the similarity between the two crops of the tracking target and the tracking result. We cannot directly compare two pictures, considering that the target is moving and the shape of the target or the angle of view may change. But at the same time, the color of the target will be roughly the same in a short time. Therefore, we consider using color histogram algorithm [21] to compare similarity between tracking result and tracking target.

In order to improve the accuracy of classification of tracking results, we abandon the traditional two-branch decision and adopt the three-way decision theory in the second layer. In 2009, Yao proposed three-way decision theory [9–11] based on decision rough set theory. Today, three decision-making theories have been widely used in many fields. So far it has been widely used in many fields, such as emotional classification [23] and image classification [24].

In general, our model could improve the accuracy of tracking results efficiently especially in some challenging cases such as fast motion, motion blur and occlusion.

The main contributions of this paper are as follows:

- Construct a long term robust pedestrian tracking model by combining the tracker and detector. Use the detector to correct the tracking result.
- Design an algorithm based on three-way decision to judge the tracking result by the feature of color histogram and decide the appropriate time to call the detector.
- Determine the appropriate threshold to get accurate tracking model.

The rest of the paper is organized as follows: In Sect. 2, we introduce the related work. In Sect. 3, we describe our model. In Sect. 4, we present the experiments and analyze the results. We make the conclusion in Sect. 5.

2 Related Work

2.1 The Faster R-CNN

Since the concept of deep learning has been introduced in many computer vision tasks, especially target detection, there were a lot of algorithms have been proposed which based on CNN. Both the R-CNN [15] and the Fast R-CNN [16] rely on the CNN to extract the feature form the proposals. But the speed of proposal generation is not good enough. Therefore, to reduce the computational burden of proposal generation, the Faster R-CNN was proposed. It could be considered as a combination of the Region Proposal Network (RPN) [17] which could extract proposals quickly and the Fast R-CNN detector whose purpose is to refine the proposals. The most important idea of the Faster R-CNN is that the RPN share the convolutional layers with the Fast R-CNN, as Fig. 1. In this way, the image could pass through the CNN only once and could extract proposals efficiently.

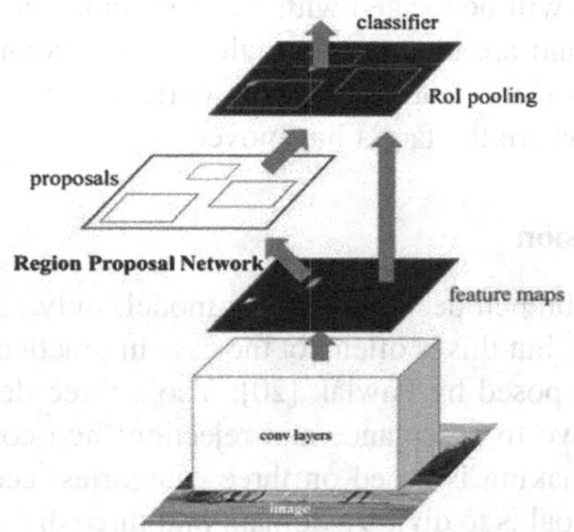

Fig. 1. Network architecture for Faster R-CNN [17].

In the Faster R-CNN model, an input image firstly passes through the Conv layers which was made up of 13 convolutional layers, 13 ReLU layers and 4 pooling layers to extract the feature map of the whole image. Then use the feature map as the input of RPN to get the region proposals. The RoI pooling gathers the proposals and the feature map and to create the proposal feature maps and sends them to the classifier to calculate the class value and process with regression to get the accurate bounding-box.

2.2 Generic Object Tracking Using Regression Network (GOTURN)

There were many algorithms of tracking a single object in a video using deep learning, but the speed of them is too difficult to be assured. So, Held proposed GOTURN which is faster than previous algorithms and can track at 100 FPS [8]. It takes offline pre-training by massive images and videos.

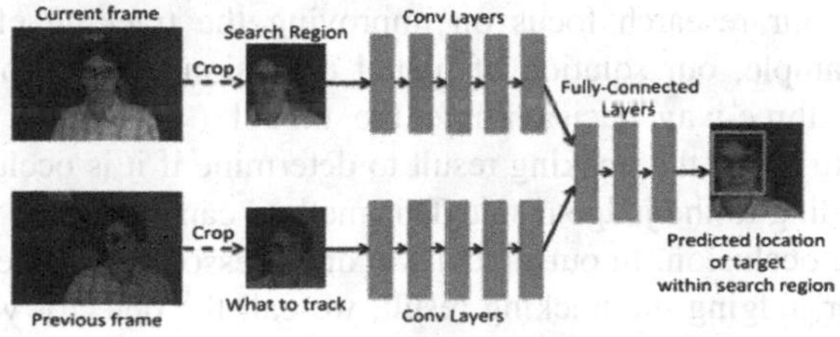

Fig. 2. Network architecture for GOTURN [8].

As seen from Fig. 2, if the target located in the bounding-box centered at c = (c_x, c_y) with a width of w and a height of h in previous frame, it takes two crops of the previous frame and the current frame at c = (c_x, c_y) with a width of $k_1 w$ and a height of $k_1 h$. Then input these crops into the convolutional layers. It supposes that the target object is not moving too quickly and will be located within this region. The outputs of the network are high-level features and are then fed through the fully connected layers. The fully connected layers compare the feature from the current frame and the feature of the previous frame to find where the target has moved.

2.3 Three-Way Decision

In the well-known two-branch decision-making model, only acceptance and rejection are generally considered, but this is often not the case in practical application. Based on the rough set theory proposed by Pawlak [20], Yao's three decision-making theories provide a third alternative to acceptance and rejection: non-commitment [9–11]. The idea of three decision-making is based on three categories: acceptance, rejection and non-commitment. The goal is to divide a domain into three disjoint parts. Positive rules acquired from positive domain are used to accept something, negative rules acquired from negative domain are used to deny something, and rules that fall on boundary domain need further observation, which called delayed decision-making. This way of decision-making describes the thinking mode of human beings in solving practical decision-making problems and has been widely used in decision tree [20] and other fields.

3 The Proposed Algorithm

3.1 Tracking by Detection Model

The tracking algorithms based on deep learning have been deeply investigated in recent years. However, the reality is rather more complicated and the trackers may occur the loss of target in the complex environment. It is because that most of trackers could use the information from previous frame to get the result in the current frame. If the target object moves too fast or it is occluded by other object, the tracker could not get the matching information in the search region and it is likely to lose the target.

In this paper, our research focus on improving the tracking effect in complex situations. For example, our solution of partial occlusion is to introduce a checking scheme based on three-way decision into the model (see as Fig. 3). We use the checking scheme to judge the tracking result to determine if it is occluded, then renew the standard according to the judge result. This method can guarantee the robustness of the standard to the occlusion. In our model, we draw lessons from the idea of tracking by detection. After judging the tracking result, we call the detector when the result is wrong. The detector will get the coordinates of all the pedestrian and select the result which is the most similar to the standard to correct the tracking result. The main notations in this paper are listed in Table 1.

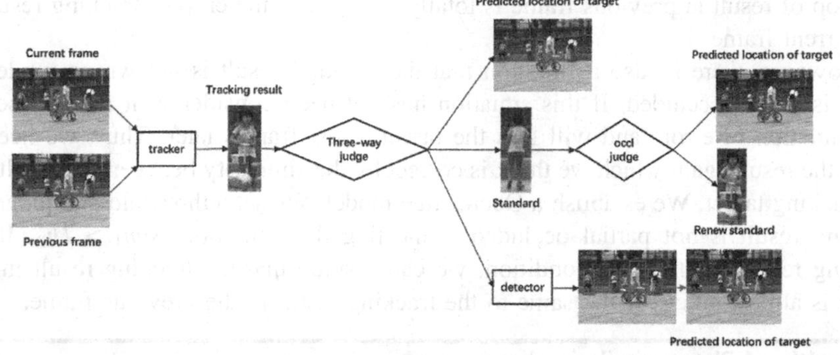

Fig. 3. Framework of our model.

Table 1. List of main notations.

Variable	Explanation
Img_{t-1}	The previous frame
Img_t	The current frame
$c_i = (c_{xi}, c_{yi})$	The center of bounding box of tracking result in frame i
(w_i, h_i)	The width and height of bounding box in frame i
$bbox_i$	The bounding box in the frames i
$crop_i$	The crop at $c_i = (c_{xi}, c_{yi})$ with $(k_1 w_i, k_1 h_i)$
st	The standard of tracking
j	The sequence number of the standard frame
sim_1	The similarity between the $bbox_t$ and $bbox_{t-1}$
$D_t = \{d_{t1}, \cdots, d_{tn}\}$	The detection results of the frame Img_t, including n results
$Th{-}corr$	The threshold to determine if the result is totally correct
$Th{-}wrong$	The threshold to determine if the result is totally wrong
$Th{-}occl$	The threshold to determine if the result is partial occluded
$Th{-}frm$	The threshold to determine if the standard has not been renewed for a long time

3.2 Picture Similarity Discriminant Model

In the long-term tracking tasks, there are many different instances of the tracking result. In the model, we need to make one of three decisions: (a) accept it if it is correct; (b) reject it if it is wrong; (c) delay judgment if it is uncertain. Since the tracker just gives the result which is the most possible, we need to judge the result by calculating the similarity between the tracking result and the tracking target. We set two threshold and judge the tracking result is correct if meeting the condition1: $sim_1 > Th{-}corr$ and is wrong if meeting the condition2: $sim_1 < Th{-}wrong$. The condition1 guarantees that the crop of result in previous frame is similar to the crop of tracking result in the current frame. We define the condition under the hypothesis that meeting the condition means that the tracking result is not wrong (see Algorithm 1). The condition2 guarantees that the crop of result in previous frame is totally different to the crop of tracking result in the current frame.

However, there is also a situation that the tracking result is not wrong while the target is partial occluded. If this situation has not been considered, it may cause the accumulation of errors and will loss the target a few frames later. Thus, we need to judge the result again which we think is correct by the similarity between the result and the tracking target. We establish a Occl-Judge model. We set a threshold and judge the tracking result is not partial occluded if meeting the condition: $sim_1 > Th_3$. If the tracking result satisfies this condition, we can assume that the tracking result in this frame is almost precisely the same to the tracking result in the previous frame.

Algorithm 1 Picture similarity discriminant

Input: $crop_{t-1}$, the current frame Img_t
Output: $bbox_t$
Input Img_t into the first layer (GOTURN), get $bbox_t$;
Send the $bbox_{t-1}$, $bbox_t$ to the discriminant layer;
Calculate the similarity between $bbox_{t-1}$ and $bbox_t$, get sim_1;
If $sim_1 > Th\text{-}corr$ **then**
  get $bbox_t$;
else if $sim_1 < Th\text{-}wrong$ **then**
  input Img_t into the third layer (Faster R-CNN), get $bbox_t$ with Formula (1);
else
  put crop c_t into the Occl-Judge model;
end if

3.3 Renew Standard Model

Under the theory of color histogram, we build a tracking by detection model based on the color histogram (see Algorithm 2). The first layer is the tracker used GOTURN, and the second layer includes judging whether the result is correct and whether the result is partial occluded. The third layer is the detector (Faster R-CNN). We put Img_t into the first layer, send the tracking result $bbox_t$ to the discriminant layer. Next, the discriminant layer calculates the similarity and make one of two decisions: (a) the result is correct and put the result to the Occl-Judge Model to renew the standard; (b) the result is wrong and call the detector.

If the result is correct, we set the crop as the bounding-box. If the result is uncertain, we calculate the similarity between $bbox_t$ and st, get the sim_2 (We initialize the st to the $bbox_1$). The model need to compare sim_2 with $Th{-}occl$ and make one of two decisions: (a) the result is almost the same to the standard; (b) the result may be partial occluded.

If the tracking result is not partial occluded, we renew the standard. If the tracking result is partial occluded, we need to return the tracking result as the final result without renewing the standard. But if we do not renew the standard for

Algorithm 2 Tracking by detection

Input: $crop_{t-1}$, the current frame Img_t, st
Output: $bbox_t$
Input Img_t into the first layer (GOTURN), get $bbox_t$;
Send the $bbox_{t-1}$, $bbox_t$ to the discriminant layer;
Calculate the similarity between $bbox_{t-1}$ and $bbox_t$, get sim_1;
if $sim_1 > Th\text{-}corr$ **then**
  get $bbox_t$;
else if $sim_1 < Th\text{-}wrong$ **then**
  input Img_t into the third layer (Faster R-CNN), detect and get $D_t = \{d_{t1}, d_{t2}, \cdots, d_{tn}\}$
  get $bbox_t$ with Formula (1);
else
  Calculate the similarity between $bbox_t$ and st, get sim_2;
  if $sim_2 > Th\text{-}occl$ **then**
    $st = bbox_t$
  else
    if $t - j > Th\text{-}frm$ **then**
      $st = bbox_t$
    end if
  end if
else
end if

a long time, that may be the object has moved too quickly or the angle of the view changed a lot. We need to force the model to update the standard.

If the result is wrong, we believe we have lost the tracking target. Then we input the Img_t to the detector layer and get all the detection results of pedestrian. We calculate the similarity between d_{ti} and st, get the sim_{ti}. We will choose the final result from D_t. Model computing see Formula (1).

$$i = index(\max(d_{t1}, d_{t2}, \cdots, d_{tn})) \tag{1}$$

We will choose the detection result which is the most similar to the standard. Return the ith detection result as the final tracking result.

To sum up, our model could handle some special cases and can avoid losing of the target efficiently.

4 Experiments

4.1 Training

In this paper, we do experiments on a deep learning framework named Caffe. We train the Faster R-CNN detector using the Caltech Pedestrian Dataset and the Pascal VOC2007. The Caltech Pedestrian Dataset consists of approximately 10 h of 640 × 480 30 Hz video taken from a vehicle driving through regular traffic in an urban environment. It has about 250000 frames with a total of 350000 bounding boxes. The Pascal VOC2007 dataset has 20 classes. It contains 9963 images and 26460 bounding boxes. We only use the labels of pedestrian in the dataset to train our model. We choose the VGG16 model and set the number of iterations to be [80000, 40000, 80000, 40000]. The mAP of the detector is 0.763 when use the Caltech Pedestrian Dataset. The mAP is 0.777 when use the Pascal VOC2007. And the further experiment shows that the model trained by the Pascal VOC2007 perform better in the deformed cases. Therefore, we select the Pascal VOC2007 pedestrian part to train the detector for the subsequent applications.

4.2 Test Set

Our test set consists of the 16 videos from the OTB-100 Dataset [10] and 14 videos from the VOT 2015 Tracking Challenge [22]. Both of the OTB-100 and the VOT 2015 are standard tracking benchmarks that allow us to compare our tracker to a wide variety of other trackers. All the videos we selected take pedestrian as tracking targets. The trackers are evaluated using two standard tracking metrics: precision and success rate, which range from 0 to 1.

Each sequence is annotated with a number of attributes and we mainly focus on occlusion and motion blur. The trackers are also compared with each other for separately from these two attributes.

4.3 Results

In order to select completely correct and incorrect tracking results from the three decision-making branches, experiments show that when the similarity of two croppers of person exceeds 0.95, they are basically identical, and when the similarity is less than 0.75, they are different. So in this experiment, we set $Th-corr$ 0.95, set $Th-wrong$ 0.75 and set $Th-occl$ 0.98. We use the tracker of GOTURN, MDNet and our model to get groups of the bounding box coordinates. Then we get the coordinates of the ground truth boxes from the corresponding data.

Calculate the center location error of every frame in the test dataset. Define the center location error as the average Euclidean distance between the center locations of the manually labeled ground truth box and the center locations of bounding box in every frame. This metric can show the performance for the whole sequence. The precision is defined as the percentage of frames whose center location error is less than the location error threshold. It can evaluate the overall tracking performance.

Calculate the bounding box overlap of every frame. The overlap is the value of the intersection of bounding box and the ground truth box divide the value of the union of these two areas. The success rate, defined as the percentage of the number of the frames whose overlap is not less than the threshold could measure the performance of the video.

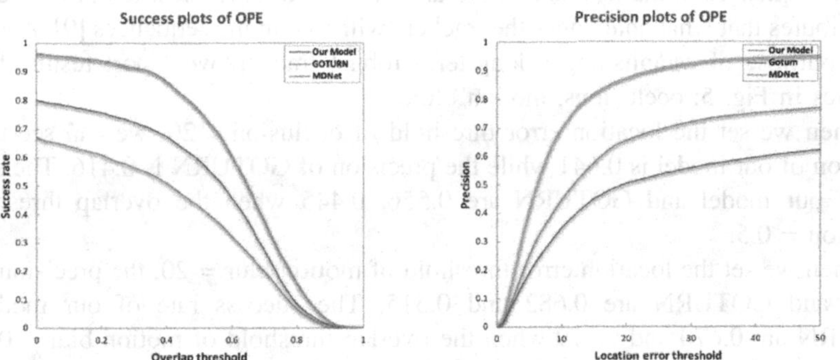

Fig. 4. Success rate plot and precision plot for all 30 sequences. Best viewed in color.

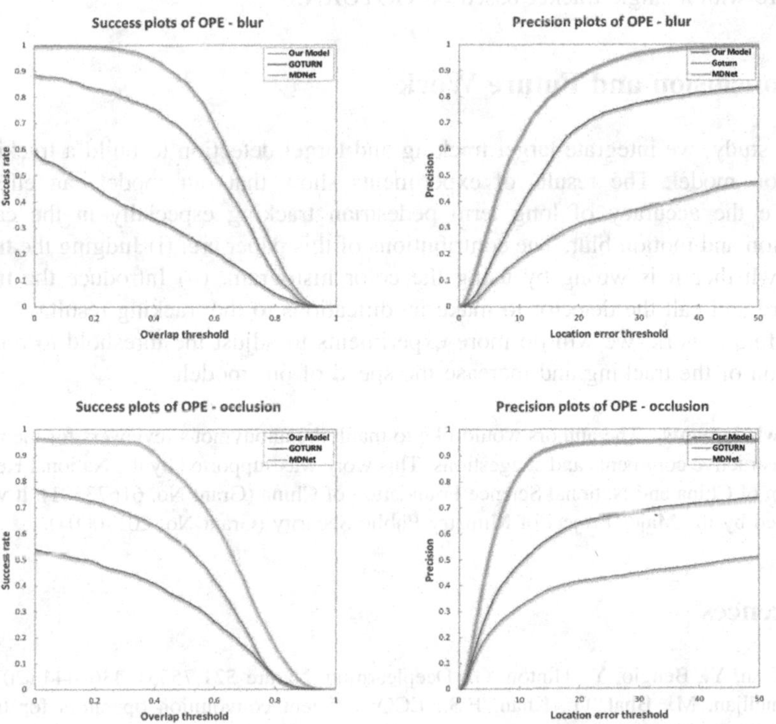

Fig. 5. Success rate plot and precision plot for sequences with attributes: occlusion, motion blur. Best viewed in color.

As can be seen in Fig. 4, both of the success rate and the precision of our model is higher than GOTURN algorithm but lower than MDNet. In the calculation results we can see that the precision of our model is 0.655 when we set the location error threshold = 20. And the success rate of our model is 0.5891 when the overlap threshold = 0.5. This suggest that certain improvement in the overall performance of our model is observed but there is still room to growth.

The sequences in the test dataset are annotated with attributes. It can be seen from the attributes that what challenges the trackers will face in the sequences [9]. According to the purpose of establishing a long-term robust tracker, we report results for two attributes in Fig. 5: occlusions, motion blur.

When we set the location error threshold of occlusion = 20, we can see that the precision of our model is 0.641 while the precision of GOTURN is 0.416. The success rate of our model and GOTURN are 0.556, 0.445 when the overlap threshold of occlusion = 0.5.

When we set the location error threshold of motion blur = 20, the precision of our model and GOTURN are 0.682 and 0.315. The success rate of our model and GOTURN are 0.670 and 0.279 when the overlap threshold of motion blur = 0.5.

Overall, the performance of our tracking-by-detection model is more stable in occlusion or motion blur situations. The experimental results show that our model can effectively improve the accuracy and the stability of long term pedestrian tracking compare with a single tracker based on GOTURN.

5 Conclusion and Future Work

In this study, we integrate target tracking and target detection to build a tracking-by-detection model. The results of experiments show that our model can effectively improve the accuracy of long term pedestrian tracking especially in the cases of occlusion and motion blur. The contributions of this paper are: (i) Judging the tracking result whether it is wrong by using the color histogram. (ii) Introduce the tracking standard and call the detector to make modifications to the tracking result.

In future work, we will do more experiments to adjust the threshold to prove the precision of the tracking and increase the speed of our model.

Acknowledgments. The authors would like to thank the anonymous reviewers for their critical and constructive comments and suggestions. This work was supported by the National Key R&D Program of China and National Science Foundation of China (Grant No. 61673301). It was also supported by the Major Project of Ministry Public Security (Grant No. 20170004).

References

1. LeCun, Y., Bengio, Y., Hinton, G.: Deeplearning. Nature **521**(7553), 436–444 (2015)
2. Danelljan, M., Bhat, G., Khan, F.S.: ECO: efficient convolution operators for tracking, pp. 6931–6939 (2016)

3. Kaaniche, M.B., Bremond, F.: Tracking HoG descriptors for gesture recognition. In: International Conference on Advanced Video and Signal Based Surveillance, pp. 140–145 (2009)
4. Wang, N., Yeung, D.Y.: Learning a deep compact image representation for visual tracking. In: International Conference on Neural Information Processing Systems, pp. 809–817 (2013)
5. Wang, N., Li, S., Gupta, A., et al.: Transferring rich feature hierarchies for robust visual tracking. Comput. Sci., arXiv:1501.04587 (preprint) (2015)
6. Shelhamer, E., Long, J., Darrell, T.: Fully convolutional networks for semantic segmentation. Trans. Pattern Anal. Mach. Intell. **39**(4), 640–651 (2014)
7. Nam, H., Han, B.: Learning multi-domain convolutional neural networks for visual tracking, pp. 4293–4302 (2015)
8. Held, D., Thrun, S., Savarese, S.: Learning to Track at 100 FPS with Deep Regression Networks. In: Leibe, B., Matas, J., Sebe, N., Welling, M. (eds.) ECCV 2016. LNCS, vol. 9905, pp. 749–765. Springer, Cham (2016). https://doi.org/10.1007/978-3-319-46448-0_45
9. Yao, Y.: Three-way decision: an interpretation of rules in rough set theory. In: Wen, P., Li, Y., Polkowski, L., Yao, Y., Tsumoto, S., Wang, G. (eds.) RSKT 2009. LNCS (LNAI), vol. 5589, pp. 642–649. Springer, Heidelberg (2009). https://doi.org/10.1007/978-3-642-02962-2_81
10. Yao, Y.Y.: Three-way decisions with probabilistic rough sets. Inform. Sci., 341–353 (2010)
11. Yao, Y.Y.: Two semantic issues in a probabilistie rough set model. Fundamenta Informatieae, Manuscript **108**(3–4), 249–265 (2011)
12. Wu, Y., Lim, J., Yang, M.H.: Online object tracking: a benchmark. In: CVPR, pp. 2411–2418 (2013)
13. Wu, Y., Lim, J., Yang, M.H.: Object tracking benchmark. IEEE Trans. Pattern Anal. Mach. Intell. **37**(9), 1834–1848 (2015)
14. Yilmaz, A., Javed, O., Shah, M.: Object tracking: a survey. CSUR **38**(4), 1–45 (2006)
15. Girshick, R., Donahue, J., Darrell, T., et al.: Rich feature hierarchies for accurate object detection and semantic segmentation. In: Conference on Computer Vision and Pattern Recognition, pp. 580–587 (2014)
16. Girshick, R.: Fast R-CNN. In: Proceedings of the IEEE International Conference on Computer Vision, pp. 1440–1448 (2015)
17. Ren, S., He, K., Girshick, R., et al.: Faster R-CNN: towards real-time object detection with region proposal networks. Trans. Pattern Anal. Mach. Intell. **39**(6), 1137–1149 (2017)
18. Redmon, J., Divvala, S., Girshick, R., et al.: You only look once: unified, real-time object detection. In: Computer Vision and Pattern Recognition, pp. 779–788 (2016)
19. Liu, W., et al.: SSD: single shot MultiBox detector. In: Leibe, B., Matas, J., Sebe, N., Welling, M. (eds.) ECCV 2016. LNCS, vol. 9905, pp. 21–37. Springer, Cham (2016). https://doi.org/10.1007/978-3-319-46448-0_2
20. Pawlak Z.: Rough sets. Int. J. Comput. Inform. Sci., 341–356 (1982)
21. Zivkovic, Z., Kröse, B.: An EM-like algorithm for color-histogram-based object tracking. In: Computer Vision and Pattern Recognition, pp. 798–803 (2004)
22. Kristan, M., et al.: The visual object tracking vot2015 challenge results. In: Proceedings of the IEEE International Conference on Computer Vision (ICCV) Workshops, pp. 1–23 (2015)
23. Zhang, Y., Miao, D., Zhang, Z.: Multi-granularity text sentiment classification model based on three-way decisions. Comput. Sci., 188–193 (2017)
24. Li, X., Zhang, Q.: Image classification algorithm based on tolerance granular model and three-way decisions. Comput. Technol. Autom., 93–96 (2014)

A Bibliometric Profile of Research on Rough Sets

Wenjie Wei[1,2], Duoqian Miao[1(✉)], and Yuxiang Li[3]

[1] College of Electronics and Information Engineering, Tongji University, Shanghai 200092, China
{weiwenjie,dqmiao}@tongji.edu.cn
[2] Siping Campus Library, Tongji University, Shanghai 200092, China
[3] School of Foreign Languages, Tongji University, Shanghai 200092, China

Abstract. Rough sets theory is a powerful mathematical tool for modelling various types of inexact, incomplete or uncertain information. Rough sets theory and its applications have attracted significant attention among researchers and extensive research has been carried out since it was first proposed by Pawlak in 1982. This paper presents a panorama of rough sets and quantitatively analyzes the developments of rough sets research by scientometrics approach. The bibliometric analysis is conducted based on 11833 Web of Science indexed papers published from 1982 to 2018. The science mapping tool, VOSviewer, is employed to cluster the documents and to assist in summarizing the important publications over the last ten years. The results are presented in the following aspects: development stages over the recent two decades, thematic structure of publications, citation distribution on subjects, core journals and conferences, international research collaboration profiles and top scholars. The results can benefit the scholars who want to go further in future research of rough sets.

Keywords: Rough sets · Bibliometric analysis · Research theme · Institutes performance · Cooperation network · Scholars distribution

1 Introduction

1.1 A Subsection Sample

Rough sets theory (briefly, RS) was proposed by Pawlak in 1982. Many complicated problems in economics, engineering, environmental science, medical science and social science may not be successfully solved because of various uncertainties arising in these problems. Motivated by the practical needs, RS models are developed to extract knowledge from incomplete, inaccurate and uncertain data sets.

The brilliant approach to classifying objects with their features and the introduction of approximation spaces can cope with large scale and diverse data easily. RS enables dealing with data granularity, which establishes the foundations of granular computing and provides an incisive approach to pattern recognition.

In the light of dealing with practical problem effectively, many researchers and practitioners have been imparted the study of hybridizations combining RS with other mathematical structures that are distinct but closely related. The RS blending with

© Springer Nature Switzerland AG 2019
T. Mihálydeák et al. (Eds.): IJCRS 2019, LNAI 11499, pp. 534–548, 2019.
https://doi.org/10.1007/978-3-030-22815-6_41

fuzzy sets, soft sets and neural network has been studied in recent years and some hybrid uncertain models occur [1]. Pawlak [2] has listed a wide range of applications of methods based on RS including machine learning, pattern recognition, data mining, knowledge discovery, bioinformatics, medicine, multicriteria decision making [3], signal processing, image processing, hierarchical learning, ontology approximation.

This paper presents a bibliometric profile of RS and quantitatively analyzes the developments of RS research by scientometrics approach. We will carry out bibliometric analysis to gain more insights in the domain of RS.

2 Data Source and Methodology

Bibliometric analysis helps to identify the influential works and to reveal some relations between academic entities. By adopting bibliometric analysis researchers will easily locate their positions in the research area and find new points for future research. The bibliographic metadata of literatures provided by publishers have abundant information for statistical treatment to evaluate the research performance of researchers, journals, countries and institutions [4–6]. Bibliographic coupling analysis is often used to outline the publications in a certain field [7, 8]. When two articles reference a common third article in their bibliographies, the two articles have bibliographic coupling, indicating that they study a related subject matter, and the similarity of their bibliographies can be defined as "coupling strength". The more citations to other articles they share the higher coupling strength they have.

The bibliometric maps in this paper are constructed by VOSviewer (www.vosviewer.com). It can be used to cluster publications and to analyze the resulting clustering solutions related to citations, co-occurrence (i.e. co-authorship and co-institute), bibliographic coupling and co-citations in bibliometric map.

The bibliographic data are obtained through the Clarivate Analytics' Web of Science™ (WoS), which contains 7 core collection databases, including SCI-EXPANDED, SSCI, A&HCI, CPCI-S, CPCI-SSH, BKCI-S, BKCI-SSH. To analyze the distribution characterizations of the literatures, the related information is extracted from particular fields of the metadata downloaded from WoS. We retrieve 11833 papers by the query as follows: TS = ("rough set$" or "rough fuzzy set$" or "rough soft set$") and PY = (1982–2018). We use the ($) in the search as a wild card character to make our search simpler and more comprehensive as it will track all possible forms of the terms used (i.e. set or sets).

3 Academic Development of RS

3.1 Development Stages

The line in Fig. 1 shows the number of RS paper publications by year from 1999 to 2018. The documents can roughly be classified into two types, proceeding papers and journal articles, as indicated by bars in Fig. 1. In fact, we used "Meeting Abstract OR Meeting Summary OR Proceedings Paper" for searching proceeding papers, and selected "Article OR Editorial Material OR Letter OR Review" to obtain journal articles, and found that few other document types left. If one paper is both labeled article and proceeding paper, it will be treated as article.

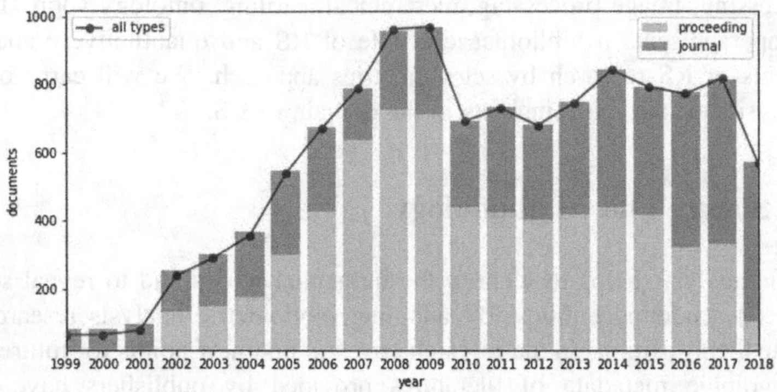

Fig. 1. RS publication year distribution in WoS.

As shown in Fig. 1, the number of articles published before 2001 is relatively small. There is a rapid growth from 2002 to 2009, and then the annual paper publication number drops below 800 and grows slowly but steadily from 2010 to 2018. Compared with that of the previous decade, the proportion of journal articles has increased and gradually exceeded that of conference papers in the 2010–2018 period. For these reasons, the period from 2002 to 2009 can be considered as a growth period of RS, and the 2010–2018 period can be interpreted as mature period. During the growth period many conferences have been held and publishing articles in journals is relatively difficult, while during the mature period journal articles have thrived.

3.2 Theme Clustering

To illustrate the general situation of original researches of RS in the last decade from 2009 to 2018 that covers the mature period, bibliographic coupling network of the documents is made by Vosviewer (Fig. 2). Bibliographic coupling of papers, as pointed out before, shows the relation between papers that citing at least one same other paper. This relationship provides the basis for topic clustering. It is proved to be an efficient approach for grasping the main themes of a research area of some scale.

As is shown in Fig. 2, every bubble shows its citations by the size. The nodes with less than 20 citations have been cut off to make the figure more readable. The articles published in the last three years have not got citations sufficiently. In order to make the newly published articles that are not sufficiently cited displayed fairly with the old ones, the papers' citations of each year are weighted by a factor calculated from the citation trends of the previous three years.

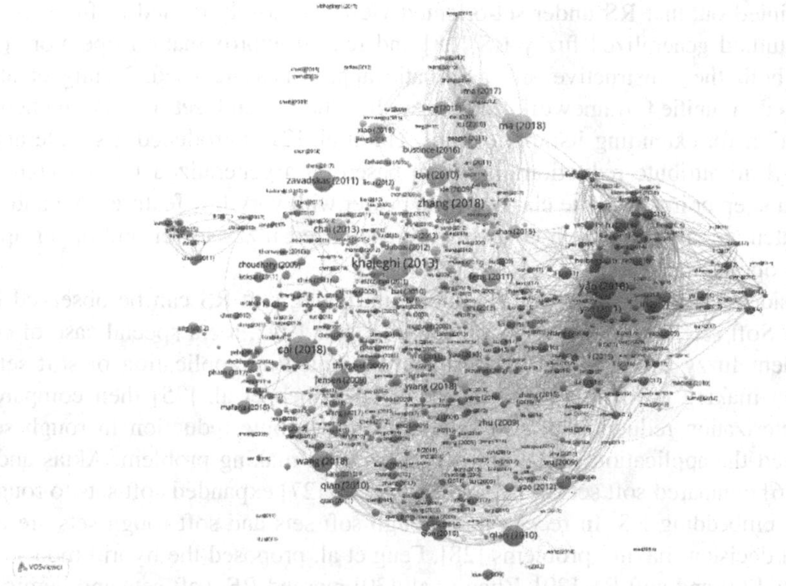

Fig. 2. Articles clustering by bibliographic coupling. (Color figure online)

Clusters with different colors reflect different themes of this area. We will explore and explain the themes of these clusters one by one by scanning the larger nodes in every cluster.

Red Cluster: Dimensionality Reduction and Its Application

One of the major limitations of the traditional rough set model in the real applications is the inefficiency in the computation of core attributes and reducts. Jensen and Shen [9] provided the approaches of fuzzy-rough feature selection for dimensionality reduction. Chen et al. [10] proposed a RS approach to solve feature selection problems successfully by using ant colony optimisation, which adopts mutual information based feature significance as heuristic information. Qian et al. [11] introduced a positive approximation framework to accelerate a heuristic process of attribute reduction. Wang et al. [12] proposed an index to characterize the discrimination of a neighborhood relation for their feature selection algorithm. He et al. [13] combined RS theory, data envelopment analysis and fuzzy artificial neural network to explore the effects of influencing factors on industrial energy efficiency. Cai et al. [14] improved the prediction of sensitive information by using RS approach to avoid inference attack for social network. Choudhary et al. [15] reviewed the multiple approaches of knowledge discovery and data mining applied in manufacturing process.

Green Cluster: Hybrids of RS

The RS hybridization with fuzzy sets has been studied much from early time because of the natural correlation between fuzzy sets and RS. Dubois and Prade [16] clarified the difference between fuzzy sets and RS and developed the concept of fuzzy RS to deal with numerical and fuzzy attributes. Yao [17] compared theories of fuzzy sets and RS and pointed out that RS under set-oriented view are closely related to fuzzy sets. Wu et al. studied generalized fuzzy RS [18] and related approximation operators [19] in which both the constructive and axiomatic approaches are used. Yeung et al. [20] presented a unified framework for fuzzy RS theory and set up its mathematical foundation for extending its applications. Hu et al. [21] introduced a simple and efficient hybrid attribute reduction algorithm based on a generalized fuzzy-rough model that can keep or improve the classification power with very few features. Mi and Zhang [22] extended approximation concepts to generalized fuzzy lower and upper approximation operators.

Besides, another kind of hybrid uncertain model, soft RS can be observed in this cluster. Soft sets, introduced by Molodtsov [23] in 1992, are a special case of context dependent fuzzy sets. Maji et al. [24] first presented an application of soft sets in a decision making problem with the help of RS. Chen et al. [25] then compared the parameterization reduction of soft sets with the attribute reduction in rough set and improved the application of a soft set in a decision-making problem. Aktaş and Çağman [26] compared soft sets to RS and Feng et al. [27] expanded soft sets to rough soft sets by embedding RS. In recent years rough soft sets and soft rough sets are mainly used in decision making problems [28]. Feng et al. proposed the hybrid models rough soft sets [27] and soft RS [29]. Zhan et al. [30] merged RS, soft sets and hemirings to provide soft rough algebraic structures. They [31] also extended the notion of soft RS and rough fuzzy sets to study roughness in hemirings.

Blue Cluster: Three-Way View Decision

The three-way decision-theoretic RS model was proposed by Yao [32]. The three-way decisions theory considers a decision-making problem as a ternary classification one. The positive, negative and boundary regions are associated with different levels of uncertainty. Yao [33] discussed the advantages of three-way decision in probabilistic rough set models.

Li and Zhou [34] proposed a three-way view decision model based on decision-theoretic RS. Herbert and Yao [35] investigated the Game-theoretic RS to reduce the boundary region in the decision problem. Sun et al. [36] constructed a multigranulation fuzzy decision-theoretic three-way group decision making method. Li et al. [37] developed an axiomatic approach to characterize three-way concepts. The three-way decision model has been used in face recognition [38] and recommender system [39]. Yu et al. [40] investigated the method for automatically determining the number of clusters by the decision-theoretic rough set model. Qi et al. [41] presented the constructing of three-way concept lattices based on classical concept lattices. Jia et al. [42] provided the minimum cost attribute reduction method for decision-theoretic RS models.

The Rest Clusters

The remaining clusters focus on some special extensive research themes which have not been studied too much. The yellow cluster studies covering-based RS. In RS theory, relation-based RS and covering-based RS are two important extensions of the classical RS. Covering-based RS [43] is a successful generalization for the Pawlak's model to make use of non-equivalence relations. Zhang et al. [44] recently established some constructive methods of rough approximation operators to make the equivalence relations in RS not too restrictive for practical applications. The pink cluster focuses on multi-granulation RS which was developed by Qian [45] to extend Pawlak's model to a multi-granulation RS model using multi-equivalence relations. In addition, a growing research interest of neighborhood RS [46] is observed in recent years for its effectiveness of dealing with data of multi-granularity [47]. Finally, the bright blue cluster is mainly concerned with formal concept analysis and RS.

3.3 Subject Distribution of Citations

The total number of citations to RS research articles is larger than 39,200 and is increasing quickly. The citations come from different subject areas in WoS. Some interesting areas are selected to reveal the extension of RS researches to the complicated world.

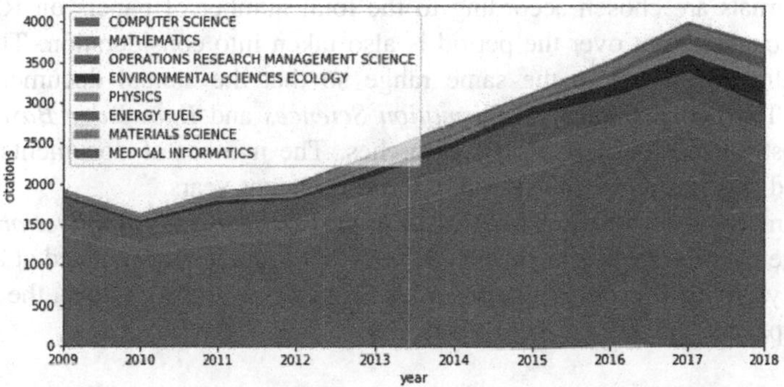

Fig. 3. Some research areas that citing RS from 2009 to 2018. (Color figure online)

Figure 3 shows some research areas with increasing RS research citations. Computer science and mathematics are the circumstances that give birth to RS. The other areas like management science and environmental science may give the application and development environments for RS. RS researches have been increasingly applied to power industry [48], natural resources sustainable utilization [49], medical diagnosis and prognosis [50] and synthetic materials design [51], etc.

4 Journals and Conferences Analysis

4.1 Core Journals

There are 4427 papers published on 778 journals in WoS from 1999 to 2018. The top 10 journals are listed in Fig. 4 and Table 1.

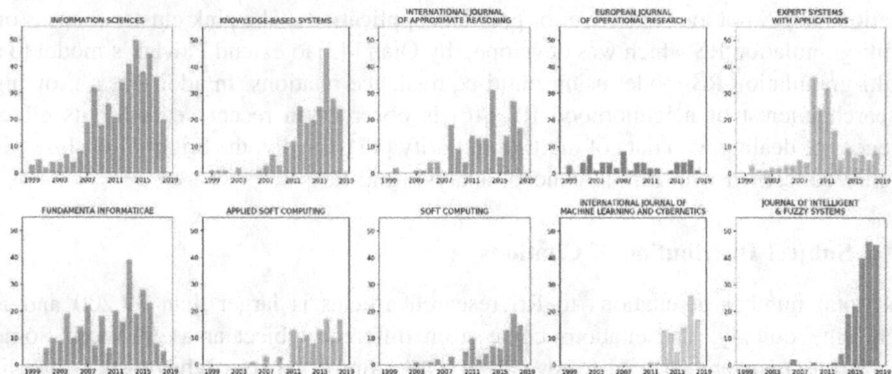

Fig. 4. Annual documents benchmarking for the top journals on RS researches.

The journals are chosen according to the total number of papers on RS, and the number of citations got over the period is also taken into consideration. The longitudinal coordinates are set to the same range so that the annual documents can be compared. The two periodicals, *Information Sciences* and *Knowledge Based Systems* are the mostly cited journals in RS researches. The number of documents on RS in these periodicals has an obvious growth trend in recent years.

The number of RS research articles in *Expert Systems with Applications* was relatively large over the particular period of 2009–2012, but it has remained at a low level in recent 6 years. In *Journal of Intelligent & Fuzzy Systems*, by contrast, the number of RS theory papers has thrived these years.

Table 1. Profile of the top journals.

Journal name	Documents	Citations	Journal IF	Quartile
INFORMATION SCIENCES	417	19156	4.305	Q1
FUNDAMENTA INFORMATICAE	279	2608	0.725	Q3
KNOWLEDGE-BASED SYSTEMS	229	4754	4.396	Q1
JOURNAL OF INTELLIGENT & FUZZY SYSTEMS	180	517	1.426	Q3
INTERNATIONAL JOURNAL OF APPROXIMATE REASONING	178	4744	1.766	Q2
EXPERT SYSTEMS WITH APPLICATIONS	172	4456	3.768	Q1
APPLIED SOFT COMPUTING	106	1825	3.907	Q1
INTERNATIONAL JOURNAL OF MACHINE LEARNING AND CYBERNETICS	89	543	2.692	Q2
SOFT COMPUTING	88	1041	2.367	Q2
EUROPEAN JOURNAL OF OPERATIONAL RESEARCH	61	4488	3.428	Q1

Table 1 gives an outline of the top journals and the last impact factors and the quartiles are listed. In addition to considering the fitness of the article to the subject of a journal, the international influence and status of a journal should also be taken into account.

4.2 Important Conferences

For computer science research, conference papers are sometimes more important than journal articles. There are about 4000 proceeding papers of RS recorded in WoS, including 2 highly cited papers in this period. The number of the proceeding papers is larger than that of the journal articles in WoS in the recent decade. But in the last decade, the conference papers are much more numerous than journal articles, which can be seen from Fig. 1.

The eight top conferences in RS research field are selected from 1999 to 2018 according to the number of papers, as shown in Table 2.

Table 2. Top conferences in RS research field from 1999 to 2018.

Conference name	Abbreviation	Papers
International conference on rough sets and knowledge technology	RSKT	377
International conference on machine learning and cybernetics	ICMLC	309
International conference on rough sets fuzzy sets data mining and granular computing	RSFDGrC	289
IEEE international conference on granular computing	GrC	226
International conference on rough sets and current trends in computing	RSCTC	216
International conference on fuzzy systems and knowledge discovery	FSKD	174
International joint conference on rough sets	IJCRS	128
IEEE international conference on fuzzy systems	FUZZ-IEEE	120

Proceeding papers have been indexed in these database: SCI-EXPANDED, SSCI, A&HCI, CPCI-S, CPCI-SSH. For some conferences, the number of papers in WoS databases might be smaller than that published actually in the conferences. The quality of papers in a conference may affect their number included in the database. The annual heatmap of papers in the top eight conferences is shown in Fig. 5. The labels of the horizontal ordinate refer to the abbreviations in Table 2.

Some conferences have changed over the past two decades. Since 2015, IJCRS has integrated the four conferences, RSKT, RSFDGrC, RSCTC and RSEISP (whose full name is *Rough Sets and Intelligent Systems Paradigms*), which are the major threads of RS conferences. FSKD has been held as part of *International Conference on Natural Computation, Fuzzy Systems and Knowledge Discovery* (ICNC-FSKD) from 2016.

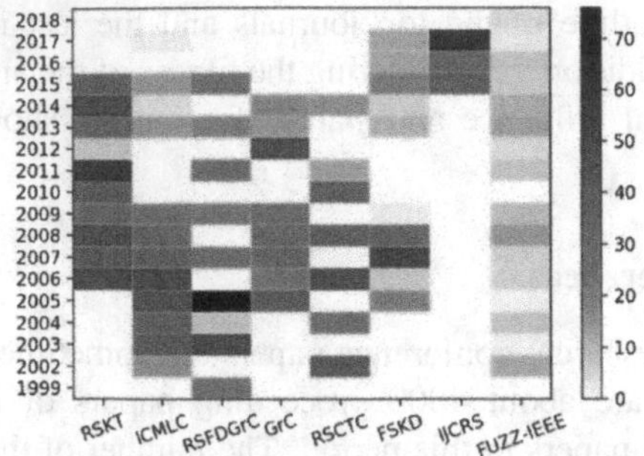

Fig. 5. Annual heatmap of the top conferences.

5 Main Institutes and Scholars of RS Research

5.1 International Collaborators

The collaboration between countries or regions is shown in Fig. 6. The line thickness of each pair of countries or regions represents the strength of the collaboration. The colors of labels represent the clusters which are calculated by their links' similarity in VOSviewer. If some countries or regions often link to each other or if they have the same links with the other nodes, they tend to be classified in the same cluster.

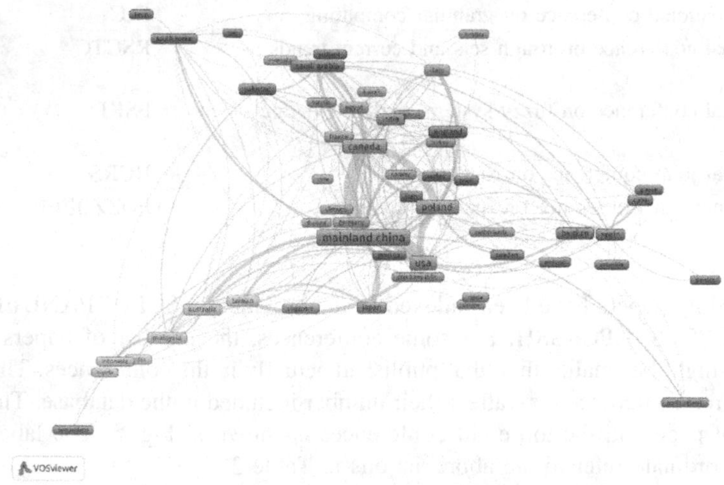

Fig. 6. Co-occurrence of country/region for RS research from 2009 to 2018.

In Fig. 6, we can see Mainland China, USA, Canada and Poland have established a tight group in the center, which features the close relationship of their cooperative RS research. Saudi Arabia England and Japan also have relatively strong partnership with China.

5.2 Main Institutes

Since more than half of the RS research articles are from China, the rankings of the institutions from other countries would be forced to sink. To avoid this, we select the top three influential institutes according to their citations in WoS for the eight top countries chosen from the map of Fig. 6, and these 24 institutes are compared in the number of documents and citations simultaneously in Fig. 7.

Fig. 7. Comparation in the number of documents and citations of the top institutes.

The main influential institutes are labeled in Fig. 7. Those organizations with less than 50 documents and less than 500 citations are too close to be labeled on the figure. It can be noticed that Southwest Jiaotong University has more papers than the others, but University of Regina has received more citations than any other institutes. Tongji University is very close to Polish Academy of Sciences both in the number of documents and citations. Poznan University of Technology and Indian Statistical Institute are about the same on citations but have different numbers of documents.

5.3 Top Authors

The cooperation networks of authors with more than 30 papers and 15 citations in the recent decade are shown in Fig. 8. The sizes of the nodes represent the amount of papers the authors have published in the RS area. To highlight some of the most productive authors and to distinguish them from their affiliations, the authors with more than or equal to 50 articles are marked with different colors according to their recent institutes. The other authors are all labeled by '_' in gray.

Min from Southwest Petr University, Liang and Qian both from Shanxi University have the broadest range of cooperation. They all have collaborated with up to 9 scholars in the co-author network. Only two of the top authors are not Chinese, Pedrycz from Univ of Alberta and Slowinski from Poznan Univ Tech. Pedrycz has cooperated much more with Chinese than Slowinski in RS area.

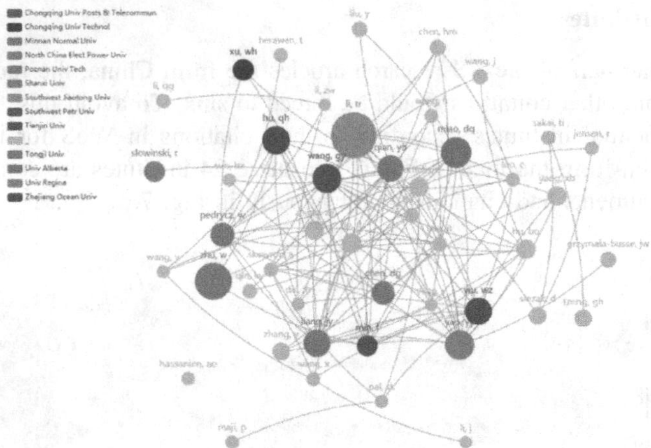

Fig. 8. Cooperation network of authors with more than 30 papers and 15 citations from 2009 to 2018. (Color figure online)

5.4 Chinese Scholars

From the above section, we see that most productive authors come from China. The RS research in China is growing very quickly in recent years. We further analyze the city distribution of Chinese RS community.

Figure 9 shows the main cities of RS researches distributed in China. The number of papers of each city is extracted from address field of bibliographic meta data downloaded from WoS. Full counting is used, i.e. if two cities co-exist in the address of scholars, the amount of papers of each city will increase by 1.

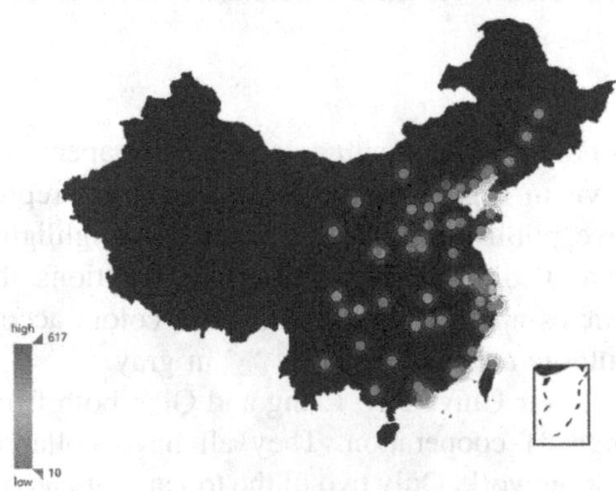

Fig. 9. The distribution of cities in China with more than 10 RS papers.

There are 4295 literatures contributed by Chinese scholars about RS indexed in WoS in recent decade. The scholars are distributed in 190 cities in China. The figure is drawn by pyecharts, and cities with fewer than 10 papers are cut out to make the main cities noticeable. Among those cities, Beijing has gathered most scholars (617 papers), followed by Chengdu, Xi'an, Shanghai and Nanjing whose numbers of papers are 417, 329, 321 and 283 respectively.

The top Chinese authors are listed in Table 3, and their institutes and their most frequently published journals are shown. These Chinese scholars are also highlighted in Fig. 9 above.

Most of the top authors prefer to publish articles in INFORMATION SCIENCES. They also favour KNOWLEDGE BASED SYSTEMS and INTERNATIONAL JOURNAL OF MACHINE LEARNING AND CYBERNETICS, etc.

Table 3. The top Chinese authors in RS in recent decade.

Author	Papers	Institute	Journal with most publications
LI, TR	110	Southwest Jiaotong Univ	INFORMATION SCIENCES
ZHU, W	88	Minnan Normal Univ	INFORMATION SCIENCES
MIAO, DQ	73	Tongji Univ	INFORMATION SCIENCES/KNOWLEDGE BASED SYSTEMS
WANG, GY	69	Chongqing Univ Posts & Telecommun	INFORMATION SCIENCES
HU, QH	67	Tianjin Univ	INFORMATION SCIENCES
LIANG, JY	66	Shanxi Univ	KNOWLEDGE BASED SYSTEMS
WU, WZ	64	Zhejiang Ocean Univ	INFORMATION SCIENCES
QIAN, YH	63	Shanxi Univ	INFORMATION SCIENCES
CHEN, DG	56	North China Elect Power Univ	INFORMATION SCIENCES/IEEE TRANSACTIONS ON FUZZY SYSTEMS
XU, WH	51	Chongqing Univ Technol	INTERNATIONAL JOURNAL OF MACHINE LEARNING AND CYBERNETICS
MIN, F	50	Southwest Petr Univ	INFORMATION SCIENCES

6 Conclusion and Discussion

This paper provides the comprehensive analysis of research landscape on research of rough sets. We use WoS databases and provide the overview of RS by conducting the bibliometric analysis of 11833 papers published from 1982 to 2018. First, from the distribution of publication, we identify the main development stages over the period The timespan of the science map covers the years from 2009 to 2018 which allows us to identify key points of RS research in recent years. We find some research areas citing

RS are broadening. Second, we identify the main journals and conferences as well as their changes over the last two decades. The third part is the analysis of research collaborations at different granularity of authorship, including the international collaboration, the research performance of top organizations in different countries, the cooperation network of authors, the distribution of cities in China and the top Chinese scholars.

We apply different approaches to visualize data in form of different illustrative graphs to make our analysis easy to read and understand. The results of the study can benefit the researchers who are ready to dive into RS research, as well as those who have launched the relevant investigation.

References

1. Ma, X., Zhan, J., Ali, M.I., Mehmood, N.: A survey of decision making methods based on two classes of hybrid soft set models. Artif. Intell. Rev. **49**(4), 511–529 (2018)
2. Pawlak, Z., Skowron, A.: Rudiments of rough sets. Inf. Sci. **177**(1), 3–27 (2007)
3. Zavadskas, E.K., Turskis, Z.: Multiple criteria decision making (MCDM) methods in economics: an overview/Daugiatiksliai sprendimu priemimo metodai ekonomikoje: apzvalga. Technol. Econ. Dev. Econ. **17**(2), 397–427 (2011)
4. Karanatsiou, D., Li, Y.H., Arvanitou, E.M., Misirlis, N., Wong, W.E.: A bibliometric assessment of software engineering scholars and institutions (2010–2017). J. Syst. Softw. **147**, 246–261 (2019)
5. Wang, X.Y., Tang, B.J.: Review of comparative studies on market mechanisms for carbon emission reduction: a bibliometric analysis. Nat. Hazards **94**(3), 1141–1162 (2018)
6. Wei, G.Y.: A bibliometric analysis of the top five economics journals during 2012–2016. J. Econ. Surv. **33**(1), 25–59 (2019)
7. Ferreira, F.A.F.: Mapping the field of arts-based management: Bibliographic coupling and co-citation analyses. J. Bus. Res. **85**, 348–357 (2018)
8. Blanco-Mesa, F., Lindahl, J.M.M., Gil-Lafuente, A.M.: A bibliometric analysis of fuzzy decision making research, pp. 1–4 (2016)
9. Jensen, R., Shen, Q.: New approaches to fuzzy-rough feature selection. IEEE Trans. Fuzzy Syst. **17**(4), 824–838 (2009)
10. Chen, Y.M., Miao, D.Q., Wang, R.Z.: A rough set approach to feature selection based on ant colony optimization. Pattern Recognit. Lett. **31**(3), 226–233 (2010)
11. Qian, Y., Liang, J., Pedrycz, W., Dang, C.: Positive approximation: an accelerator for attribute reduction in rough set theory. Artif. Intell. **174**(9), 597–618 (2010)
12. Wang, C., Hu, Q., Wang, X., Chen, D., Qian, Y., Dong, Z.: Feature selection based on neighborhood discrimination index. IEEE Trans. Neural Netw. Learn. Syst. **29**(7), 2986–2999 (2018)
13. He, Y., Liao, N., Zhou, Y.: Analysis on provincial industrial energy efficiency and its influencing factors in China based on DEA-RS-FANN. Energy **142**, 79–89 (2018)
14. Cai, Z., He, Z., Guan, X., Li, Y.: Collective data-sanitization for preventing sensitive information inference attacks in social networks. IEEE Trans. Dependable Secure Comput. **15**(4), 577–590 (2018)
15. Choudhary, A.K., Harding, J.A., Tiwari, M.K.: Data mining in manufacturing: a review based on the kind of knowledge. J. Intell. Manuf. **20**(5), 501 (2008)

16. Dubois, D., Prade, H.: Rough fuzzy sets and fuzzy rough sets. Int. J. Gen. Syst. **17**(2–3), 191–209 (1990)
17. Yao, Y.Y.: A comparative study of fuzzy sets and rough sets. Inf. Sci. **109**(1), 227–242 (1998)
18. Wu, W.-Z., Mi, J.-S., Zhang, W.-X.: Generalized fuzzy rough sets. Inf. Sci. **151**, 263–282 (2003)
19. Wu, W.-Z., Leung, Y., Mi, J.-S.: On characterizations of (I,T)-fuzzy rough approximation operators. Fuzzy Sets Syst. **154**(1), 76–102 (2005)
20. Yeung, D.S., Degang, C., Tsang, E.C.C., Lee, J.W.T., Wang, X.: On the generalization of fuzzy rough sets. IEEE Trans. Fuzzy Syst. **13**(3), 343–361 (2005)
21. Hu, Q., Xie, Z., Yu, D.: Hybrid attribute reduction based on a novel fuzzy-rough model and information granulation. Pattern Recognit. **40**(12), 3509–3521 (2007)
22. Mi, J.-S., Zhang, W.-X.: An axiomatic characterization of a fuzzy generalization of rough sets. Inf. Sci. **160**(1), 235–249 (2004)
23. Molodtsov, D.: Soft set theory - first results. Comput. Math. Appl. **37**(4–5), 19–31 (1999)
24. Maji, P.K., Roy, A.R., Biswas, R.: An application of soft sets in a decision making problem. Comput. Math. Appl. **44**(8), 1077–1083 (2002)
25. Chen, D., Tsang, E.C.C., Yeung, D.S., Wang, X.: The parameterization reduction of soft sets and its applications. Comput. Math. Appl. **49**(5), 757–763 (2005)
26. Aktaş, H., Çağman, N.: Soft sets and soft groups. Inf. Sci. **177**(13), 2726–2735 (2007)
27. Feng, F., Li, C., Davvaz, B., Ali, M.I.: Soft sets combined with fuzzy sets and rough sets: a tentative approach. Soft Comput. **14**(9), 899–911 (2010)
28. Ma, X., Liu, Q., Zhan, J.: A survey of decision making methods based on certain hybrid soft set models. Artif. Intell. Rev. **47**(4), 507–530 (2017)
29. Feng, F., Liu, X., Leoreanu-Fotea, V., Jun, Y.B.: Soft sets and soft rough sets. Inf. Sci. **181** (6), 1125–1137 (2011)
30. Zhan, J., Liu, Q., Herawan, T.: A novel soft rough set: soft rough hemirings and corresponding multicriteria group decision making. Appl. Soft Comput. **54**, 393–402 (2017)
31. Zhan, J., Zhu, K.: A novel soft rough fuzzy set: Z-soft rough fuzzy ideals of hemirings and corresponding decision making. Soft Comput. **21**(8), 1923–1936 (2017)
32. Yao, Y.: Three-way decisions with probabilistic rough sets. Inf. Sci. **180**(3), 341–353 (2010)
33. Yao, Y.: The superiority of three-way decisions in probabilistic rough set models. Inf. Sci. **181**(6), 1080–1096 (2011)
34. Li, H., Zhou, X.: Risk decision making based on decision-theoretic rough set: a three-way view decision model. Int. J. Comput. Intell. Syst. **4**(1), 1–11 (2011)
35. Herbert, J.P., Yao, J.T.: Game-theoretic rough sets. Fundam. Informat. **108**(3–4), 267–286 (2011)
36. Sun, B., Ma, W., Xiao, X.: Three-way group decision making based on multigranulation fuzzy decision-theoretic rough set over two universes. Int. J. Approx. Reason. **81**, 87–102 (2017)
37. Li, J., Huang, C., Qi, J., Qian, Y., Liu, W.: Three-way cognitive concept learning via multi-granularity. Inf. Sci. **378**, 244–263 (2017)
38. Li, H., Zhang, L., Huang, B., Zhou, X.: Sequential three-way decision and granulation for cost-sensitive face recognition. Knowl. Based Syst. **91**, 241–251 (2016)
39. Zhang, H.-R., Min, F., Shi, B.: Regression-based three-way recommendation. Inf. Sci. **378**, 444–461 (2017)
40. Yu, H., Liu, Z., Wang, G.: An automatic method to determine the number of clusters using decision-theoretic rough set. Int. J. Approx. Reason. **55**(1), 101–115 (2014). Part 2
41. Qi, J., Qian, T., Wei, L.: The connections between three-way and classical concept lattices. Knowl. Based Syst. **91**, 143–151 (2016)

42. Jia, X., Liao, W., Tang, Z., Shang, L.: Minimum cost attribute reduction in decision-theoretic rough set models. Inf. Sci. **219**, 151–167 (2013)
43. Zhu, W.: Relationship between generalized rough sets based on binary relation and covering. Inf. Sci. **179**(3), 210–225 (2009)
44. Zhang, X.H., Miao, D.Q., Liu, C.H., Le, M.L.: Constructive methods of rough approximation operators and multigranulation rough sets. Knowl. Based Syst. **91**, 114–125 (2016)
45. Qian, Y., Liang, J., Yao, Y., Dang, C.: MGRS: a multi-granulation rough set. Inf. Sci. **180** (6), 949–970 (2010)
46. Yang, X., Liang, S., Yu, H., Gao, S., Qian, Y.: Pseudo-label neighborhood rough set: measures and attribute reductions. Int. J. Approx. Reason. **105**, 112–129 (2019)
47. Wang, C., He, Q., Shao, M., Hu, Q.: Feature selection based on maximal neighborhood discernibility. Int. J. Mach. Learn. Cybern. **9**(11), 1929–1940 (2018)
48. Liu, B.H., Fu, Z.G., Wang, P.K., Liu, L., Gao, M.D., Liu, J.: Big-data-mining-based improved K-means algorithm for energy use analysis of coal-fired power plant units: a case study. Entropy **20**(9), 702 (2018)
49. Mazzorana, B., Trenkwalder-Platzer, H., Heiser, M., Hubl, J.: Quantifying the damage susceptibility to extreme events of mountain stream check dams using rough set analysis. J. Flood Risk Manag. **11**(4), e12333 (2018)
50. Juneja, A., Rana, B., Agrawal, R.K.: A novel fuzzy rough selection of non-linearly extracted features for schizophrenia diagnosis using fMRI. Comput. Methods Programs Biomed. **155**, 139–152 (2018)
51. Dey, S., Sultana, N., Dey, P., Pradhan, S.K., Datta, S.: Intelligent design optimization of age-hardenable Al alloys. Comput. Mater. Sci. **153**, 315–325 (2018)

Author Index

Printed in the United States
By Bookmasters